HANDBOOK OF NANOMATERIALS FOR SENSING APPLICATIONS

Micro and Nano Technologies

HANDBOOK OF NANOMATERIALS FOR SENSING APPLICATIONS

Edited by

CHAUDHERY MUSTANSAR HUSSAIN

SURESH KUMAR KAILASA

ELSEVIER

Elsevier
Radarweg 29, PO Box 211, 1000 AE Amsterdam, Netherlands
The Boulevard, Langford Lane, Kidlington, Oxford OX5 1GB, United Kingdom
50 Hampshire Street, 5th Floor, Cambridge, MA 02139, United States

Notices
Knowledge and best practice in this field are constantly changing. As new research and experience broaden our understanding, changes in research methods, professional practices, or medical treatment may become necessary.

Practitioners and researchers must always rely on their own experience and knowledge in evaluating and using any information, methods, compounds, or experiments described herein. In using such information or methods they should be mindful of their own safety and the safety of others, including parties for whom they have a professional responsibility.

To the fullest extent of the law, neither the Publisher nor the authors, contributors, or editors, assume any liability for any injury and/or damage to persons or property as a matter of products liability, negligence or otherwise, or from any use or operation of any methods, products, instructions, or ideas contained in the material herein.

Library of Congress Cataloging-in-Publication Data
A catalog record for this book is available from the Library of Congress

British Library Cataloguing-in-Publication Data
A catalogue record for this book is available from the British Library

ISBN: 978-0-12-820783-3

For information on all Elsevier publications
visit our website at https://www.elsevier.com/books-and-journals

Publisher: Matthew Deans
Acquisitions Editor: Simon Holt
Editorial Project Manager: Gabriela D. Capille
Production Project Manager: Kamesh Ramajogi
Cover Designer: Greg Harris

Typeset by SPi Global, India

Contents

Section III   Nano fabrication techniques—Chemical sensors

Section V   Intelligent nano sensors(INS)—Environmental applications

Section VI   Intelligent nano sensors (INS)—Electronics applications

25. Metal nanoparticles for electrochemical sensing applications   **589**

Shambhulinga Aralekallu and Lokesh Koodlur Sannegowda

Contributors

Lubna Aamir
Department of Physics, College of Science, University of Ha'il, Ha'il, Saudi Arabia

J. Ajayan
SR University, Warangal, India

B.M. Amrutha
Department of Chemistry, FMKMC College, Mangalore University Constituent College, Madikeri, Karnataka, India

Shambhulinga Aralekallu
Department of Studies in Chemistry, Vijayanagara Sri Krishnadevaraya University, Ballari, Karnataka, India

Francisco J. Arregui
Nanostructured Optical Devices Laboratory, Department of Electrical, Electronic and Communication Engineering; Institute of Smart Cities (ISC), Public University of Navarra, Pamplona, Spain

Mohsen Asadnia
School of Engineering, Macquarie University, Sydney, NSW, Australia

Jatinder Singh Aulakh
Department of Chemistry, Punjabi University, Patiala, Punjab, India

Arkaprava Bhattacharyya
SASTRA Deemed University, Thanjavur, India

N.P.G. Bhavani
Department of EEE, Meenakshi College of Engineering, Chennai, India

Rajib Biswas
Department of Physics, Tezpur University (Central University), Tezpur, Assam, India

M.M. Charithra
Department of Chemistry, FMKMC College, Mangalore University Constituent College, Madikeri, Karnataka, India

Papia Chowdhury
Department of Physics and Materials Science and Engineering, Jaypee Institute of Information Technology, Noida, Uttar Pradesh, India

Edwin S. D'Souza
Department of Chemistry, FMKMC College, Mangalore University Constituent College, Madikeri, Karnataka, India

Anita Dalal
Deenbandhu Chhotu Ram University of Science and Technology, Sonipat, India

Anand Darji
Department of Electronics Engineering, SardarVallabhbhai National Institute of Technology, Surat, Gujarat, India

B. Deepalakshmi
Department of ECE, Ramco Institute of Technology, Chennai, India

Himanshu Dehra
Monarchy of Concordia, Faridabad, Haryana, India

Sibasish Dutta
Department of Physics, Pandit Deendayal Upadhyaya Adarsha Mahavidyalaya (PDUAM) Eraligool, Karimganj, Assam, India

Marwa Farouk Elkady
Chemical and Petrochemical Engineering Department, Egypt-Japan University for Science and Technology, New Borg El-Arab City, New Borg El-Arab City; Fabrication Technology Researches Department, Advanced Technology and New Materials and Research Institute, City of Scientific Research and Technological Applications (SRTA-City), Alexandria, Egypt

Soheli Farhana
MIIT, University of Kuala Lumpur (UniKL), Kuala Lumpur, Malaysia

Amrita Ghosh
Department of Chemistry, National Institute of Technology (NIT) Kurukshetra, Kurukshetra, Haryana, India

Javier Goicoechea
Nanostructured Optical Devices Laboratory, Department of Electrical, Electronic and Communication Engineering; Institute of Smart Cities (ISC), Public University of Navarra, Pamplona, Spain

N. Hareesha
Department of Chemistry, FMKMC College, Mangalore University Constituent College, Madikeri, Karnataka, India

Hassan Shokry Hassan
Environmental Engineering Department, Egypt-Japan University for Science and Technology; Electronic Materials Researches Department, Advanced Technology and New Materials Researches Institute, City of Scientific Researches and Technological Applications (SRTA-City), New Borg El-Arab City, Egypt

Atanu Jana
Division of Physics and Semiconductor Science, Dongguk University, Seoul, South Korea

D. Amilan Jose
Department of Chemistry, National Institute of Technology (NIT) Kurukshetra, Kurukshetra, Haryana, India

Nitin Kale
NanoSniff Technologies Pvt. Ltd., IIT Bombay, Mumbai, India

J.M. Kalita
Department of Physics, Cotton University, Guwahati, India

Gauri Kalnoor
School of Engineering, Central University of Karnataka Kalaburagi, Kadaganchi, Karnataka, India

Anupreet Kaur
Basic and Applied Sciences, Punjabi University, Patiala, Punjab, India

Navneet Kaur
Department of Chemistry, Panjab University, Chandigarh, India

Ranjeet Kaur
Department of Chemistry, Panjab University, Chandigarh, India

Rahul Kaushik
Department of Chemistry, National Institute of Technology (NIT) Kurukshetra, Kurukshetra, Haryana, India

R. Krishnakumar
Department of EEE, Vels Institute of Science Technology and Advanced Studies, Chennai, India

J.G. Manjunatha
Department of Chemistry, FMKMC College, Mangalore University Constituent College, Madikeri, Karnataka, India

Mona Mittal
Department of Chemistry, Noida Institute of Engineering and Technology, Greater Noida, Uttar Pradesh, India

Hari Mohan
Centre for Medical Biotechnology, M.D. University, Rohtak, India

Subhas Chandra Mukhopadhyay
School of Engineering, Macquarie University, Sydney, NSW, Australia

Anindya Nag
DCI-CNAM Institute, Dongguan University of Technology, Dongguan, People's Republic of China

D. Nirmal
Karunya Institute of Technology and Sciences, Coimbatore, India

Ankit Kumar Pandey
Department of Applied Sciences (Physics Division), National Institute of Technology Delhi, New Delhi, India

Dimpi Paul
Department of Physics, Golaghat Engineering College, Golaghat, Assam, India

Adhithan Pon
SASTRA Deemed University, Thanjavur, India

Pooja
Department of Physics and Materials Science and Engineering, Jaypee Institute of Information Technology, Noida, Uttar Pradesh, India

Y.K. Prajapati
Department of Electronics and Communication Engineering, Motilal Nehru National Institute of Technology Allahabad, Prayagraj, Uttar Pradesh, India

Minakshi Prasad
Department of Animal Biotechnology, Lala Lagpat Rai University of Veterinary and Animal Sciences, Hisar, Haryana, India

N.S. Prinith
Department of Chemistry, FMKMC College, Mangalore University Constituent College, Madikeri, Karnataka, India

P.A. Pushpanjali
Department of Chemistry, FMKMC College, Mangalore University Constituent College, Madikeri, Karnataka, India

R. Ramesh
SASTRA Deemed University, Thanjavur, India

J.S. Rana
Deenbandhu Chhotu Ram University of Science and Technology, Sonipat, India

Meenakshi Rana
Department of Physics, School of Sciences, Uttarakhand Open University, Haldwani, Uttarakhand, India

Shweta Rana
Department of Chemistry, Panjab University, Chandigarh, India

C. Raril
Department of Chemistry, FMKMC College, Mangalore University Constituent College, Madikeri, Karnataka, India

Ravina
Centre for Medical Biotechnology, M.D. University, Rohtak, India

Pedro J. Rivero
Materials Engineering Laboratory, Department of Engineering; Institute of Advanced Materials (INAMAT), Public University of Navarra, Pamplona, Spain

Dinesh Rotake
Department of Electronics Engineering, SardarVallabhbhai National Institute of Technology, Surat, Gujarat, India

Lokesh Koodlur Sannegowda
Department of Studies in Chemistry, Vijayanagara Sri Krishnadevaraya University, Ballari, Karnataka, India

Samta Sapra
School of Engineering, Macquarie University, Sydney, NSW, Australia

Soumen Sardar
Department of Polymer Science and Technology, University of Calcutta, Calcutta, India

M.P. Sarma
Department of Physics, Cotton University, Guwahati, India

Nourwanda Mohamed Serour
Chemical and Petrochemical Engineering Department, Egypt-Japan University for Science and Technology, New Borg El-Arab City, New Borg El-Arab City, Egypt

Anuj Kumar Sharma
Department of Applied Sciences (Physics Division), National Institute of Technology Delhi, New Delhi, India

Rajni Sharma
Biosensor Technology Laboratory, Department of Biotechnology, Punjabi University, Patiala, India; School of Engineering, Macquarie University, Sydney, NSW, Australia

Ashish Kumar Singh
Biosensor Technology Laboratory, Department of Biotechnology, Punjabi University, Patiala, India

Akash Srivastava
Department of Electronics and Communication Engineering, Motilal Nehru National Institute of Technology Allahabad, Prayagraj, Uttar Pradesh, India

V. Srividhya
Department of EEE, Meenakshi College of Engineering, Chennai, India

K. Sujatha
Department of EEE, Dr. MGR Educational & Research Institute, Chennai, India

B.S. Surendra
Department of Chemistry, East West Institute of technology, Bengaluru, India

Girish Tigari
Department of Chemistry, FMKMC College, Mangalore University Constituent College, Madikeri, Karnataka, India

Aitor Urrutia
Nanostructured Optical Devices Laboratory, Department of Electrical, Electronic and Communication Engineering; Institute of Smart Cities (ISC), Public University of Navarra, Pamplona, Spain

Alka Verma
Department of Electronics Engineering, Institute of Engineering and Rural Technology, Allahabad, Uttar Pradesh, India

Neelam Verma
Division of Research and Development, Lovely Professional University, Phagwara; Biosensor Technology Laboratory, Department of Biotechnology, Punjabi University, Patiala, India

G. Wary
Department of Physics, Cotton University, Guwahati, India

Intelligent sensor—Nano candidates and their synthesis techniques

Blue phosphorene/two-dimensional material heterostructure: Properties and refractive index sensing perspectives

Ankit Kumar Pandey and Anuj Kumar Sharma
Department of Applied Sciences (Physics Division), National Institute of Technology Delhi, New Delhi, India

1. Introduction

Graphene's emergence led toward the exploration of other favorable two-dimensional (2D) materials for various optoelectronic applications. The quantum confinement effect makes 2D materials different from their bulk counterparts in terms of optical and electrical properties. Today, transition metal dichalcogenide (TMD) materials, e.g., MoS_2, $MoSe_2$, WS_2, WSe_2, $MoTe_2$ as well as their heterostructures, came into the picture for application in various devices such as phototransistors [1], solar cells [2], light-emitting diodes (LED), and sensors [3–5]. There are several techniques for synthesis of an atomically thin layer of TMDs such as liquid exfoliation method, wet chemical method, mechanical exfoliation method, chemical vapor deposition (CVD) method, etc. [6, 7]. Among the discussed synthesis techniques, the most promising is the CVD method due to its easiness. Apart from TMDs, other 2D materials such as monolayer blue phosphorene (BlueP), black phosphorus (BlackP), hexagonal boron nitride (h-BN), etc. emerged as potential materials [8]. Red, violet, black, and white are the different types of allotropes exhibited by phosphorus (P) in its bulk form [9]. BlackP possesses a puckered structure. Definite dislocation of P atoms can convert this puckered structure into a symmetric buckled structure form that is recognized as BlueP. BlueP possesses similar thermal stability as that of BlackP. Moreover, the heterostructure formation due to symmetric structure of BlueP is an advantage. Both BlueP and BlackP have intrinsic carrier mobility of the order of 10^3 cm^2/V s [10, 11]. This high carrier mobility characteristic makes them favorable candidates for application in various optoelectronic devices. However, BlueP and BlackP suffer from oxidation issues under ambient conditions, which significantly hamper their applicability in various devices. This problem can be eradicated by 2D heterostructure formation. Fig. 1 shows the potential applications of 2D heterostructure in various fields. BlueP/2D material-based heterostructures such as BlueP/TMDs [12], BlueP/graphene [13, 14], BlueP/h-BN [15], etc. have witnessed a lot of application in a few years [7]. The built-in electric field (in 2D layers) is responsible for enhanced

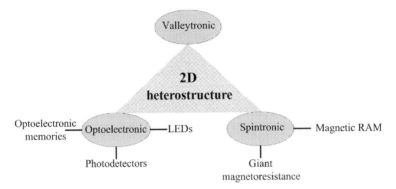

Fig. 1 Applications of 2D heterostructures in various fields.

performance in photodetectors [16]. A multibit nonvolatile optoelectronic memory was fabricated by Xiang et al. [17] based on WSe_2/h-BN heterostructure. Extremely large magnetoresistance was reported by Gopinadhan et al. in few-layer graphene/BN hetero-structure [18]. Recently, BlueP/TMD heterostructures have shown their application for performance enhancement in surface plasmon resonance (SPR)-based sensors [19, 20]. The SPR-based sensing probes are preferred over other optical sensing-based techniques owing to their precision, reliable measurement procedure, and high sensitivity behavior [21, 22]. The resonant oscillations of p-polarized (TM) waves at metal-dielectric inter-faces are commonly referred to as SPR. The inclusion of a BlueP/TMD heterostructure as sensing material is able to provide enhanced performance parameters than that of grapheme-based SPR sensors. The present chapter focuses on the properties and appli-cation of BlueP/TMD heterostructure as a sensing material in SPR-based sensors.

2. BlueP/TMDs and BlueP/2D material heterostructures

The BlueP possess a large indirect band gap of ∼2 eV [12], which is disadvantageous from an optoelectronic application point of view. On the contrary, monolayer TMDs exhibit a direct band gap nature. The direct band gap of WS_2, WSe_2, MoS_2, and $MoSe_2$ are 1.89 eV, 1.63 eV, 1.77 eV, and 1.50 eV, respectively. The formation of vdW hetero-structure causes a reduction in overall band gap that is smaller than the corresponding monolayer TMD. However, the resulting vdW heterostructures exhibit an indirect band gap nature. The transition from indirect to direct band gap can be achieved on application of external electric fields as reported by Zhang et al. in the case of the BlueP/MoS_2 het-erostructure [23]. Moreover, this transition was confirmed by Peng et al. based on density functional theory for other BlueP/TMDs heterostructures [12].

The hexagonal lattice structures of BlueP and TMDs provide an easy formation of 2D heterostructure (BlueP/TMD). For example, the lattice constant of BlueP, $MoSe_2$, and

WS$_2$ are 3.268 Å, 3.295 Å, and 3.165 Å, respectively. The corresponding lattice mismatch for BlueP/TMDs heterostructure is -0.82% and $+3.15\%$, respectively. Now, it is important to address the stability and confirmation of vdW for the BlueP/TMDs heterostructures. The information regarding formation energy (E_{form}) is required to examine the BlueP/TMDs heterostructure's stability. The E_{form} can be evaluated by the following relation [24]:

$$E_{form} = E_{total}^{BlueP/TMDs} - E_{total}^{BlueP} - E_{total}^{TMDs} \qquad (1)$$

Here, $E_{total}^{BlueP/TMDs}$, E_{total}^{BlueP}, and E_{total}^{TMDs} are the total energies in one unit cell corresponding to BlueP/TMDs, BlueP, and TMDs, respectively. Negative values of E_{form} are obtained [12] for different BlueP/TMDs heterostructures, which confirm their thermodynamic stability and experimental realizability. Furthermore, binding energy ($E_{binding}$) provides important information regarding the role of vdW force of attraction in a heterostructure. The vdW force strength in BlueP/TMDs heterostructures can be evaluated based on the binding energy considering the total energy of mutually independent monolayer BlueP and TMDs ($E_{total}^{BlueP+TMD}$) in a heterostructure as follows:

$$E_{binding} = E_{total}^{BlueP/TMDs} - E_{total}^{BlueP+TMD} \qquad (2)$$

The obtained values of $E_{binding}$ for different BlueP/TMDs heterostructures are consistent with a typical $E_{binding}$ value (20 meV/Å^2) as reported by Bjorkman et al. in a layered structure [25]. This provides confirmation about the existence of BlueP/TMDs as vdW heterostructures. Thus, it can be inferred that the individual monolayers of BlueP and TMD are strongly attached by a strong vdW force of attraction. Furthermore, the in–plane stability of the heterostructure system is provided by the strong covalent bond. Fig. 2 shows the schematic representation of BlueP/TMDs vdW heterostructures with different rotation angles of the BlueP monolayer. Here, Mo (W), S (Se), and BlueP atoms are indicated by red (dark gray in print version), green (light gray in print version), and blue (gray in print version) balls, respectively [12].

The electronic and structural properties of other BlueP/2D material-based heterostructures such as BlueP/Black phosphorus (BlueP/BP) [26], BlueP/AlN [27], BlueP/ graphene (BlueP/G), and BlueP/graphene-like gallium nitride (BlueP/g-GaN) [13, 14] have fascinated researchers. Recently, h-BN/BlueP heterostructure has been reported for application in Li/Na ion batteries [15]. Stronger adsorption energy was demonstrated in h-BN/BlueP heterostructure based on first principle computation. Li et al. proposed BlueP/G heterostructure as an anode material for Li-ion batteries [13]. Furthermore, diffusion barrier of Li on BlueP/G results in a room-temperature diffusivity of 2.61×10^{-5} cm^2/s, which is faster than that of pristine graphene (Li on G).

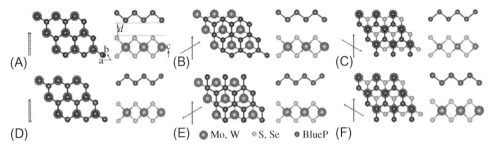

Fig. 2 Schematic representation of the BlueP/TMDs vdW heterostructures with rotation angles of the BlueP monolayer (with respect to TMDs) are (A) 0 degrees, (B) 60 degrees, (C) 120 degrees, (D) 180 degrees, (E) 240 degrees, and (F) 300 degrees. *(From Q. Peng, Z. Wang, B. Sa, B. Wu, Z. Sun, Electronic structures and enhanced optical properties of blue phosphorene/transition metal dichalcogenides van der Waals heterostructures, Sci. Rep. 6 (May) (2016) 31994. Licensed under the Creative Commons Attribution 4.0 International License.)*

3. 2D heterostructure: Promising material for SPR-based sensing application

SPR-based sensors have a large number of applications in biosensing fields such as detection of enzymes, proteins, drugs, etc. as well as in the measurement of physical quantities such as temperature, humidity, etc. [28, 29]. They also play an important role in gas detection [28], which is a crucial ingredient for overall pollution monitoring. In this section, the performance enhancement of SPR sensors is presented utilizing 2D heterostructure of BlueP/MoSe$_2$ as an analyte interacting layer. There are two advantages of having a 2D heterostructure in direct contact with an analyte: (a) increased adsorption due to strong vdW force of attraction and (b) enhancement of the field at the interface [30]. The performance is evaluated based on angular interrogation technique in terms of angular sensitivity at a wavelength of 662 nm. Effect of the number of BlueP/MoSe$_2$ layers with MgF$_2$ interlayer on sensitivity is also demonstrated.

The sensor is based on Kretschmann configuration as shown in Fig. 3. FWHM (full width at half maximum) is the width of the SPR curve. The reflectance (R) of incident light is numerically calculated using transfer matrix [31] based on an angular interrogation method. In this method, the wavelength of p-polarized (TM) incident light is fixed (here, $\lambda = 662$ nm) with variation in an incident angle. The calculations are performed on MATLAB$^{\circledR}$ platform. Coupling of incident light and SPs at metal-dielectric interface can be represented by following Eq. [32].

$$\frac{\omega}{c} n_p \sin \theta_{\mathrm{SPR}} = \mathrm{real}\left(\frac{\omega}{c} \sqrt{\frac{\epsilon_m \epsilon_s}{\epsilon_m + \epsilon_s}} \right) \tag{3}$$

where n_p is RI of the substrate medium (here, prism), ω is the angular frequency of the incident light, c is the velocity of light in vacuum, and ε_m and ε_s are the dielectric constant

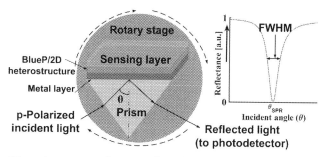

Fig. 3 Schematic of Kretschamann configuration-based SPR sensor with SPR curve. FWHM is the width of the SPR curve.

Table 1 Dielectric constant values at $\lambda = 662$ nm.

Material	Dielectric constant
BaF$_2$ [33]	2.1688
MgF$_2$ [33]	1.8950
Silver [34]	$-20.272 + 0.44527i$
Graphene (0.34 nm) [35]	$5.6503 + 7.6695i$
BlueP/MoSe$_2$ (0.75 nm) [12]	$7.3641 + 1.7097i$

of the metal and sensing (analyte) layer, respectively. θ_{SPR} is the resonance angle at which SPR curve dip is obtained. A small change in the analyte's property (i.e., ε_s) causes a shift in the angular position of the SPR dip.

The layer stacking is considered in z-direction. Fluoride glass prism, i.e., BaF$_2$ (first layer) is considered as the coupling layer. The second layer is silver (Ag) as SPR active metal. The third layer is 2D heterostructure (BlueP/MoSe$_2$), which is in direct contact with the sensing layer. This heterostructure will also prevent Ag from possible oxidation concerns. The dielectric constant values used in this study are shown in Table 1.

The minimum reflectance (R_{min}) at θ_{SPR} can be obtained only by thickness optimization of the metal layer. Fig. 4A shows the thickness optimization of Ag layer for the proposed scheme considering analyte layer refractive index (RI, i.e., n_s) of 1.33 and monolayer BlueP/MoSe$_2$. The minimum R value is obtained for Ag thickness of 53 nm. The optimized Ag layer thicknesses of 58 nm and 52 nm are considered for conventional (Ag over prism) and graphene based (prism/Ag/graphene) SPR sensors, respectively. Fig. 4B shows the variation of R with incident angle for analyte RI ranging from 1.32 to 1.37 (for monolayer BlueP/MoSe$_2$ over Ag) at incident wavelength of 662 nm based on the optimized Ag layer thickness. Thus, the proposed scheme is able to provide sensing possibilities for a wide range of analyte. The corresponding θ_{SPR}

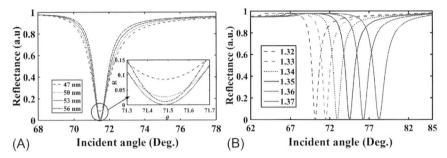

Fig. 4 (A) Variation of reflectance with incident angle for different thicknesses of Ag layer over BaF$_2$ with monolayer BlueP/MoSe$_2$ (BaF$_2$/Ag/BlueP/MoSe$_2$) considering analyte refractive index of 1.33. (B) Variation of reflectance with incident angle for different analyte refractive index (1.32 − 1.37) for monolayer BlueP/MoSe$_2$ heterostructure over Ag at $\lambda = 662$ nm.

are found to be 70.1358 degrees, 71.4822 degrees, 72.9318 degrees, 74.5132 degrees, 76.2607 degrees, and 78.2546 degrees for analyte RI of 1.32, 1.33, 1.34, 1.35, 1.36, and 1.37, respectively. The presented extent of θ_{SPR} shift is governed by the coupling equation (Eq. 3). Thus, a significant shift in the resonance angle ($\Delta\theta_{\mathrm{SPR}} = 8.1188$ degrees) is obtained for a small change in analyte RI ($\Delta n_s = 1.37 - 1.32 = 0.05$).

The sensitivity (S) in the case of angular interrogation system for a SPR–based sensor is defined as: $S = \Delta\theta_{\mathrm{SPR}}/\Delta n_s$ [36], where the change in resonance angle is $\Delta\theta_{\mathrm{SPR}}$ for a small change in RI (Δn_s) of sensing layer. Considering $n_s = 1.33$ and $\Delta n_s = 0.005$, the calculated S value for the proposed sensor is 142.10 degrees/RIU, which is higher than that of a conventional Ag-based SPR sensor (138.66 degrees/RIU).

The simultaneous effect of layer consisting of Ag and monolayer 2D material (thickness $\ll$ incident light wavelength) can be demonstrated by calculating the effective optical constant as follows:

$$n_{\mathrm{eff}} = \sqrt{\left(d_m n_m^2 + d_{2\mathrm{D}} n_{2\mathrm{D}}^2\right)/d} \tag{4}$$

$$k_{\mathrm{eff}} = \sqrt{\left(d_m k_m^2 + d_{2\mathrm{D}} k_{2\mathrm{D}}^2\right)/d} \tag{5}$$

Here, n_{eff} and k_{eff} are the real and imaginary parts, respectively. d_m, $d_{2\mathrm{D}}$, and ($d = d_m + d_{2\mathrm{D}}$) are the thickness of Ag, 2D material, and combined bilayer, respectively. Furthermore, k_m and n_m are the imaginary and real parts of RI, respectively, for the Ag layer. For 2D monolayer, $k_{2\mathrm{D}}$ and $n_{2\mathrm{D}}$ are the imaginary and real parts of RI, respectively. The effective dielectric constant value of bilayer, $\varepsilon_{\mathrm{eff}} = (n_{\mathrm{eff}} + ik_{\mathrm{eff}})^2$. ε_m will be replaced by $\varepsilon_{\mathrm{eff}}$ in SPR coupling equation (Eq. 3). Thus, a larger shift in θ_{SPR} will be provided by a multilayer system than a conventional scheme on small changes in RI of sensing layer.

Further analysis is done on the effect of prism RI on the sensitivity. Fig. 5 shows the sensitivity variation with analyte RI for three different coupling glass materials. It is evident that a BaF_2-based scheme is able to provide highest sensitivity for a wide range of analyte RI than other prisms that are widely used in SPR-based sensors. Thus, the lower the RI of prism material, the higher the obtained sensitivity [37]. However, the FWHM value will be lowest for high RI prism material.

Multilayered BlueP/$MoSe_2$ heterostructure can have slight different dielectric constant values from the corresponding monolayer. To study the effect of stacking of 2D heterostructure layers, introduction of dielectric interlayer is another approach that can be employed [37]. In the present work, a dielectric interlayer (MgF_2) is considered between two BlueP/$MoSe_2$ layers for a multilayer scheme. Fig. 6 shows the variation of R with incident angle for a different number of heterostructures with dielectric interlayer.

It is observed that there is shift in θ_{SPR} value and a decrease in R_{min} values on increasing the number of layers on Ag. Calculated value of FWHM for monolayer heterostructure is obtained as 1.0829 degrees, which increases up to 2.074 degrees for four-layer heterostructure with dielectric interlayer over Ag. This is due to a change in the effective index value of a multilayer structure and damping effect of SPs [38].

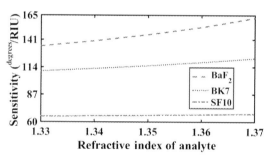

Fig. 5 Variation of reflectance with incident angle for different analyte refractive index (1.32 − 1.37) for monolayer BlueP/$MoSe_2$ heterostructure over Ag at $\lambda = 662$ nm.

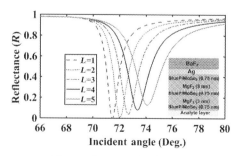

Fig. 6 Variation of reflectance with incident angle for a different number of BlueP/$MoSe_2$ heterostructures with MgF_2 interlayer at $\lambda = 662$ nm for $n_s = 1.33$.

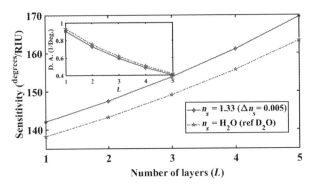

Fig. 7 Variation of sensitivity with a different number of BlueP/MoSe$_2$ heterostructure layers over Ag (with dielectric interlayer) at $\lambda = 662$ nm for two cases; H$_2$O (D$_2$O as a reference sample) as an analyte and $n_s = 1.33$, $\Delta n_s = 0.005$. Inset shows the variation of detection accuracy with a number of 2D material layers.

Sensitivity calculation for BlueP/MoSe$_2$-based schemes is made for $n_s = 1.33$ and $\Delta n_s = 0.005$ with different number of layers (with dielectric interlayer) over Ag. A sensitivity of 169.59 degrees/RIU is calculated for five-layer BlueP/MoSe$_2$ heterostructure over Ag as shown in Fig. 7 (solid line). From one- to five-layer heterostructure over Ag, the change in sensitivity is approximately 27.50 degrees/RIU. However, detection accuracy (D.A.) decreases when increasing the number of layers as shown in inset of Fig. 7. D.A. can be defined as the reciprocal of FWHM (D.A. = 1/FWHM) that provides the information regarding the preciseness of the sensor scheme. Considering a practical scenario, calculation is also performed taking H$_2$O (RI = 1.3311 at 662 nm) as an analyte and D$_2$O (RI 1.3264 at 662 nm) as a reference sample. In a previous work, it was reported that the change in sensitivity on increasing the heterostructure layer was significantly greater than increasing graphene layers. Further increase in the number of layers will cause high damping of SPs, and performance can be badly affected. Thus, optimization in the number of layers is of importance.

The effect of incident wavelength on the sensitivity and D.A. behavior of 2D heterostructure (BlueP/MoS$_2$)-based SPR sensor was demonstrated by Sharma and Pandey [19]. Fig. 8 shows the variation of sensitivity and D.A. with incident light wavelength (the figure is simulated based on the data values provided in Ref. [19]) chosen based on the requirement of sensitivity and allowable D.A. that can be compromised.

It was reported that sensitivity increases with L, which is more prominent at shorter wavelengths. The number of 2D heterostructure layers has less effect on sensitivity enhancement at longer wavelengths. Furthermore, FWHM value reduces (i.e., an increase in D.A.) when there is increasing wavelength because of a reduction in the conductive losses of metals at longer wavelengths. The damping nature of SPs in a multilayered dielectric stack is responsible for an increase in FWHM for $L > 1$. Thus, this

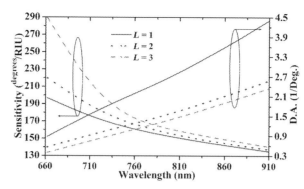

Fig. 8 Variation of sensitivity and detection accuracy with wavelength for different number of BlueP/MoS$_2$ heterostructure and graphene layers (L) over Ag : $n_s = 1.33$, $\Delta n_s = 0.005$.

Table 2 Sensitivity of some of the recently reported works based on TMD heterostructures.

Modalities	Sensitivity (degrees/RIU) ($n_s = 1.33$, $\Delta n_s = 0.005$)	λ (nm)	Ref.
BK7 + Ag + Black phosphorous + WSe$_2$	279	633	Wu et al. [30]
SF6 + Au + Black phosphorous + WSe$_2$	187.22	633	Meshginqalam and Barvestani [39]
CaF$_2$ + Ag + BlueP/MoS$_2$	432.15	662	Sharma and Pandey [19]
BK7 + Au + BlueP/MoS$_2$ + graphene	204	633	Prajapati and Srivastava [40]
BK7 + Au + Si + BlueP/MoS$_2$	230.66	632.80	Srivastava and Prajapati [41]
CaF$_2$ + Ag + black phosphorous + WS$_2$	375	633	Hasib et al. [42]

trade-off between sensitivity and D.A. should be balanced to achieve a significant overall performance. The set (L, λ) can be chosen based on required sensitivity and allowable D.A. that can be compromised. Several other works have been reported based on 2D heterostructure for sensitivity enhancement. Some of them are presented in Table 2 with maximum achieved sensitivity values.

4. Conclusion and future scope

The development in controllable CVD growth realization of layered heterostructures, and the effective modulation, various device designs, as well as unique properties of spatially separated excitons and magnetism, are some of the noteworthy advancements that have been accomplished to date in the study of 2D heterostructures. However, there are some challenges in synthesis of 2D heterostructures that require development of new controllable and scalable fabrication techniques. There is a need to explore other 2D magnetic materials that can be incorporated with BlueP enabling some new 2D heterostructures with advanced magnetic properties. The application of 2D heterostructure in SPR-based sensors is presented in this chapter along with some of their important properties. BlueP/MoSe$_2$ heterostructure-based scheme is able to provide highest sensitivity compared with conventional and graphene-based SPR sensor schemes. Apart from high sensitivity behavior, reasonable detection accuracy is also obtained. There is a lot of scope in designing 2D heterostructure-based devices. The 2D heterostructure provide stability and tunability of mechanical/optical properties exhibited by individual 2D monolayers. These properties will provide a new path to the development of 2D heterostructure-based SPR sensors (prism and fiber configurations), photodetectors, next-generation optoelectronic memories with circuit level fabrication, and integration of 2D thin-layered image sensors and other scalable applications.

Acknowledgment

The financial support from the Council of Scientific and Industrial Research (CSIR), India, is gratefully acknowledged by Dr. A.K. Sharma for research and development project grant [No. 03(1441)/18/EMR II] and Dr. A.K. Pandey for Research Associateship (RA).

References

[1] N. Huo, G. Konstantatos, Ultrasensitive all-2D MoS2 phototransistors enabled by an out-of-plane MoS2 PN homojunction, Nat. Commun. 8 (1) (2017) 1–6.

[2] M.M. Furchi, A.A. Zechmeister, F. Hoeller, S. Wachter, A. Pospischil, T. Mueller, Photovoltaics in van der Waals heterostructures, IEEE J. Sel. Top. Quantum Electron. 23 (1) (2017) 106–116.

[3] L. Wu, J. Guo, X. Dai, Y. Xiang, D. Fan, Sensitivity enhanced by MoS2-graphene hybrid structure in guided-wave surface plasmon resonance biosensor, Plasmonics (2017) 1–5.

[4] Y. Xiang, J. Zhu, L. Wu, Q. You, B. Ruan, X. Dai, Highly sensitive terahertz gas sensor based on surface plasmon resonance with graphene, IEEE Photonics J. 10 (1) (2018) 1–7.

[5] W. Li, et al., Gas sensors based on mechanically exfoliated MoS2 nanosheets for room-temperature NO2 detection, Sensors (Basel) 19 (9) (2019) 1–12.

[6] Y. Liu, S. Zhang, J. He, Z.M. Wang, Z. Liu, Recent progress in the fabrication, properties, and devices of heterostructures based on 2D materials, Nano-Micro Lett. 11 (1) (2019).

[7] N. Huo, Y. Yang, J. Li, Optoelectronics based on 2D TMDs and heterostructures, J. Semicond. 38 (3) (2017).

[8] K.S. Novoselov, A. Mishchenko, A. Carvalho, A.H.C. Neto, 2D materials and van der Waals heterostructures, Research 353 (6298) (2016).

[9] R. Gusmão, Z. Sofer, M. Pumera, Black phosphorus rediscovered: from bulk material to monolayers, Angew. Chem. Int. Ed. 56 (28) (2017) 8052–8072.

[10] H. Liu, et al., Phosphorene: an unexplored 2D semiconductor with a high hole mobility, ACS Nano 8 (4) (2014) 4033–4041.

[11] J. Xiao, M. Long, X. Zhang, J. Ouyang, H. Xu, Y. Gao, Theoretical predictions on the electronic structure and charge carrier mobility in 2D Phosphorus sheets, Sci. Rep. 5 (June) (2015) 1–10.

[12] Q. Peng, Z. Wang, B. Sa, B. Wu, Z. Sun, Electronic structures and enhanced optical properties of blue phosphorene/transition metal dichalcogenides van der Waals heterostructures, Sci. Rep. 6 (May) (2016) 31994.

[13] Y. Li, W. Wu, F. Ma, Blue phosphorene/graphene heterostructure as a promising anode for lithium-ion batteries: a first-principles study with vibrational analysis techniques, J. Mater. Chem. A 7 (2) (2019) 611–620.

[14] M. Sun, J.P. Chou, J. Yu, W. Tang, Electronic properties of blue phosphorene/graphene and blue phosphorene/graphene-like gallium nitride heterostructures, Phys. Chem. Chem. Phys. 19 (26) (2017) 17324–17330.

[15] J. Bao, et al., Hexagonal boron nitride/blue phosphorene heterostructure as a promising anode material for Li/Na-ion batteries, J. Phys. Chem. C 122 (41) (2018) 23329–23335.

[16] X. Wei, F.G. Yan, C. Shen, Q.S. Lv, K.Y. Wang, Photodetectors based on junctions of two-dimensional transition metal dichalcogenides, Chin. Phys. B 26 (3) (2017).

[17] D. Xiang, et al., Two-dimensional multibit optoelectronic memory with broadband spectrum distinction, Nat. Commun. 9 (1) (2018) 1–8.

[18] K. Gopinadhan, et al., Extremely large magnetoresistance in few-layer graphene/boron-nitride heterostructures, Nat. Commun. 6 (2015) 1–7.

[19] A.K. Sharma, A.K. Pandey, Blue phosphorene/MoS2 heterostructure based SPR sensor with enhanced sensitivity, IEEE Photon. Technol. Lett. 30 (7) (2018) 595–598.

[20] A.K. Sharma, A.K. Pandey, B. Kaur, Simulation study on comprehensive sensing enhancement of BlueP/MoS 2- and BlueP/WS 2-based fluoride fiber surface plasmon resonance sensors: analysis founded on damping, field, and optical power, Appl. Opt. 58 (16) (2019) 4518.

[21] A.K. Pandey, A.K. Sharma, R. Basu, Fluoride glass based surface plasmon resonance sensor in infrared region: performance evaluation, J. Phys. D Appl. Phys. 50 (18) (2017) 185103 (1–6).

[22] A.K. Sharma, A.K. Pandey, Self-referenced plasmonic sensor with TiO2 grating on thin Au layer: simulated performance analysis in optical communication band, J. Opt. Soc. Am. B 36 (8) (2019) F25.

[23] Z.Y. Zhang, M.S. Si, S.L. Peng, F. Zhang, Y.H. Wang, D.S. Xue, Bandgap engineering in van der Waals heterostructures of blue phosphorene and MoS2: a first principles calculation, J. Solid State Chem. 231 (2015) 64–69.

[24] J. Liao, B. Sa, J. Zhou, R. Ahuja, Z. Sun, Design of high-efficiency visible-light photocatalysts for water splitting: MoS2/AlN(GaN) heterostructures, J. Phys. Chem. C 118 (31) (2014) 17594–17599.

[25] T. Björkman, A. Gulans, A.V. Krasheninnikov, R.M. Nieminen, Van der Waals bonding in layered compounds from advanced density-functional first-principles calculations, Phys. Rev. Lett. 108 (23) (2012) 1–5.

[26] L. Huang, J. Li, Tunable electronic structure of black phosphorus/blue phosphorus van der Waals p-n heterostructure, Appl. Phys. Lett 108 (8) (2016).

[27] Q. Yang, et al., AlN/BP heterostructure photocatalyst for water splitting, IEEE Electron Device Lett. 38 (1) (2017) 145–148.

[28] S. Weng, L. Pei, C. Liu, J. Wang, J. Li, T. Ning, Double-side polished fiber SPR sensor for simultaneous temperature and refractive index measurement, IEEE Photon. Technol. Lett. 28 (18) (2016) 1916–1919.

[29] A.K. Sharma, A. Gupta, Chemical design of a plasmonic optical sensor probe for humidity-monitoring, Sens. Actuators B 188 (2013) 867–871.

[30] L. Wu, et al., Sensitivity enhancement by using few-layer black phosphorus-graphene/TMDCs heterostructure in surface plasmon resonance biochemical sensor, Sens. Actuators B 249 (2017) 542–548.

[31] Q. Ouyang, et al., Sensitivity enhancement of transition metal dichalcogenides/silicon nanostructure-based surface plasmon resonance biosensor, Sci. Rep. 6 (March) (2016) 1–13.

[32] H. Raether, Surface Plasmons on Smooth and Rough Surfaces and on Gratings, Vol. 111, Springer-Verlag, Berlin, Heidelberg, New York, 1988.

[33] H.H. Li, Refractive index of alkaline earth halides and its wavelength and temperature derivatives, J. Phys. Chem. Ref. Data 9 (161) (1980).

[34] P.B. Johnson, R.W. Christy, Optical constants of the noble metals, Phys. Rev. B 6 (12) (1972) 4370–4379.

[35] J.W. Weber, V.E. Calado, M.C.M. van de Sanden, Optical constants of graphene measured by spectroscopic ellipsometry, Appl. Phys. Lett. 97 (9) (2010) 130–132.

[36] S.A. Zynio, A.V. Samoylov, E.R. Surovtseva, V.M. Mirsky, Y.M. Shirshov, Bimetallic layers increase sensitivity of affinity sensors based on surface plasmon resonance, Sensors 2 (2002) 62–70.

[37] A.K. Pandey, A.K. Sharma, Simulation and analysis of plasmonic sensor in NIR with fluoride glass and graphene layer, Photonics Nanostruct. Fundam. Appl. 28 (2018) 94–99.

[38] L. Wu, H.S. Chu, W.S. Koh, E.P. Li, Highly sensitive graphene biosensors based on surface plasmon resonance, Opt. Express 18 (14) (2010) 14395.

[39] B. Meshginqalam, J. Barvestani, Performance enhancement of SPR biosensor based on phosphorene and transition metal dichalcogenides for sensing DNA hybridization, IEEE Sensors J. 18 (18) (2018) 7537–7543.

[40] Y.K. Prajapati, A. Srivastava, Effect of BlueP/MoS 2 heterostructure and graphene layer on the performance parameter of SPR sensor: theoretical insight, Superlattice. Microst. 129 (January) (2019) 152–162.

[41] A. Srivastava, Y.K. Prajapati, Performance analysis of silicon and blue phosphorene/MoS2 heterostructure based SPR sensor, Photonic Sens. 9 (3) (2019) 284–292.

[42] M.H.H. Hasib, J.N. Nur, C. Rizal, K.N. Shushama, Improved transition metal dichalcogenides-based surface plasmon resonance biosensors, Condens. Matter 4 (2) (2019) 49.

CHAPTER 2

Low-cost photoresponsive ITO/Ag-WO$_3$/Ag Schottky diode

Lubna Aamir
Department of Physics, College of Science, University of Ha'il, Ha'il, Saudi Arabia

1. Introduction

Tungsten oxide (WO$_3$) is a well-studied oxide semiconductor used for several applications such as a chromogenic material, sensor, and catalyst. The energy band gap of WO$_3$ corresponds to the difference between the energy levels of the valence band formed by the filled O 2p orbitals and the conduction band formed by empty W 5d orbitals, ranging from 2.6 to 3.25 eV [1]. The major important features of WO$_3$ are that it could be easily synthesized using low-cost chemical synthesis techniques, availability, good stability, easy morphological and structural control of the nanostructures, and high sensitivity [2–4]. It also possesses good electron-transport property and stability, which are important for sensing applications [5, 6], although the high rate of electron-hole recombination in oxide semiconductors such as WO$_3$ prevent their good performance in optoelectronic devices [7, 8]. To overcome this problem, nanosized metal, such as silver (Ag), gold (Au), or titanium (Ti) has been introduced in the oxide semiconductor's matrix (WO$_3$) [9–11]. The noble metals act as a sink for free charge carriers and promote interfacial charge-transfer processes, which promote charge separation. In the present research, Ag is taken as a noble metal to form Ag-WO$_3$ nanocomposite, which is already known to possess excellent gas-sensing performance along with other advantages over WO$_3$ [12].

Currently, such metal oxide-based semiconductor nanomaterials have been theoretically and experimentally studied and demonstrated their potential for electronic applications [13–15]. A Schottky diode is likely the simplest, cheapest, and most versatile diode that can be fabricated using chemically processed semiconductor nanoparticles. Also, they operate at high frequencies and produce less unwanted noise than the P-N junction diode. They are widely used as a rectifier, in radiofrequency applications, power supplies, logic circuits, and detectors.

This chapter presents the fabrication of a low-cost ITO/Ag-WO$_3$/Ag Schottky diode using a chemically synthesized p-type Ag-WO$_3$ semiconductor characterized for its current-voltage and capacitance-voltage characteristics. Mott-Schottky analysis probes the depletion capacitance at a Schottky junction and also determined the carrier

Handbook of Nanomaterials for Sensing Applications
https://doi.org/10.1016/B978-0-12-820783-3.00024-5

concentration, built-in potential, depletion width, barrier height, and maximum electric field at the Schottky junction. Indeed, the I-V characteristics of the diode is nonlinear with turn-on voltage 162 mV, and current conduction follows thermionic emission theory. The fabricated Schottky diode is also found to possess a strong photoresponsive nature and thus could be used in photodetection. Impedance spectra of the diode is also analyzed to produce the equivalent circuit of the Schottky diode. Also, the role of chemically synthesized Ag-WO$_3$ is investigated as a novel nanocomposite semiconductor candidate for solid-state device fabrication.

This chapter will be beneficial to researchers working in the field of solid-state devices such as detectors, sensors, diodes, etc. and looking for new materials and strategies to fabricate low-cost heterojunction diodes for fast switching, high frequency, detection, and sensing applications. At the end of this chapter, readers will be able to chemically synthesize the Ag-WO$_3$ nanocomposite, fabricate low-cost Schottky diodes, and be able to analyze diodes through I-V, C-V, and impedance spectroscopy. The author suggests for further analysis of a fabricated Schottky diode as a gas sensor, which would produce exciting results in the field of gas detectors or sensors.

2. Experimental

2.1 Synthesis of Ag-WO$_3$ semiconductor nanocomposite

Initially, silver nanoparticles are synthesized when 0.25 mM of silver nitrate (99% assay) is reduced by 0.1 M sodium borohydride (98.0% assay) in the presence of 0.25 mM sodium tricitrate (99.0% assay) as a surfactant. The formation of a yellow color colloid indicates the formation of Ag nanoparticles. These synthesized nanoparticles are then extracted after washing with deionized water several times and redispersed in deionized water.

Then, 34.4 nM aqueous solution of sodium tungstate (99.0% assay) under the presence of 10% polyethylene glycol (99% assay) as a stabilizer is mixed with 2 mL Ag nanoparticle dispersion and left to stir for 2 h; after this, 5 mL concentrated HCl is added and kept at 250°C until yellow precipitate of Ag-WO$_3$ nanocomposite is formed. The precipitate of Ag-WO$_3$ nanocomposite is then washed several times in deionized water to remove any unreacted HCl and by-products. The powder is then annealed at 450°C for 30 min.

2.2 Schottky junction fabrication

ITO conducting glass with resistivity 8 Ω/cm^2 is used as a transparent electrode to judge the photoresponsive nature of the fabricated Schottky junction. Before the film formation, the substrate is cleaned with acetone and 10% NaOH solution. Ag-WO$_3$ nanocomposite fine powder is dispersed in water using a sonicator to get a smooth homogeneous Ag-WO$_3$ dispersion. This dispersion is then used to form the thin film of Ag-WO$_3$

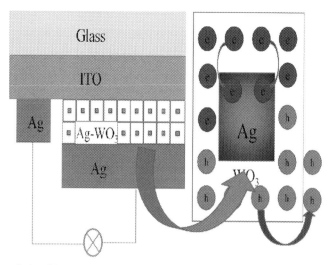

Fig. 1 Schematic of the fabricated ITO/Ag-WO₃/Ag Schottky diode indicating the mechanism of charge transfer between silver nanoparticles and WO₃ semiconductor nanoparticles; electrons will flow from Ag to WO₃.

nanocomposite over the ITO. A thin film with thickness 0.6 μm and area 1 cm² is formed using the dip-coating technique with dip rate 9 mm/s, dip time 20 s, and ret. rate 1 mm/s under consecutive drying for 99 s at 50°C. Then a silver electrode is evaporated over the Ag-WO₃ film to form the Schottky junction. The schematic of the device and mechanism of charge separation is shown in Fig. 1.

3. Characterization

Structural characterization is performed using the Rigaku Mini Flex X-ray diffractometer with Cu Kα irradiation of wavelength 1.54 Å. Microscopic characterization is performed using TEM, JEOL/EO version 1.0 and AFM, and NT-MDT Scanning Probe Microscope SOLVER NEXT. The electrical characterization of the device is performed using the I-V characterization system consisting of Keithley 6517B electrometer and Keithley 2100 6½ multimeter. C-V and Impedance analysis is performed using the Wayne Kerr 6500 B impedance analyzer. Illumination is done using a 500 W halogen lamp.

4. Results and discussion

4.1 Crystallographic and microscopic analysis

XRD pattern of synthesized Ag-WO₃ nanocomposite is given in Fig. 2A, which exhibits a crystalline nature. The XRD pattern of Ag-WO₃ nanocomposite shows the presence of elemental cubic silver (JCPDS No. 03-0931 and 01-1164) and monoclinic WO₃

(JCPDS No. 24-0747) as indexed in Fig. 2A. This indicates the formation of Ag-WO$_3$ nanocomposites. The morphologies of the obtained samples are further characterized by TEM and AFM. The TEM image is shown in Fig. 2B, which clearly confirms that Ag-WO$_3$ nanocomposite is a 3D cluster consisting of randomly oriented WO$_3$ nanoplates of lengths 100–700 nm and thickness of the order of 20–60 nm, and silver nanoparticles are attached to it either at the center or at the edge, as reported earlier [12]. AFM image of Ag-WO$_3$ nanocomposite in Fig. 2C and D also go well with the TEM result.

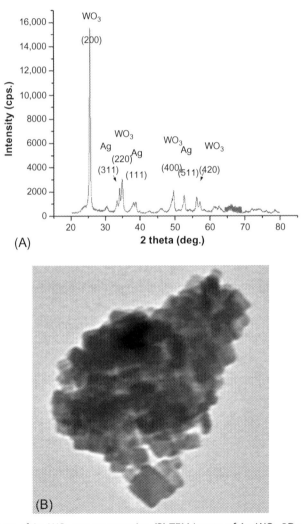

(A)

(B)

Fig. 2 (A) XRD pattern of Ag-WO$_3$ nanocomposite, (B) TEM image of Ag-WO$_3$ 3D cluster consisting of randomly oriented nanoplates, and

(continued)

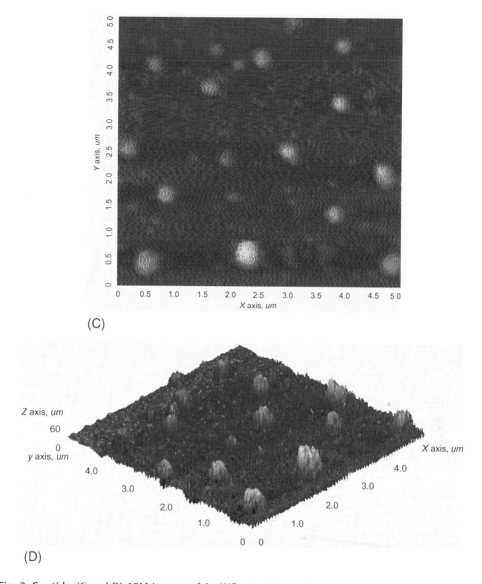

Fig. 2, Cont'd   (C and D) AFM images of Ag-WO$_3$ nanocomposite.

4.2 Current-voltage characterization

Fig. 3A shows forward and reverse characteristics of the Schottky junction, where the measurement is taken at room temperature. The device exhibits steep forward characteristics, with a turn-on voltage or knee voltage of 162 mV and depicts that the device

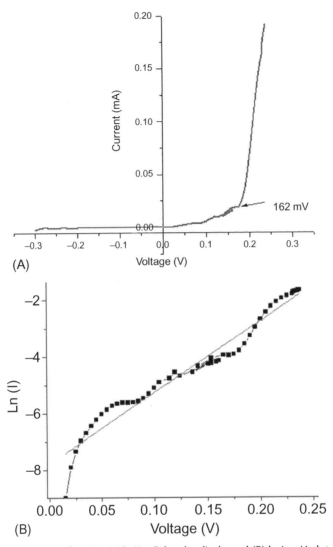

Fig. 3 (A) *I-V* characteristics of ITO/Ag-WO$_3$/Ag Schottky diode and (B) ln *I* vs *V* plot.

works as a perfect rectifier with rectification ratio at a bias of 0.22 V in dark as 219:1. The nonlinear *I-V* curve in Fig. 3A indicates that the forward current is effectively enhanced with respect to reverse current. This rectifying behavior of a Schottky barrier diode is assumed to follow the thermionic emission theory for conduction across the junction with thermal energy and expressed with the Schottky relation given in Eq. (1) [16, 17]. The logarithmic plot of ln *I* vs *V* at room temperature is shown in Fig. 3B.

In semiconductor theory, the forward Schottky I-V characteristic is given by

$$I = I_S \left[\exp \left\{ q \frac{(V - IR_s)}{nkT} \right\} - 1 \right] \qquad (1)$$

where V is the applied forward voltage, q is the electronic charge, n is the ideality factor, k is the Boltzmann constant, T is the temperature, R_S is the series resistance, and I_s is the reverse saturation current from $\ln I$ vs V plot (Fig. 3B).

$$I_S = SA^{**} T^2 \exp \left(\frac{q\Phi_b}{kT} \right) \qquad (2)$$

where S is the device area, Φ_b is the Schottky barrier height, and A^{**} is the effective Richardson constant.

The knee voltage of 162 mV is significantly less compared with the ideal silicon-based Schottky diodes (i.e., 0.2–0.5 V). This would surely be the lowest from any diode fabricated using chemically synthesized nanoparticles and suggest that this diode could be used in low voltage fast-switching and fast-sensing applications. The series resistance is found to be 3.15 Ω from the slope of the I-V curve, which is excellent for the film formed using a dip–coating technique. The ideality factor and saturation current calculated from Fig. 3B is found to be 1.48 between (0.122–0.226) V and 7.8×10^{-5} Amp, respectively. The ideality factor is one of the most important parameters that indicates the uniformity of the Schottky barrier. For an ideal diode, "n" equals 1, but in practice, it is greater than 1 for a real Schottky diode. The presence of an interfacial layer, image–force lowering, and carrier recombination due to surface states or defect levels are some of the main reasons for the ideality factor being greater than unity [18–20]. In conclusion, the I-V characteristics matches clearly with that of the ideal Schottky diode.

The author proposed that researchers working in the field of gas sensors can also measure the I-V characteristics of the fabricated Schottky diode under the presence of reducing gases. When a Schottky diode is exposed to reducing gases such as H_2, H_2S, and CO, it can change its surface conductivity and produce a lateral voltage shift in its I-V characteristics. To date, many researchers have used the forward bias diodes for gas sensing, where the change of the barrier height could describe the change in the I-V curve [21–26]. The gas-sensing mechanism in forward bias is described by the dissociation of gas at the catalytic metal, which causes an effective dipole charge to exist at the metal-semiconductor interface. The dipolar charge causes a change in barrier height, which can be calculated from the Schottky diode theory [25–27].

4.3 Analysis of photodetection capability of ITO/Ag-WO₃/Ag Schottky diode

ITO/Ag-WO₃/Ag Schottky diode is studied under visible illumination to judge its photoresponsive nature. Fig. 4 shows its I-V characteristics under the illumination of

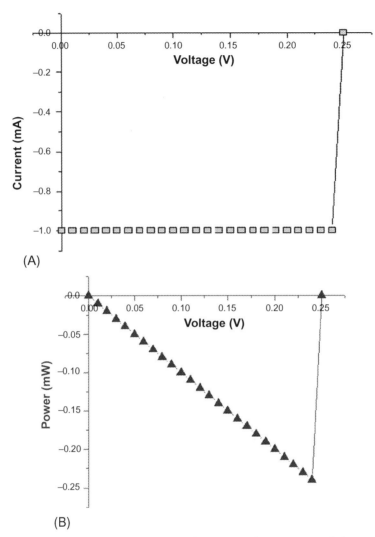

Fig. 4 (A) Current-voltage (*I-V*) and (B) power-voltage (*P-V*) characteristics of the ITO/Ag-WO$_3$/Ag Schottky diode under illumination.

0.5 mW mm^{-2} through a halogen lamp. The device shows characteristics in the fourth quadrant and hence acts as a photovoltaic device when exposed to visible light. The photovoltaic parameters of the diode are summarized in Table 1.

The calculated values mentioned in Table 1 clearly indicate that this Schottky diode has photodetection capability [28–30] with photoconversion efficiency of 0.2%, open–circuit voltage (V_{OC}) of 0.25 V, a short circuit current (I_{SC}) of 1 mA, and a fill factor of 0.99 as calculated from *I-V* graph in Fig. 4A. Also the maximum power from the power

Table 1 Photovoltaic parameters of the prepared ITO/Ag-WO$_3$/Ag Schottky diode.

	Device parameters	Calculated values
1	Open circuit voltage	0.25 V
2	Short circuit current	1 mA
3	Maximum power	0.249 mW
4	Fill factor	99.6%
5	Efficiency	0.2%
6	Ideality factor	1.48
7	Series resistance	3.15 Ω

graph in Fig. 4B is found to be 0.249 mW. Therefore, the author does not hesitate to say that the fabricated ITO/Ag-WO$_3$/Ag Schottky diode possess strong photoresponsive characteristics and hence could also be used in photodetection or photovoltaic applications [31].

4.4 Mott Schottky analysis

To determine built-in potential (Φ_{bi}), acceptor concentration (N_A), and Schottky barrier height (Φ_b) of the fabricated device, capacitance-voltage (C-V) characteristic of the device is measured at 10 MHz between (−2 to 1.5) V. The Mott-Schottky analysis probes the depletion capacitance at a Schottky junction, which is determined by the width of the bias-dependent depletion region.

The depletion capacitance, C is bias dependent and expressed by Eq. (3) [31–33].

$$C^{-2} = \frac{2}{q\varepsilon_0 \varepsilon N_A}(\Phi_{bi} - V) \tag{3}$$

where N_A is the acceptor concentration, q is the electronic charge, Φ_{bi} is the built-in potential, V is the biasing voltage, ε is the dielectric constant of semiconductor, which is 161 for Ag-WO$_3$ as determined through impedance analysis, and ε_0 is the permittivity of free space. The slope of the Mott-Schottky plot (Fig. 5) corresponds to the acceptor concentration in Ag-WO$_3$ nanocomposite layer [34]. The values for carrier concentration, built-in potential, and barrier height as determined using Eqs. (3) and (4) are depicted in Table 2.

$$\Phi_b = q|\Phi_{bi}| + \Phi_p \tag{4}$$

where Φ_p corresponds to the surface potential of p-type semiconductor (Ag-WO$_3$) and given by Eq. (5).

$$\Phi_p = \frac{kT}{q} \ln\left(\frac{N_c}{N_A}\right) = E_F - E_v \tag{5}$$

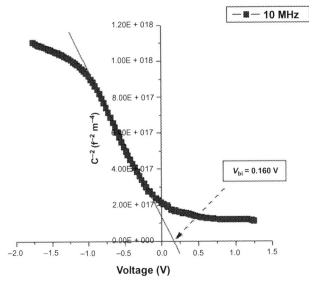

Fig. 5 Mott Schottky plot of ITO/Ag-WO₃/Ag Schottky diode at 10 MHz.

Table 2 List of Schottky junction parameters obtained from Mott-Schottky analysis at frequency 10 MHz in forward bias condition.

S. No.	Parameters	Frequency 10 MHz
1	Carrier concentration (N_A)	9.56×10^{14} cm^{-3}
2	Built-in potential (Φ_{bi})	160 mV
3	Barrier height (Φ_b)	0.563 eV

The capacitance measurement revealed the existence of a straight line with a negative slope of C^{-2} vs V plot, indicating barrier formation in p-type Ag-WO₃ nanocomposite semiconductor with acceptor concentrations N_A 9.56×10^{14} cm^{-3} [34]. P-type conductivity in Ag-WO₃ will be discussed in the author's upcoming paper.

4.5 Current conduction mechanism in Ag-WO₃/Ag Schottky diode

Ag-WO₃ is a p-type semiconductor and the forward bias direction corresponds to the situation when the front ITO electrode is positive. As shown in Fig. 6, the ITO electrode has a high work function, of 5.4 eV compared with Ag-WO₃ (work function 4.5 eV), which forms an ohmic contact with Ag-WO₃ film. The rectification observed is due to the blocking contact or a Schottky barrier that is formed between the Ag-WO₃ and Ag (work function 4.2 eV) interface, where the conduction and valence band edges bend downward at equilibrium [35–37]. After the formation of contact between Ag and Ag-WO₃ film, the Fermi level becomes constant throughout the system in thermal

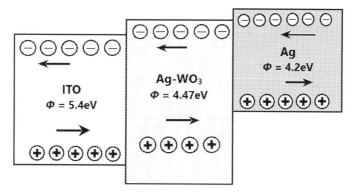

Fig. 6 Schematic showing current flow mechanism in the device.

equilibrium with the flow of holes from the semiconductor with higher work function into the metal with lower work function, as depicted in Fig. 6. This leaves negatively charged donor atoms in the semiconductor creating a space charge region, i.e., Schottky barrier between silver and Ag-WO$_3$ [35–37].

4.6 Impedance spectroscopy

Impedance spectroscopy (IS) is an excellent technique to analyze various components of a device affecting the current conduction in a device. The IS analysis of a fabricated solid-state ITO/Ag-WO$_3$/Ag Schottky diode is accomplished in the frequency range from 20 Hz to 10 MHz at a forward bias of 0.132 mV under room temperature with an amplitude of a.c. signal 0.2 mA. This technique provides a true picture of the device properties by separating the real and imaginary components of electrical parameters. The Nyquist Plot from impedance measurement is shown in Fig. 7. EIS analyzer software is used to fit the experimental spectra with the equivalent circuit model, and results are given in Table 3. A solid line (Fig. 7) represents fitted data.

The complex spectra consists of one arc with a CPE element, as seen on Nyquist plot in Fig. 7. Single arc signifies the contribution from metal and semiconductor junction (Ag-WO$_3$/Ag) with a single time constant, indicating formation of single junction, i.e., rectification is caused by Schottky junction only. [28, 38, 39]. This supports our claim for Schottky diode fabrication.

The a.c. equivalent circuit (Fig. 7) of the device consists of a resistance R_1 (due to bulk resistances) in series with a single Capacitor C_1 representing Schottky junction capacitance between Ag and Ag-WO$_3$ [28, 38–40]. The depression in the arc of the Nyquist plot (Fig. 7) indicates the presence of a constant phase element (CPE), which may be due to inhomogeneity at the junction [41–43].

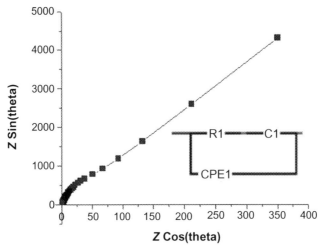

Fig. 7 Nyquist plot of the ITO/Ag-WO$_3$/Ag Schottky diode in the frequency range 20 Hz to 10 MHz and the corresponding equivalent circuit of the device.

Table 3 Table depicting the obtained values of various circuit parameters determined from equivalent circuit analysis.

S. No.	Parameters	Obtained values
1	R_1	136 (Ω)
2	C_1	25 (μF)
3	Q_O	2.93 (μS-S^{-n})
4	n	0.3

5. Conclusion

The possibilities of low-cost fabrication and analysis of photoresponsive ITO/Ag-WO$_3$/Ag Schottky diode is excellently analyzed. The Schottky diode is successfully fabricated using the dip-coating technique and characterized through *I-V*, *C-V*, and impedance spectroscopic techniques. The steep nonlinear *I-V* characteristics indicate the current flow mechanism in the Ag-WO$_3$/Ag Schottky junction, which ascribes thermionic emission theory. Various device parameters like cut-in voltage (162 mV), rectification ratio (219:1), and ideality factor (1.48) are determined through *I-V* characteristics, while other device parameters like built-in potential (160 mV) and Schottky barrier height (0.563 V) are determined at 10 MHz through *C-V* analysis. The analysis of a.c. equivalent circuit of the device supports our claim for Schottky diode facbrication, also the physical significance of individual components found in the equivalent circuit is justified. All results are enthusiastic, and the cut-in voltage of 162 mV is significantly less compared with ideal silicon based Schottky diodes (i.e., 0.2–0.5 V), which suggests its application as a fast switch, sensor, rectifier, and in radiofrequency applications and detectors. Above all, the fabricated Schottky diode is also found to possess strong photoresponsive *I–V* characteristics with conversion efficiency 0.2% and fill factor 0.99, and thus could also be used in a photosensing or detection application.

References

[1] S. Walia, S. Balendhran, H. Nili, S. Zhuiykov, G. Rosengarten, Q.H. Wang, et al., Transition metal oxides—thermoelectric properties, Prog. Mater. Sci. 58 (2013) 1443–1489.

[2] S. Adhikari, D. Sarkar, H.S. Maiti, Synthesis and characterization of WO3 spherical nanoparticles and nanorods, Mater. Res. Bull. 49 (2014) 325–330.

[3] H.G. Choi, Y.H. Jung, D.K. Kim, Solvothermal synthesis of tungsten oxide nanorod/nanowire/nanosheet, J. Am. Ceram. Soc. 88 (2005) 1684–1686.

[4] N. Prabhu, S. Agilan, N. Muthukumarasamy, C.K. Senthilkumarn, Effect of temperature on the structural and optical properties of WO$_3$ nanoparticles prepared by solvothermal method, Dig. J. Nanomater. Biostruct. 8 (2013) 1483–1490.

[5] Y. Guo, X. Quan, N. Lu, H. Zhao, S. Chen, High photocatalytic capability of self-assembled nanoporous WO3 with preferential orientation of (002) planes, Environ. Sci. Technol. 41 (12) (2007) 4422.

[6] L. Santos, J.P. Neto, A. Crespo, P. Baião, P. Barquinha, L. Pereira, R. Martins, E. Fortunato, Electrodeposition of WO$_3$ nanoparticles for sensing applications, in: M. Aliofkhazraei (Ed.), Electroplating of Nanostructures, IntechOpen, 2015, , pp. 27–47.

[7] H. Ohta, H. Hosono, Transparent oxide optoelectronics, Mater. Today 7 (2004) 42–51.

[8] Q.J. Wang, C. Pflügl, W.F. Andress, D. Ham, F. Capasso, M. Yamanishi, Gigahertz surface acoustic wave generation on ZnO thin films deposited by radio frequency magnetron sputtering on III-V semiconductor substrates, J. Vac. Sci. Technol. B 26 (2008) 1848–1851.

[9] J. Ding, Y. Chai, Q. Liu, X. Liu, J. Ren, W.-L. Dai, Selective deposition of silver nanoparticles onto WO$_3$ nanorods with different facets: the correlation of facet-induced electron transport preference and photocatalytic activity, J. Phys. Chem. 5 (2016) 10580.

[10] R. Malik, P.S. Rana, V.K. Tomer, V. Chaudhary, S.P. Nehra, S. Duhan, Nano gold supported on ordered mesoporous WO$_3$/SBA15 hybrid nanocomposite for oxidative decolorization of azo dye, Microporous Mesoporous Mater. 225 (2016) 245.

[11] C. Feng, S. Wang, B. Geng, Ti (IV) doped WO$_3$ nanocuboids: fabrication and enhanced visible-light-driven photocatalytic performance, Nanoscale 3 (2011) 3695.

[12] H. Yu, J. Li, Y. Tian, Z. Li, Gas sensing and electrochemical behaviors of Ag-doped 3D spherical WO3 assembled by nanostrips to formaldehyde. Int. J. Electrochem. Sci. 13 (2018) 9281–9291, https://doi.org/10.20964/2018.10.52.

[13] G. Ganguly, D.E. Carlson, S.S. Hegedus, D. Ryan, R.G. Gordon, D. Pang, R.C. Reedy, Improved fill factors in amorphous silicon solar cells on zinc oxide by insertion of a germanium layer to block impurity incorporation, Appl. Phys. Lett. 85 (2004) 479–481.

[14] J.R. Tuttle, et al., Accelerated publication 17.1% efficient Cu(In,Ga)Se2-based thin-film solar cell, Prog. Photovolt. Res. Appl. 3 (1995) 235.

[15] E. Delahoy, M. Cherny, Deposition schemes for low cost transparent conductors for photovoltaics, in: Materials Research Society Symposium Proceedings, Thin Films for Photovoltaic and Related Materials, vol. 426, 1996, pp. 426–467.

[16] T. Minami, et al., Group III impurity doped zinc oxide thin films prepared by RF magnetron sputtering, Jpn. J. Appl. Phys. 24 (1985) L781.

[17] S.B. Zhang, S.H. Wei, A. Zunger, A phenomenological model for systematization and prediction of doping limits in II–VI and I–III–VI$_2$ compounds, J. Appl. Phys. 83 (1998) 3192.

[18] D.J. Chadi, Doping in ZnSe, ZnTe, MgSe, and MgTe wide-band-gap semiconductors, Phys. Rev. Lett. 72 (1994) 534.

[19] S.B. Zhang, S.H. Wei, A. Zunger, Microscopic origin of the phenomenological equilibrium "doping limit rule" in -type III-V semiconductors, Phys. Rev. Lett. 84 (2000) 1232.

[20] A. Kobayashi, O.F. Sankey, J.D. Dow, Deep energy levels of defects in the wurtzite semiconductors AIN, CdS, CdSe, ZnS, and ZnO, Phys. Rev. B 28 (1983) 946.

[21] X.F. Chen, W.G. Zhu, O.K. Tan, Microstructure, dielectric properties and hydrogen gas sensitivity of sputtered amorphous $Ba0.67Sr_{0.33}TiO_3$ thin films, Mater. Sci. Eng. B 77 (2000) 177–184.

[22] E. Comini, V. Guidi, C. Malagu, G. Martinelli, Z. Pan, G. Sberveglieri, Z.L. Wang, Electrical properties of tin dioxide two-dimensional nanostructures, J. Phys. Chem. B 108 (2004) 1882–1887.

[23] S. Kandasamy, A. Trinchi, W. Wlodarski, E. Comini, G. Sberveglieri, Hydrogen and hydrocarbon gas sensing performance of $Pt/WO_3/SiC$ MROSiC devices, Sens. Actuators B 111 (2005) 111–116.

[24] K.I. Lundstrom, M.S. Shivaraman, C.M. Svensson, Hydrogen-sensitive Pd-gate MOS-transistor, J. Appl. Phys. 46 (1975) 3876–3881.

[25] T.H. Tsai, J.R. Huang, K.W. Lin, C.W. Hung, W.C. Hsu, H.I. Chen, W.C. Liu, Improved hydrogen-sensing properties of a $Pt/SiO_2/GaN$ Schottky diode, Electrochem. Solid State Lett. 10 (12) (2007) J158–J160.

[26] Y.Y. Tsai, H.I. Chen, C.W. Hung, T.P. Chen, T.H. Tsai, K.Y. Chu, L.Y. Chen, W.C. Liu, A hydrogen gas sensitive Pt–In0.5Al0.5P metal–semiconductor Schottky diode, J. Electrochem. Soc. 154 (2007) J357–J361.

[27] H. Zheng, J.Z. Ou, M.S. Strano, R.B. Kaner, A. Mitchell, K. Kalantar-Zadeh, Nanostructured tungsten oxide - properties, synthesis, and applications. Adv. Funct. Mater. 21 (2011) 2175–2196, https://doi.org/10.1002/adfm.201002477.

[28] A.F. Kohan, et al., First-principles study of native point defects in ZnO, Phys. Rev. B 61 (2000) 15019.

[29] C.G. Van de Walle, Hydrogen as a cause of doping in zinc oxide, Phys. Rev. Lett. 85 (2000) 1012.

[30] A. Krtschil, A. Dadgar, N. Oleynik, J. Bläsing, A. Diez, A. Krost, Local -type conductivity in zinc oxide dual-doped with nitrogen and arsenic, Appl. Phys. Lett. 87 (2005) 262105.

[31] T.H. Vlasenflin, M. Tanaka, P-type conduction in ZnO dual-acceptor-doped with nitrogen and phosphorus, Solid State Commun. 142 (2007) 292.

[32] J. Huang, Z. Ye, H. Chen, B. Zhao, L. Wang, Growth of N-doped p-type ZnO films using ammonia as dopant source gas, J. Mater. Sci. Lett. 22 (2003) 249–251.

[33] Y.-Y. Liu, H.-J. Jin, C.-B. Park, Analysis of photoluminescence for N-doped and undoped p-type ZnO thin films fabricated by RF magnetron sputtering method, Trans. Electr. Electron. Mater. 10 (2009) 24.

[34] M.-L. Tu, Y.-K. Su, C.-Y. Ma, Nitrogen-doped p-type ZnO films prepared from nitrogen gas radio-frequency magnetron sputtering, J. Appl. Phys. 100 (2006) 053705.

[35] M. Joseph, H. Tabata, T. Kawai, p-type electrical conduction in ZnO thin films by Ga and N co doping, Jpn. J. Appl. Phys. 38 (1999) L1205–L1207.

[36] K.R. Saravanakuma, Structural, surface morphological and electrical properties of nanostructured p-type ZnO: N films, Contemp. Eng. Sci. 4 (2011) 119–140.

[37] S. Dhara, P.K. Giri, Stable p-type conductivity and enhanced photoconductivity from nitrogen-doped annealed ZnO thin film, Thin Solid Films 520 (2012) 5000–5006.

[38] U. Ozgur, Y.I. Alivov, C. Liu, A. Teke, M.A. Reshchikov, S. Dogan, V. Avrutin, S.J. Cho, H.J. Morkoc, A comprehensive review of ZnO materials and devices, J. Appl. Phys. 98 (4) (2005) 041301.

[39] S.B. Zhang, S.H. Wei, A. Zunger, Intrinsic n-type versus p-type doping asymmetry and the defect physics of ZnO, Phys. Rev. B 63 (2001) 075205.

[40] F. Oba, S.R. Nishitani, S. Isotani, H. Adachi, I. Tanaka, Energetics of native defects in ZnO, J. Appl. Phys. 90 (2001) 824.

[41] M.C. Payne, M.P. Teter, D.C. Allan, T.A. Arias, J.D. Joannopoulos, Iterative minimization techniques for ab initio total-energy calculations: molecular dynamics and conjugate gradients, Rev. Mod. Phys. 64 (1992) 1045.

[42] E.C. Lee, Y.S. Kim, Y.G. Jin, K.J. Chang, Compensation mechanism for N acceptors in ZnO, Phys. Rev. B 64 (2001) 085120.

[43] C. Wang, Z. Ji, J. Xi, J. Du, Z. Ye, Fabrication and characteristics of the low resistive p-type ZnO thin films by DC reactive magnetron sputtering, Mater. Lett. 60 (2006) 912–914.

Nano fabrication techniques—Biosensors

CHAPTER 3

Biosensor fabrication with nanomaterials

Hari Mohan[a], Ravina[a], Anita Dalal[b], Minakshi Prasad[c], and J.S. Rana[b]
[a]Centre for Medical Biotechnology, M.D. University, Rohtak, India
[b]Deenbandhu Chhotu Ram University of Science and Technology, Sonipat, India
[c]Department of Animal Biotechnology, Lala Lagpat Rai University of Veterinary and Animal Sciences, Hisar, Haryana, India

1. Introduction

Nanobiotechnology is one of the important disciplines where biotechnology merges with nanotechnology to provide a revolutionary area for consolidation of microbiology and molecular biology to a real extent. Attention is toward the fabrication of sensors with nanomaterials for quick, specific, and sensitive real-time results [1]. Nanotechnology is a new vision pointing to the availability of room at the bottom of a material's structure level, an idea introduced by Richard P. Faymen in 1959 [2]. Researchers around the world are developing methods for the fast detection, fast response, fast cure, and fast measurements in biological or other sciences. Every single reaction in science occurs at a molecular level, even at an atomic level. Calculating or detecting these single changes, or even thousands of them, can give us large decision-making data. Leland C. Clark Jnr. presented the concept of a biosensor and is considered the father of this field. In 1956, he published a paper on the oxygen electrode. Clark's idea became commercial reality in 1975 with the successful launch of a glucose analyzer based on amperometric detection of hydrogen peroxide [3]. This was the first attempt to make a laboratory-based analyzer from many of the biosensor-based laboratory analyzers produced by companies at that time [4]. Then a new phase started in the field of chemistry, physics, and material science to make a reliable, sophisticated biosensor for application in various fields of science such as agriculture, medicine, military, and biotechnology. A biosensor is a synergic combination of bioelectrochemistry and microelectronics, which enable the signals produced by specific biochemical reactions to be recorded, registered, and quantified. A number of factors influence the performance of a sensor like transducers, electrochemical behaviors of the materials (mediators, stabilizers) used for development of sensor. It also has some limitations such as less stability, interferences in electrochemically activity, and complex electron-transfer pathways, but these can be sorted out with the help of nanosensors. Nanosensors are small devices or setups that can calculate or sense changes at the nanolevel at a nanosize. It can be applied in different fields like agriculture, environment, pharmaceutical, chemistry, information technology, medicine, etc. Recently many types of nanosensors have been developed in different field such as optical biosensors, biological nanosensors, chemical nanosensors, and physical nanosensors [5]. Nanodiagnostic methods are very specific, fast, and easy to handle at early detection [6]. The basic structure of a nanosensor has a base surface of metal, glass, paper, and plastic on which recognition elements such as DNA, RNA, protein, antigen, and antibodies are attached and a transducer interprets the signal in different forms such as optical, amperometric, potentiometric, piezoelectric, calorimetric, and acoustic [7]. A lot of work has been done to develop noninvasive and easy target-specific delivery systems that can work at the molecular level with the help of nanomaterials. A range of nanomaterials of different dimensions are present that can be used for fabrication of sensors, in drug delivery systems, for cancer therapy, as a label, a virocidal agent, and antitumor drugs, and they also show antigenic activity [8–10]. Nanomaterials help in a certain way either in transducers or the biorecognition element of sensors. It can increase the biorecognition area or interacting site for analytes of sensors by increasing the surface-to-volume ratio, which helps increase the sensitivity of the biosensor. The biorecognition area is a major factor in determining the sensitivity of a sensor because it is the area that interacts with the analyte. Thus, nanomaterials have provided a vast range of options to develop sensors that can detect analytes even at small concentrations.

Hence, nanotechnology is an area of huge opportunities that can enhance stability and sensitivity, and also decrease the interference in an electrochemical system. It is a discipline where materials are handled at a nanoscale level to improve the performance of sensors, and a number of novel materials are used for fabrication that enhances the performance of biosensors. This chapter describes the major nanomaterials and fabrication techniques onto the surface.

2. Nanomaterials

A nanomaterial is a structure having a size up to 1–100 nm nanolevel. These are classified as 0D that have dimensions in x, y, z plane (nanoparticles [NPs]); 1D have dimensions in x, y plane (rods, wire, and tube); 2D have dimensions in x plane (nanofilms, nanocoatings); and 3D have no confinement barrier (nanocomposites) [7]. These are present in many structural forms such as NPs, nanowire, nanofilms, nanotubes, nanocrystals, nanobelts, quantum dots (QDs), and self-assembled nanomaterials [11]. Different nanomaterials have different specific properties that contribute to the development of nanobiosensors, which have more sensitivity with an amplified signal. These materials can be used for surface modification and as a label for signal amplification to overcome the signal-to-noise (S/N) ratio. The size dimensions of the metal NPs makes them structurally compatible to proteins, enzymes, nucleic acids, and antibodies [12,13].

Ding et al. [14] have described some useful aspects of metal NPs as labels in the development of electrochemical sensors: (1) these nanomaterials can enhance the signal of electroactive species; (2) can act as an ultramicroelectrode array for the electrolysis reaction of large amounts of analytes; (3) can act as a surface for deposition; and (4) can act as a mediator for the deposition of electrocatalysts. There are a number of NPs used for the development of biosensors, for example, Au, Ag, QDs, carbon nanotubes (CNTs), magnetic particles, etc. CNTs are a center of attraction due to their superior characteristics in electron transfer reaction, physical characteristics, electrochemical activities, stability, and natively good mechanical properties. They create special interest due to their novel electronic, metallic, and structural characteristics [15]. Carbon has allotropes with different bond arrangements and shows important chemical and physical properties. These days, metal-organic NPs and metal NPs of a number of metals are in use (MNPs) having different beneficial properties required in the development of transducers of biosensors. Some noble metals such as gold, platinum, and silver are used and extensively studied as nanomaterials, which have been the most popular in nanosensor development. Noble metals are chemically inert in nature but show some unique physiochemical properties at the nanoscale level. These NPs provide an immobilizing surface, can increase electron transfer, help in catalyzing the reaction of chemiluminescents (CLs) with their substrates, can change their mass, and make differences in the refractive index (RI). Also MNPs act as "electron wires" in electrochemical biosensors, which helps in transportation of electrons produced in a bioreaction to the sensing electrodes and convert physiochemical changes to a measurable signal in proportion to the concentration of the analyte [16,17]. Different properties of NPs are shown in Fig. 1.

2.1 Gold nanoparticles

Gold NPs are used due to their size availability, easy modification, stability in different conditions, nontoxic nature, optical, biocompatibility, simple modification, easy production, and electronic conductive behaviors [16–18]. AuNPs exhibit extraordinary catalytic capability compared with bulk gold. A number of studies show that gold NP has an interface-dominated property and high surface-to-volume ratio, which makes these NPs capable of reducing the overall potential of electrochemical reactions and accelerating the chemical reaction, leading to an improvement of detection sensitivity. Different shapes of AuNPs make surface modification easy and also make them easily accessible to develop detection techniques.

Optical behavior is an important property of gold NPs. Metal particles were irradiated with light of a specific wavelength, responsible for an oscillation in the conduction band of electrons, and formed resonant surface plasmons. In the case where size of an NP is less than the incident light, then propagation of oscillating electrons along the surface becomes difficult. Then, electron density will become polarized on one side of the gold

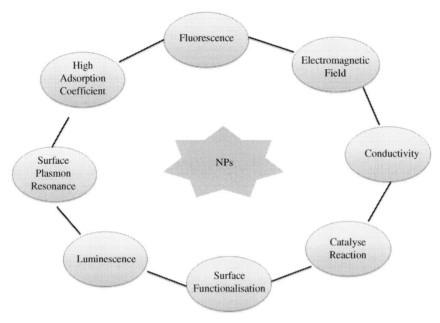

Fig. 1 A number of properties are shown by nanoparticles.

particle where the plasmons show oscillation with the light frequency according to the Mie theory [19,20], and the characteristics are strongly dependent on the dielectric constant, shape, and size of the NP [21]. Recognition event in sensing devices cause a change in the oscillation frequency, which results in a color change of the gold NPs, which can be observed with the naked eye. But in 1850, Michael Faraday's work already proved that the color produced was due to the size of the gold particles [22]. This application was recorded for first time with the extremely small gold particles used in biosensing [23]. This change in color during aggregation (or dispersion of aggregates) provides a new way of using gold NPs as transducers in colorimetric detection methods [24,25]. A structural change will occur in the substrate composition when ligand binds with a target, and it will produce a change in the environment of the NP, resulting in a color change with changed aggregation status of NPs. Nanomaterials can be easily fabricated to develop platforms with improved S/N ratios by miniaturization of the sensor device [26]. Gold NPs are conductive in nature, transferring electrons, and can be used in a wide range of electroactive biological species. This property can be used in redox reactions of enzymes, different dyes, etc. In the case of an enzyme that catalyzes the reaction oxidation and reduction occurs at the bioreceptor unit with the analyte, this increase in electron transfer further increases the amount of current and helps increase the sensitivity. Actually, Au NPs act as electron shuttles by transferring the electrons involved in the redox reaction to the electrode in the redox center of the enzyme. These gold NPs act as connecting wire and provide an increase in the electrochemical signal because it stops the need for enzymatically formed redox species in an enzymatic reaction. Thus, AuNPs can be used as fluorescence quenchers, catalysts, immobilization platforms, and colorimetric NPs as well as surface plasmon resonance (SPR) and surface-enhanced Raman scattering (SERS) enhancers in optical biosensors.

Gold NP's electrochemical properties provide an advantage in SPR transduction. SPR is a phenomenon that happens when a polarized light falls on a metal surface at the interface of different RI of media. Localized surface plasmon resonances (LSPRs) is a term used when light is used for excitation on metal NPs, and they show electron's oscillations with increased near-field amplitude at the resonance wavelength. Near-field is highly localized at the metal NP but decays quickly away from the particle/dielectric interface. Due to resonance, far-field scattering of particles also happens. LSPR increases the light intensity and also has high spatial resolution but is influenced with the size of the NPs [27]. Optical detection methods require large and expensive instruments such as SPR/SERS instruments and fluorescent spectrometers, but nonoptical methods like piezoelectric and electrochemical do not require these instruments. In piezoelectric biosensors, AuNPs are used as labels to increase the mass changes, which enhances the sensitivity of the detection methods [28,29]. It proves that, by using AuNPs, the sensitivity of optical biosensors can be enhanced. AuNPs also increase the surface area for probe attachment so that more probes can hybridize with the target DNA. Thus, it increases the sensitivity of a sensor better than without AuNP's immobilization. In a study, AuNPs of 50 nm size were used, which increased the frequency of signals of quartz crystal microbalance (QCM) to get a detection limit of 10^{-14} M for the target DNA [30]. A QCM–DNA sensor was developed for dengue virus (DENV) using an AuNP layer–by–layer method for DNA hybridization to get a low detection limit of 2 plaque-forming units (PFU)/mL. To further increase the frequency shift, Chen et al. developed a QCM–DNA sensor with a layer–by–layer AuNP structure by DNA hybridization, which provided even lower detection limit of 2 PFU/mL for DENV [31]. Kerman et al. [32] reported an electrochemical sensor for DNA detection by the direct oxidation of AuNPs without acid treatment, which provided a detection limit of 2.17 pM for target DNA. A different way of modification for AuNPs on the surface of electrodes is the direct immobilization of AuNPs. In a study, 2.2 pg mL^{-1} LOD was achieved for a target protein with the help of direct immobilization. Jarocka et al. immobilized AuNPs on the surface of a gold electrode as the electron migration enhancer, achieving an LOD of 2.2 pg·mL^{-1} for target protein [33]. Another study developed an electrochemical sensor by immobilizing methylene blue-labeled DNA probes on AuNPs. These NPs increase the surface area for labeled probes and enhance the signal with a detection range of 10^{-13} to 10^{-8} M; LOD was 50 fM [34]. An amperometric DNA sensor was developed based on the same principle of immobilization of gold NPs for the detection of *S. pyogenes* in a throat swab sample of patients. Single-stranded DNA was used as a target after isolation from bacteria.

DNA probe hybridize with target DNA with a sensitivity of 951.34 $(\mu A/cm^2)$/ng DNA, 130 fg/6 μL samples [35]. In a study, AuNPs were used as labels to achieve an ultrahigh sensitivity for detection of human immunodeficiency virus (HIV)-1 p24 antigen using inductively coupled plasma–mass spectroscopy (ICP-MS) technology. Diluted HNO_3 was used to dissociate AuNPs from the immunoassay complex, with a detection limit of 1.49 pg·mL^{-1} by ICP-MS measurement [36]. A piezoelectric biosensor for detection of *E. coli* O157:H7 was developed for foodborne disease detection. In this sensor, bacteria-specific gene thiolated probe conjugated with AuNP. Here, these particles are enhancing mass as well as the sensitivity of the sensor. They obtained a LOD 1.2×10^2 CFU/mL with a working range of 10^2–10^6 CFU/mL [37]. Another example is colorimetric sensor based on AuNP conjugated with anti–Salmonella antibody to selectively detect *S. typhimurium* [38]. In the food industry, aflatoxin B1 was detected using an AuNP aptasensor developed by Hosseini et al. [39]. According to World Health Organization (WHO), Zika virus is considered one of the most dreadful viral diseases these days, and new diagnostic methods are developed for its diagnosis along with DENV. Even though there are many tests for its diagnosis, those methods are slow, costly, and nonspecific. This is a reason we need a novel biosensor for detection. A three-contact configuration was developed by thermal evaporation of a disposable electrode on polyethylene terephthalate substrates and then covered by a layer of nanometric gold. The biosensor detected the viral sequence and DNA sequences in a drop of sample by using its three-contact electrode; it allowed direct reading with a response time of 1.5 h without labeling on the disposable electrode. The biosensor showed Zika selectivity in synthetic DNA assays with a limit of detection of 25.0 $\pm$ 1.7 nM [40]. An electrochemical immunosensor was developed using a functionalized interdigitated microelectrode of gold (IDE-Au) for Zika virus (ZIKV) protein detection. A miniature immunosensing chip was prepared using IDE-Au by immobilizing ZIKV-specific envelope antibodies (Zev-Abs) onto dithiobis(succinimidyl propionate) (DTSP)-functionalized IDE-Au to measure the current response of the sensing chip. Electrochemical impedance spectroscopy (EIS) was performed as a function of ZIKV-protein concentrations. EIS results demonstrate that the detection range and limit of detection of the chip was 10 pM–1 nM and 10 pM, respectively, along with a high sensitivity of 12 kΩM^{-1}, and it detected ZIKV-protein selectively. Such a miniature chip can be used for early diagnosis at a point-of-care application by integrating with a miniature potentiostat (MP) that is interfaced with a smartphone for rapid diagnosis of infection [41].

Thus, gold NPs have been used in a number of techniques for different applications. All these properties of gold NPs can be modified and adjusted according to the desired methodology after using an appropriate synthesis technique to get any desired size and shape of these NPs. Thus, gold NP properties made them a promising candidate not only in the field of sensor technology but also in many other research fields. These different morphologies of gold NPs provide different catalytic, optical, and electronic behaviors [42].

2.2 Silver NPs

AgNPs are also providing a promising area for nanosensor development that is more affordable than AuNPs. A TiO_2 nanotube array base modified with AgNPs amperometric biosensor was developed that provided a sensitivity of 1151.98 $\mu A\ mM^{-1}\ cm^{-2}$ [34]. Nanospheres of Ag and Au were also used for LSPR optical fiber-based biosensor development to detect gastric cancer biomarkers [43]. AgNPs show virocidal activity and was used to cure viral infections. To date, the efficiency of AgNPs as an antiviral has been checked in HIV-1; herpes simplex virus (HSV)-1, 2, 3; feline calicivirus, poliovirus type-1; Peste des petits ruminants virus (PPRV); murine norovirus-1; avian influenza A, virus subtype H7N3; coxsackievirus B3; infectious bursal disease virus; and *S. cerevisiae* dsRNA viruses. These particles bind with surface molecules of a virus or host cell and inhibit the entry of viruses. In the case of HIV-1, AgNPs bind with gp120 to inhibit CD4-based virion binding and blending, acting as a barrier to virus entry, also as a virocidal agent [44]. AgNPs bind with HA molecule of H7N3 subtype of avian influenza A virus and block the HA function and various interactions required for viral replication and virus entry [45]. In the case of poliovirus type-1, silver particles interact with viral proteins resulting in compromised interaction with the host cell, inhibiting the viral internalization [46]. An AgNP-based CL sensor based on triple-channel properties of the luminal-functionalized silver NP (Lum-AgNP) and H_2O_2 CL system was developed for identification of organophosphate and carbamate pesticides in agriculture. The triple-channel properties will change after interaction with pesticides and produce different patterns of CL response, as the CL response is like a "fingerprint" for every specific pesticide. This sensor can detect five different organophosphate and carbamate pesticides, including chlorpyrifos, dipterex, dimethoate, carbaryl, and carbofuran at a concentration of 24 $\mu g/$ mL. They have 95% accuracy to identify 20 unknown pesticides in samples [47]. Due to damage to the body, high death rate, and various complications, acute myocardial infarction (AMI) is considered one the severe cardiovascular diseases. Recently, Xiong et al. [48] developed an immunosensor based on EIS composed of AgNPs and anti–cTNI antibody. AgNPs were synthesized from green algae *Stoechospermum marginatum* using algal extract and Ag salt. Self-assembly method was employed for sensor fabrication in which indium tin oxide (ITO) substrate was attached to 3-Aminopropyl triethoxysilane. Then, 3-Aminopropyl triethoxysilane was conjugated to ITO along with AgNPs (with mercaptopropionic acid (MPA)-N-hydroxysuccinimide (NHS)). Using a coupling reaction after this process, anti–cTNI antibodies were immobilized onto this modified surface. While doing this, Bovine-serum albumin (BSA) was applied to antibody/AgNPs/ITO electrode to prevent physical–adsorption immobilization, and fericynanide was used for redox reaction. The limit of detection was reported to be 0.001 g/mL, which meets the AMI determination value. Several other EC biosensors were fabricated for TN detection, which includes nanostructures such as gold nanodumbbells, gold NPs, CNT-based screen-printed electrodes, and ZnO nanostructures [49–52].

2.3 Aluminum nanoparticles

Al is also providing a promising option for optical biosensor development in an ultraviolet range where Au- and Ag-like metals cannot be used at wavelengths less than 500 nm. In a study, it was reported that RI sensitivity increases Al arrays with the nanoconcave shape due to an increase in the pitch [53].

2.4 Carbon nanostructures

In 1991, CNTs were discovered for the first time. These nanotubes have carbon atoms in hexagonal shapes in which every carbon atom is covalently bonded to three other carbon atoms. CNTs have a small diameter of 1 nm and lengths up to several centimeters. CNTs having only one cylinder are known as single-walled CNTs, and multiple-wall concentric cylinders are known as multiwalled CNTs (MWCTs). These nanostructured carbons have many structural variations such as CNTs or graphene making them a widely used material as electronic or electrochemical transducers in biosensor devices [54,55]. In particular, CNTs possess an outstanding combination of nanowire, electronic properties, morphology, and biocompatibility [56]. Moreover, CNT films have a highly porous 3D network providing good electroactive surface areas due to the natural formation of highly porous three-dimensional networks, suitable for the anchoring of a high amount of bioreceptor units, leading consequently to high sensitivities [57,58]. CNTs have attracted much attention as an electrode material for electrochemical biosensors because of their excellent electrochemical properties, rapid electron kinetics, semi- and superconducting electron transport, and high strength composites [59]. Metal NPs and CNTs amplified the conductance of direct transfer of electrons between the electrode and enzyme. There is a salient role of direct electrochemistry of enzymes in the instigation of biofuel cells, biosensors, and biomedical devices. CNTs encourage the transfer of electrons between electrode, substrate, and enzyme due to its extraordinary quality of electron transport. CNTs act as both immobilization matrices and mediators for the development of third-generation biosensor systems with good analytical characteristics. The invention of CNT is a valuable turning point in the history of carbon research; their unique structural, physical, chemical, and electronic properties yield a broad range of implementation such as field emission devices, DNA biosensors, scanning probe microscopy tips, chemical sensors, gas sensors, potential hydrogen storage materials, nanoelectronic devices and batteries, etc. [60]. The applications and properties of CNT has been a thread of many studies. It relies on diameter, structure, and helicity of CNT, and either it will act as a semiconductor or exhibit metallic behavior [61]. Open end of CNTs result in electrocatalytic properties.

In recent times, CNTs have been employed for the development of electrodes to enhance the kinetics of electron transfer. Thus, CNTs can act as an electrode material, and a number of sensors have been developed using theses carbon-based structures in every field of science. Functionalization of CNTs ameliorates their selective binding to biotargets and solubility in physiological solutions. During the process of functionalization of CNTs with different acids, some defects arise on CNT walls such as breaks and caps in the sidewall due to introduction of certain functional groups [62]. The fundamental structure of a CNT is formed by a layer of sp^{2-} bonded carbon atoms, in which every atom in the x–y plane is attached to three other carbon atoms by a weakly delocalized electron cloud in the z-axis. This configuration of CNT is accountable for blocking dispersion and solubility of carbon NPs due to formation of strong Van der Waal's forces. A major drawback of CNT is their pivotal solubilization. Organic solvents like DMF or DMSO and Nafion's aqueous solution can be used control this problem. CNT holds high S/V ratio that offers high enzyme loading and also provides effectual contact between the electrode and deeply buried active sites of enzymes, enhancing the performance of the sensor. For the fabrication of a desirable sensor, two utmost requirements are transformation of CNT's hydrophobic surface to a hydrophilic surface with the assistance of chemically active functional groups by treating them with nitric acid or sulfuric acid; the other is to attain the required connection between the enzyme and CNT with the distribution of CNTs within a matrix of a polymer such as Nafion polyelectrolyte or other polymers. A biosensor was successfully fabricated using chitosan-doped multi-walled CNTs for the detection of gonorrhea in which methylene blue was used as a DNA indicator. It was noticed that CNTs can increase the electroactive surface by three-fold ($0.093 + 0.060$ and $0.28 + 0.03$ cm^2 for chitosan-modified electrodes and chitosan-CNT, respectively). The response time, limit of detection, and concentration range of the biosensor was observed to be 60 s, 1×10^{-16} M, and 1×10^{-16}-1×10^{-17} M, respectively. The stability of the biosensor was found to be 4 months when stored at 4°C [63].

The fruit industry is a worldwide business, and losses occur every year due to the absence of rapid detection methods. One such sensor was developed by Dalal et al. [64] for the detection of malic acid, which is a fruit-ripening indicator. Tomato ripening before transport to market creates huge losses to farmers. A screen-printed carboxylated multiwall CNT (MWCNT) electrode was used for immobilization of NADP-malate dehydrogenase (malic enzyme) covalently through EDC-NHS chemistry. An LOD of 0.01 mM was achieved through differential pulse voltammetry analysis.

2.5 Quantum dots

QDs are semiconducting luminescent materials that provide a very prominent area for nanosensor development. This nanomaterial shows size-dependent optical and electronic properties that makes it favorable for fluorometric-based sensor development [65], and cadmium selenide (CdSe) is the most commonly used QDs [66]. A QD-based fluorescent biosensor was developed to detect S. *typhimurium* in chicken carcass wash water. In this sensor, anti-salmonella antibody-coated magnetic beads are used to bind another set of anti-salmonella antibodies labeled with biotin. Then fluorescence happens when QDs coated streptavidin interact with biotin. A LOD of 10^3 CFU/mL was achieved, which represents the number of bacteria present in a sample [67]. A number of nanosensors were developed using CdSe QDs for detection of Shiga toxin, Staphylococcal enterotoxin A, and cholera. QDs of different sizes provide a large range of emission wavelengths, which helps in multiplex analysis [68,69]. Some structural defects are reported in the crystal lattice of QDs, which can cut off the exited electrons leading to nonradiative relaxation [70]. But this problem can be overcome by using another semiconductor material that has wider band gap range (generally ZnS) to increase the quantum yields and photostability [71]. A high photochemical stability of core/shell QDs provide a good alternative to other organic fluorophores [72]. This material can be attached to any kind of biomolecule without affecting photophysical recombination. Fluorescence resonance energy transfer (FRET) is a nonradiative energy transfer between the exited QD (donor) and a quencher (acceptor), and fluorescence happens again when the quencher separates. Organic fluorophores show this activity and are used to develop optical DNA sensors [73,74]. A number of studies are reported where hybridization between QD-tagged receptor DNA and a short complementary DNA that is attached with a gold NP was performed because QDs are a good quencher. Complementary DNA hybridizes with the analyte DNA, which will detach from gold NP, and the QDs' fluorescence reappears where its intensity is directly proportional to analyte concentration. In a study, a light-producing protein was used as a label to transfer the energy to the QDs and eradicate the requirement of an external light source for excitation. QDs can also interact with surface plasmons present on the gold surface, which helps light emission from QDs [75]. An electrochemical biosensor has been reported based on the use of semiconductor QDs for the detection of avian influenza virus. Glassy carbon electrode (GCE) was modified using ss-DNA/CdSe for measuring DNA hybridization. QDs shows greater selectivity and sensitivity for virus DNA sequence detection [76].

2.6 Magnetic nanoparticles

Using magnetic NPs as labels are quite interesting and provide an alternate option for sensing applications because biological molecules do not have any magnetic properties; therefore, the chances of any interference or noise during signal capturing is zero [77]. These particles provide a large surface area for immobilization for functionalized groups of molecules [78]. Bulk magnetic domains do not show superparamagnetic behaviors because of a reduced number of magnetic domains compared with magnetic NPs. This property determines that magnetization can change the direction of electrons randomly within a very short time (Neel's relaxation time) and, without an external magnetic field, magnetization will be zero. Magnetic NPs have been used for sensor development for detection and removal of contaminants from food. These particles functionalized with d-mannose were used for detection of *E. coli* cells. These modified NPs also help in separation and visualization of cells through a magnetic effect when incubated with fluorescently labeled concanavalin [79]. Antibodies and MNPs are conjugated for the development of a detection tool against salmonella present in milk. These immobilized Abs are enabled for entrapping bacteria and later segregated by applying a magnetic field. These segregated cells are then pregnable by antibody-immobilized nanocrystals of TiO_2. Then, the nanocrystals of Ab-MNP-TiO_2 are segregated magnetically, and the remaining unbound nanocrystals of TiO_2 are governed by using UV–visible spectrophotometer. The LOD of 100 CFU/mL was observed in milk [80]. Another study reported the fast detection of both Gram–positive and Gram–negative bacteria in food matrices and water with the help of amine-functionalized MNPs. Some organisms such as *E. coli*, *Sarcina lutea*, *P. vulgaris*, *B. cereus*, *P. aeruginosa* and *Salmonella* showed greater adsorption affinity. It was noticed that ionic strength of buffer and amine quantity of functionalized magnetic NPs was pivotal for effective and fast exchange [81]. *Salmonella* were detected in a sample of skimmed milk having a detection limit of 1 CFU/mL with the help of magneto-gene-sensing setup [82]. An additional advantage of magnetic NPs is the possibility to transfer the analytes onto the transduction platform in microfluidic organization, which enable the simultaneous detection of complex and multiplexed analytes [83].

2.7 New approaches for other nanoparticle-based sensors

In a study, two different metal NPs (Pd-Co alloy NPs) were used with synergistic effect to enhance electron transfer by using two different kinds of noble metal NPs to develop electrochemical biosensors. In an experiment, NPs of Pd-Co alloy were embedded in carbon nanofibers, which provided a superior analytical ability for sensing H_2O_2 and nitrite [84]. A novel nanomaterial supramolecular ionic liquid grafted on nitrogen–doped graphene aerogels (SIL-*g*-(N)GAs) was prepared for the development of a biosensor for the detection of BRCA1 gene, a marker of breast cancer. The analytical measurement was recorded using differential pulse voltammetry (DPV) and cyclic voltammetry (CV). A LOD of 2 pM was achieved with reproducible results without use of any label, enzyme, or equipment [85]. Another study reported an ampermetric glucose sensor prepared with NPs of Rhodium (Rh) and gold. Improved electrocatalytic activity was achieved using Rh NP-modified Pt electrode as it provides a large surface area, and AuNPs catalyzed the oxidation reaction of peroxide molecules present in the active regions of an enzyme. It increases the sensitivity of the glucose sensor [86]. ZnO nanostructures also provide high crystallinity, larger surface area, and better optoelectrical properties. It makes them highly desirable for various applications including biosensor development, tissue engineering, drug delivery, and cancer detection. ZnO nanostructure is synthesized by dry and wet chemical method. A number of biomolecules have been immobilized onto 1D ZnO NPs such as bovine serum albumin, human serum albumin, DNA and angiotensin II, cancer, fluorescence detection, mammalian cell biosafety, etc., and DNA [87]. A thrombin electrochemical biosensor was developed by immobilizing platinum NPs (PtNPs) on DNA thrombin aptamer. Its limit of detection was as low as 1 nm [88]. An ultrasensitive bio-barcode assay was used to measure the amyloid-β-derived diffusible ligands (ADDLs) concentration, which is a soluble pathogenic Alzheimer's disease (AD) marker present in cerebrospinal fluid (CSF) of 30 individuals. Because of the low concentrations of ADDL or any other pathogenic AD markers, their study is not possible [89]. *M. tuberculosis* detection using a specific amperometric nucleic acid biosensor was developed by Malhotra and his coworkers immobilizing 21-m oligonucleotide probe (ss-DNA) onto zirconium oxide (ZrO_2) nanostructured film that was deposited onto a gold electrode. The sensor displayed excellent electrocatalytic response due to the presence of nanostructured metal oxide film with a LOD of 0.065 ng/μL within 60 s. Fabrication of nanoporous CeO_2/chitosan composite films as an immobilization matrix was reported for colorectal cancer DNA sequence [90]. In this matrix, the affinity of nano CeO_2 toward oxygen present in DNA and its nontoxicity, electronic conductivity, and good biocompatibility was combined with the excellent film-forming property of chitosan. After hybridization, DPV was recorded using methylene blue as a DNA hybridization indicator. This response was used to identify and determine the amount of target DNA sequence of colorectal cancer. The working range of the sensor was 1.59×10^{-11} to 1.16×10^{-7} M.

3. Fabrication of biosensor with nanomaterials

A number of methods and new experiments have been designed to make better combinations of nanomaterials with sensors to achieve greater sensitivity, so the method of fabrication can determine or affect the basic advantage of the sensor. Two basic methods of fabrication are top-down and bottom-up, which are explained in the following section with further classification as seen in Fig. 2.

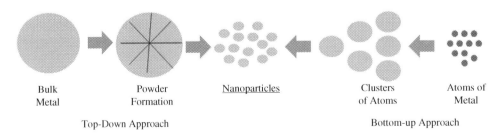

| Bulk Metal | Powder Formation | Nanoparticles | Clusters of Atoms | Atoms of Metal |

Top-Down Approach   Bottom-up Approach

Fig. 2 A schematic diagram for nanoparticle formation.

3.1 Top-down method

This method involves fabrication of solid material on the surface by slicing a large substrate or removal of bulk substrate into NPs. Due to involvement of glass and bulk silicon processing in device development, this methodology is extremely popular among sensor fabrication applications. The first step in every sensor fabrication is channel formation on a blank surface, and this step is done by lithography methods. Lithography is a process of printing and is followed by an etching process, which helps in formation of actual features of the surface. Afterward, functionality to the device is provided by addition of other materials such as metals. There are many types of lithography method. UV lithography involves a photoresist, which is a light-sensitive material, for removal of extra solvent from the surface. Photoresist is gyrated onto a clean surface followed by warm heating for activation of photoinitiators in photoresist and also for solvent removal. Next step is UV exposure on the surface, which is done by placing a desired photomask on the photoresist. The dissolution of the exposed area takes place when placed in developer in the case of a positive photoresist but dissolution of the unexposed area takes place in the case of a negative photoresist [91]. This process is used for sensor surface development, but certain problems occur due to uneven coverage of the photoresist. There are a few methods such as detachment lithography [92] and photoresist spin-step alterations [93] to overcome this problem. This method is usually employed in fabrication of fluidic channels in nano$-$/microfluidic-based biosensors and also in the silicon nanowire (having a 2–5 mm size limit) fabrication [94]. Other types of lithography use a highly focused H + 2 ion beam to form a pattern in a polymer substrate; this type of lithography is known as focused proton beam lithography. This method is highly advantageous over UV lithography as we can get high aspect ratio structures such as pillars and a feature size down to 22 nm [95,96]. Above all, the advantages of this method is costly compared with UV lithography. In nanoimprint lithography (another type of lithography) a prepatterned mold of silicon is pressed into a flat sheet of polymer, which is often made of PDMS or PMMA, followed by heating up to 30 mins. After removing the mold, the device can be used for further immobilization procedure with other molecules [97]. Many researchers have used nanoimprint lithography for the fabrication of polydimethylsiloxane (PDMS) fluidic channels in different biosensors. Along with UV, detachment lithography is also a photoresist-based technique used for development of three-dimensional structures. In this method, a photoresist-coated flat PDMS stamp is brought into contact with a prepatterned silicon substrate. After rapid peeling and annealing, the extended areas of silicon substrate become coated with the photoresist. For further modification into more complicated structures, other lithography methods can be employed.

Lithography is followed by an etching process used for pattern formation. Depending on the type of etch medium (liquid or vapor), the process of etching is classified into dry or wet techniques. Hydrofluoric acid (HF) can be used in buffered oxide etch [98] for glass or wet etching technique [99,100] and KOH for silicon, respectively. HF undercuts the masking layer by providing isotropic wet etching, thus resulting in downward tapering or semicircular sidewalls. Duration time of etching and continuous exposure of HF is essential for depth features. Gold or aluminum metal masking layers are used to protect the substrate with an intermediate chrome adhesion layer [101]. On the other hand, mirror-like sidewalls are obtained in the case of silicon, because KOH provides an anisotropic etch. Reactive ion etching (RIE) can be used to transfer a pattern onto the sensor surface. RIE is more beneficial than wet etching because it is anisotropic [98,102]. CF_4 or CHF_3 is typically used for the glass etching process, while Cl_3/BCl_3 plasma is used for the silicon etching process [103].

3.2 Bottom-up approaches

In this approach, material is manipulated to make building blocks one at a time, atom by atom, which leads to self-assembly of atoms in a controlled manner. Thermodynamic energy minimization is used in self-assembly to make polymers of molecules based on phase segregation. This approach is divided into gas-phase and liquid-phase approaches. Plasma arcing and chemical vapor deposition comes under gas-phase approach. A number of methods are possible in liquid-phase approach like sol-gel synthesis, molecular beam epitaxy, and molecular self-assembly. Alkanethiols are used for self-assembly to form organized monolayer formation on a sensor surface that makes surface modification possible [104]. There are certain methods also discussed by some researchers using advanced tools including pick-and-place approaches like optical tweezers and AFMs [105,106].

Plasma arcing is a common method for fabrication of sensors, where an ionized gas (plasma) is used. When a certain potential difference is applied between the electrodes, ionization of the gas will take place, and the first electrode vaporizes due to the potential difference. In the case of CNT formation, a carbon electrode is used for vaporization to make cations of carbon and takes electron. These electrons are deposited on the second electrode to make CNT.

Chemical vapor deposition is used to make carbide and oxides. In this method, vacuum is used and material is heated to a gas before its deposition on a flat surface, and the deposited material crystallizes as a layer, which grows vertically.

Molecular beam epitaxy is when a molecular beam interacts with a heated crystalline substrate under ultrahigh vacuum conditions to produce a single crystal form, one atomic layer at a time. All processes are performed in a controlled manner to avoid every contamination. This process is mainly used in the semiconductor industry for development of computer chips. Various surface analysis techniques are used to analyze the fabrication at each step. In a study, bismuth selenide (Bi_2Se_3) QDs were made by molecular beam epitaxy through self-assembled process on gallium arsenide (GaAs) substrate. These QDs were formed after acclimatization of bismuth droplets under selenium flux. This method of NP formation provides a convenient way to develop topological insulator QDs in a reproducibly controlled way, with singular spin properties [107].

Sol-gel synthesis is a highly explored technique for synthesis of metal oxides. It is a low temperature-based, highly controllable, and cost-effective method for production of homogeneous, highly stoichiometric and high-quality ultrafine nanostructures. Low temperature chemistry, reproducibility, and high S/V ratio of obtained products are other features that add merit to this technology [108]. A solid material is suspended in a solution in a homogenous manner to form a colloid of solid. Then gelation is performed to make a network of colloid on the surface [109]. Various metal ions, metal alkoxides, and alkoxysilanes are used for synthesis of colloids such as terramethoxysilane and tetra-ethoxysilane, which are the most widely used alkoxysilanes. Sol-gel method is performed at low temperatures compared with other methods. Therefore, it is favorable for many bio-organic molecules fabrication. This technique was used for CuO NP formation to develop a highly sensitive gas sensor [110]. Various NPs of metal oxides are prepared from a sol-gel method such as ZnO, Fe_2O_3, CuO, etc. [16,17,111,112].

Molecular self-assembly is a process to assemble molecules into a well-organized structure with the help of various intermolecular forces like hydrogen bonding, weak noncovalent interactions, electrostatic interactions, π-π stacking, Van der Waal forces, coordination, and ion–dipole interactions [113,114]. Molecular self-assembly is a promising field providing opportunities of research in nanotechnology to get nanostructures with accurate and controlled application of intermolecular forces. There are many examples where molecular self-assembly is used; formation of colloids, lipid bilayers, molecular crystals, and self-assembled monolayers are all examples of the molecular self-assembly. Molecular self-assembly is divided into two types; static and dynamic [12]. In static self-assembly, ordered structures are formed and the system does not disperse energy and reaches an energy minimum. Examples of static self-assembly are NPs, nanorod-structured NPs, nanorods, liquid crystals, block copolymers [115,116] and hierarchical supramolecular systems [117–119]. In dynamic self-assembly, molecules assemble in a complex way only if all components in the system are dissipating energy. It is further sub-divided into two types; template and biological self-assembly. Templated particle assembly involves particle adsorption and arrangement in a specific manner on surfaces unlike the classical adsorption process where adsorption happens in a nonspecific manner. Also the interaction between component and environmental features determine the structure that will form on the surface, for example, crystallization of colloids in three-dimensional optical fields on the surface [120–122]. Biological self-assembly is a hybrid methodology that provides variety and complexity in the organization of molecules. Biological nano-materials based on the self-assembly of biomolecules exhibited wide applications from nanotechnology to biomedical engineering due to their unique properties such as con-trollable self-assembly, good biocompatibility, high thermal stability and flexibility, easy functionality, and facile synthesis. A number of examples are present in nature where self-assembly occurs such as in viruses, DNA, RNA, protein-protein, and enzymes-substrate. With the help of internal interactions such as hydrophobic interaction, hydrogen bonds, π–π interaction, electrostatic interaction, DNA/RNA base pairing, and ligand-receptor interactions play important roles for mediating the self-assembly of biomolecules and promoting the formation of biological nanostructures and nanomaterials. A number of factors affects self-assembly of biomolecules like pH, ionic strength, temperature of solution, and organic stimulators (ethanol, surfactants, and others), and adding NPs or enzymes stimulators have been proven to be effective to promote the self-assembly of biomolecules [123]. The interactions between biomolecules (peptide, protein, DNA, etc.) and material (metal NPs, solid substrates, graphene, CNTs, and others) can guide and inspire the synthesis of hybrid nanomaterials [124,125]. According to the type of binding, biomolecular self-assembly is classified into types such as molecule-molecule interaction, molecule-material recognition, molecule-mediated nucleation and growth, molecule-mediated reduction/oxidization, and various hybrid bionanomaterials by combining with corresponding building blocks like NPs, CNTs, graphene, and others as type 1, 2, 3, and 4, respectively [123]. Graphene oxide was used to develop a fluorescent turn-on sensor to detect lipopolysaccharide (LPS) in clinical injectable fluids, and also it can detect which have this endotoxin in different strains. This sensor was developed by assembling a LPS-binding peptide labeled with tetramethylrhodamine dye on graphene oxide (GO) surface. Quenching happens when this dye-labeled peptide interacts with GO. But a specific binding of peptide with LPS enables the release of the peptide-LPS complex from GO, which results in fluorescence recovery [126]. Another example of self-assembled monolayers (SAM) of 11-mercaptoundecanoic acid was used to adsorb on a gold surface of a screen-printed electrode. EDC-NHS chemistry was used to link this 11-mercaptoundecanoic acid with antitubulin antibody. This biosensor is capable of detecting binding and unbinding events depending upon redox reactions on an electrode surface in a controlled manner [127]. Thus, biological self-assembly provides a vast area of work by using biological molecules. As you can see in Fig. 3, all fabrication methods of sensors are mentioned.

4. Summary

Nanomaterials is prominent in developing a number of methods for their use in agriculture field, medical science as labels, in drug delivery systems, as virocidal agents and surface modifying agents, for sensor development, and they also have anticancerous properties. NPs are improving our daily life by augmenting the efficiency, output, and performance of diagnostic devices, and their use in medicine, diagnostics, and drug delivery is set to rapidly increase. Due to their size ranging from few nanometers to 100 nm, the surface area of NPs increases enormously, and this property makes them

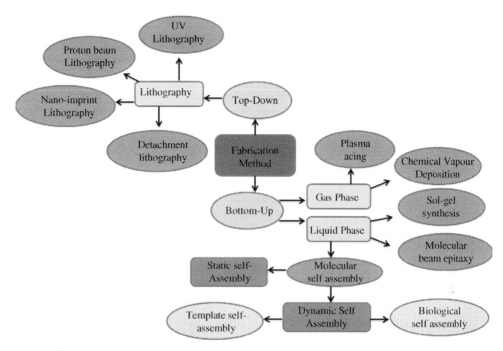

Fig. 3 Different types of fabrication methods for surface modification.

suitable for various advanced applications. In this chapter, we have given a detailed overview of types of NPs, synthesis, characterization, and biosensor fabrication techniques using nanomaterials. These NPs are replacing organic dyes due to their higher photostability and multiplexing features. Nanomaterial-based biosensors are applied in almost every field of science as you can see in Table 1. These "magic bullets" have many interesting features for novel research, and use of nanoproducts is expectedly increasing worldwide. Advanced nanomaterials have been prepared by adapting new fabrication techniques, and many government and private sectors are focusing toward this technology. Incorporation of NPs in a biosensor will help in early disease diagnosis, and improve imaging of disease progression and drug delivery systems with lesser side effects. Biomarkers present in a very minute quantity in body fluids could be easily detected with these NP-based biosensors. Point of care (POC) clinical monitoring devices based on NPs are also under clinical trial for in-vivo use; however, health hazards due to NPs toxicity need to be kept in consideration. Future strategies should focus on utilizing the novel properties of biofilms and to synthesize next-generation biosensors that will increase their use in-vivo clinical, environmental surveillance, and food testing.

Table 1 Use of different nanoparticles in different fields of sensor development.

Sr. no	Nanoparticle	Organism	Limit of detection	Sensitivity	Advantage	References
1	Ag nanoparticles	*M. tuberculosis*	0.03 fM	—	Effective in low concentration	[128]
2.	ZnO nanoparticles	*N. gonorrhea*	0.000704 fmol	—	Effective in low concentration	[129]
3	Platinum and gold flower-like nanoparticles (PtAuNPs)	Glucose	1 µM	17.85 µA/mM cm^2	Provide synergistic effect of both metals	[130]
4	AuNPs	Foot and Mouth Disease Virus (FMDV)	1 copy number of the RNA standard in the rRT-PCR	—	Stable, nontoxic, biocompatiblility	[131]
5	Cadmium–tellurium quantum dots	Human T-lymphotropic virus–1 (HTLV-1)	19.5 pg/µl	—	Enhanced electron transfer	[132]
6	Carboxyl–derivatized magnetic beads	Hepatitis A virus	100 pfu/25 g of green onions	—	Enhanced usability	[133]
7	silver nanoparticles (AgNPs)	microRNA let-7a	10^{-3} f molar	—	Cheaper than gold nanoparticles	[134]
8	SiO$_2$ 20–60 nm	Effective against *Callosobruchus maculatus*	1, 1.5, 2, and 2.5 g/kg	—	Kills insect and larvae in a dose–dependent manner; useful in protection of stored grain	[135]
9	Si–NPs	Effective against *Spodoptera littoralis*	200, 300, 400, and 500 ppm	—	Kills larvae in a dose–dependent manner; increases plant longevity and number of leaves per plant after 15 days of application	[136]
10	Rhodamine B-doped silica nanoparticles	Cu^{2+} in real tap water	35.2 nM	—	Fluorescence imaging of Cu2+ in cells and determination of Cu2+ in real tap water	[134]
11	Silica NPs	Numerous bacterial pathogens	—	—	Multiplexed monitoring of bacterial pathogens within 30 min	[95,96]
12	Glycoconjugate-specific antibody-bound gold NanoWire arrays (GNWA)	*E. coli*	10 cells/0.173 cm^2	—	A rapid method for UTI detection in very low bacteria concentration	[137]
13	Quantum dot	*S. enteritidis*	10^5 CFU/mL	—	Can detect in <3 min	[138]
14	Magnetic nanoparticles	*S. saprophyticus*	7 × 10^4 CFU/mL	—	Selectively traps Gram-positive pathogens	[139]
15	Au nanoparticle	HIV-1 Virus	600 fg/mL to 375 pg mL^{-1}	—	Sensitive virus sensor, label-free system	[140]
16	Ag nanoparticle	HIV-1 p24 antigen	10 and 1000 pg mL^{-1}	—	First time, we have engineered streptavidin-labeled fluorescent silver nanoparticles	[141]

References

[1] B.S. Sekhon, Nanotechnology in agri-food production: an overview, Nanotechnol. Sci. Appl. 7 (2014) 31–53.

[2] R.P. Feynman, There's Plenty of Room at the Bottom, California Institute of Technology, Engineering and Science Magazine, 1960.

[3] L.C. Clark Jr., C. Lyons, Electrode systems for continuous monitoring in cardiovascular surgery, Ann. N. Y. Acad. Sci. 102 (1) (1962) 29–45.

[4] M. Pohanka, P. Skládal, Electrochemical biosensors—principles and applications, J. Appl. Biomed. 6 (2) (2008) 57–64.

[5] L. Liao, S. Wang, J. Xiao, X. Bian, Y. Zhang, M. Canlon, X. Hu, Y. Tang, B. Liu, H. Girault, A nanoporous molybdenum carbide nanowire as an electrocatalyst for hydrogen evolution reaction, Energy Environ. Sci. 7 (1) (2014) 387–392.

[6] H. Nguyen, S.A. El-Safty, Meso-and macroporous Co_3O_4 nanorods for effective VOC gas sensors, J. Phys. Chem. C 115 (17) (2011) 8466–8474.

[7] A. Munawar, Y. Ong, R. Schirhagl, M.A. Tahir, W.S. Khan, S.Z. Bajwa, Nanosensors for diagnosis with optical, electric and mechanical transducers, RSC Adv. 9 (12) (2019) 6793–6803.

[8] J.V. Jokerst, A.J. Cole, D. Van de Sompel, S.S. Gambhir, Gold nanorods for ovarian cancer detection with photoacoustic imaging and resection guidance via Raman imaging in living mice, ACS Nano 6 (11) (2012) 10366–10377.

[9] H.F. Liang, C.T. Chen, S.C. Chen, A.R. Kulkarni, Y.L. Chiu, M.C. Chen, H.W. Sung, Paclitaxel-loaded poly (γ-glutamic acid)-poly (lactide) nanoparticles as a targeted drug delivery system for the treatment of liver cancer, Biomaterials 27 (9) (2006) 2051–2059.

[10] J. Malmo, A. Sandvig, K.M. Vårum, S.P. Strand, Nanoparticle mediated P-glycoprotein silencing for improved drug delivery across the blood-brain barrier: a siRNA-chitosan approach, PLoS One 8 (1) (2013) e54182.

[11] H. Liu, A.T. Neal, Z. Zhu, Z. Luo, X. Xu, D. Tománek, P.D. Ye, Phosphorene: an unexplored 2D semiconductor with a high hole mobility, ACS Nano 8 (4) (2014) 4033–4041.

[12] G.M. Whitesides, B. Grzybowski, Self-assembly at all scales, Science 295 (5564) (2002) 2418–2421.

[13] C.A. Mirkin, et al., A DNA-based method for rationally assembling nanoparticles into macroscopic materials. Nature 382 (6592) (1996) 607–609, https://doi.org/10.1038/382607a0 12.

[14] L. Ding, et al., Utilization of nanoparticle labels for signal amplification in ultrasensitive electrochemical affinity biosensors: a review. J. Anal. Chim. Acta 797 (2013) 1–12, https://doi.org/10.1016/j.aca.2013.07.035 14.

[15] T.W. Odom, J.L. Huang, P. Kim, C.M. Lieber, Atomic structure and electronic properties of single-walled carbon nanotubes, Nature 391 (6662) (1998) 62.

[16] Y. Li, H. Schluesener, S. Xu, Gold nanoparticle-based biosensors, Gold Bull. 43 (2010) 29–41.

[17] Y. Li, L. Xu, X. Li, X. Shen, A. Wang, Effect of aging time of ZnO sol on the structural and optical properties of ZnO thin films prepared by sol–gel method, Appl. Surf. Sci. 256 (14) (2010) 4543–4547.

[18] V. Biju, Chemical modifications and bioconjugate reactions of nanomaterials for sensing, imaging, drug delivery and therapy, Chem. Soc. Rev. 43 (2014) 744–764.

[19] E. Hao, G.C. Schatz, J.T. Hupp, Synthesis and optical properties of anisotropic metal nanoparticles, J. Fluoresc. 14 (4) (2004) 331–341.

[20] P. Mulvaney, Surface plasmon spectroscopy of nano sized metal particles, Langmuir 12 (1996) 788–800.

[21] K.L. Kelly, E. Coronado, L.L. Zhao, G.C. Schatz, The optical properties of metal nanoparticles : The influence of size, shape, and dielectric environment, J. Phys. Chem. B 107 (2002) 668–677.

[22] M. Faraday, X. The Bakerian lecture.—experimental relations of gold (and other metals) to light, Phil. Trans. R. Soc. London 147 (1857) 145–181.

[23] C.G. Grulee, A.M. Moody, The Lange gold chlorid reaction on the cerebrospinal fluid of infants and young children, Am. J. Dis. Child. 9 (1) (1915) 17–27.

[24] H. Li, L. Rothberg, Colorimetric detection of DNA sequences based on electrostatic interactions with unmodified gold nanoparticles, Proc. Natl. Acad. Sci. 101 (39) (2004) 14036 14039.

[25] R.A. Reynolds, C.A. Mirkin, R.L. Letsinger, Homogeneous, nanoparticle-based quantitative color-imetric detection of oligonucleotides, J. Am. Chem. Soc. 122 (15) (2000) 3795–3796.

[26] P.E. Sheehan, L.J. Whitman, Detection limits for nanoscale biosensors, Nano Lett. 5 (4) (2005) 803–807.

[27] P. Jiang, Y. Wang, L. Zhao, C. Ji, D. Chen, L. Nie, Applications of gold nanoparticles in non-optical biosensors, Nanomaterials 8 (12) (2018) 977.

[28] L. Lin, H.Q. Zhao, J.R. Li, J.A. Tang, M.X. Duan, L. Jiang, Study on colloidal Au-enhanced DNA sensing by quartz crystal microbalance, Biochem. Biophys. Res. Commun. 274 (3) (2000) 817–820.

[29] X.C. Zhou, S.J. O'Shea, S.F.Y. Li, Amplified microgravimetric gene sensor using au nanoparticle modified oligonucleotides, Chem. Commun. 11 (2000) 953–954.

[30] H.Q. Zhao, L. Lin, J.R. Li, J.A. Tang, M.X. Duan, L. Jiang, DNA biosensor with high sensitivity amplified by gold nanoparticles, J. Nanopart. Res. 3 (4) (2001) 321–323.

[31] S.H. Chen, Y.C. Chuang, Y.C. Lu, H.C. Lin, Y.L. Yang, C.S. Lin, A method of layer-by-layer gold nanoparticle hybridization in a quartz crystal microbalance DNA sensing system used to detect dengue virus, Nanotechnology 20 (21) (2009) 215501.

[32] K. Kerman, Y. Morita, Y. Takamura, M. Ozsoz, E. Tamiya, Modification of Escherichia coli single-stranded DNA binding protein with gold nanoparticles for electrochemical detection of DNA hybrid-ization, Anal. Chim. Acta 510 (2) (2004) 169–174.

[33] U. Jarocka, R. Sawicka, A. Góra-Sochacka, A. Sirko, W. Zagórski-Ostoja, J. Radecki, H. Radecka, An immunosensor based on antibody binding fragments attached to gold nanoparticles for the detec-tion of peptides derived from avian influenza hemagglutinin H5, Sensors 14 (9) (2014) 15714–15728.

[34] W. Wang, Y. Xie, C. Xia, H. Du, F. Tian, Titanium dioxide nanotube arrays modified with a nano-composite of silver nanoparticles and reduced graphene oxide for electrochemical sensing, Micro-chim. Acta 181 (11 − 12) (2014) 1325–1331.

[35] S. Singh, A. Kaushal, S. Khare, A. Kumar, DNA chip based sensor for amperometric detection of infectious pathogens, Int. J. Biol. Macromol. 103 (2017) 355–359.

[36] Q. He, Z. Zhu, L. Jin, L. Peng, W. Guo, S. Hu, Detection of HIV-1 p24 antigen using streptavidin–biotin and gold nanoparticles based immunoassay by inductively coupled plasma mass spectrometry, J. Anal. At. Spectrom. 29 (8) (2014) 1477–1482.

[37] S.H. Chen, V.C. Wu, Y.C. Chuang, C.S. Lin, Using oligonucleotide-functionalized Au nanoparti-cles to rapidly detect foodborne pathogens on a piezoelectric biosensor, J. Microbiol. Methods 73 (1) (2008) 7–17.

[38] S. Wang, A.K. Singh, D. Senapati, A. Neely, H. Yu, P.C. Ray, Rapid colorimetric identification and targeted photothermal lysis of Salmonella bacteria by using bioconjugated oval-shaped gold nanopar-ticles, Chem. Eur. J. 16 (19) (2010) 5600–5606.

[39] M. Hosseini, H. Khabbaz, M. Dadmehr, M.R. Ganjali, J. Mohamadnejad, Aptamer-based colorimet-ric and chemiluminescence detection of aflatoxin B1 in foods samples, Acta Chim. Slov. 62 (3) (2015) 721–728.

[40] H.A.M. Faria, V. Zucolotto, Label-free electrochemical DNA biosensor for zika virus identification, Biosens. Bioelectron. 131 (2019) 149–155.

[41] A. Kaushik, A. Yndart, S. Kumar, R.D. Jayant, A. Vashist, A.N. Brown, C.Z. Li, M. Nair, A sensitive electrochemical immunosensor for label-free detection of Zika-virus protein, Sci. Rep. 8 (1) (2018) 9700.

[42] S. Eustis, M.A. El-Sayed, Why gold nanoparticles are more precious than pretty gold: noble metal surface plasmon resonance and its enhancement of the radiative and nonradiative properties of nano-crystals of different shapes, Chem. Soc. Rev. 35 (3) (2006) 209–217.

[43] B. Sciacca, T.M. Monro, Dip biosensor based on localized surface plasmon resonance at the tip of an optical fiber, Langmuir 30 (3) (2014) 946–954.

[44] H.H. Lara, N.V. Ayala-Nuñez, L. Ixtepan-Turrent, C. Rodriguez-Padilla, Mode of antiviral action of silver nanoparticles against HIV-1, J. Nanobiotechnol. 8 (1) (2010) 1.

[45] M. Fatima, N.U.S.S. Zaidi, D. Amraiz, F. Afzal, In vitro antiviral activity of Cinnamomum cassia and its nanoparticles against H7N3 influenza a virus, J. Microbiol. Biotechnol. 26 (1) (2016) 151–159.

[46] T.Q. Huy, N.T.H. Thanh, N.T. Thuy, P. Van Chung, P.N. Hung, A.T. Le, N.T.H. Hanh, Cyto-toxicity and antiviral activity of electrochemical–synthesized silver nanoparticles against poliovirus, J. Virol. Methods 241 (2017) 52–57.

[47] Y. He, B. Xu, W. Li, H. Yu, Silver nanoparticle-based chemiluminescent sensor array for pesticide discrimination, J. Agric. Food Chem. 63 (11) (2015) 2930–2934.

[48] M. Xiong, X. Wang, Y. Kong, B. Han, Development of cardiac troponin I electrochemical imped-ance immunosensor, Int. J. Electrochem. Sci. 12 (2017) 4204–4214.

[49] A.S. Ahammad, Y.H. Choi, K. Koh, J.H. Kim, J.J. Lee, M. Lee, Electrochemical detection of cardiac biomarker troponin I at gold nanoparticle–modified ITO electrode by using open circuit potential, Int. J. Electrochem. Sci. 6 (6) (2011) 1906–1916.

[50] M. Negahdary, M. Behjati-Ardakani, N. Sattarahmady, H. Yadegari, H. Heli, Electrochemical apta-sensing of human cardiac troponin I based on an array of gold nanodumbbells-applied to early detec-tion of myocardial infarction, Sensors Actuators B Chem. 252 (2017) 62–71.

[51] N.R. Shanmugam, S. Muthukumar, A.P. Selvam, S. Prasad, Electrochemical nanostructured ZnO biosensor for ultrasensitive detection of cardiac troponin-T, Nanomedicine 11 (11) (2016) 1345–1358.

[52] B.V. Silva, I.T. Cavalcanti, M.M. Silva, R.F. Dutra, A carbon nanotube screen-printed electrode for label-free detection of the human cardiac troponin T, Talanta 117 (2013) 431–437.

[53] M. Norek, M. Włodarski, P. Matysik, UV plasmonic-based sensing properties of aluminum nanocon-cave arrays, Curr. Appl. Phys. 14 (11) (2014) 1514–1520.

[54] F. Valentini, M. Carbone, G. Palleschi, Carbon nanostructured materials for applications in nano-medicine, cultural heritage, and electrochemical biosensors, Anal. Bioanal. Chem. 405 (2–3) (2013) 451–465.

[55] V. Vamvakaki, N.A. Chaniotakis, Carbon nanostructures as transducers in biosensors, Sensors Actu-ators B Chem. 126 (1) (2007) 193–197.

[56] A. Battigelli, C. Ménard-Moyon, T. Da Ros, M. Prato, A. Bianco, Endowing carbon nanotubes with biological and biomedical properties by chemical modifications, Adv. Drug Deliv. Rev. 65 (15) (2013) 1899–1920.

[57] A. Le Goff, K. Gorgy, M. Holzinger, R. Haddad, M. Zimmerman, S. Cosnier, Tris (bispyrene-bipyridine) iron (II): a supramolecular bridge for the biofunctionalization of carbon nanotubes via π-stacking and pyrene/β-cyclodextrin host–guest interactions, Chem. Eur. J. 17 (37) (2011) 10216–10221.

[58] J. Wang, H. Ji, W. Hou, Carbon-nanotube based electrochemical biosensors: a review, Electroanalysis 17 (1) (2005) 7–14.

[59] S. Iijima, Helical microtubules of graphitic carbon, Nature 354 (6348) (1991) 56–58.

[60] S.-P. Germarie, et al., Vertical attachment of DNA–CNT hybrids on gold. J. Electroanal. Chem. 606 (1) (2007) 47–54, https://doi.org/10.1016/j.jelechem.2007.04.010 58.

[61] S. Laschi, E. Bulukin, I. Palchetti, C. Cristea, M. Mascini, Disposable electrodes modified with multi-wall carbon nanotubes for biosensor applications, IRBM29 (2–3) (2008) 202–207.

[62] S. Goyanes, G.R. Rubiolo, A. Salazar, A. Jimeno, M.A. Corcuera, I. Mondragon, Carboxylation treatment of multiwalled carbon nanotubes monitored by infrared and ultraviolet spectroscopies and scanning probe microscopy, Diam. Relat. Mater. 16 (2) (2007) 412–417.

[63] R. Singh, G. Sumana, R. Verma, S. Sood, K.N. Sood, R.K. Gupta, B.D. Malhotra, Fabrication of Neisseria gonorrhoeae biosensor based on chitosan–MWCNT platform, Thin Solid Films 519 (3) (2010) 1135–1140.

[64] A. Dalal, J.S. Rana, A. Kumar, Ultrasensitive Nanosensor for detection of malic acid in tomato as fruit ripening indicator, Food Anal. Methods 10 (11) (2017) 3680–3686.

[65] X. Michalet, F.F. Pinaud, L.A. Bentolila, J.M. Tsay, S.J.J.L. Doose, J.J. Li, G. Sundaresan, A.M. Wu, S.S. Gambhir, S. Weiss, Quantum dots for live cells, in vivo imaging, and diagnostics, Science 307 (5709) (2005) 538–544.

[66] L. Yang, Y. Li, Quantum dots as fluorescent labels for quantitative detection of Salmonella typhimur-ium in chicken carcass wash water, J. Food Prot. 68 (6) (2005) 1241–1245.

[67] K.S. Yao, S.J. Li, K.C. Tzeng, T.C. Cheng, C.Y. Chang, C.Y. Chiu, C.Y. Liao, J.J. Hsu, Z.P. Lin, Fluorescence silica nanoprobe as a biomarker for rapid detection of plant pathogens, Adv. Mater. Res. 79 (2009) 513–516.

[68] D. Geißler, L.J. Charbonnière, R.F. Ziessel, N.G. Butlin, H.G. Löhmannsröben, N. Hildebrandt, Quantum dot biosensors for ultrasensitive multiplexed diagnostics, Angew. Chem. Int. Ed. 49 (8) (2010) 1396–1401.

[69] E. Petryayeva, W.R. Algar, Multiplexed homogeneous assays of proteolytic activity using a smartphone and quantum dots, Anal. Chem. 86 (6) (2014) 3195–3202.

[70] C.J. Murphy, Peer reviewed: optical sensing with quantum dots, Anal. Chem. 74 (2002) 520A–526A.

[71] J.K. Jaiswal, H. Mattoussi, J.M. Mauro, S.M. Simon, Long-term multiple color imaging of live cells using quantum dot bioconjugates, Nat. Biotechnol. 21 (1) (2003) 47–51.

[72] U. Resch-Genger, M. Grabolle, S. Cavaliere-Jaricot, R. Nitschke, T. Nann, Quantum dots versus organic dyes as fluorescent labels, Nat. Methods 5 (9) (2008) 763–775.

[73] R. Freeman, J. Girsh, I. Willner, Nucleic acid/quantum dots (QDs) hybrid systems for optical and photoelectrochemical sensing, ACS Appl. Mater. Interfaces 5 (8) (2013) 2815–2834.

[74] C.Y. Zhang, H.C. Yeh, M.T. Kuroki, T.H. Wang, Single-quantum-dot-based DNA nanosensor, Nat. Mater. 4 (11) (2005) 826–831.

[75] H. Wei, D. Ratchford, X. Li, H. Xu, C.K. Shih, Propagating surface plasmon induced photon emission from quantum dots, Nano Lett. 9 (12) (2009) 4168–4171.

[76] H. Fan, P. Ju, S. Ai, Controllable synthesis of CdSe nanostructures with tunable morphology and their application in DNA biosensor of avian influenza virus, Sensors Actuators B Chem. 149 (1) (2010) 98–104.

[77] C.R. Tamanaha, S.P. Mulvaney, J.C. Rife, L.J. Whitman, Magnetic labeling, detection, and system integration, Biosens. Bioelectron. 24 (1) (2008) 1–13.

[78] A. Hayat, C. Yang, A. Rhouati, J. Marty, Recent advances and achievements in nanomaterial-based, and structure switchable aptasensing platforms for ochratoxin A detection, Sensors 13 (11) (2013) 15187–15208.

[79] K. El-Boubbou, D.C. Zhu, C. Vasileiou, B. Borhan, D. Prosperi, W. Li, X. Huang, Magnetic glyco-nanoparticles: a tool to detect, differentiate, and unlock the glyco-codes of cancer via magnetic resonance imaging, J. Am. Chem. Soc. 132 (12) (2010) 4490–4499.

[80] Y.F. Huang, Y.F. Wang, X.P. Yan, Amine-functionalized magnetic nanoparticles for rapid capture and removal of bacterial pathogens, Environ. Sci. Technol. 44 (20) (2010) 7908–7913.

[81] J. Joo, C. Yim, D. Kwon, J. Lee, H.H. Shin, H.J. Cha, S. Jeon, A facile and sensitive detection of pathogenic bacteria using magnetic nanoparticles and optical nanocrystal probes, Analyst 137 (2012) 3609–3612.

[82] S. Liebana, A. Lermo, S. Campoy, J. Barbe, S. Alegret, M. Pividori, Electrochemical magneto-immunosensing of Salmonella based on nano and micro-sized magnetic particles, Anal. Chem. 81 (2009) 5812–5820.

[83] T. Konry, S.S. Bale, A. Bhushan, K. Shen, E. Seker, B. Polyak, M. Yarmush, Particles and microfluidics merged: perspectives of highly sensitive diagnostic detection, Microchim. Acta 176 (3–4) (2012) 251–269.

[84] D. Liu, Q. Guo, X. Zhang, H. Hou, T. You, PdCo alloy nanoparticle–embedded carbon nanofiber for ultrasensitive nonenzymatic detection of hydrogen peroxide and nitrite, J. Colloid Interface Sci. 450 (2015) 168–173.

[85] H. Kazerooni, B. Nassernejad, A novel biosensor nanomaterial for the ultraselective and ultrasensitive electrochemical diagnosis of the breast cancer-related BRCA1 gene, Anal. Methods 8 (15) (2016) 3069–3074.

[86] X. Guo, B. Liang, J. Jian, Y. Zhang, X. Ye, Glucose biosensor based on a platinum electrode modified with rhodium nanoparticles and with glucose oxidase immobilized on gold nanoparticles, Microchim. Acta 181 (5–6) (2014) 519–525.

[87] S.S. Bhat, A. Qurashi, F.A. Khanday, ZnO nanostructures based biosensors for cancer and infectious disease applications: perspectives, prospects and promises, TrAC Trends Anal. Chem. 86 (2017) 1–13.

[88] R. Polsky, R. Gill, L. Kaganovsky, I. Willner, Nucleic acid-functionalized Pt nanoparticles: catalytic labels for the amplified electrochemical detection of biomolecules, Anal. Chem. 78 (7) (2006) 2268–2271.

[89] D.G. Georganopoulou, L. Chang, J.M. Nam, C.S. Thaxton, E.J. Mufson, W.L. Klein, C.A. Mirkin, Nanoparticle-based detection in cerebral spinal fluid of a soluble pathogenic biomarker for Alzheimer's disease, Proc. Natl. Acad. Sci. 102 (7) (2005) 2273–2276.

[90] K.J. Feng, Y.H. Yang, Z.J. Wang, J.H. Jiang, G.L. Shen, R.Q. Yu, A nano-porous CeO2/chitosan composite film as the immobilization matrix for colorectal cancer DNA sequence-selective electrochemical biosensor, Talanta 70 (3) (2006) 561–565.

[91] M.J. Madou, Fundamentals of Microfabrication: The Science of Miniaturization, CRC Press, 2002.

[92] J. Yeom, M.A. Shannon, Detachment lithography of photosensitive polymers: a route to fabricating three-dimensional structures, Adv. Funct. Mater. 20 (2) (2010) 289–295.

[93] M. Pinti, S. Prakash, A two-step wet etch process for the facile fabrication of hybrid micro-nanofluidic devices, in: ASME 2011 International Mechanical Engineering Congress and Exposition, American Society of Mechanical Engineers Digital Collection, 2011, pp. 647–651 January.

[94] D.R. Kim, C.H. Lee, X. Zheng, Probing flow velocity with silicon nanowire sensors, Nano Lett. 9 (5) (2009) 1984–1988.

[95] L. Wang, W. Zhao, M.B. O'Donoghu, W. Tan, Fluorescent nanoparticles for multiplexed bacteria monitoring, Bioconjug. Chem. 18 (2) (2007) 297–301.

[96] L.P. Wang, P.G. Shao, J.A. van Kan, K. Ansari, A.A. Bettiol, X.T. Pan, T. Wohland, F. Watt, Fabrication of nanofluidic devices utilizing proton beam writing and thermal bonding techniques, Nucl. Instrum. Methods Phys. Res., Sect. B 260 (1) (2007) 450–454.

[97] P. Abgrall, L.N. Lowb, N.T. Nguyen, Fabrication of planar nanofluidic channels in a thermoplastic by hot-embossing and thermal bonding, Lab Chip 7 (2007) 520–522.

[98] L. Gervais, M. Hitzbleck, E. Delamarche, Capillary-driven multiparametric microfluidic chips for one-step immunoassays, Biosens. Bioelectron. 27 (2011) 64–70.

[99] N.F. Durand, P. Renaud, Label-free determination of protein–surface interaction kinetics by ionic conductance inside a nanochannel, Lab Chip 9 (2) (2009) 319–324.

[100] S.H. Roper, M.A. Feldman, E.L. Hughes, Hewlett, T.J. Merkel, J.P. Ferrance, J.P. Landers, A fully integrated microfluidic genetic analysis system with sample-in-answer-out capability, Proc. Natl. Acad. Sci. U. S. A. 103 (2006) 19272–19277.

[101] H. Zhu, M. Holl, T. Ray, S. Bhushan, D.R. Meldrum, Characterization of deep wet etching of fused silica glass for single cell and optical sensor deposition, J. Micromech. Microeng. 19 (6) (2009) 065013.

[102] A.A. Yanik, M. Huang, A. Artar, T.Y. Chang, H. Altug, Integrated nanoplasmonic-nanofluidic biosensors with targeted delivery of analytes, Appl. Phys. Lett. 96 (2) (2010) 021101.

[103] Z.L. Zhang, N.C. MacDonald, A RIE process for submicron, silicon electromechanical structures, J. Micromech. Microeng. 2 (1) (1992) 31.

[104] S. Prakash, M.B. Karacor, S. Banerjee, Surface modification in microsystems and nanosystems, Surf. Sci. Rep. 64 (7) (2009) 233–254.

[105] C. Stavis, T.L. Clare, J.E. Butler, A.D. Radadia, R. Carr, H. Zeng, W.P. King, J.A. Carlisle, A. Aksimentiev, R. Bashir, R.J. Hamers, Surface functionalization of thin-film diamond for highly stable and selective biological interfaces, Proc. Natl. Acad. Sci. 108 (3) (2011) 983–988.

[106] K. Svaboda, S.M. Block, Biological applications of optical forces, and references therein, Annu. Rev. Biophys. Biomol. Struct. 23 (1994) 247–285.

[107] M.S. Claro, I. Levy, A. Gangopadhyay, D.J. Smith, M.C. Tamargo, Self-assembled bismuth selenide (Bi 2 Se 3) quantum dots grown by molecular beam epitaxy, Sci. Rep. 9 (1) (2019) 3370.

[108] L.C. Klein (Ed.), Sol-Gel Optics: Processing and Applications, In: 259 Springer Science & Business Media, 2013.

[109] D. Levy, M. Zayat, The sol-Gel Handbook: Synthesis, Characterization and Applications, Wiley-VCH Verlag GmbH & Co, Germany, 2015.

[110] F. Wang, H. Li, Z. Yuan, Y. Sun, F. Chang, H. Deng, L. Xie, H. Li, A highly sensitive gas sensor based on CuO nanoparticles synthetized via a sol–gel method, RSC Adv. 6 (83) (2016) 79343–79349.

[111] V. Dhanasekaran, S. Anandhavelu, E.K. Polychroniadis, T. Mahalingam, Microstructural properties evaluation of Fe2O3 nanostructures, Mater. Lett. 126 (2014) 288–290.

[112] T. Shrividhya, G. Ravi, Y. Hayakawa, T. Mahalingam, Determination of structural and optical parameters of CuO thin films prepared by double dip technique, J. Mater. Sci. Mater. Electron. 25 (9) (2014) 3885–3894.

[113] J.A. Elemans, A.E. Rowan, R.J. Nolte, Mastering molecular matter. Supramolecular architectures by hierarchical self-assembly, J. Mater. Chem. 13 (11) (2003) 2661–2670.

[114] J.M. Lehn, Toward self-organization and complex matter, Science 295 (5564) (2002) 2400–2403.

[115] D. Chapman, M.N. Jones, M.N. Jones, Molecular self-assembly, in: Micelles, Monolayers, and Biomembranes, Wiley-Liss, 1995, p. 37 (Chapter 1).

[116] C. De Rosa, C. Park, E.L. Thomas, B. Lotz, Microdomain patterns from directional eutectic solidification and epitaxy, Nature 405 (6785) (2000) 433.

[117] A. Kumar, N.A. Abbott, E. Kim, H.A. Biebuyck, G.M. Whitesides, Patterned self-assembled monolayers and meso-scale phenomena, Acc. Chem. Res. 28 (5) (1995) 219–226.

[118] M.L. Schmidt, A. Fechtenkotter, K. Mullen, E. Moons, R.H. Friend, J.D. Mackenzie, Self-organized discotic liquid crystals for high-efficiency organic photovoltaics, Science 293 (5532) (2001) 1119–1122.

[119] K.E. Schwiebert, D.N. Chin, J.C. MacDonald, G.M. Whitesides, Engineering the solid state with 2-benzimidazolones, J. Am. Chem. Soc. 118 (17) (1996) 4018–4029.

[120] F.S. Bates, M.A. Hillmyer, T.P. Lodge, C.M. Bates, K.T. Delaney, G.H. Fredrickson, Multiblock polymers: panacea or Pandora's box? Science 336 (6080) (2012) 434–440.

[121] A.V. Ruzette, L. Leibler, Block copolymers in tomorrow's plastics, Nat. Mater. 4 (1) (2005) 19.

[122] Y. Xia, T.D. Nguyen, M. Yang, B. Lee, A. Santos, P. Podsiadlo, Z. Tang, S.C. Glotzer, N.A. Kotov, Self-assembly of self-limiting monodisperse supraparticles from polydisperse nanoparticles, Nat. Nanotechnol. 6 (9) (2011) 580.

[123] C. Gong, S. Sun, Y. Zhang, L. Sun, Z. Su, A. Wu, G. Wei, Hierarchical nanomaterials via biomolecular self-assembly and bioinspiration for energy and environmental applications, Nanoscale 11 (10) (2019) 4147–4182.

[124] Y. Cheng, L.D. Koh, D. Li, B. Ji, Y. Zhang, J. Yeo, G. Guan, M.Y. Han, Y.W. Zhang, Peptide–graphene interactions enhance the mechanical properties of silk fibroin, ACS Appl. Mater. Interfaces 7 (39) (2015) 21787–21796.

[125] L. Miao, J.S. Han, H. Zhang, L.L. Zhao, C.Y. Si, X.Y. Zhang, C.X. Hou, Q. Luo, J.Y. Xu, J.Q. Liu, Quantum-dot-induced self-assembly of cricoid protein for light harvesting, ACS Nano 8 (4) (2014) 3743–3751.

[126] S.K. Lim, P. Chen, F.L. Lee, S. Moochhala, B. Liedberg, Peptide-assembled graphene oxide as a fluorescent turn-on sensor for lipopolysaccharide (endotoxin) detection, Anal. Chem. 87 (18) (2015) 9408–9412.

[127] X. Hu, A. Guiseppi-Elie, C.Z. Dinu, Biomolecular interfaces based on self-assembly and self-recognition form biosensors capable of recording molecular binding and release, Nanoscale 11 (11) (2019) 4987–4998.

[128] Y. He, D. Liu, X. He, H. Cui, One-pot synthesis of luminol functionalized silver nanoparticles with chemiluminescence activity for ultrasensitive DNA sensing, Chem. Commun. 47 (38) (2011) 10692–10694.

[129] A.A. Ansari, R. Singh, G. Sumana, B.D. Malhotra, Sol–gel derived nano-structured zinc oxide film for sexually transmitted disease sensor, Analyst 134 (5) (2009) 997–1002.

[130] M.F. Hossain, J.Y. Park, Amperometric glucose biosensor based on Pt-Pd nanoparticles supported by reduced graphene oxide and integrated with glucose oxidase, Electroanalysis 26 (5) (2014) 940–951.

[131] M.E. Hamdy, M. Del Carlo, H.A. Hussein, T.A. Salah, A.H. El-Deeb, M.M. Emara, G. Pezzoni, D. Compagnone, Development of gold nanoparticles biosensor for ultrasensitive diagnosis of foot and mouth disease virus, J. Nanobiotechnol. 16 (1) (2018) 48.

[132] M. Norouzi, M. Zarei Ghobadi, M. Golmimi, S.H. Mozhgani, H. Ghourchian, S.A. Rezaee, Quantum dot-based biosensor for the detection of human T-lymphotropic virus-1, Anal. Lett. 50 (15) (2017) 2402–2411.

[133] Y. Zheng, Y. Hu, Development of a fast and efficient method for hepatitis A virus concentration from green onion, J. Virol. Methods 249 (2017) 161–164.

[134] H.K. Elhakim, S.M. Azab, A.M. Fekry, A novel simple biosensor containing silver nanoparticles/propolis (bee glue) for microRNA let-7a determination, Mater. Sci. Eng. C 92 (2018) 489–495.

[135] M. Rouhani, M.A. Samih, S. Kalamtari, Insecticidal effect of silica and silver nanoparticles on the cowpea seed beetle, Callosobruchus maculatus F. (Col.: Bruchidae), J. Entomol. Res. 4 (2012) 297–305.

[136] A.A. El-Helaly, H.M. El-Bendary, A.S. Abdel-Wahab, M.A.K. El-Sheikh, S. Elnagar, The silica-nano particles treatment of squash foliage and survival and development of Spodoptera littoralis (Bosid.) larvae, Pest Control 4 (1) (2016) 175–180 (Biswal et al., 2012, Brennan 2012 and (Elbendary and El-Helaly 2013).

[137] M. Basu, S. Seggerson, J. Henshaw, J. Jiang, R. del A Cordona, C. Lefave, P.J. Boyle, A. Miller, M. Pugia, S. Basu, Nano-biosensor development for bacterial detection during human kidney infection: use of glycoconjugate-specific antibody-bound gold NanoWire arrays (GNWA), Glycoconj. J. 21 (8–9) (2004) 487–496.

[138] G. Kim, A.S. Om, J.H. Mun, Nano-particle enhanced impedimetric biosensor for detedtion of food-borne pathogens, J. Phys. Conf. Ser. 61 (1) (2007) 555 IOP Publishing.

[139] Y.S. Lin, P.J. Tsai, M.F. Weng, Y.C. Chen, Affinity capture using vancomycin-bound magnetic nanoparticles for the MALDI-MS analysis of bacteria, Anal. Chem. 77 (6) (2005) 1753–1760.

[140] J.H. Lee, B.K. Oh, J.W. Choi, Electrochemical sensor based on direct electron transfer of HIV-1 virus at Au nanoparticle modified ITO electrode, Biosens. Bioelectron. 49 (2013) 531–535.

[141] A.D. Kurdekar, L.A. Chunduri, S.M. Chelli, M.K. Haleyurgirisetty, E.P. Bulagonda, J. Zheng, I. K. Hewlett, V. Kamisetti, Fluorescent silver nanoparticle based highly sensitive immunoassay for early detection of HIV infection, RSC Adv. 7 (32) (2017) 19863–19877.

CHAPTER 4

Effect of 2D, TMD, perovskite, and 2D transition metal carbide/nitride materials on performance parameters of SPR biosensor

Akash Srivastava[a], Alka Verma[b], and Y.K. Prajapati[a]
[a]Department of Electronics and Communication Engineering, Motilal Nehru National Institute of Technology Allahabad, Prayagraj, Uttar Pradesh, India
[b]Department of Electronics Engineering, Institute of Engineering and Rural Technology, Allahabad, Uttar Pradesh, India

1. Introduction

Infectious diseases cause 40% of all deaths worldwide [1]. Particularly in many developing countries, bacterial diseases constitute the major causes of death. Thus, the important demands for a bacterial detection sensor are high sensitivity and fast response. Those sensors that are actively working in a commercial purpose are very expensive [2]. Moreover, they need consumable sensor chips, which must include some necessary specifications focused to less size, less thickness, effective sensing area, and so on. Since 1990, prestigious healthcare company General Electric Healthcare (former name Biacore) provided a range of models of surface plasmon resonance (SPR)-based instruments that were only compatible with Biacore accessories [3–5]. Moreover, they were very costly. Thus, developing a low-cost SPR-based multipurpose optical sensor and to perform theoretical modeling of optical biosensors that have higher sensitivity and faster response than existing biosensors is the key motivation for constant research in the biosensing field [6]. A biosensor determines the presence and concentration of a specific biological substance in a biological analyte, and the bioreceptor and transducer are the key components of a biosensing device [7]. A bioreceptor or ligand can be some enzymes, otherwise proteins or antibodies are also used as a bioreceptor; a transducer converts biochemical activity into electrical energy. Professor Leland C Clark Jnr (1918–2005) is known as the "father of biosensors," and today glucose sensors, used daily by millions of diabetic patients, are based on his research [8]. The basic block diagram of a biosensor is given in Fig. 1.

Biosensor performance depends upon parameters like sensitivity, precision, cost-effectiveness, utility, simplicity, ruggedness, reliability, reproducibility, speed accuracy, stability, ease of calibration, etc. [9]. In the last 20 years, plasmonics has played a very

Handbook of Nanomaterials for Sensing Applications
https://doi.org/10.1016/B978-0-12-820783-3.00005-1

57

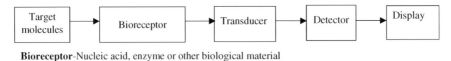

Bioreceptor-Nucleic acid, enzyme or other biological material
Transducer –Metal nano film coated on some substrate (prism)

Fig. 1 Basic block diagram of a biosensor.

important role in nanophotonics as well as in sensing [10], because it provides high sensitivity and reliability [11]. Even though there are various efficient detection schemes available, SPR is the most preferred biosensing method because of its adorable sensing capability, ease of multiplexing, and reliability in various bioapplications like DNA hybridization [12], enzyme detection [13], protein-protein hybridization, protein DNA hybridization [14], and cell- or virus-protein hybridization, and most important, it's a label-free detection method (Fig. 2).

Plasmon are a combined oscillation of those electrons presented at the metal surface and are excited when treated with an electromagnetic source of light, refer to Fig. 2 [15]. Surface plasmon are propagating in nature [16]. When a monochromatic light is incident on the metal dielectric interface, the energies carried by the photons is transferred to the SPs; this phenomenon is called surface plasmon resonance (SPR) or surface plasmon polariton (SPP). Two conditions are necessary for SPR; first, the incident light must be a TM wave or P polarized wave [17], and the energy as well as the momentum of the incident light and SPW wave must be matched [10]. This matching is responsible for a sharp dip in SPR angle or simple decrement of reflected light intensity. The sensing performance of a SPR biosensor is highly dependent upon the reflectance curve; moreover, the incident light angle, wavelength, and dielectric function of metal and dielectric also cause a major effect on resonance condition [18]. In the entire study included in this chapter, the angle interrogation method is proposed where the wavelength of incident life keeps constant while the angle of incidence tends to change [19].

The attachment of analyte on antibodies is a very important phenomenon for a biosensor because the shift in SPR angle takes place when antigen–antibodies come in contact. Fig. 3A and B is dedicated to explaining the basic idea of an SPR-based biosensing device. SPR biosensors are so sensitive that they can detect a very small change in the refractive index of a dielectric medium up to an order of 10^{-7} [20]. Moreover, it's a label-free detection technique, meaning it is able to respond in real time for biomolecular

Fig. 2 Collective oscillation of free electrons.

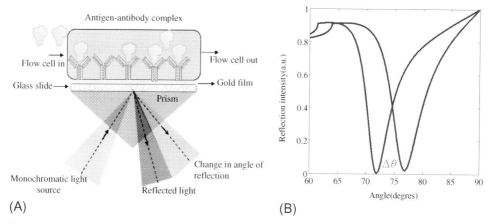

Fig. 3 (A) The scheme of the possible experimental setup for surface plasmon resonance-based instrument. (B) Change in the refractive index of the sensing layer affected by any biomolecules leads to shifting in the SPR angle.

interactions between proteins, DNA/RNA, and small molecules. It also has the capacity to handle small sample sizes, and one can use less expensive materials [21]. SPR also has the ability to reuse sensor chips. These qualities make the SPR technique very cost-effective. Its ability to handle complex samples make it a suitable candidate for testing a variety of complicated sample matrices including for serum analysis.

2. Possible experimental arrangements that realize the idea of SPR plasmon resonance

SPR phenomenon was first observed in 1902 by Wood [22] when he was experimenting. Later in 1941, Fano [23] explained the concept of Wood's observation by using a surface electromagnetic wave propagating along with the interface, which is known as SPR phenomenon. In 1957, it was the combined effect of Powell, and Swan [24] who discovered the idea that fast-moving electrons lost their energy when passing through a thin metal film. Otto (1968) [16,21] and Kretschmann (1971) [10,11] developed a light prism-coupling configuration to excite SPR using the concept of "total internal reflection" (TIR) [22]. A prism–based SPR biosensor is mainly based on two different configurations: Otto configuration and Kretschmann configuration. In Otto configuration, there is an air gap between the metal and the TIR surface, as seen in Fig. 4A. Although, to study SPR in solid-phase media, this is a suitable method, but maintaining distance between the metal and TIR surface is a very tedious task, as it reduces the SPR efficiency so it is less useful. In the Kretschmann configuration, the metal layer is in immediate contact on top of the TIR surface, which makes it an efficient plasmon generator configuration. Current

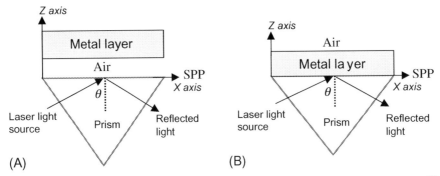

Fig. 4 Experimental arrangements that realize the idea of SPR plasmon resonance. (A) Otto configuration and (B) Kretschmann configuration.

research is based on Kretschmann configuration [24] due to its simplified, user-friendly, metal-dielectric interface. Fig. 4B refers to the Kretschmann configuration.

Metal has a negative dielectric constant (ε) as well as a complex refractive index, which leads to the strong imaginary part of the refractive index ($n + ik$); this property of metal prevents light from penetrating very much [25]. In SPR-based biosensor, metallic film plays a very significant role so its selection should be given specific attention [26].

Silver is better than gold in terms of Full-Width Half-Maximum (FWHM) and shows a sharp curve compared with gold [5]. By comparing several thicknesses of the gold film in the construction of the SPR sensor surface, 45–50 nm metal film thickness shows better sensitivity and the best SPR angle shift [27]. Curve sharpness directly influences the signal-to-noise (S/N) ratio [28], and from Fig. 5 it's concluded that Ag-based biosensing

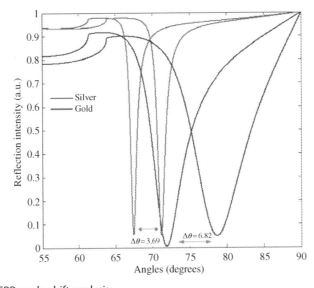

Fig. 5 Au vs Ag SPR angle shift analysis.

device S/N is always greater than Au-based biosensing device [29]. In addition to all these facts, adhesion of thin film of noble metals to common optical glasses is not sufficiently strong, and that's why a thin metallic adhesion layer is deposited onto a glass structure before the deposition of noble metals [30–33]. This adhesion layer [34] binds strongly to the glass substrate, and a subsequently deposited thin film of noble metals also adheres well to the adhesion layer. Titanium (Ti) [35] thin-film 2-nm thick and chromium (Cr) thin film [36] is ideal for the adhesion layer. It's observed that, for BK7 prism, 2-nm adhesion layer of Ti is better as it's less reactive than Cr [13].

3. Mathematical concept and performance parameters

In this section, a brief mathematical overview of SPR condition with a relevant mathematical formula to estimate the performance parameter is given.

3.1 Mathematical analysis and conceptual overview of surface plasmon resonance

When a TM polarized light hits a dielectric–metal interface, SPR occurs under phase-matching condition [37–39], seen in Fig. 4. This internal reflectance is responsible for evanescent wave generation. The generated evanescent wave [40,41] extends beyond the metal film into the applied sample and is attenuated in the region of the infrared spectrum where the sample absorbs energy [42,43]. To get total internal reflectance, incident light angle must be greater than the critical angle; the critical angle depends upon the refractive indices of the sample and ATR crystal and is defined as:

$$\theta_C = \sin^{-1} \frac{n_2}{n_1}$$

where n_1 and n_2 are the refractive index of the ATR crystal (like prism) and sample (like water or biosamples), respectively. A high refractive index is preferred for the ATR crystal to get the minimum critical angle [44]. The intensity of the evanescent wave decays exponentially with distance from the surface of the ATR crystal (refer to the inset of Fig. 6) [45], and distance is in the order of microns [46]. Evanescent wave extension into the sample is known as penetration depth (d_p) [47], and it is the distance from the crystal-sample interface where the intensity of the evanescent wave decays to $1/e$ (~37%) from its original value [48]. It is calculated by:

$$d_p = \frac{\lambda}{\sqrt{2\pi(n_1^2 \sin^2\theta - n_2^2)}} \tag{1}$$

where λ is the wavelength of the incident light, and θ is the angle of incidence. Depending upon the experimental conditions, typical depth of penetration is from 0.5 to 5 μ [49]. As we have discussed earlier, the propagation constant of evanescent wave is made to match that of surface plasmon by adjusting the incident angle.

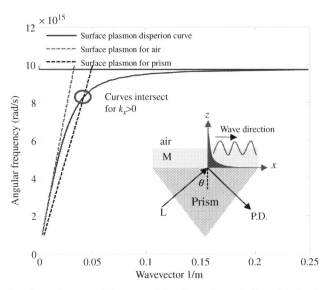

Fig. 6 Excitation of surface plasmons: Close-up of the dispersion relation with the free-space light line and the tilted light line in glass. For air $\omega = ck_x$ (*red* dotted line; *gray* in print version) and glass $\omega = {ck_x}/{N_{glass}}$ (*black* dotted line), *blue* solid line (*light gray* in print version) is dedicated to the surface plasmon dispersion curve.

$$\frac{2\pi}{\lambda} n_p \sin\theta = \mathrm{Re}\left\{\beta_{sp}\right\} \qquad (2)$$

Here, θ is the incident angle, n_p is the refractive index of prism, and β_{sp} denotes the propagation constant of surface plasmon; its mathematical representation is $\left(\beta_{sp} = \frac{2\pi}{\lambda}\sqrt{\frac{\varepsilon_m \varepsilon_s}{\varepsilon_m + \varepsilon_s}}\right)$. Binding of an antibody to a free analyte onto the sensor's surface affects SPR conditions. The refractive index is changed, and a shift (following Beer–Lambert law) of SPR angle occurs. It is a key point to note that antibodies make a coordination bond with a metal layer by sharing its lone pair bond with a central metal atom. Second, for SPR resonance, the surface plasmon generates at the metal–dielectric interface, and permittivity of metal (negative permittivity) is always greater than the permittivity of dielectric (positive permittivity). A popular 2D material, graphene, works as a biorecognition element (BRE) [10] and helps increase the adsorption property of the sensor. The involvement of graphene in numerous optoelectronic devices explores new fields for other two-dimensional materials to handshake with existing and upcoming technologies. Strong absorption of biomolecules on a graphene layer is due to π stacking interaction between the honeycomb lattice structure and carbon-based ring structure, which are proliferating in biomolecules [50]. The size of 2D particles are too small comparison with the wavelength, so due to the quantum confinement effect, the 2D materials show different physical, chemical, and optical properties compared with its bulk form. For the last few years, researchers are continuously working on the application of 2D materials and their applications in the field of SPR-based sensors. In more advanced research, some more pure 2D materials like Black Phosphorene, Germanene, Silicene, Mxenes, and Stanine [18] are the hot topic of research as well.

Before going into the depth, it should be noted that ε_1 is the real and positive quantity and is independent of ω, which is true for air, $\varepsilon_1 = 1$. For all values of $k_x > 0$, the dispersion curve for surface plasmons propagating along the metal-air boundary lies to the right of the dispersion curve for electromagnetic waves in air, $\omega = ck_x$. These features are shown in Fig. 6. Because the dispersion curve for propagation in a vacuum does not intersect the dispersion curve for surface plasmons, it is not possible to match the frequency (ω) and wave vector ($\tilde{k}$) of the surface plasmons to the frequency and wave vector of incident electromagnetic radiation (light, for our purposes) in air [51]. (This is why these surface plasmons are called nonradiative-they cannot propagate into empty space.) But if, for a given (ω), the magnitude of ($\tilde{k}$) of the light incident on the metal-air boundary can be increased, the dispersion curves will intersect. Physically, this can be accomplished by passing the incident light through a medium such as glass, on which a metal film is deposited [52]. The glass has the effect of multiplying the wave-vector by N_{glass} (refractive index of glass) then, the (sufficiently) thin metal film allows some of the incident light to be transmitted through to the metal-air boundary [53], where the surface plasmons are excited. Dielectric constant of the metal layer calculated through Drude-Lorentz model is given by [54]:

$$\varepsilon_m(\lambda) = -\varepsilon_{mr} + i\varepsilon_{mi} = 1 - \frac{\lambda^2 \lambda_c}{\lambda_p^2(\lambda_c + i\lambda)} \tag{3}$$

Here, λ_p (1.68×10^{-7} m) and λ_c(8.93×10^{-6} m) are the plasma wavelength and the collision wavelength of Au, respectively. If a sensing device structure contains a monolayer or two layers, its reflection and transmission value can be estimated through Fresnel's formula. But for a multilayer thin film setup, it can be proven to be a troublesome technique. In this chapter, we have used the transfer matrix method [51], which is a well-suited method for the N-layer model [55]. The tangential field between the boundaries is related by:

$$\begin{bmatrix} A_1 \\ B_1 \end{bmatrix} = M_2 M_3 M_4 \dots M_{N-1} \begin{bmatrix} A_{N-1} \\ B_{N-1} \end{bmatrix} = M \begin{bmatrix} A_{N-1} \\ B_{N-1} \end{bmatrix} \tag{4}$$

Here, A_1 and B_1 are the tangential components of electric and magnetic fields, respectively, at the boundary of the first layer; similarly, A_{N-1} and B_{N-1} are the tangential components of electric and magnetic fields, respectively, at the boundary of the Nth layer. M is the characteristic matrix of the combined structure, which is given by:

$$M = \prod_{K=2}^{N-1} M_K = \begin{bmatrix} M_{11} & M_{12} \\ M_{21} & M_{22} \end{bmatrix} \tag{5}$$

Here,

$$M_K = \begin{bmatrix} \cos\beta_k & -\sin\left(\beta_k/q_k\right) \\ -iq_k\sin\beta_k & \cos\beta_k \end{bmatrix} \tag{6}$$

$$q_k = \left(\frac{\mu_k}{\varepsilon_k}\right)^{1/2} \cos\theta_k$$

and

$$\beta_k = \frac{2\pi}{\lambda} n_k \cos\theta_k(d_k)$$

Going through some calculations, the amplitude reflection coefficient (r) and reflectivity (R) for p-polarized light [56] is obtained as:

$$R = |r|^2 = \frac{(M_{11} + M_{12}q_N)q_1 - (M_{21} + M_{22}q_N)}{(M_{11} + M_{12}q_N)q_1 + (M_{21} + M_{22}q_N)} \tag{7}$$

3.2 Performance parameters

The important parameters that are directly responsible for SPR biosensor performance are sensitivity (it must be high for ideal sensing), figure of merit (FoM) (must be high), detection accuracy (must be high), resolution (must be high), FWHM (must be minimum), and minimum reflectance (R_{min}) (must be minimum showing complete transfer of energy of EM wave). Although each parameter has its significance in sensing device performance, sensitivity and figure of merit are always at the center, which must be high. Sensitivity (S) is defined as:

$$S = \frac{\delta\theta_{res}}{\delta n_c} \; (^\circ/_{RIU}) \tag{8}$$

Where ($\delta\theta_{res}$) is shift in SPR angle, and δn_c is change in the refractive index of the sensing region. FWHM stands for full-width at half-maximum of reflectance dip, expressed as:

$$FWHM = \Delta\theta_{0.5} \tag{9}$$

The value of the FOM is inversely proportional to FWHM [57] and can be calculated by:

$$FOM = \frac{S}{FWHM} \left(^1/_{RIU}\right) \qquad (10)$$

For reliable sensing, SPR curve must be sharp and narrower to get a better S/N ratio. SPR sensing is nothing; it is the detection of refractive index changes in the sample in the direct contact to the sensor's analyzing surface. There is a tradeoff between detection accuracy $(DA = \frac{1}{FWHM})$ and sensitivity of the SPR sensor. It should be noted that at near-infrared regions such as 785 nm, the FWHM tends to be smaller than that of 633 nm under the same condition [58].

3.3 An attractive method for improvement of sensitivity of SPR biosensors

The major issue with conventional SPR sensors is that the metal layer absorption property toward biomolecules is very little, which limits the sensitivity of the biosensor. To find enhanced sensitivity, the metal film must be functionalizing with BRE [59]. A BRE layer plays a very important role to enhance the adsorption of biomolecules on the metal surface. In a BRE layer, the following materials and methods are applicable:

(a) Applying 2D materials like graphene [31,32], black phosphorus [5], or phosphorene
(b) Transition metal dichalcogenides (TMDs) [24, 25] like MoS_2, $MoSe_2$, WS_2, and WSe_2
(c) Heterostructure of 2D/TMD [8,13,60]
(d) Applying perovskite element [61–64] like $BaTiO_3$ [65]
(e) 2D transition metal carbide/nitride materials like MXene [40–45,66]

Two-dimensional nanomaterials are one of the most attractive research topics from the last two decades due to their outstanding potential application in many fields. If we generalize the term *nano,* we get one that's 100,000 times thinner than a strand of wire and 20 times tougher/harder than steel. 2D nanomaterials have a high S/V ratio, less electron scattering, and lower toxicity for biomolecules than its bulk counterpart. Among 2D materials, graphene is outstanding as it has the highest charge carrier mobility, high transparency, high thermal conductivity, high Young's modulus, more stability, and a high specific area compared with others. Moreover, every material has some of its demerits like bandgap absent in graphene, which makes its application limited; MoS_2 has tunable bandgap but low mobility and less stability as a BRE layer; and monolayer black phosphorus or phosphorene have a large tunable bandgap compared with MoS_2 but has the issue of instability [34]. Currently, 2D Van der Waals heterostructure is used to resolve these issues [13,37]. The advantage of these materials is that they preserve the individual properties of each layer and adjust and control the property of the final heterostructure. Following the research paper published by Peng et al. [60], we came to know that the lattice constant of BlueP and MoS$_2$ is 3.268 and 3.164 Å, respectively, which shows that the lattice mismatch is very less (3.18%) and is acceptable. Because MoS$_2$ monolayer and BlueP both have hexagonal lattice structures, heterostructure can be accomplished easily by stacking BlueP on top of MoS$_2$ [67]. Van der Waals force [36] of attraction is responsible for holding both monolayers (with AB or AB$'$ stacking), and in-plane stability is assured by the strong covalent bond between them. Phosphorene is degraded when it comes in contact with an external agent. Heterostructure of two layers of BlueP/MoS$_2$ are able to prevent this unwanted degradation [13]. In the upcoming section of this chapter, we have briefly described the merit and demerits of the optical and electrical properties of these 2D materials, as mentioned earlier, and the effect on the performance parameters of SPR-based biosensors as a BRE layer.

4. Design background and theoretical analysis of the proposed structure

In this section, we present a short introduction of each layer used in the proposed biosensor. SPR sensor structures in Fig. 7 are based on Kretschmann configuration where a metal film is in immediate contact of the base of the prism. A p-polarized (TM Polarized) light of wavelength 633 nm is launched on the base of a prism and, after getting reflected, the light is detected at the other side using a photodetector. To get brief information about each layer, the comparative analysis of different 2D materials, TMD materials, perovskite materials, and 2D transition metal carbide/nitride materials and how can they influence the performance parameters of the biosensor is given with a relevant graph in upcoming sections.

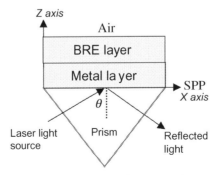

Fig. 7 A simple Kretschmann configuration with BRE layer to enhance evanescent wave strength.

4.1 Analysis of the major components in an SPR-based biosensor

(a) *Prism selection*

Prism is used to increase the wavevector of incident light to match the wavevector of surface plasmon. Among various kinds of prisms (like SF10, SF11, SF5, BAK1, BK7, 2S2G, BAF10) BK7 prism (also called a right-angle prism) is preferred due to its low refractive index ($n = 1.51$). As a result, it proves to be a good substrate for high sensitivity [67]. The property of BK7 contains excellent transmission from 350 to 2000 nm wavelength and has good thermal expansion, low cost, and hard glass that is robust to handle with good chemical resistance.

(b) *Selection of metal layer and optimization of thickness*

Strongest evanescent field is observed at the metal/dielectric interface, so it's very essential to select a suitable metal layer as well as optimize its thickness. Light doesn't penetrate very far into a metal, typically on the order of a nanometer (skin-depth). Thus, the thickness of metal should be kept in the nanometer range. The silver-based configuration shows narrower reflection intensity vs SPR angle curve compared with gold; consequently, the value of FWHM is less as well as S/N ratio is high, therefore high FOM ($FOM \propto 1/FWHM$), respectively [28] (refer to Fig. 5). Despite these favorable conditions, silver is not chemically stable whereas gold is chemically more inert and exhibits strong dispersive characteristics [4]. Moreover, gold-based sensing configuration exhibits much higher sensitivity [29] due to the larger shift in SPR angle. Metal layer thickness must be optimized because if a metal film has a thickness less than optimal value, it will cause the broadening of the SPR curve with greater minimum reflectance near the resonance angle. Similarly, greater thickness of metal layer than optimal value is responsible for SPR curve with smaller resonance angle with minimum greater reflectance. An experiment that reviewed information found that the penetration depth of 50-nm-thick gold film is about 164 nm, whereas for silver it's 219 nm at 633 nm wavelength of the incident light. Even though it has more penetration depth, silver is not preferred because of its stability; the degradation of Ag layer is due to the aging effect, which happens due to the growth of the oxide layer on Ag as AgO_2. It's observed from X-ray photoelectron-spectroscopy that oxide layer thickness in Au is ($\sim$0–0.1 nm), which is very much less compared with another nano-thick metal layer like Ag ($\sim$0.3 nm), Cu ($\sim$1–1.6 nm), and Al ($\sim$1.3–3.3 nm) [49]. The metal layer has a complex refractive index ($n + ik$), where n refers to the real part and k is the extinction coefficient, which represents the amount of attenuation when an EM wave propagates through it. A large value of imaginary part of the complex refractive index of material shows higher damping, which results in reduction in detection accuracy [48]. One more very important reason to prefer Ag and Au over other metal layers is that both metals have 20 times lower imaginary dielectric values than other supporting metal layers.

(c) *Selection of BRE layer*

   Functionalization of the metal film with BRE influences the sensitivity by increasing the adsorption of biomolecules on the metal surface [11]. As a BRE layer, the following materials, monolayer or combined heterostructures, are applicable:

- Applying 2D materials like graphene, black phosphorus, or phosphorene
- Transition metal dichalcogenides (TMDs) like MoS_2, $MoSe_2$, WS_2, and WSe_2
- Heterostructure of 2D/TMD
- Applying perovskite element like $BaTiO_3$
- 2D transition metal carbide/nitride materials like MXene

4.2 Functionalization of metal film with BRE elements

This section is the most important segment of this chapter because here we have briefly described the effect of the previously mentioned materials as a BRE layer step-by-step and investigated the performance parameters. We have also tried to put a brief description of the optical and electronics property of each layer and their behavior with biomolecules. Some graphs reveal the heterostructure of two layers and their combined effect on SPR biosensor performance parameters. In the conventional SPR biosensor configuration, a thin metallic film is coated on the hypotenuse (base) of a rectangular prism, working as a separation layer between the sample (sensing medium) and dielectric (prism). As discussed earlier, the Au layer has an advantage as a metal film, even though the major drawback is that the biomolecules adsorb poorly on gold [48]. This drawback limits the sensitivity of the conventional SPR biosensor; to overcome this issue, we functionalized the metal film with BRE elements.

4.2.1 Analysis with graphene monolayer and multilayers

The discovery of graphene as a 2D material revolutionized the world of sensors. Novoselov and Geim won the Nobel Prize in physics for graphene in 2010. In this study, we investigate the effect of a graphene monolayer as a BRE layer on the performance parameters of SPR-based biosensors. To do so, a monolayer of graphene is coated on the metal surface (Au in this case) in the conventional SPR biosensor setup. From experiments and the literature, it's observed that graphene-on-Au (111) is shown to stably adsorb biomolecules (e.g., ssDNA) [9, 22] due to π stacking interaction between its hexagonal cells and carbon-based ring structures that are wide presented in biomolecules. This advantageous property of graphene performs a greater refractive index change near the graphene-sensing medium interface compared with conventional SPR biosensors (Table 1; Fig. 8).

   Here all the refractive indexes of biosamples are taken as 1.33 and changed in refractive index $\Delta\theta = 0.03$ means, after a change in concentration of sample, the refractive

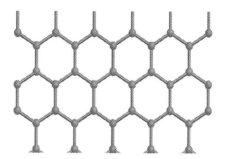

Honeycomb lattice structure of graphene monolayer.

Table 1 Performance parameter analysis of graphene-based sensor than the conventional structure.

Performance parameter	Graphene Au/prism-based sensor	Conventional SPR sensor
Sensitivity $(°/RIU)$ at $\Delta\theta = 0.03$	169.33	164.66
$\Delta\theta_{SPR}$	5.08	4.94
FWHM	6.38	5.87
FOM	26.54	28.05
Detection accuracy	0.15	0.17

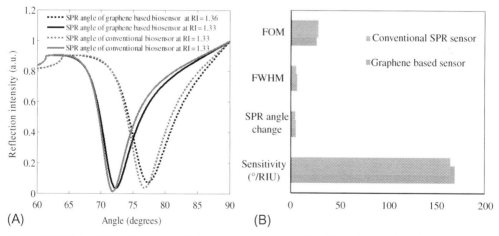

Fig. 8 (A) Shift in SPR angle observed after a change in refractive. The *red* curve (*gray* in print version) for a conventional SPR device *black* curve for graphene-based SPR biosensor. (B) Performance parameter analysis for both structures

index is 1.36. We are restricting the increase of the refractive index beyond a certain limit because, if so, it indicates that the presence of biomolecules is very high, then the sampling layer behaves like a separate layer with a particular thickness. It is not a favorable condition, and it may change the reflectance property of the sensor. Complex refractive index of graphene is calculated by the direct formula:

$$n + ik = 3 + \frac{C_1 \lambda}{n} \tag{11}$$

where $C_1 = 5.446 \; \mu\mathrm{m}^{-1}$ that are necessarily implied by the opacity, and $\lambda = 0.633 \; \mu\mathrm{m}$ (in the present case). Eq. (11) is applicable for any refractive index proposed for graphene in a visible region; above the visible region, Kubo formula is applied.

Graphene layer coating on a metal layer is responsible to modify the propagation constant of SPP; thereby, enhancement in sensitivity concerns the change in refractive index. Addition of the graphene layer, although it increases adsorption of biomolecules, which causes an increase in sensitivity at the same time, it also affects detection accuracy (DA) due to the broadness of the SPR curve. The supporting graph for SPR angle vs. reflection intensity and effect on performance parameters after an increase in the graphene monolayer in the form of $(0.34 \times L$, where $L = 1, 2, 3 \ldots)$ is given in Fig. 9A and B. One more fact should also be noted: By increasing the number of layers up to five, the maximum sensitivity reaches up to 183°/RIU, but minimum reflectance is also an important parameter to optimize the number of the graphene layer. Graphene monolayer provides deeper penetration of surface plasmon compared with other 2D materials like MoS_2. Because the refractive index of graphene monolayer is complex ($n + ik$ form), the smaller the thickness of 2D material's layer, the less effective the damping of surface plasmons, which leads to greater penetration depth. However, if the graphene layers increase more than a single layer, it will produce more damping in the surface plasmon due to large imaginary dielectric constant for higher graphene layers [24] and hence result in a decrement of detection accuracy (FWHM $\propto$ damping and DA $\propto$ 1/FWHM). It's concluded from the data shown in Table 2 and from Fig. 9A and B that sensitivity increases linearly with inclusion of several layers. However, as described earlier, sensitivity increment occurs due to π stacking interaction between honeycomb lattice structure of graphene and biomolecules' carbon ring structure, although the increment in sensitivity is a linear function of several graphene layers up to a certain limit and linearity breaks down beyond a certain number of graphene layers. The graphene has zero bandgap, and valance band and conduction band touch each other at a conical point. Graphene thin film can produce micromechanical cleavage or exfoliation of graphene from graphite using the scotch tape method, and by using chemical vapor deposition (CVD) [68] synthesis of graphene for a large area is possible. Transfer of graphene on a metal layer can be done by using wet etching or floating method where transfer method possible by using polymethyl methacrylate (PMMA) [25,67].

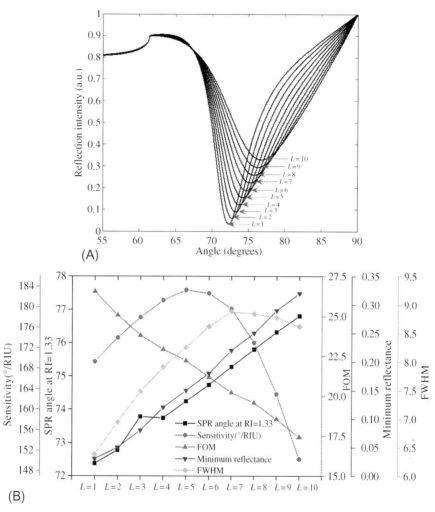

Fig. 9 Analysis of performance parameter by increasing the number of the graphene layer. (A) Shift in SPR angle from $L = 1$ to $L = 10$. (B) Effect on the performance parameter.

4.2.2 Analysis of graphene and MoS$_2$-based combined structure

Graphene has zero bandgap, which limits its application, whereas the presence of band-gap, work function, and high absorption efficiency, compared with graphene, MoS$_2$ achieves an extra advantage for use in biosensing devices [5]. Moreover, if we focus on the optical property of single-layer graphene, light absorption efficiency is only 2.3%, whereas the absorption efficiency of MoS$_2$ is higher, up to 5% [32]. To integrate the advantages of both materials, graphene as well as MoS$_2$, we have incorporated both in a single biosensor to get increased sensitivity. MoS$_2$ is a semiconductor that possesses the

Table 2 Performance parameter analysis by increasing the graphene layer in the proposed sensor.

Number of graphene layer	SPR angle at RI = 1.33	Sensitivity (°/RIU) at $\Delta\theta = 0.03$	FOM	R_{min}	FWHM	DA
$L = 1$	72.38	169.33	26.54	0.03	6.38	0.15
$L = 2$	72.78	174	25.07	0.05	6.94	0.14
$L = 3$	73.78	178	23.79	0.08	7.48	0.13
$L = 4$	73.74	181.33	22.92	0.12	7.91	0.12
$L = 5$	74.24	183.33	22.22	0.15	8.25	0.12
$L = 6$	74.74	182.66	21.19	0.18	8.62	0.11
$L = 7$	75.28	179.66	20.23	0.22	8.88	0.11
$L = 8$	75.8	173	19.57	0.25	8.84	0.11
$L = 9$	76.32	163	18.56	0.29	8.78	0.11
$L = 10$	76.81	150.33	17.43	0.32	8.62	0.11

property of direct bandgap of 1.8 eV [69] that will largely compensate for the weakness of gapless graphene. Molybdenite is easily available in nature, thus MoS_2 is one of the most studied layered transition metal dichalcogenides (TMDCs). The combined structure of graphene and MoS_2 is possible due to the same lattice structure. The front and side view of the crystal structure diagram of monolayer MoS_2 shown in the inset indicate a layer of molybdenum atoms (blue) sandwiched between two layers of sulfur atoms (yellow). To use the combined property of graphene and MoS_2, monolayers of both materials are placed side-by-side between the metal film and sensing medium. The results show that sensitivity is enhanced 13.56% compared with a conventional sensor. This sensitivity enhancement is calculated using $\frac{S-S_0}{S_0} \times 100\%$ [70], where S is modified sensitivity of MoS_2/graphene sensor and S_0 is the sensitivity of the conventional sensor. Comparative analysis regarding the performance parameter is given in Table 3. In this structure, the base of a prism is coated with Au thin film having a thickness of 50 nm, then a MoS_2 layer is applied in immediate contact with a gold layer having the thickness of

Table 3 Effect of 2D and TMD materials on SPR biosensor's parameter.

Performance parameter	Graphene/MoS₂/Au/prism-based sensor	Conventional SPR sensor
Sensitivity (°/RIU) at $\Delta\theta = 0.03$	187	164.66
θ_{SPR} at RI = 1.33	74.27	72
$\Delta\theta_{SPR}$	5.6	4.94
FWHM	8.51	5.87
FOM	21.97	28.05
Detection accuracy	0.11	0.17
R_{min}	0.11	0.01

$M \times 0.65$ nm, where M is the number of MoS_2 layers. Finally there is a BRE layer at the top of the sensor graphene layer ($L \times 0.34$ nm, where L is the number of graphene layers). The reflection intensity versus SPR angle analysis based on a four-layer structure (prism/Ag/graphene-MoS_2/sensing medium) is briefly given in Fig. 10A and B. It is observed in the previous case that, by applying the graphene layer over an Au layer, sensitivity increased up to 169.33 (°/RIU) from its conventional structure's sensitivity of 164.66 (°/RIU), but this increment in sensitivity is relatively low and not sufficient.

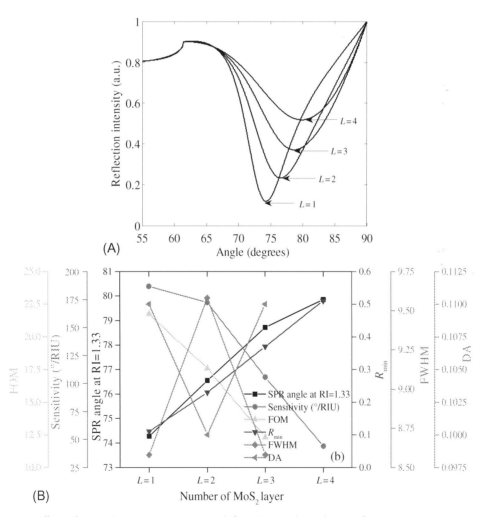

Fig. 10 Effect of MoS_2, layer variation on (A) shift in SPR angle and (B) performance parameters.

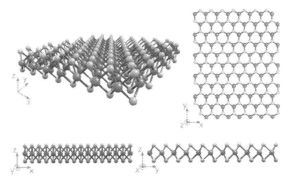

This image remains the copyright of Ossila Ltd. Taken with permission from Ossila.com.

An appreciable enhancement in the sensitivity of up to 187(°/RIU) was obtained by implementing only one layer of MoS_2. Relevant graph for SPR angle modification compared with graphene-based and conventional structure-based device as well as performance parameter analysis is given in Fig. 11. MoS_2 have an indirect bandgap of 1.2 eV in its bulk form, whereas 1.8 eV for monolayer have direct bandgap (Quantum Confinement effect) so tunable bandgap is the advantage of MoS_2, but absorption efficiency toward biomolecules is low and charge carrier mobility is also less. Moreover, due to indirect bandgap in its bulk form limiting the use of more layers for further improvement in its sensitivity, that's why the combined structure of limited MoS_2 and graphene layer-based structure plays a significant role in sensitivity enhancement. Table 3 shows the effect of 2D and TMD materials on SPR biosensor's parameter. The effect of MoS_2, layer variation on performance parameter, and shift in SPR angle is shown in Table 4. In this subsection, the performance of the proposed SPR design is checked by changing several MoS_2 layers, keeping one of graphene layers fixed as a monolayer. Fig. 10A presents the reflectance curve for a varying number of MoS_2 layers up to $L = 4$, and Fig. 10B shows the effect on performance parameters.

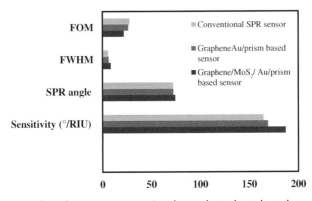

Fig. 11 Comparative analysis between conventional, graphene-based, and graphene with MoS_2-based biosensor's performance parameters.

Table 4 Comparative study of performance parameters.

Number of MoS$_2$ layer	SPR angle at RI = 1.33	Sensitivity (°/RIU) at $\Delta\theta = 0.03$	FOM	R_{min}	FWHM	DA
$L = 1$	74.27	187	21.75	0.11	8.58	0.11
$L = 2$	76.55	172.33	17.58	0.23	9.58	0.10
$L = 3$	78.72	105.66	12.31	0.37	8.58	0.11
$L = 4$	79.85	44	–	0.51	–	–

It reflects that resonance angle increases on the increasing trend for several MoS$_2$ layers. When calculating sensitivity in a similar way, as we have calculated it in Section 4.2.1, we get a sensitivity of 187°/RIU ($L = 1$), 172.3°/RIU ($L = 2$), 105.6°/RIU ($L = 3$), and 44°/RIU ($L = 4$). Thus, sensitivity decreases due to the limitation of the angular range. Finally, in Fig. 11, a comparative analytical graph between conventional, graphene-based, and graphene/MoS$_2$-based SPR biosensor is given. Monolayer synthesis of MoS$_2$ is possible on quartz substrate by CVD process where molybdenum trioxide (MoO$_3$) powder and sulfur powder used as a reactant and for transfer of MoS$_2$ on metal [69] with ultrasonic bubbling transfer method [35] is most suitable and in fashion technique.

4.2.3 Analysis with WS$_2$ as a TMDC material

Tungsten disulfide (WS$_2$) [22] is a well-known competitor of molybdenum disulfide (MoS$_2$) and graphite as observed in various literature and sensing applications. It's a matter of fact that molybdenum and tungsten belong to the same chemical family as in the periodic table. Tungsten disulfide is thermally heavier and more stable than molybdenum disulfide (50°C–100°C) [22]. Numerous features of MoS$_2$ such as cheaper price, easier availability, and strong and innovative marketing makes MoS$_2$ extremely popular, but WS$_2$ is not a new chemical and has been around as long as MoS$_2$ [71] and is widely used by NASA, in various military devices and weapons, and the aerospace and automotive industry. The thermal stability of tungsten disulfide is one more property that makes WS$_2$ a promising candidate, as it has a higher density 7500 kg/m^3 compared with MoS$_2$, 5060 kg/m^3 [21,68,72].

It appears from Table 5 that the structure contains WS$_2$ as a TMDC material with the highest sensitivity up to 201.33°/RIU compared with the sensitivity of 187°/RIU for the same type of structure containing MoS$_2$ as a TMDC material. SPR angle vs. reflection intensity graph by varying refractive index from 1.33 to 1.36 is given in Fig. 12C.

4.2.4 Analysis of black phosphorus and WS$_2$ based combined structure

Black phosphorus (BP) falls in the family of 2D material having unique electrical, mechanical, and optical properties like thickness-dependent tunable bandgap (from

Table 5 Performance parameter analysis between conventional, WS₂, and MoS₂-based sensor.

Performance parameter	Graphene/WS$_2$/Au/ prism-based sensor	Graphene/MoS$_2$/Au/ prism-based sensor	Conventional SPR sensor
Sensitivity (°/RIU) at $\Delta\theta = 0.03$	201.33	187	164.66
θ_{SPR} at $RI = 1.33$	74.78	74.27	72
$\Delta\theta_{SPR}$	6.04	5.6	4.94
FWHM	7.84	8.51	5.87
FOM	25.67	21.97	28.05
Detection accuracy	0.12	0.11	0.17
R_{min}	0.054	0.11	0.01

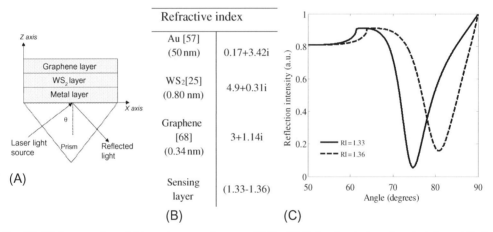

Fig. 12 (A) Proposed model based on graphene and WS₂. (B) Refractive index of materials and their optimized thickness used in the sensor. (C) SPR angle variation.

0.3 to 2 eV), high work function, high carrier mobility, and greater adsorption energy than graphene and MoS₂ and 40 times faster fast response time, making it suitable for efficient binding of biomolecules and improvising its sensing ability [8,31,73]. Monolayer of BP possesses good biocompatibility, good renal clearance, and low toxicity rather than bulk crystal, which is more toxic and does not possess good renal clearance. One of the major issues with BP is that it is degraded easily underexposure with air or water, although several methods have been recently suggested and experimentally verified to synthesize stable BP monolayers [10]. One such method was presented by Avsar et al. [74] where he suggested encapsulation of a BP-based biosensing device with hexagonal boron nitride (h-BN) in a layer-by-layer manner under inert argon and water in oxygen-less environment (Fig. 13).

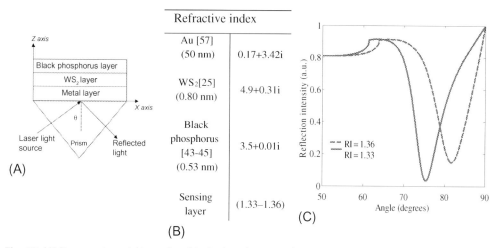

Fig. 13 (A) Proposed model based on black phosphorus and WS$_2$. (B) Refractive index of materials and their optimized thickness used in the SPR sensor. (C) SPR angle variation.

Table 6 Performance parameter analysis between graphene and black phosphorus-based sensor.

Performance parameter	Graphene/WS$_2$/Au/prism-based sensor	Black phosphorus/WS$_2$/Au/prism-based sensor
Sensitivity (°/RIU) at $\Delta\theta = 0.03$	201.33	214
θ_{SPR} at $RI = 1.33$	74.78	75.40
$\Delta\theta_{SPR}$	6.04	6.42
FWHM	7.84	7.95
FOM	25.67	26.97
Detection accuracy	0.12	0.12
R_{min}	0.054	0.03

This kind of encapsulation provides equally good performances under ambient as well as in vacuum conditions as used in heterostructures. In this subsection, we have applied black phosphorus in place of graphene as a 2D material and got very appreciable outcomes like high sensitivity 214°/RIU, high FOM 26.97, and minimum reflectance (R_{min}) 0.03. A comparative study for performance parameter analysis between graphene and black phosphorus-based sensor is given in Table 6 and a pictorial overview to elaborate the results are shown in Fig. 14.

4.2.5 Analysis of MXenes, black phosphorus, and WS$_2$-based combined structure

MXene is two-dimensional transition metal carbide, nitride, or carbonitride that shows highly accessible hydrophilic surfaces in contrast to the hydrophobic nature of graphene

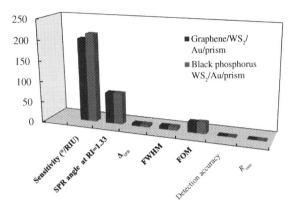

Fig. 14 Comparison graph between graphene/WS$_2$/Au/prism and black phosphorus/WS$_2$/Au/prism-based sensor's performance parameters.

and most of the other 2D materials. With a large surface area, high biocompatibility, and long-term stability, MXene sensors/biosensors offer high reproducibility of results over a long period of time. 2D transition metal carbide and nitride is known as MXene with general formula $M_{n+1}AX_n$ ($n = 1, 2, 3$). Researchers coined these ceramics as MAX phase.

  M is a d block transition metal (Ti, Sc, Cr, etc.)

  A is a main group element

  X is either C or N atoms.

MXenes are molecular sheets obtained from carbides and nitrides of transition metals like titanium. Until recent research, titanium carbide ($Ti_3C_2T_x$) was the only scrutinized material that falls in the family of MXene, as well as in the field of sensing. According to recent reports, other transition metal MXene-based detection systems are very rare. The simplest way to synthesize MXene is by removing the "A" element from MAX phase compounds. In MAX phase or $M_{n+1}X_n$, layers are packed with X atoms and M_{n+1} interleaved with layers of A. Here, it should be noted that M is strongly bonded with X with covalent/metallic/ionic bond, whereas M-A layers are weakly bonded. Treating with high temperature causes the decomposition of M-A bonds. This process converts the MAX phase into $M_{n+1}X_n$ form, and after going through the process of recrystallization, a 3D $M_{n+1}X_n$-like structure is obtained. To obtain a 2D MXene sheet, the MAX phase powder is stirred in aqueous hydrofluoric acid (HF) for a certain period followed by centrifugation and washing several times with distilled water until pH level reaches between 4 and 6. Consequently, loosely packed, exfoliated, graphite-layered structures are formed and named as MXene. A block diagram related to the systematic process of preparation of graphite-like 2D MXenes from 3D MAX phase is given in Fig. 15 [43].

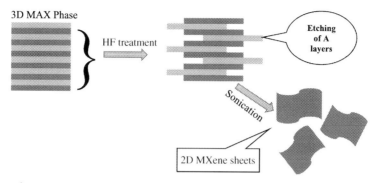

Fig. 15 Pictorial overview to obtain 2D MXenes from 3D MAX phase crystal.

MXene (Ti$_3$C$_2$T$_x$) exhibits plasmonic property, shows 77% transmittance in the visible range, is s useful material as 2D for SPR sensors, and has surface-enhanced Raman spectroscopy applications [5, 10]. There are plenty of surface plasmons in multilayered MXene (Ti$_3$C$_2$T$_x$) possessing energy from 0.3 to 1 eV compared with bulk form. So, it is a new 2D-layered biosensing material having extraordinary properties of tunable bandgap, tunable work function that can be modified as per surface terminations, chemical and physical stability, and larger specific area to attach biomolecules. Larger surface area and its hydrophilic nature enhance the adsorption of biomolecules and, due to this, it can be used as a BRE layer for biosensing. Exploiting these novel properties of MXene, exploration of vdW heterostructure is possible to design new electronic and optoelectronic devices for future applications. Recent studies show that, using MXene, a 2DLM can be assembled layer-by-layer with MXene itself or other 2DLMs resulting in vdW heterostructures. The refractive index and monolayer thickness for MXene (Ti$_3$C$_2$T$_x$) is 2.39 + 1.33i and 0.993 nm in the visible range at 633 nm operating wavelength [10]. Authors have done a study of the effect of nanomaterial layers (i.e., multilayers performance) on the sensitivity of sensors according to the state of the art performance with multilayers like the published research paper written by Wu, You, et al. [46] and Xu, Ang, et al. [47]. Here, the authors presented the performance of the proposed sensor using different metal layers like copper (Cu), silver (Ag), and gold (Au) along with the effect of nanomaterial layer as explained by Wu, You, et al. and Xu, Ang, et.al.

Fig. 16A–C shows the sensitivity analysis for the proposed SPR biosensor based on different metals (Cu, Ag, and Au) at λ = 633 nm. Effect on the sensitivity of biosensing device was tested by increasing one particular layer and keeping the rest of the layers fixed at the monolayer. In Fig. 16A, the sensing device is based on the Cu layer, the highest sensitivity 172°/RIU of the sensor obtained when a monolayer of each layer MXene, BP, and WS$_2$ was used. Furthermore, in Fig. 16B, it's observed that by using silver as a metal layer, sensitivity is reduced up to 146°/RIU at the monolayer-based structure,

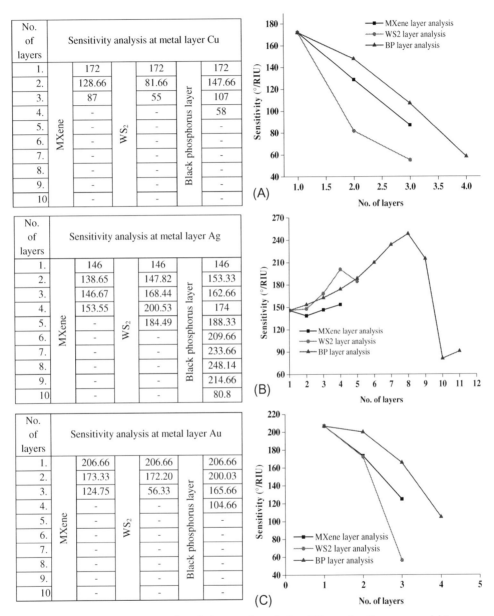

Fig. 16 Sensitivity analysis at incident light wavelength $\lambda = 633$ nm for the proposed biosensors. (A) Cu as a metal layer. (B) Ag as a metal layer. (C) Au as a metal layer.

but by an increasing the number of black phosphorus layers, sensitivity tends to increase. The corresponding sensitivity enhancement analysis tables are attached to a concerned figure. Finally in another case shown in Fig. 16C, we have taken the structure based on Au as a metal layer, using the monolayer of MXene/WS$_2$/BP and got

Table 7 Analysis of performance parameter SPR biosensor based on 2D/TMD/2D transition metal carbide, nitride, or carbonitride-based material.

Performance parameter RI = 1.33–1.36	Black phosphorus/WS$_2$/MXene Au/prism-based sensor
Sensitivity (°/RIU) at $\Delta\theta = 0.03$	206.66
θ_{SPR} at RI = 1.33	75.98
$\Delta\theta_{SPR}$	6.2
FWHM	8.65
FOM	23.89
Detection accuracy	0.11
R_{min}	0.10

much-improved sensitivity up to 206.66°/RIU than conventional metal-based structure sensitivity 164.40°/RIU (sensitivity of conventional metal-based structure), the enhancement to sensitivity is 25.70% (Table 7).

It's clear from Fig. 16B and related table data that, by using silver as a metal layer, the device gives the highest sensitivity up to 248°/RIU using multilayers of black phosphorus $L = 8$. But it's also a matter of fact that when one opts the monolayer of each layer used in the device, the gold-based sensing device as a metal layer shows the highest sensitivity up to 206.66°/RIU. In the upcoming section, we present a performance parameter analysis table based on black phosphorus/WS$_2$/MXene Au/prism combined structure.

4.2.6 Analysis of perovskite element-based SPR biosensor

Perovskite materials are those that have the generic form ABX$_3$, and they have the same crystallographic structure [22]. There is some important material that falls in this category like barium titanate (BaTiO$_3$) [75], lead titanate (PbTiO$_3$), potassium niobate (KNbO$_3$), methylammonium lead halide (CH$_3$NH$_3$PbI$_3$), etc. [23]. If we take the example of BaTiO$_3$ then the simplest way to describe a perovskite material [76] structure is a cubic unit cell with titanium (Ti) atoms at the corners, oxygen (O) atoms are at the midpoint of the edges, and barium (Ba) atom at the center [24]. In this segment, we have taken barium titanate (BaTiO$_3$), also known as BTO, for device structure due to its excellent properties like ferroelectricity, piezoelectricity, and high dielectric constant value. The high dielectric constant is directly responsible for increasing the electric flux density [25,67], if all other physical parameters remain unchanged. This enables the sensing device to hold their electric charge for a long period of time, which is the key requirement of the biosensor. BaTiO$_3$ nanoparticles are synthesized using various chemical methods such as coprecipitations, hydrothermal reaction, and alkoxide sol–gel process [8]. Silver-barium titanate (Ag-BaTiO$_3$)-based SPR biosensor proposed by Fouad, Sabri et al. [77] in 2016 investigated the sensitivity of 280°/RIU. BaTiO$_3$ was discovered during World War II in 1941, and it shows a ferroelectric behavior and is a photorefractive material. This material exhibits a sequence of ferroelectric phase transition and has an indirect bandgap. At high temperature, $BaTiO_3$ is paraelectric with a cubic structure. On cooling, this material undergoes a successive structural phase transition. In 1968, Wemple et al. [78] calculated the refractive index of $BaTiO_3$ at room temperature using a single-term Sellmeier equation:

$$n = \sqrt{1 + \frac{S_0 \lambda_0^2}{1 - \frac{\lambda_0^2}{\lambda^2}}} \tag{12}$$

where S_o is an average interband oscillator strength, and λ_0 is an average interband oscillator position in wavelength units. After inputting these values directly from this chapter, we have obtained generalized formula as:

$$n = \sqrt{1 + \frac{4.187}{1 - \left(\frac{0.223}{\lambda}\right)^2}} \tag{13}$$

Where λ is the wavelength of incident light (in this chapter, we have taken $\lambda = 0.633$ μm). In pure form, $BaTiO_3$ is insulating in nature, however, doping with small amounts of metals, most notably with scandium, yttrium, neodymium, samarium, etc., converts its property in a semiconductor material. BTO films with different layer numbers were highly transparent, and the optical transmittance of the films with one layer is above 80% when the wavelength is between 350 and 1000 nm. High transparency of BTO shows that it has small surface roughness and good homogeneity (Table 8).

Second, the optical transmittance of the BTO films decreases with the increasing of the layer number, which indicates that the optical transmittance of the BTO films decreases with the increasing of the thickness. Moreover, in the ultraviolet wavelength

Table 8 Comparative study between graphene/MoS$_2$ and black phosphorus/WS$_2$-based SPR biosensor including BaTiO$_3$ as a perovskite element

Performance parameter RI = 1.33–1.34	Graphene/BaTiO₃/MoS₂/Ag/ prism-based sensor (Structure-1)	Black phosphorus/BaTiO₃/WS₂/Ag/ prism-based sensor (Structure-2)
Sensitivity (°/RIU) at $\Delta\theta = 0.01$	268.66	371
θ_{SPR} at RI = 1.33	80.12	82.14
$\Delta\theta_{SPR}$	2.68	3.71
FWHM	6.89	5.92
FOM	38.99	62.66
Detection accuracy	0.14	0.16
R_{min}	0.07	0.00033

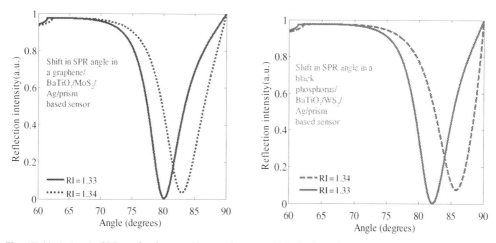

Fig. 17 Variation in SPR angle observed in graphene and black phosphorus-based SPR biosensor using BaTiO$_3$.

region around 350 nm, the transmission of the BTO films decreases sharply, which shows interband transitions [79]. In 2019, Sun, Wang et al. [80] found the sensitivity to be 294°/RIU using double metal layers Ag and Au with a combination of BaTiO$_3$ and graphene as a BRE layer (Fig. 17).

After going through these research papers, we can claim that BaTiO$_3$ plays a very significant role in enhancing biosensor sensitivity [16,77,81]. To investigate the effect on the performance parameter of a biosensor, we have applied a monolayer of BaTiO$_3$ first between graphene and MoS$_2$ and a second time between black phosphorus and WS$_2$. Both of the analyses were attempted on an Ag layer. For the comparative analysis, with respect to the advantage of a black phosphorus/WS$_2$-based sensor concerning the graphene/MOS$_2$-based sensor, a brief study of the performance parameter is given in Table 9 and a graphical overview shown in Fig. 18. Sensitivity analysis of black phosphorus-based biosensor concerning the surrounding materials is discussed with the help of Fig. 19. The figure indicates the change in sensitivity concerning the change

Table 9 Refractive index and optimized thickness of the materials used in the proposed perovskite element-based SPR biosensor.

Type of material/optimized thickness	Refractive index
Ag (46 nm)	0.056206 + 4.2776i
WS$_2$ (0.80 nm)	4.9 + 0.31i
MoS$_2$ (0.65 nm)	5.08 + 1.17i
Black phosphorus (0.53 nm)	3.5 + 0.01i
Graphene (0.34 nm)	3 + 1.1491i
BaTiO$_3$ (10 nm)	2.4043

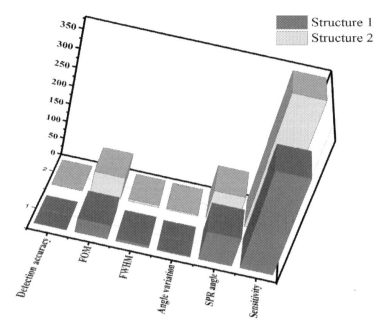

Fig. 18 Comparison graph between structure-1 and structure-2 sensor's performance parameters.

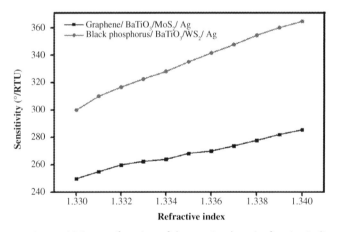

Fig. 19 Enhancement in sensitivity as a function of the sensing layer' refractive indices for structure-1 and structure-2.

in RI of the sensing layer. The change in RI of the sensing layer is fixed by $\Delta n = 0.01$ from 1.33 to 1.34 and enhancement in sensitivity observed from 268.66°/RIU (conventional) to 371°/RIU (the total increment is 38.09%).

At the end of this chapter, we provide a summary of the materials we have used in this chapter by which we have observed the change or modification occurring in the performance parameters of SPR-based biosensors (Table 10).

Table 10 Electronic and optical properties of the materials used in the chapter.

Graphene (2D material)	1. Semimetal, bandgap zero
	2. Hybridized form SP^2
	3. Crystal system hexagonal
	4. Electrical conduction type ambipolar
	5. Thermal conductance 2000–5000 W/m k
	6. Optical adsorption-fast photoresponse, large bandwidth, low photoresponsivity
	7. Carrier mobility 200,000 cm^2/V/s
	8. Effective mass ∼0
	9. Light absorption efficiency 2.3%
	10. hydrophobic surfaces nature
	11. stable in ambient environment and water/oxygen contact
Black phosphorus (2D material) [82]	1. Effective mass ∼0.146 m_e
	2. 0.3–2 eV tunable bandgap depend upon thickness
	3. Thermal conductance 52 W m/k
	4. Electrical conduction type ambipolar
	5. Thermal conductance 36 W/m k
	6. Carrier mobility ∼1000 cm^2/V/s
	7. Hydrophobic surfaces nature
	8. Not stable in ambient environment
MoS_2 (TMD material)	1. Semiconductor, 1.2–1.8 eV direct bandgap
	2. Thermal conductance 52 W m/k
	3. High photoresponsivity
	4. Carrier mobility 2000 cm^2/V/s
	5. Penetration depth at 800 nm, 5.33 μm
	6. Effective mass ∼0.47–0.6 m_e
	7. Light absorption efficiency 5%
	8. Hydrophobic surfaces nature
	9. Not stable in ambient environment
$Ti_3C_2T_x$ (MXene)	1. 2D transition metal carbide, nitride or carbonitride
	2. Hydrophilic surfaces nature
	3. Large surface area high biocompatibility
	4. Long term stability
	5. 2D transition metal carbide and nitride known as MXene with general formula $M_{n+1}AX_n$ (n = 1, 2, 3)
	6. Exhibits plasmonic property, show 77% transmittance in visible range
	7. Possess plenty of surface plasmons in multilayered MXene ($Ti_3C_2T_x$) possessing energy from 0.3 to 1 eV in comparison to bulk form
$BaTiO_3$ (Perovskite material)	1. Generic form ABX_3 have same crystallographic structure [83]
	2. Cubic unit cell with titanium (Ti) atoms at the corners, oxygen(O) atoms are at midpoint of edges and barium (Ba) atom at the center

Continued

Table 10 Electronic and optical properties of the materials used in the chapter—cont'd

	3. Excellent properties like ferroelectricity, piezoelectricity, and high dielectric constant value, photorefractive material [84] 4. T exhibits a sequence of ferroelectric phase transition and have indirect bandgap 5. At high temperature $BaTiO_3$ is Para electric with cubic structure 6. On cooling this material undergoes successive structural phase transition

5. Conclusion

In this chapter, the theoretical and mathematical concept of surface plasmons and significant condition for SPR is discussed. A short overview is given about the application of SPR-based sensors in the biosensing field as well as its commercial benefits. Angle interrogation-based Kretschmann configuration is taken throughout this chapter and, according to this performance, parameter analysis has been done. Attractive methods for improvement of sensitivity of SPR biosensor have been discussed and appreciable outcomes than conventional structures metal layer functionalized with various BRE elements. Role of various 2D nanomaterials in biosensing applications, their advantages, and drawbacks were discussed step-by-step. The demerit of the monolayer of 2D materials like graphene and how 2D/TMD materials' heterostructure overcome the drawbacks related to 2D monolayer were also shown. An analytical discussion about TMD/perovskite/2D material-based combined structure of SPR biosensor was also discussed. At the end of this chapter, the property of 2D transition metal carbide/nitride materials like MXene were discussed, and its various advantages to increase the performance parameter of SPR biosensor was given briefly. At the end of this chapter, we have discussed significant enhancement in sensitivity up to 371°/RIU by using black phosphorus/$BaTiO_3$/WS_2/Ag/prism-based sensor, which is 125% greater than conventional SPR biosensor's sensitivity (164°/RIU).

Acknowledgment

The author acknowledges Project No. 34/14/10/2017-BRNS/34285 entitled "Role of Graphene and MoS_2 on Performance of Surface Plasmon Resonance-Based Sensors: An Application to Biosensing" funded by BRNS, Department of Atomic Energy for fellowship and Department of Electronics and Information Technology (DEITy), Government of India.

References

[1] C.-T. Li, H.-F. Chen, I. Wai, H. C. Lee, T.-J. Yen, Study of optical phase transduction on localized surface plasmon resonance for ultrasensitive detection, Optics Exp. 20 (2012) 3250–3260.

[2] J. Hamola, S.S. Yee, G. Gauglitzand, Surface plasmon resonance sensor: review, Sens. Actuators B Chem. 54 (1999) 3–15.

[3] P. Sarika, Y.K. Prajapati, J.P. Saini, V. Singh, Resolution enhancement of optical SPR sensor using metamaterial, Photonics Sensors 5 (4) (2015) 330–338.

[4] F.S. Ligler, C.R. Taitt, L.C. Shiver-Lake, K.E. Sapsford, Y. Shubin, J.P. Golden, Array bio-sensors for detection of toxins, Anal., Bioanal., Chem. 377 (2003) 469–477.

[5] L. Wu, Y. Jia, L. Jiang, J. Guo, X. Dai, Y. Xiang, D. Fan, Sensitivity improved SPR biosensor based on the MoS2/graphene–aluminum hybrid structure, J. Light. Technol. 35 (1) (2017) 82–87.

[6] J.B. Maurya, Y.K. Prajapati, A comparative study of different metal and prism in the surface plasmon resonance biosensor having MoS_2-graphene, Opt. Quantum Electron. 48 (5) (2016) 1–12.

[7] J.B. Maurya, A. François, Y.K. Prajapati, Two-dimensional layered nanomaterial based one-dimensional photonic crystal refractive index sensor, Sensors 18 (3) (2018) 857.

[8] S. Pal, A. Verma, Y.K. Prajapati, J.P. Saini, Influence of black phosphorous on performance of surface plasmon resonance biosensor, Opt. Quantum Electron. 49 (12) (2017) 403.

[9] S. Pal, A. Verma, S. Raikwar, Y.K. Prajapati, J.P. Saini, Detection of DNA hybridization using black phosphorus-graphene coated surface plasmon resonance sensor, Appl. Phys. A 124 (2018) 394.

[10] J.B. Maurya, S. Raikawar, Y.K. Prajapati, J.P. Saini, A silicon-black phosphorous based surface plasmon resonance sensor for the detection of NO_2 gas. Optik 160 (2018) 428–433, https://doi.org/10.1016/j.ijleo.2018.02.002.

[11] J.B. Maurya, Y.K. Prajapati, V. Singh, J.P. Saini, Sensitivity enhancement of surface plasmon resonance sensor based on graphene-MoS_2 hybrid structure with TiO_2-SiO_2 composite layer, Appl. Phys. A 121 (2) (2015) 525–533.

[12] Q. Ouyang, S. Zeng, L. Ji, L. Hong, G. Xu, X.Q. Dinh, J. Qian, S. He, J. Qu, P. Coquet, K.T. Yong, Sensitivity enhancement of transition metal dichalcogenides/silicon nanostructure-based surface plasmon resonance biosensor, Sci. Rep. 6 (2016) 28190.

[13] L. Wu, J. Guo, Q. Wang, S. Lu, X. Dai, Y. Xiang, D. Fan, Sensitivity enhancement by using few-layer black phosphorus-graphene/TMDCs heterostructure in surface plasmon resonance biochemical sensor, Sens. Actuators B Chem. 249 (2017) 542–548.

[14] S. Cui, P. Haihui, S.A. Wells, Z. Wen, S. Mao, J. Chang, M.C. Hersam, J. Chen, Ultrahigh sensitivity and layer-dependent sensing performance of phosphorene-based gas sensors, Nat. Commun. 6 (2015) 8632.

[15] N. Liu, S. Zhou, Gas adsorption on monolayer blue phosphorus: implications for environmental stability and gas sensors, Nanotechnology 28 (17) (2017) 175708.

[16] J.B. Maurya, Y.K. Prajapati, S. Vivek, J.P. Saini, R. Tripathi, Performance of graphene-MoS_2 based surface plasmon resonance sensor using silicon layer, Opt. Quantum Electron. 47 (11) (2015) 3599–3611 Springer publication.

[17] I. Pockrand, J.D. Swalen, J.G. Gordon, M.R. Phllpott, Surface plasmon spectroscopy of organic monolayer assemblies, Surf. Sci. 74 (1977) 237–244.

[18] M. Bruna, S. Borini, Optical constants of graphene layers in the visible range, Appl. Phys. Lett. 94 (3) (2009) 4–13.

[19] S. Cheon, K. David Kihm, H.G. Kim, G. Lim, J.S. Park, J.S. Lee, How to reliably determine the complex refractive index (RI) of graphene by using two independent measurement constraints, Sci. Rep. 4 (1) (2014) 6364.

[20] I. Pockrand, J.D. Swalen, Anomalous dispersion of surface plasma oscillations, J. Opt. Soc. Am. 68 (8) (1978) 1147–1151.

[21] P.K. Maharana, R. Jha, S. Palei, Sensitivity enhancement by air mediated graphene multilayer based surface plasmon resonance biosensor for near infrared, Sensors Actuators B Chem. 190 (2014) 494–501.

[22] R.W. Wood, On a remarkable case of uneven distribution of light in a diffraction grating spectrum, Philos. Mag. Ser. 6 (4) (1902) 396–402. https://doi.org/10.1080/14786440209462857.

[23] U. Fano, The theory of anomalous diffraction gratings and of quasi-stationary waves on metallic surfaces (Sommerfeld's waves), J. Opt. Soc. Am. 31 (3) (1941) 213–222.

[24] C.J. Powell, J.B. Swan, Origin of the characteristic electron energy losses in aluminum, Phys. Rev. 115 (1959) 869.

[25] J.B. Maurya, Y.K. Prajapati, Influence of dielectric coating of metal layer in surface plasmon resonance sensor, J. Plasmonics 12 (4) (2017) 1121–1130.

[26] J. Homola, M. Piliarik, Surface Plasmon Resonance Based Sensors, 4 Springer, 2006, pp. 46–47.

[27] Q. Ouyang, S. Zeng, L. Jiang, L. Hong, G. Xu, X.Q. Dinh, J. Qian, S. He, J. Qu, P. Coquet, et al., Sensitivity enhancement of transition metal dichalcogenides/silicon nanostructure-based surface plasmon resonance biosensor, Sci. Rep. 6 (2016) 1–13.

[28] Y. Xu, C.Y. Hsieh, L. Wu, L.K. Ang, Two-dimensional transition metal dichalcogenides mediated long range surface plasmon resonance biosensors, J. Phys. D Appl. Phys. 52 (6) (2019) 1–8.

[29] J. Zhu, B. Ruan, Q. You, J. Guo, X. Dai, Y. Xiang, Terahertz imaging sensor based on the strong coupling of surface plasmon polaritons between PVDF and graphene, Sens. Actuators B 264 (2018) 398–403.

[30] B. Ruan, Q. You, J. Zhu, L. Wu, J. Guo, X. Dai, Y. Xiang, Fano resonance in double waveguides with graphene for ultrasensitive biosensor, Opt. Express 26 (13) (2018) 16884–16892.

[31] L. Wu, Q. Wang, B. Ruan, J. Zhu, Q. You, X. Dai, Y. Xiang, High performance lossy-mode resonance sensor based on few-layer black phosphorus. J. Phys. Chem. C (2018), https://doi.org/10.1021/acs.jpcc.7b12549.

[32] L. Wu, J. Guo, X. Dai, Y. Xiang, D. Fan, Sensitivity enhanced by MoS₂–graphene hybrid structure in guided-wave surface plasmon resonance biosensor, Plasmonics 13 (2018) 281–285.

[33] Z. Lin, L. Jiang, L. Wu, J. Guo, X. Dai, Y. Xiang, D. Fan, Tuning and sensitivity enhancement of surface plasmon resonance biosensor with graphene covered Au-MoS₂-Au films, IEEE Photonics J. 8 (6) (2016).

[34] Y. Xiang, J. Zhu, L. Wu, Q. You, B. Ruan, X. Dai, Highly sensitive terahertz gas sensor based on surface plasmon resonance with graphene, IEEE Photonics J. 10 (1) (2018).

[35] D. Ma, J. Shi, Q. Ji, K. Chen, et al., A universal etching-free transfer of MoS₂ films for applications in photodetectors, Nano Res. 8 (11) (2015) 3662–3672.

[36] A. Srivastava, Y.K. Prajapati, Performance analysis of silicon and blue phosphorene/MoS₂ heterostructure based SPR sensor, Photonic Sens. 9 (33) (2019) 1–9.

[37] J.D. Wood, S.A. Wells, D. Jariwala, K.S. Chen, E. Cho, V.K. Sangwan, et al., Effective passivation of exfoliated black phosphorus transistors against ambient degradation, Nano Lett. 14 (2014) 6964–6970.

[38] Y. Zhao, H. Wang, H. Huang, Q. Xiao, Y. Xu, Z. Guo, et al., Surface coordination of black phosphorus for robust air and water stability, Angew Chem. 128 (2016) 5087–5091.

[39] Y. Zhang, A.L. Wang, B.N. Zhanga, Z. Zhoua, Adsorptive environmental applications of MXene nanomaterials: a review, RSC Advance 8 (2018) 19895–19905.

[40] X. Sang, Y. Xie, M.W. Lin, M. Alhabeb, K.L. VanAken, Y. Gogotsi, P.R. Kent, K. Xiao, R.R. Unocic, Atomic defects in monolayer titanium carbide (Ti₃C₂Tₓ) MXene, ACS Nano. 10 (2016) 9193–9200.

[41] A. Lipatov, M. Alhabeb, M.R. Lukatskaya, A. Boson, Y. Gogotsi, A. Sinitskii, Effect of synthesis on quality, electronic properties and environmental stability of individual monolayer Ti₃C₂Tₓ MXene flakes, Adv. Electron. Mater. 2 (2016) 1–9.

[42] B. Anasori, M.R. Lukatskaya, Y. Gogotsi, 2D metal carbides and nitrides (MXenes) for energy storage, Nat. Rev. Mater. 2 (2017) 1–17.

[43] F. Wang, C. Yang, M. Duan, Y. Tang, J. Zhu, TiO2 nanoparticle modified organlike Ti3C2 MXene nanocomposite encapsulating hemoglobin for a mediator free biosensor with excellent performances, Biosens. Bioelectron. 74 (2015) 1022–1028.

[44] H. Lin, Y. Chen, J. Shi, Insights into 2D MXenes for versatile biomedical applications: current advances and challenges ahead, Adv. Sci. 5 (10) (2018) 1–20.

[45] Y. Fang, X. Yang, T. Chen, G. Xu, M. Liu, J. Liu, Y. Xu, "Two-dimensional titanium carbide (MXene)-based solid-state electrochemiluminescent sensor for label-free single-nucleotide mismatch discrimination in human urine" Sens, Actuators B Chem. 263 (2018) 400–407.

[46] L. Wu, Q. You, Y. Shan, S. Gan, Y. Zhao, X. Dai, Y. Xiang, Few-layer Ti₃C₂Tₓ MXene: a promising surface plasmon resonance biosensing material to enhance the sensitivity, Sens. Actuators B Chem. 277 (2018) 210–215.

[47] Y. Xu, Y.S. Ang, L. Wu, L.K. Ang, High sensitivity surface plasmon resonance sensor based on two-dimensional MXene and transition metal dichalcogenide: a theoretical study, Nanomaterials 9 (2) (2019) 2–11.

[48] A. Srivastava, A. Verma, R. Das, Y.K. Prajapati, A theoretical approach to improve the performance of spr biosensor using MXene and black phosphorus, Optik 203 (2020) 1–9.

[49] E. Palik, Handbook of Optical Constants of Solids, vol. 1, Academic press, 1985.

[50] Z. Chen, X. Zhao, C. Lin, S. Chen, L. Yin, Y. Ding, Figure of merit enhancement of surface plasmon resonance sensors using absentee layer, Appl. Opt. 55 (2016) 6832–6835.

[51] X. Wang, Y. Deng, Q. Li, Y. Huang, Z. Gong, et al., Excitation and propagation of surface plasmon polaritons on a non-structured surface with a permittivity gradient, Light Sci. Appl. 5 (2016) 1–6.

[52] B. Richard, M. Schasfoort, A.J. Tudos, Handbook of Surface Plasmon Resonance, Royal Society of Chemistry, 2008, pp. 1–14.

[53] A. Shalabney, I. Abdulhalim, Figure-of-merit enhancement of surface plasmon resonance sensors in the spectral interrogation, Opt. Lett. 37 (7) (2012) 1175–1177.

[54] Y.K. Prajapati, A. Yadav, A. Verma, V. Singh, J.P. Saini, Effect of metamaterial layer on optical surface plasmon resonance sensor, Optik 124 (18) (2013) 3607–3610.

[55] E. Kretschmann, The determination of the optical constants of metals by excitation of surface plasmons, J. Phys. 241 (1971) 313–324.

[56] S. Zeng, D. Baillargeat, et al., Nanomaterials enhanced surface plasmon resonance for biological and chemical sensing applications, Chem. Soc. Rev. 43 (2014) 3426–3452.

[57] S. Nivedha, P.R. Babu, K. Senthilnathan, Surface plasmon resonance: physics and technology, Curr. Sci. 115 (1) (2018).

[58] S.A. Han, R. Bhatia, S.W. Kim, Synthesis, properties and potential applications of two-dimensional transition metal dichalcogenides. Nano Converg. (2015), https://doi.org/10.1186/s40580-015-0048-4.

[59] A.K. Geim, I.V. Grigorieva, Van der Waals heterostructure, Nature 499 (2013) 419–425.

[60] Q. Peng, et al., Electronic structures and enhanced optical properties of blue phosphorene/transition metal dichalcogenides van der Waals heterostructures, Sci. Rep. 6 (2016) 1–10.

[61] M.A. Green, A. Ho-Baillie, H.J. Snaith, The emergence of perovskite solar cells, Nat. Photonics 8 (7) (2014) 506–514.

[62] A. Munawar, Y. Ong, R. Schirhagl, M.A. Tahir, W.S. Khanae, S.Z. Bajwa, Nanosensors for diagnosis with optical, electric and mechanical transducers, RSC Adv. 9 (2019) 6793–6803.

[63] N.J. Jeon, J.H. Noh, W.S. Yang, Y.C. Kim, S. Ryu, J. Seo, S.I. Seok, Compositional engineering of perovskite materials for high-performance solar cells, Nature 517 (2015) 476–480.

[64] W.J. Merz, The electric and optical behavior of $BaTio_3$ single-domain crystals, Phys. Rev. 76 (1949) 1221.

[65] S.L. Simon, A. Hajjaji, Y. Emziane, B. Guiffard, D. Guyomar, Re-investigation of synthesis of $BaTiO_3$ by conventional solid-state reaction and oxalate coprecipitation route for piezoelectric applications, Ceram. Int. 33 (2007) 35–40.

[66] S.J. Kim, H.J. Koh, C.E. Ren, O. Kwon, et al., Metallic Ti_3C_2Tx MXene Gas Sensors with Ultrahigh Signal-to-Noise Ratio, ACS Nano 12 (2) (2018) 986–993.

[67] Y.K. Prajapati, A. Srivastava, Effect of blueP/MoS_2 heterostructure and graphene layer on the performance parameter of SPR sensor: theoretical insight, Superlattice Microstruct. 129 (2019) 152–162.

[68] Z. Liu, L. Lin, H. Ren, X. Sun, Chapter-2 CVD synthesis of graphene, in: Micro and Nano Technologies2017, pp. 19–56.

[69] U. Krishnan, M. Kaura, K. Singha, M. Kumarb, A. Kumara, A synoptic review of MoS_2: synthesis to applications, Superlattices Microstruct. 128 (2019) 274–297.

[70] A. Verma, A. Prakash, R. Tripathi, Performance analysis of graphene based surface plasmon resonance biosensors for detection of pseudomonas-like bacteria, Opt. Quant. Electron. 47 (2015) 1197–1205.

[71] M.S. Anower, M.S. Rahman, K.A. Rikta, Performance enhancement of graphene coated surface plasmon resonance biosensor using tungsten disulfide, Opt. Eng. 57 (1) (2018) 1–8.

[72] R.H. Holma, E.I. Solomonb, A. Majumdar, A. Tenderholt, Comparative molecular chemistry of molybdenum and tungsten and its relation to hydroxylase and oxo transfer as e enzymes, Coord. Chem. Rev. 255 (2011) 993–1015.

[73] J.B. Maurya, Y.K. Prajapati, Comparative analysis of silicon and black phosphorous as an add-layer in nanomaterial based plasmonic sensor, Opt. Quantum Electron. 51 (4) (2019) 91–126.

[74] A. Avsar, I.J. Vera-Marun, J.Y. Tan, K. Watanabe, T. Taniguchi, A.H.C. Neto, B. Ozyilmaz, Air-stable transport in graphene-contacted, fully encapsulated ultrathin black phosphorus-based field effect transistors, ACS Nano. 9 (4) (2015) 4138–4145.

[75] R. Sonia, K. Patel, P. Kumar, C. Prakash, D.K. Agrawal, Low temperature synthesis and dielectric, ferroelectric and piezoelectric study of microwave sintered $BaTiO_3$ ceramics, Ceram. Int. 38 (2012) 1585–1589.

[76] A. Djurisic, F. Liu, H. Tam, et al., Perovskite solar cells—an overview of critical issues, Prog. Quantum Electron. 53 (2017) 1–37.

[77] S. Fouad, N. Sabri, Z.A.Z. Jamal, P. Poopalan, Enhanced sensitivity of surface plasmon resonance sensor based on bilayers of silver-barium titanate, J. Nano- Electron. Phys. 8 (2) (2016) 4085–4090.

[78] S.H. Wemple, M. Didomenico Jr., I. Camlibel, Dielectric and optical properties of melt-grown $BaTio_3$, J. Phys. Chem. Solids 29 (1968) 1797–1803.

[79] W. Cai, et al., Preparation and optical properties of barium titanate thin films, Physica B Condens. Matter 406 (2011) 3583–3587.

[80] P. Sun, M. Wang, et al., Sensitivity enhancement of surface plasmon resonance biosensor based on graphene and barium titanate layers, Appl. Surf. Sci. 475 (2019) 342–347.

[81] Q. Meng, X. Zhao, C. Lin, S. Chen, Y. Ding, Z.-Y. Chen, Figure of merit enhancement of a surface plasmon resonance sensor using a low-refractive-index porous silica film, Sensors 17 (8) (2017) 1846.

[82] T. Srivastava, R. Jha, Black phosphorus: a new platform for gaseous sensing based on surface plasmon resonance, IEEE Photon. Technol. Lett. 30 (4) (2018).

[83] Z. Shi, A.H. Jayatissa, Perovskites-based solar cells: a review of recent progress, materials and processing methods, Materials (Basel) 11 (5) (2018) 729–785.

[84] L. Liu, M. Wang, L. Jiao, et al., Sensitivity enhancement of a graphene–barium titanate-based surface plasmon resonance biosensor with an Ag–Au bimetallic structure in the visible region, J. Opt. Soc. Am. B 36 (4) (2019).

SECTION III

Nano fabrication techniques—Chemical sensors

CHAPTER 5

Jatropha extract mediated synthesis of ZnFe$_2$O$_4$ nanopowder: Excellent performance as an electrochemical sensor, UV photocatalyst, and antibacterial activity

B.S. Surendra
Department of Chemistry, East West Institute of technology, Bengaluru, India

1. Introduction

Currently, nanosized substances are more attractive due to their environmentally sustainable and unique physicochemical properties. Nanoparticles have attracted more attention from researchers around the world, increasing the search for new and innovative research applications for modern civilization. The overdosage of medicines causes many hazards for humans. However, this problem can be resolved using nanoparticles, which are involved in sensing such types of medicines. Presently, some nanomaterials are used in sensor applications in various disciplines such as gas sensors (LPG) (sensing leakage and preventing accidents that might be caused when gas leaks out accidentally or by mistake) and chemical sensors (H$_2$S, ethanol, Cl$_2$), etc. [1–5]. The Spinel ferrites are increasing interest in research activity due to their versatile technical applications [6–11]. Among these, ZnFe$_2$O$_4$ nanomaterials have gained increasing interest for many applications in different areas of industry. However, its CV analysis was carried out to characterize the modified graphite–ZnFe$_2$O$_4$ electrode paste and further applied to find sensor activity with different chemicals such as Paracetamol, Dolo-650, and Mifepristone-Misoprostol, which was tested by EIS with 0.1 M KCl analyte. Equivalent circuit-fitting characteristics of the sensors were developed and discussed.

Currently, there has been increasing potential for energy conservation and hence, the search for new materials as energy storage devices with high capacity and densities continues. Today, researchers are looking into modern electrode materials for energy variations such as high storage capacity, power density, cycle life, etc. Ferrites are prominent substances, which fulfills the gap between capacitor and battery [4]. Developing sustainable materials have attracted much attention for removal of hazardous, toxic chemicals in wastewater released from various industries and laboratories [12–15]. Mainly, organic dyes that are released from various sectors like textile, paper, cosmetic industries, etc., into water is very toxic and causes environmental problems [16–20]. Most nanomaterials were used as photocatalysts for wastewater treatment to protect the environment from such contaminants. Recently, it was found that $ZnFe_2O_4$ is one of the most efficient nanomaterials for the degradation of organic dyes released from various sectors into harmless chemicals (CO_2, NO_2, H_2O, etc.). Green combustion synthesis of spinel $ZnFe_2O_4$ photocatalysts and their catalytic performance on MG dye have been studied. The authors concluded that the $ZnFe_2O_4$ photocatalyst shows highly active photocatalytic activity in pollutant degradation (MG dye) because of its capability to absorb UV light.

The objective of the current work is to find the chemical-sensing properties, nature of supercapacitors, and photocatalytic activities of prepared zinc ferrite nanoparticles using Jatropha oil extracts and well-characterized by powder X-ray diffraction (P-XRD), Fourier-transform infrared spectroscopy (FT-IR), and ultraviolet-visible (UV-Vis) spectroscopy.

2. Materials and methods

2.1 Synthesis of $ZnFe_2O_4$ NPs

In green–assisted synthesis, a stoichiometric amount of analytic grade Zn $(NO_3)_2 \cdot 3H_2O$ and $Fe(NO_3)_3 \cdot 9H_2O$ were dissolved with a small quantity of distilled water in a silica crucible and 10 mL of Jatropha oil extract (green fuel) was added. This was mixed thoroughly to attain homogeneity and subjected to combustion in a muffle furnace at 500°C. The mixture boiled and caught fire at one point due to fatty acids present in the oil extract that slowly spread throughout the reaction mixture (Flowchart 1). Finally, the obtained $ZnFe_2O_4$ powder sample was collected and its structural characterizations were studied.

3. Results and discussion

3.1 P-XRD analysis

The phase formation and purity of green–assisted $ZnFe_2O_4$ nanoparticles were characterized by P-XRD analysis (Fig. 1). The 2θ values of Jatropha oil extract-synthesized $ZnFe_2O_4$ nanoparticles were 29.95, 35.21, 36.78, 42.81, 53.03, 56.56, 62.14, and 73.75 degrees and their corresponding diffraction peaks were (220), (311), (222), (400), (422), (511), (440), and (533), respectively. The intensity of peaks were observed and well matched with JCPDS No: 74-2397, which confirms the cubic spinel structure of $ZnFe_2O_4$ nanoparticles [21]. The sharp intensity of the crystal plane (311) shows high

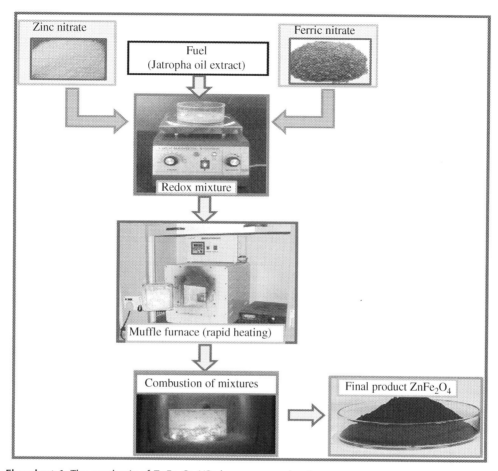

Flowchart 1 The synthesis of ZnFe$_2$O$_4$ NPs by green combustion process.

crystallinity of the prepared nanomaterials, and its average crystallite size is ∼18 nm, calculated by the following Debye-Scherrer's formula:

$$d = \frac{(K \times \lambda)}{(\beta \times \cos\theta)} \tag{1}$$

3.2 Scanning electron microscope studies

Scanning electron microscope (SEM) photograph of ZnFe$_2$O$_4$ nanomaterial was prepared by green-assisted combustion method as shown in Fig. 2. The morphological changes of the prepared nanosample shows the porous flakes and agglomeration in nature due to the presence of fatty acids in the oil extract that coordinates with zinc ions (Zn^{2+}), forming complex coordination bonds with hydroxyl (OH$^-$) groups from different fatty acids chains. Furthermore, it forms a superstructure by reacting with proteins at a low

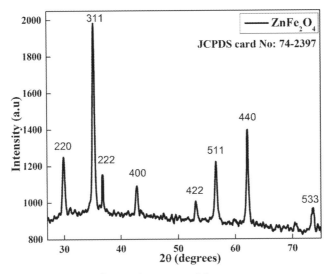

Fig. 1 P-XRD studies of the prepared ZnFe$_2$O$_4$ nanoparticles.

Fig. 2 SEM photographs of the synthesized ZnFe$_2$O$_4$ nanoparticles.

temperature (Fig. 3). Finally, this complex network underwent slow decomposition during the combustion process, which was responsible in producing different structures of ZnFe$_2$O$_4$ nanomaterial [22].

3.3 FT-IR studies

Fig. 4 shows the FT-IR spectra of prepared ZnFe$_2$O$_4$ nanoparticles, and its functional groups was ascertained in the range 400–4000 cm^{-1}. The vibrations of Zn–O and

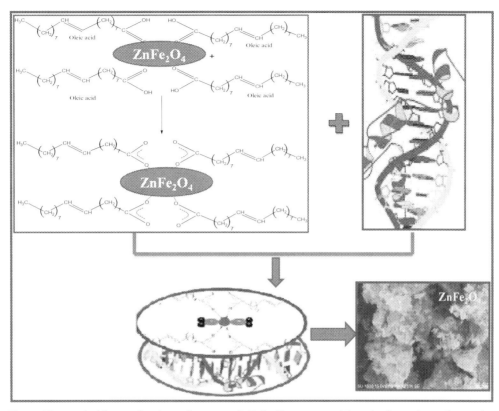

Fig. 3 The probable mechanism of prepared ZnFe$_2$O$_4$ nanoparticles via formation of complex coordination bonds.

Fe-O metal ions occur in the range 400–600 cm^{-1}. The tetrahedral and octahedral complexes are the two prominent vibrational peaks of ZnFe$_2$O$_4$ nanoparticles, which were observed at ∼552 and ∼437 cm^{-1}, respectively. Furthermore, these results show the high-frequency peak attributed to stretching vibration of tetrahedral complexes Zn-O (552 cm^{-1}) and the low-frequency peak to octahedral complexes Fe-O (437 cm^{-1}). Finally, the broad band appears at 3442 cm^{-1} indicating the confirmation of hydroxyl (–OH) group of adsorbed water molecules and its corresponding bending vibration observed peak at ∼1672 cm^{-1} [23].

4. Electrochemical measurements

The redox peak's current (reversible anodic and cathodic current) and charge efficiency of prepared electrode was studied by electrochemical measurements (CV and EIS analysis). These investigations are very important for current research work to study the nature of reduction–oxidation reactions taking place in a prepared ZnFe$_2$O$_4$ electrode in the presence of platinum (counter electrode) and Ag/AgCl reference electrodes in

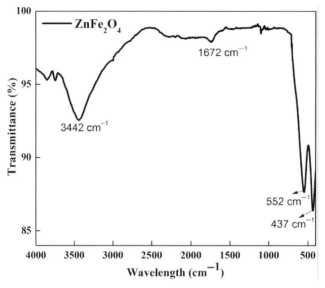

Fig. 4 FT-IR spectra of synthesized ZnFe$_2$O$_4$ nanoparticles.

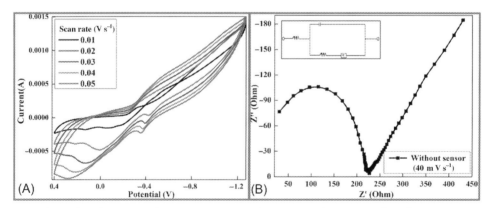

Fig. 5 (A) CV plots for prepared ZnFe$_2$O$_4$ nanoparticles; (B) Impedance plot of synthesized ZnFe$_2$O$_4$ nanoparticles (insight: its equivalent circuit).

0.1 M KCl electrolyte in potential cycled between -1.24 and 0.4 V at the scan rates of 10–50 mV s^{-1} as shown in Fig. 5. The increasing scan rate shows significant changes in increasing current and hysteresis of the anodic and cathodic peaks. The prepared ZnFe$_2$O$_4$ electrode showed a good redox peak and hence its good photocatalytic activity is noteworthy. The CV study for a prepared ZnFe$_2$O$_4$ electrode in 0.1 M KCl electrolyte with 6 drops of different sensor chemicals (Paracetamol, Dolo-650, and mifepristone-misoprostol) was discussed (Fig. 6).

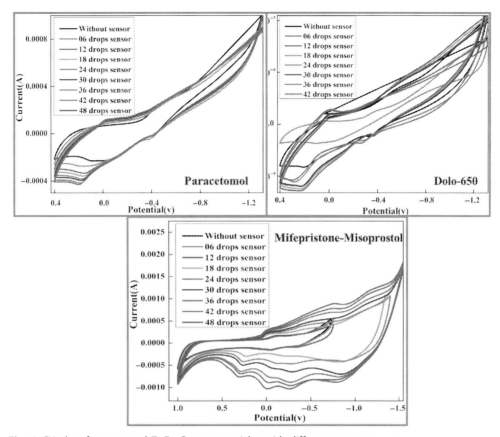

Fig. 6 CV plots for prepared ZnFe$_2$O$_4$ nanoparticles with different sensors.

EIS measurements show the specific capacitance and charge transfer resistance between the electrons-holes of the ZnFe$_2$O$_4$ electrode. The specific capacitance of these electrodes can be measured by the appearance of a semicircle at the high-frequency region and a line at the low-frequency region in the Nyquist curve [24]. The capacitance values of the ZnFe$_2$O$_4$ electrode were observed for electrolyte without a sensor and with sensors of every six drops of Dolo-650, Paracetamol, and Mifepristone-Misoprostol sensor chemicals, which were 235, 208, 269, and 91 Ω, respectively, at scan rate of 40 mV (Fig. 7). The Nyquist curve was discussed based on an equivalent Randles circuit that includes resistance (R_s), capacitance (C_{dl}), charge transfer resistance (R_{ct}), and Warburg impedance [25]. The errors measured by the equivalent circuit of the ZnFe$_2$O$_4$ electrode and its activities in the presence of sensors like Dolo-650, Paracetamol, and Mifepristone-Misoprostol are 0.0358%, 0.0177%, 0.02933%, and 0.01593% error, respectively, as shown in Fig. 8. The present work reported that the electrochemical studies of prepared

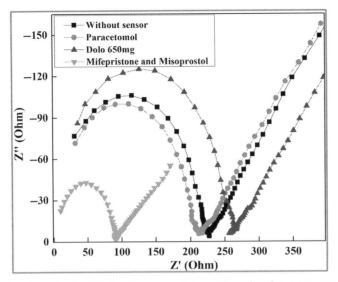

Fig. 7 Nyquist plot of synthesized $ZnFe_2O_4$ nanoparticles with and without sensors.

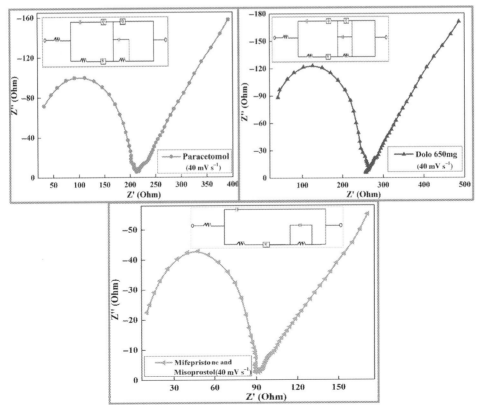

Fig. 8 Impedance plot of synthesized $ZnFe_2O_4$ nanoparticles (insight: mentions its equivalent circuit).

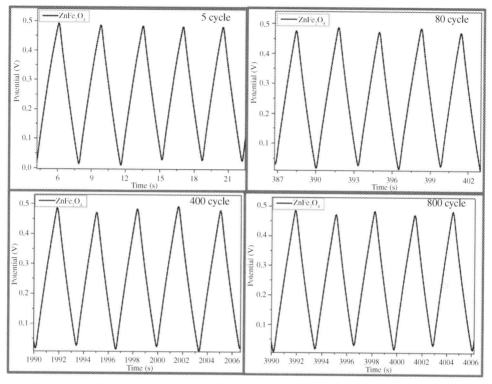

Fig. 9 The galvanostatic charge-discharge plots of synthesized ZnFe$_2$O$_4$ nanomaterial.

ZnFe$_2$O$_4$ electrode conducted with different chemicals. Among these chemicals, Mifepristone-Misoprostol chemical showed excellent sensing activity.

Fig. 9 shows the galvanostatic charge–discharge plots for electrodes of synthesized ZnFe$_2$O$_4$ nanomaterial at different current densities. Generally, the integrated area of current potential axis increased due to increasing in current density, representing a good storage rate capacity [26]. The obtained plots indicated that the curves of charge-discharge do not conform with that of a true electric double-layer capacitor and hence it is a pseudocapacitor due to its agreement with the resulting CV analysis. The graph (Fig. 9) shows the life cycle stability of ZnFe$_2$O$_4$ electrode measurement during charge-discharge plots at a current density of 5 Ag^{-1} for 4000 cycles within the potential window of 0–0.5 V [27].

5. Photocatalytic degradation

The photocatalytic degradation of MG dye under UV light was carried out using ZnFe$_2$O$_4$ samples prepared by green-assisted combustion route (Fig. 10A). The photo-catalytic reaction of MG was also conducted in the absence of light showing less degra-dation (12%). However, this reaction showed 98% photo-decolorization of MG under

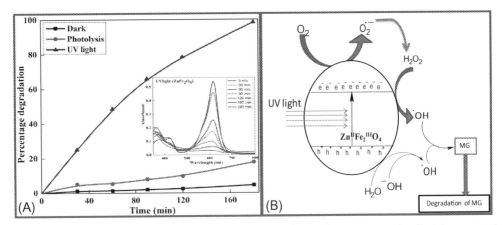

Fig. 10 (A) Percentage decomposition of MG from synthesized nanocomposite (insight: spectral absorbance for photocatalytic decomposition of MG with the variation of irradiation time under UV light); (B) Mechanism for the photocatalytic decomposition of MG under UV light irradiation.

UV radiation due to small particle size and more active sites. Generally, the ferrite compounds are capable of capturing UV light and thus, prepared $ZnFe_2O_4$ showed good photocatalytic activity for degradation of MG.

The probable mechanisms of photocatalytic degradation reactions for MG dye in the presence of $ZnFe_2O_4$ nanoparticles were discussed taking this into account (Fig. 10B). The $ZnFe_2O_4$ photocatalysts are able to absorb the incident light with sufficient energy and generate electron-holes (h^+), which facilitates the redox reactions of MG dye in an aqueous solution. The photoexcitation reaction occurs due to absorption of oxygen (O_2), water (H_2O), and dye molecules on the surface of the $ZnFe_2O_4$ photocatalyst, creating holes (h^+) by electrons' (e^-) jump from the conduction band (CB) into the valence band (VB) (Eqs. 1, 2). During the photoexcitation mechanism of MG molecules in the presence of UV light, the surface Fe(III) could first be reduced into Fe(II), and then Zn(II) on the surface might also be reduced into Zn(0). A number of superoxide anion-free radicals ($O^{\bullet-}$) are formed due to more oxygen vacancies, which are active sites for O_2 molecules' adsorption on the conduction band of photocatalyst surface (Eq. 3), which acts as a strong oxidant. But it is not involved directly in the reduction process to occur. Furthermore, these superoxide radicals are neutralized by combining with proton and released OH^-, $OH^{\bullet}$ radicals with strong oxidation ability (Eq. 4). These produced active radicals are directly involved in degradation by reacting with the dye molecules converted into the harmless chemicals (H_2O, CO_2, NO^{2-}, NO^{3-}, etc.) [28].

The probable mechanisms of photocatalytic degradation reactions for MG dye in presence of $ZnFe_2O_4$ NPs.

$$ZnFe_2O_4 \xrightarrow{\text{UV light irradiation}} ZnFe_2O_4{}^*(\text{Energy}) \qquad (1)$$

$$ZnFe_2O_4^*(Energy) \rightarrow ZnFe_2O_4^*(h^+ + e^-) \tag{2}$$

$$ZnFe_2O_4(e^-) + O_2 \rightarrow O^{2\cdot-} (superoxide \ radical) \tag{3}$$

$$O^{2\cdot-} + H_2O \rightarrow OH^\cdot + OH^- \tag{4}$$

6. Conclusions

This chapter has presented the green synthesis of $ZnFe_2O_4$ nanoparticles by a simple combustion method using Jatropha as a fuel. These synthesized nanomaterials are characterized by P-XRD, FT-IR, and UV-Vis spectroscopy and confirmed as an electrode material for all solid-state supercapacitors. CV analysis of prepared electrodes showed high reversible redox reaction and excellent performance in electrochemical sensor activity for the detection of mifepristone-misoprostol and Dolo-650. The photodegradation activity of $ZnFe_2O_4$ nanosamples showed 98% MG dye degradation activity under UV light irradiation. Hence, the synthesized $ZnFe_2O_4$ powder is effectively used as a supercapacitor, sensor, and energy storage devices. It is a nonhazardous, ecofriendly, prominent material for organic dye degradation that can be reused up to 6–10 cycles.

References

[1] C.V.G. Reddy, S.V. Manorama, V.J. Rao, Preparation and characterization of ferrites as gas sensor materials, J. Mater. Sci. Lett. 19 (2000) 775–778.

[2] Z. Tianshu, P. Hing, Z. Jiancheng, K. Lingbing, Ethanol-sensing characteristics of cadmium ferrite prepared by chemical coprecipitation, Mater. Chem. Phys. 61 (1999) 192–198.

[3] K.M. Reddy, L. Satyanarayana, S.V. Manorama, R.D.K. Mishra, A comparative study of the gas sensing behavior of nanostructured nickel ferrite synthesized by hydrothermal and reverse micelle techniques, Mater. Res. Bull. 39 (2004) 1491–1498.

[4] N. Iftimie, E. Rezlescu, P.D. Popa, N. Rezlescu, Gas sensitivity of nanocrystalline nickel ferrite, J. Optoelectron. Adv. Mater. 8 (2006) 1016–1018.

[5] Y.L. Liu, H. Wang, Y. Yang, Z.M. Liu, H.F. Yang, G.L. Shen, R.Q. Yu, Hydrogen sulfide sensing properties of NiFe₂O₄ nanopowder doped with noble metals, Sensors Actuators B Chem. 102 (2004) 148–154.

[6] G. Concas, G. Spano, C. Cannas, A. Musinu, D. Peddis, G. Piccaluga, Inversion degree and saturation magnetization of different nanocrystalline cobalt ferrites, J. Magn. Magn. Mater. 321 (2009) 1893–1897.

[7] M. Siddique, N.M. Butt, Effect of particle size on degree of inversion in ferrites investigated by Mossbauer spectroscopy, Physica B 405 (2010) 4211–4215.

[8] Y.W. Jun, J.H. Lee, J. Cheon, Chemical design of nanoparticle probes for high-performance magnetic resonance imaging, Angew. Chem. Int. Edit. 47 (2008) 5122–5135.

[9] Y. Lee, J. Lee, C.J. Bae, J.G. Park, H.J. Noh, J.H. Park, T. Hyeon, Large-scale synthesis of uniform and crystalline magnetite nanoparticles using reverse micelles as nanoreactors under reflux conditions, Adv. Funct. Mater. 15 (2005) 503–509.

[10] L. Nalbandian, A. Delimitis, V.T. Zaspalis, E.A. Deliyanni, D.N. Bakoyannakis, E.N. Peleka, Hydrothermally prepared nanocrystalline Mn–Zn ferrites: synthesis and characterization, Microporous Mesoporous Mater. 114 (2008) 465–473.

[11] K.E. Sickafus, J.M. Wills, N.W. Grimes, Structure of spinel, J. Am. Ceram. Soc. 82 (1999) 3279–3292.

[12] F. Y, Q. Chen, M. He, Y. Wan, X. Sun, H. Xia, X. Wang, A multifunctional hetero architecture for photocatalysis and energy storage, Ind. Eng. Chem. Res. 51 (2012) 11700–11709.

[13] X. Shu, J. He, D. Chen, Visible-light-induced photocatalyst based on nickel titanate nanoparticles, Ind. Eng. Chem. Res. 47 (2010) 4750–4753.

[14] R.Q. Guo, L.A. Fang, W. Dong, F.G. Zheng, M.R. Shen, Enhanced photocatalytic activity and ferromagnetism in Gd doped $BiFeO_3$ nanoparticles, J. Phys. Chem. C 114 (2010) 21390–21396.

[15] M.A. Al-Ghouti, M.A. Khraisheh, S.J. Allen, M.N. Ahmad, The removal of dyes from textile wastewater: a study of the physical characteristics and adsorption mechanisms of diatomaceous earth, J. Environ. Manag. 69 (2003) 229–238.

[16] P.H. Borse, J. Hwichan, S.H. Choi, S.J. Hong, J.S. Lee, Phase and photoelectrochemical behavior of solution-processed Fe_2O_3 nanocrystals for oxidation of water under solar light, Appl. Phys. Lett. 93 (2008) 173103-3.

[17] Y. Tamaura, Y. Ueda, J. Matsunami, N. Hasegawa, M. Nezuka, T. Sano, M. Tsuji, Solar hydrogen production by using ferrites, Sol. Energy 65 (1999) 55–57.

[18] S.B. Han, T.B. Kang, O.S. Joo, K.D. Jung, Water splitting for hydrogen production with ferrites, Sol. Energy 81 (2007) 623–628.

[19] S.H. Yu, M. Yoshimura, Ferrite/metal composites fabricated by soft solution processing, Adv. Funct. Mater. 12 (2002) 9–15.

[20] C.W. Nan, M.I. Bichurin, S. Dong, D. Viehland, G. Srinivasan, Multiferroicmagnetoelectric composites: historical perspective, status, and future directions, J. Appl. Phys. 103 (2008) 031101.

[21] B.S. Surendra, Green engineered synthesis of Ag-doped $CuFe_2O_4$: Characterization, cyclic voltammetry and photocatalytic studies, J. Sci. Adv. Mater. Dev. 3 (2018) 44–50.

[22] K. Kombaiah, J.J. Vijaya, L.J. Kennedy, M. Bououdina, Studies on the microwave assisted and conventional combustion synthesis of Hibiscus rosa-sinensis plant extract based $ZnFe_2O_4$ nanoparticles and their optical and magnetic properties, Ceram. Int. 42 (2016) 2741–2749.

[23] K.J. Rao, S. Paria, Green synthesis of silver nanoparticles from aqueous Aegle marmelos leaf extract, Mater. Res. Bull. 48 (2013) 628–634.

[24] R. Ashok Kumar, M. Ramaswamy, Phytochemical screening by FT-IR spectroscopic analysis of leaf extracts of selected Indian medicinal plants, Int. J. Curr. Microbiol. App. Sci. 3 (2014) 395–406.

[25] C.R. Ravikumar, M.R.A. Kumar, H.P. Nagaswarupa, S.C. Prashantha, A.S. Bhatt, M.S. Santosh, D. Kuznetsov, CuO embedded β-$Ni(OH)_2$ nanocomposite as advanced electrode materials for supercapacitors, J. Alloys Compd. (2017).

[26] B.S. Surendra, M. Veerabhdraswamy, K.S. Anantharaju, H.P. Nagaswarupa, S.C. Prashantha, Green and chemical-engineered $CuFe_2O_4$: characterization, cyclic voltammetry, photocatalytic and photoluminescent investigation for multifunctional applications, J. Nanostruct. Chem. 8 (2018) 45–59.

[27] N.M. Deraz, A. Alarifi, Microstructure and magnetic studies of zinc ferrite nano-particles, Int. J. Electrochem. Sci. 7 (2012) 6501–6511.

[28] P.M. Prithviraj Swamy, S. Basavaraja, A. Lagashetty, N.V. Srinivas Rao, R. Nijagunappa, A. Venkataraman, Synthesis and characterization of zinc ferrite nanoparticles obtained by self-propagating low-temperature combustion method, Bull. Mater. Sci. 34 (7) (2011) 1325–1330.

CHAPTER 6

Fabrication, characterization and application of poly(acriflavine) modified carbon nanotube paste electrode for the electrochemical determination of catechol

J.G. Manjunatha, C. Raril, N.S. Prinith, P.A. Pushpanjali, M.M. Charithra, Girish Tigari, N. Hareesha, Edwin S. D'Souza, and B.M. Amrutha
Department of Chemistry, FMKMC College, Mangalore University Constituent College, Madikeri, Karnataka, India

1. Introduction

Catechol (CA), generally known as pyrocatechol or benzene-1,2-diol and belonging to the simple phenolic compounds, has a predominant significance in biological and ecological systems like bearing antiviral and antifungal properties, which affects some of the enzyme activities and root growth of the plant [1]. The main source of CA is basically from nature via plants like tobacco, tea, coffee, and the tannin layer of mycorrhiza of the douglas fir, and a small quantity of CA is present in some fruits and vegetables like apples and potatoes [2–4]. CA has a major role in various industries such as pharmacy; in the photography industry as a developing agent in photographic negative film; in rubber and oil industries as an intermediate for the antioxidants of rubber, olefins, and polyolefins, and in the industries of dyes, perfumes, detergents, and pesticides [5, 6]. Even with all these benefits, it is a hazardous compound and toxic to human beings, flora, fauna, and aquatic species at low concentrations [7]. Moreover, it cannot be degraded easily and hence it is considered as pollutants according to the European Union and US Environmental Protection Agency [8]. CA in contact with skin brings about eczematous dermatitis [9]. Exposure to CA also affects the central nervous system causing depression, hampering DNA replication, and causing chromosomal abnormalities [10, 11]. Therefore, it's essential to determine CA by establishing sensitive and consistent techniques.

Recently, numerous techniques have been applied for the determination of CA involving spectrophotometry [12], fluorimetry [13], gas chromatography [14], high-performance liquid chromatography [15], flow-injection chemiluminescence [16], and electroanalytical techniques [17, 18]. Contrary to the electroanalytical techniques, all

Handbook of Nanomaterials for Sensing Applications
https://doi.org/10.1016/B978-0-12-820783-3.00022-1

105

the other techniques are highly complicated to operate, require a pretreatment process, more expensive, consume time and process, which in turn limits their application for analysis [19]. Therefore, electroanalytical techniques were chosen as pertinent for the determination of CA for their low expenditure, and simple and quick response.

Carbon nanotubes have enthralled many researchers after their discovery due to their peculiar structure, and splendid electrical, mechanical, and chemical characteristics [20–22]. Sensors with carbon nanotube materials are superior over other traditional carbon sensors for their ability in improving electron transfer reactions inhibiting surface fouling, and greater surface area offering high sensitivity, broad potential window, and good biocompatibility [23–25].

At present, electrodes modified with conducting polymer films by electropolymerization method have impressed many scientists with their physical and chemical features and their applications in electrocatalysis. Numerous works have been reported based on polymer film coated on carbon nanotube electrodes for the determination of various biological and electrochemical active substances [26–28]. They exhibited excellent stability, enhancing the redox behavior, reproducibility, transducing, and permselective. Dyes have been used as mediators to investigate the electrochemical behavior of bioactive molecules [29–32]. Usually, dye molecules covalently bonded to the aromatic ring forming the derivatives of dye, which minimize their ability to donate their protons and enhance their adsorptive action. Acriflavine (AF) is a fluorescent dye commonly used as an antiseptic and as a photosensitizer for chemotherapy [33–35].

In our recent study, we have developed an electrochemical sensor by fabricating AF by electropolymerization process on the surface of the carbon nanotube paste electrode for determining a trace quantity of CA. The electrochemical response of PAFMCNTPE was studied with the interference of caffeine (CF) as it is present in tea. The application of the electrode was studied in tap water samples.

2. Materials and methods

2.1 Instrumentation

CV and differential pulse voltammetry (DPV) experiments were operated with the aid of electrochemical workstation CHI-6038E (United States) attached to a conventional three-electrode system. Bare carbon nanotube paste electrode (BCNTPE) and PAFMCNTPE electrodes served as working electrodes, aqueous saturated calomel electrode performed as the reference electrode, and a Pt wire operated as an auxiliary electrode. The instrument was connected to a personal computer to store and extract data.

2.2 Chemicals and reagents

CA was acquired from Sisco Research Laboratories, India. AF and CF were attained from Molychem, India. Multiwalled carbon nanotube (MWCNT) with a length of 10–30 μm and possessing optical density of 30–50 nm were purchased from Sisco Research Laboratory Pvt. Ltd., Maharashtra, India, and silicone oil was bought from Nice Company, Cochin, Kerala, India. Disodium hydrogen phosphate (Na_2HPO_4) and monosodium dihydrogen phosphate (NaH_2PO_4) were purchased from Himedia, India, and 25×10^{-4} M CA, 25×10^{-4} M AF, and 25×10^{-4} M CF stock solutions were prepared by dissolving in distilled water. The 0.2 M PBS was prepared by mixing the acquired volume of prepared 0.2 M NaH_2PO_4 solution and 0.2 M Na_2HPO_4 solution.

3. Results and discussion

3.1 Preparation of poly(acriflavine) coated carbon nanotube paste electrode

CV is a suitable technique to prepare a conducting polymer at the surface of the electrode through an effortless electropolymerization technique. The PAFMCNTPE was prepared by 5 cycle scans in a potential domain of −0.5 to 1.0 V in aqueous media containing 0.1 mM AF as revealed in Fig. 1A. Since the polymerization curves inclined as cyclic period increases, this indicates the formation of poly(acriflavine) on carbon nanotube paste electrode, and it is then washed with distilled water to remove the unreacted monomer. The number of cycles were optimized by varying cycles from 3 to 15 (Fig. 1B), the highest sensitivity was achieved at 5 cycles. Then, a further increase in the number of cycles' current sensitivity decreases due to increases in the thickness of the film. So, the 5 multiple cycles were chosen as an optimized condition for further electrochemical measurements.

3.2 Electrocatalytic characterization of BCNTPE and PAFMCNTPE

Before starting with the electroanalysis of the desired analyte, the performance of BCNTPE and PAFMCNTPE were tested by CV in 0.1 M KCl containing standard

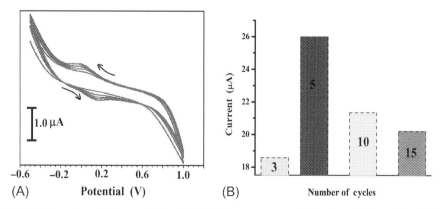

Fig. 1 (A) Electropolymerization of AF on BCNTPE in aqueous media for 5 multiple cycles. (B) Plot of number of scans vs peak current sensitivity.

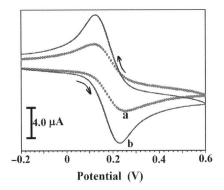

Fig. 2 Electrochemical characterization of BCNTPE *(curve a)* and PAFMCNTPE *(curve b)*.

1 mM $K_4[Fe(CN)_6]$, and the generated voltammogram is depicted in Fig. 2. PAFMCNTPE (curev b) detects oxidation and reduction of $K_4[Fe(CN)_6]$ at 0.221, 0.138 V with ΔE_p 0.083 V. The unmodified, i.e., BCNTPE (curve a) traces the diminished $K_4[Fe(CN)_6]$ oxidation and reduction at 0.123, 0.250 V with ΔE_p 0.127 V. So, after polymer modification, significant improvement in current response was observed with lower ΔE_p. The current obtained at PAFMCNTPE is twice the current obtained at BCNTPE. The effective surface area of BCNTPE and PAFMCNTPE can be calculated by using the Randles-Sevcik equation [36]:

$$I_p = 2.69 \times 10^5 \, n^{3/2} \, A \, D^{1/2} \, C_0 v^{1/2}$$

where I_p is anodic/cathodic peak current in A, C_o is electroactive species concentration (mol cm^{-3}), n is the number of electrons transferred, D is the coefficient of diffusion (cm^2 s^{-1}), v is the potential scan rate (V s^{-1}), and A is the effective surface area (cm^2). The active surface area for BCNTPE and PAFMCNTPE were calculated as 0.012 and 0.022 cm^2, respectively.

3.3 FESEM characteristics of BCNTPE and PAFMCNTPE

In electroanalysis, surface topography is a very important aspect that affects the performance of electrode and electrode processes. So, it is necessary to study the surface nature of the electrode. The BCNTPE and PAFMCNTPE field emission scanning electron microscopy (FE-SEM) characterizations are shown in Fig. 3A and B. The BCNTPE shows the random arrangement of carbon nanotubes with a rough surface. But after polymer modification, i.e., PAFMCNTPE shows a homogenous surface with uniform alignment. This indicates that the BCNTPE is modified with a poly(acriflavine) polymer layer.

3.4 Electrochemical sensing of CA with BCNTPE and PAFMCNTPE

The electrocatalytic behavior of CA was examined by CV in 0.2 M PBS (pH 7.0), at a sweep rate of 0.1 V s^{-1} as revealed in Fig. 4. BCNTPE (curve a) traces the CA oxidation

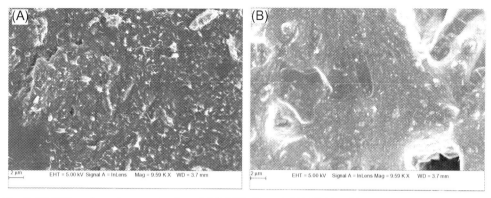

Fig. 3 FE-SEM surface characterization of BCNTPE (A) and PAMCNTPE (B).

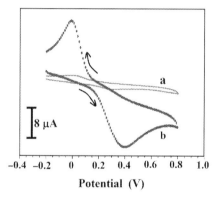

Fig. 4 CVs of CA (0.1 mM) sensing at BCNTPE (curve a) and PAFMCNTPE (curve b) in 0.2 M PBS (pH 7.0), with 0.1 V s^{-1} sweep rate.

at 0.321 V and reduction at 0.026 V with current responses of 1.9 and 1.75 μA, respectively. In similar conditions, the PAFMCNTPE (curve b) recognizes the CA oxidation and reduction at 0.399 and −0.009 V with current responses of 17.72 and 16.71 μA, respectively. The results show that current produced at PAFMCNTPE was enhanced nine times higher than BCNTPE. So, the conducting polymer on the electrode creates an electrocatalytic effect for sensing CA.

3.5 Impact of solution pH

The solution pH has a marked effect on the electrochemical performance of the electrode and mechanism of organic molecules. Therefore, it is essential to study the influence of pH in sensing analytes at the fabricated electrode. The effect of pH variation was investigated for CA at PAFMCNTPE over a range of 5.5–8.0 in PBS (0.2 M) at a sweep rate 0.1 V s^{-1}, which is represented in Fig. 5A. The plot of oxidation peak current vs solution pH (Fig. 5B) confirms the highest current sensitivity was observed at pH 7.0. So, it was

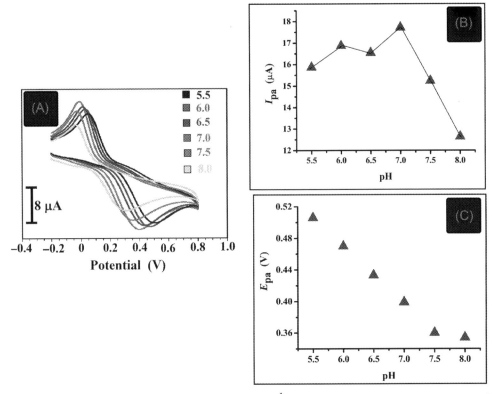

Fig. 5 (A) CVs of CA at different pH (5.5–8.0) at 0.1 V s^{-1} scan rate. (B) Graph of pH vs anodic peak current. (C) Plot of oxidation potential current response vs different pH values.

chosen as fine-tuned pH for all voltammetric experiments. At this pH, the strong electrochemical interaction occurs between electrode and analyte.

The graphical plot of anodic peak potential vs solution pH (Fig. 5C) proves that, as the pH increases, the oxidation peak is shifted toward the negative side by validating linear regression equation E_{pa} (V) $= -0.064$ pH $+ 0.85$, the slope 0.064 V/pH unit is close to the ideal value 0.059, which specifies the electrons (e$^-$) and protons (H$^+$) involved in the electrocatalytic reaction is in the proportion 1:1. So the passage of reduction/oxidation of CA comprises two electrons (2 e$^-$) and two protons (2 H$^+$) as revealed in Scheme 1.

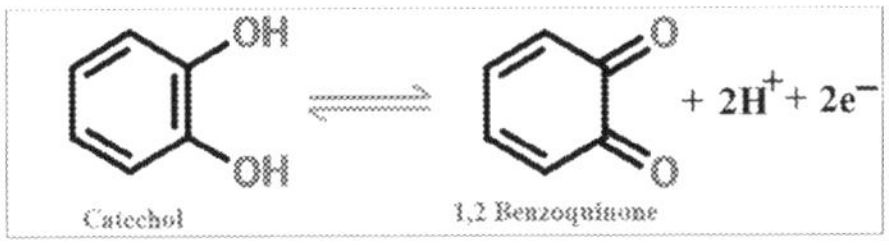

Scheme 1 Electron transfer mechanism of CA.

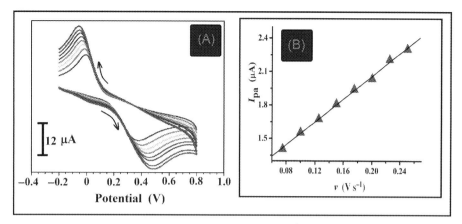

Fig. 6 (A) CA (0.1 mM) scan rate studies from 0.75 to 0.250 V s^{-1} in 0.2 PBS (pH 7.0). (B) linear plot of v vs I_{pa}.

3.6 Potential scan rate effect

The sweep rate studies reveal the process occurring at the electrode surface-solution interface. The impact of potential scan rate (0.75–0.250 V s^{-1}) on the oxidation of CA (0.1 mM) at PAFMCNTPE in PBS (0.2 M; pH 7.0) was deliberated by CV technique as portrayed in Fig. 6A. As the sweep rate increases, the anodic and cathodic current signals also increase, and the cathodic and anodic potentials shift opposite to each other with an increase in ΔE_p values, which shows the CA redox process is quasireversible. The linear corelation was established between v vs I_{pa} (Fig. 6B) by aggregating linear equation I_{pa} (μA) = 1.03 + 5.12 v (V s^{-1}) with $R = 0.99$. This illustrates the process of electrode solution interface was controlled by adsorption of CA.

3.7 Electrochemical detection of CA using DPV

DPV is used to analyze the electrocatalytic behavior of PAFMCNTPE toward the CA. As seen in Fig. 7, in a blank (curve a) PBS (0.2 M, pH 7.0), no peak was observed over a working potential range at PAFMCNTPE (curve c). After addition of 0.1 mM CA to 0.2 M PBS of pH 7.0, a well-defined peak appeared at potential 0.268 V with high current response. At BCNTPE (curve b), CA peak was sensed at 0.172 V with a diminished current signal. So, after modification with polymer layer, the CA sensing was upgraded.

3.8 Analytical performance

DPV measurements were carried out for the sensitive estimation of CA in the concentration range of 20–800 μM at the PAFMCNTPE electrode as shown in Fig. 8A and B. Well-defined oxidation peak at 0.268 V is observed and, on increasing CA concentration, the oxidation peak current increases. Plot of peak current vs CA concentration shows (Fig. 8B) linear range (LR) from 20 to 800 μM by validating linear equation I_p

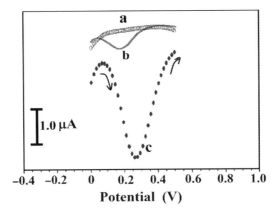

Fig. 7 DPVs of Blank *(curve a)*, CA sensing at BCNTPE *(curve b)*, and CA sensing at PAFMCNTPE in 0.2 M PBS (pH 7.0), with 0.1 V s^{-1} sweep rate.

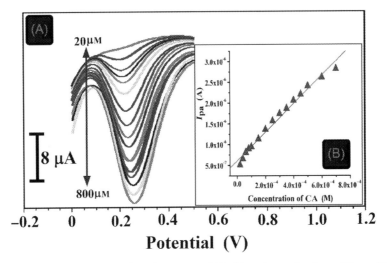

Fig. 8 (A) DPV curves for variation of CA from 20 to 800 µM under optimal conditions. (B) Standard calibration graph for CA concentration against peak currents.

(A) $= 6.12 \times 10^{-6} + 0.0034$ C (M), ($r^2 = 0.999$). The limit of detection and limit of quantification were determined as 3 σ/m and 10 σ/m, respectively; here, "σ" is determined as the standard deviation of 5 DPV measurements carried out in blank solution, and "m" is the slope of the calibration curve. The LOD and LOQ for CA estimation at PAFMCNTPE were found to be 2.441×10^{-7} and 8.153×10^{-7} M, respectively. Table 1 shows an assessment of the PAFMCNTPE with other electrodes described in the literature for estimation of CA [37–42]. The fabricated sensor works at neutral pH (pH 7.0) without using toxic materials, has extensive linear range (20–800 µM), and a

Table 1 Comparison of the proposed electrochemical sensor with previously reported sensor for CA analysis.

Electrode	LOD (µM)	LR (µM)	Reference
PEDOT/GO	1.6	5–400	37
AuNPS-MPS/CPE	1.1	30–1000	38
OM-MnFeOx/GCE	0.48	1–150, 150–400	39
RGO-MWNTS	1.8	5.5–540	40
CNx/GCE	2.71	20–1000	41
CG/poly-L-Asn	0.132	3.3–20 and 20–66.7	42
CNTPE/poly(acriflavine)	0.2441	20–800	Present work

AuNPS-MPS, gold nanoparticles mesoporous silica modified carbon paste electrode; *CG/poly-L-Asn*, carboxylated graphene/poly-lasparagine modified electrode; *CNx/GCE*, nitrogen-doped carbon nanotubes; *GCE*, glassy carbon electrode; *OM-MnFeOx*, oxidized multiwalled carbon nanotubes, manganese dioxide and manganese ferrite; *PEDOT/GO*, graphene oxide doped poly(3,4ethylenedioxythiophene); *RGO-MWNTS*, reduced graphene oxide and multiwall carbon nanotubes.

low detection limit (0.24 µM) with high stability. These all factors make the equipped sensor analogous with earlier studies.

3.9 PAFMCNTPE stability, repeatability, and reproductivity

The stability of PAFMCNTPE was studied by running 30 continuous cycles for CA in PBS (0.2 M, pH 7.0) using CV; 90% of the initial current signal for CA is retained even after 30 cycles. This shows the steadiness of a proposed electrode. Good repeatability with 2.4 RSD % was observed for repetitive measurements of five separate CA solutions with an unchanged electrode. Reproductivity was performed at optimal conditions for five renewals of an electrode with a constant analyte solution. The excellent reproductivity of CA detection was achieved for five distinct measurements with 2.8 RSD %.

3.10 CA estimation in water sample

The devised sensor was applied for CA analysis in a tap water sample at optimal conditions. The water sample was diluted with PBS (0.2 M, pH 7.0). The standard addition method was adopted for the estimation of CA in the water sample. The recovery for CA estimation in the water sample was found as 97%–100.1% with RSD less than 2%. This shows the analytical applicability of the proposed sensor.

3.11 Instantaneous analysis of CA with CF

The concurrent analysis of CA with CF is significant because CA and CF coexist in tea samples. So, voltammetric separation of CA and CF was performed at PAFMCNTPE (curve b) and BCNTPE (curve a) at calibrated conditions as shown in Fig. 9. At BCNTPE, the CA and CF were detected at a potential 0.249, 1.357 V with current response 2.9 and 10.9 µA, respectively. But at PAFMCNTPE, the clear and sharp

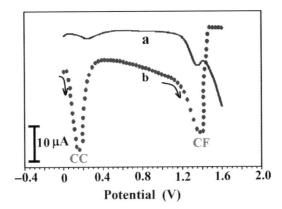

Fig. 9 Concurrent separation of CA and CF at BCNTPE *(curve a)* and PAFMCNTPE *(curve b)* under calibrated conditions.

separations of CA and CF were noticed at 0.164 and 1.387 V with current responses of 35 and 30 μA, respectively. So, the PAFMCNTPE can be applied for the simultaneous separation of CA and CF.

4. Conclusion

In the present work, the simple and active electrochemical sensor PAFMCNTPE was developed through the electropolymerization technique. The devised sensor was characterized by FESEM and voltammetric methods. The electrocatalytic behavior of PAFMCNTPE toward CA detection was studied with CV and DPV techniques. The polymer-capped electrode exhibits higher effective surface area and conducting track through a polymer layer, which has a high affinity for determination of CA. The devised sensor provides a lower detection limit with good linear range. The stable sensor is effectively tested for analysis of CA in a water sample with good recovery. Moreover, the electrode exhibits good repeatability and reproductivity. By these detailed analyses, we can conclude the proposed sensor is another alternative for CA sensing.

References

[1] B. Gigante, C. Santos, A.M. Silva, M.J.M. Curto, M.S.J. Nascimento, E. Pinto, M. Pedro, F. Cerqueira, M.M. Pinto, M.P. Duarte, et al., Catechols from abietic acid: synthesis and evaluation as bioactive compounds. Bioorg. Med. Chem. 11 (8) (2003) 163, https://doi.org/10.1016/S0968-0896(03)00063-4.
[2] M.C. Chang, H.H. Chang, T.M. Wang, C.P. Chan, B.R. Lin, S.Y. Yeung, C.Y. Yeh, R.H. Cheng, J. H. Jeng, Antiplatelet effect of catechol is related to inhibition of cyclooxygenase, reactive oxygen species, ERK/P38 signaling and thromboxane A2 production. PLoS One 9 (2014) 1, https://doi.org/10.1371/journal.pone.0104310.

[3] F.A.L. Clowes, The structure of mycorrhizal roots of fagus sylvatica. New Phytol. 50 (1) (1951) 1, https://doi.org/10.1111/j.1469-8137.1951.tb05166.x.

[4] C. Kaur, H.C. Kapoor, Inhibition of enzymatic browning in apples, potatoes and mushrooms, J. Sci. Ind. Res. (India) 59 (5) (2000) 389.

[5] P.K. Dhar, M.A. Zubaer, M.K. Hasan, M.S. Hossain, M.K. Amin, Unusual solubility behavior of catechol-arginine derivative, J. Basic Appl. Chem. 8 (2) (2018) 14.

[6] A. Weissberger, D.S. Thomas, Oxidation processes. XIV'. The effect of silver on the autoxidation of some photographic developing agents. J. Am. Chem. Soc. 64 (7) (1942) 1561, https://doi.org/10.1021/ja01259a021.

[7] C. Gu, X. Li, H. Liu, Simultaneous determination of hydroquinone, and catechol using a multi-walled carbon nanotube/GC electrode modified by electrodeposition of carbon nanodots. J. Nanopart. Res. 54 (2018) 42, https://doi.org/10.4028/www.scientific.net/JNanoR.54.42.

[8] Y.H. Huang, J.H. Chen, X. Sun, Z.B. Su, H.T. Xing, S.R. Hu, W. Weng, H.X. Guo, W.B. Wu, Y. S. He, One-pot hydrothermal synthesis carbon nanocages-reduced graphene oxide composites for simultaneous electrochemical detection of catechol and hydroquinone. Sensors Actuators B Chem. 212 (2015) 165, https://doi.org/10.1016/j.snb.2015.02.013.

[9] K. Kim, Influences of environmental chemicals on atopic dermatitis. Toxicol. Res. 31 (2) (2015) 89, https://doi.org/10.5487/TR.2015.31.2.089.

[10] J.M. McCue, K.L. Link, S.S. Eaton, B.M. Freed, Exposure to cigarette tar inhibits ribonucleotide reductase and blocks lymphocyte proliferation. J. Immunol. 165 (12) (2000) 6771, https://doi.org/10.4049/jimmunol.165.12.6771.

[11] M. Do Ceu Silva, J. Gaspar, I.D. Silva, D. Leão, J. Rueff, Induction of chromosomal aberrations by phenolic compounds: possible role of reactive oxygen species. Mutat. Res. 540 (1) (2003) 29, https://doi.org/10.1016/S1383-5718(03)00168-2.

[12] D.W. Barnum, Spectrophotometric determination of catechol, epinephrine, dopa, dopamine and other aromatic vic-diols. Anal. Chim. Acta 89 (1) (1977) 157, https://doi.org/10.1016/S0003-2670(01)83081-6.

[13] M.F. Pistonesi, M.S. Di Nezio, M.E. Centurion, M.E. Palomeque, A.G. Lista, B.S. Fernández Band, Determination of phenol, resorcinol and hydroquinone in air samples by synchronous fluorescence using partial least-squares (PLS). Talanta 69 (5) (2006) 1265, https://doi.org/10.1016/j.talanta.2005.12.050.

[14] O. Grodnick, J. S, G.D. Dupre, B.J. Gulizia, S.H. Blake, Determination of benzene, phenol, catechol, and hydroquinone in whole blood of rats and mice. J. Chromatogr. Sci. 21 (7) (1983) 289, https://doi.org/10.1093/chromsci/21.7.289.

[15] C.H. Risne, S.L. Cash, A high-performance liquid chromatographic determination of major phenolic compounds in tobacco smoke. J. Chromatogr. Sci. 28 (5) (1990) 239, https://doi.org/10.1093/chromsci/28.5.239.

[16] L. Zhao, B. Lv, H. Yuan, Z. Zhou, D. Xiao, A Sensitive chemiluminescence method for determination of hydroquinone and catechol. Sensors 7 (4) (2007) 578, https://doi.org/10.3390/s7040578.

[17] S.J. Caballero, M.A. Guerrero, L.Y. Vargas, C.C. Ortiz, J.J. Castillo, J.A. Gutiérrez, S. Blanco, Electroanalytical determination of catechol by a biosensor based on laccase from aspergillus oryzae immobilized on gold screen-printed electrodes. J. Phys. Conf. Ser. 1119 (2018) 1, https://doi.org/10.1088/1742-6596/1119/1/012009.

[18] J.G. Manjunatha, Electrochemical polymerised graphene paste electrode and application to catechol sensing. Open Chem. Eng. J. 13 (1) (2019) 81, https://doi.org/10.2174/1874123101913010081.

[19] H. Beitollahi, Z. Dourandish, M.R. Ganjali, S. Shakeri, Voltammetric determination of dopamine in the presence of tyrosine using graphite screen-printed electrode modified with graphene quantum dots. Ionics (Kiel). 24 (12) (2018) 4023, https://doi.org/10.1007/s11581-018-2489-3.

[20] P.A. Pushpanjali, J.G. Manjunatha, Electroanalysis of sodium alizarin sulfonate at surfactant modified carbon nanotube paste electrode : a cyclic voltammetric study, J. Mater. Environ. Sci. 10 (10) (2019) 939.

[21] G. Tigari, J.G. Manjunatha, C. Raril, N. Hareesha, Determination of riboflavin at carbon nanotube paste electrodes modified with an anionic surfactant. ChemistrySelect 4 (7) (2019) 2168, https://doi.org/10.1002/slct.201803191.

[22] J.G. Manjunatha, M. Deraman, N.H. Basri, N.S.M. Nor, I.A. Talib, N. Ataollahi, Sodium dodecyl sulfate modified carbon nanotubes paste electrode as a novel sensor for the simultaneous determination of dopamine, ascorbic acid, and uric acid. C. R. Chim. 17 (5) (2014) 465, https://doi.org/10.1016/j.crci.2013.09.016.

[23] B.M. Amrutha, J.G. Manjunatha, A.S. Bhatt, C. Raril, P.A. Pushpanjali, Electrochemical sensor for the determination of alizarin red-S at non-ionic surfactant modified carbon nanotube paste electrode. Phys. Chem. Res. 7 (3) (2019) 523, https://doi.org/10.22036/pcr.2019.185875.1636.

[24] N.S. Prinith, J.G. Manjunatha, C. Raril, Electrocatalytic analysis of dopamine, uric acid and ascorbic acid at poly(adenine) modified carbon nanotube paste electrode: a cyclic voltammetric study, Anal. Bioanal. Electrochem. 11 (6) (2019) 742.

[25] J.G. Manjunatha, Surfactant modified carbon nanotube paste electrode for the sensitive determination of mitoxantrone anticancer drug. J. Electrochem. Sci. Eng. 7 (1) (2017) 39, https://doi.org/10.5599/jese.368.

[26] M.M. Charithra, J.G. Manjunatha, Poly (l-proline) modified carbon paste electrode as the voltammetric sensor for the detection of estriol and its simultaneous determination with folic and ascorbic acid. Mater. Sci. Energy Technol. 2 (3) (2019) 365, https://doi.org/10.1016/j.mset.2019.05.002.

[27] J.G. Manjunatha, Electroanalysis of estriol hormone using electrochemical sensor. Sens. Bio-Sensing Res. 16 (2017) 79, https://doi.org/10.1016/j.sbsr.2017.11.006.

[28] J.G. Manjunatha, A novel voltammetric method for the enhanced detection of the food additive tartrazine using an electrochemical sensor. Heliyon 4 (11) (2018) e00986, https://doi.org/10.1016/j.heliyon.2018.e00986.

[29] J.G. Manjunatha, M. Deraman, N.H. Basri, Electrocatalytic detection of dopamine and uric acid at poly (basic blue b) modified carbon nanotube paste electrode, Asian J. Pharm. Clin. Res. 8 (5) (2015) 48.

[30] J.G. Manjunatha, Fabrication of poly (solid red A) modified carbon nano tube paste electrode and its application for simultaneous determination of epinephrine, uric acid and ascorbic acid. Arab. J. Chem. 11 (2) (2018) 149, https://doi.org/10.1016/j.arabjc.2014.10.009.

[31] N. Hareesha, J.G. Manjunatha, C. Raril, G. Tigari, Sensitive and selective electrochemical resolution of tyrosine with ascorbic acid through the development of electropolymerized alizarin sodium sulfonate modified carbon nanotube paste electrodes. ChemistrySelect 4 (15) (2019) 4559, https://doi.org/10.1002/slct.201900794.

[32] C. Raril, J. Manjunatha, Cyclic voltammetric investigation of caffeine at methyl orange modified carbon paste electrode. Biomed. J. Sci. Tech. Res. 9 (3) (2018) 7149, https://doi.org/10.26717/bjstr.2018.09.001804.

[33] P. Taylor, T. Meinelt, A. Rose, M. Pietrock, Effects of calcium content and humic substances on the toxicity of acriflavine to juvenile zebrafish Danio rerio. J. Aquat. Anim. Health 37 (2015), https://doi.org/10.1577/1548-8667(2002)014 No. May.

[34] C.H. Browning, R. Gulbransen, L.H.D. Thornton, The antiseptic properties of acriflavine and proflavine, and brilliant green. Br. Med. J. 2 (2951) (1917) 70, https://doi.org/10.1136/bmj.2.2951.70.

[35] M. Broekgaarden, R. Weijer, M. Krekorian, B. van den IJssel, M. Kos, L.K. Alles, A.C. van Wijk, Z. Bikadi, E. Hazai, T.M. van Gulik, et al., Inhibition of hypoxia-inducible factor 1 with acriflavine sensitizes hypoxic tumor cells to photodynamic therapy with zinc phthalocyanine-encapsulating cationic liposomes. Nano Res. 9 (6) (2016) 1639, https://doi.org/10.1007/s12274-016-1059-0.

[36] C. Raril, J.G. Manjunatha, Sensitive electrochemical analysis of resorcinol using polymer modified carbon paste electrode. a cyclic voltammetric study, Anal. Bioanal. Electrochem. 10 (2018) 488.

[37] W. Si, W. Lei, Y. Zhang, M. Xia, F. Wang, Q. Hao, Electrodeposition of graphene oxide doped poly (3,4 ethylenedioxythiophene) film and its electrochemical sensing of catechol and hydroquinone. Electrochim. Acta 85 (2012) 295, https://doi.org/10.1016/j.electacta.2012.08.099.

[38] J. Tashkhourian, M. Daneshi, F. Nami-Ana, M. Behbahani, A. Bagheri, Simultaneous determination of hydroquinone and catechol at gold nanoparticles mesoporous silica modified carbon paste electrode. J. Hazard. Mater. 318 (2016) 117, https://doi.org/10.1016/j.jhazmat.2016.06.049.

[39] S. Chen, R. Huang, J. Yu, X. Jiang, Simultaneous voltammetric determination of hydroquinone and catechol by using a glassy carbon electrode modified with a ternary nanocomposite prepared from

oxidized multiwalled carbon nanotubes, manganese dioxide and manganese ferrite. Microchim. Acta 186 (2019) 643, https://doi.org/10.1007/s00604-019-3750-9.

[40] F. Hu, S. Chen, C. Wang, R. Yuan, D. Yuan, C. Wang, Study on the application of reduced graphene oxide and multiwall carbon nanotubes hybrid materials for simultaneous determination of catechol, hydroquinone, p-cresol and nitrite. Anal. Chim. Acta 724 (2012) 40, https://doi.org/10.1016/j.aca.2012.02.037.

[41] J. Dong, X. Qu, L. Wang, C. Zhao, J. Xu, Electrochemistry of nitrogen-doped carbon nanotubes (CNx) with different nitrogen content and its application in simultaneous determination of dihydroxybenzene isomers. Electroanalysis 20 (18) (2008) 1981, https://doi.org/10.1002/elan.200804274.

[42] L. Yanrong, R. Jin, Q. Yina, L. Jingmin, W. Xiaojian, W. Kun, W. Chaoqi, Simultaneous electrochemical determination of hydroquinone and catechol using a carboxylated graphene/poly-lasparagine modified electrode. Int. J. Electrochem. Sci. 14 (2019) 10043, https://doi.org/10.20964/2019.11.20.

CHAPTER 7

Nanofabrication techniques for semiconductor chemical sensors

Mona Mittal[a], Soumen Sardar[b], and Atanu Jana[c]
[a]Department of Chemistry, Noida Institute of Engineering and Technology, Greater Noida, Uttar Pradesh, India
[b]Department of Polymer Science and Technology, University of Calcutta, Calcutta, India
[c]Division of Physics and Semiconductor Science, Dongguk University, Seoul, South Korea

1. Introduction

Demand for information has grown enormously in every prospect of daily life owing to the technological advancement of our civilization. Chemical sensors are at the forefront of the information acquisition chain about the environment in which we live. They have applications in various fields such as healthcare-genetics, diagnostics, drug discovery, quality control, and environmental and industrial monitoring [1–3]. There has been a strong demand for the development of highly selective, sensitive, responsive, and cost-effective sensors. Chemical sensors are broadly divided into classes based on the transduction principle. The classes employed in wireless chemical sensors are specifically electrochemical, optical, electrical, and mass sensitive. Electrochemical sensors transduce the electrochemical interaction of an analyte at a modified electrode into a voltage or current signal. Major groups within electrochemical sensors are potentiometric sensors, which measure the potential of an indicator electrode (ion-selective electrode) against a reference electrode, and voltammetric sensors, which measure current flow at a constant or varying potential. Amperometric sensors, which measure current at a constant potential, are very commonly used for electrochemical biosensors. Optical chemical sensors are based on measuring optical phenomena caused by analyte-receptor interaction. Optical sensors can measure the absorbance of the analyte directly (or an analyte-sensitive indicator molecule) or luminescence intensity emitted by an indicator molecule in the receptor layer [4–6]. More complex optical sensors employing measurement of refractive index change, surface plasmon resonance (SPR), and light scattering are often encountered in the chemical sensor literature but are generally not suitable for use in wireless chemical sensors due to the need for laser sources and specialty optics that result in large power demand, size, and cost of such systems. Electrical chemical sensors are generally somewhat simpler, where the analytical signal arises from the change in electrical properties of a material, such as the conductivity of a metal oxide semiconductor (MOS) or organic semiconductor, caused by a chemical interaction with the analyte. Mass sensitive

Handbook of Nanomaterials for Sensing Applications
https://doi.org/10.1016/B978-0-12-820783-3.00023-3

119

chemical sensors transform a mass change, or surface energy change, caused by the accumulation of an analyte on a chemically modified surface into a change of property of the supporting material [7, 8]. The field of chemical sensing is wide and endless. The challenge of the scientific world is to develop new technology that will move from expensive and large instrumentation to small and inexpensive methods and technologies that can be operated by anyone. This chapter is dedicated to the chemical sensors and their various fabrication techniques.

2. Chemical sensors

A chemical sensor is a device or instrument that transforms different types of chemical information such as concentration, chemical activity, partial pressure, composition, and presence of metal ions or gases into a measurable and distinguishable signal [4–6] (Fig. 1).

The chemical information arises from either a chemical reaction or a physical property of the system under investigation. They have gained more attention in various practical applications in automotive, mining, environmental monitoring, industrial process monitoring, gas compositional analysis, and pharmaceutical as well as personal safety and homeland security [1–3] owing to their excellent properties like small size, acceptable sensitivity, larger dynamic range, low cost, easy to realize automatic measurement, and in situ and continuous detection. Hence, the chemical sensor becomes one of the most dynamic and effective directions of modern sensor technology.

2.1 Composition of chemical sensor

A chemical sensor contain two basic components that are connected in series: a receptor (a chemical or molecular recognition system) and a physicochemical transducer (Fig. 1). The receptor interacts selectively with the analyte molecules at their interface. The interaction processes for chemical sensors are adsorption/desorption of analyte, ion exchange, and liquid-liquid extraction. These interactions result in a change of physical properties of the receptor in such a way that the appending transducer can amplify an electrical signal. The transducer transforms the actual concentration value, a nonelectric quantity, into an electric quantity, voltage, current, or resistance.

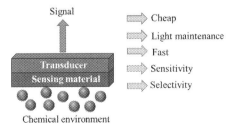

Fig. 1 Chemical sensor and detection mechanism.

2.2 Classification of chemical sensors

Chemical sensors can be classified into many types such as electrochemical, mass, optical, magnetic, and thermal sensors depending upon their working principle [1]. The working principle of an electrochemical sensor depends on the electrochemical effect among the analyte molecules and featured electrodes. The mass sensor depends on the quality change induced by the mass loading from the adsorption toward the analyte by the special modification of the sensor surface. The optical chemical sensor is based on the changes in optical properties, which results from the interaction between the analyte and the receptor. The magnetic device is based on the changes in the magnetic properties during analyte adsorption, whereas the thermal sensor is based on the thermal effect generated by the specific chemical reaction or adsorption process between analyte and receptor surface.

Chemical sensors can also be classified into various categories depending upon the object that has to be analyzed such as gas, ion, humidity, and biosensors. These sensors are further categorized into many types depending upon the working principle. Gas sensors are represented by a semiconductor, an electrochemical, a solid electrolyte, a contact combustion, a photochemical, and a polymer gas sensor.

2.3 Common examples of chemical sensors

The commonly used chemical sensors are carbon monoxide detector, glucose detector, mosquito, and pregnancy test strips. *Carbon monoxide detector* detects the presence of carbon monoxide (CO) gas. CO is a colorless and odorless gas, and it is also known as a "silent killer" because it is virtually undetectable without detection technology. Elevated levels of CO are dangerous to humans depending on the amount present and period of exposure. CO detectors are based on a chemical reaction causing a color change, an electrochemical reaction, which produces a current to trigger an alarm, or a semiconductor sensor that changes its electrical resistance in the presence of CO. *Glucose detector* measures the concentration of glucose/sugar within a person's blood using a complex chemical process. In this process, blood is mixed with glucose oxidase, which reacts with the glucose in the blood sample to create gluconic acid. Ferricyanide reacts with the gluconic acid to create ferrocyanide. The electrode within the test strip then runs a current through the blood sample and the ferrocyanide influences this current in such a way that the concentration of blood glucose within the sample can be accurately measured. *Mosquitoes* have a battery of sensors in their antennas, and one of them is a chemical sensor. They can sense carbon dioxide, lactic acid, and chemicals present in sweat. Therefore, people who sweat more easily will tend to attract more mosquitoes. *Pregnancy test strips* detect the presence of human chorionic gonadotropin (hcG) hormone, which is produced by the placenta and can be found in a woman's system as soon as implantation of a fertilized egg has occurred.

2.4 The parameters of chemical sensors

The performances of chemical sensors are usually characterized by various parameters such as sensitivity, selectivity, response and recovery times, stability, and reproducibility.

2.4.1 Sensitivity

Sensitivity of a sensor is the minimum perturbation of physical parameters, which produce a detectable output signal [1, 7, 8]. The sensitivity, S (%), of a chemical sensor is defined as the ratio of the difference in amplitude of a sensor signal in the presence (R_i) and in the absence (R_0) of the sample, and the amplitude of sensor signal in the absence of sample (R_0). The signal could be the resistance, capacitance, current, voltage, or conductance.

$$S(\%) = (R_i - R_0)/R_0 \times 100$$

2.4.2 Response and recovery time

Response time is defined as the time required by a chemical sensor to achieve 90% of the total signal change or stable output reading upon exposure to the analyte [9, 10]. Recovery time is defined as the time taken by a chemical sensor to achieve 90% of its original signal state. A chemical sensor should have a short response and recovery time. Reversibility is the extent to which the signal is restored to its initial state prior to analyte exposure, which is important paramount for practical applications of sensors for continuous sample monitoring [11].

2.4.3 Detection limit

A high-performance chemical sensor should be able to detect a tiny amount of chemical species, and the lowest amount of the object that the sensor could have a response to is called the detection limit. The low detection limit is determined by exposing a sensing device to a known concentration of analyte, which will generate a calibration curve. The detection limit is obtained by doing a comparison of sensor resolution with its sensitivity [8, 12, 13].

2.4.4 Selectivity and specificity

The selectivity of a chemical sensor means the extent to which a chemical sensor can detect a particular analyte in the mixture of analyte molecules without any interference from other components [7, 14]. Specificity is the ability of a chemical sensor to discriminate the analyte of interest from other compounds, which are present within the analyzed sample and is known as the ultimate selectivity [15].

2.4.5 Stability and reproducibility

The stability of a sensor is evaluated by the change of sensing behavior after numerous times of switching between "ON" state and "OFF" state. The stability is good when the sensing performance shows little change after numerous tests over a period of time [8]. Reproducibility is the ability of the chemical sensors to produce the same signal output after the alteration of experimental conditions [14].

3. Nanomaterials-based chemical sensors

Researchers have introduced nanomaterials with different sizes and shapes for chemical sensing applications owing to their unique and captivating properties. Nanomaterials such as nanocrystals, nanowires, nanorods, nanobelts, nanorings, and nanosheets are becoming attractive candidates for various chemical sensing devices owing to their high surface-to-volume ratio, very small grain and pore size, and superior optical, electrical, mechanical, chemical, catalytic, and interfacial properties compared with bulk counterparts. Their novel properties are further explored by controlling them at the atomic and molecular scales. These properties impart better sensitivity, selectivity, stability, and reproducibility. Their diameters are comparable to the size of chemical species and hence, few chemical species are sufficient to change their electrical characteristics like resistance or capacitance after absorption [16]. Hence, nanomaterials-based chemical sensors produce analytical signals by detecting very small concentrations of a sample. Therefore, nanomaterials-based chemical sensors could revolutionize many aspects of sensing and detection of chemical species like gases, metal ions, and small biomolecules. They have various applications in the field of transportation, communications, building and facilities, medicine, safety, and national security, including both homeland defense and military operations. There is a great demand for the development of efficient and cost-effective sensors.

3.1 Sensors based on nanoparticles

Semiconductor metal oxide nanoparticles such as SnO_2, ZnO, TiO_2, and WO_3 nanoparticles have certain specific advantages such as large specific surface area, less sensitive toward moisture and temperature, biocompatible, highly robust, high electron mobility, and piezoelectricity. They have ease of preparation and scope of developing a large variety of nanoparticles with good control on their size and availability of a larger number of active centers. Hence, they can be used to design gas (reductive or oxidative) sensors, chemical sensors, biosensors, UV sensors, and pH sensors depending upon their sensing mechanisms. These show simple interface electronics, and faster response as well as recovery time. Their sensing properties are highly influenced by a number of parameters including size, morphology, phase composition, structure, and type of dopants. However, their high-temperature synthesis procedure, large energy consumption, and poor selectivity must be minimized for their practical application in chemical sensing [17–21].

3.2 Sensors based on nanowires

The one-dimensional systems show promising sensing capabilities for gas molecules and biological species. Based on the mechanism of detection, various systems have been configured as chemical sensors for the detection of toxic and flammable gases such as NO_2, CO, NH_3, and ethanol [22–30].

3.3 Sensors based on nanotubes

Carbon nanotubes have novel electronic properties, which are confirmed by experimental as well as theoretical investigation. Hence, they can be utilized as an electric wire between two electrodes and their electronic properties (conductance or resistance) can be measured easily. The atomic structure of nanotubes, chemical or mechanical deformations in their structure, and chemical doping in them influence their electronic properties and hence, induce strong change in conductance or resistance, which can be detected by electron current signals. These properties make nanotubes extremely small sensors sensitive to their chemical and mechanical environments. Carbon nanotubes possess low operating temperatures and high sensitivity but suffer from long response and recovery times, poor stability, degradation, and difficult processing [31–34].

3.4 Sensors based on 2D nanomaterials

Two-dimensional nanomaterials have gained tremendous attention owing to their unique chemical and physical properties, which make them suitable and improve sensitivity for chemical sensing. They possess a high surface-to-volume ratio and ultrahigh surface sensitivity toward environment, which are key characteristics for their applications in chemical sensing. Furthermore, their superior electrical and optical properties, combined with their excellent mechanical characteristics (robustness and flexibility), make them ideal components for the fabrication of a new generation of high-performance chemical sensors. Depending on the particular device, they can be tailored to interact with various chemical species at the noncovalent level. These modifications make them powerful platforms for fabricating devices, which exhibits a high sensitivity and selectivity toward the detection of various analytes such as gases, metal ions, and small biomolecules.

Among 2D nanomaterials, graphene and graphene oxide are particularly appealing owing to their low resistivity and electrical noise, which enable the detection of small changes in the intrinsic resistance resulting from the interaction of graphene with chemical species [35–37]. Transition metal dichalcogenides nanosheets (MoS_2, WS_2, $MoSe_2$, and WSe_2) [38–40], layered metal oxides (MoO_3, SnO_2) [41, 42], phosphorene (black phosphorous) [43, 44], h-BN [45, 46], and metal-organic frameworks (MOFs) [47, 48] have also been employed for chemical sensing by exploiting their thickness-dependent optoelectronic properties. The band gap of 2D materials are modified after interaction with chemical species, which enhanced the sensitivity of chemical sensors based on them. Furthermore, their semiconducting properties render them suitable for integration as active components in field–effect transistors (FETs), which allow the realization of low power consumption and miniaturized devices.

4. Applications of chemical sensors

4.1 Detection of gases

A gas sensor senses the presence and quantify the concentration of a specific gas in the atmosphere such as CO, CO_2, H_2S, SO_x, NH_3, NO, N_2O, N_2O_4, etc. These gases originate from a wide range of natural and anthropogenic sources and act as targets for chemical sensing. Their exceeding concentrations have significance on human health and safety, industrial process, and environmental monitoring, as well as emissions and production control, air quality management, and medical diagnostics [49]. Many of these gases are produced in large amount by combustion of fuels. For example, complete combustion of fossil fuels can lead to generation of CO_2 gas. CO and NO_x compounds are by-products of hypothermic combustion, whereas H_2S and SO_x compounds are by-products of the combustion of fuels containing sulfur. NH_3 gas is industrially produced by the reaction of N_2 and H_2 through the Haber–Bosch process; it is a good fertilizer and hence, acts as the backbone of the world's agricultural infrastructure.

The monitoring of these gases for biological purposes requires a unique set of sensor design criteria besides them being cost-effective, small size, and reusable. Sensors, which can monitor *in-vivo* processes and detect small reactive gases, are of great interest. The concentration of analyte is controlled using mass-flow controllers and inert gas streams. The strong reducing gases (NH_3) and oxidizing gases (NO_2) can easily induce observable electronic changes in many conductive materials. However, less reactive gases such as H_2, CO_2, and C_2H_4 require more specialized materials or material hybrids to induce observable electronic changes and provide robust detection. Hence, the demand for easily produced, robust, cost-effective, and highly integrated devices and design strategies is increasing exponentially for accomplishing the need for industry, environment, human health, and safety.

In a gas sensing device, receptors are integrated as an active component in FETs, and the adsorption of gases on the receptors results in a change of their conductivity, which can be detected and measured as a variation in the drain current [50]. Another common gas sensing devices are chemiresistors and chemicapacitors, in which the sensing material is interposed between two electrodes, and the gas molecules adsorbed on the surface of the material induce a change in the resistance or capacitance, respectively, which can be directly quantified [51, 52].

4.2 Detection of metal ions

A metal ion sensor detects the presence and quantifies the concentration of metal ions especially alkali, heavy metal ions, and ionic electrolytes. Seawater and wastewater contain alkali metal ions. The discharge of metallic contaminants from industrial processes contains heavy metal ions, which lead to extensive contamination of drinking water and agricultural products [53]. Metal ions in aqueous environments have caused various diseases and have been a serious threat to ecosystems and public health with the rapid development of industry in recent years. Hence, there is a commercial demand to manufacture rapid, sensitive, cost-effective, portable, and simple analytical platforms for the detection and monitoring of metallic contaminants in water and soil [54]. Furthermore, ionic electrolytes are essential for various bodily functions such as cell functioning and cell signaling. Their imbalance can result in various life-threatening conditions including cardiac arrest, neurological disorders, or kidney failure [55]. Hence, it is important to develop portable diagnostic devices that can provide real-time information about electrolyte concentration in biological samples. Metal sensors detect heavy metal ions and ionic electrolytes via electrical and optical outputs based on physiosorption or chemisorption. They rely on the capturing of the pollutant (analyte) by an adsorbent (receptor).

4.2.1 The adsorption process

The development of specific receptors for the detection of heavy metal ions or ionic electrolytes allows for exploitation of the reversible processes of adsorption and desorption as extremely versatile strategies toward sensing in a variety of environments. Moreover, by relying on the reversible nature of noncovalent interactions, the sensor can exhibit a quick response, fast recovery rate, and facile regeneration to enable its use multiple times. Nanomaterials have a high surface-to-volume ratio, which helps in metal ion adsorption. The presence of functional groups containing oxygen atoms on nanomaterials enhances the occurrence of adsorption events as oxygen atoms act as reactive sites for further covalent functionalization and can interact via dipole-dipole or strong electrostatic interactions with metal ions. In addition, chalcogenides can act as potential coordination sites for certain heavy metal ions.

The adsorption capabilities of receptors are evaluated through the maximum adsorption capacity (q_{max}), which is defined as the ratio between the maximum loaded mass of the analyte and the mass of the receptor. The conventional methods for the quantification of heavy metal ions include plasma mass spectrometry (ICP-AES), atomic absorption/emission spectroscopy (AAS/AES), and polarography. The concentration of the remaining heavy metal ions after adsorption is the difference between the initial (C_0) and equilibrium (C_e) concentrations. The equilibrium sorption capacity and time-dependent capacity were determined using the following equation:

$$q_e - (C_0 - C_e) \times V/m_{adsorbant}$$

Where q_e is the equilibrium amount of heavy metal ions adsorbed per unit mass (m) of adsorbent, and V is the volume of the metal ion solution. Among various adsorption isotherm models, the Freundlich and Langmuir models are most commonly used to estimate the maximum adsorption capacity (q_{max}) of metal ions on nanomaterials [56–58].

4.2.2 Fluorescence-based metal sensors

Fluorescence-based metal sensors or colorimetric sensors for detection of heavy metal ions comprise two key features; first, a metal chelating or binding (coordination) pocket and second, at least one fluorophore capable of absorbing and/or emitting light. Fluorescence sensing is basically based on analyte-induced changes in the physicochemical properties of fluorophores such as fluorescence intensity, lifetimes, and anisotropy, which are related to either charge or energy transfer processes. The electronic structure of the sensor must be altered upon the metal binding, which lead to changes in either intensity or wavelength of light absorption or emission. However, changes in the molecular structure lead to modification of the distance or alignment between a pair of fluorophores, which serve as a donor–acceptor pair [59, 60].

4.2.3 Field-effect transistors (FETs)-based metal sensors

The integration of nanomaterials as an active component into FETs has recently revealed their enormous potential for the detection of heavy metals. The working principle of a FET sensor is based on the changes in their critical parameters such as field–effect mobility, threshold voltage, and I_{on}/I_{off} ratio upon adsorption of targeted heavy metal ions. Chemical sensors based on FETs can overcome the obstacles of previous detection methods [61, 62]. The use of FET sensors enables the rapid label-free detection of metal ions in real-time by monitoring the resistance or the Dirac point shift caused by the adsorption of target analyte molecules. Such devices can also be characterized by low power consumption and can be miniaturized for the development of portable sensors, eventually supported on flexible foils.

4.2.4 Electrochemical-based metal sensors

Electrochemical sensing of metal ions is based on the use of sensing electrodes, which are employed for passing the current to an aqueous solution. Then, the sensing electrode generates an electrical signal, corresponding to the electrochemical reaction within the solution owing to the presence of metal ions. Common experimental setups for electrochemical detection of heavy metal ions consist of an electrolytic cell containing an electrolyte (a solution of heavy metal ions) in contact with an electrode. The cell potential is measured at the interface of the electrode with the electrolyte solution. As heavy metal ions have defined redox potentials, the selectivity toward specific metal ions could be achieved by using bare electrodes without the need for a molecular recognition probe. Numerous techniques have been employed in electrochemical sensing, including potentiometry, voltammetry, impedimetry, amperometry, and conductometry. Anodic stripping voltammetry (ASV) method has been explored widely for the detection of heavy metal ion. ASV analysis involves two steps, i.e., deposition of metals onto the electrode surface and stripping or dissolution of the deposited analyte molecule from the electrode. Several electrochemical sensors for bioanalysis and environmental analyses have been developed [63].

4.3 Detection of volatile compounds

The chemical sensors are also utilized to detect and quantify the concentration of volatile inorganic and organic compounds, which are easily vaporized. Among various volatile inorganic compounds such as water, volatile acids, halogen, and mercury, sensing of humidity is important. Humidity affects physiological activities, building construction, climate, storage conditions for foods and medicines, chemical refineries, and degradation of metal surface and instruments [64]. Therefore, humidity sensors with high sensitivity, selectivity, repeatability, long-term stability under ambient conditions, corrosion resistance to pollutants, and low cost of manufacturing are in great demand. Volatile organic compounds (VOCs) are composed of a variety of chemicals such as gasoline, alcohols, formaldehyde, and aromatic solvents (benzene, toluene and xylene, styrene, and perchloroethylene). VOCs also constitute an important fraction of gaseous pollutants over urbanized areas, which originate from exhaust gases, evaporation of petroleum products, and utilization of organic solvents. Short-term exposure leads to eye and respiratory tract irritation, headaches, dizziness, visual disorders, fatigue, loss of coordination, allergic skin reactions, nausea, and memory impairment. Long-term exposure to VOCs can cause damage to the liver, kidneys, and central nervous system. Therefore, detection and monitoring their concentration are really important in environmental protection, chemical process control, and personal safety [65, 66].

4.4 Detection of biomolecules

Several nanomaterials are used for the detection of biomolecules such as glucose, urea, cholesterol, carbohydrate antigen, and DNA sequence owing to their biocompatibility and stability, as well as high isoelectric point. These biomolecules are immobilized on the surface of nanomaterials, and the transducer transforms the analyte signal into fluorescence, electrochemical, SPR, or field emitter signal. Electrochemical sensors have been widely reported to detect vitamins, metabolites, neurotransmitters, and biomarkers. These sensors improve heterogeneous electron transfer rates after their integration on electrode surfaces as functional material. Ultrafast sensing of chemically and biologically active molecules at low concentrations is critical in a wide range of research fields [67, 68].

5. Nanofabrication techniques of chemical sensors

5.1 Lithography

Lithography is the process where we can transfer a computerized pattern over a surface (Fig. 2) [69]. To create a pattern in a material, a micro- or nanofabrication technique is mostly used. The most commonly used nanofabrication technique is the photolithography technique. A common lithography technique contains three-step coating with sensitive polymer, illuminated light, electron, ion beam, and devolved resist material by suitable chemical. In this photolithography technique, a pattern on the material surface is created by UV light irradiation. When UV light is illuminated over a surface of positive photoresist material, the surface of it breaks down whereas a negative photoresist material is crosslinked. So, we can make a pattern on the surface of the material by selecting an opaque and positive or negative photoresist material. After completing the pattern over a device, the photoresist materials are removed usually by sonication method. The most commonly used photoresist material is SU-8 (negative photoresist), which was developed by IBM. SU-8 is a negative photoresist material, as it is cross-linked when UV is illuminated (350–450 nm). Another technique of lithography is soft lithography, similar to photolithography, which is also used to create a pattern over a surface. In soft lithography, a molding polymer, poly(dimethyl siloxane) (PDMS), is used in biomedical purposes for its biocompatibility, good thermal, and optical properties.

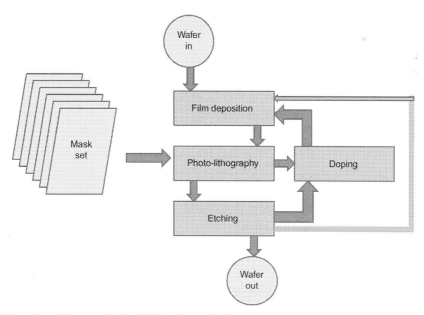

Fig. 2 Flow chart diagram of microfabrication techniques.

The soft lithography technique has great advantages over other lithography techniques. It is reusable and less expensive. Soft lithography process is done by microstamping, microfluidic patterning, and stencil patterning. There are other techniques of lithography such as X-ray lithography, electron-beam lithography, focused ion beam lithography, optical projection lithography, electron and ion projection lithography, nanoimprint lithography, proximity probe lithography, and near-field optical lithography. The lithography technique is shown in Fig. 2 by a flow chart diagram.

5.2 Electroplating

Electrodeposition or electroplating is a process for metallic coating over a base material [70]. Electroplating over a substrate material enhances heat tolerance and corrosion resistance. Electroplating is an electrocrystallization process because electrodeposition is performed by a crystalline mechanism. Electroplating needs an electrolyte solution that is made by a molten salt solution of desired metal, containing both positive, and negative ions. An electrolyte solution is made mostly by a water solvent; for special cases, an organic and ionic solvent may be used. For electroplating, two electrodes are required; a cathode (working electrode) and anode (counterelectrode). To start the electroplating process, the electrodes are connected to a power source where the anode is connected with the positive terminal and the cathode is connected with the negative terminal of the battery.

In the electroplating process, metal ions are reduced to metal atoms and deposited over the electrode. The thickness of the electroplating layer depends on time; the longer the time, the thicker the plating will be. The advantage of this electroplating process is its simplicity as you only need an electrolyte solution and two electrodes. Constituents of the electrolytic cell are: (i) the outer circuit with a power supply source; (ii) two electrodes, a cathode (working electrode) and an anode (counterelectrode) immersed into the solution; and (iii) the electrolytic plating solution (Fig. 3). The electrolyte plating solution for alkaline solution is plain mild steel and for acidic solution, lined steel is used.

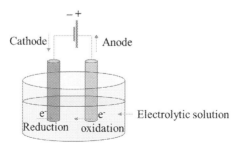

Fig. 3 Electroplating process.

5.3 Surface treatment by plasma

Generally, plasma is related to the process when energy is applied to a gas that is converted into a mixture of electrons, ions, radicals, and neutral species [71]. This energy may be two types; one is directly thermal and the other is derived from an electric current, i.e., electromagnetic radiation. In many chemical laboratories, plasmas are synthesized from electrical energy. The mechanism associated with it is the transformation of electrical energy to the electrons, which in turn passes a little part of this energy to the neutral gas molecules by collisions. The collision may be two types: elastic and inelastic. For the first, the energy transmitted resulting only in an increase of the kinetic energy of the neutral molecules. For the last one, i.e., inelastic, the electronic structure of the gas molecules changes. In recent times, plasma activation has demonstrated to be very promising in a plethora of processes required for the modification of the surface of polymer materials. These processes involve cleaning and etching for the removal of contaminants and polymer material. Plasmas are probably the most frequently used method for surface modification of polymers. Capsulation of a polymer to plasma results in the immediate formation of radical active sites on its surface, typically up to a depth of a few nanometers, and an increase of roughness and formation of nanosized pores. By changing the parameters like pressure, power, process time, gas flow and composition, and distance from the substrate surface, the actual effect of the plasma is clearly observed. Plasma treatment also leads to a variation of the zeta potential. For industrial application, the less expensive option is plasma processes using air or nitrogen atmospheres but the use of noble gases with eventual admixture of reactive gases (O_2, steam, NH_3, etc.). Plasma treatments are able to alter the surface characteristics of polymers by substitution of chemical groups. The surface treatment by plasma can introduce various functional groups such as alcohol (O—H), carbonyl (C=O), carboxylic (O—C=O), or amino (NH_2) on the polymer surface and/or insert reactive oxygen or nitrogen species. The addition of O_2 enhances the generation of polar, oxygen-containing groups on the treated surface. However, some etching and damage of the surface layer may also occur.

Plasmas can be generated using a variety of discharge types, including but not limited to dc glow, radiofrequency (RF), microwave (MW), and electron cyclotron resonance (ECR) discharges. The configurations for the plasma generation have also been continuously developed in the past years; the simplest method is two parallel electrodes and more complicated systems include dielectric barrier, capacitively coupled plasma (CCP), inductively coupled plasma (ICP), atmospheric pressure microplasmas, filtered cathodic vacuum arc (FCVA), and atmospheric pressure plasma jet.

5.4 Chemical vapor deposition

CVD is a technique where a solid material is deposited from a vapor by some chemical reaction occurring on or in the vicinity of a normally heated substrate surface (Fig. 4).

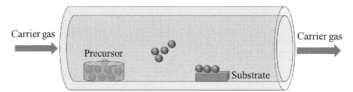

Fig. 4 Chemical vapor deposition process.

CVD is an example of vapor-solid reaction. The process is often used in the semiconductor industry to produce thin films. Microfabrication processes widely use CVD to deposit materials in various forms. There are different types of chemical vapor deposition methods. These processes generally differ in the means by which chemical reactions are initiated. CVD has been classified by three parameters: 1. operating conditions: (i) atmospheric pressure CVD (APCVD), (ii) low-pressure CVD (LPCVD), and (iii) Ultrahigh vacuum CVD (UHVCVD); 2. physical characteristics of vapor: (i) aerosol-assisted CVD (AACVD) and (ii) direct liquid injection CVD (DLICVD); and 3. substrate heating: (i) hot wall CVD and (ii) cold wall CVD.

5.5 Doping of silicon

The most common method for fabrication of the MOS sensors is the silicon micromachining technique [72]. Herein, the fabrication steps include preparation of silicon substrate with insulation layer, a microheater, and deposition of sensing layer along the electrodes. An insulating layer of SiO_2 is grown by thermal oxidation process to a couple of micrometers thickness. The substrate is then etched to make it more thinner by chemical etching or plasma etching process. The initiation reaction at the sensor surface requires high temperature, which can be achieved by the use of microheaters. Due to this, a microheater is employed at the bottom of the device to achieve high temperature. The fabrication of a microheater can be done by DC magnetron sputtering of a 10/60-nm thick Ti/Pt metal layer over a microheater-patterned mask in combination with a lift-off process. The microheater is coated with a layer of SiO_2, which is 300-nm thick to avoid a short circuit via the electrical connection between the metal oxide-sensing layer and the microheater. The same method as previously described is followed for the MOS electrode deposition. Generally, electrodes are made of Cr/Au layer with a thickness of 10/60 nm. Various methods can be applied for the deposition of an MOS electrode layer over the sensing layer. As the sensor dimension decreases and matches that of the analyte, the sensitivity increases. Carbon nanotubes and silicon nanowires have been successfully used as single molecule biosensors, but their fabrication technique is quite different from that of manufacturing typical semiconductors, and it is quite problematic. This can be now achieved by a microfabrication technique that can be employed for the designing of the nanowire-like structure. Top-down synthesis of silicon nanowires is more pronounced in this arena. The confined lateral selective epi-taxial growth process is used for producing single crystal silicon nanoplates with only 7 nm thickness and nanowires with 40 nm in diameter. This method paves the way for the realization of a truly integrated dense array of sensors which showed an excellent sensitivity toward oxygen indicating the possibility of utilizing these sensors for both chemical and biological detection.

5.6 Micro-electro-mechanical systems (MEMS)

MEMS is a process for the fabrication of tiny devices integrated with mechanical and electrical components [73]. Using integrated circuit (IC) batch processing techniques, MEMS can be developed in various sizes ranging from a few micrometers to millimeters. These devices can sense, control, and actuate on the microscale creating effects on the macro scale. These techniques are highly promising technologies that have the potential to revolutionize industrial products. Not only that, MEMS has a potential impact on our day-to-day lives. If semiconductor microfabrication is considered as the first microma-nufacturing revolution technique, MEMS can be considered as the second revolution technique.

5.7 Cracking

Cracking should be avoided during the nanofabrication process. However, cracks aren't always bad. If cracking is performed in a controlled way, this technique will be really helpful for making definite patterns. Cracking is a cost-effective and time-saving tech-nique, and highly beneficial for electronics and microfluidics research areas. Controlled cracking provides a cheaper and easier way of making nanosized patterns. Generally, lithography and etching are used for controlled nanofabrication, which are time-consuming as well as costly. However, cracking should be compatible with lithography and etching techniques in semiconductor fabrication which is often used for the fabri-cation of future generations of silicon chips in a controlled way. Cracking can be also useful in the field of microfluidics, in which microchannels are essential to manipulate biomolecules.

6. Conclusions and perspectives on the future

Nanomaterials are well-known candidates for sensing applications. The smart use of such nano-objects synthesized by various nanofabrication techniques result in improved sen-sitivities and lowered detection limits. The high specific surface of nanomaterials and the unique size-dependent physical and chemical properties of nanomaterials make them very attractive for achieving very high sensitivity. Various nanoparticles such as inorganic nanoparticles, semiconductor quantum dots, polymer nanoparticles, and carbon-based nanomaterials have been used as chemical sensors for detection of gases, metal ions, VOCs, and biomolecules. Carbon nanomaterials such as carbon black, carbon nanofibers, carbon nanotubes, graphene, and diamond have been extensively employed for the development of gas-sensitive chemical sensors. But there are many other allotropes of carbon that still need to be investigated for sensing gas as well as other environmental hazardous achieving high sensitivity, stability, and improved limit of detection in the presence of interfering species.

The interaction between two-dimensional materials and molecules/ions occur through noncovalent interactions, or through the chemical adsorption to form covalent bonds onto their basal planes. During sensing, noncovalent interactions are favorable when a quick response and a fast recovery rate are required. However, the weakness of the supramolecular forces is unfavorable when biomolecules need to be immobilized on the surface and thus covalent linkages will be more suitable. So more suitable nano-fabrication techniques are required in the future for realizing better sensitivity.

Among the different types of sensors, metal oxides semiconductors have shown excel-lent properties. The sensing mechanisms of these materials are complex and yet to explore. The different surface properties such as the nanostructure, morphology, and crystallinity play an important role in its sensing mechanism and performance. Three crit-ical challenges still remain for nanofabrication techniques: (i) how to enhance the sensor sensitivity? (ii) how to achieve the required selectivity?, and (iii) how to miniaturize the sensors? We believe that most of the present challenges will be solved by various new fabrication techniques and this will pave the way for the commercialization of new chem-ical sensors for various applications.

Acknowledgments

Dr. Atanu Jana acknowledges Dongguk University for providing the research facilities. Soumen Sardar acknowledges the University Grant Commission under UGC-JRF scheme (Ref. No. 22/12/2013(ii) EU-V) for providing the financial assistant to carry out this project work.

References

[1] F.G. Bănică, Chemical Sensors and Biosensors: Fundamentals and Applications, (2012, Wiley).
[2] T.C. Pierce, S.S. Schiffman, H.T. Nagle, J.W. Gardner, Chemical sensing in humans and machines, in: Hand Book of Machine Olfaction: Electronic Nose Technology, Wiley-VCH Verlag GmbH and Co, Weinheim, Germany, 2003, pp. 33–53 (Chapter 4).
[3] J. Janata, A. Bezegh, Chemical sensors, Anal. Chem. 60 (1988) 62–74.
[4] M.A. McEvoy, N. Correll, Materials that couple sensing, actuation, computation, and communication, Science 347 (2015) 1261689.
[5] R. Paolesse, S. Nardis, D. Monti, M. Stefanelli, C. Di Natale, Porphyrinoids for chemical sensor appli-cations, Chem. Rev. 117 (2017) 2517–2583.
[6] J. Watson, K. Ihokura, Gas-sensing materials, MRS Bull. 24 (1999) 14–17.

[7] P. Li, D. Zhang, C. Jiang, X. Zong, Y. Cao, Ultra-sensitive suspended atomically thin-layered black phosphorus mercury sensors, Biosens. Bioelectron. 98 (2017) 68–75.

[8] J.R. Vig, F.L. Walls, A review of sensor sensitivity and stability, in: Proceedings of the 2000 IEEE/EIA International Frequency Control Symposium & Exhibition, 2000, pp. 30–33.

[9] X. Yan, Y. Wu, R. Li, C. Shi, R. Moro, Y. Ma, L. Ma, High-performance UV-assisted NO$_2$ sensor based on chemical vapor deposition Graphene at room temperature, ACS Omega 4 (10) (2019) 14179–14187.

[10] O.K. Varghese, D. Gong, M. Paulose, K.G. Ong, C.A. Grimes, Hydrogen sensing using titania nanotubes, Sensors Actuators B Chem. 93 (2003) 338–344.

[11] V.N. Mishra, R.P. Agarwal, Sensitivity, response and recovery time of SnO$_2$ based thick-film sensor array for H$_2$, CO, CH$_4$ and LPG, Microelectron. J. 29 (1998) 861–874.

[12] H.P. Loock, P.D. Wentzell, Detection limits of chemical sensors: applications and misapplications, Sensors Actuators B Chem. 173 (2012) 157–163.

[13] A. Shrivastava, V. Gupta, Methods for the determination of limit of detection and limit of quantitation of the analytical methods, Chron. Young Sci. 2 (2011) 21–25.

[14] D.C. Harris, Quantitative Chemical Analysis, W. H. Freeman, 2016.

[15] W.J. Peveler, M. Yazdani, V.M. Rotello, Selectivity and specificity: pros and cons in sensing, ACS Sens. 1 (2016) 1282–1285.

[16] K. Manzoor, S.R. Vadera, N. Kumar, K. Trn, Synthesis and photoluminescent properties of ZnS nanocrystals doped with copper and halogen, Mater. Chem. Phys. 82 (2003) 718–725.

[17] G.S. Devi, V.B. Subrahmanyam, S.C. Gadkari, S.K. Gupta, NH$_3$ gas sensing properties of nanocrystalline ZnO based thick films, Anal. Chim. Acta 568 (2006) 41–46.

[18] G. Zhang, M. Liu, Effect of particle size and dopant on properties of SnO$_2$-based gas sensors, Sensors Actuators B Chem. 69 (2000) 144–152.

[19] C.M. Carney, S. Yoo, S.A. Akbar, TiO$_2$–SnO$_2$ nanostructures and their H$_2$ sensing behavior, Sensors Actuators B Chem. 108 (1–2) (2005) 29–33.

[20] A.M. Ruiz, A. Cornet, K. Shimanoe, J.R. Morante, N. Yamazoe, Effects of various metal additives on the gas sensing performances of TiO2 nanocrystals obtained from hydrothermal treatments, Sensors Actuators B Chem. 108 (1–2) (2005) 34–40.

[21] R. Mathur, D.R. Sharma, S.R. Vadera, S.R. Gupta, B.B. Sharma, N. Kumar, Room temperature synthesis of nanocomposites of Mn-Zn ferrites in a polymer matrix, Nanostruct. Mater. 11 (5) (1999) 677–686.

[22] C.S. Rout, S.H. Krishna, S.R.C. Vivekchand, A. Govindaraj, C.N.R. Rao, Hydrogen and ethanol sensors based on ZnO nanorods, nanowires and nanotubes, Chem. Phys. Lett. 418 (4–6) (2006) 586–590.

[23] C. Xiangfeng, W. Caihong, J. Dongli, Z. Chenmou, Ethanol sensor based on indium oxide nanowires prepared by carbothermal reduction reaction, Chem. Phys. Lett. 399 (4–6) (2004) 461–464.

[24] A.S. Zuruzi, A. Kolmakov, N.C. MacDonald, M. Moskovits, Highly sensitive gas sensor based on integrated titania nanosponge arrays, Appl. Phys. Lett. 88 (10) (2006) 102904–102913.

[25] I. Raible, M. Burghard, U. Schlecht, A. Yasuda, T. Vossmeyer, V$_2$O$_5$ nanofibres: novel gas sensors with extremely high sensitivity and selectivity to amines, Sensors Actuators B Chem. 106 (2) (2005) 730–735.

[26] K.M. Sawicka, A.K. Prasad, P.I. Gouma, Metal oxide nanowires for use in chemical sensing applications, Sens. Lett. 3 (1) (2005) 31–35.

[27] V. Dobrokhotov, D.N. McIlroy, M.G. Norton, A. Abuzir, W.J. Yeh, I. Stevenson, R. Pouy, J. Bochenek, M. Cartwright, L. Wang, J. Dawson, M. Beaux, C. Berven, Principles and mechanisms of gas sensing by GaN nanowires functionalized with gold nanoparticles, J. Appl. Phys. 99 (10) (2006) 104302–104306.

[28] M.W. Shao, Y.Y. Shan, N.B. Wong, S.T. Lee, Silicon nanowire sensors for bioanalytical applications: glucose and hydrogen peroxide detection, Adv. Funct. Mater. 15 (9) (2005) 1478–1482.

[29] T.I. Kamins, S. Sharma, A.A. Yasseri, Z. Li, J. Straznicky, Metal-catalysed, bridging nanowires as vapour sensors and concept for their use in a sensor system, Nanotechnology 17 (11) (2006) S291–S297.

[30] J. Kong, N.R. Franklinn, C. Zhou, M.G. Chapline, S. Peng, K. Cho, H. Dai, Nanotube molecular wires as chemical sensors, Science 287 (2000) 622–625.

[31] P. Young, Y. Lu, R. Terriu, J.J. Li, High-sensitivity NO_2 detection with carbon nanotube–gold nanoparticle composite films, J. Nanosci. Nanotechnol. 5 (9) (2005) 1509–1513.

[32] B.Y. Wei, M.C. Hsu, P.J. Su, H.M. Lin, R.J. Wu, H.J. Lai, A novel SnO2 gas sensor doped with carbon nanotubes operating at room temperature, Sensors Actuators B Chem. 101 (2004) 81–89.

[33] M. Peneza, M.A. Tagliente, P. Aversa, J. Cassano, Organic-vapor detection using carbon-nanotubes nanocomposite microacoustic sensors, Chem. Phys. Lett. 409 (4) (2005) 349–354.

[34] J. Zhu, D. Yang, Z. Yin, Q. Yan, H. Zhang, Graphene and graphene-based materials for energy storage applications, Small 10 (2014) 3480–3498.

[35] M. Pumera, A. Ambrosi, A. Bonanni, E.L.K. Chng, H.L. Poh, Graphene for electrochemical sensing and biosensing, TrAC Trends Anal. Chem. 29 (2010) 954–965.

[36] K.S. Novoselov, A.K. Geim, S.V. Morozov, D. Jiang, M.I. Katsnelson, I.V. Grigorieva, S.V. Dubonos, A.A. Firsov, Two-dimensional gas of massless Dirac fermions in graphene, Nature 438 (2005) 197–200.

[37] Q.H. Wang, K. Kalantar-Zadeh, A. Kis, J.N. Coleman, M.S. Strano, Electronics and optoelectronics of two-dimensional transition metal dichalcogenides, Nat. Nanotechnol. 7 (2012) 699–712.

[38] M. Chhowalla, H.S. Shin, G. Eda, L.J. Li, K.P. Loh, H. Zhang, The chemistry of two-dimensional layered transition metal dichalcogenide nanosheets, Nat. Chem. 5 (2013) 263–275.

[39] C. Tan, H. Zhang, Two-dimensional transition metal dichalcogenide nanosheet-based composites, Chem. Soc. Rev. 44 (2015) 2713–2731.

[40] K. Kalantar-zadeh, J.Z. Ou, T. Daeneke, A. Mitchell, T. Sasaki, M.S. Fuhrer, Two dimensional and layered transition metal oxides, Appl. Mater. Today 5 (2016) 73–89.

[41] S. Balendhran, S. Walia, H. Nili, J.Z. Ou, S. Zhuiykov, R.B. Kaner, S. Sriram, M. Bhaskaran, K. Kalantar-zadeh, Two-dimensional molybdenum trioxide and dichalcogenides, Adv. Funct. Mater. 23 (2013) 3952–3970.

[42] A. Khandelwal, K. Mani, M.H. Karigerasi, I. Lahiri, Phosphorene—the two-dimensional black phosphorous: properties, synthesis and applications, Mater. Sci. Eng. B 221 (2017) 17–34.

[43] H. Liu, A.T. Neal, Z. Zhu, Z. Luo, X. Xu, D. Toma'nek, P.D. Ye, Phosphorene: an unexplored 2D semiconductor with a high hole mobility, ACS Nano 8 (2014) 4033–4041.

[44] Y. Lin, T.V. Williams, J.W. Connell, Soluble, exfoliated hexagonal boron nitride nanosheets, J. Phys. Chem. Lett. 1 (2010) 277–283.

[45] L.H. Li, Y. Chen, Atomically thin boron nitride: unique properties and applications, Adv. Funct. Mater. 26 (2016) 2594–2608.

[46] L. Sun, M.G. Campbell, M. Dincă, Electrically conductive porous metal-organic frameworks, Angew. Chem. Int. Ed. 55 (2016) 3566–3579.

[47] M. Ko, L. Mendecki, K.A. Mirica, Conductive two-dimensional metal–organic frameworks as multifunctional materials, Chem. Commun. 54 (2018) 7873–7891.

[48] J.S.G. Dos Santos-Alves, R.F. Patier, The environmental control of atmospheric pollution. The framework directive and its development. The new European approach, Sensors Actuators B Chem. 59 (1999) 69–74.

[49] B. Liu, L. Chen, G. Liu, A.N. Abbas, M. Fathi, C. Zhou, High-performance chemical sensing using Schottky-contacted chemical vapor deposition grown monolayer MoS2 transistors, ACS Nano 8 (2014) 5304.

[50] W. Li, X. Geng, Y. Guo, J. Rong, Y. Gong, L. Wu, X. Zhang, P. Li, J. Xu, G. Cheng, M. Sun, L. Liu, Reduced graphene oxide electrically contacted graphene sensor for highly sensitive nitric oxide detection, ACS Nano 5 (2011) 6955.

[51] N.L. Teradal, S. Marx, A. Morag, R. Jelinek, Porous graphene oxide chemi-capacitor vapor sensor array, J. Mater. Chem. C 5 (2017) 1128.

[52] M.S. Mauter, M. Elimelech, Environmental applications of carbon-based nanomaterials, Environ. Sci. Technol. 42 (2008) 5843–5859.

[53] F. Perreault, A.F. de Faria, M. Elimelech, Environmental applications of graphene-based nanomaterials, Chem. Soc. Rev. 44 (2015) 5861–5896.

[54] A. Ambrosi, C.K. Chua, A. Bonanni, M. Pumera, Electrochemistry of graphene and related materials, Chem. Rev. 114 (2014) 7150–7188.

[55] I.˙. Duru, D. Ege, A.R. Kamali, Graphene oxides for removal of heavy and precious metals from wastewater, J. Mater. Sci. 51 (2016) 6097.

[56] R. Sitko, P. Janik, B. Feist, E. Talik, A. Gagor, Suspended aminosilanized graphene oxide nanosheets for selective preconcentration of lead ions and ultrasensitive determination by electrothermal atomic absorption spectrometry, ACS Appl. Mater. Interfaces 6 (2014) 20144.

[57] F. Zhang, B. Wang, S. He, R. Man, Preparation of graphene-oxide/polyamidoamine dendrimers and their adsorption properties toward some heavy metal ions, J. Chem. Eng. Data 59 (2014) 1719.

[58] K.P. Carter, A.M. Young, A.E. Palmer, Fluorescent sensors for measuring metal ions in living systems, Chem. Rev. 114 (2014) 4564.

[59] C. Zhu, D. Du, Y. Lin, Graphene and graphene-like 2D materials for optical biosensing and bioimaging: a review, 2D Mater. 2 (2015) 32004.

[60] K. Chen, G. Lu, J. Chang, S. Mao, K. Yu, S. Cui, J. Chen, Hg(II) ion detection using thermally reduced graphene oxide decorated with functionalized gold nanoparticles, Anal. Chem. 84 (2012) 4057.

[61] Y. Huang, J. Guo, Y. Kang, Y. Ai, C.M. Li, Two dimensional atomically thin MoS2 nanosheets and their sensing applications, Nanoscale 7 (2015) 19358.

[62] Z.-Q. Zhao, X. Chen, Q. Yang, J.-H. Liu, X.-J. Huang, Selective adsorption toward toxic metal ions results in selective response: electrochemical studies on a polypyrrole/reduced graphene oxide nanocomposite, Chem. Commun. 48 (2012) 2180.

[63] H. Farahani, R. Wagiran, M.N. Hamidon, Humidity sensors principle, mechanism, and fabrication technologies: a comprehensive review, Sensors 14 (2014) 7881–7939.

[64] A. Mirzaei, S.G. Leonardi, G. Neri, Detection of hazardous volatile organic compounds (VOCs) by metal oxide nanostructures-based gas sensors: a review, Ceram. Int. 42 (2016) 15119–15141.

[65] B. Li, G. Sauve, M.C. Iovu, M. Jeffries-El, R. Zhang, J. Cooper, S. Santhanam, L. Schultz, J.C. Revelli, A.G. Kusne, T. Kowalewski, J.L. Snyder, L.E. Weiss, G.K. Fedder, R.D. McCullough, D.N. Lambeth, Volatile organic compound detection using nanostructured copolymers, Nano Lett. 6 (2006) 1598–1602.

[66] Y. Chen, C. Tan, H. Zhang, L. Wang, Two-dimensional graphene analogues for biomedical applications, Chem. Soc. Rev. 44 (2015) 2681–2701.

[67] D. Chimene, D.L. Alge, A.K. Gaharwar, Two-dimensional nanomaterials for biomedical applications: emerging trends and future prospects, Adv. Mater. 27 (2015) 7261–7284.

[68] M.M. Khin, A.S. Nair, V.J. Babu, R. Murugan, S. Ramakrishna, A review on nanomaterials for environmental remediation, Energy Environ. Sci. 5 (2012) 8075–8109.

[69] Z. Meng, R.M. Stolz, L. Mendecki, K.A. Mirica, Electrically-transduced chemical sensors based on two-dimensional nanomaterials, Chem. Rev. 119 (2019) 478–598.

[70] C. Anichini, W. Czepa, D. Pakulski, A. Aliprandi, A. Ciesielski, P. Samorì, Chemical sensing with 2D materials, Chem. Soc. Rev. 47 (2018) 4860–4908.

[71] H. Nazemi, A. Joseph, J. Park, A. Emadi, Advanced micro- and nano-gas sensor technology: a review, Sensors 19 (2019) 1285.

[72] S.Y. Lee, H.C. Jeon, S.-M. Yang, Unconventional methods for fabricating nanostructures toward high-fidelity sensors, J. Mater. Chem. 22 (2012) 5900.

[73] S.A.M. Al-Bat'hi, Electroplating of Nanostructures, Electrodeposition of Nanostructure Materials, InTech, 2015, pp. 3–26.

CHAPTER 8

Design and fabrication of new microfluidic experimental platform for ultrasensitive heavy metal ions (HMIs) sensing

Dinesh Rotake[a], Anand Darji[a], and Nitin Kale[b]
[a]Department of Electronics Engineering, SardarVallabhbhai National Institute of Technology, Surat, Gujarat, India
[b]NanoSniff Technologies Pvt. Ltd., IIT Bombay, Mumbai, India

1. Introduction

Today's world faces many environmental problems of which water contamination is proving to be very hazardous, according to many health organizations, and needs to be solved as early as possible. Approximately 15 million children under the age of 5 die each year because of diseases caused by water pollution, out of which heavy metal ions (HMIs) are the most harmful water contaminants. Most of the HMIs are transition metals that can persuade environmental pollution because of their high residues, and they are very difficult to remove [1]. By definition, heavy metals are metals (HMIs) with a density higher than 5 g/cm^3 and atomic weights varying between 63.5 and 200.6 g/mol. Most of the industrial wastes and processes produce the problematic HMI concentrations in discharged water from various factories that are harmful to humans and also contaminate the agricultural land. Testing for heavy metals at part per billion (ppb) levels is essential to meet internationally established limits, which is one of the significant challenges in micro electro mechanical systems (MEMS), and detected amounts as little as part per trillion (ppt) is today needed. There are various HMIs, out of which mercury (Hg^{2+}), cadmium (Cd^{2+}), lead (Pb^{2+}), arsenic (Ar^{2+}), and chromium (Cr^{2+}) are more toxic, and the World Health Organization (WHO) has provided a limit (μg/L) of 6, 3, 10, 10, and 50, respectively [2]. So from these limits, we can understand that they are very hazardous and cause many harmful diseases, and even minimal concentrations below the WHO limit are also extremely dangerous and toxic to human health.

We have investigated various research articles in the field of HMIs detection such as fluorescent- or colorimetric-based detection; a one-step synthetic approach was proposed by the authors [3] using fluorescent gold nanoclusters (AuNCs) with trypsin as a ligand. A one-pot synthetic method approach was also proposed by the authors [4] using

Handbook of Nanomaterials for Sensing Applications
https://doi.org/10.1016/B978-0-12-820783-3.00014-2

fluorescent gold nanoclusters (AuNCs) with bovine serum albumin (BSA) and bromelain as templates, and ZnSe quantum dots (QDs) modified with mercaptoacetic acid (MAA) were used for selective target Cu^{2+} ions [5]. A DNA molecular logic gate with ECL signals as outputs for increased versatility, low background, and simplified optical setup based on the T-rich or C-rich oligonucleotides for the selective analysis of Hg^{2+} and Ag^+ ions was proposed [6]. A novel colorimetric and ratiometric fluorescent chemosensor with 2-(2-mercaptophenyl)-1H-benzo[de]isoquinoline-1,3-(2H)–dione (L3) is reported for Hg^{2+} ions [7].

Some of the authors used an electrochemical sensor for HMIs detection; a commercial screen-printed carbon electrode (SPCE) was modified with a NiO-PPS-CnP composite material to selectively detect mercury (II) ions [8]. A one-shot disposable sensor was proposed with Calixarene bulk–modified screen-printed electrodes (SPCCEs) used to detect Pb^{2+}, Cu^{2+}, and Hg^{2+} ions simultaneously [9]. Bismuth nanoparticles were decorated with graphenated carbon nanotubes modified with screen-printed electrodes for Hg^{2+} ions detection [10].

Many authors proposed flexible sensors for HMIs detection:; flexible MoS2 sensor arrays for high performance label-free HMIs sensing [11]; a flexible superhydrophilic-superhydrophobic tape introduced by the authors for on-site HMIs monitoring and detection [12]; a flexible micropatterned reduced graphene oxide (rGO) and a carbon nanotube (CNT) composite-based electrochemical sensor for Cd^{2+} and Pb^{2+} detection was reported [13]; a flexible polyethylene terephthalate (PET)-based electrochemical sensor introduced for cadmium sulfide (CdS) [14]; and a flexible electrochemical sensor using a direct laser engraving process that transforms commercial polyimide (PI) films into a laser-engraved graphene electrode (LEGE) for detection of Pb^{2+} ions [15].

Surface plasmon resonance (SPR)–based detection is also a very popular method for HMIs sensing. A reusable SPR is used for fast detection of Cu^{2+} based on optimized Ag/Au composite thin film with modified-chitosan (MCS) film as an active layer [16]. A reflection-based localized SPR fiber-optic sensor was introduced to detect Pb^{2+} ions [17], and a combination of valinomycin–doped chitosan–graphene oxide (C–GO–V) thin film and SPR system was used for detection of K^+ ions [18]. Localized SPR–based U-shaped optical fiber sensor was proposed for detection of Pb^{2+} ions in water [19].

Many authors also used surface-enhanced Raman scattering (SERS) for HMIs detection such as ultrasensitive SERS-based sensing of Hg^{2+} ions was proposed using the assembled gold nanochains [20]; novel ratiometric SERS aptasensor for selective and reproducible sensing of Hg^{2+} ions [21]; an ultrathin and isotropic metal sulfide wrapping on plasmonic metal nanoparticles for SERS-based detection of trace HMIs [22]; and a facile Ag–film-based SERS using DNA molecular switch for ultrasensitive Hg^{2+} ions detection [23].

So from a literature survey, and considering the various benefits of MEMS, we have designed a system that will overcome the disadvantages of other systems because most of the systems studied required costly instruments for characterization and processing, with interference noise when more than one HMI was present. Many of the systems are capable of suitable heavy metal sensing but incapable of selective ion sensing, and others required setup for detection and hence can't be a portable device. So there is a need for designing a portable system for on-site detection of HMIs in the water at the picomolar range.

2. HMIs detection system overview

We proposed a portable, highly sensitive, easy-to-use, continuous, and fast HMI detection system, which can be accessible to remote areas because most of the remote areas across the country will face a water contamination problem. The microfluidic platform can be used for sensing HMIs using a piezoresistive microcantilever-based sensor fabricated by using MEMS technology. The proposed system is divided into two main parts; the first one includes a microfluidic chamber for handling HMI samples in which the sensing material with the appropriate buffer can be integrated to improve selectivity, and the second one uses an array of piezoresistive microcantilever-based sensors with different thiols for selective HMIs detection. This type of sensor, called a surface stress–based biosensor, is used for effectively solving the problem of underwater detection. Surface stress–based biosensors, as one new technology of microscale and label-free systems, have immense potential to satisfy the demand for better sensor quality and have been investigated extensively in recent years.

3. Block diagram of proposed system

A biosensor is one kind of device used for the detection of a targeted analyte, which can be heavy metals in our case. So the first two blocks, i.e., analyte and selective layer, combine to form the biosensor part of the system. The selective layer can be designed employing principles of molecular and biomolecular recognition, for example, antigen–antibody binding; specific surface selectivity can be achieved using molecular self-assembled monolayer's (SAMs) as the sensing layers assembled on the surface of biosensor sensitive element. Once the targeted analyte is detected, i.e., binded with the microcantilever top surface, microcantilever starts bending due to the mass of HMIs, and respective deflection can be measured in terms of change of resistance (ΔR). So the transducer will convert the mass to a respective change in resistance, and in this case, we use the MEMS-based piezoresistive sensor for this purpose. Now, to extend the lower detection limit of microcantilever up to part per trillion ranges will produce a change in resistance (1–250 Ohm)

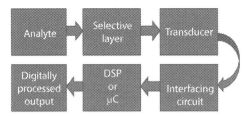

Fig. 1 Block diagram of proposed system.

range. So our main challenge is to design the interfacing circuit for very effectively measuring the small change in resistance. The last two blocks are used to process the data coming from the interfacing circuit and displaying the appropriate result depending on the concentration of different HMIs. As we know, that minimum concentration of all toxic HMIs is not known to the average person using the system, hence it is necessary to convert the output concentration into digital form, making the comparison and producing a report that can be understood by any nontechnical person for simplicity of design. Deionization is the process in which detected HMIs can be removed from the microcantilever surface to be reused again. The biosensor was able to regenerate using various materials, which is discussed in detail under the selective layer block. The block diagram of the proposed system shown in Fig. 1 explains the function of each every block in detail.

3.1 Analyte

The analytes in biosensors are always geometrically constrained to a small size, typically submillimeter. It has some features, small volume [nanoliter (nl), picoliter (pl)] and characteristic sizes, low energy consumption, and microdomain effects. Analytes are those who need to be detected by the selective layer. In our case, highly toxic heavy metals such as lead (Pb^{2+}), mercury (Hg^{2+}), cadmium (Cd^{2+}), arsenic (As^{2+}), and manganese (Mn^{2+}) can be consider as analytes.

3.2 Selective layer

The sensor's selectivity, i.e., the selective layer of the biosensors, determines the specificity of the sensor to a particular analyte. The selective layer can be designed employing principles of molecular and biomolecular recognition, for example, antigen–antibody binding (i.e., any chemicals, bacteria, viruses, or pollen binding to a specific protein). Otherwise, precise surface selectivity can be achieved using SAMs as the sensing layers assembled on the surface of the biosensor sensitive element. In our case, different thiols can be used as a selective layer to bind the HMIs.

3.3 Transducer

In this case, a microcantilever beam is used as a transducer, which will convert mechanical energy to the desired change in resistance. A surface stress-based transducer structure can be made by silicon nitride or polysilicon and a contact layer, i.e., the gold layer.

The microsensing techniques for MEMS technology can be divided into five basic categories depending on the principle of detection, which are listed here:

Piezoelectric sensing is based on the piezoelectric effect of piezoelectric materials. The electrical charge change is generated when a force is applied across the face of a piezo-electric film or vice versa [24], i.e., the internal generation of electrical charge resulting from an applied mechanical force or the internal generation of a mechanical strain resulting from an applied electrical field. A piezoelectric microcantilever sensor (PEMS) measures the mass of the target molecules on the basis of the microbalance technique, which uses the change in the resonant frequency of the PEMS due to the added mass of the target molecules. Environmental and operating conditions should be controlled to improve the frequency stability of a PEMS. Notably, the calibration or frequency stabilization using preactuation to correct for the frequency drift is required by Refs. [25–27]. The standard piezoelectric material lead–zirconate-titanate (PZT) is mostly used, but there are several issues with fabrication using PZT [28]. The resolution can be improved by increasing the thickness at the cost of bandwidth [29].

Optical detection is the most common method used for the microcantilever-based biosensor, and is the simplest way to measure the beam deflection resulting from surface stress, with regard to the position sensing (photo) detection method and fiber interferometry.

However, optical detection always has the problem of heat drift [30]. Thermal effects are clearly evident and occur slowly (over a time scale of hours) for the duration of the experiment. In some cases, radiative heating can increase the microcantilevers temperature up to $130 \pm 20°C$. Thermal effects are particularly troublesome for cantilever/membrane sensors where a gold layer is evaporated on one surface, creating sensitive bimetallic. Optical detection systems are still typically large, and so they are more suited for bench-top applications than for portable handheld use.

Resonant sensing is easily understood as the natural frequency of a spring changing as a result of tensile force. In the development of a resonant microsensor, the strain caused pressure on the diaphragm leading to the natural frequency of a resonator varying. By picking up the natural frequency variation of the resonator, the physical information that caused the strain will be sensed. Subatomic resolution force and displacement measurements in air and liquid were demonstrated using rotational mode disk resonant structures acting as stress/strain sensors [31].

Capacitive sensing utilizes the diaphragm or cantilever surface deformation-induced capacitance change to convert the information of pressure, force, etc., into an electrical signal. The capacitive microsensors can be used for pressure, force, acceleration, flow rate, displacement, position, orientation measurement, etc. The capacitive sensor for differential pressure measurement due to HMIs and enhancement in sensitivity by introducing different shape, stress concentration region (SCR), changing the dimension of the microcantilever is already proposed with the design of capacitance to digital converter (CDC) circuit for 1–10 fF (femtofarad) range [32, 33]. For capacitive microsensors, the capacitance change is not linear with respect to the diaphragm deformation, and, also, the small capacitance (generally 1 to 3 pF) requires the measurement circuit to be integrated on the chip.

Piezoresistive sensing utilizes the phenomenon of the piezoresistive effect in which the resistance of piezoresistor varies through external pressure or force. Readout systems are one of the critical factors that affect the sensitivity of microcantilever systems (MCSss). The readout of piezoresistive systems has advantages over optical ones in many aspects. The best sensor performance is expected to achieve when the Si piezoresistor is a thin, narrow, low doping level, and when the piezoresistor length is approximately two–fifths of the SiO2 cantilever length [34].

Piezoresistivity is the most commonly used phenomenon for micro–nano film technology and MEMS-based sensors. The piezoresistive effect exhibited a relative change in electrical resistance under mechanical strain/stress. This phenomenon is the primary operating principle for all types of piezoresistive sensors used in engineering measurements. The fabrication of a polymeric cantilever with an encapsulated polysilicon piezoresistor using a five-mask process was reported the first time [35]. SiO_2-based microcantilevers were fabricated to improve the sensitivity of cantilever sensors [36]. The piezoresistive sensor is gathering attention as a biosensing device whose initial resistance value (R) changes by a few parts per million (ΔR) in deflected mode [37].

We have already investigated the stiffness and sensitivity of the microcantilever-based piezoresistive sensor for bio-MEMS application to target the low-pressure range using COMSOL Multiphysics 5.3 software [38]. In this chapter, we have proposed the MEMS-based portable piezoresistive experimental platform for on-site HMIs detection and focus our aim toward lower-order detection (1 to 50 ng/mL).

The piezoresistive MEMS sensor experimental setup for microfluidic application design and manufactured from NanoSniff Technologies Pvt. Ltd., IIT-Bombay, is shown in Fig. 2, and technical specifications for liquid phase analysis are shown in Table 1. Each sensor PCB has an eight-microcantilever shown in Fig. 2, out of which one microcantilever is blocked with acetyl chloride material using a nanodispenser. The block microcantilever is used as reference cantilever, which makes it possible to minimize the effect of drift directly and also avoid the thermal drift and other parametric effects in the measurement. The surface-stress change due to HMI binding is measured in terms of change in resistance of these microcantilevers using the piezoresistive MEMS sensor-based experimental platform.

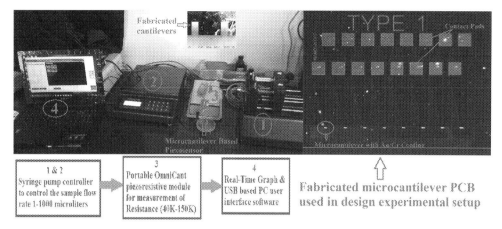

Fig. 2 Piezoresistive MEMS sensor experimental setup for microfluidic application.

Table 1 Technical specifications for: piezoresistive MEMS microcantilever sensor based experimentation platform for liquid phase analysis.

Sr. no.	Features
1.	Channels: The system has 4–8 individual cantilever sensor channels
2.	Electrical Resistance: Cantilever resistance range 40 to 150 kΩ
3.	Detection Capability of the Electronics: Minimum 5 µV signal is detected
4.	Electrical Isolation: Analyte fluid flow path is isolated from electronics and contact pads of the cantilever
5.	Surface Functionalization: Cantilever is coated with SAM or appropriate receptor compounds
6.	Liquid Cell: A dedicated flow cell houses for the cantilever die
7.	Liquid Flow: Configurable liquid flow rates 1–1000 µL/min for analyte compound
8.	Operation modes: Full-featured PC user interface mode using PC software, the stand-alone mode through on-system graphical display and buttons
9.	User Interface: One-touch experiment start/stop. One-touch channel Enable/Disable for individual cantilever channels. One-touch to start the flow of analyte. Individual dedicated buttons for setting fluid flow rate
10.	Logging and memory: Internal memory for logging experiment data and activity logs. Logs available in CSV format through USB interface software
11.	PC software: USB based PC user interface software. Real-time graphs and system parameter. Complete instrument control including setting of parameters view, download, and erase experiments

4. Results and discussion

To investigate the performance of piezoresistive MEMS sensors for microfluidic application, we have used the modified OmniCant setup for liquid phase analysis. The microcantilevers used in our experiments are 220 μm long, 80 μm wide, and 0.625 μm thick.

Experimental procedure:

- Prepared the stock solution with a concentration of 10 mM/10 mL of different thiols (4-mercaptobenzoic acid, homocysteine, cysteamine) to create the SAM on microcantilever surface.
- Immerse microcantilever dies in a thiol solution.
- Store the sample for 24–48 h.
- Prepare the stock solution with a concentration of 1 mM/10 mL of different HMIs (Al, Mn, Cr, Cd, Pb, Hg).
- The constant flow rate of 30 μL/minute is set for all the experiments.
- Initially, DI water started to stabilize microcantilevers with the liquid flow, and then all the HMI is injected.

4.1 Experimental results of microcantilever based piezoresistive sensor for SAM of cysteamine (Cys)

We have prepared the stock solution of 20 mL for manganese (II) chloride ($MnCl_2$) with a concentration of 1 mM/20 mL to check the experimental platform for HMIs detection. Initially, we have to test the designed portable experimental platform for lower detection limits and simplicity; we have used the Mn^{2+} ions as an HMI to test the experimental setup.

The complete process flow of biosensor for Mn^{2+} ions detection is shown in Fig. 3 with SAM of cysteamine.

4.2 Experimental results of microcantilever based piezoresistive sensor for SAM of 4-mercaptobenzoic acid (4-MBA)

We have used the same stock solution of 20 mL for manganese (II) chloride ($MnCl_2$) with a concentration of 1 mM/20 mL to check the experimental platform for HMIs detection with different thiol solutions. Hence, we have to test the designed portable experimental platform for lower detection limits using 4-mercaptobenzoic acid. The complete process flow of biosensor for Mn^{2+} ions detection is shown in Fig. 4 with SAM of 4-mercaptobenzoic acid.

The change in resistance of unblock cantilever with respect to block cantilever is shown in Figs. 3 and 4. Also, it is observed that change in resistance of microcantilever with respect to block cantilever after the injection of Mn^{2+} molecules is 80 Ohm and 18 Ohm for SAM of cysteamine and 4-mercaptobenzoic acid, respectively. The change

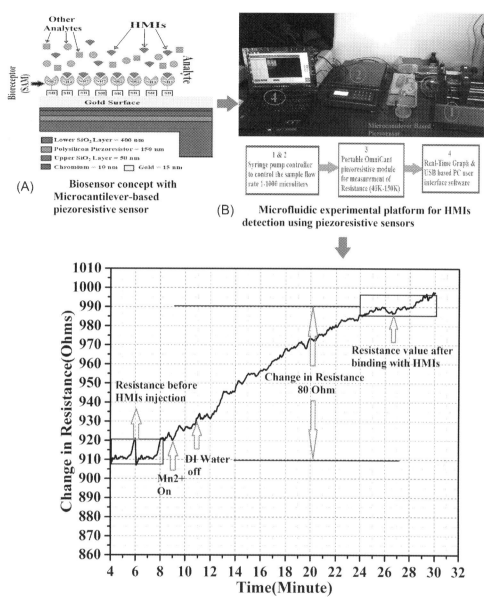

Fig. 3 Complete process flow of biosensor for Mn^{2+} ions detection with SAM of cysteamine (Cys). (A) Biosensor concept with microcantilever-based piezoresistive sensor. (B) Microfluidic experimental platform for HMIs detection using piezoresistive sensor. (C) The change in resistance of unblocked cantilever with respect to the resistance of block cantilever using SAM of Cysteamine.

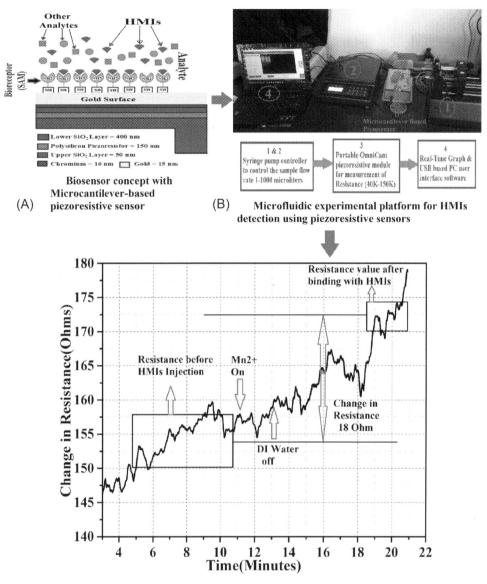

Fig. 4 Complete process flow of biosensor for Mn^{2+} ions detection with SAM of 4-mercaptobenzoic acid. (A) Biosensor concept with microcantilever-based piezoresistive sensor. (B) Microfluidic experimental platform for HMIs detection using piezoresistive sensor. (C) The change in resistance of unblock cantilever with respect to the block cantilever of acetyl chloride using SAM of 4-mercaptobenzoic acid.

Change in Resistance (ΔR) = [Change in Resistance of Block
cantilever–Change in Resistance of sensor cantilever (Unblock)]

Fig. 5 Change in resistance formula.

in resistance is calculated with respect to block cantilevers using a formula, as shown in Fig. 5.

4.3 Experimental results of microcantilever based piezoresistive sensor for SAM of homocysteine (HCys)

There are various HMIs, out of which mercury (Hg^{2+}), cadmium (Cd^{2+}), lead (Pb^{2+}), arsenic (Ar^{2+}), and chromium (Cr^{2+}) are most toxic according to WHO. In this chapter, we tried to selectively detect the mercury (Hg^{2+}) ions using SAM of homocysteine with pyridinedicarboxylic acid (PDCA) as a chelating ligand. In this approach, we tried to combine the advantages of three technologies, namely thin-film, nanoparticles (NPs), and MEMS to selectively detect the mercury ions at the picomolar range. The excessive commercialization of nanomaterials and NPs is quickly expanding their toxic impact on health and the environment [39–41]. Here, we try to transfer the AuNPs-based sensors technology [42–46] for selective detection of Hg^{2+} ions using PDCA as a chelating ligand to the remarkably sensitive MEMS-based sensor platform.

There are two forms of mercury available; one is inorganic and the other is organic. The inorganic Hg is generally found in batteries and also used in chemical industries. This is produced from elemental Hg using an oxidation process. The inorganic Hg mostly present in ground water is not considered very toxic, but the accumulation of this results in kidney damage. The organic mercury (methyl mercury) is highly toxic to human health and the ecosystem, produced from the inorganic Hg by specific bacteria. Both forms of mercury have a significant effect on children as they quickly accumulate in their body and pose a significant threat to their life. Volcanoes, fossil fuels (coal and petroleum), and forest fire are the natural sources of mercury in the environment. The levels of mercury are increasing day by day because of discharge from the pulp, hydroelectric, mining, and paper industries. The most dominant contribution is through medical waste and coal-fired power plant emissions. The mercury emitted by power plants in the air is then deposited in the water of lakes, rivers, and ocean sediments, contaminating sources of drinking water. Some of the necklaces and glass pendants imported from Mexico contain mercury that can release metallic mercury in water if broken. Gold and silver mining are also one of the leading contributors to metallic mercury pollution.

One of the most crucial facts that mercury is a volatile organic compound (VOC) is that is can evaporate at relatively lower temperatures in the atmosphere and then contaminate sources of water. So, there are many sources of mercury available to contaminate the ground and drinking water. Excessive exposure to mercury can cause damage to many human organs such as the brain, lungs, stomach, and kidneys, which may lead to congenital disabilities and mental retardation, and lead to harmful diseases such as acro-dynia and Hunter-Russell syndrome. Methylmercury (MeHg) is an organic form of mercury that can damage the developing brains of human fetuses [47]. Women who consume methyl mercury during pregnancy can bear children who have neurological issues because methyl mercury has toxic effects on the nervous system during embryonic development. Also, it may have an impact on vision, hearing, etc., and mercury is extremely harmful compared with other HMIs according to WHO limits. Hence, we have tried to selectively detect the Hg ions in the picomolar range using a microcantilever-based piezoresistive sensor.

Experimental Procedure for selective mercury detection:
- The stock solution of 10 mL is prepared with a concentration of 10 mM/10 mL to develop the SAM of homocysteine (Hcys) on the microcantilever surface.
- Immerse microcantilever dies in a container containing thiol solution.
- Store the sample for 24–48 h. Longer assembly time results in better monolayer packing.
- The stock solution of 10 mL is prepared with a concentration of 1 mM (1 mM/10 mL) of different HMIs (Al, Mn, Cr, Cd, Pb, Hg).
- The flow rate of 30 μL/minute is set for all the experiments.
- Initially, DI water started, and then all the HMI is injected one by one after 07:00 min.
- The flow of DI water is stopped at 07:00 min, and only HMIs + PDCA ligands molecules are injected through the syringe pumps. Here, PDCA acts as a masking agent and suppressing other HMIs interaction with SAM layer.
- At last, $HgCl_2$ + PDCA is injected to check the selectivity and sensitivity of homocysteine (Hcys) with pyridinedicarboxylic acid (PDCA) as a chelating ligand.

The complete process flow of biosensor for Hg^{2+} ions detection is shown in Fig. 6 with SAM of homocysteine with pyridinedicarboxylic acid.

The change in resistance of unblock cantilever with respect to block cantilever is shown in Fig. 6. Initially, when DI water is used to stabilize the microcantilever in a wet environment for the first 7 min, the change in resistance is stable around a particular value. When different HMIs injected after 7 min except for Hg, the change in resistance is in the same range for all other HMIs (50 Ω). When Hg is injected after 7 min, the change in resistance is different compared with other HMIs. This variation is around 250–300 Ω for the microcantilever-based piezoresistive sensor. From the experimental results, it is clear that the proposed experimental platform required a maximum of 23 min for selective detection of mercury ions. So, with the constant flow rate of 30 μL/minute, the total sample volume is 0.69 mL for 23 min in the experiments (see appendix for calculation). Hence, the equivalent concentration for Al, Mn, Cr, Cd, Pb, and Hg is 0.67 ng, 0.55 ng, 0.74 ng, 0.56 ng, 0.77 ng, and 0.75 ng, respectively. It is clear that this technique is capable of detecting HMIs as low as 0.75 ng/mL (3.73 pM/mL). Many researchers have developed a colorimetric sensor that shows the

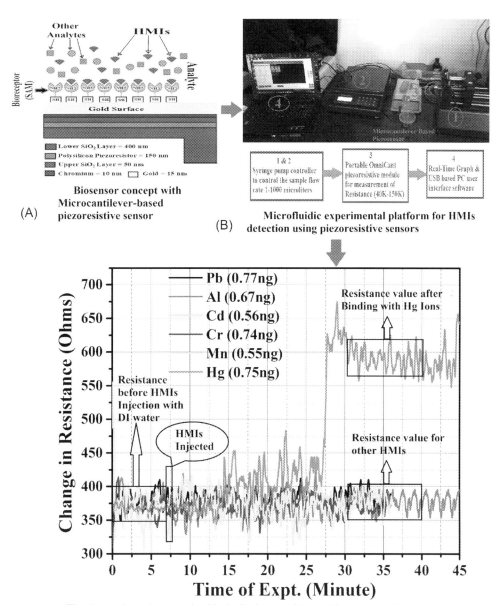

Biosensor concept with Microcantilever-based piezoresistive sensor

(A)

(B) **Microfluidic experimental platform for HMIs detection using piezoresistive sensors**

(C) **The change in resistance of unblocked microcantilever with respect to block cantilever of Acetyl Chloride using SAM of Homocysteine (Hcys)-Pyridinedicarboxylic acid (PDCA)**

Fig. 6 Complete process flow of biosensor for Hg^{2+} ions detection with SAM of homocysteine with pyridinedicarboxylic acid. (A) Biosensor concept with microcantilever-based piezoresistive sensor. (B) Microfluidic experimental platform for HMIs detection using piezoresistive sensor. (C) The change in resistance of unblocked microcantilever with respect to block cantilever of acetyl chloride using SAM of Homocysteine (Hcys)-Pyridinedicarboxylic acid (PDCA).

output in terms of color change, but this may lead to confusion between two nearby ranges and also required costly lab equipment for analysis.

5. Conclusion and future scope

The proposed microcantilever-based piezoresistive sensor is capable of selectively detecting mercury ions, the most toxic HMIs according to the WHO limits. So this system can be used in household water purification systems, real-time water quality check systems, water purification plants, and industrial wastewater treatment plants as an early warning system. The proposed system used as a real-time quality check of DI water used in the microfabrication facility, which ultimately improves the quality of fabricated devices. The real-time accumulation of HMIs on the surface was used as a filter for the removal of selective HMIs from the water sample. This approach is the best option because most of the RO-based water purifiers can remove 99% of contaminants but unfortunately also removes and filters out essential minerals. The pore size of the RO membrane is 100 pm, so small that it only allows pure water molecules to escape it. [Atomic Radii of Hg (150 pm) and that of essential minerals is Ca (197 pm), K (227 pm), Na (186 pm), Mg (160 pm)]. Real-time detection of HMIs from the water sample can be used as an early warning system in water purifiers to indicate that the RO filter is not working correctly and the need to replace it. This proposed system used as an on-field instrument in rural areas to test the groundwater quality due to its portability. As the system is designed using MEMS, performance is at the optimum level and satisfies every need today.

The experimental analysis shows an excellent response to the selective detection of Hg^{2+} ions in the presence of other HMIs. The change resistance (ΔR) is around 250–300 Ω for Hg^{2+} ions with Hhomocysteine (Hcys)-pyridinedicarboxylic acid (PDCA). From the experimental results, it is clear that this technique has the potential and capability to selectively detect Hg^{2+} ions as low as 0.75 ng/mL (3.73 pM/mL) and outperform compared with other methods that require sophisticated analytical tools for measurements. Also, the proposed method can be used for on-site selective detection of HMIs.

Acknowledgments

The authors would like to thank IIT, Bombay for the support of AFM facility under INUP and "Visvesvaraya Ph.D. Scheme for Electronics and IT" funded by the Ministry of Electronics and Information Technology (MeitY) and FESEM facility under PUMP, NCPRE funded by the Ministry of New and Renewable Energy (MNRE), Government of India.

References

[1] H.C. Chang, Y.L. Hsu, C.Y. Tsai, Y.H. Chen, S.L. Lin, Nanofiber-based brush-distributed sensor for detecting heavy metal ions, Microsyst. Technol. 23 (2) (2017) 507–514.

[2] T. Thompson, J. Fawell, S. Kunikane, D. Jackson, S. Appleyard, P. Callan, J. Bartram, P. Kingston, S. Water, World Health Organization, Chemical Safety of Drinking Water: Assessing Priorities for Risk Management, (2007).

[3] S. Ghosh, J.R. Bhamore, N.I. Malek, Z.V.P. Murthy, S.K. Kailasa, Trypsin mediated one-pot reaction for the synthesis of red fluorescent gold nanoclusters: sensing of multiple analytes (carbidopa, dopamine, $Cu2+$, $Co2+$ and $Hg2+$ ions), Spectrochim. Acta A Mol. Biomol. Spectrosc. 215 (2019) 209–217.

[4] J.R. Bhamore, S. Jha, H. Basu, R.K. Singhal, Z.V.P. Murthy, S.K. Kailasa, Tuning of gold nanoclusters sensing applications with bovine serum albumin and bromelain for detection of $Hg2+$ ion and lambda-cyhalothrin via fluorescence turn-off and on mechanisms, Anal. Bioanal. Chem. 410 (11) (2018) 2781–2791.

[5] D. Wu, Z. Chen, G. Huang, X. Liu, ZnSe quantum dots based fluorescence sensors for $Cu2+$ ions, Sensors Actuators A Phys. 205 (2014) 72–78.

[6] X. Li, L. Sun, T. Ding, Multiplexed sensing of mercury (II) and silver (I) ions: a new class of DNA electrochemiluminescent-molecular logic gates, Biosens. Bioelectron. 26 (8) (2011) 3570–3576.

[7] M. Bahta, N. Ahmed, A novel 1, 8-naphthalimide as highly selective naked-eye and ratiometric fluorescent sensor for detection of $Hg2+$ ions, J. Photochem. Photobiol. A Chem. 373 (2019) 154–161.

[8] M.A. Armas, R. María-Hormigos, A. Cantalapiedra, M.J. Gismera, M.T. Sevilla, J.R. Procopio, Multiparametric optimization of a new high-sensitive and disposable mercury (II) electrochemical sensor, Anal. Chim. Acta 904 (2016) 76–82.

[9] P.S. Adarakatti, C.W. Foster, C.E. Banks, A.K. NS, P. Malingappa, Calixarene bulk modified screen-printed electrodes (SPCCEs) as a one-shot disposable sensor for the simultaneous detection of lead (II), copper (II) and mercury (II) ions: application to environmental samples, Sensors Actuators A Phys. 267 (2017) 517–525.

[10] N. Jeromiyas, E. Elaiyappillai, A.S. Kumar, S.T. Huang, V. Mani, Bismuth nanoparticles decorated graphenated carbon nanotubes modified screen-printed electrode for mercury detection, J. Taiwan Inst. Chem. Eng. 95 (2019) 466–474.

[11] P. Li, D. Zhang, Z. Wu, Flexible MoS2 sensor arrays for high performance label-free ion sensing, Sensors Actuators A Phys. 286 (2019) 51–58.

[12] X. He, T. Xu, W. Gao, L.P. Xu, T. Pan, X. Zhang, Flexible superwettable Tapes for on-site detection of heavy metals, Anal. Chem. 90 (24) (2018) 14105–14110.

[13] X. Xuan, J.Y. Park, A miniaturized and flexible cadmium and lead ion detection sensor based on micropatterned reduced graphene oxide/carbon nanotube/bismuth composite electrodes, Sensors Actuators B Chem. 255 (2018) 1220–1227.

[14] D. Maddipatla, B. Narakathu, V. Turkani, B. Bazuin, M. Atashbar, A gravure printed flexible electrochemical sensor for the detection of heavy metal compounds, in: Multidisciplinary Digital Publishing Institute Proceedings, vol. 2, No. 13, 2018, p. 950.

[15] Z. Lu, X. Lin, J. Zhang, W. Dai, B. Liu, G. Mo, J. Ye, J. Ye, Ionic liquid/poly-L-cysteine composite deposited on flexible and hierarchical porous laser-engraved graphene electrode for high-performance electrochemical analysis of lead ion, Electrochim. Acta 295 (2019) 514–523.

[16] W. Wang, X. Zhou, S. Wu, S. Li, W. Wu, Z. Xiong, W. Shi, X. Tian, Q. Yu, Reusable surface plasmon resonance sensor for rapid detection of $Cu2+$ based on modified-chitosan thin film as an active layer, Sensors Actuators A Phys. 286 (2019) 59–67.

[17] P. Dhara, R. Kumar, L. Binetti, H.T. Nguyen, L.S. Alwis, T. Sun, K.T. Grattan, Optical fiber-based heavy metal detection using the Localized Surface Plasmon Resonance technique, IEEE Sensors J. 19 (19) (2019) 8720–8726.

[18] A.A. Zainudin, Y.W. Fen, N.A. Yusof, S.H. Al-Rekabi, M.A. Mahdi, N.A.S. Omar, Incorporation of surface plasmon resonance with novel valinomycin doped chitosan-graphene oxide thin film for sensing potassium ion, Spectrochim. Acta A Mol. Biomol. Spectrosc. 191 (2018) 111–115.

[19] B.S. Boruah, R. Biswas, Localized surface plasmon resonance based U-shaped optical fiber probe for the detection of Pb2+ in aqueous medium, Sensors Actuators B Chem. 276 (2018) 89–94.

[20] L. Xu, H. Yin, W. Ma, H. Kuang, L. Wang, C. Xu, Ultrasensitive SERS detection of mercury based on the assembled gold nanochains, Biosens. Bioelectron. 67 (2015) 472–476.

[21] Y. Wu, T. Jiang, Z. Wu, R. Yu, Novel ratiometric surface-enhanced raman spectroscopy aptasensor for sensitive and reproducible sensing of Hg2+, Biosens. Bioelectron. 99 (2018) 646–652.

[22] H. Bao, H. Zhang, L. Zhou, H. Fu, G. Liu, Y. Li, W. Cai, Ultrathin and isotropic metal sulfide wrapping on plasmonic metal nanoparticles for surface enhanced ram scattering-based detection of trace heavy-metal ions, ACS Appl. Mater. Interfaces 11 (31) (2019) 28145–28153.

[23] X. Liu, M. Liu, Y. Lu, C. Wu, Y. Xu, D. Lin, D. Lu, T. Zhou, S. Feng, Facile Ag-film based surface enhanced Raman Spectroscopy using DNA molecular switch for Ultra-Sensitive Mercury Ions Detection, Nanomaterials 8 (8) (2018) 596.

[24] E.S. Kim, CMOS-Compatible Piezoelectric Microphone, Hawaii University at Manoa Honolulu Department of Electrical Engineering, 1999.

[25] S. Lee, Y. Lee, J.S. Yang, J. Kim, W. Moon, Experimental analysis on frequency stability of piezoelectric microcantilever sensor under varying environmental and operational conditions, Procedia Eng. 25 (2011) 1517–1520.

[26] S. Lee, J. Cho, Y. Lee, S. Jeon, H.J. Cha, W. Moon, Measurement of hepatitis B surface antigen concentrations using a piezoelectric microcantilever as a mass sensor, J. Sens. 2012 (2012).

[27] R. Littrell, K. Grosh, Modeling and characterization of cantilever-based MEMS piezoelectric sensors and actuators, J. Microelectromech. Syst. 21 (2) (2012) 406–412.

[28] R. Andosca, T.G. McDonald, V. Genova, S. Rosenberg, J. Keating, C. Benedixen, J. Wu, Experimental and theoretical studies on MEMS piezoelectric vibrational energy harvesters with mass loading, Sensors Actuators A 178 (2012) 76–87.

[29] D. Zhu, S. Beeby, J. Tudor, N. White, N. Harris, Improving output power of piezoelectric energy harvesters using multilayer structures, Procedia Eng. 25 (2011) 199–202.

[30] S. Sang, Y. Zhao, W. Zhang, P. Li, J. Hu, G. Li, Surface stress-based biosensors, Biosens. Bioelectron. 51 (2014) 124–135.

[31] E. Mehdizadeh, M. Rostami, X. Guo, S. Pourkamali, Atomic resolution disk resonant force and displacement sensors for measurements in liquid, IEEE Electron Device Lett. 35 (8) (2014) 874–876.

[32] D. Rotake, A.D. Darji, Heavy metal ion detection in water using MEMS based sensor, Mater. Today Proc. 5 (1) (2018) 1530–1536.

[33] D.R. Rotake, A.D. Darji, CMOS based capacitance to digital converter circuit for MEMS sensor, in: IOP Conference Series: Materials Science and Engineering, vol. 310, No. 1, IOP Publishing, 2018, p. 012030.

[34] V. Chivukula, M. Wang, H.-F. Ji, A. Khaliq, J. Fang, K. Varahramyan, Simulation of SiO2-based piezoresistive microcantilevers, Sensors Actuators A 125 (2006) 526–533.

[35] N.S. Kale, S. Nag, R. Pinto, V.R. Rao, Fabrication and characterization of a polymeric microcantilever with an encapsulated hotwire CVD polysilicon piezoresistor, J. Microelectromech. Syst. 18 (1) (2008) 79–87.

[36] Y. Tang, J. Fang, X. Yan, H.F. Ji, Fabrication and characterization of SiO2 microcantilever for microsensor application, Sensors Actuators B Chem. 97 (1) (2004) 109–113.

[37] S. Nag, N.S. Kale, V.R. Rao, D.K. Sharma, An ultra-sensitive ΔR/R measurement system for biochemical sensors using piezoresistive micro-cantilevers, in: 2009 Annual International Conference of the IEEE Engineering in Medicine and Biology Society, IEEE, 2009, pp. 3794–3797.

[38] D.R. Rotake, A.D. Darji, Stiffness and sensitivity analysis of microcantilever based piezoresistive sensor for Bio-MEMS application, in: 2018 IEEE SENSORS, IEEE, 2018, pp. 1–4.

[39] P. Asharani, Y. Lianwu, Z. Gong, S. Valiyaveettil, Comparison of the toxicity of silver, gold and platinum nanoparticles in developing zebrafish embryos. Nanotoxicology 5 (1) (2011) 43–54, [Online]. Available: https://doi.org/10.3109/17435390.2010.489207.

[40] M. Kovochich, T. Xia, J. Xu, J.I. Yeh, A.E. Nel, Principles and procedures to assess nanomaterial toxicity, in: Environmental Nanotechnology: Applications and Impacts of Nanomaterials, McGraw Hill, New York, 2007, pp. 205–229.

[41] S.G. Royce, D. Mukherjee, T. Cai, S.S. Xu, J.A. Alexander, Z. Mi, L. Calderon, G. Mainelis, K. Lee, P.J. Lioy, et al., Modeling population exposures to silver nanoparticles present in consumer products. J. Nanopart. Res. 16 (11) (2014) 2724, [Online]. Available: https://doi.org/10.1007/s11051-014-2724-4.

[42] C.C. Huang, H.T. Chang, Selective gold-nanoparticle-based "turn-on" fluorescent sensors for detection of mercury (II) in aqueous solution, Anal. Chem. 78 (24) (2006) 8332–8338.

[43] C.J. Yu, T.L. Cheng, W.L. Tseng, Effects of Mn2 + on oligonucleotide-gold nanoparticle hybrids for colorimetric sensing of Hg2 +: improving colorimetric sensitivity and accelerating color change, Biosens. Bioelectron. 25 (1) (2009) 204–210.

[44] C. Lai, L. Qin, G. Zeng, Y. Liu, D. Huang, C. Zhang, P. Xu, M. Cheng, X. Qin, M. Wang, Sensitive and selective detection of mercury ions based on papain and 2, 6-pyridinedicarboxylic acid functionalized gold nanoparticles, RSC Adv. 6 (4) (2016) 3259–3266.

[45] J. Du, L. Jiang, Q. Shao, X. Liu, R.S. Marks, J. Ma, X. Chen, Colorimetric detection of mercury ions based on plasmonic nanoparticles. Small 9 (9–10) (2013) 1467–1481, [Online]. Available: https://doi.org/10.1002/smll.201200811.

[46] G.K. Darbha, A.K. Singh, U.S. Rai, E. Yu, H. Yu, P.C. Ray, Selective detection of mercury (ii) ion using nonlinear optical properties of gold nanoparticles. J. Am. Chem. Soc. 130 (25) (2008) 8038–8043, [Online]. Available: https://doi.org/10.1021/ja801412b.

[47] M. Minai, Embryo Project Encyclopedia, http://embryo.asu.edu/handle/10776/11335, 2016.

SECTION IV

Nano fabrication techniques—Physical sensors

CHAPTER 9

Synthesis and development of solid-state X-ray and UV radiation sensor

J.M. Kalita, M.P. Sarma, and G. Wary
Department of Physics, Cotton University, Guwahati, India

1. Introduction

Since the early days of radiation testing by Roentgen and Becquerel, scientists have sought ways to measure and observe the radiation given off by materials. One of the earliest means of studying radioactivity was a "photographic plate." A photographic plate would be placed in the path/vicinity of a radioactive beam or material. The plate was so developed that it would have spots or be fogged from the exposure to the radiation. Henri Becquerel used a method similar to this to demonstrate the existence of radiation in 1896. Another common early detector was the "electroscope." These used a pair of gold leaves that would become charged by the ionization caused by radiation and repel each other. This provided a means of measuring radiation with a better level of sensitivity than photographic plates. Depending on the arrangement of the device, they could be configured to measure alpha or beta particles and were a valuable tool for early experiments involving radiation measurement.

Another interesting early device developed for measurement of a radioactive field was the "spinthariscope." The spinthariscope was developed by William Crookes, who had also invented the Crookes Tube used by Wilhelm Roentgen to discover X-rays. A spinthariscope uses a zinc sulfide screen at the end of a tube with a lens at the other end with a small amount of a radioactive substance near the zinc sulfide screen. The zinc sulfide would react with the alpha particles emitted and each interaction would result in a tiny flash of light. This was one of the first means of counting a rate of decay, albeit a very tedious one, as it meant researchers had to work in shifts watching and literally counting the flashes of light. Therefore, the spinthariscope was not very practical as a long-term solution for radiation detection. However, the tendency of certain materials to give off light when exposed to radiation later proved to be valuable for future radiation detection technologies.

These early devices and many others, such as cloud chambers, were valuable in developing an understanding of the basic principles of radiation and conducting important

Handbook of Nanomaterials for Sensing Applications
https://doi.org/10.1016/B978-0-12-820783-3.00021-X

experiments that set the stage for later developments such as G–M tubes, ion chambers, scintillators, and solid-state radiation sensors, etc.

Depending on the purpose for use, all radiation-sensing instruments can be broadly categorized into three different categories: measurement, protection, and search. Radiation measurement tasks are for situations where there is a known presence of radioactive materials that need to be monitored. The goal of this type of detection is awareness of a radiation field. Radiation protection is similar to radiation measurement applications in the sense that it is usually in a setting where radiation is expected to be found. However, the goals of radiation protection are different. In radiation protection, the goal is to monitor people in the radiation field. The most common example of radiation protection is dosimetry. The importance of radiation protection is that it provides information and protection from the most harmful effects of radiation exposure through awareness. In contrast, radiation search differs from the other two basic categories of radiation detection applications. Primarily the goal of radiation security personnel, first responders, or groups such as customs and border inspectors, radiation search has a different set of requirements to monitor. In this case, the detectors need to be highly sensitive with the concern being more about smaller, concealed radioactive sources or materials.

Depending on the working principle, there are three basic types of radiation detectors. These are gas-filled detectors, scintillators, and solid-state sensors. Each has various strengths and weaknesses that recommend them to their own specific roles. In a gas-filled detector when the gas in the detector comes in contact with radiation, it reacts with the gas, becoming ionized, and the resulting electronic charge is measured by a meter. Various types of gas-filled detectors are ionization chambers, proportional counters, and Geiger–Mueller tubes. In contrast, the scintillators are based on the principle of scintillation. Scintillation is the act of giving off light when some materials are exposed to ionizing radiation. Each photon of radiation that interacts with a scintillator material will result in a distinct flash of light, meaning that, in addition to being highly sensitive, scintillation detectors are able to capture specific spectroscopic profiles for the measured radioactive materials. Scintillation detectors work through the connection of a scintillator material with a photomultiplier (PM) tube. The PM tube uses a photocathode material to convert each pulse of light into an electron and then amplifies that signal significantly to generate a voltage pulse that can then be read and interpreted. The number of pulses that are measured over time indicates the strength of the radioactive source being measured, whereas the information on the specific energy of the radiation, as indicated by the number of photons of light being captured in each pulse, gives information on the type of radioactive material present.

In contrast to gas-filled detectors and scintillators, the solid-state sensor uses semiconductor materials such as silicon to detect radiation. The silicon-based solid-state sensors are composed of two layers of n-type and p-type silicon material. Electrons from the n-type migrate across the junction between the two layers to fill the holes in the p-type, creating a depletion zone. This depletion zone acts like the detection area. Radiation interacting with the atoms inside the depletion zone causes them to re-ionize and create an electronic pulse that can be measured. The small scale of the detector and the depletion zone itself means that the electron-hole pairs can be collected quickly, meaning that instruments utilizing this type of sensor can have a particularly quick response time. Due to their smaller sizes, this type of solid-state sensor is very useful for electronic dosimetry applications. They are also able to withstand a much higher amount of radiation over their lifetime than other detector types such as G-M tubes, meaning that they are also useful for instruments operating in areas with particularly strong radiation fields.

This chapter is confined to highlight the basic theory associated with solid-state radiation sensors, synthesis and design of semiconductor-based radiation sensors, and various associated characteristics of such sensors for the detection of X-ray and ultraviolet (UV) radiation only.

2. Theoretical background of solid-state radiation sensor

When radiation incident on the surface of the material, it gets absorbed as it penetrates into the material. If P_o represents the power of the incident radiation and P_x is the power of the radiation after traveling to a distance x through the material, then P_x is given by

$$P_x = P_o[1 - \exp(-\alpha x)] \tag{1}$$

where α is the absorption coefficient of the material at wavelength λ [1].

The basic theory behind the radiation sensor is that when the energy of an incident photon is greater than or equal to the bandgap of a material, electron-hole pairs generate. These electron-hole pairs separate and are swept away by the electric field developed by the applied potential and thereby produce current in the external circuit. The maximum wavelength or the cut-off wavelength (λ_c) that can be detected by a device is governed by the relation as $\lambda_c = hc/E_g$ where h is the Planck's constant, c is the speed of light, and E_g is the bandgap of the material. For the radiation of wavelength greater than λ_c, the energy of the photon is not sufficient to excite electrons from the valence band (VB) to the conduction band (CB).

For a p-n junction-based radiation sensor, if the width of the depletion layer is d, the total power absorbed in the depletion layer is given by Eq. (1) as

$$P(d) = P_o[1 - \exp(-\alpha d)] \tag{2}$$

If we consider the reflectivity of the surface is R_f, the total power absorbed will be

$$P'(d) = (1 - R_f)P(d) = P_o(1 - R_f)[1 - \exp(-\alpha d)] \tag{3}$$

Since each photon has an energy equal to $h\nu$, the number of photon incidents on the surface will be $n = P'(d)/h\nu$. Therefore, the current I through the external circuit is given by

$$I = ne = \frac{eP_o}{h\nu}(1 - R_f)[1 - \exp(-\alpha d)] \tag{4}$$

where e is the charge of an electron, ν is the frequency of the incident radiation, and the rest of the parameters are as defined earlier [1].

The generation of electron–hole pairs in a material depends on various factors including the bandgap of the material, the doping concentration (in case of extrinsic semiconductor), and photon flux of the incident radiation [1]. The ratio of the number of electron–hole pairs generated in the material to the number of photons incident per unit time is called "quantum efficiency" of the device [1]. If r_p is the incident photon per second and r_e is concentration of corresponding electron–hole pair generated per second, then the quantum efficiency is defined as $\eta = r_e/r_p$. But $r_e = I/e$ and $r_p = P_o/h\nu$; therefore,

$$\eta = \frac{I/e}{P_o/h\nu} = \frac{I\,h\nu}{e\,P_o} \tag{5}$$

or

$$\eta = (1 - R_f)[1 - \exp(-\alpha d)] \quad \text{(by Eq.4)} \tag{6}$$

The higher the quantum efficiency of a material, the higher the radiation generated current through the external circuit will be. Quantum efficiency for most materials varies from 30% to 95% depending on the physical characteristics as well as design structures of the material. It can be improved by increasing the thickness of the depletion layer and thereby allowing more photons to be absorbed. However, the wider depletion width creates longer drifting time of the charge carriers, which slow down the response speed of the sensor. Therefore, a compromise has to be made between the response speed and quantum efficiency.

The performance of a radiation sensor is determined by a parameter called "responsivity." It is defined as the amount of current generated per unit radiation power incident on the sensor [1]. The responsivity (R) is given by

$$R = \frac{I}{P_o} = \frac{e}{h\nu}(1 - R_f)[1 - \exp(-\alpha d)] \tag{7}$$

or

$$R = \frac{e\eta}{h\nu} = \frac{e\eta\lambda}{hc} \quad \text{(by Eq.6)} \tag{8}$$

The detection sensitivity of a solid-state radiation sensor can also be studied from the "signal-to-noise" ratio of the device [2]. However, the signal-to-noise ratio of a device can vary depending on several factors including the distance from the radiation source to the device, the orientation of the sensing area of the device, and the surrounding environment. When radiation is allowed to incident perpendicularly to the sensing area of the device and the distance between the radiation source and the device is constant, the sensitivity (S) of the device at a constant bias voltage for a constant incident power can be determined using

$$S(V) = (I - I_d)/I_d \tag{9}$$

where I is the current at a bias voltage V when the radiation incident on the surface and I_d is the dark current under the same bias voltage [3].

In addition to the previously mentioned parameters, another important parameter of radiation sensor is the "response time." It is the time the sensor takes for the output to change in response to the change in input radiation intensity [1]. Ideally, the response time of a sensor should be very short. It depends on the transit time of the charge carriers and their mobility, width of the depletion layer, etc.

3. Materials for X-ray and UV radiation sensor

In the initial stage, most of the photosensing devices were silicon based. The silicon-based sensors possess a good sensitivity over 800 to 900 nm, low dark current, and long-term stability. However, the response time of a silicon-based sensor is relatively high. On the other hand, a germanium-based sensor has relatively high dark current owing to its lower bandgap.

For high-energy (X-ray and UV) radiation sensor, it is preferable to use high bandgap semiconductor materials [3]. The higher bandgap materials result in lower dark current and thereby improves the sensitivity of the device [3]. Literature shows that various high bandgap semiconductors including ZnO [4, 5], ZnS [3, 6, 7], TiO_2 [8, 9], Y_2O_3 [10] etc. give excellent properties suitable for UV and X-ray radiation sensors. However, the characteristics of those materials highly depend on the physical structures and synthesis process. For example, for high-performance UV photodetectors, the one-dimensional nanostructure ZnS was found to be most effective among the other nanostructures [11–15]. The large surface-to-volume ratios and the deep-level surface trap states of a one-dimensional nanostructure not only extend the lifetime of photocarriers, but also the reduced dimensionality of the active area in low-dimensional devices shortens the carrier transit time [16, 17]. However, the UV photodetectors fabricated with 1D ZnS nanostructures show a weak photocurrent and very poor stability [18–21]. Apart from this type of photodetector, some other ZnS-based heterostructure photodetectors also show some problems in the stability of the photocurrent [22–24]. To overcome this serious issue, recently the highest performance UV photodetectors were fabricated via a facile process based on the hybrid structure of solution-grown ZnS nanobelts and chemical vapor-deposited graphene [25]. Three kinds of UV photodetectors, namely ZnS spin-coated on the surface of double-layer graphene, ZnS sandwiched between two graphene layers, and multiply sandwiched graphene with ZnS, were synthesized and their UV photosensitivity was studied in ambient conditions at a low bias voltage. Among these three types, the photodetector with ZnS sandwiched between two graphene layers showed the optimum photocurrent of 37 μA, which was 10^6 times higher than the reported values of graphene-free UV photodetectors based on ZnS nanobelts [19] and ZnS–ZnO nanowires [14].

In this chapter, we confined our discussion within the synthesis of ZnS and TiO_2 thin film and their performance as X–ray and UV radiation sensors.

4. Synthesis and physical characteristics

There are various methods such as chemical bath deposition (CBD), chemical vapor deposition, thermal evaporation, etc. for the deposition of thin films. Among those, CBD is a simple and cost-effective method of deposition of thin film. It is a controlled process of precipitation of a chemical compound on a solid substrate from a solution of chemical precursors. Using this method, stable, adherent, uniform, and hard films can be deposited. It is a useful method for large-scale deposition of films. In CBD, the growth and quality of the films strongly depend on the growth conditions. The concentration, composition, and temperature of the precursor solution; pH of the solution; duration of deposition; and topographical and chemical nature of the substrate influence the deposition process. In addition, a chemical complexing agent used in the precursor solution also affects the quality of the films.

ZnS and TiO_2 thin films synthesized by CBD technique were found to be very effective for X–ray and UV radiation sensors. Sarma et al. [3] reported the physical characteristics of ZnS films synthesized by the CBD method. During synthesis, at first, zinc acetate [$Zn(CH_3COO)_2$] (Merck) and thiourea [$SC(NH_2)_2$] (Merck) solutions were prepared separately at 0.10, 0.15, and 0.20 M. Then 30 mL of zinc acetate solution was added to the 30 mL of equimolar thiourea solution, and the reaction bath temperature was simultaneously increased up to 343 K. The pH of the solution was kept constant within 8 to 9 by adding NH_3 solution (Merck). At this stage, some chemically cleaned glass substrates were immersed vertically into the solution. The bath temperature was maintained for 40 min and then the whole system was kept at room temperature. After 24 h of deposition, the glass substrates were taken out and rinsed with distilled water and dried in air. The films were polycrystalline, and the grains were hexagonal rod-like shape and uniformly deposited on the substrate [3]. The bandgap of the 0.10, 0.15, and 0.20 M films were found to be ~3.50, 3.47, and 3.38 eV, respectively [3].

A similar process was followed to deposit TiO_2 film. To synthesize the film at say, 3.0 M, at first 3.0 M solutions of titanium (IV) isopropoxide ($C_{12}H_{28}O_4Ti$) and 2-propanol ($CH_3CHOHCH_3$) are prepared separately. The two solutions are then mixed in a beaker and stirred for 1 h at 353 K. To maintain the pH within 2.0–2.3, glacial acetic acid (CH_3COOH) is added drop-wise. After stirring, a glass substrate is immersed in the solution and the temperature is maintained until 2 h. The glass substrate is taken out after 24 h and dried in air. A thin layer of the film can be seen on the substrate. The films were polycrystalline, uniform, and had good adherent to the surface [9]. The bandgap of the 1.5, 2.0, 2.5, and 3.0 M films were found to be ~3.38, 3.36, 3.33, and 3.26 eV, respectively [9].

To improve the sensitivity of the films, ZnS and TiO_2 were later doped with Cr. The doping of Cr in both ZnS and TiO_2 was found to reduce the bandgap of the host [7].

5. Current-voltage (I-V) characteristics

Sarma et al. [3] reported an experimental setup as shown in Fig. 1 to study the current-voltage (I-V) characteristics of ZnS and TiO_2 films. Here, the current through the film can be measured as a function of bias-voltage applied across the film under dark condition and X-ray/UV illumination condition.

Fig. 2 shows I-V characteristics of ZnS and TiO_2 films. For the ZnS film, the current under X-ray illumination is significantly high, whereas that under the UV illumination is moderate compared with the dark current [3]. Moreover, the minimum voltage for a typical dark current (~9.02×10^{-12} A) is 0.5 V; however, under UV radiation, approximately the same amount of current is found to flow through the external circuit at 0.3 V. Furthermore, the same amount of current is found to flow at 0.03 V under X-ray radiation [3].

The electrical conductivity under UV and X-ray radiation is calculated from the linear region of the I-V characteristic, and it is found to be 3.26×10^{-6} and 4.13×10^{-5} Siemens/cm, respectively. In contrast, the electrical conductivity under the dark condition is 2.06×10^{-6} Siemens/cm [3]. Similar behavior is also observed for the TiO_2 film. The minimum bias-voltage for a start-up dark current (~1×10^{-8} A) is 0.2 V; however, under UV radiation, approximately the same amount of current was found to flow at 0.08 V. Furthermore, the same amount of current flowed at 0.02 V under X-ray radiation [9]. The electrical conductivity under the X-ray illumination was ~2.22×10^{-1}

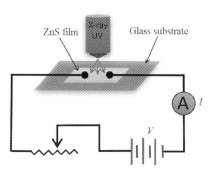

Fig. 1 An experimental setup used for UV and X-ray sensitivity analysis. *(Reproduced from M.P. Sarma, J.M. Kalita, G. Wary, Chemically deposited ZnS thin film as potential X-ray radiation sensor, Mater. Sci. Semicond. Process. 61 (2017) 131–136, https://doi.org/10.1016/j.mssp.2017.01.013 with permission from Elsevier.)*

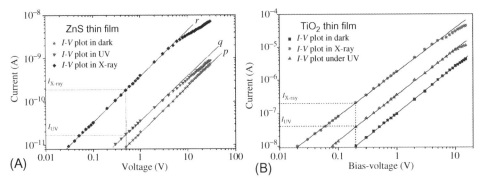

Fig. 2 I-V characteristics of (A) 0.20 M ZnS film and (B) 3.0 M TiO$_2$ film recorded under UV (line q) and X-ray (line r) radiation. The line p shows the same features but was recorded under dark condition. *(Reproduced from M.P. Sarma, J.M. Kalita, G. Wary, Chemically deposited ZnS thin film as potential X-ray radiation sensor, Mater. Sci. Semicond. Process. 61 (2017) 131–136, https://doi.org/10.1016/j.mssp.2017.01.013 and M.P. Sarma, J.M. Kalita, G. Wary, Chemical bath deposited nanocrystalline TiO$_2$ thin film as x-ray radiation sensor, Mater. Res. Express 4 (4) (2017) 045005, https://doi.org/10.1088/2053-1591/aa6a91 with permission from Elsevier and IOP Science, respectively.)*

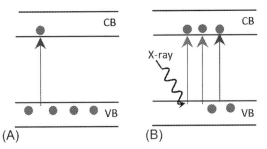

Fig. 3 Two energy band models to explain the current (A) under dark condition and (B) under UV or X-ray radiation. *(Reproduced from M.P. Sarma, J.M. Kalita, G. Wary, Chemically deposited ZnS thin film as potential X-ray radiation sensor, Mater. Sci. Semicond. Process. 61 (2017) 131–136, https://doi.org/10.1016/j.mssp.2017.01.013 with permission from Elsevier.)*

Siemens/cm whereas that under the UV illumination ~4.62 × 10^{-2} Siemens/cm. In comparison, the conductivity under dark condition was 1.11 × 10^{-2} Siemens/cm. This shows the conductivity under X-ray and UV irradiation was significantly higher than that under the dark condition [9].

It was also noted that the electrical conductivity of Cr-doped ZnS and TiO$_2$ films under dark condition as well as X-ray and UV illumination conditions are found to be 10 times higher than the undoped films [3, 7].

The enhancement of the currents under UV and X-ray radiation in both ZnS and TiO$_2$ films have been explained on the basis of simple energy band scheme as shown in Fig. 3 [3].

Under dark condition when the voltage across the film is increased, a few electrons excite from the VB to the CB and drift through the external circuit. Further increase in voltage excites more electrons to the CB. Hence, the current through the external circuit is found to increase linearly as shown by the straight line p (Fig. 2). During the UV illumination, a few electrons excite from VB to the CB, hence, the same amount of current is found to flow through the external circuit at relatively lower bias voltage. On the other hand, since the energy of X-ray is very high compared with UV radiation, a significant number of electrons excite to the CB during X-ray radiation. As a result, current is found to flow through the external circuit at relatively low bias voltage. The I-V characteristic under X-ray radiation is identical to that under the dark and UV radiation. This signifies that the current contributed by the electrons under X-ray radiation are of the same origin as those under the UV radiation or in dark condition.

6. Radiation sensitivity

Sarma et al. [3, 7, 9] reported the X-ray radiation detection sensitivity of the films as a function of the voltage applied across the films. The detection sensitivity (more precisely, the signal-to-noise ratio) at a particular energy of X-ray radiation has been calculated using Eq. (9). Here, the X-ray or UV radiation was incident perpendicularly on the surface of the film and the distance between the source and the film was kept constant throughout the experiment. Therefore, any dependence of the angle of incidence of the radiation and distance between the source and the film on detection sensitivity was ignored.

Fig. 4 shows a plot of normalized sensitivity of ZnS and TiO_2 films under UV and X-ray radiation as a function of applied voltage. For ZnS film, the sensitivity to X-ray radiation is very high compared with the UV radiation (Fig. 4A (inset)). It is very

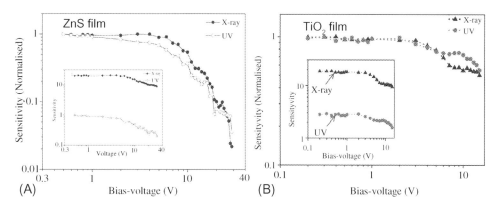

Fig. 4 Change of normalized sensitivity under the X-ray and UV radiation with bias-voltage for (A) ZnS films and (B) TiO_2 films, respectively. The inset in each figure shows the same features without normalization for the ZnS and TiO_2 films, respectively. *(Reproduced from M.P. Sarma, J.M. Kalita, G. Wary, Chemically deposited ZnS thin film as potential X-ray radiation sensor, Mater. Sci. Semicond. Process. 61 (2017) 131–136, https://doi.org/10.1016/j.mssp.2017.01.013 and M.P. Sarma, J.M. Kalita, G. Wary, Chemical bath deposited nanocrystalline TiO_2 thin film as x-ray radiation sensor, Mater. Res. Express 4 (4) (2017) 045005, https://doi.org/10.1088/2053-1591/aa6a91 with permission from Elsevier and IOP Science, respectively.)*

significant that the normalized sensitivity at very low voltage (<0.8 V) is the same under the UV and X–ray radiation. However, as the applied voltage increases, the normalized sensitivity under UV radiation decreases more quickly than the normalized sensitivity under X–ray radiation. The normalized sensitivity under X–ray radiation is found to be constant up to 6 V and then it decreases gradually.

On the other hand, the variation of normalized sensitivity with bias voltage under the X–ray and UV illuminations are similar for TiO_2 film (Fig. 4B). The inset in Fig. 4B shows the un-normalized sensitivity under the X–ray and UV illuminations. It can be observed that the sensitivity under X–ray radiation was significantly higher than that under the UV radiation. The sensitivity under the X–ray radiation was consistent within 0.2 to 3.0 V and thereafter decreased gradually. On the other hand, the sensitivity under the UV radiation was constant within 0.2 to 2.0 V and then decreased with the bias-voltage. It is also noted that the effect of doping on the sensitivity is not very prominent. The doped films show similar sensitivity as that of the undoped ones [7].

7. Response time

The response time for a radiation sensor is the measure of how quickly a device can sense the radiation. Sarma et al. [7] reported the response time of ZnS- and Cr-doped ZnS films for radiation detection. In this experiment, a voltage of 1.00 V was applied across the films and X–ray having a constant energy allowed to fall on the films for 2.00 s. The real-time current flowing through the films was measured by the electrometer (Keithley, 6517B, United States). The electrometer was set to collect only 40 data points at a resolution of 0.12 s/channel as soon as the X–ray fell on the surface of the films. For each film, the whole experiment was carried out three times and the average of the three independent measurements was taken for the analysis. Fig. 5 shows the current as a function of time. Here the solid lines through the data points are the experimentally measured current flowing through the films. The dotted lines represent the dark currents corresponding to the voltage of 1.00 V applied across the films. Fig. 5B shows the X–ray radiation-induced currents that are digitally mustered by subtracting the dark currents from the corresponding experimentally measured currents.

Figs. 5A and B clearly show the rise and decay of the radiation-induced currents. It is evident that the current under X–ray radiation rises very quickly, reaches a steady state, and as soon as the X–ray is switched off, the current decreases to the dark level current. It is also observed that the net radiation-induced currents increase after doping of Cr (see Fig. 5B).

When X–ray photons fall on the films, electrons from the VB excite to the CB and electron-hole pairs generate. Due to the applied voltage, the conduction electrons drift through the external circuit and produce the radiation-induced current. It is reported that in polycrystalline thin films, there may be built-in potential barriers at the grain boundaries [26]. In such a case, a part of the photogenerated charge carriers may recombine with

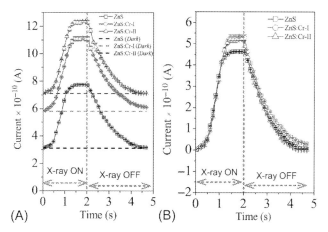

Fig. 5 (A) The real-time current through ZnS- and Cr-doped ZnS films as a function of time when the X-ray is ON and OFF. Here, the solid lines through the data points are the experimentally measured current flowing through the films. The dotted lines represent the dark currents corresponding to the voltage of 1.00 V applied across the films. (B) The X-ray radiation-induced currents that are digitally mustered by subtracting the dark currents from the corresponding experimentally measured currents. *(Reproduced from M.P. Sarma, J.M. Kalita, G. Wary, X-ray sensing performance of ZnS:Cr film synthesised by a chemical bath deposition technique, Sens. Actuators A 276 (2018) 328–334, https://doi.org/10.1016/j.sna.2018.04.044 with permission from Elsevier.)*

their respective opposite charges confined at the depletion region of the grain boundaries and eventually reduce the potential barriers of the grain boundaries. On the other hand, the remaining part of the photogenerated charge carriers take part in the conduction process. The reduction of potential barriers enhances the effective mobility of the charge carriers. As a result, conductivity of the sample increases. In ZnS- and Cr-doped ZnS films, it can be considered that the increase in radiation-induced currents is mainly due to the resultant increase in the X-ray-generated charge carriers and enhancement of their effective mobility. As soon as the X-ray falls on the films, the number of X-ray-generated charge carriers increase rapidly. However, as the time elapses, the concentration of the X-ray-generated charge carriers reaches a saturation state and causes a constant radiation-induced current through the external circuit. When the X-ray is switched off, a fraction of conduction electrons recombines with the holes whereas the remaining part of the conduction electrons drift through the external circuit due to the influence of voltage applied across the films. As no more charge carriers are excited to the CB, the concentration of conduction electron reduces and thus the radiation-induced current falls to the dark level.

The rise and decay times of the radiation-induced current corresponding to the doped and undoped films have been calculated by drawing two tangents on the rise and decay parts of the plots. For ZnS film, the rise and decay times are estimated as 0.87 s and 1.61 s, respectively [7]. In contrast, the rise and decay times of 0.05 M Cr-doped ZnS film (ZnS:Cr-I) are estimated as 0.83 s and 1.65 s, respectively, whereas those of 0.10 M Cr-doped ZnS film (ZnS:Cr-II) are found as 0.80 s and 1.67 s, respectively [7]. This implies that the rise time in any film is shorter than the decay time. The rise time in ZnS:Cr-II film is the shortest whereas the decay time is the longest. On the other hand, the rise time in ZnS is the longest whereas the decay time is the shortest. In ZnS:Cr-II, the concentration of doping is high. Therefore, under X-ray radiation, the trivalent Cr impurity offers more excess electrons in the ZnS matrix and thereby causes more electron-hole pairs to form. Some of these radiation-induced charge carriers recombine with their respective opposite charges confined at the depletion region of the grain boundaries and thereby reduce the potential barriers of the grain boundaries. The reduction of potential barriers enhances the effective mobility of the charge carriers. Due to the enhancement of charge mobility, the electrons take less time to reach the saturation state and thus cause the shortest rise time in ZnS:Cr-II film. On the other hand, owing to the greater concentration of Cr in ZnS:Cr-II film, the number of electrons at the CB might be higher than the undoped and ZnS:Cr-I films. Therefore, when the X-ray is switched off, the conduction electrons take more time to drift through the external circuit. Thus, the decay time is the longest in ZnS:Cr-II film. From these analyses, although some differences in the rise and decay times have been observed between the undoped and doped films but in terms of practical applications, the magnitude of differences are not too high. It means, in terms of response speed, the effect of doping is not very significant [7].

The TiO_2-film and Cr-doped TiO_2 films show similar nature in response speed as that of ZnS films. The rise and decay times for the undoped film are estimated to be 0.91 s and 1.18 s, respectively. On the other hand, the same parameters for 0.05 M Cr-doped TiO_2 film are estimated as 0.89 s and 1.22 s, respectively; and for 0.10 M Cr-doped TiO_2 film 0.84 s and 1.32 s, respectively. In all the films, the rise time is found to be shorter than the decay time. The 0.10 M Cr-doped TiO_2 film shows the shortest rise time however the longest decay time. In contrast, the undoped film shows the longest rise time but the shortest decay time. This behavior of response time between the doped and undoped films is somewhat related to concentration of charge carriers and thus can be explained in line with the similar discussion as ZnS films.

8. Summary

We briefly summarized the development of solid-state X-ray and UV radiation sensors and their advantages over the other radiation sensors. We confined our discussion on the X-ray and UV radiation sensing characteristics of two high bandgap semiconductor materials, namely ZnS and TiO_2. Synthesis and experimental arrangements for studying radiation sensitivity of thin film-based X-ray and UV sensors have been included. Sensing characteristics such as radiation sensitivity and response time of the sensor have been discussed for the ZnS and TiO_2 films.

References

[1] G. Keiser, Optical Fiber Communication, McGraw-Hill, Inc, Singapore, 1986.

[2] G.H. Rieke, Detection of Light From the Ultraviolet to the Submillimeter, second ed., Cambridge University Press, New York, 2003.

[3] M.P. Sarma, J.M. Kalita, G. Wary, Chemically deposited ZnS thin film as potential X-ray radiation sensor. Mater. Sci. Semicond. Process. 61 (2017) 131–136, https://doi.org/10.1016/j.mssp.2017.01.013.

[4] K. Liu, M. Sakurai, M. Aono, ZnO-based ultraviolet photodetectors. Sensor 10 (9) (2010) 8604–8634, https://doi.org/10.3390/s100908604.

[5] R. Khokhra, B. Bharti, H.-N. Lee, R. Kumar, Visible and UV photo-detection in ZnO nanostructured thin films via simple tuning of solution method. Sci. Rep. 7 (2017) 15032, https://doi.org/10.1038/s41598-017-15125-x.

[6] S. Premkumar, D. Nataraj, G. Bharathi, S. Ramya, T.D. Thangadurai, Highly responsive ultraviolet sensor based on ZnS quantum dot solid with enhanced photocurrent. Sci. Rep. 9 (2019) 18704, https://doi.org/10.1038/s41598-019-55097-8.

[7] M.P. Sarma, J.M. Kalita, G. Wary, X-ray sensing performance of ZnS:Cr film synthesised by a chemical bath deposition technique. Sens. Actuators A 276 (2018) 328–334, https://doi.org/10.1016/j.sna.2018.04.044.

[8] Z. Liu, F. Li, S. Li, C. Hu, W. Wang, F. Wang, F. Lin, H. Wang, Fabrication of UV photodetector on TiO_2/diamond film. Sci. Rep. 5 (2015) 14420, https://doi.org/10.1038/srep14420.

[9] M.P. Sarma, J.M. Kalita, G. Wary, Chemical bath deposited nanocrystalline TiO_2 thin film as x-ray radiation sensor. Mater. Res. Express 4 (4) (2017) 045005, https://doi.org/10.1088/2053-1591/aa6a91.

[10] P. Praveenkumar, T. Subashini, G.D. Venkatasubbu, T. Prakash, Crystallite size effect on low-dose X-ray sensing behaviour of Y_2O_3 nanocrystals. Sens. Actuators A 297 (2019) 111544, https://doi.org/10.1016/j.sna.2019.111544.

[11] X.D. Gao, X.M. Li, W.D. Yu, Morphology and optical properties of amorphous ZnS films deposited by ultrasonic-assisted successive ionic layer adsorption and reaction method. Thin Solid Films 468 (1–2) (2004) 43–47, https://doi.org/10.1016/j.tsf.2004.04.005.

[12] I.K. Sou, Z.H. Ma, Z.Q. Zhang, G.K.L. Wong, Temperature dependence of the responsivity of ZnS-based UV detectors. J. Cryst. Growth 214/215 (2000) 1125–1129, https://doi.org/10.1016/S0022-0248(00)00287-6.

[13] I.K. Sou, Z.H. Ma, G.K.L. Wong, ZnS-based visible-blind UV detectors: effects of isoelectronic traps. J. Electron. Mater. 29 (2000) 723–726, https://doi.org/10.1007/s11664-000-0213-2.

[14] W. Tian, C. Zhang, T. Zhai, S.L. Li, X. Wang, J. Liu, X. Jie, D. Liu, M. Liao, Y. Koide, D. Golberg, Y. Bando, Flexible ultraviolet photodetectors with broad photoresponse based on branched ZnS-ZnO heterostructure nanofilms. Adv. Mater. 26 (19) (2014) 3088–3093, https://doi.org/10.1002/adma.201305457.

[15] Y. Liang, H. Liang, X. Xiao, S. Hark, The epitaxial growth of ZnO nanowire arrays and their applications in UV-light detection. J. Mater. Chem. 22 (2012) 1199–1205, https://doi.org/10.1039/C1JM13903G.

[16] J.S. Jie, W.J. Zhang, Y. Jiang, X.M. Meng, Y.Q. Li, S.T. Lee, Photoconductive characteristics of single-crystal CdS nanoribbons. Nano Lett. 6 (9) (2006) 1887–1892, https://doi.org/10.1021/nl060867g.

[17] C. Soci, A. Zhang, B. Xiang, S.A. Dayeh, D.P.R. Aplin, J. Park, X.Y. Bao, Y.H. Lo, D. Wang, ZnO nanowire UV photodetectors with high internal gain. Nano Lett. 7 (4) (2007) 1003–1009, https://doi.org/10.1021/nl070111x.

[18] X. Wang, Z. Xie, H. Huang, Z. Liu, D. Chen, G. Shen, Gas sensors, thermistor and photodetector based on ZnS nanowires. J. Mater. Chem. 22 (14) (2012) 6845–6850, https://doi.org/10.1039/C2JM16523F.

[19] X. Fang, Y. Bando, M. Liao, U.K. Gautam, C. Zhi, B. Dierre, B. Liu, T. Zhai, T. Sekiguchi, Y. Koide, D. Golberg, Single-crystalline ZnS nanobelts as ultraviolet-light sensors. Adv. Mater. 21 (20) (2009) 2034–2039, https://doi.org/10.1002/adma.200802441.

[20] X. Fang, Y. Bando, M. Liao, U.K. Gautam, C. Zhi, B. Dierre, B. Liu, T. Zhai, T. Sekiguchi, Y. Koide, D. Golberg, An efficient way to assemble ZnS nanobelts as ultraviolet-light sensors with enhanced

photocurrent and stability. Adv. Funct. Mater. 20 (3) (2010) 500–508, https://doi.org/10.1002/adfm.200901878.

[21] Y. Yu, J. Jie, P. Jiang, L. Wang, C. Wu, Q. Peng, X. Zhang, Z. Wang, C. Xie, D. Wu, Y. Jiang, High-gain visible-blind UV photodetectors based on chlorine-doped n-type ZnS nanoribbons with tunable optoelectronic properties. J. Mater. Chem. 21 (34) (2011) 12632–12638, https://doi.org/10.1039/C1JM11408E.

[22] D.I. Son, H.Y. Yang, T.W. Kim, W.I. Park, Photoresponse mechanisms of ultraviolet photodetectors based on colloidal ZnO quantum dot-graphene nanocomposites. Appl. Phys. Lett. 102 (2) (2013) 021105, https://doi.org/10.1063/1.4776651.

[23] A.V. Babichev, H. Zhang, P. Lavenus, F.H. Julien, A.Y. Egorov, Y.T. Lin, L.W. Tu, M. Tchernycheva, GaN nanowire ultraviolet photodetector with a graphene transparent contact. Appl. Phys. Lett. 103 (20) (2013) 201103, https://doi.org/10.1063/1.4829756.

[24] Z. Zhan, L. Zheng, Y. Pan, G. Sun, L. Li, Self-powered, visible-light photodetector based on thermally reduced graphene oxide–ZnO (rGO–ZnO) hybrid nanostructure. J. Mater. Chem. 22 (6) (2012) 2589–2595, https://doi.org/10.1039/C1JM13920G.

[25] Y. Kim, S.J. Kim, S.-P. Cho, B.H. Hong, D.-J. Jang, High-performance ultraviolet photodetectors based on solution-grown ZnS nanobelts sandwiched between graphene layers. Sci. Rep. 5 (2015) 12345, https://doi.org/10.1038/srep12345.

[26] K.C. Sarmah, H.L. Das, Effect of substrate temperature on photoconductivity in CdTe thin films. Thin Solid Films 198 (1–2) (1991) 29–34, https://doi.org/10.1016/0040-6090(91)90321-N.

CHAPTER 10

Nanosensors for checking noise of physical agents

Himanshu Dehra
Monarchy of Concordia, Faridabad, Haryana, India

1. Introduction

A noise sensor is a technological device or biological organ that detects and senses a noise signal or noise due to physical agents of light, sound, heat, electricity, fluid, fire, and sun. Because a noise signal is a form of energy, sensors can be classified according to the type of energy they detect, for example, electrical power watt-hour meters, voltmeter, ammeter, multimeter, photocells, image sensor, thermocouples, thermistors, bolometer, calorimeter, speed indicator, anemometer, flow meter, speedometer, tachometer, position sensor, magnetometer, etc. This chapter presents a brief overview of nanosensors for evaluating noise of physical agents with a description of its noise-measurement sensing system along with its characteristics.

There are two types of nanosensors: mechanical nanosensors and chemical nanosensors, which both have different sensing mechanisms.

Nanosensors that detect chemicals work by measuring the change in the electrical conductivity of the nanomaterial once an analyte has been detected. Many nanomaterials have a high electrical conductivity, which will reduce upon binding or adsorption of a molecule. It is this detectable change that is measured; 1D materials, such as nanowires and nanotubes, are excellent examples of chemical nanosensors, as their electrically confined structure can act as both the transducer and the electronic wires once an analyte has been detected.

At present, the best examples of nanosensors, as far as applications for sensing noise of physical agents, are associated with information science and environmental industry. The development of nanostructures from nanoparticles has been accelerated over the past decade because of unexpected breakthroughs in the synthesis and assembly of nanometer-scale structures, as well as the capabilities of manipulating matter from the top-down during this period. Some development areas include:

- discovery and controlled production of carbon nanotubes (CNTs) and the use of proximal probe and lithographic schemes;
- success in placing engineered individual molecules onto appropriate electrical contacts and in measuring transport through these molecules;

Handbook of Nanomaterials for Sensing Applications
https://doi.org/10.1016/B978-0-12-820783-3.00015-4

- availability and use of proximal probe techniques for fabrication of nanometer-scale structures;
- development of chemical synthetic methods to realize nanocrystals and implantation technique for putting these crystals into a variety of larger organized structures and its assembly making;
- incorporation of biological motors into nonbiological environments;
- integration of nanoparticles into gas sensors;
- fabrication of photonic level bandgap structures;
- development of quantum effect, single electron memory, and logic elements that can operate at room temperature.

One area in sensing noise of physical agents is the development of nanometer-scale objects that can manipulate and develop other nanometer-scale objects economically and efficiently.

1.1 Monitoring the environment

Nanosensors have a great ability to monitor and analyze micro-organisms and toxic-chemical compounds found in environmental samples. Nanomaterials can be used to enhance the sensitivity of electrochemical sensors and ion-selective electrodes (ISEs), which are the conventional techniques used for the detection of trace amount of metals, nitrates, phosphates, and pesticides in waterborne samples. Nanosensors also have the ability to measure in real time, which is a highly valued property for environmental monitoring applications.

1.2 Light sensing

Many applications focus on the detection of various molecules in a certain environment. However, nanosensors can also be used to detect electromagnetic radiation. One example is using zinc oxide nanorods, or zinc oxide nanowires, to detect UV radiation at low levels. Nanowires are often used in electromagnetic radiation-sensing applications because they change their resistive state and invoke a measurable response to electromagnetic rays. Nanowires can also be used in parallel where the electrons cascade across all the nanowires and provide a quick and effective response.

1.3 Combining nanosensors with other technologies

There is a growing trend of combining nanosensors with other useful technologies, such as microelectromechanical systems (MEMS) and microfluidic devices. Examples of where this has been useful include: depositing nanoparticles onto silicon substrates for more efficient chemical and gas sensing applications, gold nanowires in microfluidic devices to detect cholesterol in blood samples, using CNTs on silicon to detect harmful traces of ammonia, and in fluid-based MEMS devices to detect trace amounts of micro-organisms in a fluidic sample.

Sensors based on vibrating elements have also been produced by bonded silicon-on-insulator (BSOI) process to measure many physical and chemical parameters. The technique is responsible for reduction in size of the resonant sensors by process of miniaturization of the sensors, producing them on silicon substrate and, in the process, making them resistant to shock. Internal losses can also be reduced through use of high-quality materials with single crystal or the polysilicon type; in these cases, high Q-factor is obtainable.

Gas-sensing systems: For gas sensing, a silicon-based metal oxide chemical sensor chip is used. The chip is fabricated from the oxide, is in the thin film form, and has an integrally produced polysilicon microheater buried in the system. The gas-sensitive layer converts gas concentration into electrical conductivity when exposed to air. The heater and the sensing element are built on a precision-generated diaphragm micromachined from a silicon oxynitride ceramic. This material has thermal properties that minimize both response time and sensor power.

Operation of integrated sensor: Ambient oxygen molecules chemisorb onto the film surface of the sensor only to dissociate into oxygen ions. This requires the heater to be at a temperature of around $450°C$. Formation of these oxygen ions removes free electrons from the tin oxide grain surfaces. This process increases the sensing film resistance. If carbon monoxide is now present in the gas, it is acting as a reducing agent, reacting with oxygen to produce CO_2, and releasing electrons such that the conductance of the sensor is restored. As CO concentration increases, the film resistance further decreases. This change is called the "detecting principle." If instead an oxidizing agent like NO_2 is present in the gas, the effect reverses and the resistance of the film sensor increases. The response of the sensor can be made dependent on the constituents of the gas by appropriate control of the heater temperature. This phenomenon has been utilized to optimize the sensor response to a specific gas against a background of interfering gases. Thus, a single sensor can be used to speciate a mixture of gases such as hydrogen, CO_2, and methane. It must be remembered, however, that the gas mixture cannot be arbitrary. The important thing is the sensitivity of the metal oxide film depends on the grain size as well as the structure. Smaller grain size provides greater surface area and increased film porosity, factors that improve sensitivity and long-term stability of the film. The nanostructured metal oxide films with 20-nm grain size have shown fourfold improvement in sensitivity.

While progressing toward the development of fast and miniaturized memory structures, giant magnetoresistance structures have been produced using the Thomson effect. These giant magnetoresistors (GMR) consist of layers of magnetic and nonmagnetic metal films wherein the critical layers have thickness of the order of nanometers. They are used as extremely sensitive magnetic field sensors. The spin–polarized electrons are transported between the magnetic layers on the nanometer-length scale, and this process, during operation, allows the structure to sense a magnetic field. Organic nanostructures have been developed combining chemical self-assembly with a mechanical device. The organic sample is reduced to a size that consists of a single molecule, and this is connected by two gold leads. This structure has been successfully used to measure the electrical conductivity of a single molecule.

Sensors for monitoring biomedical activities: A mix of conventional, micro-, and nanosensors are being adopted more and more now in biomedical activities. Many of these are based on:

(i) radiation—electromagnetic and acoustic;
(ii) force and pressure;
(iii) temperature;
(iv) electromagnetic variables;
(v) chemical and electrochemical principles;
(vi) variables related to blood flow;
(vii) kinematic and geometric.

2. Theory of noise characterization

A unified theory for stresses and oscillations is proposed by the author [1]. The following standard measurement equations are derived and adopted from standard definitions for sources of noise interference as developed by the author [1–26].

Noise of sol: For a pack of solar energy waves, the multiplication of solar power storage and the velocity of light gives solar power intensity I. A logarithm of two intensities of solar power, I_1 and I_2, provides intensity difference. It is mathematically expressed as:

$$Sol = \log(I_1)(I_2)^{-1} \qquad (1)$$

Whereas logarithmic unit ratio for noise of sol is expressed as *Sol*. The oncisol (oS) is more convenient for solar power systems. The mathematical expression by the following equality gives an oncisol (oS), which is 1/11th unit of a *Sol*:

$$oS = \pm 11 \log(I_1)(I_2)^{-1} \qquad (2)$$

Noise of therm: For a pack of heat energy waves, the multiplication of total power storage and the velocity of light gives heat power intensity I. The pack of solar energy waves and heat energy waves (for same intensity I) have same energy areas, therefore their units of noise are same as *Sol*.

Noise of photons: For a pack of light energy beams, the multiplication of total power storage and the velocity of light gives light power intensity I. The pack of solar energy waves and light energy beams (for same intensity I) have same energy areas, therefore their units of noise are same as Sol.

Noise of electrons: For a pack of electricity waves, the multiplication of total electrical storage and the velocity of light gives electrical power intensity I. The pack of solar energy waves and electricity waves (for same intensity I) have same energy areas, therefore their units of noise are same as Sol.

Noise of scattering: For a pack of fluid energy waves, the multiplication of total power storage and the velocity of fluid gives fluid power intensity I. A logarithm of two intensities of fluid power, I_1 and I_2, provides intensity difference. It is mathematically expressed as:

$$Sip = \log(I_1)(I_2)^{-1} \tag{3}$$

Whereas logarithmic unit ratio for noise of scattering is Sip. The oncisip (oS) is more convenient for fluid power systems.

The mathematical expression by the following equality gives an oncisip (oS), which is 1/11th unit of a Sip:

$$oS = \pm 11 \log(I_1)(I_2)^{-1} \tag{4}$$

For energy area determination for a fluid wave, the water with a specific gravity of 1.0 is the standard fluid considered with power of ± 1 W m^{-2} for a reference intensity I_2.

Noise of scattering and lightning: For a pack of fire waves, the intensity, I, of fire flash with power of light is the multiplication of total power storage and the velocity of light. Whereas for a pack of fire waves, the intensity, I, of fire flash with power of fluid is the multiplication of total power storage capacity and velocity of fluid.

- For a noise due to fire flash, the collective effect of scattering and lightning is obtained by superimposition principle.
- For same intensity I, the pack of solar energy waves and a fire flash with light power have same energy areas, therefore their units of noise are same as Sol. The therm power may also be included in fire flash with power of light.
- For same intensity I, the pack of fluid energy waves and a fire flash with fluid power have same energy areas, therefore their units of noise are same as Sip. In determining the areas of energy for the case of fluids other than water, a multiplication factor in specific gravity has to be evaluated.

Noise of elasticity: For a pack of sound energy waves, the product of total power storage and the velocity of sound gives sound power intensity I. A logarithm of two intensities of sound power, I_1 and I_2, provides intensity difference. It is mathematically expressed as:

$$Bel = \log(I_1)(I_2)^{-1} \tag{5}$$

Whereas logarithmic unit ratio for noise of elasticity is *Bel*. The oncibel (oB) is more convenient for sound power systems. The mathematical expression by the following equality gives an oncibel (oB), which is 1/11th unit of a *Bel*:

$$oB = \pm 11 \log (I_1)(I_2)^{-1} \tag{6}$$

There are following elaborative points on choosing an *onci* as 1/11th unit of noise [24]:

(i) Reference value used for I_2 is -1 W m^{-2} on a positive scale of noise and 1 W m^{-2} on a negative scale of noise. In a power cycle, all types of wave form one positive power cycle and one negative power cycle [2]. Positive scale of noise has 10 positive units and 1 negative unit, whereas, negative scale of noise has 1 positive unit and 10 negative units.

(ii) Each unit of sol, sip and bel is divided into 11 parts, 1 part is 1/11th unit of noise.

(iii) The base of logarithm used in noise-measurement equations is 11.

(iv) Reference value of I_2 is -1 W m^{-2} with I_1 on positive scale of noise, which should be taken with negative noise-measurement expression (see Eqs. 2, 4, 6), therefore it gives positive values of noise.

(v) Reference value of I_2 is 1 W m^{-2} with I_1 on negative scale of noise, which should be taken with positive noise-measurement expression (see Eqs. 2, 4, 6), therefore it gives negative values of noise.

The choosing of *onci* in noise units is done so as to have separate market product and system of noise scales and their units distinguished from prevailing *decibel* unit (which has its limitations) in the International System of Units. More discussions on energy conversion, noise characterization theory, and choice of noise scales and its units are presented in many papers by the author [1–26].

2.1 Estimating changes in noise power and noise pressure levels

In some cases, it is difficult to measure intensity of a power source; therefore, pressure p can be measured. The relationship between pressure and intensity is given by:

$$I = \frac{p_{rms}^2}{\rho c} \tag{7}$$

where p = root mean-square (rms) pressure, N m^{-2}, ρ = density of wave medium, kg m^{-3}, c = speed of wave, m s^{-1}.

Eqs. (3), (5), (7) are rewritten in the form:

$$\text{Intensity difference } \Delta I = \pm 22 \log \frac{p}{po} \tag{8}$$

The addition of two equal pressures results in an increase of:

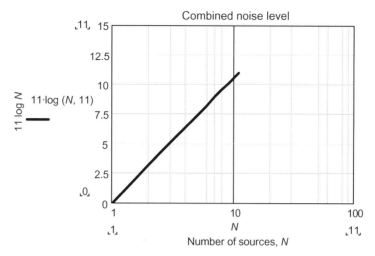

Fig. 1 Predicting the combined noise level of identical sources.

22 log 112 = 6.4 onci sol (oS) and addition of two equal powers results in an increase of 3.2 onci sol. When two equal pressure levels are added, we are in effect adding two equal power levels, therefore:

$$Lp1 + Lp2 = 11 \log 11 \left(\frac{p}{pref}\right)^2 + 3.2 oS \tag{9}$$

Similarly, it can be shown that, when N identical noise sources are added,

$$Lp(\text{total}) = Lp1 + 11 \log 11 N \tag{10}$$

11 Log N is plotted as a function of N in Fig. 1.

Table 1 shows how to add two unequal noise levels and Fig. 2 presents Table 1 graphically.

Fig. 3 presents a double-sided hexagonal slide rule with seven edges for noise measurement representing seven sources of noise. Reference value used for I_2 is −1 W m^{-2} on the positive scale of noise and 1 W m^{-2} on the negative scale of noise. Positive scale of noise has 10 positive units and 1 negative unit, whereas negative scale of noise has 1 positive unit and 10 negative units. Each unit of sol, sip, and bel is divided into 11 parts, 1 part is 1/11th unit of noise. The base of logarithm used in noise-measurement equations is 11.

The results of noise filtering using various noise-measurement equations for an outdoor duct exposed to solar radiation are tabulated in Tables 2–6. Table 7 presents noise calculation charts based on intensity and pressure differences so as to calculate onci sol, onci sip, and onci bel.

Table 1 Addition of unequal noise levels.

Difference between two levels, oncisol, oncisip, and oncibel	Add to the higher level, oncisol, oncisip, and oncibel
0	3.18
1	1.86
2	1.319
3	1.024
4	0.836
5	0.708
6	0.612
7	0.54
8	0.484
9	0.437
10	0.399
11 or more	0

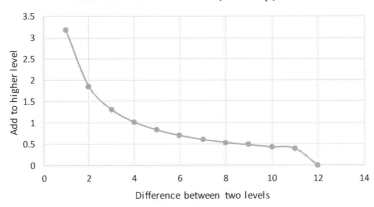

Fig. 2 Noise addition.

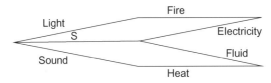

Fig. 3 A double-sided hexagonal scales of noise with seven edges (S denotes Sun).

Table 2 Properties of physical domain.

Property	Value	Property	Value
Solar irradiation	650 W m^{-2}	Width of air gap	0.025 m
Ambient heat transfer coefficient	$13.5 \text{ W m}^{-2} \text{ K}^{-1}$	Thermal conductivity of air	$0.02624 \text{ W m}^{-1} \text{ K}^{-1}$
Ambient air temperature	$-5°C$	Specific heat of air (cp)	$1000 \text{ J kg}^{-1} \text{ K}^{-1}$
Building space temperature	$20°C$	Density of air	1.1174 kg m^{-3}
Height of duct	3.0 m	Kinematic viscosity of air	$15.69 \times 10^{-6} \text{ m}^2 \text{ s}^{-1}$
Width of duct	1.0 m	Prandtl number of air	0.708
Thickness of outer wall of duct	0.0025 m	Air velocity for obtaining mass flow rate	0.75 m s^{-1}
Absorptance of outer wall with flat black paint	0.95	Stefan Boltzmann constant for surface of duct walls	$5.67 \times 10^{-8} \text{ W m}^{-2} \text{ K}^{-4}$
Thermal conductivity of aluminum alloy for HVAC duct	$137 \text{ W m}^{-1} \text{ K}^{-1}$	Emissivity of back surface of duct walls	0.95
RSI value	$1.0 \text{ m}^2 \text{ K W}^{-1}$	Number of nodes in x-direction	$Nx = 3$
Thickness	0.04 m	Number of nodes in y-direction	$Ny = 10$, $\Delta y = 0.3$ m

Table 3 Temperature difference and noise of sol with solar irradiation (air velocity: 0.75 ms^{-1}).

Solar irradiation (W m^{-2})	Air temperature difference (ΔT) °C	Noise of sol oS (oncisol)
450	15.50	28
550	18.90	28.93
650	22.40	29.7
750	25.90	30.36
850	29.40	30.91

2.2 Real-time noise informatics: Sensors for sensing noise of physical agents

Noise sensors can be classified as per type of physical agent (light, sound, heat, fluid, electricity, fire, and sun) for which monitoring is required. To ensure proper environmental control, the climatic parameters within an enclosure should be periodically checked and measured. Noise monitoring data of cities and their habitants can be collected in a

Table 4 Temperature difference and noise of scattering with air velocity ($S = 650$ W m^{-2})

Air velocity (ms^{-1})	Fluid power (W m^{-2})	Air temperature difference (ΔT) °C	Noise of scattering oS (oncisip)
1.35	47.62	15.28	17.72
1.05	37.0	18.22	16.50
0.75	26.45	22.40	15.02
0.45	15.87	28.15	12.65
0.15	05.29	29.80	07.64

Table 5 Mass flow rate and noise of therm with (ΔT) °C.

(ΔT) °C	Mass flow rate (Kg s^{-1})	Thermal power (W m^{-2})	Noise of therm oS (oncisol)	(ΔT) °C	Mass flow rate (Kg s^{-1})	Thermal power (W m^{-2})	Noise of therm oS (oncisol)
15.50	0.01376	71.09	19.5602	15.28	0.0231	117.65	21.868
18.90	0.01275	80.325	20.119	18.22	0.0171	103.85	21.296
22.40	0.0120	89.6	20.614	22.40	0.0120	89.6	20.614
25.90	0.0115	99.2833	21.043	28.15	8.1×10^{-3}	76.0	19.866
29.40	0.0111	108.78	21.505	29.80	6.2×10^{-3}	61.59	18.898

Table 6 Noise of elasticity with air particle velocity (impedance $Z_0 = 413$ N s m^{-3} at 20°C).

Air velocity (m s^{-1})	Fluid Power (W m^{-2})	Noise of scattering oS (oncisip)	Sound pressure (N m^{-2})	Sound power intensity (W m^{-2})	Noise of elasticity oB (oncibel)
1.35	47.62	17.72	557.5	752.7	30.36
1.05	37.0	16.50	433.65	455.33	28.05
0.75	26.45	15.02	309.75	232.31	24.97
0.45	15.87	12.65	185.85	83.63	20.24
0.15	05.29	07.64	61.94	09.29	10.12

real-time domain with the aid of computerized monitor and control distributed systems at a master location. The system is called Supervisory Control and Data Acquisition (SCADA), and control may be automatic or initiated by operator commands. The data acquisition is accomplished first by the remote terminal units (RTUs) scanning the field inputs connected to the programmable logic controller (PLC). This is usually done at a fast rate. The central host will scan the RTUs usually at a slower rate. The data is processed to detect alarm conditions, and if an alarm is present, it will be displayed on special

Table 7 Noise calculation chart estimating Onci Sol, Onci Sip, and Onci Bel.

a	b	Intensity ratio (11^a)	Pressure ratio (11^b)	←oSol→ ←oSip→ ←oBel→	Pressure ratio (1/11)^b	Intensity ratio (1/11)^a
0	0	1	1	0	1	1
1/11	1/22	1.244	1.115	±01	0.897	0.804
2/11	2/22	1.546	1.244	±02	0.804	0.647
4/11	4/22	2.392	1.546	±04	0.647	0.418
6/11	6/22	3.699	1.923	±06	0.520	0.270
8/11	8/22	5.720	2.392	±08	0.418	0.175
10/11	10/22	8.845	2.974	±10	0.336	0.113
12/11	12/22	13.679	3.699	±12	0.270	0.073
14/11	14/22	21.155	4.599	±14	0.217	0.047
16/11	16/22	32.715	5.720	±16	0.175	0.031
18/11	18/22	50.594	7.113	±18	0.141	0.020
20/11	20/22	78.242	8.845	±20	0.113	0.013
22/11	22/22	121.000	11.000	±22	0.091	8.264×10^{-3}
24/11	24/22	187.124	13.679	±24	0.073	5.344×10^{-3}
26/11	26/22	289.383	17.011	±26	0.059	3.456×10^{-3}
28/11	28/22	447.525	21.155	±28	0.047	2.235×10^{-3}
30/11	30/22	692.089	26.308	±30	0.038	1.445×10^{-3}
32/11	32/22	1070	32.715	±32	0.031	9.343×10^{-4}
34/11	34/22	1655	40.684	±34	0.025	6.042×10^{-4}
36/11	36/22	2560	50.594	±36	0.020	3.907×10^{-4}
38/11	38/22	3959	62.917	±38	0.016	2.526×10^{-4}
40/11	40/22	6122	78.242	±40	0.013	1.633×10^{-4}
42/11	42/22	9467	97.300	±42	0.010	1.056×10^{-4}
44/11	44/22	14640	121.0	±44	8.264×10^{-3}	6.830×10^{-5}
46/11	46/22	22640	150.47	±46	6.646×10^{-3}	4.417×10^{-5}
48/11	48/22	35020	187.12	±48	5.344×10^{-3}	2.856×10^{-5}
50/11	50/22	54150	232.70	±50	4.297×10^{-3}	1.847×10^{-5}
66/11	66/22	1.772×10^6	1331	±66	7.513×10^{-4}	5.645×10^{-7}
77/11	77/22	1.949×10^7	4414	±77	2.265×10^{-4}	5.132×10^{-8}
88/11	88/22	2.144×10^8	14640	±88	6.830×10^{-5}	4.665×10^{-9}
99/11	99/22	2.358×10^9	48560	±99	2.059×10^{-5}	4.241×10^{-10}
110/11	110/22	2.594×10^{10}	161100	±110	6.209×10^{-6}	3.855×10^{-11}

Example: To find oSol corresponding to a pressure ratio of 363
Ratio of 363 = 11 × 33
In oSol = + 22 + 32 oSol
          = + 54 oSol

alarm lists. Data can be of three main types. Analogue data (i.e., real numbers) will be trended on data analytics software (i.e., placed in graphs). Digital data (on/off) may have alarms attached to one state or the other. Pulse data (e.g., counting revolutions of a meter or counter) is normally accumulated or counted.

A typical SCADA system includes remote sensors, controllers, or alarms located at various facilities and places, as well as a central processing system situated in an appropriate location. SCADA systems integrate data acquisition systems with data transmission systems and graphical software to provide a centrally located monitor and control system for numerous process inputs and outputs. SCADA systems are designed to collect information, transfer it back to a central computer, and display the information to the operators, thereby allowing the operator to monitor and control the entire noise system parameters from a central location in real time.

2.2.1 Components of SCADA system

An SCADA system is composed of the following:
- Central Monitoring Station;
- Remote Terminal Units (RTUs);
- Field Instrumentation;
- Communications Network

The Central Monitoring Station (CMS) refers to the location of the master or host computer. Several workstations may be configured on the CMS. It uses a Man–Machine Interface (MMI) program to monitor various types of data needed for the operation. The Remote Station is installed at the remote points in the facilities being monitored and are controlled by the central host computer. This can be an RTU or a PLC. Field instrumentation refers to the sensors and actuators that are directly interfaced to the remote locations in the various facilities. They generate the analog and digital signals that will be monitored by the Remote Station. Signals are also conditioned to make sure they are compatible with the inputs/outputs of the RTU or PLC at the Remote Station. The Communications Network is the medium for transferring information from one location to another. This can be via telephone line, radio, or cable.

2.3 Checking human noise behavior

Noise behavior is checked by identifying a source and a sink of noise, i.e., a person making noise in the environment and a person affected by such noise in the environment. Behavior of human beings is controlled by the central nervous system. Neurology is the study of the nervous system. The behavior of human beings is studied in Psychology. Psychophysics is a study in which physical stimuli are perceived by human beings. Psychophysiology is the field of study of the interaction between environmental stimuli and physiological functions of the body. The study of abnormal behavior is dealt with by a psychiatrist. Abnormal human noise behavior interferes in proper functioning and wellness of the individual and society. With proper psychophysiological measurements, it is possible to detect the source and sink of noise, i.e., a person making noise in the environment and a person affected by such noise in the environment.

Biomedical instrumentation for measurement and monitoring of human noise behavior from human systems of indoor environment require real-time informatics capabilities. Sensing and actuating capabilities as well as measurement systems for noise can be "in vivo" and "in vitro". The signals can be classified as bioelectric, biosound sample, biomechanical, biochemical, biomagnetic, bio-optic, and bioimpedance depending upon origin of "stresses and oscillations" in a noise system.

Use of Biological Sensors: All living organisms contain biological sensors with functions similar to those of the electro- and mechanical sensors. Most of these are specialized cells that are sensitive to light, motion, temperature, magnetic fields, gravity, humidity, vibration, pressure, electrical fields, sound, and other physical aspects of the external environment, and also physical aspects of the internal environment such as stretch motion of the organism and position of appendages (proprioception) an enormous array of environmental molecules including toxins, nutrients, and pheromones, and many aspects of the internal metabolic milieu, such as glucose level, oxygen level, or osmolality, an equally varied range of internal signal molecules such as hormones, neurotransmitters, and cytokines, and even the differences between proteins of the organism itself and the environment or alien creatures. The human senses are examples of specialized neuronal sensors.

2.4 Indoor air quality and noise

With the increased interest in indoor air quality and the need to monitor potentially dangerous gases, gas concentration measurements have become increasing more prevalent in building DDC system design. Many gas measuring devices are currently available for use in HVAC applications.

There are many types of gas-measuring devices available for use with DDC systems. Currently, the three most common gases measured in HVAC applications are carbon monoxide, carbon dioxide, and refrigerant gases. These gas-measuring devices can be modified for measuring noise of scattering (oncisip) based on concentration and fluid power of these gases. Some examples illustrating noise of scattering (oncisip) are mentioned in previous sections.

Carbon monoxide: Carbon monoxide is a poisonous gas that is most commonly generated as the by-product of the incomplete combustion of carbon-based fuels. Carbon monoxide is generated by all fuel-burning equipment including internal combustion engines. Carbon monoxide detectors are used to operate ventilation equipment to prevent carbon monoxide levels from becoming unsafe. They are also used to warn facility owners and occupants of unsafe levels in garages, loading docks, tunnels, and other areas where vehicles are operated. Solid-state sensing technology is most commonly used. Simple or multiple sensing point versions are available that can provide contact closures at one or more set levels and/or analog signals that are proportional to carbon monoxide concentration.

Carbon dioxide: Carbon dioxide is a nontoxic gas produced by the respiration of living organisms, by the complete combustion of carbon, and by photosynthesis in green plants. Carbon dioxide exists in the air in the amount of 320–350 ppm. Carbon dioxide concentration inside buildings has been related to general ventilation adequacy and is commonly monitored by DDC control systems as a measure of indoor air quality and ventilation adequacy. It is also measured by building DDC systems and used to control outdoor air fans and dampers to keep concentration below set levels. The most commonly used sensing technology is Nondispersive Infrared (NDIR). This is based on the principle that carbon dioxide gas absorbs infrared radiation at the 4.2 μm wavelength. Attenuation of an infrared source can be related to the gas concentration in the air in the range of 0–5000 ppm with a general accuracy of ±150 or 50 ppm over narrower ranges.

Refrigerant gas: Refrigerant gas detectors have been in widespread use since safety codes for mechanical refrigeration required their use in the operation of emergency ventilation systems to evacuate hazardous concentrations of refrigerant gas in machinery rooms and other applicable enclosed areas. Detectors broadly sensitive to families of CFC and HCFC gases are commonly used, as refrigerants are available. Gas-specific detectors are also available to detect individual refrigerant gases including CFC, HFC, HCFC, and ammonia specific to the equipment in use. The most commonly used are infrared (IR), photoacoustic, and solid-state sensing technologies. Single or multiple sensing point versions are available that can provide contact closures at one or more set levels and/or analog signals that are proportional to refrigerant concentration.

2.5 Noise-sensing system

Noise sensors convert an input phenomenon to an output in a different form. This transformation relies upon a manufactured noise-sensing system with limitations and imperfections. As a result, noise sensor limitations are characterized with utility of a noise slide rule by the following points:

(a) *Accuracy:* This is the maximum difference between the indicated and actual reading on a noise scale. For example, if a sensor reads a noise of 100 oncisol with a ±1% accuracy, then the noise could be anywhere from 99 oncisol to 101 oncisol.

(b) *Resolution:* It's used for a noise-sensing system that steps through readings. This is the smallest increment that the noise sensor can detect, and this may also be incorporated into the accuracy value. For example, if a noise sensor measures up to 11 oncibel of noise, and it outputs a number between 0 and 110, then the resolution of the noise-measuring device is 0.1 oncibel.

(c) *Repeatability:* When a single noise sensor condition is made and repeated, there will be small variation for that particular reading. If statistical range is taken for repeated readings (e.g., ±3 standard deviations), this will be the repeatability. For example, if a flow rate sensor measures oncisip with a repeatability of 0.5 oncisip, readings with actual flow noise of 100 oncisip should rarely be outside 99.5 oncisip to 100.5 oncisip.

(d) *Linearity:* In a linear noise sensor, the input phenomenon has a linear relationship with the output signal. In most noise types sensors, this is a desirable feature. When the relationship is not linear, the conversion from the sensor output (e.g., voltage) to a calculated quantity (e.g., intensity) becomes more complex.

(e) *Precision:* Precision of a noise-measuring device considers accuracy, resolution, and repeatability relative to another noise-measuring device.

(f) *Range:* There are natural limits for the noise-measuring sensor. For example, a sensor based on noise-measuring slide rule system can measure noise range from 0 to 110 oncisol, oncisip, and oncibel units.

(g) *Dynamic response:* Dynamic response gives the frequency range for regular operation of the noise-measuring sensor. Typically, noise-measuring sensors will have an upper operation frequency, and occasionally there will be lower frequency limits. For example, our ears hear best between 10 Hz and 16 kHz.

(h) *Environmental:* All noise-measuring sensors have some limitations over factors such as temperature, humidity, dirt, oil, corrosives, and pressures. For example, many sensors will work in relative humidity (RH) from 10% to 80%.

(i) *Calibration:* When manufactured or installed, many noise types sensors will need some calibration to determine or set the relationship between the input phenomena and output. For example, a temperature sensor for measurement of noise due to heat may need to be zeroed or adjusted so that the measured temperature matches the actual temperature. This may require a special device and needs to be performed frequently.

The term "noise-measurement sensing system" includes the noise-sensing element, variable conversion and manipulation elements, data transmission, and data presentation. These are realized with suitable sensors, signal conditioners, and recording equipment. The performance characteristics of a noise-measurement sensing system are judged by how faithfully the system measures the desired input and how thoroughly it rejects the undesirable inputs. Quantitatively, it relates to the degree of approach to perfection. The noise-measurement sensing system operation is defined in terms of static and dynamic characteristics. The former represents the nonlinear and statistical effects and the latter generally represents the dynamic behavior of the system. The static characteristic is represented by precision and accuracy. The accuracy may be influenced by the sensitivity, working range, nonlinearity, hysteresis, and other properties of the sensor. The dynamics of a real system is reflected by the time constant, damping coefficient, and natural frequency. The response of cascade elements processing individual dynamic response can be ascertained from the characteristic of the dominant elements or combined thereof.

2.6 Classification of noise-measurement errors

A good noise-measurement sensor applies to the following rules: (1) the sensor should be sensitive to be measured property; (2) the sensor should be insensitive to any other property; and (3) the sensor should not influence the measured property. In an ideal situation, the output signal of a sensor is exactly proportional to the value of the measured property. The gain is then defined as the ratio between output signal and measured property. For example, if a sensor measures temperature and has a voltage output, the gain is a constant with the unit ($V\,K^{-1}$).

If the noise measurement sensor is not ideal, several types of deviations can be observed:

- **(j)** The gain may in practice differ from the value specified. This is called a gain error. Because the range of the output signal is always limited, the output signal will eventually clip when the measured property exceeds the limits. The full-scale range defines the outmost values of the measured property where the sensor errors are within the specified range.
- **(k)** If the output signal is not zero when the measured property is zero, the sensor has an offset or bias. This is defined as the output of the sensor at zero output.
- **(l)** If the gain is not constant, this is called nonlinearity. Usually this is defined by the amount the output differs from ideal behavior over the full range of the sensor, often noted as a percentage of the full range.
- **(m)** If the deviation is caused by a rapid change of the measured property over time, there is a dynamic error. Often, this behavior is described with a bode plot showing gain error and phase shift as a function of the frequency of a periodic input signal.
- **(n)** If the output signal slowly changes independent of the measured property, this is defined as drift.
- **(o)** Long-term drift usually indicates a slow degradation of sensor properties over a long period of time.
- **(p)** Sensor or instrumentation noise is a random deviation of the signal that varies in time.
- **(q)** Hysteresis is an error caused by the fact that the sensor does not instantly follow the change of the property of physical agents being measured and therefore involves the history of the measured property.
- **(r)** If the noise sensor has a digital output, the signal is discrete and is essentially an approximation of the measured property. The approximation error is also called digitization error.
- **(s)** If the signal is monitored digitally, limitation of the sampling frequency also causes a dynamic error.
- **(t)** The sensor may, to some extent, be sensitive for other properties than the property being measured. For example, most sensors are influenced by the temperature of their environment.

(u)   All these deviations can be classified as systematic errors or random errors. Systematic errors can be sometimes be compensated for by means of some kind of collaboration strategy. Sensor noise is a random error that can be reduced by signal processing, such as filtering, usually at the expense of the dynamic behavior of the sensor.

3. Conclusion

The chapter has presented an overview of nanosensors for their applications in sensing noise of physical agents. To monitor noise and human noise behavior, these noise characterization techniques are elaborated. For example, monitoring of an improved transportation system or built environment for sensing integrated noise parameters would not only result in energy conservation and economical affordability but also result in generating less noise pollution of the indoor pollutants mentioned earlier. With integrated environmental control, these environmental parameters can be characterized and checked for comfort and wellness, and controlled through various environmental monitoring sensors. The effect of these human behavioral parameters is characterized on a logarithmic noise scale. A theory on noise-measurement equations due to noise of physical agents is devised. A slide rule for noise measurement is also presented. Noise calculations charts are presented for calculating noises of oncisol, oncisip, and oncibel based on pressure and intensity ratios. A noise-measurement sensing system along with its brief description is also elaborated.

References

[1] H. Dehra, A unified theory for stresses and oscillations, in: Proceedings from CAA Conf. Montréal. 2007, Canada, Canadian Acoustics. vol. 35. No. 3, September 2007, pp. 132–133.

[2] H. Dehra, Power transfer and inductance in a star connected 3-phase RC circuit amplifier, in: Proc. AIChE 2008 Spring Meeting, New Orleans, LA, USA, April 6-10, session 96a, 2008.

[3] H. Dehra, A novel theory of psychoacoustics on noise sources, noise measurements and noise filters, in: INCE Proc. NoiseCon16 Conf., Providence, Rhode Island, USA, 13-15 June, 2016, pp. 933–942.

[4] H. Dehra, The noise scales and their units, in: Proc. CAA Conf., Vancouver 2008, Canada, Canadian Acoustics, vol. 36 (3), 2008, pp. 78–79.

[5] H. Dehra, A multi-parametric PV solar wall device, in: Proceedings from IEEE International Conference on Power, Control, Signals and Instrumentation Engineering (ICPCSI-2017), Chennai, India on Sep 21-22, 2017, pp. 392–401.

[6] H. Dehra, Characterization of noise in power systems, in: Proceedings from IEEE International Conference on Power Energy, Environment & Intelligent Control (PEEIC2018), Greater Noida, India on April 13-14, 2018, pp. 320–329.

[7] H. Dehra, A paradigm of noise interference in a wave, in: Internoise-2018, 47th International Congress and Exposition on Noise Control Engineering, Chicago, Illinois, USA on Aug 26-29, 2018, pp. 451–462.

[8] H. Dehra, A paradigm for characterization and checking of a human noise behavior, Int. J. Psychol. Behav. Sci. 11 (5) (2017) 317–325 WASET (scholar.waset.org/1307-6892/10007615) doi.org/10.5281/zenodo.1131589 (http://www.waset.org/publications/10007615).

[9] H. Dehra, Acoustic filters for sensors and transducers. in: ICAE2018, Energy Procedia, vol. 158, February, Elsevier, 2019, pp. 4023–4030, https://doi.org/10.1016/j.egypro.2019.01.837.

[10] H. Dehra, A benchmark solution for interference of noise waves, in: Proc. AIChE 2009 Spring Meeting, Tampa, FL, USA, session 67c, 2009.

[11] H. Dehra, A theory of acoustics in solar energy, Nat. Res. 4 (1A) (2013) 116–120.

[12] H. Dehra, On sources and measurement units of noise, in: Proc. International Conference on Innovation, Management and Industrial Engineering (IMIE 2016), Kurume, Fukuoka, Japan, 05-07 August, 2016, pp. 219–227 ISSN: 2412-0170.

[13] H. Dehra, A slide rule for noise measurement, in: 10th International Conference on Sustainable Energy Technologies (SET 2011), Istanbul, Turkey, September 4-7, 5 p, 2011.

[14] H. Dehra, A guide for signal processing of sensors and transducers, in: Proc. AIChE 2009 Spring Meeting, Tampa, FL, USA, session 6b, 2009.

[15] H. Dehra, Solar energy absorbers, (Chapter 6) in: R. Manyala (Ed.), Solar Collectors and Panels, Theory and Applications, InTech Publication, 2010, pp. 111–134.

[16] H. Dehra, Acoustic filters, (Chapter 5) in: V.A. Romano, A.S. Duval (Eds.), Ventilation: Types, Standards and Problems, Nova Publishers, 2012, pp. 135–154.

[17] H. Dehra, A heat transmission model for a telephone line, in: Proc. of 21st CANCAM, Department of Mechanical and Industrial Engineering, Ryerson University, Toronto, Ontario, Canada, June 3-7, 2007, pp. 356–357.

[18] H. Dehra, Solar energy conversion and noise characterization in photovoltaic devices with ventilation, invited chapter in book, in: Recent Developments in Photovoltaic Materials and Devices, IntechOpen, 2019, pp. 1–20, https://doi.org/10.5772/intechopen.79706. ISBN 978-953-51-6690-0, edited by Dr. Natarajan Prabaharan, Dr. Marc A. Rosen and Dr. Pietro Elia Campana, Chapter 1. Available from: https://www.intechopen.com/chapter/pdf-download/62929.

[19] H. Dehra, Acoustic Filters for Sensors and Transducers: Energy Policy Instrument for Monitoring and Evaluating Holy Places and Their Habitants, in book 'Energy Policy' edited by Dr. Tolga Taner, InTech Publication, 2019 IntechOpen, DOI:10.5772/intechopen.81949, 23 pages (Available from: https://www.intechopen.com/chapter/pdf-download/64590).

[20] H. Dehra, Noise Calculation Charts and Indoor Environmental Quality for Evaluating Industrial Indoor Environment and Health, in book 'Indoor Environment and Health' edited by Dr. Orhan Korhan, InTech Publication, 2019. IntechOpen, DOI:10.5772/intechopen.84993, ISBN 978-1-78984-374-3, 25 pages, (Available from https://www.intechopen.com/chapter/pdf-download/67373.

[21] H. Dehra, Acoustic signal processing and noise characterization theory via energy conversion in a PV solar wall device with ventilation through a room, Adv. Sci. Technol. Eng. Syst. J. 3 (4) (2018) 130–172.

[22] H. Dehra, Principles of energy conversion and noise characterization in air ventilation ducts exposed to solar radiation, Appl. Energy 242 (2019) 1320–1345.

[23] H. Dehra, Integrated acoustic and thermo-fluid insulation modeling of an airflow window with a photovoltaic solar wall, in: Building Simulation 2019 (Session: Building Acoustics), Rome, Italy, on Sep 2-4, (http://buildingsimulation2019.org/), IBPSA, pp. 2-9, 2020, 2019 http://www.ibpsa.org/proceedings/BS2019/BS2019_210178.pdf.

[24] H. Dehra, Cooling load and noise characterization modeling for photovoltaic driven building integrated thermoelectric cooling devices. in: Proc. XII International Conference on Computational Heat, Mass and Momentum Transfer, 3-6 September 2019, Rome, Italy, E3S Web of Conferences 128, 01019, 2019, https://doi.org/10.1051/e3sconf/201912801019.

[25] H. Dehra, Monarchy of concordia: a globalized society on maintaining peace and harmony in the world by controlling human noise behavior. Int. J. Soc. Sci. IX (1) (2020), https://doi.org/10.20472/SS2020.9.1.001 ISSN 1804-980X (online) The International Institute of Social and Economic Sciences, https://www.iises.net; Publication Detail: https://www.iises.net/international-journal-of-social-sciences/publication-detail-25601 (Download from: https://www.iises.net/international-journal-of-social-sciences/publication-detail-25601?download=1).

[26] H. Dehra, Data for Monarchy of Concordia: A Globalized Society on Maintaining Peace and Harmony in the World by Controlling Human Noise Behavior (Version v1) [Data set]. (2019), https://doi.org/10.5281/zenodo.3515734Zenodo.

CHAPTER 11

Optical fiber-based localized surface plasmon resonance for volatile liquid sensing for different probe geometry

Dimpi Paul[a], Rajib Biswas[b], and Sibasish Dutta[c]
[a]Department of Physics, Golaghat Engineering College, Golaghat, Assam, India
[b]Department of Physics, Tezpur University (Central University), Tezpur, Assam, India
[c]Department of Physics, Pandit Deendayal Upadhyaya Adarsha Mahavidyalaya (PDUAM) Eraligool, Karimganj, Assam, India

1. Introduction

To date, optical fiber probes have been utilized by researchers owing to their design flexibility and compactness. It is well known that there are a large number of sensing probes available, but optical fiber is considered to be one of the handier platforms out of the rest. Starting from 1979 to 1984, a number of researchers reviewed the progress and performance of optical fiber when exposed to an unknown environment [1, 2]. Likewise, a number of scientists reviewed the effect of radiation on an exposed portion of optical fiber as a probe [3–5]. With time, the optical fiber probe was utilized by a number of researchers for varying sensitivity analyses. In 1991, Mullen et al. confirmed the possibility of high-quality surface-enhanced Raman spectra (SERS) with deposition of gold onto the rough surface of a optical fiber [6]. Later, for the first time in 1992, fiber optic surface plasmon resonance sensor was reported by Jorgenson et al. [7]. Eventually, a number of fiber optic SPR sensors have been reported by different groups of researchers in several chemical and biochemical sensing applications utilizing multimode fibers [8–10]. Although an SPR sensor has certain advantages over other sensing techniques, it requires very expensive components to establish the sensing head. To construct an SPR sensor, thin films of noble metals are coated onto the exposed portion of an optical fiber, which is considered an expensive technique compared with other sensing techniques. In the case of fiber optic sensors, expensive metals with a sophisticated environment to control the films is required. To optimize this expensive technique, study of the plasmonic behavior has been improved by introducing noble metal nanoparticles (NPs) coating on a probe instead of using metal films. Harnessing the plasmonic signal of noble metal NPs is considered to be localized surface plasmon resonance (LSPR)-based sensing. LPSR phenomenon basically takes place only when the incident photons undergo resonant oscillation of conduction electrons in metallic NPs [11, 12]. This technique not only reduces the cost of the sensing schemes but also reduces the complexity of the sensing schemes. On a similar note, incorporation of noble NPs onto the exposed portion of an optical fiber can execute sensing within a very small dimension [11]. Nowadays, optical fiber-based LSPR sensing is utilized in several sensing schemes such as remote sensing and in vivo techniques [11]. To optimize the design of a sensor, LSPR-based optical fiber sensor was reported by a number of researchers using broadband source as well as light-emitting diode (LED) sources to ease the complexity, making a handy platform to record the changes with incorporation of noble metal NPs [11, 12]. One of the fine techniques was introduced by Mitsui et al. [12] in which a new LSPR-based fiber optic sensor used a red LED as a source instead of white light, and the changes were recorded using a spectrometer. By obviating the conventional light source, they reported a very important aspect regarding LSPR sensing and thereby using bulk refractive indices of water-glycerol mixtures with variable compositions were demonstrated [12]. Thus, it led to the initiation of studying the optical properties of noble metal NPs onto the exposed portion of optical fibers and, to date, several fiber optic sensors based on LSPR have grabbed the attention of many researchers. By considering the advantages of LSPR phenomenon in combination with conventional optical fibers, ethanol vapor sensing has been studied based on the influence of different probe geometry.

2. Different types of fiber optic LSPR sensors based on variable probe geometry

2.1 Tapered shaped optical fiber sensor

Tapered structured probe was fabricated by removing the plastic jacket and cladding of the fiber at the central part. The evanescent field spread throughout the exposed portion of the probe, which was chemically modified with NPs that led to evanescent wave coupling between the NPs and evanescent wave to detect adjacent changes in the refractive index (RI) [13–15]. This technique was utilized by a number of researchers in different types of chemical, biochemical, as well as physical sensing applications such as temperature sensing, gas sensing, and detection of chemical agents and biomolecules [16–22]. In a tapered probe, the propagation modes are considered along the beam waist, which is exposed at the central part of the fiber [22]. The distinction in refractive indices of core and cladding leads to propagation of light through the core following the principle of total internal reflection (TIR). The light guidance through the core results in exponential decay of evanescent field in the cladding and reaches $1/e$ of its value to a certain distance at the core-cladding interface [22]. The working principle of tapered probe has been mathematically described here by penetration depth "d_p":

$$d_p = \frac{\lambda}{2\pi\sqrt{n_{co}^2 \sin^2\theta - n_{cl}^2}} \qquad (1)$$

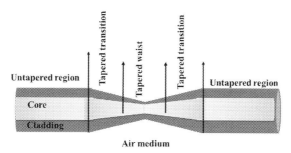

Fig. 1 Variable segments in the case of a tapered optical fiber.

where λ is the wavelength of the source, and θ is the angle of incidence at the core and cladding interface, whereas n_{co} and n_{cl} are the refractive indices of the core and cladding, respectively.

Optical fiber provides low loss during communication due to smaller penetration depth d_p, which is much smaller than the thickness of the cladding. When light propagates, there is no interaction between the optical field and the adjacent medium. However, the exponentially decaying evanescent field of decladded optical fiber exposed to the adjacent medium can be utilized for certain sensing applications [22]. In the case of tapered optical fiber, the decladded portion leads to interaction between the evanescent fields with the surrounding media [22]. The tapered optical fiber probe is composed of a tapered waist along with untapered transition regions as depicted in Fig. 1. Besides, interference depends upon several parameters of the tapered waist, i.e., the tapered diameter, length, and the untapered transition region shape, respectively.

2.2 D-typed/side-polished

As mentioned earlier, the sensitivity of a fiber optic sensor has a direct dependence on the geometry, shape, and core radii of the fiber probe [23–25]. Thus, D-type geometry also has a huge effect on altering the sensing performance of a sensing setup. In this case, only half of the central part of a fiber has been removed, i.e., only a half-portion is exposed to interact with the coated NPs, which tend to couple with the evanescent wave. In a quest for sensitivity enhancement of fiber optic sensors, D-type fiber optic sensors were also fabricated. The working principal of D-type optical fiber follows the attenuated total internal reflection (ATIR) as only half of the central part is exposed. Basically, the electromagnetic wave coming out from the cladding to the exposed core was monitored by ATR owing to its half-removed core. The propagation of light through a conventional optical fiber depends on angle of incidence as well as the critical angle of the incident light. While propagating from core to cladding region, the evanescent wave is absorbed to a certain penetration depth (d_p) in these kinds of probes. Fig. 2 illustrates the propagation characteristics of D-type or side-polished optical fiber probe. As the sensing

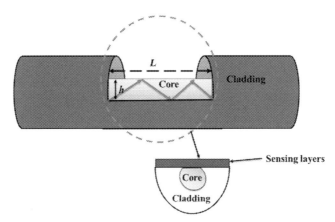

Fig. 2 Electromagnetic field propagation in the case of a D-type optical fiber.

principle is based on ATR, the sensing performance depends upon the number of reflections, say "m" of the sensor. The value of m can be determined as follows [23–26]:

$$m = \left(\frac{L}{2h\tan\theta} \right) \qquad (2)$$

Here, L is the length of the probe with h as the height of the sensing region. Additionally, the sensing mechanism of D-type optical fiber is dependent on power transmission through the probe, which has a direct dependence on the effective reflectance R and the reflectance for a single reflection R'. The effective reflectance in given by [23–26]:

$$R = R^{'m} \qquad (2.1)$$

2.3 U-bent optical fiber sensor

U-bent fiber optic probe has also been widely reported for sensing applications owing to their compact size. Here, the dependence of the bending radius, core diameter, and the refractive index of the analytes were used as a sensing entity. The wide utilization of the U-bent fiber probe can be ascribed to its compactness, high sensitivity, and ease in fabrication during sensing mechanisms [27–34]. The design flexibility in U-bent optical fiber probe has certain merits over other shaped fiber probes. The central portion of this fiber probe is exposed with a bend, thus the working principal is different than that of the other two. The sensing mechanism of U-bent optical fiber is also based on modified attenuated total reflection (ATR) spectroscopy [27]. Besides, the transmitted power of the sensing scheme depends on the absorbing fluid that leads to absorption of the evanescent wave field into cladding to a certain penetration depth (say d_p). While propagation light confined only through the plane of the bending portion, i.e., the sensing region of the fiber with meridional rays, decrease in skewness of the incoming electromagnetic

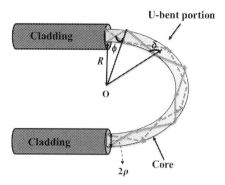

Fig. 3 Electromagnetic field propagation in the case of a U-bent optical fiber.

wave increases the sensitivity of a U-bent optical fiber. The conventional light propagation method through the exposed portion of a fiber is dependent on the critical angle of light rays that strike the core–cladding interface, which should be greater than the critical angle of the sensing region of the probe. On the contrary, in the case of the U-bent optical fiber probe, the propagation characteristics follow two different prospectives (Fig. 3). In the figure, the continuous line shows the path of the light rays undergoing TIR only at the outer surface, whereas the dotted lines show the optical path of the light rays undergoing TIR at both the inner and outer surface of the probe. The guided rays follows Eq. (3) for guiding light from straight to the bending portion of the U-bent fiber and is given by [27–34]:

$$\sin\phi = \left[\frac{R+h}{R+2\rho}\right]\sin\theta \tag{3}$$

Here, θ is the angle of guidance, which is normal to the core and cladding interface in a straight portion of the fiber, and ϕ is the angle at the outer surface of the bent fiber. Here, h represents the distance of the light ray, which is incident onto the exposed bent portion of the probe. The bent portion acts as the sensing region of the sensor with R as bending radius. Accordingly, the angle δ at the inner surface of the bending region is given by [27–34]:

$$\sin\delta = \left[\frac{R+h}{R}\right]\sin\theta \tag{3.1}$$

Thus, optical fiber-based LSPR sensors have been developed by modulating the sensing probe into different shapes such as tapered, U-bent, and D-shaped. Such modifications of the fiber probes have been done to enhance the effective coupling of the evanescent wave in the probe region and thereby increase the sensitivity of the designed sensor.

3. Results and discussion

This piece of work has been introduced to analyze volatile liquid (viz., alcohol) sensitivity using a very low-cost sensing setup with optical fiber probe having three variable geometries. Also, to test the output signal response of the designed sensor, ethanol has been chosen as a volatile liquid sample. Volatile liquid tends to vaporize and thereby change the effective refractive index of the medium adjacent to the probe coated with NPs. Gold NPs adhered to the decladded portion of the proposed probes interacts with the evanescent field of the light propagating along the fiber. The change in the refractive index of the surrounding medium of the NPs leads to a change in output signal of the optical fiber that has been recorded using the photodetector [35]. The schematic of the proposed chamber has been illustrated in Fig. 4. Utilizing three different shapes of decladded optical fibers coated with noble metal NPs (gold NPs), the change in the effective refractive index in the medium has been detected in terms of change in output voltage.

Before initiation, each probe has been cleaned and kept inside a desiccator to avoid attachment of unwanted particles. Thereafter, each probe has been coated with noble gold NPs. At first, the NP-coated tapered optical fiber is exposed to the chamber containing ethanol vapors and allows the evanescent light wave of the broadband source to interact with the NPs. With time, the effective refractive index of the medium changes due to an increase in density of the ethanol vapor inside the chamber, which correspondingly changes the output response. Thus, the change in voltage response has been recorded by a photodetector each time to observe the variation at the output. To ensure change in the presence of the volatile liquid inside the chamber, the blank voltage response has been noted before initiating in each case. The variation in voltage response in the presence

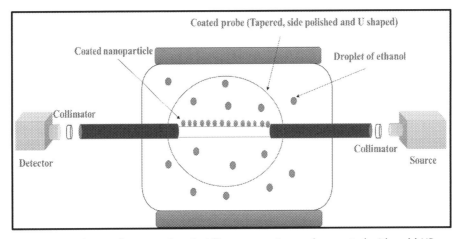

Fig. 4 Experimental setup for tapered optical fibers as sensing probes coated with gold NPs.

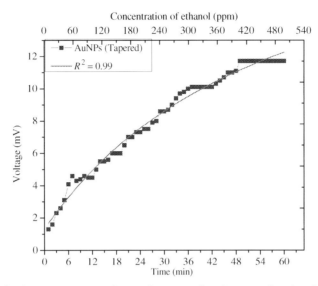

Fig. 5 Variation of voltage response with time for tapered probe coated with gold NPs.

of ethanol confirms the changes in refractive index inside the chamber. For the tapered probe coated with AuNPs, the initial output voltage in the presence of ethanol is found to be ~1.3 mV. As the time increases, the output voltage registered by the photodetector is found to be increasing, which is due to an increase in ethanol vapors and thereby an increase in the refractive index of the surrounding medium of the probe. As the ethanol vapor reaches a saturation value, the output voltage recorded by the photodetector attains a steady value. This change in output voltage with time is found to follow a sigmoid response with R-squared value of 0.992. The sensitivity of this probe design is found to be 0.02328 mV/ppm and having a working domain range of 0–244.84 ppm. The curve followed by the sensing setup with a tapered fiber is illustrated in Fig. 5.

The fitting equation for the sigmoid pattern followed by the tapered probe for the sensing setup is:

$$Y = 22.98 - \frac{21.63}{1 + \left(\dfrac{X}{58.81}\right)} \tag{4}$$

Similarly, D–shaped and U–bent optical fiber probes coated with gold NPs have been exposed to the ethanol vapor medium, and the corresponding change in voltage response of the photodetector has been recorded accordingly.

In the D-shaped probe, only half of the cylindrical central part is exposed to the vapor medium, and the coupling between evanescent waves with NPs occurs only on that

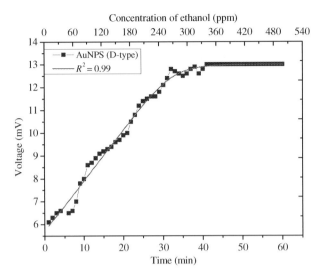

Fig. 6 Variation of voltage response with time for D-type probe coated with gold NPs.

exposed portion. The principal followed by D-shaped probe is ATIR as already mentioned in the introduction section. Unlike the tapered probe, initially a dip in the output voltage response has been observed. This dip is attributed to the coupling between the evanescent wave and the NPs only at the cylindrical half of the probe due to ATIR. Initial response of this probe is found to be ∼6.1 mV. Similar to that of the tapered probe after complete vaporization of ethanol inside the chamber, the output voltage response ceases and reaches a constant value. The variation in voltage response with time for the given probe follows a sigmoid curve as shown in Fig. 6 From the graph, the sensitivity obtained is 0.0204 mV/ppm and having its working domain in the range 0–244.84 ppm.

With an R-squared value of 0.991, the sigmoid fitted regression equation for the given D-type probe is given as:

$$Y = 13.72 - \frac{7.28}{1 + \left(\dfrac{X}{18.47}\right)^{2.3}} \tag{5}$$

Similarly, for the U-bent probe, coupling between the evanescent wave and the NPs with ethanol vapor as a medium produces a progressive change in the output voltage response. This variation of voltage response in the presence of ethanol with respective change in refractive index using U-bent fiber has been found to be following a sigmoid response (Fig. 7).

Upon doing sigmoid fitting, the R-squared value is found to be 0.993 and the corresponding regression equation for the given probe is given as:

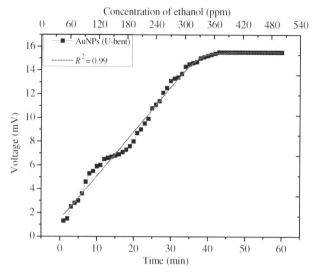

Fig. 7 Variation of voltage response with time for U-bent probe coated with gold NPs.

Table 1 Figure of merits (FOM) for the three different probes.

Sl. no.	Probe geometry	Sensitivity (mV/ppm)	Regression (R^2)	Working domain (ppm)
1	Tapered-type	0.02328	0.992	0–244.84
2	D-type	0.0204	0.991	0–244.84
3	U-type	0.04084	0.993	0–244.84

$$Y = 17.66 - \frac{15.85}{1 + \left(\dfrac{X}{20.65}\right)^{1.9}} \qquad (6)$$

From the graph, the sensitivity of the probe is found to be 0.04084 mV/ppm and having a working domain in the range 0–244.84 ppm.

The results of each proposed optical fiber probes shows variation which confirms the change in response with respective change in geometry (Table 1).

In all the three cases, it has been found that output voltage increases with an increase in the vapor concentration inside the chamber. In each case, the changes have been observed due to the change in plasmonic response of the NPs upon interaction of the evanescent field with the ethanol vapors in the vicinity of the NPs inside the chamber. Geometry does have an effect on the LSPR response that can be clearly understood from

Table 1. It has been found that the U-bent probe shows highest sensitivity value in the same range of working domain compared with the tapered probe and D-shaped probe.

4. Conclusion

In summary, this chapter deals with the plasmonic response of optical fiber sensors based on variable geometry. Also, the use of optical fibers reduces the cost as well as the size of the sensor setup. Based on different geometry of the probe region, namely tapered, D-shaped, and U-bent type, the sensitivity is found to be different. As a whole, the U-bent probe is showing highest sensitivity, which indicates its future implications for developing other types of volatile liquid sensing. Furthermore, the low-cost detection technique demonstrated in this work can open a myriad of applications in the field of developing environment and healthcare sensors.

References

[1] E.J. Friebele, Optical fiber waveguides in radiation environments, Opt. Eng. 18 (6) (1979) 186552.
[2] E.J. Friebele, C.G. Askins, M.E. Gingerich, K.J. Long, Optical fiber waveguides in radiation environments, II, Nucl. Instrum. Methods Phys. Res., Sect. B 1 (2–3) (1984) 355–369.
[3] P.B. Lyons, Review of high bandwidth fiber optics radiation sensors, in: Fiber Optic and Laser Sensors III, vol. 566, International Society for Optics and Photonics, 1986, January, , pp. 166–171.
[4] E.M. Dianov, L.S. Kornienko, E.P. Nikitin, A.O. Rybaltovskiĭ, V.B. Sulimov, P.V. Chernov, Radiation-optical properties of quartz glass fiber-optic waveguides, Sov. J. Quantum Electron. 13 (3) (1983) 274.
[5] W. Schneider, Radiation sensitivity of thick core all-glass fibers in reactor radiation fields, Appl. Phys. A 28 (1) (1982) 45–51.
[6] K.I. Mullen, K.T. Carron, Surface-enhanced Raman spectroscopy with abrasively modified fiber optic probes, Anal. Chem. 63 (19) (1991) 2196–2199.
[7] R.C. Jorgenson, S.S. Yee, A fiber-optic chemical sensor based on surface plasmon resonance, Sensors Actuators B Chem. 12 (3) (1993) 213–220.
[8] H.P. Uranus, H.J.W.M. Hoekstra, Modelling of microstructured waveguides using a finite-element-based vectorial mode solver with transparent boundary conditions, Opt. Express 12 (12) (2004) 2795–2809.
[9] J.C. Knight, J. Arriaga, T.A. Birks, A. Ortigosa-Blanch, W.J. Wadsworth, P.S.J. Russell, Anomalous dispersion in photonic crystal fiber, IEEE Photon. Technol. Lett. 12 (7) (2000) 807–809.
[10] R. Slavík, J. Homola, J. Čtyroký, Single-mode optical fiber surface plasmon resonance sensor, Sensors Actuators B Chem. 54 (1–2) (1999) 74–79.
[11] S.F. Cheng, L.K. Chau, Colloidal gold-modified optical fiber for chemical and biochemical sensing, Anal. Chem. 75 (1) (2003) 16–21.
[12] K. Mitsui, Y. Handa, K. Kajikawa, Optical fiber affinity biosensor based on localized surface plasmon resonance, Appl. Phys. Lett. 85 (18) (2004) 4231–4233.
[13] H.Y. Lin, C.H. Huang, G.L. Cheng, N.K. Chen, H.C. Chui, Tapered optical fiber sensor based on localized surface plasmon resonance, Opt. Express 20 (19) (2012) 21693–21701.
[14] A. Leung, P.M. Shankar, R. Mutharasan, A review of fiber-optic biosensors, Sensors Actuators B Chem. 125 (2) (2007) 688–703.
[15] H.S. MacKenzie, F.P. Payne, Evanescent field amplification in a tapered single-mode optical fibre, Electron. Lett. 26 (2) (1990) 130–132.
[16] T.A. Birks, W.J. Wadsworth, P.S.J. Russell, Supercontinuum generation in tapered fibers, Opt. Lett. 25 (19) (2000) 1415–1417.

[17] D.A. Akimov, A.A. Ivanov, A.N. Naumov, O.A. Kolevatova, M.V. Alfimov, T.A. Birks, … A. M. Zheltikov, Generation of a spectrally asymmetric third harmonic with unamplified 30-fs Cr: forsterite laser pulses in a tapered fiber, Appl. Phys. B 76 (5) (2003) 515–519.

[18] S. Lacroix, F. Gonthier, J. Bures, All-fiber wavelength filter from successive biconical tapers, Opt. Lett. 11 (10) (1986) 671–673.

[19] J. Zhang, P. Shum, X.P. Cheng, N.Q. Ngo, S.Y. Li, Analysis of linearly tapered fiber Bragg grating for dispersion slope compensation, IEEE Photon. Technol. Lett. 15 (10) (2003) 1389–1391.

[20] Z.Z. Feng, Y.H. Hsieh, N.K. Chen, Successive asymmetric abrupt tapers for tunable narrowband fiber comb filters, IEEE Photon. Technol. Lett. 23 (7) (2011) 438–440.

[21] J. Villatoro, D. Monzón-Hernández, D. Luna-Moreno, In-line optical fiber sensors based on cladded multimode tapered fibers, Appl. Opt. 43 (32) (2004) 5933–5938.

[22] Y. Tian, W. Wang, N. Wu, X. Zou, X. Wang, Tapered optical fiber sensor for label-free detection of biomolecules, Sensors 11 (4) (2011) 3780–3790.

[23] M.H. Chiu, S.F. Wang, R.S. Chang, D-type fiber biosensor based on surface-plasmon resonance technology and heterodyne interferometry, Opt. Lett. 30 (3) (2005) 233–235.

[24] M.F. Ubeid, M.M. Shabat, Numerical investigation of a D-shape optical fiber sensor containing graphene, Appl. Phys. A 118 (3) (2015) 1113–1118.

[25] P.C. Chiu, M.H. Chiu, M.C. Wang, Y.H. Kuo, C.G. Wang, M.T. Chen, C.W. Yang, Portable D-type optical fiber sensor based on ATR effect in temperature detection, in: International Conference on Smart Materials and Nanotechnology in Engineering (Vol. 6423, p. 64232R). International Society for Optics and Photonics, 2007, November.

[26] S.F. Wang, M.H. Chiu, R.S. Chang, New idea for a D-type optical fiber sensor based on Kretschmann's configuration, Opt. Eng. 44 (3) (2005) 030502.

[27] B.D. Gupta, H. Dodeja, A.K. Tomar, Fibre-optic evanescent field absorption sensor based on a U-shaped probe, Opt. Quant. Electron. 28 (11) (1996) 1629–1639.

[28] V.V.R. Sai, T. Kundu, S. Mukherji, Novel U-bent fiber optic probe for localized surface plasmon resonance based biosensor, Biosens. Bioelectron. 24 (9) (2009) 2804–2809.

[29] B.D. Gupta, N.K. Sharma, Fabrication and characterization of U-shaped fiber-optic pH probes, Sensors Actuators B Chem. 82 (1) (2002) 89–93.

[30] P.K. Choudhury, T. Yoshino, On the pH response of fiber optic evanescent field absorption sensor having a U-shaped probe: an experimental analysis, Optik 114 (1) (2003) 13–18.

[31] S.K. Khijwania, B.D. Gupta, Maximum achievable sensitivity of the fiber optic evanescent field absorption sensor based on the U-shaped probe, Opt. Commun. 175 (1–3) (2000) 135–137.

[32] S.K. Srivastava, V. Arora, S. Sapra, B.D. Gupta, Localized surface plasmon resonance-based fiber optic U-shaped biosensor for the detection of blood glucose, Plasmonics 7 (2) (2012) 261–268.

[33] R. Bharadwaj, S. Mukherji, Gold nanoparticle coated U-bend fibre optic probe for localized surface plasmon resonance based detection of explosive vapours, Sensors Actuators B Chem. 192 (2014) 804–811.

[34] D. Paul, S. Dutta, R. Biswas, LSPR enhanced gasoline sensing with a U-bent optical fiber, J. Phys. D. Appl. Phys. 49 (30) (2016) 305104.

[35] S.K. Khijwania, B.D. Gupta, Fiber optic evanescent field absorption sensor with high sensitivity and linear dynamic range, Opt. Commun. 152 (1998) 4259–6262.

CHAPTER 12

Colorimetric and fluorescent nanosensors for the detection of gaseous signaling molecule hydrogen sulfide (H$_2$S)

Rahul Kaushik, Amrita Ghosh, and D. Amilan Jose
Department of Chemistry, National Institute of Technology (NIT) Kurukshetra, Kurukshetra, Haryana, India

1. Introduction

Gaseous signaling molecule hydrogen sulfide (H$_2$S) is endogenously produced and plays a key role in many biological processes [1]. The significance of these signaling molecules may vary depending upon the different expression level of availability in biological systems. Reports suggest that the level of H$_2$S in the blood varies in the range of 10–100 μM [2]. Various biological processes are mediated by H$_2$S such as vasodilation, neuromodulation, and anti-inflammation. Changes in H$_2$S level or concentration in the human biological system have been correlated with hypertension, Down syndrome, diabetes, and many other health problems. Owing to the significance in the physiological system, research related to the recognition and release of H$_2$S is progressing very quickly [3].

It is important to have efficient analytical methods and tools that estimate, quantify, and monitor the H$_2$S activity in various biological systems. Chemical sensors based on small molecules, polymers, nanoparticles, and smart materials have been reported to detect a trace amount of H$_2$S in biological and environmental systems [3–7]. Methods based on colorimetric and fluorescent systems are advantageous compared with other analytical methods such as electrochemical, chromatography, NMR, and other sophisticated analytical instruments. Colorimetric and fluorescent methods are advantageous because, using these methods, real-time analysis could be performed even by a nonexpert and it may not need any special sample preparation [6].

Nanoprobes or nanomaterials with controlled sizes are advantageous for the detection of various analytes compared to small molecule-based probes due to their excellent optical behaviors, promising immobilization matrices, and transduction platforms [8]. Nanoprobes have also addressed problems such as sensitivity, selectivity, low solubility,

Handbook of Nanomaterials for Sensing Applications
https://doi.org/10.1016/B978-0-12-820783-3.00008-7

photostability, and biocompatibility that most of the small molecular probes suffer from, expanding their application in physiological fluids. Furthermore, the optical properties of nanosensors can be easily modulated by surface modification, ligand substitution, varying size, and morphology. Considering the advantages of nanoprobes, recently several colorimetric and fluorescent nanoprobes based on inorganic nanoparticles such as silver nanoparticles (AgNPs), gold nanoparticles (AuNPs), upconversion nanoparticles (UCNPs), silicon-based nanomaterials, and other metal nanoparticles were used as H_2S sensors. In an effort to highlight the advantages of nanotechnology, in this chapter, we discuss the recent achievements for the development of colorimetric and fluorescent nanoprobes for the selective detection of H_2S.

2. Colorimetric sensing of hydrogen Sulfide (H_2S)

The signaling processes by H_2S are mediated most likely by $H\bar{S}$, but the role of H_2S cannot be eliminated. These active species are controlled by a prototrophic equilibrium defined by pK_1 of 6.9 and pK_2 of 19 of H_2S at a physiological pH of 7.4. For the colorimetric sensing of H_2S, several nanoprobes have been developed. The commonly used nanoprobes are metal nanoparticles composed of metals such as silver (Ag), gold (Au), manganese (Mn), mercury (Hg), and copper (Cu). These metal nanoparticles are mainly involved in metal-etching mechanisms induced by sulfide anions. The chemical reactions involved in the detection of H_2S at the surface of different metal nanoparticles are shown in Scheme 1.

Ag-NPs based nanoprobes:

$$4\,Ag + 2\,H_2S + O_2 \longrightarrow 2\,Ag_2S + 2\,H_2O$$
$$4\,Ag + 2\,HS^- + O_2 \longrightarrow 2\,Ag_2S + 2\,HO^-$$
(A) $$4\,Ag + 2\,S^{2-} + O_2 + 2\,H_2O \longrightarrow 2\,Ag_2S + 4\,HO^-$$

Mn-NPs based nanoprobes:

(B) $$MnO_2 + 2\,H_2S \longrightarrow MnSO_4 + S + 2\,H_2O$$

Cu-NPs based nanoprobes:

$$2\,Cu + 2\,H_2S + O_2 \longrightarrow 2\,CuS + 2\,H_2O$$
(C) $$4\,Cu + 2\,HS^- + O_2 \longrightarrow 2\,Cu_2S + 2\,HO^-$$

Au-HgNPs based nanoprobes:

$$Au\text{-}Hg + S^{2-} + 1/2\,O_2 + 2\,H^+ \longrightarrow Au\text{-}HgS + H_2O$$
$$Au\text{-}Hg + HS^- + 1/2\,O_2 + H^+ \longrightarrow Au\text{-}HgS + H_2O$$
(D) $$Au\text{-}Hg + H_2S + 1/2\,O_2 \longrightarrow Au\text{-}HgS + H_2O$$

Scheme 1 General reactions for the interaction of H_2S with metal nanoparticles composed of Ag, Au, Mn, Hg, and Cu.

As shown in Scheme 1, formation of stable metal sulfides such as Ag_2S, CuS, and HgS is responsible for the detection of H_2S.

In the following discussion, colorimetric nanoprobes are discussed in three subsections, such as silver-based nanoprobes, gold-based nanoprobes, and other metal (mercury, manganese, and copper)-based nanoprobes for the sensitive detection of H_2S through color change. Their structural compositions, mechanism, spectral or optical properties, working conditions, limit of detection (LOD), sensitivity, and applications in various fields are discussed.

2.1 Silver-based nanoprobes

AgNPs are easily synthesized via designing different ligands as stabilizing agents. Using surfactant-free synthetic route, Wang and co-workers [9] synthesized positively charged silver nanoparticles (PPF-AgNPs) caged with cross-linked polyhedral oligomeric silsesquioxane under microwave irradiation. Polyhedral oligomeric silsesquioxane is an organic–inorganic hybrid with cubic silica cage and eight functional groups. In particular, octa-amino polyhedral oligomeric silsesquioxane exhibits good water solubility with abundant reactive sites, which could enhance the sensing properties of the probes. PPF-AgNPs displayed outstanding selectivity and colorimetric response to H_2S among other thiols and anions. PPF-AgNPs exhibited absorbance band at 400 nm and showed decreased absorbance in the presence of H_2S. The response time was less than 3 min, and the LOD was calculated as 0.2 μM. H_2S detection in real samples such as human serum, calf serum, egg white, urine, whey, and milk samples showed the practical applicability of the PPF-AgNPs.

Ilanchelian and co-workers [10] have synthesized a yellow colored chitosan-capped silver nanoparticle (chit-AgNPs) that exhibits strong surface plasmon band at 404 nm. Only in the presence of sulfide to chit-AgNPs solution, the SPR band gradually decreased and finally disappeared with the color change from yellow to colorless. The sulfide anions react with metallic Ag and generate Ag_2S at RT, which is also confirmed by HR-TEM and Zeta potential studies. The LOD was calculated as 0.35 μM, and real-time applications were confirmed in tap water, drinking water, and sea water.

Singh and co-workers [11] reported silver nanorods arrays that displayed silver-white to blackish color change in the presence of H_2S. The change in color intensity was recorded in the presence of 1% of H_2S exposure for 5 min, and the image was captured with mobile cameras and processed using MATLAB software for further analysis. The AgNRs were employed on aged wool fabric textile, and significant color change was recorded within 3 days. Therefore, colorimetric response using AgNRs proved to be better compared with the conventional method for material testing such as Oddy test, which requires 28 days for measurements and could be useful for art conservation applications.

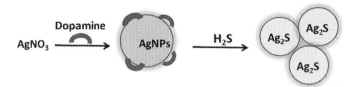

Fig. 1 Silver nanoparticles stabilized with dopamine for colorimetric H₂S sensor.

Silver nanoparticles stabilized with dopamine (Fig. 1) for the colorimetric detection of sulfide was reported by Zhao et al. [12]. The plasmon absorbance band at 400 nm decreased with a bright yellow to dark brown naked-eye detectable color change. This could be due to the decomplexation of dopamine from AgNPs and formation of Ag_2S. The LOD was calculated to be 0.03 μM, but other bio-thiols also showed change in absorbance spectrum. The change with other bio-thiols could be due to interaction of –SH group with AgNPs. These nanoparticles were also employed for the detection of DTT due to presence of two proximal sulfhydryl groups that can cross-link monodispersed AgNPs to form aggregated AgNPs. The application of Ag/dopamine nanoparticles were investigated in the fetal bovine serum samples, and average S^{2-} concentration in serum was found to be 16.3 μM. The spiked-recovery assays confirmed the applications of the nanoparticles in real biological samples.

Veerappan and co-workers [13] prepared ciprofloxacin (CIP)-capped silver nanoparticles (CIP-AgNPs) for selective colorimetric detection of H_2S. Ciprofloxacin and silver are known for their antimicrobial behavior. Therefore, the CIP-AgNPs offered more resistance to microbial contamination even after storing for a long time. The UV-vis absorbance maximum of 410 nm was due to typical surface plasmon resonance (SPR) with a yellow color. Only in the presence of H_2S, the absorbance peak at 410 nm decreases and color change from yellow to colorless was observed. The LOD of H_2S was reported as 0.112 nM by UV-vis change and 16 nM through naked eyes. The silver nanoparticles were suitable for versatile pH range and for assaying sulfide in environmental water samples. Furthermore, silver nanoparticle-coated cotton swabs were used for colorimetric detection of H_2S in gaseous form as well as aqueous solutions for easy on-spot sensing applications (Fig. 2).

Silver nanoparticles are known to oxidize organic dyes such as 3,3′,5,5′-tetramethylbenzidine (TMB), which consequently induced blue color with absorbance band at 652 nm. However, this catalytic oxidation could be enhanced in the presence of Ag_2S. This chemical phenomenon was utilized by Li and co-workers [14] for the colorimetric sensing of sulfide anions (Fig. 3).

In the presence of S^{2-}, the Ag will oxidize to Ag_2S and thereby enhance the catalytic oxidation of the TMB dye, inducing a blue color. The enhancement in absorbance peak was observed at 652 nm at pH 4 with incubation time of 20 min. The LOD was found to

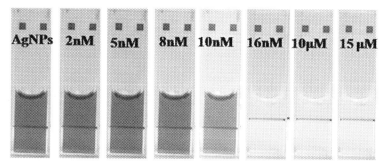

Fig. 2 Colorimetric sensing AgNPs in the presence of different concentrations of H$_2$S. *Reproduced with a permission from reference Sensors and Actuators B: Chemical, 2017, 244, 831–836.*

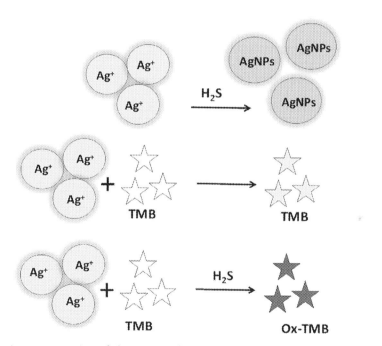

Fig. 3 Schematic representation of the proposed colorimetric sensing of sulfide ions using silver nanoparticle and TMP.

be 0.2 nM using absorbance spectral change. The colorimetric change could be detectable at 8 nM of S^{2-}, which is better than the previously discussed Ag nanoparticles-based probes. The application of silver nanoparticles was explored to detect sulfide anions in environmental water samples.

Gellen gum-silver nanoparticle-based bionanocomposite was prepared and utilized for colorimetric H$_2$S detection by Zou and co-workers [15]. Gellen gum could act as both a reducing agent as well as capping agent for the AgNPs synthesis. The prepared GG-AgNPs solution showed absorbance peak at 420 nm and displayed a yellow color. For colorimetric detection, H_2S gas was produced by treating sodium sulfide with HCl solution and allowed to react with gellen gum silver nanoparticles. The yellow color of the solution disappeared and peak at 420 nm decreased with the addition of H_2S at pH 7. The LOD was calculated as 0.81 μM and application in monitoring meat spoilage was explored by preparing a sensory hydrogel of the GG-AgNPs. The sensor presented visible color changes from yellow to colorless in situ and nondestructively detected the H_2S generated from chicken breast and silver carp in a packaging system. It exhibited excellent selectivity toward H_2S against other volatile components generated from chicken breast and silver carp during spoilage. The low toxicity and economically cheaper hydrogel showed potential application in intelligent food packaging.

2.2 Gold-based nanoprobes

Gold nanoparticles (AuNPs) are ideal for developing sensitive and selective sensing systems for the detection of various environmental and biological important analytes [16]. AuNPs strong local SPR and color-tunable behaviors have great advantages for the development of sensors. The color change during AuNPs aggregation or redispersion provides a general platform for the colorimetric detection of analytes. AuNPs were also used for the development of nanoprobes for H_2S detection.

Yang and co-workers [17] have developed a Cu@AuNP colorimetric probe-based competition assay for S^{2-} sensing. The Cu@Au initially displayed red color (0, 1 μM of S^{2-}) changed to purple reddish (3, 5 μM of S^{2-}) and purple (7, 10 μM of S^{2-}) to blue (50, 100 μM of S^{2-}). The clear spectral and color change indicated the stronger binding affinity of sulfide. Although the same particles are known for binding with iodide, it showed better sensitivity with H_2S at the surface of the Cu@AuNP. The UV–vis spectral change at 650 nm and 520 nm were recorded after 15 min of incubation time with LOD calculated to be 0.3 μM. The temperature-dependent studies revealed that the Cu@AuNPs stability decreases at higher temperature; the working temperature was kept at about 20 degree Celsius with pH 5.2. However, the cysteine showed absorbance as well as color change that could be disadvantageous while working with biological samples.

AuNPs modified with GSH were also used for the colorimetric sensing of sulfide [18]. Interestingly, the sulfide sensing could not be achieved by AuNPs or AuNPs barely functionalized with other thiols such as cysteine, 3-mercaptopropionic acid, mercapto-succinic acid, and 11- mercaptoundecanoic acid. The color of solution selectively turned from red through purple reddish and purple to blue with sulfide anions due to exchange reaction between GSH and sulfide anions. The LOD was calculated as 7 μM and 3 μM through naked ayes and UV–vis spectral change, respectively. The NaCl dependent test was also carried out due to its ability to minimize the electrostatic repulsion force for the particle aggregation.

The etching property of gold nanorods was utilized by Lin and co-workers [19] for the colorimetric detection of H_2S. The H_2S detection relied on the catalytic oxidation of 3,3',5,5'-tetramethylbenzidine (TMB) via horseradish peroxidase (HRP) to produce TMB^{2+}. The TMB^{2+} etches the gold nanorods and a change in color was observed. The pH used for the studies was 6.5 and response time was calculated about 2.5 min with LOD of 0.0190 μM. The LSPR shifts toward higher wavelength with vivid colors (pink, purple, blue, green, gray, and reddish-brown) detectable with naked eyes with an increase in H_2S concentration. The validity of gold nanorods in biological samples was demonstrated by in vivo microdialysate in rat brain, including hippocampus, striatum, and cortex.

Taking advantage of the reducing nature of H_2S, Chang and co-workers [20] developed active ester cofunctionalized gold nanoparticles (AE-AuNPs) composed with azide group. Only with H_2S conversion of azide group to primary amine takes place and consequent cross-linking reaction with N-hydroxysuccinimide ester to give acylamide compounds, resulting in aggregation among active ester cofunctionalized AuNPs. The ratiometric change in absorbance spectra was monitored at 520 nm and 720 nm at pH 9; 20 mM of NaCl was used to avoid self-aggregation of AE-AuNPs, and the LOD was calculated as 0.2 μM with an incubation time of 15 min. The real-life practical application of this probe was tested in lake water samples.

Yin and co-workers [21] have developed dimeric nanoparticles based on Au/AgI for selective and sensitive colorimetric sensing of sulfide anions. Au/AgI dimeric NPs were obtained by deposition of Ag shells onto AuNPs by using the Tollens reagent, followed by oxidation of Ag shell in the presence of I_2 to form AgI. LSPR peak of Au@Ag core-shell NPs fall between 383 nm and 501 nm. In the presence of S^{2-}, the peak at 421 nm vanished, and the peak at 527 nm showed a bathochromic shift to 565 nm (Fig. 4).

The sulfide sensing relies on the reaction with AgI shell to form Ag_2S, thus changing the composition and core/shell ratio of the NPs and inducing color and spectrum changes. The LOD calculated by UV-vis spectral change was 0.5 μM and through naked eyes was 5 μM. The real water analysis was carried out to confirm the applicability of Au/

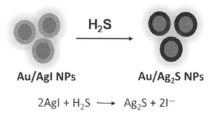

Fig. 4 Au/AgI dimeric nanoparticles for the sensing of H_2S.

AgI dimeric NPs for environmental samples. Furthermore, the Au/AgI dimeric NPs were immobilized in agarose gel as solid test strips. With S^{2-}, these strips showed purplish red to blue to green color change, similar to the naked eye change observed in the solution and found to be better than well-known commercially available lead acetate test strips.

The authors have further investigated the potential applications of Au/AgI dimeric NPs for gaseous H_2S detection and analysis of H_2S released by cells. The test strips to detect H_2S generated during cell cultivation play a vital role in physiology and pathophysiology.

Fluorosurfactant-functionalized gold nanorods (FSN-AuNRs) have been proposed to act as selective colorimetric nanoprobes for H_2S. With the combination of strong gold–thiol interactions and small FSN bilayer interstices, FSN-AuNRs demonstrate favorable selectivity and sensitivity toward H_2S over other anions and small biological molecules. The LOD for H_2S was 0.2 μM. The practical application in biological H_2S detection was validated with human and mouse serum samples [22].

Halder and co-workers [23] have discussed the photoreduction of $AgNO_3$ to generate silver nanoparticles by using synthetic azo food colorant dyes (carmoisine, cochineal red A, orange G, coumarin314 (C314), rhodamine 6G). The AgNPs prepared by the photoinduced reduction in the presence of coumarin314 dye showed absorbance peak at about 400 nm, which decreased with increase in S^{2-} concentration. However, to enhance the selectivity toward S^{2-}, Cu^{2+} was used as masking agents. The LOD was calculated as 2.0 μM with naked eye detectable transformation of yellow to colorless solution. Along with the practical applicability of prepared AgNPs in detecting S^{2-} in real water samples, the photocatalytic dye degradation could also be achieved and is advantageous toward environment healing compared with other silver nanoparticles-based probes.

2.3 Other metal nanoparticles

Jiang and co-workers [24] described simple, stable, and economically cheaper nanocrystalline cellulose (NCC)-based Pb nanoparticles for H_2S sensing (Fig. 5A). The approach is based on well-known lead acetate method for H_2S detection. Color change from colorless to light yellow then to brown was observed with H_2S. The change in absorbance was monitored at three different wavelengths, i.e., 300 nm, 350 nm, and 400 nm with the LOD of 2.4 μM, 2.5 μM, and 2.6 μM, respectively. The repeatability of the system was checked by pumping H_2S gas to the mixture of nanocrystalline cellulose and lead acetate, and the mixture was stored for 24 months. The absorbance at 350 nm did not change even after 24 months, and this proved that the system is better than other nanoprobes in long-term storage and assaying H_2S gas.

(A)

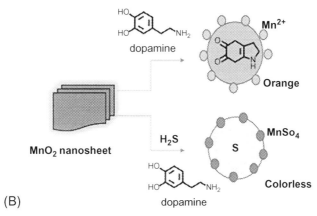

(B)

Fig. 5 (A) Nanocrystalline cellulose (NCC)-based lead Pb nanoparticles for H_2S sensor. (B) Manganese dioxide (MnO_2) nanosheets and dopamine-based nanosensor.

Liu and co-workers [25] have described Au–Hg alloy nanorods (Au–HgNRs) as probes for H_2S. The prepared alloy nanoparticles were stabilized with surfactant. The LSPR absorbance peak of the composition Au–HgNRs–(I, II, and III) was prepared by adding 50, 100, and 150 μmol/L of Hg^{2+} found at 680 nm, 640 nm, and 620 nm, respectively. In the presence of H_2S, Au–HgNRs showed a color change from cyan to green gray along with a significant redshift in longitudinal wavelength of SPR (LSPR) peak and LOD was determined as 2.5 μM with high selectivity toward S^{2-}. However, the more acidic medium with working pH 3.5 could be a disadvantage for S^{2-} sensing. The application of Au–HgNPs was explored in tannery waste water samples and the results were consistent with the standard methylene blue method.

Manganese dioxide (MnO_2) nanosheets and dopamine-based nanosensors were developed by Cai and co-workers [26] for colorimetric detection of H_2S (Fig. 5B). MnO_2 nanosheets induce the oxidation of dopamine, producing an orange color. The addition of H_2S into the reaction system will make the color fade, which is due to the competitive reactions of H_2S and dopamine. The reaction system of MnO_2 nanosheets and dopamine is employed for colorimetric sensing of H_2S. The presence of H_2S inhibits the redox process of dopamine to aminochrome induced by MnO_2

nanosheets. The nanosheets showed absorbance maximum at 470 nm and detection limit of 0.08 μM in buffer solution of pH 7. This system proved to be simple and low-cost and avoided multistep synthesis as compared to expensive Ag- and Au-based nanoparticle probes. The application of MnO_2 nanosheets-based nanoprobe was explored in natural water samples using spiking method.

Yang and co-workers [27] have explored a versatile sensing platform composed of Cu@Ag core shell nanoparticles for the colorimetric sensing of different ions including sulfide anions. The SPR and naked eye color change was easily detected with an increasing concentration of S^{2-} anions. The absorbance peak was significantly decreased at 400 nm with brown to reddish orange color change when Cu@Ag NPs were treated with S^{2-} for 5 min at pH 5.2. However, the lack of selectivity among various anions could be disadvantageous for the sensing of H_2S.

Hatamie and co-workers have reported copper nanoparticles for the colorimetric detection of sulfide ions and explored their application in spiked water samples [28]. Cu is less expensive than Au and Ag, and displays a localized plasmon band in the visible part of the spectrum. The colloidal plasmon exhibit absorbance band at 570 nm and red color is typically due to SPR absorption band. The red color turned to yellow in about 5 min when treated with sulfide anions. The etching of CuNPs due to the formation of CuS or Cu_2S was confirmed by decrease in size of NPs in TEM measurements. The LOD was calculated as 8.1 μM, and selectivity was confirmed in the presence of 500-fold high concentration of other anions. The advantages of CuNPs are easy and cost-effective synthesis compared with expensive Ag and AuNPs synthesis.

3. Fluorescent nanoprobes

Fluorescent H_2S nanoprobes mainly work on three important sensing methods [29] such as (a) nucleophilic reaction: where nanoparticles functionalized with nitro or azide group are converted to an amine via a reduction reaction; this chemical conversion leads to the change in the emission properties (Fig. 6A); (b) Displacement of metal ions such as Cu^{2+}, Au^{3+}, and Ag^+ from the nanosystem owing to their high affinity with S^{2-} (Fig. 6B); (c) Fluorescence quencher removed from the functionalized nanoprobes due to thiolysis reaction by H_2S and consequently recovering the original fluorescence of the nanoprobes (Fig. 6C).

3.1 H_2S-induced conversion azide to amine

H_2S-induced chemical reactions such as reduction of nitro and azides groups, nucleophilic addition, and metal-sulfide precipitation are mostly used for the detection of H_2S by fluorescence method. Inspired by the one of the earlier reports by Chang's group [3] for the H_2S fluorescence switch-on sensor by reduction of azides to amines, many small molecules and few nanoprobes have been reported.

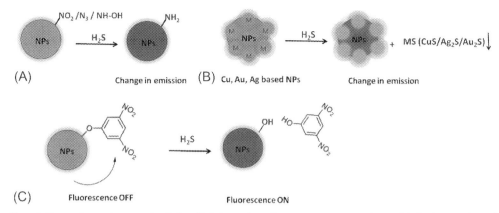

Fig. 6 General sensing method for H$_2$S by using fluorescent nanoprobes. (A) Reduction-based nanoprobes, (B) metal displacement-based nanoprobes, and (C) thiolysis reaction-based nanoprobes. *Reproduced with permission from reference Methods 2019, 168, 62–75.*

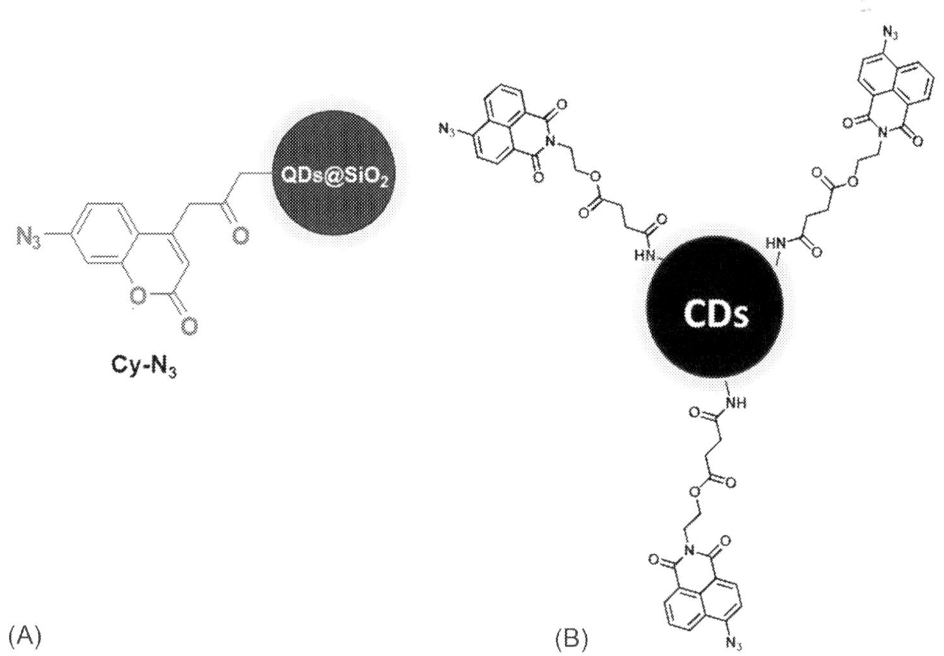

(A)   (B)

Fig. 7 (A) CdTe @SiO2 for H$_2$S sensor. (B) Carbon dots-based nanosensor for H$_2$S.

Yan et al. described organic dye molecules Cy-N$_3$ functionalized on CdTe quantum dots stabilized on SiO$_2$ (CdTe @SiO2) for H$_2$S detection (Fig. 7B) [30]. Fluorescent turn-ON was observed only with H$_2$S via azide to amine group reduction using this nanohybrid material. The electron withdrawing Cy-N$_3$ group responsible for the weak

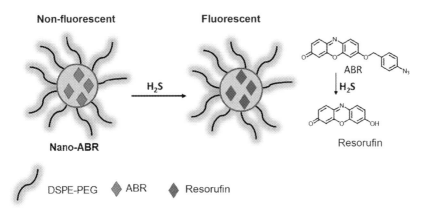

Fig. 8 Schematic representation of nano-ABR for H_2S sensor.

emission of this nanomaterial but in presence of H_2S of azide group of Cy-N_3 converted to the amine (Cy-NH_2). Therefore, fluorescence signal become strong, and good LOD of 7.0 nM was observed with H_2S in water.

FRET-based carbon nanodot (CDs) nanosensor was reported by Yu et al. [31]. The naphthalimide azide moiety was appended on the surface of the CDs, which also served as an energy donor. The conversion of naphthalimide-azide to naphthalimide-amine in the presence of H_2S was accountable for FRET between CDs and naphthalimide moiety. The probe displayed precise detection of H_2S in water and serum with a LOD of 10 nM. The probe was also useful to measure trace levels of H_2S in live cells with good biocompatibility.

In another approach, fluorogenic nanoreactor (nano-ABR, Fig. 8) composed of amphiphilic lipid 1-iodooctadecane, DSPE-PEG, and azidobenzylresorufin (ABR) was reported by Kim et al. [26]. This method allowed use of the hydrophobic molecular probe in an aqueous environment for sensing purpose without any chemical modification. In this, hydrophobic molecular probe was encapsulated into the inner core of nanoreactors. ABR is nonfluorescent and undergoes self-immolative cleavage upon reaction with HS to release resorufin with fluorescence recovery. Probe nano-ABR displayed change in emission and color within a minute in the presence of H_2S, with a low LOD of 18 nM. In addition, the nanoprobe nano-ABR has the ability to detect endogenous H_2S in living cells and the serum samples of mice by fluorescence response.

3.2 Nucleophilic attack-based H_2S probes

H_2S prefer to exist as $\overline{HS}$ in aqueous solution at neutral pH, which is a good nucleophile. Nucleophilic nature $\overline{HS}$ was used to design many small probes. Li et al. [32] reported graphene quantum dots (GQDs) functionalized with electron-withdrawing (2,4-dinitrophenoxy)tyrosine for H_2S sensor (Fig. 9A). Initially the fluorescent is in off stage due to

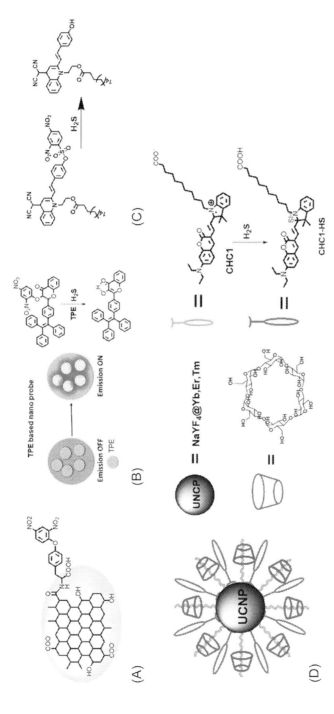

Fig. 9 (A) Quantum dots functionalized with electron-withdrawing (2,4-dinitrophenoxy) tyrosine for H$_2$S sensor. (B) Schematic illustration for the TPE-based nanoprobe and its fluorescence response to H$_2$S. (C) H$_2$S responsive organic dye used for the preparation of nanoprobe (D) UCNPs-based nanoprobe for H$_2$S sensor.

PET process between GQD and dinitrophenoxyl moiety. Only in the presence of H_2S, dinitrophenoxyl group was removed the PET process was inhibited, and the original fluorescence of GQD was recovered. This quantum dot-based nanoprobe was highly selective, and LOD was reported as 2 nM. Trace level of H_2S present in the living cell was also detected by using this probe.

Zhang et al. [33] reported a nanoprobe that was prepared by precipitation of phospholipid, 2,4-dinitrophenyl (DNP) group-modified tetraphenylethene (TPE) derivative and peptide. Weak fluorescence (at 550 nm) of this nanoaggregate became strong in the presence of H_2S. 2,4-dinitrophenyl (DNP) ether moiety was cleaved by H_2S to regenerate fluorophores and recover its original fluorescence emission (Fig. 9B). The emission is due to excited-state intramolecular proton transfer (ESIPT). LOD was determined as 0.09 μM with good linear relationship established in the range of 0–40 μM, which is in the limit of physiological levels of H_2S (10–600 μM).

A mixture of amphiphilic block copolymer PEO113-*b*-PS46 and H_2S responsive dye were used for the preparation of aggregation-induced emission dots (AIED) by Zhang et al. [34] (Fig. 9C). Due to the PET process, initially, the emission of AIED was in OFF stage, but H_2S induces an enhancement in the fluorescence spectrum with large stokes shift (~150 nm). AIED exhibited good water dispersibility, high sensitivity (LOD = 43.8 nM), and good biocompatibility.

Upconversion nanoparticles (UCNPs) have the ability to convert near-infrared radiations with lower energy into visible radiations with higher energy via a nonlinear optical process [35]. UCNPs are also called new generation fluorophores. Recently, UCNPs have evolved as alternative fluorophores with a great potential for designing chemical sensors and bioimaging and biological assays. First H_2S sensor based on UCNPs was reported by Loo et al. [36]. UCNPs incorporated with the H_2S responsive long carboxylic-terminated Coumarin-hemicyanine dye (CHC1) was prepared, and cyclodextrin (CD) was assembled on the surface to yield a water-soluble nanoparticle CD-UCNPs (Fig. 9D). Only with H_2S, the nanoprobe CD-UCNPs displays ratiometric upconversion luminescence (UCL). Nucleophilic addition of H_2S to the dye molecule is responsible for the fluorescence change.

Cationic near-infrared cyanine dye (Cy7-Cl) loaded with UCNPs was reported by Wang et al. [37] for monitoring H_2S. Here, NIR-absorbing cyanine dye act as an energy acceptor, and UCNPs act as energy donors (Fig. 10). Due to the energy transfer (ET) between UCNPs and dye Cy7-Cl, the nanoprobe displayed weak fluorescence at 800 nm. But, the turn-*on* luminescence of UCNPs at 800 nm was observed only with H_2S due to nucleophilic thiolation reaction between Cy7-Cl and H_2S to form Cy7-SH. The LOD was reported as 510 nM; endogenous H_2S detection and bioimaging studies were also performed by using this probe.

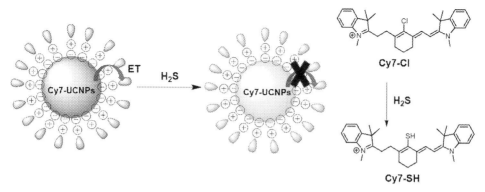

Fig. 10 UCNPs modified with Coumarin-hemicyanine dye (Cy7-Cl) for fluorescent H_2S sensor.

3.3 Metal displacement approach

Compared with the reaction-based detection of H_2S, metal displacement approach (MDA) shows quick response time and reusability. Kaushik et al. [38] developed a metal indicator displacement assay (MIDA) for the detection of H_2S. Amphiphilic DPA-based copper complex was mixed with phospholipid to form nanoscale vesicular receptor. Copper complex-incorporated vesicles quenched the fluorescence of indicator dye eosin (EY) dye due to binding of copper with the EY. Only in the presence of H_2S, the copper is displaced as CuS, instantaneously EY dye displaced from nanovesicles, and the original fluorescence of EY was restored. Fluorescence emission changes, and the naked eye detection through color change confirmed the selectivity, sensitivity, and LOD (4.06 μM) of the nanoscale vesicular receptor for H_2S.

In an another approach, mesoporous silica nanoparticle (MSN)-based nanosensor (**Ru@FITC-MSN**) was prepared by coupling fluorescent Ru(II) complex and fluorescein isothiocyanate (FITC) (Fig. 11). The probe displayed two emissions at 520 nm and 600 nm responsible for FITC and Ru(II) complex, respectively. In presence of Cu(II), the red emission was quenched due to complex formation with the DPA unit attached to the Ru(II) complex. In situ generated **Ru–Cu@FITC-MSN** responded to H_2S rapidly and selectively, with a LOD of 0.36 μM. Formation of CuS in the presence of H_2S was responsible for the emission change. The use of the **Ru–Cu@FITC-MSN** for imaging of H_2S in live cells and to monitor different H_2S levels in cancer cells was also explored [39].

4. Conclusion

In this chapter, we summarized the recent development of colorimetric and fluorescent nanosensors for the detection of gaseous signaling molecule H_2S. Nanoprobes' structural compositions, working mechanism, spectral or optical properties, conditions, LOD,

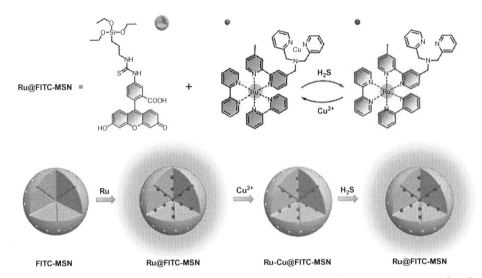

Fig. 11 Schematic representation of the H_2S responsive **Ru@FITC-MSN** nanosensor. *Reproduced with permission from reference Analytica Chimica Acta, 2019. doi.org/10.1016/j.aca.2019.12.056.*

sensitivity, and applications in different areas were discussed. From the examples discussed in the chapter, it is clear that nanoprobes are advantageous over the traditional small molecule-based probes in hydrophobicity, biocompatibility, sensitivity, and photostability, which are beneficial to the long-time tracking of H_2S. Most of the nanoprobes offer a sensitive detection of H_2S in living cells for bioimaging application, but their potential applications in vivo are insufficient. For further improvement, the nanomaterials should not only be used as a platform to carry the organic fluorophores, more attention must be paid on the functionalization of nanomaterials to fulfill the requirement of more sensitive and rapid detection of H_2S.

References

[1] H. Kimura, Signaling by hydrogen sulfide (H_2S) and polysulfides (H_2Sn) in the central nervous system, Neurochem. Int. 126 (2019) 118–125.
[2] C. Szabó, Hydrogen sulphide and its therapeutic potential, Nat. Rev. Drug Discov. 6 (2007) 917.
[3] V.S. Lin, C.J. Chang, Fluorescent probes for sensing and imaging biological hydrogen sulfide, Curr. Opin. Chem. Biol. 16 (2012) 595–601.
[4] R. Kaushik, A. Ghosh, D. Amilan Jose, Recent progress in hydrogen sulphide (H_2S) sensors by metal displacement approach, Coord. Chem. Rev. 347 (2017) 141–157.
[5] H. Peng, W. Chen, S. Burroughs, B. Wang, Recent advances in fluorescent probes for the detection of hydrogen sulfide, Curr. Org. Chem. 17 (2013) 641–653.
[6] N. Kumar, V. Bhalla, M. Kumar, Recent developments of fluorescent probes for the detection of gasotransmitters (NO, CO and H2S), Coord. Chem. Rev. 257 (2013) 2335–2347.
[7] S.K. Pandey, K.-H. Kim, K.-T. Tang, A review of sensor-based methods for monitoring hydrogen sulfide, TrAC Trends Anal. Chem. 32 (2012) 87–99.

[8] X. Huang, J. Song, B.C. Yung, X. Huang, Y. Xiong, X. Chen, Ratiometric optical nanoprobes enable accurate molecular detection and imaging, Chem. Soc. Rev. 47 (2018) 2873–2920.

[9] Y. Zhang, H.-Y. Shen, X. Hai, X.-W. Chen, J.-H. Wang, Polyhedral oligomeric silsesquioxane polymer-caged silver nanoparticle as a smart colorimetric probe for the detection of hydrogen Sulfide, Anal. Chem. 89 (2017) 1346–1352.

[10] K. Shanmugaraj, M. Ilanchelian, Colorimetric determination of sulfide using chitosan-capped silver nanoparticles, Microchim. Acta 183 (2016) 1721–1728.

[11] S.K. Gahlaut, K. Yadav, C. Sharan, J.P. Singh, Quick and selective dual mode detection of H_2S gas by mobile app employing silver nanorods array, Anal. Chem. 89 (2017) 13582–13588.

[12] L. Zhao, L. Zhao, Y. Miao, C. Liu, C. Zhang, A colorimetric sensor for the highly selective detection of sulfide and 1,4-dithiothreitol based on the in situ formation of silver nanoparticles using dopamine, Sensors 17 (2017) 1–13.

[13] K.B. Ayaz Ahmed, M. Mariappan, A. Veerappan, Nanosilver cotton swabs for highly sensitive and selective colorimetric detection of sulfide ions at nanomolar level, Sensors Actuators B Chem. 244 (2017) 831–836.

[14] P. Ni, Y. Sun, H. Dai, J. Hu, S. Jiang, Y. Wang, Z. Li, Z. Li, Colorimetric detection of sulfide ions in water samples based on the in situ formation of Ag_2S nanoparticles, Sensors Actuators B Chem. 220 (2015) 210–215.

[15] X. Zhai, J. Shi, X. Huang, Z. Sun, D. Zhang, Y. Sun, J. Zhang, Z. Li, X. Zou, M. Holmes, Y. Gong, M. Povey, S. Wang, A colorimetric hydrogen sulfide sensor based on gellan gum-silver nanoparticles bionanocomposite for monitoring of meat spoilage in intelligent packaging, Food Chem. 290 (2019) 135–143.

[16] K. Saha, S.S. Agasti, C. Kim, X. Li, V.M. Rotello, Gold nanoparticles in chemical and biological sensing, Chem. Rev. 112 (2012) 2739–2779.

[17] J. Zhang, X. Xu, Y. Yuan, C. Yang, X. Yang, A Cu@Au nanoparticle-based colorimetric competition assay for the detection of sulfide anion and cysteine, ACS Appl. Mater. Interfaces 3 (2011) 2928–2931.

[18] J. Zhang, X. Xu, X. Yang, Highly specific colorimetric recognition and sensing of sulfide with glutathione-modified gold nanoparticle probe based on an anion-for-molecule ligand exchange reaction, Analyst 137 (2012) 1556–1558.

[19] Z. Chen, C. Chen, H. Huang, F. Luo, L. Guo, L. Zhang, Z. Lin, G. Chen, Target-induced horseradish peroxidase deactivation for multicolor colorimetric assay of hydrogen sulfide in rat brain microdialysis, Anal. Chem. 90 (2018) 6222–6228.

[20] Z. Yuan, F. Lu, M. Peng, C.-W. Wang, Y.-T. Tseng, Y. Du, N. Cai, C.-W. Lien, H.-T. Chang, Y. He, E.S. Yeung, Selective colorimetric detection of hydrogen sulfide based on primary amine-active ester cross-linking of gold nanoparticles, Anal. Chem. 87 (2015) 7267–7273.

[21] J. Zeng, M. Li, A. Liu, F. Feng, T. Zeng, W. Duan, M. Li, M. Gong, C.-Y. Wen, Y. Yin, Au/AgI dimeric nanoparticles for highly selective and sensitive colorimetric detection of hydrogen sulfide, Adv. Funct. Mater. 28 (2018) 1800515.

[22] X. Zhang, W. Zhou, Z. Yuan, C. Lu, Colorimetric detection of biological hydrogen sulfide using fluorosurfactant functionalized gold nanorods, Analyst 140 (2015) 7443–7450.

[23] N. Mahapatra, S. Datta, M. Halder, A new spectroscopic protocol for selective detection of water soluble sulfides and cyanides: use of Ag-nanoparticles synthesized by Ag(I)–reduction via photo-degradation of azo-food-colorants, J. Photochem. Photobiol. A Chem. 275 (2014) 72–80.

[24] Y. Jia, Y. Guo, S. Wang, W. Chen, J. Zhang, W. Zheng, X. Jiang, Nanocrystalline cellulose mediated seed-growth for ultra-robust colorimetric detection of hydrogen sulfide, Nanoscale 9 (2017) 9811–9817.

[25] X. Zhu, C. Liu, J. Liu, Colorimetric detection of sulfide anions via redox-modulated surface chemistry and morphology of Au-Hg nanorods, Int. J. Anal. Chem. 2019 (2019) 8961837.

[26] M. Kim, Y.H. Seo, Y. Kim, J. Heo, W.-D. Jang, S.J. Sim, S. Kim, A fluorogenic molecular nanoprobe with an engineered internal environment for sensitive and selective detection of biological hydrogen sulfide, Chem. Commun. 53 (2017) 2275–2278.

[27] J. Zhang, Y. Yuan, X. Xu, X. Wang, X. Yang, Core/Shell Cu@Ag nanoparticle: a versatile platform for colorimetric visualization of inorganic anions, ACS Appl. Mater. Interfaces 3 (2011) 4092–4100.

[28] A. Hatamie, B. Zargar, A. Jalali, Copper nanoparticles: a new colorimetric probe for quick, naked-eye detection of sulfide ions in water samples, Talanta 121 (2014) 234–238.

[29] D. Amilan Jose, N. Sharma, R. Sakla, R. Kaushik, S. Gadiyaram, Fluorescent nanoprobes for the sensing of gasotransmitters hydrogen sulfide (H$_2$S), nitric oxide (NO) and carbon monoxide (CO), Methods 168 (2019) 62–75.

[30] Y. Yan, H. Yu, Y. Zhang, K. Zhang, H. Zhu, T. Yu, H. Jiang, S. Wang, Molecularly engineered quantum dots for visualization of hydrogen sulfide, ACS Appl. Mater. Interfaces 7 (2015) 3547–3553.

[31] C. Yu, X. Li, F. Zeng, F. Zheng, S. Wu, Carbon-dot-based ratiometric fluorescent sensor for detecting hydrogen sulfide in aqueous media and inside live cells, Chem. Commun. 49 (2013) 403–405.

[32] N. Li, A. Than, J. Chen, F. Xi, J. Liu, P. Chen, Graphene quantum dots based fluorescence turn-on nanoprobe for highly sensitive and selective imaging of hydrogen sulfide in living cells, Biomater. Sci. 6 (2018) 779–784.

[33] P. Zhang, X. Nie, M. Gao, F. Zeng, A. Qin, S. Wu, B.Z. Tang, A highly selective fluorescent nanoprobe based on AIE and ESIPT for imaging hydrogen sulfide in live cells and zebrafish, Mater. Chem. Front. 1 (2017) 838–845.

[34] P. Zhang, Y. Hong, H. Wang, M. Yu, Y. Gao, R. Zeng, Y. Long, J. Chen, Selective visualization of endogenous hydrogen sulfide in lysosomes using aggregation induced emission dots, Polym. Chem. 8 (2017) 7271–7278.

[35] B. Gu, Q. Zhang, Recent advances on functionalized upconversion nanoparticles for detection of small molecules and ions in biosystems, Adv. Sci 5 (2018) 1700609.

[36] Y. Zhou, W. Chen, J. Zhu, W. Pei, C. Wang, L. Huang, C. Yao, Q. Yan, W. Huang, J.S.C. Loo, Q. Zhang, Inorganic–organic hybrid nanoprobe for NIR-excited imaging of hydrogen Sulfide in cell cultures and inflammation in a mouse model, Small 10 (2014) 4874–4885.

[37] F. Wang, C. Zhang, X. Qu, S. Cheng, Y. Xian, Cationic cyanine chromophore-assembled upconversion nanoparticles for sensing and imaging H2S in living cells and zebrafish, Biosens. Bioelectron. 126 (2019) 96–101.

[38] R. Kaushik, R. Sakla, A. Ghosh, G.T. Selvan, P.M. Selvakumar, D.A. Jose, Selective detection of H2S by copper complex embedded in vesicles through metal Indicator displacement approach, ACS Sens. 3 (2018) 1142–1148.

[39] J. Liu, C. Duan, W. Zhang, H.T. Ta, J. Yuan, R. Zhang, Z.P. Xu, Responsive nanosensor for ratiometric luminescence detection of hydrogen sulfide in inflammatory cancer cells, Anal. Chim. Acta 1103 (2020) 156–163.

SECTION V

Intelligent nano sensors(INS)—Environmental applications

CHAPTER 13

Study of carbon quantum dots as smart materials for environmental applications

Anupreet Kaur[a] and Jatinder Singh Aulakh[b]
[a]Basic and Applied Sciences, Punjabi University, Patiala, Punjab, India
[b]Department of Chemistry, Punjabi University, Patiala, Punjab, India

1. Introduction

Carbon quantum dots (CQDs) have been a focal point of interest by world researchers these days due to their unique properties such as small size, biocompatibility, photoluminescence (PL) properties, high temperature stability, chemically inert structure, and easy routes of functionalization. QCDs are discrete, quasispherical particles with size below 10 nm [1–4]. These generally possess a sp^2 conjugated core with suitable oxygen content in the form of multiple oxygen-containing groups such as carboxyl, hydroxyl, and aldehyde group [5–7]. In 2004, during the electrophoretic fractionation of arc-discharge soot, fluorescent carbon nanoparticles with vertical sizes of about 1 nm were isolated with the assertion from discoveries that they promise good nanomaterials [8]. But 2 years later, Sun's group then showed interest in these carbon nanomaterials, which were afterward coined as CQDs or carbon nanodots or carbogenic quantum dots [9]. Since 2006, research become a burning topic on the carbon nanodots due to several merits of CQDs, including simple synthesis, low cost, low toxicity, high photostability, and excellent biocompatibility. The CQDs have water solubility, acceptable biocompatibility, high fluorescence emission intensity, resistance to photobleaching, neglectable cytotoxicity, and are chemically inert. The CQDs may have functional groups such as hydroxyl, carboxyl, carbonyl, and epoxy that results in their water solubility. The main applications of these CQDs are sensing, bioimaging, nanomedicine, catalysis, optoelectronics, and energy conversion/storage. CQDs have many desirable and tunable properties such as upconversion of PL and biocompatibility, rendering them promising candidates for wide applications in remediation of environmental pollution, bioimaging, and optoelectronics (Fig. 1). In this chapter, we studied the uses of CQDs for the removal of emerging environmental pollutants.

Handbook of Nanomaterials for Sensing Applications
https://doi.org/10.1016/B978-0-12-820783-3.00019-1

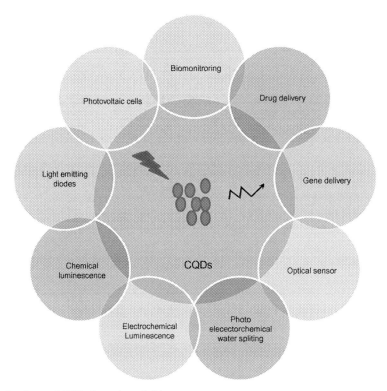

Fig. 1 Applications of CQDs in various fields.

2. Properties

The structures and components of CQDs determine their diverse properties. Based on carbon, CQDs possess such properties as good conductivity, benign chemical composition, and photochemical stability. By surface passivation, the fluorescence properties as well as physical properties of CQDs are enhanced (Fig. 2).

2.1 Optical properties

The optical absorption of CQDs is in the near-UV region, and the absorption intensity is lowered in the visible as well as near-infrared region. There are many shoulder peaks in the UV region due to the transitions from π-π^* of benzene rings and C$=$C bonds [10, 11]. The n-π^* transition is due to C$=$O bonds. CQDs have rich surface functionalities mainly consisting of oxygen-based groups such as hydroxyl, carboxyl, and carbonyl groups that play an important role in their luminescence properties. The degree of oxidation/reduction and the functional groups play important roles in tuning the optical properties.

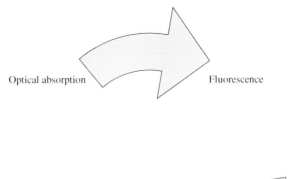

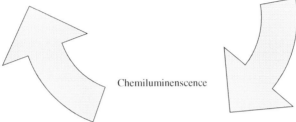

Fig. 2 Various properties of CQDs.

2.2 Fluorescence

The most important feature of CQDs is fluorescence due to the broad size distribution, different surface defects, and different surface states. The functional groups such as N—H and O—H on the surface of CQDs can form intermolecular and intramolecular hydrogen bond, which endows the CQDs with different surface states. When pH increases or decreases, deprotonation or protonation of functional groups can lead to variation of surface states and fluorescence. In addition to normal fluorescence, there is upconversion of fluorescence emission.

2.3 Photoluminescence

PL is one of the most fascinating behaviors of carbon dots. The PL spectra of CQDs have much variation in excitation wavelength (300–380 nm and 400–490 nm). The classic signature of carbon dots is their emission wavelength and size-dependent photoluminescent behavior [12, 13]. The emission peak shifted to higher wavelengths with an increase in excitation wavelength, but also a decrease in the intensity was also observed in CQDs.

Two main reasons for the PL behavior of CQDs are the presence of different sizes and the distribution of the different surface energy traps on the particles. Like semiconductor quantum dots, the energy gap increases with an decrease in size of the carbon dots and vice versa due to the quantum confinement effect. Thus, the particles with a smaller size are excited at a lower wavelength, whereas those with a larger size are excited at higher wavelengths. The intensity of the PL depends on the number of particles excited at a particular wavelength. The highest PL intensity of carbon dots was observed at an excitation wavelength of 360 nm, because the largest number of particles being excited at that wavelength and the reason for the excitation-dependent PL behavior of carbon dots is the nature of their surface. The presence of various functional groups on the surface of the carbon dots may result in a series of emissive traps between π and π^* of C=C. On illuminating the carbon dots at a certain excitation wavelength, a surface energy trap dominates the emission. As the excitation wavelength changes, other corresponding surface state emissive traps become dominant. Hence the PL mechanism is controlled by both the size effects and surface defects.

3. Surface passivation and functionalization operation

Surface passivation is the process in which the CQDs are protected as well as increasing the optoelectric property. So, the protection layer is vital for the stability and long-life usage of CQDs. Surface passivation forms a thin insulating capping layer that shields CQDs from adhesive of impurities and further improves their fluorescence intensity [14]. Various polymers or organic compound molecules have been and may be used as surface passivation agents as long as they do not contain visible or near-UV chromophores and therefore have nonemission at visible wavelength, leaving the observed colorful emission of passivated CQDs intact. As the CQDs are prepared by various chemical oxidation, generally fluorescence is low; to increase this, surface modification is done by various process. Surface modification includes capping the CQDs with organic polymer, with inorganic salts, as well as compounds having heteroatoms. This also increases the hydrophobicity and specific chemical's reactivity. One of the major advantages of passivation of CQDs is that it increases the cross-linking, and this improves the fluorescence emissions because cross-linking stabilizes the surface functionalization by strengthening the soft shell surrounding the CQDs. CQDs have also been "passivated" by oligomeric polymers (e.g., oligomeric PEG, molecular weight 1500) to improve PL.

Functionalization with heteroatoms like oxygen or nitrogen also increases the quantum yield. To get the specificity and selectivity, CQDs are functionalized with small organic compounds having functional groups such as amine, carboxylic groups, and hydroxyl groups. Commonly, two strategies have been employed: heteroatom doping and surface modification. Heteroatoms (e.g., nitrogen, boron, and sulfur) can be facilely incorporated into CQDs by utilizing heteroatom-containing reactants. For example, water-soluble nitrogen- or selenium-doped CQDs have been prepared from the reactants comprising nitrogen or selenium groups. CQDs can also be co-doped by more than one type of heteroatom type, such as nitrogen-sulfur-doped CQDs prepared through hydrothermal treatment of cysteine. Although heteroatom doping is a facile method for CQDs modification, the resulting CQD's chemical structure is commonly unclear because of a complicated reaction mechanism, severely limiting the rational structural design. Controlled doping with well-defined molecular structures must be further developed. Conversely, surface functionalization via well-established organic reactions is more controllable. CQDs functional groups can be selectively formed, eliminated, or converted. For example, hydroxyl groups have been converted into trifluoroacetate groups by esterification with trifluoroacetic anhydride, and carboxylic acid/ketone groups have been transformed into hydroxamic acid/oximes groups after treatment with hydroxylamine. CQDs can also be decorated with molecules or oligomeric polymers through specific or nonspecific interaction. CQDs have been covalently modified through condensation reactions between carbonyl and amino groups by several aromatic molecules, including o-phenylenediamine, 2,3-diaminonaphthalene, and 1,8-diaminonaphthalene [15]. These polyaromatic molecules alter the conjugated domain of CQDs so as to rationally regulate the resulting bandgap.

4. Synthesis techniques

Controllable size and other parameters of CQDs make them more useful in applications, and CQDs are prepared by two methods. Currently, CQDs can be prepared by two potential approaches: bottom-up [16–21] and top-down [6, 8, 22–24]. The former utilizes organic monomer/polymer precursors, whereas the latter employs carbon materials such as graphite or carbon nanotubes (CNTs). Carbohydrates, organic acids, and polymers have been widely explored in the bottom-up approach as precursors to synthesize CQDs through the partial dehydration and dehydrogenation assisted by microwaves, solvothermal reaction, or thermal decomposition. Often, the requisite harsh reaction conditions (e.g., high temperature and pressure) result in multistep organic reactions and varied intermediates that severely complicate the formation mechanism. The type of precursors, reaction time, solvent type, and temperature are the key factors affecting the CQD's chemical structure and size [6, 18, 20, 21]. Thus, the CQD size and composition can be controlled through the careful optimization of reaction parameters. In the top-down approach, carbon materials such as CNTs, graphite, and soot are "cut" into CQDs via arc-discharge, laser ablation, or chemical oxidation. In particular, electrochemical oxidation of CNTs or graphite has been heavily explored because of the facile manipulation and mild experimental conditions required (e.g., no toxic gas exhausted). The size and composition of CQDs can be controlled by careful optimization of reaction parameters.

5. Detection of metal ions by CQDs
5.1 Mercury ion

Mercury is the most toxic heavy metal ion because of its maximum use in industries as well as agricultural use. The bioaccumulation as well as strong toxicity of this metal ion causes various diseases even at low concentration such as headaches, hyperspasmia, expiratory dyspnea, stomach perforation, renal failure, and so on. Zhang et al. synthesized a nanosensor for the selective determination of Hg(II) ions having limit of detection (LOD) of 4 ppb [25]. N–doped CQDs were used for the determination of mercury in the aqueous phase having LOD of 0.23 μM [26]. Silva et al. synthesized the CQDs immobilized on the silica optical fiber for the detection of Hg(II) [27]. Methyl mercury was detected by CQD-based detector synthesized by Bendicho's group [28]. An ultrasensitive sensor was developed by Kang et al. for the detection of Hg(II), and its LOD value was 1 fM [29]. Nandi et al. synthesized the Thiol-functionalized N-CQDs for detection of Hg(II) having LOD of 18 pM [30, 31]. Wang et al. synthesized the CQDs from apple juice for the determination of mercury ions having LOD of 2.3 nM [32]. Shao et al. synthesized the O,S-co-doped CQDs for Hg(II) ions and the LOD is 0.37 nM [33]. CQDs were synthesized from citric acid used for the determination of Hg(II) ions having LOD of 0.57 nM/0.55 nM by Chen et al. Mobin et al. used the CQDs for the detection of Hg(II), and the LOD of this method is 100 nM [34]. N-CQDs were synthesized from citric acid having LOD of 7.3 nM by Huang et al. for Hg(II) ions [35]. $Eu^{3+}/$CQDs-MOF-253 were used for the determination of HG(II), and the LOD is 13 nM [36]. Tang et al. synthesized the N-rich CQDs for mercury ions, and the LOD is 0.63 μM [37].

5.2 Zinc ion

It is a major trace element for human beings and plays an important role in the process of growth, inheritance, immunity, and incretion. As such, analysis of zinc content is inevitable. Yi et al. synthesized the Quinoline-modified CQDs for the determination of Zn(II) ions, and a color change visible with the naked eye is possible; the LOD is 6.4 nM [38]. Raju et al. used the CQDs for the determination of Zn(II). Kang et al. synthesized CQDs from glucose for the sensitive detection of Zn(II). In this system, the CQDs serve as the energy donor and Zn(II)-QCT(3,3,4,5,7-pentahydroxyflavone) as an acceptor having a detection limit of 2 μM [39].

5.3 Cadmium ions

Cadmium ions as a class of heavy and toxic metal ions have attracted much attention due to their chronic toxicity in humans. Sharma et al. synthesized the heteroatom CQDs from cysteine for the detection of Cd(II) [40]. This shows the high sensitivity of CQDs toward this metal and LOD to 0.0061 $μg^{-1}$ and 2.0 $μg^{-1}$. CQDs were synthesized from camphor for sensing Cd(II) ions by Raju et al. [41]. CQDs/Au cluster nanohybrids were used for the sensing of Cd(II) [42].

5.4 Copper ions

It is an essential element for animals as well as human beings as an integral part of proteins and enzymes, so it is widely used in pharmaceutical industry and medicines. Salinas-Castillo et al. reported CQDs-based sensor for detection of Cu(II), and these CQDs showed the upconversion and downconversion of fluorescence giving a sensitive detection of copper ion by inner filter effect [43]. CQDs–alginate gel was prepared by the Chang et al., and LOD is 5 ppm. N-CQDs were synthesized by Vallvey et al., and LOD of this method is 10 nM [44]. Agarose/CQDs was used for the Cu(II) by Chowdhary et al [45]. The LOD for Cu(II) using Agarsose/CQDs is 0.5 μM. CQDs-TPEA was used the detection of copper ions, and LOD is 1–100 nM by Tian et al. Polyamine-CQDs were used for he copper ions, and this facile methodology can offer a rapid, reliable, and selective detection of Cu^{2+} with a detection limit as low as 6 nM. Silica modified with CQDs were used for the detection of Cu(II) in brain cells [46].

A facile, rapid, and frugal approach was developed for the preparation of highly fluorescent CQD by pyrolysis of Finger millet ragi (*Eleusine coracana*) as a carbon source. CQDs have been applied in sensing of Cu^{2+} in real water samples with a LOD of 10 nM [47]. A facile one-step hydrothermal method was developed to synthesize nitrogen-doped CQDs (N-CQDs) by utilizing hexamethylenetetramine as the carbon and nitrogen source. This N-CQD fluorescent probe has been successfully applied to selectively determine the concentration of copper ion (Cu^{2+}) with a linear range of 0.1–40 μM and a LOD of 0.09 μM [48].

6. Detection of drugs by CQDs

ZnO-doped CQDs were prepared by the chemical oxidation of activated carbon followed by zinc acetate impregnation and precipitation. The ZnO-doped CQDs had a stable fluorescence and a high fluorescence quantum yield. The fluorescence of the ZnO-doped CQDs can be quenched by reduction with metronidazole, allowing a flow-injection chemiluminescence method to be used to determine the metronidazole content in tablets. The detection limit is 1.08×10^{-10} g/mL.

7. Detection of anions by CQDs

Lin et al. used CQDs for the detection of nitrite ions. In their work, the CL properties of CQDs in the presence of peroxynitrous acid were studied. Peroxynitrous acid is formed by the online mixing of nitrite ions and acidified hydrogen peroxide. Lin et al. discovered the CL response of PEG-capped CQDs for the detection of nitrite from peroxynitrous acid with a detection limit of 53 nM [48]. Peroxynitrite is a powerful oxidizing agent of biological importance, which exists in many forms, such as *trans*-peroxynitrous acid (ONOOH), *cis*-ONOOH, activated form of ONOOH, and peroxynitrous anion. ONOOH, as a weak acid with a pKa of 6.8, is an important form of peroxynitrite. In acid solution, ONOOH is an unstable compound with a lifetime less than 1 s. In basic solution, ONOOH could be converted to peroxynitrous anion ($ONOO^-$) that is fairly stable and can decompose slowly to nitrite and superoxide anion radical. The oxidation reaction by peroxynitrite is beneficial for destroying invading organisms. However, peroxynitrite is also involved in protein and lipid nitration, which has been found to be an early event in the etiology of some pathologies. The study concerning peroxynitrite has attracted much attention. We first found that carbon dots had an enhancing effect on the ONOOH system. In the present work, we further found that carbon dots could greatly amplify the CL from ONOOH-carbonate system. Furthermore, the carbon dots-ONOOH-carbonate CL system has higher CL intensity than that from carbon dots-ONOOH system. More interestingly, a linear relationship between the nitrite for the formation of ONOOH and CL signal produced from the carbon dots-ONOOH-Na_2CO_3 system was found. Hence, the proposed method can be employed for the determination of nitrite with improved sensitivity. The CL-enhanced mechanism of carbon dots on ONOOH-Na_2CO_3 system was also illustrated in detail, and its role as energy acceptor was further discussed. Phosphate off-on by Eu^{3+}-CQDs with 51 nM [49]. Yin et al. synthesized CQDs that showed a down- and upconversion fluorescent biosensor for superoxide anion (O_2^-).

7.1 Anion detection

Zhao et al. used CQDs functionalized with carboxyl groups to detect PO_4^{3-} ions. The working principle is that the coordination of Eu^{3+} ions with carboxyl groups on the surface of the CQDs leads to fluorescence quenching. The ability of PO_4^{3-} ions to form a complex with Eu^{3+} ions is stronger than that of carbon, and the CQDs are separated from the Eu^{3+} ions to restore their fluorescence. Liu et al. designed a novel fluorescent probe (Zr(CQDsCOO)$_2$EDTA) for the detection of F^- content based on competitive ligand reactions that occurred between carboxylate groups (–COOH) on the surface of the luminescent CQDs and F^- ions coordinated to Zr(H$_2$O)$_2$EDTA. The strong and stable fluorescence signal that was generated was quenched upon the addition of F^- ions as a result of the formation of the non-fluorescent complex Zr(F)$_2$EDTA, because the affinity of F^- ions for Zr(IV) ions was stronger than that of COOH groups in the CQDs. The change in fluorescence (DF) in this process displayed a linear correlation with the F^- content in a range from 0.10 mM to 10 mM. Hou et al. applied the electrochemical luminescence properties of CQDs in the detection of S^{2-} ions [50]. They reported a fluorescence sensor based on CQDs, which displayed excellent water solubility, low cytotoxicity, and a short response time. The sensor was based on a ligand/Cu^{2+} approach so as to achieve rapid sensing of sulfide anions. The CQDs serve as the fluorophore as well as an anchoring site for ligands that bound to copper ions. In this CQDs-based system, as copper ions bind to ligands that are present on the surface of the CQDs, the paramagnetic copper ions efficiently quench the fluorescence of the CQDs, which enables the system to act as a turn-off sensor for copper ions. More importantly, the subsequent addition of sulfide anions, which can extract Cu$_2$C from the system to form very stable CuS, results in fluorescence enhancement and enables the system to act as a turn-on sensor for sulfide anions [51]. This rapidly responding and selective sensor can operate in an entirely aqueous solution or in a physiological medium with a low detection limit of 0.78 mM.

Other applications of CQDs in anion sensing include the detection of I^- [52], $C_2O_4^{2-}$ [53], CN^- [54], O_2^- [55] ions, and so on. Barman et al. synthesized graphitic carbon nitride QDs (g-CNQDs) with strong blue fluorescence by a simple microwave-mediated method from formamide ($HCONH_2$). These QDs are highly sensitive and selective fluorescent probes for Hg^{2+} ions in aqueous media owing to the "superquenching" of fluorescence. The addition of I^- ions abstracts bound Hg^{2+} ions to form HgI_2 and restores the fluorescence characteristics of g-CNQDs. Thus, g-CNQDs can play a dual role in the selective and sensitive detection of Hg^{2+} ions, as well as I^- ions, in aqueous media via an ON-OFF-ON fluorescence response. Gao et al. developed a ratiometric fluorescence biosensor for O_2^- ions by employing CQDs as a reference fluorophore and hydroethidine (HE) as an organic molecule that is specific for O_2^- ions, which plays the role of both a specific recognition element and a response signal [55].

8. Detection of other pollutants by CQDS

Tartrazine, also known as lemon yellow, is a water-soluble synthetic pigment widely used as a coloring in ice cream, jelly, yogurt, beverages, canned foods, candy coatings, and so on. This dye revealed that tartrazine may cause adverse health effects such as changes in hepatic and renal parameters and reproductive toxicity, as well as neurobehavioral toxicity when consumed in excess [56, 57]. Therefore, the food industry must strictly control and regulate the content of tartrazine in foods, which necessitates an interest in the development of measurement techniques for determining tartrazine in foods that are efficient in terms of rapidity, simplicity, and sensitivity. Xu et al. found that tartrazine could result in strong quenching of the fluorescence of CQDs that were synthesized from aloe via a hydrothermal process [58]. The linear range was 0.25–32.50 mM, and the detection limit was 73 nM. Cysteine (Cys) is the only thiol-containing amino acid among the amino acids that constitute proteins in the body. It is unstable, easily oxidized, and can be converted into cysteine [59]. It can be used for detoxification, the prevention and treatment of radiation injury, and the inhibition of cancer cells and bacterial growth. Therefore, cysteine has important physiological functions and is widely used in the medical, food, and other industries. The early diagnosis of certain diseases can be achieved by monitoring the concentration of cysteine in the organism. In addition, it can be used in food products and human serum samples for the determination of cysteine. Wu et al. synthesized CQDs co-doped with nitrogen and sulfur via a facile hydrothermal method using citric acid and cysteine as precursors. The N,S-CQDs acted as a highly sensitive and selective "turn-off-on" probe for the determination of Fe^{3+} ions or cysteine [60]. The limits of detection were 14 nM for Fe^{3+} and 0.54 mM for cysteine. Glucose is the most widely distributed and most important monosaccharide in nature and is a type of polyhydroxyaldehyde. Glucose has an important position in biology as the energy source of living cells and metabolic intermediates; that is, it provides the main energy supply for biological substances. Plants can produce glucose by photosynthesis. It has a wide range of applications in candy manufacturing and pharmaceuticals. Shan et al. found that the fluorescence of boron-doped CQDs can be quenched by hydrogen peroxide, and they achieved the rapid and highly sensitive detection of glucose based on the oxidation of glucose catalyzed by glucose oxidase to produce gluconate and hydrogen peroxide. When glucose oxidase and glucose were present, the degree of quenching of the fluorescence of CQDs was linearly related to the glucose concentration. The linear range was 8.0–80.0 mM, and the detection limit was 8.0 mM. Phytic acid is the main form of phosphorus in plant tissues (cereals, fruits, and vegetables) and is mainly present in the seeds, roots, and stems of plants. In the food industry, phytic acid can be used as a food additive; in the wine industry, it can be used as a metal-removing agent. Gao et al. reported a facile, one-step pyrolytic synthesis of photoluminescent CQDs using citric acid as the carbon source and lysine as a surface passivation agent [61]. The fluorescence of the CQDs was found to be effectively quenched by ferric (Fe^{3+}) ions with high selectivity via a photoinduced electron transfer (PET) process. Upon the addition of phytic acid (PA) to a dispersion of the CQDs/Fe^{3+} complex, the fluorescence of the CQDs was significantly recovered as a result of the release of Fe^{3+} ions from the CDs/Fe^{3+} complex, because PA has a higher affinity for Fe^{3+} ions in comparison with CQDs. This analytical method is simple, low cost, and sensitive (linear range 0.68–18.69 mM, detection limit 0.36 mM) with high selectivity, stability, and recovery ratios for standard and actual samples. Vitamins are a class of trace organic substances that humans and animals must obtain from food to maintain normal physiological functions. They play important roles in the body in growth, metabolism, and development. Wang et al. reported a novel fluorescence resonance energy transfer sensor based on thermally reduced CQDs for the determination of vitamin B12 (VB12) in aqueous solutions [62]. When the concentration of VB12 was increased, the fluorescence intensity of the CQDs decreased. The CQDs were used to detect VB12 at concentrations ranging from 1 to 12 mg/mL, and the LOD was as low as 0.1 mg/mL. Furthermore, highly sensitive detection of VB2 and VB9 was also achieved [63, 64].

9. Detection of explosives by CQDs

In recent years, there has been much international concern over the reliable and accurate detection of explosives [65]. In addition, the detection of explosives for preventing terrorist crimes, safeguarding people's lives and property, maintaining national security, and protecting the environment and health is of great significance [66, 67]. Owing to the wide variety of explosives, together with the fact that the vapor pressure of most explosives is very low, the detection of explosives has always been a challenging problem [68, 69]. Nitroaromatic explosives are major environmental pollutants with high toxicity and are difficult to degrade [70]. They pose a serious threat to the ecological balance, social stability, and national security. Nitroaromatic explosives in the environment are mainly derived from the industrial production of explosives, and wastewater is generated during the processing of bombs and military activities. At present, most explosives contain nitroaromatic compounds such as 2,4,6-trinitrotoluene (TNT), dinitrotoluene (DNT), and 2,4,6-trinitrophenol (PA, picric acid), etc. Technology for the ultratrace detection of explosives is mainly used to detect vapor from explosives, which adheres to the surface of any object that has been exposed to explosives. Fluorescence methods are considered to be among the most appropriate techniques because of their high sensitivity, acquisition of multiple parameters, and relatively mature instrument designs. The fluorescence detection of explosives has therefore aroused widespread interest in recent years and has developed rapidly. One of the most notable experts in this field is Professor Swager of the Massachusetts Institute of Technology (MIT), who achieved breakthroughs in the detection of explosives by utilizing conjugated organic polymer films [71]. However, the syntheses of such conjugated organic polymer materials are time-consuming and costly, and problems arise such as the photobleaching and chemical degradation of fluorescent organic materials. Therefore, the development of new fluorescent materials with high resistance to photobleaching and degradation ability has become a focus of research on fluorescence detection of nitroaromatic explosives. Recently, CQDs have attracted extensive attention owing to their unique properties, such as green synthesis, excellent solubility, easy functionalization, and high biocompatibility, etc. As a green substitute for QDs, CQDs have found further promising applications in the trace detection of explosives. During the World Wars, huge amounts of TNT were released into the soil, lakes, and rivers. As an environmental contaminant, the explosive nitroaromatic compound TNT has aroused social concern. Currently, TNT continues to be a major component of ordnance used by the military and terrorists, which causes environmental damage and threatens public security [72–74]. Therefore, the development of practicable analytical platforms for monitoring ultratrace levels of TNT is urgently necessary. Zhang et al. successfully synthesized CQDs via a microwave-assisted pyrolysis method and employed them in both the fluorescence and the electrochemical determination of TNT. The fluorescence-sensing platform was based on the strong interaction between TNT and amino groups, which can quench the PL of CQDs via charge transfer. The resulting linear detection range extended from 10 nM to 1.5 mM, with a short response time of 30 s. A glassy carbon electrode modified with CDs exhibited high ability for the reduction of TNT, with a linear detection range extending from 5 nM to 30 mM, which was greater than that obtained by the fluorescence method. Xu et al. developed a facile strategy for preparing a fluorescence sensor based on CQDs capped by molecularly imprinted polymers with mesoporous structures (M-MIPs@CQDs) for the highly sensitive and selective determination of TNT [75]. The strategy, which used amino-CQDs directly as a "functional monomer" for imprinting, simplified the imprinting process and provided highly accessible recognition sites. The as-prepared M-MIPs@CQDs sensor, which used periodic mesoporous silica as the imprinting matrix and amino-CQDs directly as a "functional monomer", exhibited excellent selectivity and sensitivity toward TNT, with a detection limit of 17 nM. The detection process could be repeated 10 times with no obvious decrease in efficiency. The feasibility of the developed method for real sample analysis was successfully demonstrated by the analysis of TNT in soil and water samples, with satisfactory recoveries of 88.6%–95.7%. TNP, also known as picric acid, which is a nitroaromatic explosive with higher explosive power and thermal expansion than other nitroaromatic compounds such as TNT, has been widely used in the preparation of matches and fireworks [76, 77] and has become a great threat to the environment, homeland security, and public safety. With regard to the significance of TNP, it is highly desirable to develop simple analytical methods, in particular an effective and selective sensor, for detecting trace amounts of TNP. Chen et al. prepared CQDs doped with rare earths with multifunctional properties by simply keeping a mixture of terbium(III) nitrate pentahydrate and citric acid at 190 °C for 30 min [78]. The as-prepared terbium–doped CQDs (Tb-CQDs), which were synthesized via a rapid and simple direct carbonization route, had a size of about 3 nm. They exhibited blue fluorescence emissions that depended on the excitation wavelength, were stable, and were employed for the selective colorimetric detection of TNP in the range of 500 nM to 100 mM with a LOD of 200 nM. Liu et al. developed a selective and sensitive method for the detection of TNP with a LOD of 28 nM using amorphous photoluminescent CQDs, which were prepared via a simple hydrothermal route using spermine and m-phenylenediamine as precursors [79]. The as-prepared CQDs were found to exhibit blue-green PL that was independent of the excitation wavelength and excellent chemical and optical stability. Zearlenone (ZEA) contaminates cereals and derived products across the world. Toxic effects of ZEA include infertility, teratogenesis, neurotoxicity, carcinogenicity, and abortion. Manyu et al. synthesized CQDs that encapsulated molecularly imprinted fluorescence-quenching particles in a detection limit of 0.02 mg/L [80].

10. Concluding remarks

Nanotechnology is one of the most important technologies in the 21st century. Currently, nanotechnology has been widely applied in numerous fields. In recent years, fluorescent nanomaterials have garnered much interest. Therefore, over the past decade, many CQDs of varying size and chemistry have been successfully prepared. The designability of both the carbon bulk and surface functional groups affords extensive property tunability, significantly extending potential applications from traditional fields such as sensing, light-emitting diodes, bioimaging, and photocatalysts, to new areas such as tumor therapy, antibacterial agents, self-healing materials, and analytical chemistry. Although CQDs have been studied extensively, there are still several significant challenges, including controlling and quantifying CQD chemical structures, as well as quantitative analysis of structure-property relationships. This chapter has highlighted recent progress in the field of CQDs in terms of their rational synthesis, tunable optical properties, and analytical applications (Table 1).

Table 1 Synthesis of CQDs by various techniques.

#	Method	Size	QY	Advantages	Disadvantages
1.	Arc discharge method	—	—	Most attainable method	Harsh conditions, possesses low quantum yield and composite method
2.	Laser ablation method	5 nm	QY ranges between 4%–10%	Effortless, effective technique, different-sized nanoparticles can be prepared	Large amount of carbon matter is required, poor control over sizes, low quantum yield
3.	Electrochemical method	6–8 nm	QY ranges between 2.8%–8.9%	Stable method, extent of carbon dots can be managed by changing current density, water-soluble carbon dot can also be prepared	Complex method
4.	Thermal route	2–6 nm	QY ranges between 0.1%–3%	Easy straight method	Low quantum yield
5.	Microwave-assisted method	4.51 nm	QY ranges between 3.1%–6.3%	Simple and convenient method	Poor control over size
6.	Hydrothermal and aqueous-based method	—	—	Highly water dispersible carbon dots	Poor control over size
7.	Template method	—	—	Carbon dots have biocompatibility and colloidal stability	Time consuming and expensive method
1.	Arc discharge method	—	—	Most attainable method	Harsh conditions, possesses low quantum yield and composite method
2.	Laser ablation method	5 nm	QY ranges between 4%–10%	Effortless, effective technique, different-sized nanoparticles can be prepared	Large amount of carbon matter is required, poor control over sizes, low quantum yield
3.	Electrochemical method	6–8 nm	QY ranges between 2.8%–8.9%	Stable method, extent of carbon dots can be managed by changing current density, water-soluble carbon dot can also be prepared	Complex method
4.	Thermal route	2–6 nm	QY ranges between 0.1%–3%	Easy straight method	Low quantum yield
5.	Microwave-assisted method	4.51 nm	QY ranges between 3.1%–6.3%	Simple and convenient method	Poor control over size
6.	Hydrothermal and aqueous-based method	—	—	Highly water dispersible carbon dots	Poor control over size
7.	Template method	—	—	Carbon dots have biocompatibility and colloidal stability	Time consuming and expensive method

References

[1] M. Bruchez Jr., M. Moronne, P. Gin, S. Weiss, A.P. Alivisatos, Semiconductor nanocrystals as fluorescent biological labels, Science 281 (1998) 2013–2016.

[2] X. Michalet, F.F. Pinaud, L.A. Bentolila, J.M. Tsay, S. Doose, J.J. Li, G. Sundaresan, A.M. Wu, S. S. Gambhir, S. Weiss, Quantum dots for live cells, in vivo imaging, and diagnostics, Science 307 (2005) 538–544.

[3] J. Peng, W. Gao, B.K. Gupta, Z. Liu, R. Romero-Aburto, L. Ge, L. Song, L.B. Alemany, X. Zhan, G. Gao, S.A. Vithayathil, B.A. Kaipparettu, A.A. Marti, T. Hayashi, J.J. Zhu, P.M. Ajayan, Graphene quantum dots derived from carbon fibers, Nano Lett. 12 (2012) 844–849.

[4] Y. Li, Y. Hu, Y. Zhao, G. Shi, L. Deng, Y. Hou, L. Qu, An electrochemical avenue to green-luminescent graphene quantum dots as potential electron-acceptors for photovoltaics, Adv. Mater. 23 (2011) 776–780.

[5] L. Cao, S. Sahu, P. Anikumar, C.E. Bunker, J.A. Xu, K.A.S. Fernando, P. Wang, E.A. Guliants, K. N. Tackett, Y.P. Sun, Effect of UV irradiation on photoluminescence of carbon dots, J. Am. Chem. Soc. 133 (2011) 4754–4757.

[6] L. Cao, X. Wang, M.J. Meziani, F. Lu, H. Lou, P.G. Wang, Y. Lin, B.A. Harruff, L.M. Veca, D. Murray, S.Y. Xie, Y.P. Sun, Carbon dots for multiphoton bioimaging, J. Am. Chem. Soc. 129 (2007) 11318–11319.

[7] P. Anilkumar, X. Wang, L. Cao, S. Sahu, J.H. Liu, P. Wang, L.L. Tian, K.W. Sun, M.A. Bloodgood, Y.P. Sun, Carbon "quantum" dots for optical bioimaging, Nanoscale 3 (2011) 2023–2027.

[8] X.Y. Xu, R. Ray, Y.L. Gu, H.J. Ploehn, L. Gearheart, K. Raker, W.A. Scrivens, Electrophoretic analysis and purification of fluorescent single-walled carbon nanotube fragments, J. Am. Chem. Soc. 126 (2004) 12736–12737.

[9] Y.P. Sun, B. Zhou, Y. Lin, W. Wang, K.A.S. Fernando, P. Pathak, M.J. Meziani, B.A. Harruff, X. Wang, H.F. Wang, P.G. Luo, H. Yang, M.E. Kose, B.L. Chen, L.M. Veca, S.Y. Xie, Preparation of carbon dots and their application in food analysis as signal probe, J. Am. Chem. Soc. 128 (2006) 7756–7757.

[10] Y.H. Deng, D.X. Zhao, X. Chen, F. Wang, H. Song, D.Z. Shen, Long lifetime pure organic phosphorescence based on water soluble carbon dots, Chem. Commun. 49 (2013) 5751–5753.

[11] B. Liao, P. Long, B.Q. He, S.J. Yi, B.L. Ou, S.H. Shen, J. Chen, Reversible fluorescence modulation of spiropyran-functionalized carbon nanoparticles, J. Mater. Chem. C 1 (2013) 3716–3721.

[12] Z. Lin, W. Xue, H. Chen, J.M. Lin, Classical oxidant induced chemiluminescence of fluorescent carbon dots, Chem. Commun. 48 (2012) 1051–1053.

[13] C.X. Li, Z.Y. Hou, Y.L. Dai, D.M. Yang, Z.Y. Cheng, P.A. Ma, J. Lin, A facile fabrication of upconversion luminescent and mesoporous core–shell structured β-NaYF$_4$:Yb^{3+}, Er^{3+}@mSiO$_2$ nanocomposite spheres for anti-cancer drug delivery and cell imaging, Biomater. Sci. 1 (2013) 213–223.

[14] Z.L. Wu, Z.X. Liu, Y.H. Yuan, Carbon dots: materials, synthesis, properties and approaches to long-wavelength and multicolor emission, J. Mater. Chem. B 5 (2017) 3794–3809.

[15] X. Gao, C. Du, Z. Zhuang, W. Chen, Carbon quantum dot-based nanoprobes for metal ion detection, J. Mater. Chem. C 4 (2016) 6927–6945.

[16] H.P. Liu, T. Ye, C.D. Mao, Fluorescent carbon nanoparticles derived from candle soot, Angew. Chem. Int. Ed. 46 (2007) 6473–6475.

[17] Y.F. Sha, J.Y. Lou, S.Z. Bai, D. Wu, B.Z. Li, Y. Ling, Hydrothermal synthesis of nitrogen-containing carbon nanodots as the high-efficient sensor for copper(II) ions, Mater. Res. Bull. 48 (2013) 1728–1731.

[18] S. Liu, J.Q. Tian, L. Wang, Y.W. Zhang, X.Y. Qin, Y.L. Luo, A.M. Asiri, A.O. Al-Youbi, X.P. Sun, Green synthesis of fluorescent carbon quantum dots for detection of Hg^{2+}, Adv. Mater. 24 (2012) 2037–2041.

[19] B. Zhang, C.Y. Liu, Y. Liu, A novel one-step approach to synthesize fluorescent carbon nanoparticles, Eur. J. Inorg. Chem. 2010 (2010) 4411–4414.

[20] J. Li, B. Zhang, F. Wang, C.Y. Liu, Silver/carbon-quantum-dot plasmonic luminescent nanoparticles, New J. Chem. 35 (2011) 554–557.

[21] Y.H. Yang, J.H. Cui, M.T. Zheng, C.F. Hu, S.Z. Tan, Y. Xiao, Q. Yang, Y.L. Liu, One-step synthesis of amino-functionalized fluorescent carbon nanoparticles by hydrothermal carbonization of chitosan, Chem. Commun. 48 (2012) 380–382.

[22] S.T. Yang, X. Wang, H.F. Wang, F.S. Lu, P.G.J. Luo, L. Cao, M.J. Meziani, J.H. Liu, Y.F. Liu, M. Chen, Y.P. Huang, Y.P. Sun, Gd(iii)-doped carbon dots as a dual fluorescent-MRI probe, J. Phys. Chem. C 113 (2009) 18110–18114.

[23] H.T. Li, X.D. He, Z.H. Kang, H. Huang, Y. Liu, J.H. Liu, S.Y. Lian, C.H.A. Tsang, X.B. Yang, S. T. Lee, Full-colour carbon dots: integration of multiple emission centres into single particles, Angew. Chem. Int. Ed. 122 (2010) 4532–4536.

[24] H. Ming, Z. Ma, Y. Liu, K. Pan, H. Yu, F. Wang, Z.H. Kang, Full-colour carbon dots: integration of multiple emission centres into single particles, Dalton Trans. 41 (2012) 9526–9531.

[25] R. Zhang, W. Chen, Nitrogen-doped carbon quantum dots: facile synthesis and application as a "turn-off" fluorescent probe for detection of Hg^{2+} ions, Biosens. Bioelectron. 55 (2014) 83–90.

[26] C. Yuan, B.H. Liu, F. Liu, M.Y. Han, Z.P. Zhang, Dual-colored carbon dot ratiometric fluorescent test paper based on a specific spectral energy transfer for semiquantitative assay of copper ions, Anal. Chem. 86 (2014) 1123–1130.

[27] H.M.R. Goncalves, A.J. Duarte, J.C.G.E. da Silva, Optical fiber sensor for Hg(II) based on carbon dots, Biosens. Bioelectron. 26 (2010) 1302–1306.

[28] I. Costas-Mora, V. Romero, I. Lavilla, C. Bendicho, In situ building of a nanoprobe based on fluorescent carbon dots for methylmercury detection, Anal. Chem. 86 (2014) 4536–4543.

[29] R.H. Liu, H.T. Li, W.Q. Kong, J. Liu, Y. Liu, C.Y. Tong, X. Zhang, Z.H. Kang, High-bright fluorescent carbon dots and their application in selective nucleoli staining, Mater. Res. Bull. 48 (2013) 2529–2534.

[30] A. Gupta, A. Chaudhary, P. Mehta, C. Dwivedi, S. Khan, N.C. Verma, C.K. Nandi, Nitrogen-doped, thiol-functionalized carbon dots for ultrasensitive Hg(ii) detection, Chem. Commun. 51 (2015) 10750–10753.

[31] Y. Xu, C.J. Tang, H. Huang, C.Q. Sun, Y.K. Zhang, Q.F. Ye, A.J. Wang, Green synthesis of fluorescent carbon quantum dots for detection of Hg^{2+}, Chin. J. Anal. Chem. 42 (2014) 1252–1258.

[32] Y.C. Lu, J. Chen, A.J. Wang, N. Bao, J.J. Feng, W.P. Wang, L.X. Shao, Facile synthesis of oxygen and sulfur co-doped graphitic carbon nitride fluorescent quantum dots and their application for mercury(ii) detection and bioimaging, J. Mater. Chem. C 3 (2015) 73–78.

[33] Y. Liang, H. Zhang, Y. Zhang, F. Chen, Simple hydrothermal preparation of carbon nanodots and their application in colorimetric and fluorimetric detection of mercury ions, Anal. Methods 7 (2015) 7540–7547.

[34] V. Sharma, A.K. Saini, S.M. Mobin, Multicolour fluorescent carbon nanoparticle probes for live cell imaging and dual palladium and mercury sensors, J. Mater. Chem. B 4 (2016) 2466–2476.

[35] Z.H. Gao, Z.Z. Lin, X.M. Chen, H.P. Zhong, Z.Y. Huang, Carbon quantum dot-based nanoprobes for metal ion detection, Anal. Methods 8 (2016) 2297–2304.

[36] X.Y. Xu, B. Yan, Fabrication and application of ratiometric and colorimetric fluorescent probe for Hg^{2+} based on dual-emissive metal-organic framework hybrids with carbon dots and Eu^{3+}, J. Mater. Chem. C 4 (2016) 1543–1549.

[37] Z.S. Wu, M.K. Feng, X.X. Chen, X.J. Tang, N-dots as a photoluminescent probe for the rapid and selective detection of Hg^{2+} and Ag^{+} in aqueous solution, J. Mater. Chem. B 4 (2016) 2086–2089.

[38] Z.M. Zhang, Y.P. Shi, Y. Pan, X. Cheng, L.L. Zhang, J.Y. Chen, M.J. Li, C.Q. Yi, Quinoline derivative-functionalized carbon dots as a fluorescent nanosensor for sensing and intracellular imaging of Zn^{2+}, J. Mater. Chem. B 2 (2014) 5020–5027.

[39] M.M. Yang, W.Q. Kong, H. Li, J. Liu, H. Huang, Y. Liu, Z.H. Kang, Carbon dots serve as an effective probe for the quantitative determination and for intracellular imaging of mercury(II), Microchim. Acta 182 (2015) 2443–2450.

[40] Cadmium:P. Karfa, E. Roy, S. Patra, S. Kumar, A. Tarafdar, R. Madhuri, P.K. Sharma, Amino acid derived highly luminescent, heteroatom-doped carbon dots for label-free detection of Cd^{2+}/Fe^{3+}, cell imaging and enhanced antibacterial activity, RSC Adv. 5 (2015) 58141–58153.

[41] R.R. Gaddam, D. Vasudevan, R. Narayan, K.V.S.N. Raju, Controllable synthesis of biosourced blue-green fluorescent carbon dots from camphor for the detection of heavy metal ions in water, RSC Adv. 4 (2014) 57137–57143.

[42] W.J. Niu, D. Shan, R.H. Zhu, S.Y. Deng, S. Cosnier, X.J. Zhang, Dumbbell-shaped carbon quantum dots/AuNCs nanohybrid as an efficient ratiometric fluorescent probe for sensing cadmium (II) ions and l-ascorbic acid, Carbon 96 (2016) 1034–1042.

[43] A. Salinas-Castillo, M. Ariza-Avidad, C. Pritz, M. Camprubi-Robles, B. Fernandez, M.J. Ruedas-Rama, A. Megia-Fernandez, A. Lapresta-Fernandez, F. Santoyo-Gonzalez, A. Schrott-Fischer, L. F. Capitan-Vallvey, Carbon dots for copper detection with down and upconversion fluorescent properties as excitation sources, Chem. Commun. 49 (2013) 1103–1105.

[44] Y. Dong, R. Wang, G. Li, C. Chen, Y. Chi, G. Chen, Polyamine-functionalized carbon quantum dots as fluorescent probes for selective and sensitive detection of copper ions, Anal. Chem. 84 (2012) 6220–6224.

[45] N. Gogoi, M. Barooah, G. Majumdar, D. Chowdhury, Carbon dots rooted agarose hydrogel hybrid platform for optical detection and separation of heavy metal ions, ACS Appl. Mater. Interfaces 7 (2015) 5–10.

[46] N. Murugan, M. Prakash, M. Jayakumar, A. Sundaramurthy, A.K. Sundramoorthy, Green synthesis of fluorescent carbon quantum dots from *Eleusine coracana* and their application as a fluorescence 'turn-off' sensor probe for selective detection of Cu^{2+}, Appl. Surf. Sci. 476 (2019) 468–480.

[47] Z. Yan, X. Hua, Z. Liu, J. Chen, Carbon dots rooted agarose hydrogel hybrid platform for optical detection and separation of heavy metal ions, New Carbon Mater. 29 (2014) 216–224.

[48] Z. Lin, W. Xue, H. Chen, J.M. Lin, Peroxynitrous-acid-induced chemiluminescence of fluorescent carbon dots for nitrite sensing, Anal. Chem. 83 (2011) 8245–8251.

[49] J.M. Liu, L. Lin, X.X. Wang, J. Li, M. Cui, S. Jiang, W. Cai, L. Zhang, Z. Zheng, $Zr(H_2O)_2EDTA$ modulated luminescent carbon dots as fluorescent probes for fluoride detection, Analyst 138 (2013) 278–283.

[50] X.F. Hou, F. Zeng, F.K. Du, S.Z. Wu, Carbon-dot-based fluorescent turn-on sensor for selectively detecting sulfide anions in totally aqueous media and imaging inside live cells, Nanotechnology 24 (2013) 335502.

[51] S. Barman, M. Sadhukhan, Facile bulk production of highly blue fluorescent graphitic carbon nitride quantum dots and their application as highly selective and sensitive sensors for the detection of mercuric and iodide ions in aqueous media, J. Mater. Chem. 22 (2012) 21832–21837.

[52] S.R. Zhang, Q. Wang, G.G. Tian, H.G. Ge, Hyperbranched polyaniline: a new conductive polyaniline with simultaneously good solubility and super high thermal stability, Mater. Lett. 115 (2014) 233–236.

[53] Y.Q. Dong, R.X. Wang, W.R. Tian, Y.W. Chi, G.N. Chen, Fluorescent carbon dots: rational synthesis, tunable optical properties and analytical applications, RSC Adv. 4 (2014) 3701–3705.

[54] X. Gao, C. Ding, A. Zhu, Y. Tian, Carbon-dot-based ratiometric fluorescent probe for imaging and biosensing of superoxide anion in live cells, Anal. Chem. 86 (2014) 7071–7078.

[55] T. Tanaka, Reproductive and neurobehavioural toxicity study of erythrosine administered to mice in the diet, Food Chem. Toxicol. 44 (2006) 179–187.

[56] K.A. Amin, H.A. Hameid, A.H. Abd Elsttar, Effect of food azo dyes tartrazine and carmoisine on biochemical parameters related to renal, hepatic function and oxidative stress biomarkers in young male rats, Food Chem. Toxicol. 48 (2010) 2994–2999.

[57] H.P. Xu, X. Yang, G. Li, C. Zhao, X.J. Liao, J. Agric, Green synthesis of fluorescent carbon dots for selective detection of tartrazine in food samples, Food Chem. 63 (2015) 6707–6714.

[58] X.Y. Shan, L.J. Chai, J.J. Ma, Z.S. Qian, J.R. Chen, H. Feng, B-doped carbon quantum dots as a sensitive fluorescence probe for hydrogen peroxide and glucose detection, Analyst 139 (2014) 2322–2325.

[59] H. Wu, J. Jiang, X. Gu, C. Tong, Highly crystalline graphitic carbon nitride quantum dots as a fluorescent probe for detection Fe(III) via an inner filter effect, Microchim. Acta (2017) 1–8.

[60] Z. Gao, L.B. Wang, R.X. Su, R.L. Huang, W. Qi, Z.M. He, A carbon dot-based "off-on" fluorescent probe for highly selective and sensitive detection of phytic acid, Biosens. Bioelectron. 70 (2015) 232–238.

[61] J.L. Wang, J.H. Wei, S.H. Su, J.J. Qiu, Novel fluorescence resonance energy transfer optical sensors for vitamin B_{12} detection using thermally reduced carbon dots, New J. Chem. 39 (2015) 501–507.

[62] Z.B. Chen, J. Wang, H. Miao, L. Wang, S. Wu, X.M. Yang, Fluorescent carbon dots derived from lactose for assaying folic acid, Sci. China Chem. 59 (2016) 487–492.

[63] A. Kundu, S. Nandi, P. Das, A.K. Nandi, Facile and green approach to prepare fluorescent carbon dots: Emergent nanomaterial for cell imaging and detection of vitamin B2, J. Colloid Interface Sci. 468 (2016) 276–283.

[64] S.S.R. Dasary, D. Senapati, A.K. Singh, Y. Anjaneyulu, H. Yu, P.C. Ray, Highly Sensitive and selective dynamic light-scattering assay for TNT detection using p-ATP attached gold nanoparticle, ACS Appl. Mater. Interfaces 2 (2010) 3455–3460.

[65] Y. Yang, G.A. Turnbull, D.W. Samuel, Sensitive explosive vapor detection with polyfluorene lasers, Adv. Funct. Mater. 20 (2010) 2093–2097.

[66] Y. Wang, A. La, Y. Ding, Y.X. Liu, Y. Lei, Nanostructure-based optoelectronic sensing of vapor phase explosives—a promising but challenging method, Adv. Funct. Mater. 22 (2012) 3547–3555.

[67] S. Kumar, N. Venkatramaiah, S. Patil, Fluoranthene based derivatives for detection of trace explosive nitroaromatics, J. Phys. Chem. C 117 (2013) 7236–7245.

[68] W.E. Tenhaeff, L.D. McIntosh, K.K. Gleason, Chemical vapor deposition of conformal, functional, and responsive polymer films, Adv. Funct. Mater. 20 (2010) 1144–1151.

[69] K.S. Ju, R.E. Parales, Nitroaromatic compounds, from synthesis to biodegradation, Microbiol. Mol. Biol. Rev. 74 (2010) 250–272.

[70] J.D. Rodgers, N.J. Bunce, Treatment methods for the remediation of nitroaromatic explosives, Water Res. 35 (2001) 2101–2111.

[71] S. Zahn, T.M. Swager, Three-dimensional electronic delocalization in chiral conjugated polymers, Angew. Chem. Int. Ed. 41 (2002) 4225–4230.

[72] T.S. Tsai, Hazardous waste treatment technologies, Hazard. Waste Hazard. Mater. 8 (1991) 231–244.

[73] M. Emmrich, Kinetics of the alkaline hydrolysis of 2, 4, 6-trinitrotoluene in aqueous solution and highly contaminated soils, Environ. Sci. Technol. 33 (1999) 3802–3805.

[74] A. Mills, A. Seth, G. Peters, Alkaline hydrolysis of trinitrotoluene, TNT, Phys. Chem. Chem. Phys. 5 (2003) 3921–3927.

[75] L.L. Zhang, Y.J. Han, J.B. Zhu, Y.L. Zhai, S.J. Dong, Simple and sensitive fluorescent and electrochemical trinitrotoluene sensors based on aqueous carbon dots, Anal. Chem. 87 (2015) 2033–2036.

[76] S.F. Xu, H.Z. Lu, Ion imprinted dual reference ratiometric fluorescence probe for respective and simultaneous detection of Fe^{3+} and Cu^{2+}, Biosens. Bioelectron. 85 (2016) 950–956.

[77] S.J. Toal, W.C. Trogler, Mesoporous structured MIPs@CDs fluorescence sensor for highly sensitive detection of TNT, Biosens. Bioelectron. 85 (2016) 950–956.

[78] B.B. Chen, Z.X. Liu, H.Y. Zoub, C.Z. Huang, Highly selective detection of 2,4,6-trinitrophenol by using newly developed terbium-doped blue carbon dots, Analyst 141 (2016) 2676–2681.

[79] M.L. Liu, B.B. Chen, Z.X. Liub, C.Z. Huang, Highly selective and sensitive detection of 2,4,6-trinitrophenol by using newly developed blue–green photoluminescent carbon nanodots, Talanta 161 (2016) 875–880.

[80] M. Sha, M. Yao, S. de Sarger, L. Yan, S. Song, Carbon quantum dots encapsulated molecularly imprinted fluorescence quenching particles for sensitive detection of zearalenone in corn sample, Toxins 10 (2018) 1–10.

CHAPTER 14

Polymer and bionanomaterial-based electrochemical sensors for environmental applications

Navneet Kaur, Ranjeet Kaur, and Shweta Rana
Department of Chemistry, Panjab University, Chandigarh, India

1. Introduction

Environmental pollution is a worldwide problem posing an alarming situation. The ever-increasing demand of industries as a result of a rise in population has laid environmental sources at risk. The release of pollutants from industries is becoming a major issue among all sets of environmental components like air, water, and soil [1–3]. Among these pollutants, pesticides, antibiotics, and heavy metal ions are of major concern. Most of these sources continue to stay for a long tenure and enter water resources via soil leaching or absorption of contaminants and hence pose a great threat to human health [4, 5]. So, the control and monitoring of such pollutant's level in soil and water resources is of major concern.

In light of these emerging concerns from overuse of environmental pollutants, several analytical approaches including spectrophotometry [6], liquid chromatography-mass spectrometry (LC-MS) [7], and high-performance liquid chromatography (HPLC) [8], surface-enhanced Raman scattering (SERS) [9, 10], etc. have been put forward to determine the levels of their residues in real-time assays. But again, expensive instrumentation and long analysis times limit their uses. Hence, as a matter of urgency, particularly in underdeveloped areas, there is a need of designing compact, high-throughput, and low-cost detection systems and, for this, electroanalytical techniques offer practical solutions to trace level detection of antibiotics. Also, it includes the use of pulse techniques that offer ultratrace level determination of these environmental pollutants [11, 12]. However, to improve the sensitivity of a sensor and specificity toward a particular analyte, various electrode materials have been put forward. Among them, polymeric and bionanomaterials have emerged as efficient materials as they not only impart amplification and nanolevel sensitivity but also provide specificity toward the analyte [5, 13].

Polymeric nanomaterials and, in particular, conducting polymers pose diversity in electrochemical amplification strategies including the intrinsically advantageous properties of linearity over a wide range, sensitivity, and ease of fabrication [14]. They impart stability and tremendous conductivity to an electrode. The list includes mainly polypyrrole (PPy), polyaniline (PANI), and polythiophene (PT) that can be obtained via various chemical or electrochemical methods [15–18]. PPy enhances electronic conductivity and also acts as a binder to offer better interactions toward nanomaterials to develop polymeric nanocomposites (PNCs), whereas PANIs offer ease of fabrication along with good thermal stability and electric conductivity offered by protonation [17]. Moreover, PTs and PTNCs are the most promising candidates as electrode materials due to their environmental-friendly nature, cost-effectiveness, and high electric conductivity [16]. But the applications of these PNCs are limited as they do not offer specificity. So, a new class of polymeric nanomaterials was put forward, i.e., molecularly imprinting polymers (MIPs). MIPs are basically developed based on lock-key theory wherein an analyte works as a template and noncovalent interactions between the template and functional monomer generate sites for specific recognition of an analyte [19]. This will help to attain trace level determination as they retain molecular memory after elution of the template. MIPs are rather stable polymers obtained from polymerization in the presence of a template that is removed afterward [20]. The use of MIPs in electrochemical sensor fabrication is rather recent, which provides excellent selectivity and sensitivity. This technique has directed researchers toward synthetic polymerization with high specificity using complementary binding sites. MIPs have the power to recognize pesticides or antibiotics by utilizing molecularly imprinting technology (MIT). Ion imprinting is the class of MIT where ions act as template molecules thus establishing a new approach for heavy metal ions sensing [21]. The general mechanism of MIT is shown in Scheme 1.

Apart from this, bionanomaterial is another class of electrochemical sensors that has also been successfully employed in electrochemical sensors used for environmental pollutants. In particular, biosensors integrate the biological moieties with nanostructures for a combined effect of nanomaterial characteristics and biomolecular catalysis [5]. In this

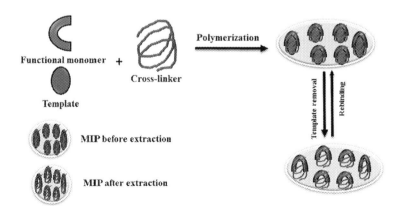

Scheme 1 The general diagram showing the fabrication of molecularly imprinted polymers (MIPs).

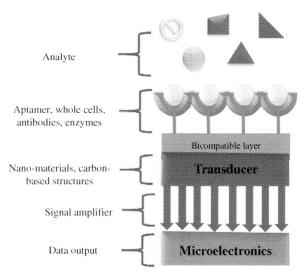

Analyte

Aptamer, whole cells, antibodies, enzymes

Biocompatible layer

Transducer

Nano-materials, carbon-based structures

Signal amplifier

Data output

Microelectronics

Scheme 2 The schematic diagram showing the components of a biosensor.

regard, different receptor elements like enzymes, aptamers, and antibodies with various nanostructures have emerged as a new class of functional bionanomaterials for environmental applications [22]. Biosensors are portable devices comprised of enzymes, microbes, nucleic acids, aptamers, whole cells, and micro-organisms assimilated into physiochemical signal transducers that, in this case, are an electrochemical workstation where current produced at certain oxidation or reduction potential corresponds to analyte concentration (Scheme 2). So, on the basis of a biomaterial fixed onto the modified electrode surface, the electrochemical biosensors can be enzymatic sensor, aptasensor, and immunosensors [22]. The nature of the recognition element depends on the target analyte. In recent years, biosensing has witnessed applications to various fields due their specificity, remarkable selectivity, and sensitivity. However, to widen the applicability of biosensors, the diversity in their functionality is there.

Henceforth, in this chapter, we will briefly outline the various types of polymeric and bionanomaterials that have been exploited for environmental applications. Sensing of different pollutants like pesticides, antibiotics, and heavy metal ions using these materials will be discussed along with detailed electrochemical approaches in designing sensor systems in recent years.

2. Detection of pesticides

In an effort to eradicate yield loss and maintain the level of product quality in agriculture, pesticides are commonly used to control pests, diseases, and weeds along with other herb pathogens. With thriving agricultural practices, pesticide residues are being found in water, sludge, soil, and aquatic biota. Today, the accumulation of pesticides residues is becoming a serious concern toward human health and environmental problems. Thus, the quantity of these pesticides in groundwater and food samples must be controlled [23–25], and to address this, development of high sensitivity sensors is the first and foremost step.

2.1 Polymeric-based electrochemical sensors for pesticide detection

2.1.1 Conducting polymers for pesticide detection

In the case of pesticides, conducting polymers are the most widely used material owing to their excellent conductivity and film-forming ability. Zhang et al. [26] designed a novel electrochemical sensor based on SnO_2@PANI/Au nanocomposite for the detection of ethephon using electrochemical impedance spectroscopy (EIS). The amine functionality present in PANI is hydrolyzed to ammonium cations, while that of phosphates of ethephon are hydrolyzed to phosphate anions leading to electrostatic interactions between the two. While the absorption of both phosphate groups and its anions can occur in SnO_2, more absorption happens in the modified electrode due to synergistic effects of the two. The changes in electron transfer resistance ($\triangle R_{ct}$) were used to produce a calibration plot between the range from 0.01 to 5.0 ng/mL and produced limit of detection (LOD) of 4.76 pg/mL followed by stability and selectivity studies with application to practical samples. Combination of conducting polymers with other nanomaterials augments their sensitivity toward the analyte. Wang et al. [27] highlighted a novel approach for the detection of methyl parathion (MP) using a zirconia/ordered macroporous polyaniline (ZrO_2/OMP) electrode. The modified electrode was designed by first electropolymerizing aniline in the presence of colloidal suspension of SiO_2 nanospheres, which is removed using HF after the polymerization. Here, PANI was grown between the interstitial spaces of the SiO_2 nanospheres resulting in a porous morphology with increased surface area. Then, ZrO_2 was electrodeposited onto the electrode using cyclic voltammetry (CV), which had further increased the roughness parameter in the electrode leading to better sensitivity in the ZrO_2/OMP electrode. The high affinity of ZrO_2 toward phosphate groups and benefits of good conducting properties and high catalytic performance led to low LOD of 2.28×10^{-10} mol/L (Table 1).

PEDOT is another highly stable conducting polymer endorsed with high conductivity extensively used in the detection of pesticides [30]. Zamora et al. [31] fabricated an electrochemical sensor combining the excellent properties of two potent materials, PEDOT and multiwalled carbon nanotubes (MWCNTs), for the determination of mancozeb (MCZ) filtrates in edible water samples. The proposed electrode and electrochemical methodology gave a wide linear working range from 25 to 150 μM and lower LOD of 10 μM. Migliorini et al. [33] fabricated nanofibers of polyamide 6 and ppy (PA6-Ppy) using electrospinning where the surface of an electrode was modified with chemically reduced graphene oxide (CRGO) as well as electrochemically reduced graphene oxide (ERGO). The effect of the two on the electrical conductivity was compared using CV

Table 1 Conducting polymer-based electrochemical sensors for the detection of OPs.

Sensor	Characterization techniques	Analytical technique used	Pesticide detected	LOD (M)	Linear range for detection (M)	Refs.
SnO_2@PANI/Au	XRD, FTIR, SEM, XPS	EIS	Ethephon	3.29×10^{-11}	0.069×10^{-9} to 0.034×10^{-7}	[26]
ZrO_2/OMP/GCE	SEM, XRD, CV, EIS	SWV	MP	2.28×10^{-10}	5.96×10^{-10} to 1.02×10^{-8}	[27]
PANI-ES-SWCNTs graphite	FTIR, SEM, XRD	DPV	Malathion	2.0×10^{-7}	2.0×10^{-7} to 14.0×10^{-7}	[28]
ITO/MnPc-TA/ N_3^- PANI	SEM	SWV	Eserine, diazinon, fenitrothion	0.049×10^{-6}, 0.088×10^{-6}, 0.062×10^{-6}	$(0.1-5) \times 10^{-6}$, $(0.2-7.5) \times 10^{-6}$, $(0.1-15) \times 10^{-6}$	[29]
PEDOT:PSS	UV, CV	LSV	DEDNPP	4.74×10^{-4}	$(10-50) \times 10^{-4}$	[30]
PEDOT/MWCNTs	AFM, SEM	CV	MCZ	10×10^{-6}	$(25-150) \times 10^{-6}$	[31]
Ni NPs–modified poly(p–aminophenol)/GCE	SEM, XPS, CV	DPV	DCF	0.08×10^{-6}	$(0.83-30.7) \times 10^{-6}$	[32]
FTO/PA6/PPy/CRGO	SEM, AFM, CV, EIS	DPV	Malathion	0.0024×10^{-6}	$(1.51-60) \times 10^{-6}$	[33]
SPE/PHA/mPEG	FESEM, CV, EIS	CA	Paraoxon–ethyl, fenitrothion, CPS	0.36×10^{-6}, 0.61×10^{-6}, 0.83×10^{-6}	$(1-10) \times 10^{-6}$	[34]

and EIS, which shows high conductivity of CRGO toward detection of malathion (LOD ~ 0.8 ng/mL) credited to the higher degree of roughness in the surface morphology provided by the nanomaterial. Sgobbi et al. [34] came up with a completely new approach wherein derivatized polyacrylamide polyhydroxamicalkanoate (PHA) was used as a recognition element. The authors have proposed PHA as moiety, which mimics the enzyme acetylcholinesterase (AChE) in its activity, hence resembling its behavior in performance toward pesticide detection (Scheme 3). In this biomimetic electrode, protonation of immobilized PHA leads to a nucleophillic attack from the hydroxamic acid group followed by the formation of a disulfide dimer from the oxidation of thiocholine. On incubation with the pesticide, deprotonation of the acidic group present on the polymer chain occurs. Thus mimicking the enzyme's activity, they rule out the difficulties in handling biological moieties. After necessary optimization, the sensor was able to detect paraoxon–ethyl, fenitrothion, and chlorpyrifos (CPS) with LOD as low as 0.36, 0.61, and 0.83 μmol/L, respectively.

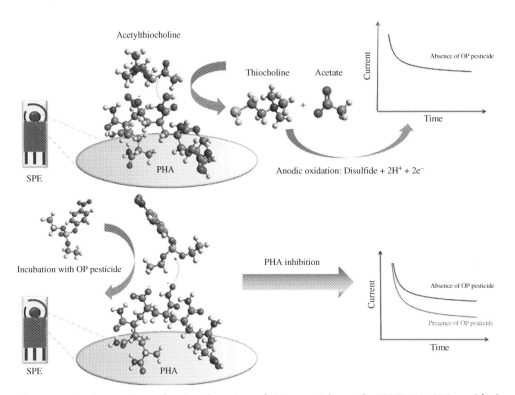

Scheme 3 Sensing pathway for the detection of OP pesticides with SPE/PHA/mPEG-modified electrode. *(Reproduced from L.F. Sgobbi, S.A.S. Machado, Functionalized polyacrylamide as an acetylcholinesterase-inspired biomimetic device for electrochemical sensing of organophosphorus pesticides, Biosens. Bioelectron. 100 (2018) 290–297.)*

2.1.2 Molecularly imprinted polymers

The detection mode employed for electrochemical sensing of pesticides is the result of performance of an analyte, i.e., pesticides under the effect of potential. Thus based on this, there are two categories in which MIPs can be divided [35]:

(a) Direct detection: In these MIPs, electrochemical response generated increases with an increase in the concentration of pesticides. This technique involves the use of analyte, i.e., pesticide as substrate and current response obtained is directly proportional to the pesticide concentration. This technique involves the use of pesticide as an electroactive compound, thus it can be exploited for use of electroactive pesticides only.

(b) Indirect detection: This technique involves the quantification and sensing of electroactive as well as nonelectroactive pesticides. It involves the use of redox probe generally $[Fe(CN)_6]^{3-/4-}$ or Prussian blue (PB) as redox probes. The electrochemical signal thus generated is the result of redox behavior of the probe. The current response is inversely proportional to pesticide concentration and is used for quantification of pesticides (Scheme 4).

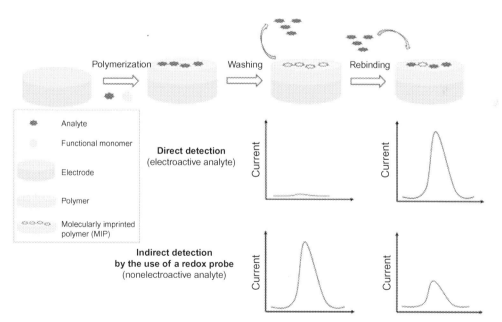

Scheme 4 Scheme showing the fabrication of electrochemical sensors based on molecularly imprinted polymer using electrochemical methodology and direct and indirect approach for analyte sensing. (Reproduced from D. Capoferri, F. Della Pelle, M. Del Carlo, D. Compagnone, Affinity sensing strategies for the detection of pesticides in food, Foods 7 (2018) 148.)

Direct detection MIPs

Various MIPs have been reported under this category for the detection of pesticides. Motaharian et al. [36] fabricated an electrochemical sensor using this approach utilizing methylacrylic acid (MAA) as a monomer where they have tried to get improved sensitivity toward diazinon detection by forming MIP nanoparticles (nano-MIPs) of higher adsorption capacity. These MIPs were further mixed with graphite powder to get an enhanced surface area resulting in an LOD of 7.9×10^{-10} M. To meet the demand of rapid and sensitive detection of pesticides, an iso-proturon-imprinted MIP was developed by Sadriu et al. [37] using pyrrole as a monomer. The electrochemical imprinting methodology has resulted in highly selective artificial recognition sites. The improved LOD of 0.5 μg/L with wide linearity highlights the better performance of the sensor thus hinting at the role of choice of material in getting higher outputs (Table 2).

Although high selectivity and sensitivity of MIPs and their role in electrochemical sensing is greatly appreciated by researchers, these types of electrochemical sensors suffer from limited conductivity and low surface-to-volume ratios leading to their limited applications. However, the use of carbon structures has paved a way in achieving sufficiently improved detection limits. Khadem et al. [38] developed MIP-based electrochemical sensors using MWCNTs to detect diazinon with LOD of 1.3×10^{-10} M. The sensing studies were performed using square wave voltammetry (SWV) after carrying out necessary optimization in the analyte specific potential range and then exploited in environmental and biological samples. GO, another carbon-based structure, has exciting properties such as high conductivity, large surface area, thermal stability, etc. To make use of these attributes, Khalifa et al. [39] developed an electrochemical sensor for the detection of profenofos (PFF) by loading GO on the surface of a glassy carbon electrode (GCE) before drop-casting MIPs. The electrochemical studies were thoroughly investigated using CV and EIS studies. After the necessary

Table 2 Various electrochemical sensor designs for the detection of pesticides utilizing direct MIP approach.

Sensor	Characterization techniques	Analytical technique used	Pesticide detected	LOD (M)	Linear range for detection (M)	Refs.
Nano-MIP/CP	SEM	SWV	Diazinon	7.9×10^{-10}	2.5×10^{-9} to 1.0×10^{-7}, 1.0×10^{-7} to 2.0×10^{-6}	[36]
MIP-GC electrode	CV	SWV	Iso-proturon	2.76×10^{-9}	2.5×10^{-9} to 5×10^{-6} M	[37]

optimization, the LOD obtained was 5 nM over a wide linear range from 5×10^{-8} M to 35×10^{-4} M. The system has been applied to detect PFF in a real sample of vegetables and spiked water samples.

In view of creating functionalities onto the carbon structures in a simplified manner, Xie et al. [40] chose vinylbenzoic acid as a functional monomer that helped add recognition units onto graphene via π-π interactions (Scheme 5). The elution of template was done using CV, whereas the performance of a sensor was evaluated using LSV. The reported electrochemical sensor shows LOD of 0.04 mM with excellent results for the detection of thiamethoxam in grain samples. Zhang et al. [41] fabricated an electrochemical sensor based on graphene for selective and highly sensitive detection of imidacloprid (IDP) residues in rice samples. In this, the first step involves the use of p-vinyl benzoic

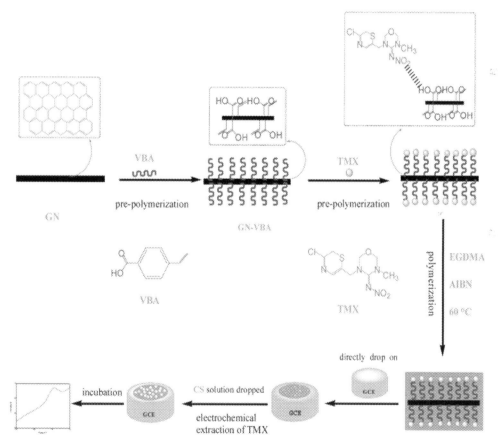

Scheme 5 Fabrication procedure for MIP-GN/GCE and mechanism behind sensing. *(Reproduced from T. Xie, M. Zhang, P. Chen, H. Zhao, X. Yang, L. Yao, H. Zhang, A. Dong, J. Wang, Z. Wang, A facile molecularly imprinted electrochemical sensor based on graphene: application to the selective determination of thiamethoxam in grain, RSC Adv. 7 (2017) 38884–38894.)*

Table 3 Different electrochemical sensors based on carbon-based structures utilizing direct MIP technique for pesticide detection.

Sensor	Characterization techniques	Analytical technique used	Pesticide detected	LOD (M)	Linear range for detection (M)	Refs.
MIP/ CP	SEM, FTIR	SWV	Diazinon	4.1×10^{-10}	5×10^{-10} to 1.0×10^{-6}	[38]
MIP/ GO/ GCE	SEM, FTIR	SWV	PFF	5×10^{-9}	5×10^{-8} to 35×10^{-4}	[39]
MIP- GN/ GCE	SEM, AFM, EIS	LSV	Thiamethoxam	0.04×10^{-8}	5×10^{-7} to 2.0×10^{-5}	[40]
MIP/ GN/ GCE	UV–Vis	LSV	IDP	1.0×10^{-7}	5×10^{-7} to 1.5×10^{-5}	[41]
MIP/ CP	–	SWV	Dichloran	4.8×10^{-10}	1×10^{-9} to 1.0×10^{-6}	[42]

acid (VBA) as a functional monomer that was absorbed onto the surface via π-π interactions. The studies of adsorption isotherms indicated the single-layer formation along with homogeneity with LOD of 0.10 μM (Table 3).

More promising outcomes with an increased number of imprinted sites come with the use of metal nanoparticles (MNPs) and metal oxide nanoparticles (MONPs) that exhibit catalytic activity apart from high electronic conductivity. Li et al. [43] developed a novel supramolecular-imprinted electrochemical sensor utilizing Pt-In as a surface modifier. Here, bromophenol blue-doped o-aminophenol was utilized as a functional monomer while inclusion complex of 4-*tert*-butylcalix[6]arene-imidacloprid served as a template for the preparation of a supramolecular-imprinted electrochemical sensor. Catalytic amplification was observed owing to the combination of Pt-In and bromophenol, and electricity generated by the combination resulted in the oxidation of IDP. The proposed sensor exhibited a wide linear potential range from 1×10^{-10} to 5×10^{-8}M with LOD of 1.2×10^{-11} M. The sensor thus developed has paved a way to the determination of pesticides in environmental and food products. Wang et al. [44] proposed a MIP-based GCE modified with reduced AuNPs and MP as an analyte as well as a template molecule. Modification with the AuNPs has resulted in an increased surface area making available a higher number of active sites for facile electron transfer resulting in effective quantification of MP with LOD of 0.01 μM. The real-time applicability of the sensor was monitored using quantification in different matrices with excellent recoveries pointing toward practical applications. Kumar et al. [45] fabricated a cost-effective

Table 4 Different electrochemical sensors based on nanostructures utilizing direct MIP technique for pesticide detection.

Sensor	Characterization techniques	Analytical technique used	Pesticide detected	LOD (M)	Linear range for detection (M)	Refs.
MIP/Pt-In/GCE	SEM, XPS	DPV	IDP	1.2×10^{-11}	1×10^{-10} to 5×10^{-8}	[43]
MIP/Au/GCE	SEM, EDX	CV	MP	1.0×10^{-8}	5×10^{-8} to 1.5×10^{-5}	[44]
MISP-modified SPIONs/PGE	FE-SEM, XRD, BET	SWSV	MCZ	1.77×10^{-9}	1.1×10^{-8} to 4.75×10^{-7}	[45]

and novel electrochemical sensor for the detection of MCZ for vegetable and soil samples where electrode modification was done using molecular imprinting of star polymer's (MISP) surface imprinted on superparamagnetic iron oxide nanoparticles (SPIONs). The synthesis of SPIONs was done using a hydrothermal approach followed by surface modification with vinyl silanes. The growth of MIPs on this surface was then carried out using surface-imprinted methodology. The resulting MISPs offer good recognition capacity along with high adsorption capacity. Low LODof 1.77×10^{-9} M highlights the potential for detection and removal of pesticides even at trace levels (Table 4).

Indirect determination-based MIPs

Indirect use of the redox moieties is also one of the prevalent approaches in detecting pesticides. Li et al. [46] described a hybrid electrochemical sensor based on Ag and N co-doped ZnO (Ag-N@ZnO/CHAC) for selective determination of cypermethrin (CYP) with reliable accuracy (Scheme 6). The surface of Ag-N@ZnO/CHAC-modified GCE was coated with a layer of MIP using dopamine and resorcinol as functional monomers resulting in diversity in blotting sites in the final design of the MIP thereby increasing the binding capacity. Availability of multiple functional groups using bifunctional monomers has dramatically enhanced the specific recognition ability. The optimization of parameters followed by low LOD of 6.7×10^{-14} M was achieved. The present method poses great potential for determination of drug residues, food safety, and environmental monitoring.

The same research group has also developed novel surface molecularly imprinted microspheres (SMIPMs) using distillation precipitation polymerization on the carbon paste electrode (CPE) for the detection of MP [47]. These SMIPMs have high adsorption and binding capacity compared with nonsurface-imprinted ones that has also been reflected in their electrochemical response. For enhanced surface area and higher conductivity, Feng et al. [48] used nitrogen and sulfur-doped hollow Mo_2C/C spheres (N,S-Mo_2C) with MIP to sense carbendazim (CBD) using high temperature

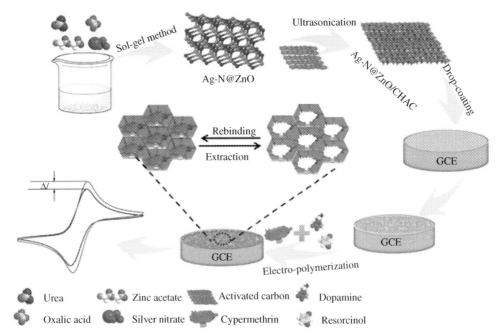

Scheme 6 Schematic showing the fabrication of Ag-N@ZnO/CHAC-modified GCE. *(Reproduced from Y. Li, L. Zhang, Y. Dang, Z. Chen, R. Zhang, Y. Li, B.C. Ye, A robust electrochemical sensing of molecularly imprinted polymer prepared by using bifunctional monomer and its application in detection of cypermethrin, Biosens. Bioelectron. 127 (2019) 207–214.)*

carbonization. The electrochemical performance was monitored using CV and EIS analysis showing better electrochemistry of the modified electrode owing to the combinatorial effect of the doping of heteroatom and high surface area via carbonization treatment. Under optimized parameters, LOD of 6.7×10^{-13} M has been achieved with application to fruit and vegetable samples (Table 5).

MNPs have always acted as the signal enhancers in the designing of sensors. Qi et al. [49] developed an electrochemical sensor for the quantification of carbofuran (CBF) in which AuNPs in the form of electron wires acted as transducers for signal monitoring while MIP served as recognition element. The electrochemical analysis was carried out using CV and DPV resulting in LOD as low as 2.0×10^{-8} M. The other necessary parameters like reproducibility, sensitivity, and stability were evaluated systematically indicating the potential of prepared sensors for the detection of CBF in vegetable samples. An electrochemical assay has been developed for the determination of glyphosate (GLY) in food and environment samples where the composites of PB with urchin-like AuNPs had served as electron mediators onto the surface of indium titanium oxide (ITO)–coated glass slides [50]. The response of analyte, measured with PB as a redox mediator, has offered a lower LOD of 5.44×10^{-7} M. An MIP has also been developed for fingerprinting of cyhexatin (CYT) in a pear sample using AuNPs onto the surface of a screen-printed

Table 5 Different electrochemical sensors based on nanostructures utilizing MIP technique for indirect pesticide detection (by redox probe).

Sensor	Characterization techniques	Analytical technique used	Pesticide detected	LOD (M)	Linear range for detection (M)	Refs.
DM-MIP-Ag-N@ZnO/CHAC	SEM, BET, XRD, XPS	CV, $[Fe(CN)_6]^{3-/4-}$	CYP	6.7×10^{-14}	2×10^{-13} to 8×10^{-9} M	[46]
SMIPMs/CPE	SEM, FTIR, BET	CV, $[Fe(CN)_6]^{3-/4-}$	MP	3.40×10^{-13}	1.0×10^{-12} to 8.0×10^{-9}	[47]
MIP/N,S–Mo$_2$C/GCE	SEM, TEM, FTIR, XRD	CV, $[Fe(CN)_6]^{3-/4-}$	CBD	6.7×10^{-13}	1×10^{-12} to 8×10^{-9}	[48]

carbon electrode (SPCE) by exploiting the hydrogen-bonding interaction between the template and functional monomer [51]. After the optimization of required conditions, the imprinted sensor imparted satisfactory linearity with LOD of 0.20 ng/mL. Roushani et al. [52] have utilized Au nanorods to modify the electrode surface as they promote electron transfer rate along with aptasensor science. After elution of the template molecule, i.e., CPS, the double-imprinting sites achieved more efficiently compared with conventional imprinted or aptasensor technology (Table 6).

2.2 Bionanomaterials for the detection of pesticides

Based on their biorecognition elements, these pesticide sensors can be categorized as:

2.2.1 Enzymatic biosensors for pesticide detection

The mechanistic approach of enzymatic biosensors for the determination of pesticides also involves two basic pathways (Scheme 7).

Direct determination

In this, the pesticides are determined directly by treating them as a substrate rather than as inhibitors. The pesticides are catalyzed by enzymes, and signal collection is done with a physiochemical signal transducer. Organophosphorus hydrolase (OPH), organophosphorus acid hydrolase and methyl parathion hydrolase (MPH) are the most commonly used enzymes for direct determination of pesticides, and in this regard different nanomaterials have been used for the immobilization of these enzymes.

(a) OPH-based assays: These have evolved as an emerging biorecognition element for direct detection of OPs. OPH catalyzes the hydrolysis of commonly used OPSs as follows [54]:

$$OP + Water \longrightarrow OP\text{-}Acid + R\text{-}OH (\text{-}SH) \quad Eq. (1)$$

Table 6 Different electrochemical sensors based on MNPs utilizing MIP technique for indirect pesticide detection (by redox probe).

Sensor	Characterization techniques	Analytical technique used	Pesticide detected	LOD (M)	Linear range for detection (M)	Refs.
MIP/ AuNPs/ GCE	FESEM	DPV, (K$_3$[Fe (CN)$_6$]/ K$_4$[Fe(CN)$_6$])	CBF	2.4×10^{-8}	5×10^{-8} to 4×10^4	[49]
AuNP– PB– MIP/ ITO	SEM, EDS, CV	DPV, (PB)	GLY	5.4×10^{-7}	2.37×10^{-6} to 7.1×10^{-6}	[50]
MIP– AuNPs/ ERGO/ SPCE	SEM, AFM, FTIR, LC–MS/MS	DPV, (K$_3$[Fe (CN)$_6$])	CYT	0.5×10^{-9}	2.5×10^{-9} to 1.2×10^{-6}	[51]
MIP– aptamer/ AuNR/ GCE	SEM, EIS, CV	DPV, ([Fe (CN)$_6$]$^{3-/4-}$)	CPS	0.35×10^{-15}	1.0×10^{-15} to 0.4×10^{-12}	[52]
MIPPy/ Au	SEM	DPV, (K$_3$[Fe (CN)$_6$])	GLY	1.6×10^{-9}	2.96×10^{-8} to 4.73×10^{-6}	[53]

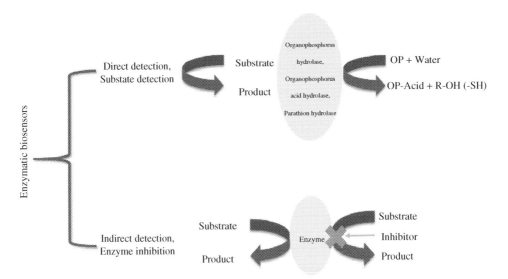

Scheme 7 Classification of enzymatic biosensors based on the mechanism involved in the detection of pesticides.

The electroactive product thus generated can be monitored with the help of electrochemical signal transducers making it feasible for direct detection of pesticides. These biorecognition elements are combined with various transducer components to enhance the sensitivity and selectivity of sensors. Utilizing elastin-like polypeptide organophosphate hydrolase (ELP-OPH) as a biorecognition element, an electrochemical biosensor has been developed for sensitive and selective determination of parathion and MP [55]. The nanocomposite of bovine serum albumin (BSA), titanium dioxide nanofibers (TiO_2NF), and carboxylic acid functionalized MWCNTs (c-MWCNTs) were utilized to enhance the adsorption of OPs attributed to its strong affinity toward phosphates. The principle for electrochemical sensing involves the electrochemical oxidation of paranitrophenol (PNP) formed as a result of hydrolysis of MP catalyzed by ELP-OPH carried out using chronoamperometry. The biosensor thus fabricated offers a wide linear range and low LOD of 10 and 12 nM for parathion and MP, respectively. This has been further exploited for detection of OPs in spiked lake water samples, indicating its excellent performance for real-time monitoring.

(b) MPH-based assays: These have evolved as an emerging biosensing element offering specificity toward MP with several platforms designed for MPH immobilization and capturing for biosensor fabrication.

Chen et al. [56] offered a nanocomposite film composed of AuNPs onto the surface of silica NPs mixed with $MWCNT_S$ for the immobilization of MPH. The transducer material offered high surface-to-volume ratio along with excellent conductivity. The biosensing element, i.e., MPH carries out the hydrolysis of MP in an effective manner resulting in the release of PNP, which undergoes oxidation-reduction and is detected by CV and SWV (Scheme 8). A synergistic effect of prepared nanocomposite is seen on the performance of MPH showing a wide linear range from 0.001 to 5 μg/mL with LOD of 0.3 ng/mL. Similarly, Liu et al. [57] designed an electrochemical biosensor to provide interface stability between biomolecules and the electrode by using aryldiazonium salt monolayers and an AuNPs-modified electrode interacted through Au-C chemistry. The MPH has been covalently immobilized on the surface of AuNPs. The sensing performance was evaluated using amperometry wherein the MPH carries out the hydrolysis of MP resulting in the formation of 4-nitrophenol. LOD is 0.07 ppb with a linear range from 0.2 to 100 ppb with excellent reproducibility.

(c) Lipases-based assays: Lipases, generally called triacylglycerol ester hydrolases, also offer catalysis of hydrolysis of esters in aqueous/organic media that is quite economical. Moreover, the designed biosensors are reusable, which has directed researchers toward their presentation as a cost-effective approach for the detection of Ops, especially MP. Ma et al. [58] developed a quick and sensitive electrochemical biosensor based on lipase@ZIF-8/CS/GCE and lipase@amine-modified ZIF-8/CS/GCE where modified materials ZIF-8 and An-ZIF-8 have offered a large surface area and biocompatible

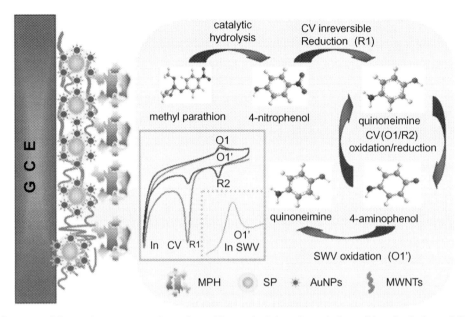

Scheme 8 Schematic representation of working principle of methylparathion hydrolase (MPH) biosensor for determination of methylparathion (MP). *(Reproduced from S. Chen, J. Huang, D. Du, J. Li, H. Tu, D. Liu, A. Zhang, Methyl parathion hydrolase based nanocomposite biosensors for highly sensitive and selective determination of methyl parathion, Biosens. Bioelectron. 26 (2011) 4320–4325.)*

environment for the immobilization of lipase (Scheme 9). After optimizing the various parameters, the lipase@amine-modified ZIF-8/CS/GCE and lipase@ZIF-8/CS/GCE modified electrodes have offered a wide linear range of 0.1–25 and 0.1–38 μM with LOD of 0.51 and 0.28 μM, respectively. The performance is attributed to lipase activity that favors the formation of PNP from MP and hence undergoes electrochemical oxidation.

Wang et al. [59] focused on developing a direct, sensitive, and highly reliable method using *Burkholderiacepacia* lipase as a biorecognition material with MOF nanofibers as a transducer for detection of MP residues in real-time samples. The sensor displayed similar electrochemical behavior to the already reported mechanism involving the formation of PNP with a wide linear range of 0.1–38 μM with LOD of 0.067 μM (Table 7).

Indirect determination

In this phenomenon, the determination of analyte is done by considering them as inhibitors. The pesticide attacks the active site of the enzyme leading to phosphorylation. This decreases the activity of the enzyme toward the electroactive substrate formed as a result of hydrolysis. This, in turn, is used to monitor the concentration of pesticide by using the formula:

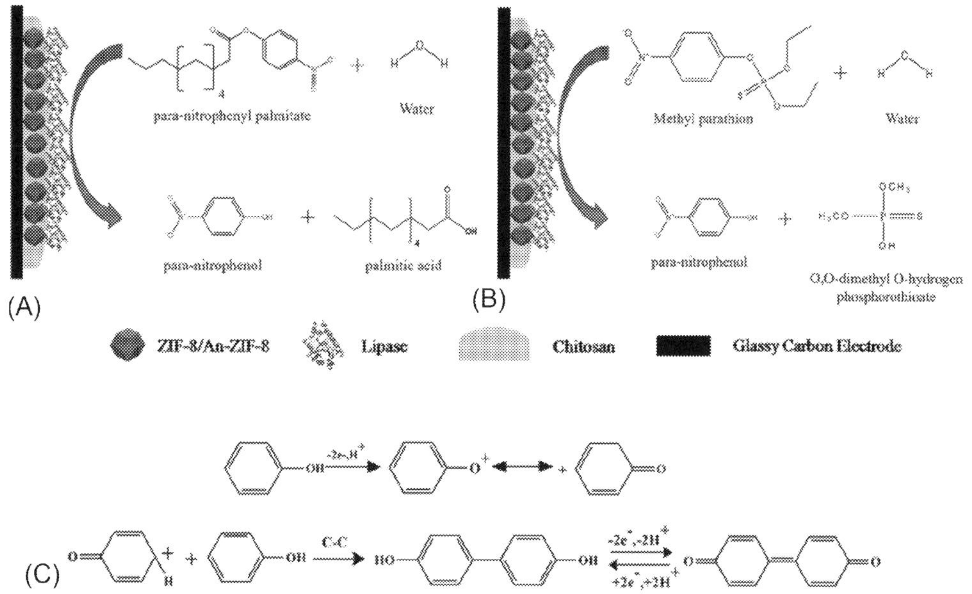

Scheme 9 Schematic illustration of sensing principles based on lipase activity and also the phenolic oxidation in sensing. *(Reproduced from B. Ma, L.Z. Cheong, X. Weng, C.-P. Tan, C. Shen, Lipase@ZIF-8 nanoparticles-based biosensor for direct and sensitive detection of methyl parathion. Electrochim. Acta 283 (2018) 509–516.)*

Table 7 Various biological sensors developed for determination of pesticides using direct approach.

Sensor	Characterization techniques	Analytical technique used	Pesticide detected	LOD (M)	Linear range for detection (M)	Refs.
ELP-OPH/ BSA/ TiO$_2$NFs/c– MWCNTs	SEM, RAMAN, FTIR, XRD	CA	MP, parathion	12×10^{-9} and 10×10^{-9}	–	[55]
SP@AuNPs/ MWNTs/ GCE	SEM, EIS	SWV	MP	1.1×10^{-9}	$3.8 \times \times 10^{-9}$ to 19.0×10^{-6}	[56]
MPH/AuNP	SEM, XPS	SWV	MP	0.2×10^{-9}	0.7×10^{-9} to 0.38×10^{-6}	[57]
BCL@An– ZIF-8/CS/ GCE	SEM	CA	MP	0.51×10^{-6}	$(0.1–25) \times 10^{-6}$	[58]
BCL@MOF nanofibers/ CS/GCE	XRD, BET	DPV	MP	0.067×10^{-6}	$(0.1–38) \times 10^{-6}$	[59]

$$I(\%) = \left[1 - \frac{I_1}{I_0}\right] \times 100 \tag{2}$$

The resulting inhibition percentage calculated using current measurements of the electrochemical sensor is then plotted against the concentration of inhibitor, and the concentration of pesticide in the samples can be calculated from the calibration plot thus generated [60, 61].

(a) Cholinesterase (ChE)-based assays: The most commonly used recognition element for inhibition-based biosensors is ChE enzyme where generally AChE is the one that combines with various transducer components to enhance the sensitivity and selectivity of the sensor [62]. The AChE consists of two active sites, serine hydroxyl group and histidine, that mainly participate in the whole phenomenon. The positively charged pockets in enzymes attract the substrate acteylthiocholine (ATCl) and subsequently hydrolyze it to thiocholine and acetic acid. The species thus produced, i.e., choline is electroactive and hence undergoes oxidation. This oxidation process is suppressed in presence of pesticide, i.e., that blocks the active sites of the enzyme and hence is used for the quantification of pesticides (Scheme 10) [63].

One of the extensively used matrixes used with AChE-based sensors is chitosan (CS). It offers good film-forming ability and imparts biocompatibility along with nontoxic behavior [64]. CS also imparts stability to enzyme performance along with good adhesive properties and is susceptible for chemical modifications [65]. Most of the researchers prepared dispersion of AChE in CS solution and used it for the fabrication of a biosensor on the surface of electrode. The biosensor is prepared by directly drop-casting the prepared

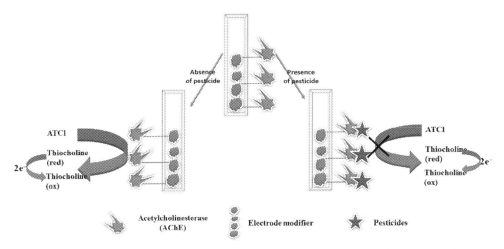

Scheme 10 Scheme showing the mechanism of AChE biosensor and its electrochemical-sensing performance.

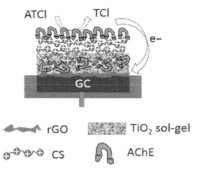

Scheme 11 Scheme showing AChE biosensor fabrication and chitosan as adhesion. *(Reproduced from H.F. Cui, W.W. Wu, M.M. Li, X. Song, Y. Lv, T.T. Zhang, A highly stable acetylcholinesterase biosensor based on chitosan-TiO2-graphene nanocomposites for detection of organophosphate pesticides, Biosens. Bioelectron. 99 (2018) 223–229.)*

dispersion on the surface of transducer followed by incubation at 4°C for almost 10–12 h. This provides stability to the prepared sensor as it avoids leakage from the electrode surface via good adhesion properties. Cui et al. [64] developed an electrochemical biosensor by adsorption of AChE on the surface of CS to deal with the enzyme leakage problems from electrode surface, wherein TiO_2 sol–gel and rGO were utilized as an immobilization matrix (CS@TiO_2-CS/rGO). The assimilation of CS and electrodeposition of CS layers into/on TiO_2 sol–gel has imparted mechanical strength to the sensor (Scheme 11). The catalytic activity of the prepared samples was higher and LOD was as low as 29 Nm toward Dichlorvos (DDVP) sensing. Zhou et al. [65] fabricated a novel electrochemical biosensor utilizing transition metal carbide nanosheets (MXenes) and CS as AChE an immobilization matrix (CS-$Ti_3C_2T_x$) for sensing of malathion. The good adhesion, excellent film-forming nature, and nontoxic behavior of CS with large surface area of $Ti_3C_2T_x$ nanosheets imparted good reproducibility and stability to the developed biosensor. The list of some biosensors utilizing CS as a biocompatible layer are shown in Table 8.

Amongst carbon materials, graphene-based structures are also one of the emerging materials used for electrochemical-sensing analysis. These materials impart high surface area along with excellent electrocatalysis for sensing applications [72]. Also, the loading of desired biomolecules is favored as they offer easy adsorption on their surface. Mahmoudi et al. [73] developed an electrochemical biosensor utilizing Ce metal incorporated into UiO-66-template to form Ce/UiO-66 and mixed it with different carbonaceous forms like GO, carbon black (CB), and MWCNTs. The electrochemical characterization of differently loaded carbon structures showed an excellent electron transfer rate with more adsorption of AChE onto the porous material of Ce (7%) and MWCNTs (30%) (Scheme 12). The oxophilicity of Ce imparted good electrocatalytic performance with Michaelis–Menten constant of 0.258 mM. This leads to low LOD of approximately 0.004 nM and hence can be exploited for on-field analysis.

Table 8 Different AChE-based electrochemical biosensors utilizing CS as binder.

Sensor	Characterization techniques	Analytical technique used	Pesticide detected	LOD (M)	Linear range for detection (M)	Refs.
AChE/CS@TiO$_2$-CS/rGO/GCE	SEM	DPV	DDVP	29×10^{-9}	$0.036–22.6 \times 10^{-6}$	[64]
AChE/CS-Ti$_3$C$_2$T$_x$/GCE	SEM, TEM, XRD	CA	Malathion	0.3×10^{-14}	1×10^{-14} to 1×10^{-8}	[65]
AChE-CS/3DG-CuO NFs/GCE/GCE	FESEM, EIS	SWV	Malathion	0.92×10^{-12}	3×10^{-12} to 46.66×10^{-9}	[66]
AChE-Chit/Pd@AuNWsN/GCE	TEM	DPV	Malathion	0.037×10^{-12}	0.1×10^{-12} to 100×10^{-9}	[67]
AChE-Cs/Pd-Cu NWs/GCE	TEM, XRD	DPV	Malathion	4.5×10^{-12}	$15–3000 \times 10^{-12}$ and $1500–9000 \times 10^{-9}$	[68]
NA/AChE-CS/Pd@Au NRs/GCE	HRTEM	DPV	Paraoxon	3.6×10^{-12}	33.6×10^{-12} to 100×10^{-9}	[69]
AChE-Chit/PdNi NWs/m-MoS2/GCE	HRTEM, XRD, XPS	DPV	Omethoate	0.05×10^{-12}	$10^{-13} – 10^{-7}$	[70]
Pt/ZnO/AChE/Chitosan bio-electrode	FESEM, XRD	CV	Carbosulfan	0.24×10^{-9}	$(5–30) \times 10^{-9}$	[71]

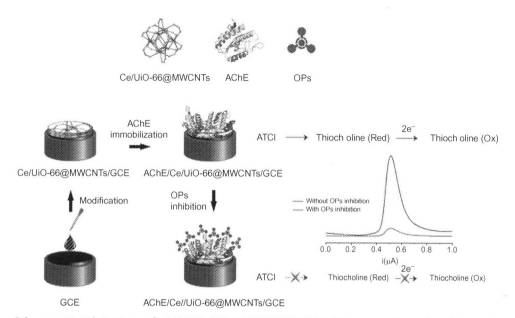

Scheme 12 Fabrication of AChE/Ce/UiO-66@MWCNTs/GCE biosensor. *(Reproduced from E. Mahmoudi, H. Fakhri, A. Hajian, A. Afkhami, H. Bagheri, High-performance electrochemical enzyme sensor for organophosphate pesticide detection using modified metal-organic framework sensing platforms, Bioelectrochemistry 130 (2019) 107348.)*

A highly sensitive electrochemical biosensor based on AChE utilizing the electrochemically polymerized 4,7-di(furan-2-yl)benzothiadiazole and nanocomposite Ag-rGO-NH$_2$ has been reported by Zhang et al. [74]. The composite provides the hydrophilic surface for AChE and offers excellent conductivity and biocompatibility. After necessary optimization, a wide linear range with LOD of 0.032 μg/L has been obtained. Rana et al. [75] developed an electrochemical biosensor based on PEDOT/ ILRGO utilizing AChE as an enzyme for quantification of OPs. The high surface area and good electronic conductivity offered by synergistic interactions has resulted in enhanced current response. The generation of electroactive thiocholine forms the basis of quantification leading to low LOD of 0.04, 0.117, and 0.108 ng/mL for CPS, malathion, and MP, respectively. The sensor offered good reactivation with real-time applications for detection of OPs in beverages (Table 9).

Apart from use of a single enzyme for sensing purposes, bienzymatic sensors have also been developed for the detection of pesticides [83]. For this purpose, AChE is often coupled with choline oxidases (ChOx) for the determination of pesticides. The product resulted from oxidation of substrate of AChE supplied to ChOx, and the electroactive species thus generated is exploited for analysis of pesticides [84]. For the very first time, AChE was coupled with OPH for selective detection and discrimination between

Table 9 Various AChE-based electrochemical biosensors based on carbon-based nanostructures.

Sensor	Characterization techniques	Analytical technique used	Pesticide detected	LOD (M)	Linear range for detection (M)	Refs.
AChE/Ce/UiO-66-/MWCNTs/GCE	XRD, FTIR, SEM	DPV	Paraoxon	0.004×10^{-9}	$(0.01\text{--}150) \times 10^{-12}$	[73]
poly(FBThF)/Ag-rGO-NH$_2$/AChE/GCE	SEM, TEM, XRD, FTIR	CV	Malathion, richlorfon	0.09×10^{-9}, 3.9×10^{-12}	$(0.0003\text{--}0.03) \times 10^{-6}$, $(0.082\text{--}0.008) \times 10^{-9}$	[74]
AChE/PEDOT/RGO/IL/FTO	FTIR, XRD, FE-SEM	DPV	Malathion, CPS, MP	0.35×10^{-9}, 0.11×10^{-9}, 0.37×10^{-9}	—	[75]
AChE/NA/AuNPs/rGO/GCE	FTIR, SEM, TEM, EDS	DPV	Malathion, MP	8.4×10^{-14}, 8.24×10^{-14}	1.0×10^{-6} to 1.0×10^{-10}	[76]
Au/VNSWCNTs/AuNPs/AChE	RAMAN, AFM	DPV	MP, Malathion, CPS	0.01×10^{-6}, 5.9×10^{-9}, 5.8×10^{-12}	3.7×10^{-14} to 3.7×10^{-9}, 3.03×10^{-14} to 3.03×10^{-9}, 2.8×10^{-14} to 2.8×10^{-9}	[77]
AuNPs/DAR/AChE	UV, TEM	CV	Malathion, MP	1.5×10^{-15}, 2.22×10^{-13}	0.038×10^{-9} to 0.038×10^{-13}, 0.038×10^{-9} to 0.038×10^{-13}	[78]
IL$_1$-MWCNTs/AChE/GC	TEM, TGA	DPV	OPs	3.3×10^{-11}	1.0×10^{-10} to 5.0×10^{-7}	[79]
AChE/Fe$_3$O$_4$@MHCS/GCE	SEM, TEM, EDS, XRD, BET	DPV	Malathion	0.044×10^{-9}	0.03×10^{-9} to 0.30×10^{-6}	[80]
NF/AChE/NF-NiCo$_2$S$_4$/CPE	SEM, TEM, XRD, EDS	DPV	MP, Malathion	0.015×10^{-10}, 0.010×10^{-11}	$(3.7 \times 10^{-12}$ to $3.7 \times 10^{-8})$, $(3.03 \times 10^{-13}$ to $3.03 \times 10^{-10})$	[81]
AChE/PEDOT-MWCNTs/FTO	FTIR, FE-SEM	DPV	Malathion	1×10^{-15}	1×10^{-15} to 1×10^{-6}	[82]

pesticides [85] using layer-by-layer assembly of MWCNTs/OPH and MWCNTs/ AChE with MWCNTs/PEI and MWCNTs/DNA bilayers. The sensor detects paraoxon by catalytic and inhibition approach yet carbaryl by inhibition phenomenon. The remarkable distinction between OPs and non-OPs has been achieved with potential for practical applications.

(b) Glutathion-S-transferase (GST)-based assays: Attempts have also been made for the development of an electrochemical biosensor using GST as a biorecognition element as it is capable of binding with hydrophobic moieties and undergoes changes in electrocatalytic activity [86]. Borah et al. [87] put forward a sensitive biological sensor for quantification of organothiophosphate and carbamate pesticides (temephos, dimethoate, and fenobucarb). The sensor works on the inhibition principle wherein the residual catalytic activity of GST was monitored via CV studies. The mechanism of inhibition by three different pesticides was studied and corresponding K_i value was compared. After optimization of necessary parameters, LOD of fenobucarb, temephos, and dimethoate were 2, 4, and 5 ppb, respectively.

(c) Others: Apart from these, alkaline phosphatase- [88], peroxidase- [89], and tyrosinase- [90] based enzymatic biosensors have also been used for the detection of pesticides [91], but the use of these enzymes is limited these days. Lipase-mobilized sensors have also been investigated in this category using an inhibition approach like ChE. They can hydrolyze p-nitrophenyl acetate to p-nitrophenol which, being electroactive, undergoes oxidation. This forms the basis for pesticide detection [92, 93].

2.2.2 Aptasensors for pesticide detection

The use of aptasensors for pesticide detection has also been explored to check the efficiency and sensitivity for pesticide detection [94–96]. These are synthetic ssDNA or RNA that can be utilized in in vitro detection process with the advantage of high stability. In comparison with other biological moieties like enzymes and antibodies, aptamer-based electrochemical sensors have attracted remarkable attention in electrochemical sensing as they offer exceptional specificity and affinity with significant conformational changes on target binding owing to the generation from automated nucleic acids, leading to high detection sensitivity and selectivity, thus having the ability to respond with variety of ligands.

Despite their high sensitivity, the process of amplification in these systems is a tedious and expensive process that creates the demand for various materials that can enhance their loading onto the electrode surface. With aptamers as well, carbon nanostructures have gained considerable attention because of their large surface area, good electronic and electrical conductivity, and large charge transfer rate, which adds to the sensor properties in terms of sensitivity and LOD [97–105]. Xu et al. [98] put forward an efficient electrochemical aptasensor for the selective and sensitive detection of CPS where the sensor surface was modified with copper oxide nanoflowers (CuONFs) and carboxyl group-

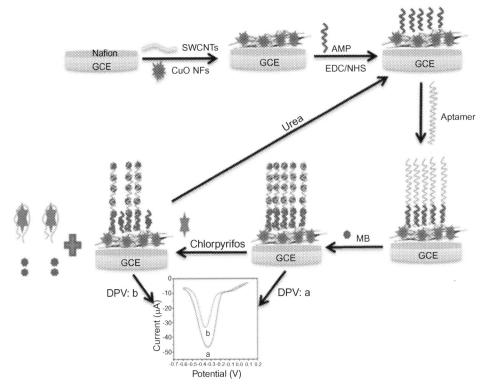

Scheme 13 Detailed formation of Apt/AMP/CuO NFs-SWCNTs/Nafion/GCE sensor for selective detection of CPS. *(Reproduced from G. Xu, D. Huo, C. Hou, Y. Zhao, J. Bao, M. Yang, H. Fa, A regenerative and selective electrochemical aptasensor based on copper oxide nanoflowers-single walled carbon nanotubes nanocomposite for chlorpyrifos detection, Talanta 178 (2018) 1046–1052.)*

derivatized SWCNTs (CuO NFs and c-SWCNTs). Here, the presence of c-SWCNTs has supported covalent bonding between the surface of the sensor and aminated probe allowing easy hybridization of probes and aptasensor. After grafting it with the aminated probe, the current response decreases, which is further suppressed in the presence of an aptamer, indicating its successful formation (Scheme 13). The regeneration can be achieved via incubation with urea as almost flat curves were obtained after incubation due to a break in aptamer from the surface. The sensor surface posed a wide linear range of 0.1–150 ng/mL with LOD of 70 pg/mL (Table 10).

Jiao et al. [99] designed an aptasensor based on a novel GO@Fe$_3$O$_4$ composite film consisting of CB. The CB has been functionalized with CS providing large surface area, good dispensability, and hence high efficacy for the capture of GO@Fe$_3$O$_4$. The composite offered synergistic effects and hence provided large electron transfer rate that favors aptamer immobilization. The sensor thus fabricated offers 0.1–10 ng/mL linear range and wide LOD of 0.033 ng/mL. Yi et al. [104] developed a label-free approach where a

Table 10 Various aptasensors based on carbon-based nanostructures for the detection of pesticides.

Sensor	Characterization techniques	Analytical technique used	Pesticide detected	LOD (M)	Linear range for detection (M)	Refs.
BSA/Apt/Fc@MWCNTs/OMC/GCE	SEM	CV	CPS	0.94×10^{-9}	2.8×10^{-9} to 2.8×10^{-4}	[97]
Apt/AMP/CuO NFs-SWCNTs/NA/GCE	SEM, FTIR	DPV	CPS	0.20×10^{-9}	0.2×10^{-9} to 0.42×10^{-6}	[98]
BSA/Apt/GO@Fe$_3$O$_4$/CB–CS/GCE	TEM, SEM	CV	CPS	0.09×10^{-9}	0.28×10^{-9} to 0.28×10^{-3}	[99]
Apt/PEDOT–c–MWCNTs/FTO	FTIR, FESEM	DPV	Malathion	3.03×10^{-12}	3.03×10^{-12} to 3.03×10^{-8}	[100]
BSA/Apt/rGO-CuNPs/SPCE	SEM, CV	DPV	Profenofos, phorate, isocarbophos, and omethoate	0.003×10^{-9}, 0.3×10^{-9}, 0.03×10^{-9}, 0.3×10^{-9}	$(0.01–100) \times 10^{-9}$, $(1-1000) \times 10^{-9}$, $(0.1–1000) \times 10^{-9}$, $(1–500) \times 10^{-9}$	[101]
rGO-AgNPs/PB-AuNPs/Apt/BSA/GCE	TEM, SEM	CV	Acetamiprid	0.30×10^{-12}	1×10^{-12} to 1×10^{-6}	[102]
MCH/aptamer/Glu–GQD/Au/GCE	XRD, FTIR, SERS	DPV	Acetamiprid	0.0003×10^{-12}	$(0.001–1000) \times 10^{-12}$	[103]
S2/S1/Apt/Au/3D–CS/rGO/GCE	SEM	SWV	Acetamiprid	71.2×10^{-15}	0.1×10^{-12} to 0.1×10^{-6}	[104]
BSA/aptamer/NiHCF NPs/rGO	TEM, EDX, XRD, RAMAN	DPV	PCBs	0.75×10^{-12}	$(3.4–340) \times 10^{-12}$	[105]

three-dimensional porous framework design based on 3D-CS/rGO/GCE has been proposed leading to high loading of acetamiprid aptamer. The determination of acetamiprid was done in tea samples with LOD 71.2 fM. Fan et al. [105] developed electrochemical sensor for detection of polychlorinated biphenyl (PCBs) based on nickel hexacyanoferrates nanoparticles (NiHCF NPs)/rGO hybrids. NiHCF NPs have been utilized as signal probes, and graphene materials have been utilized for attaining large surface area for good electrical conductivity. The sensor has displayed good efficiency for real-time monitoring.

Apart from these, electrochemical sensors utilizing NPs with or without carbon-based materials have also been reported [106–110]. Zhu et al [106] reported impedimetric aptasensor based on CBD using CNHS/AuNP-based composites for detection at pictogram levels. Compared with nanotubes, CNHs exhibits better adsorption behavior owing to more defects in the surface morphology. Its modification with 1-amino pyrene helped in affixing AuNPs, which is further covalently bound to the thiol-terminated CBD aptamer via Au–S interactions (Scheme 14). The sensor possesses better electrocatalytic response displaying no interference form atrazine and thiomethoxam in its sensing performance. The excellent recoveries have been obtained for the detection and are in good agreement with HPLC-MS reference method.

Eissa et al. [107] developed an impedimetric aptasensor for CBD detection in food matrices. The development of DNA aptamers was done using SELEX, and thiol-modified aptamer was self-assembled onto Au surface. The conformational changes were used to bring redox changes allowing specific detection of CBZ with LOD of 8.2 pg/mL. The sensor displayed no interferences from fenamiphos, thiamethoxam, trifluralin, isoproturon, linuron, methyl-parathion carbaryl, and atrazine. Acetamiprid (a neoictinoid), a widely used insecticide that has replaced Ops, was detected utilizing AuNPs as surface

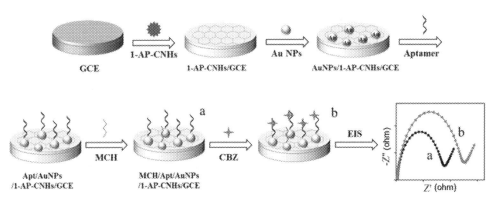

Scheme 14 Fabrication of CBD aptasensor and its electrochemical response. *(Reproduced from C. Zhu, D. Liu, Z. Chen, L. Li, T. You, An ultra-sensitive aptasensor based on carbon nanohorns/gold nanoparticles composites for impedimetric detection of carbendazim at picogram levels, J.Colloid Interf Sci. 546 (2019) 92–100.)*

modifiers in an aptasensor [109]. Taghdisi et al. also [110] reported electrochemical apta-sensor for detection of acetamiprid using silica nanoparticle-coated streptavidin (SiNP-Streptavidin) and methylene blue (MB). The sensing mechanism was based on the release of redox probe MB from dsDNA resulted from aptamer and complementary strand at the site of acetamiprid. This develops an electrochemical signal at the electrode surface and hence offers a low LOD of 153 pM (Table 11).

2.2.3 Immunosensors for pesticide detection

These are basically antibodies (Ab) or antigens (Ag)-based affinity biosensors where binding specificity is the focal point in the sensing mechanism. In the market, antibodies for different classes of pesticides are already available which have been explored for the detection process [111]. But again, for the augmentation of a signal, there is a need for conductive matrix with Ab probe. In this regard, immunosensors have been developed by utilizing carbon-based structures as they impart large surface area, easy functionalization, and immobilization of a large amount of Ab onto them. Various electrochemical immunosensors based on carbon-based structures have been reported in the literature [112–116] as shown in Table 12.

Mehta et al. [112] have utilized GQDs to modify the surface of SPEs, which was further modified using 2-aminobenzylamine (2-ABA) via electrochemical approach (Scheme 15). After this, the derivatized graphene was interfaced with Ab (anti-parathion Ab), and the resulting sensor provides selectivity for parathion in the presence of other pesticides like malathion, fenitrothion, and paraoxon with LOD 52 pg/L. Chuc et al. [113] developed a layer-by-layer assembly composite comprised of PANI/graphene for the detection of atrazine. The graphene layer was developed by using thermal chemical vapor deposition methodology and was shifted onto the surface of predeposited PANI microelectrode. The system provides a suitable microenvironment and also provides enhanced electron transfer rate and hence imparts good sensitivity and LOD of 43 pg/mL.

Pérez-Fernández et al. [117] developed a competitive immunosensor for the electrochemical quantification of IDP using SPE modified with AuNPs followed by specific monoclonal Ab immobilization. Mechanism here involves the competition between free IDP and IDP conjugated with horseradish peroxidase (IDP-HRP) for antibody recognition. The current response initiated by redox process of $3',5,5'$-tetramethylbenzidine (TMB) was inhibited by IDP, which indirectly corresponds to an amount of IDP (Scheme 16). After optimization, the excellent LOD of 22 pM has been obtained for determination in tap water and watermelon samples.

Similarly, Wang et al. [118] also developed Co_3O_4 nanoparticle-based immunoassay for the detection of CPS residues in environmental samples. The artificial antigen CPF-BSA was coupled to the surface of an electrode with a layer of Co_3O_4-Pan, and the competitive reaction between bound and free CPS leads to an electrochemical response. The detection of CPS was attained up to 0.01 μg/mL with high precision (Table 13).

Table 11 Various nanoparticle-based electrochemical aptasensors for detection of OPs.

Sensor	Characterization techniques	Analytical technique used	Pesticide detected	LOD (M)	Linear range for detection (M)	Refs.
MCH/Apt/AuNPs/1-AP-CNHs/GCE	UV, Zetapotential, TEM	EIS	CBD	2.6×10^{-12}	$(5.2–5200) \times 10^{-12}$	[106]
Aptamer/Au	–	EIS	CBD	0.042×10^{-9}	52.3×10^{-9} to 52.3×10^{-6}	[107]
DZN/thiolatedaptamer/AuNP/SPGEs	FESEM	DPV	Diazinon	0.0169×10^{-12}	$(0.1–1000) \times 10^{-12}$	[108]
PANI/AuNPs/GSPE	–	DPV	Acetamiprid	0.086×10^{-6}	$(0.25–2.0) \times 10^{-6}$	[109]
MB-dsDNA-modified SiNP streptavidin	–	DPV	Acetamiprid	153×10^{-12}	–	[110]

Table 12 Various immunosensors for the detection of pesticides utilizing carbon-based structures.

Sensor	Characterization techniques	Analytical technique used	Pesticide detected	LOD (M)	Linear range for detection (M)	Refs.
Ab/NH$_2$GQD-SPE	TEM, AFM	EIS	Parathion	0.15×10^{-12}	0.03×10^{-12} to 3.4×10^{-6}	[112]
Pt/PANi/Gr/α-ATZ	HRTEM, Raman	SWV	Atrazine	0.2×10^{-12}	9.3×10^{-9} to 9.3×10^{-5}	[113]
CNTFETs	–	–	Atrazine	4.6×10^{-12}	4.6×10^{-12} to 46×10^{-9}	[114]
Parathion/ab-fG-SPE	FESEM, FTIR, UV	EIS	Parathion	0.17×10^{-12}	0.34×10^{-12} to 3.4×10^{-9}	[115]
Ab/AuNPs-PANABA/ MWCNTs-SPE	CV, EIS	EIS	2,4-Dichlorophenoxy acetic acid	1.3×10^{-9}	4.5×10^{-12} to 4.5×10^{-10}	[116]

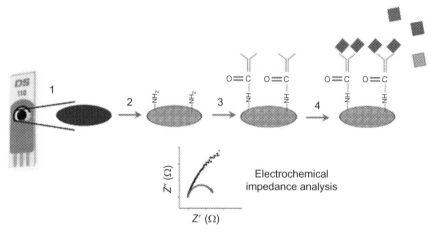

Step 1: Drop-casting of graphene on carbon screen printed electrode
Step 2: Electro-catalyzed amine (—NH$_2$) functionalization of graphene with 2-aminobenzylamine
Step 3: Immobilization of anti-parathion antibodies on —NH$_2$ functionalized graphene SPE
Step 4: Immunosensing of parathion with above sensor

Antibody ⧫ Parathion ▪ ▪ ▪ Non-specific pesticides

Scheme 15 Fabrication of graphene-based immunosensor for the detection of parathion. *(Reproduced from J. Mehta, P. Vinayak, S.K. Tuteja, V.A. Chhabra, N. Bhardwaj, A.K. Paul, K.H. Kim, A. Deep, Graphene modified screen printed immunosensor for highly sensitive detection of parathion, Biosens. Bioelectron. 83 (2016) 339–346.)*

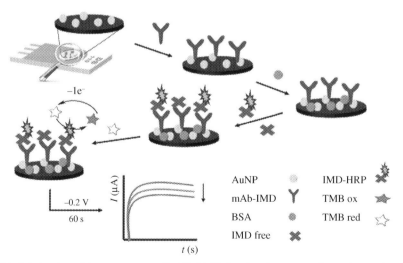

Scheme 16 Fabrication of immunosensor (AuNP-SPCE) for the detection of IDP. *(Reproduced from B. Perez-Fernandez, J.V. Mercader, A. Abad-Fuentes, B.I. Checa-Orrego, A. Costa-Garcia, A. Escosura-Muniz, Direct competitive immunosensor for imidacloprid pesticide detection on gold nanoparticle-modified electrodes, Talanta 209 (2020) 120465.)*

Table 13 Various immunosensors for the detection of pesticides utilizing different nanostructures.

Sensor	Characterization techniques	Analytical technique used	Pesticide detected	LOD (M)	Linear range for detection (M)	Refs.
AuNPs–SPCE	SEM, DLS	CA	IDP	22×10^{-12}	$(50–10{,}000) \times 10^{-12}$	[117]
Co_3O_4/Pan/CPF antigen	TEM, SEM	CV	CPS	0.028×10^{-6}	0.028×10^{-6} to 0.028×10^{-3}	[118]
FTO–AuNPs–chl–Ab	UV, DLS, TEM	DPV	CPS	10×10^{-15}	1×10^{-15} to 1×10^{-6}	[119]

3. Detection of antibiotics

Antibiotic is a class of compounds that has been utilized in animal foods since the 1940s. In the market, the range of available antibiotics include β-Lactams, aminoglycosides (KANA, streptomycin, tobramycin), CAP, glycopeptides, quinolones (NFX, CIP, enrofloxacin), oxazolidinones, sulfonamides (sulfadiazines), TC, macrolides (azithromycin), streptogramins, lipopeptides, and penicillins (Cloxacillin, AMP, AMX) [120, 121]. These antibiotics are often employed to resolve health issues and stimulate growth in animals. They have also been employed for agriculture purposes to deal with various infections. However, the uninterrupted use of antibiotics has resulted in their release in water bodies through run-off or during purification of irrigation water [122, 123]. Liu et al. [124] reviewed 234 cases of suspected antibiotics and, in total, 32 antibiotics were reported in the detected marine products with quinolones and sulfonamides as top raters followed by CIP, sulfisoxazole, and NFX. Their residues have exceeded the minimum residual limit (MRL) set by the World Health Organization (WHO), making it one of the emerging environmental pollutants [125]. Moreover, with increasing antibiotic pollution, the major concern is with the growth of antimicrobial resistance (AMR) engraving numerous public health problems [126] shown as a worldwide concern even by the WHO [127].

3.1 Polymeric-based electrochemical sensors for antibiotics

To obtain improved electrochemical properties, the integration of polymeric nanomaterials with nanoparticles or carbon-based structures is always an effective approach [128]. Jalal et al. [129] reported the synthesis MWCNTs, magnetite nanoparticles, and conducting polymer polyethyleneimine (PEI)-based nanocomposites for sensitive determination of ciprofloxacin (CIP) using electrochemical methods. Owing to the high protonation present in PEI, a strong interactive matrix was available for the modification, which resulted in the exceptional electrocatalytic effect of the composite on the oxidation of CIP. The modified electrode exhibited LOD 3 nmol/L with excellent recoveries (Table 14).

Table 14 Conducting polymer-based electrochemical nanosensors for the detection of antibiotics.

Sensor	Characterization techniques	Analytical technique used	Antibiotics detected	LOD (M)	Linear range for detection (M)	Refs.
QDs-P6LC-PEDOT: PSS/GCE	XRD, TEM, confocal microscopy	SWV	AMX	0.05×10^{-6}	$(0.9–69) \times 10^{-6}$	[128]
PEI@Fe$_3$O$_4$@CNTs/ GCE	FESEM, TEM, XRD, EDX	DPV	CIP	0.003×10^{-6}	$(0.03–70) \times 10^{-6}$	[129]
AMT–Ag/CPE	FTIR, XRD, XPS, FESEM, EDX	DPV	CIP	0.005×10^{-6}	$(18–180) \times 10^{-6}$	[130]

A new class of polymeric sensors, nanocrystalline coordination polymers (NCCPs) has also come forward as emerging electrodes in the detection field of antibiotics [130]. An ultrasensitive novel, portable, and low–cost AMT-Ag NCCP-based electrode material was developed utilizing electron tunneling for effective detection of CIP hydro–chloride. Various interactions comprised of π-π and hydrophobic ones, as well as avail–ability of anchoring points in the designed electrodes, have provided an easy pathway for the oxidation of CIP (Scheme 17). A LOD of 20 and 50 nM has been noticed in eye drops and urine samples with least intervention.

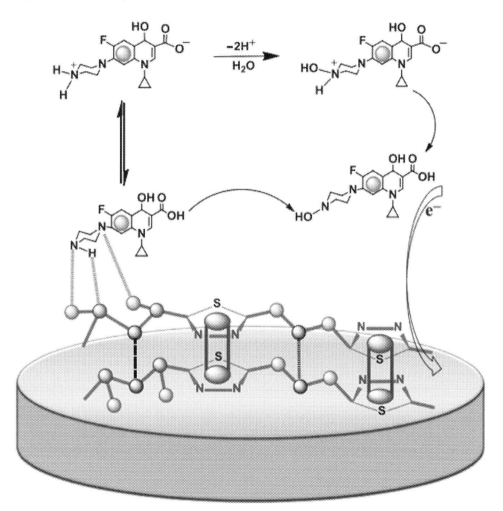

Scheme 17 The schematic diagram showing the electro-oxidation of CIP at AMT-Ag/CPE surface. *(Reproduced from M. Tiwari, A. Kumar, U. Shankar, R. Prakash, The nanocrystalline coordination polymer of AMT-Ag for an effective detection of ciprofloxacin hydrochloride in pharmaceutical formulation and biological fluid, Biosens. Bioelectron. 85 (2016) 529–535.)*

Significant efforts have been directed toward scheming of various electrode materials providing better amplification paths for achieving ultrahigh sensitivity. In the modern era, MIP stands out as an effective methodology for highly selective electrochemical detection of antibiotics using artificial recognition sites [131–133]. Liu et al. [132] developed a novel one-step fabrication protocol without the use of any eluting agents for the sensing of anticancer antibiotic mitoxantrone (MTX) where electrochemical polymerization of β-CD was used to prepare MIP onto GCE. The imprinting technique imparts cavities complementary to MTX due to binding-extraction procedure resulting in a sensor equipped with high selectivity and sensitivity. The linear response of 6×10^{-8} M to 1×10^{-5} M using DPV with low LOD of 3×10^{-8} M has made this an apt candidate for the detection of anticancer drugs (Table 15).

However, the performance of conventional MIPs is limited by their low surface area, slow diffusion rate, long response time, etc. Here arises an increasing demand of additional supporters like carbon structures, MNPs, etc., that can resolve these issues by increasing the binder capacity. Liu et al. [134] synthesized MIP along with 3D framework of derivatized fMWCNTs where synergistic interactions between the two offer high electrocatalytic behavior toward norfloxacin (NFX) (a class of quinolones) (Scheme 18). The use of fMWCNTs ensures high surface area, whereas tailor-made recognition sites offered by MIP ensure selectivity toward the NFX. After necessary optimizations, wide linearity was observed in the range from 0.003 to 0.391 µM and 0.391 to 3.125 µM from DPV with appreciably lower LOD of 1.58 nM. The sensor is highly selective as inferred from possible analysis of interferences and is offering fast quantification in pharmaceutical formulations and rat plasma samples.

Moro et al. [135] devised MWCNTs-modified graphite SPEs (MWCNTs-G-SPE) array to achieve high surface-to-volume ratio for the detection of β–lactams-based antibiotics in milk. The polymer of aminobenzoic acid (ABA) has been employed as a conductive film on substrate for MIPs formation to fingerprint the target analyte, i.e., cefquinome (CFQ). After successful polymerization on MWCNTs-G-SPE array, two different protocols were tested to improve selectivity and gain high adsorption. Appreciably low LOD of the order of 50 nM has been achieved leading to their extension in

Table 15 MIP-based electrochemical nanosensors for the detection of antibiotics.

Sensor	Characterization techniques	Analytical technique used	Antibiotics detected	LOD (M)	Linear range for detection (M)	Refs.
β–CD/ MIP/ GCE	SEM	DPV	MTX	0.03×10^{-6}	$(0.06–10) \times 10^{-6}$	[132]
MIP/ GCE	SEM	DPV	Metroindazole (MNZ)	3.33×10^{-9}	$(1–10) \times 10^{9}$	[133]

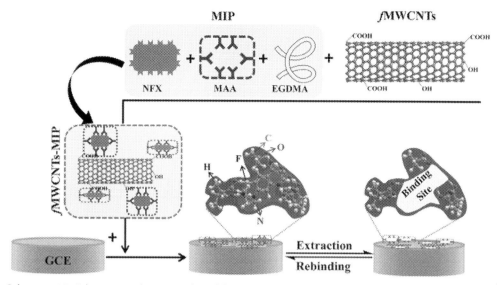

Scheme 18 Schematic showing the fabrication of MWCNTs-MIP-modified electrode and corresponding adsorption-desorption mechanism of NFX in the imprinted cavity. *(Reproduced from Z. Liu, M. Jin, H. Lu, J. Yao, X. Wang, G. Zho, L. Shui, Molecularly imprinted polymer decorated 3D-framework of functionalizedmulti-walled carbon nanotubes for ultrasensitive electrochemical sensing of norfloxacin in pharmaceutical formulations and rat plasma, Sens. Actuat. B Chem. 288 (2019) 363–372.)*

field analysis. Munawar et al. [136] synthesized electrochemical nanosensors by utilizing 3D-imprinted nanostructures. The surface of the working electrode was modified with CuNPs decorated onto amino-functionalized CNTs followed by layering of MIPs. The material formed a good model compound for electrochemical detection of chloramphenicol (CAP) (Table 16).

Besides these carbon nanomaterials, MNPs are also one of the leading high performance supporters strongly backed by their large surface area, high conductivity and electrocatalytic performance leading to fast equilibration in response to an analyte. Simple synthetic modes make them more attractive in the use of sensor technology. Devkota et al. [139] designed an electrochemical sensor based on molecular imprinting employed over oxidized pyrolle (MIOPPy) and AuNPS as modifiers on SPEs. The optimizations of each parameter including electropolymerization cycles, accumulation time, and pH were done to gain high sensitivity and better performance for the detection of tetracyclines (TC) namely CAP and AMX with LOD 0.65 μmol/dm^3. The proposed method was applied to determine TC in shrimp samples and hence can be modified to detect TCs in food and environmental applications. Bougrini et al. [140] developed an electrochemical sensor for the detection of TC wherein the modifier consists of a MIP-based microporous MOF (Scheme 19). The imprinting consists of p-aminothiophenol as a functional monomer in the presence of TC as templates, and electropolymerization was carried out

Table 16 Carbon structure-based MIPs as electrochemical nanosensors for the detection of antibiotics.

Sensor	Characterization techniques	Analytical technique used	Antibiotics detected	LOD (M)	Linear range for detection (M)	Refs.
fMWCNTs/MIP/GCE	FESEM, FTIR, XRD, UV	DPV	NFX	1.5×10^{-6}	$(0.003-391) \times 10^{-6}$	[134]
MIP-MWCNTs-G-SPE	SEM, EIS, CV	SWV	CFQ	50×10^{9}	$(50-1000) \times 10^{9}$	[135]
3DCNTs@CuNPs@MIP-modified GCE	SEM, AFM, XRD, zetapotential	CV	CAP	10×10^{-6}	$0.01-05) \times 10^{-3}$	[136]
AgDs/MIP/cMWCNTs/GCE	SEM, CV	ASDPV	CFQ	1×10^{-9}	$(0.01-600) \times 10^{-6}$	[137]
MIP/MWCNTs/GCE	CV	DPV	Doxycycline–hyclate (DC)	13×10^{-9}	$(50-500) \times 10^{9}$	[138]

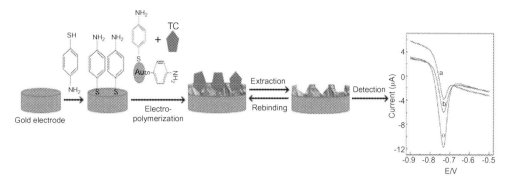

Scheme 19 Schematic showing fabrication of MIP for determination of TC and its voltammetric behavior. *(Reproduced from M. Bougrini, A. Florea, C. Cristea, R. Sandulescu, F. Vocanson, A. Errachid, B. Bouchikhi, N. El Bari, N. Jaffrezic-Renault, Development of a novel sensitive molecularly imprinted polymer sensor based on electropolymerization of a microporous-metal-organic framework for tetracycline detection in honey, Food Control 59 (2016) 424–429.)*

in the presence of AuNPs. Higher current densities observed in the MIP indicate the specific cavities available to bind TC provided by the unique structure of microporous MOFs. The electrochemical characterization was done using LSV for the optimization of several factors under which the sensor exhibited good linearity in the range from 224 to 22.4 fM with low LOD of 0.22 fM. The system was also investigated to quantify TC in honey and showed excellent recoveries.

A combination of both carbon structures and MNPs in MIP has also been reported for the detection of antibiotics. Jafari et al. [141] developed an electrochemical nanosensor based on fabrication of MIP using GO and AuNPs and evaluated its efficiency for selective absorption of cloxacillin (CLO). The quantification of CLO includes two steps: (a) the incubation of MIP with samples containing CLO and (b) removal of MIP followed by collection of an incubated sample, which is then monitored by electrochemical nanosensor using DPV. The sensor exhibited linearity in the range of 110 to 750 nM with LOD as low as 36 nM. The removal of CLO has been achieved up to 92%, and its performance for CLO detection in milk samples extends its use for real-time monitoring. Dehghani et al. [142] used GO and Au nanowires as modifiers on electrode surface of GCE, and imprinting of cefixime (CEF) was achieved on the modified surface using PANI. Owing to the synergic combination of GO and Au nanowires encompassing the selectivity of MIP, the sensor worked well for biological samples with excellent recovery exhibiting a low LOD of 7.1 nM (Table 17).

Roushani et al. [145] combined this technology further with aptamer chemistry to achieve an excellent sensitivity with LOD of 0.3 pM for the selective detection of CAP. Interaction of Ag NPs and —NH₂ groups of aptamer and Ag NPs and —NH of functionalized GO has resulted in a stable hybrid system that was followed by the electropolymerization of resorcinol around the aptamer. The design has resulted in double recognition imprinting cavities showing better performance than an aptamer.

Table 17 Carbon structures and NPs as electrode modifiers in MIPs for the electrochemical detection of antibiotics.

Sensor	Characterization techniques	Analytical technique used	Antibiotics detected	LOD (M)	Linear range for detection (M)	Refs.
MIOPPy/AuNP/SPCE	SEM	DPV	TC	0.65×10^6	$(1-20) \times 10^6$	[139]
PATP-functionalized AuNPs	–	LSV	TC	0.22×10^{-15}	224×10^{-15} to 22.4×10^{-9}	[140]
SPCE/GO/AuNPs	SEM, FTIR, TGA	DPV	CLO	36×10^{-9}	$(110-750) \times 10^{-9}$	[141]
EMIP/GNU/GO/GCE	FESEM	DPV	CEF	7.1×10^{-9}	$(20-950) \times 10^{-9}$	[142]
MIP/GNU/GO/GCE	FESEM	DPV	Azithromycin	0.1×10^{-9}	$(0.3-920) \times 10^{-9}$	[143]
MIP/AuNP-PEDOT:PSS/SPCE	SEM	DPV	Nitrofurantoin	0.1×10^{-9}	$(1-1000) \times 10^{-9}$	[144]

3.2 Bionanomaterials for detection of antibiotics

In the fabrication of nanosensor arrays for antibiotics detection, biosensors using biona-nomaterials are promising candidates that can accurately apprehend different targets and then consecutively alter them to discern electrochemical signals. Li et al. [146] fabricated vertically aligned multisegment Au-Pt NW/NP hybrid electrodes for the simultaneous detection of penicillin and TC using CV. The biological system penicillinase was immobilized onto the surface of a Au-Pt segment while L-cysteine worked as an efficient bioreceptor. The fabricated hybrid array showed good sensitivity of penicillin and TC of the order of 41.2 μA μM^{-1} cm^{-2} and 26.4 μA μM^{-1} cm^{-2}, respectively, with appreciable recovery from chicken and beef extracts. Similarly, a low LOD of 4.5 nM was observed for penicillin using penicillinase enzyme affixed on the Au electrode via cysteine self-assembled monolayer using CA tool [147].

Besides most commonly used enzymes, immunosensors, are one of the eye-catching tools comprising recognition of antigens on the surface of an antibody at molecular level for various types of analytes including antibiotics [148–150]. A wide range of immunosensors supported on a variety of nanomaterials are available for detecting different classes of antibiotics. Liu et al. [151] put forward an immunosensor by modifying Au electrode with magnetic NPs and utilizing CS as a binder for the detection of TC. The covalent bonding between anti–TC antibody and carboxyl-Fe$_3$O$_4$ MNPs played a significant role in immunosensor sensitivity. The quantification of TC was done using DPV, and linearity in the range of 0.08–1 ng/mL with LOD of 0.0321 ng/mL was obtained. The validation of the proposed method was done by quantification of TC in milk samples after spiking. The comparison of proposed methods was also done with ELISA, and both were in good agreement with each other.

A label-free immunosensor utilizing SPCE modified with poly(vinyl alcohol-*co*-ethylene) (PVA-*co*-PE) membrane covalently modified with antibody (anti-CAP) has been reported (Scheme 20) for the sensing of CAP in spiked milk samples [152]. The current was monitored using CA with LOD 4.7 pg/mL (Table 18).

NFX was detected using AuNPs encapsulated into poly(amidoamine) dendrimer by immobilizing anti–NFX. Here, signal labels in the form of HRP-labeled antigen were introduced to catalyze the reaction of substrate to get the LOD of 0.3837 ng/mL [155]. The very first sulfonamide detected using immunosensor methodology was sulfapyridine. In this regard, Hassani et al. [156] proposed a novel fully integrated bio-microelectromechanical system (Bio-MEMS) with Au microelectrodes. The surface of the electrode was covered with poly(pyrrole-*co*-pyrrole-2-carboxylic acid) (Py/Py-COOH and the detection analysis involves competitive methodology with 5-[4-(amino)phenylsulfonamide]-5-oxopentanoic acid-BSA antigens with polyclonal antibody. The selectivity was monitored in presence of other sulfonamides, and validity was checked using spiked honey samples. Another polyether ionophore antibiotic, monensin, was detected using electrochemical immunosensor comprising of AuNPs/Zn/Ni-ZIF-8-800@graphene composites. The synergy of highly porous nature of MOFs

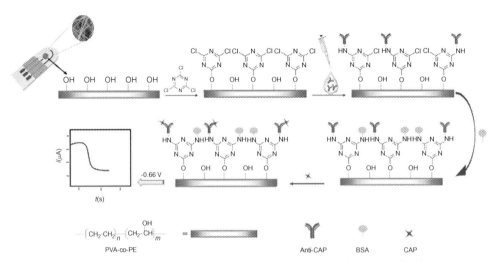

Scheme 20 Fabrication of PVA-co-PE NFM/Anti-CAP/SPCE for the detection of chloramphenicol. *(Reproduced from A.Y. El-Moghazy, C. Zhao, G. Istamboulie, N. Amaly, Y. Si, T. Noguer, G. Sun, Ultrasensitive label-free electrochemical immunosensor based on PVA-co-PE nanofibrous membrane for the detection of chloramphenicol residues in milk, Biosens. Bioelectron. 117 (2018) 838–844.)*

and large surface area provided by graphene has greatly enhanced the performance of the sensor [160]. Monoglycoside polyether ionophore–type antibiotics have been successfully detected using electrochemical immunosensor in egg samples by a newly developed multiple amplification system. The Ag-Ab (MD/BSA) was immobilized onto AuNPs-modified GCE, and the amplification strategy was achieved using hemin-encapsulated Fe-MIL-88-NH$_2$-designed frameworks (Table 19). Immobilization of Ab2-HRP has effectively catalyzed the performance of the sensor with LOD of 0.045 ng/mL [161].

Utilizing the surface chemistry of nanomaterials, excellent redox sensing of antibiotics like ampicillin (AMP) [162, 163], TC [164–166], streptomycin [167, 168], etc., have been done using aptamer chemistry. Generally, these are obtained by an in vitro methodology named SELEX resulting in binding of a variety of targets [169]. A few researchers have also tried to develop electrochemical biosensors based on multiple recycling amplification initiated by binding of a target–aptamer as a new approach toward the highly sensitive and specific detection of antibiotics [170]. In an electrochemical sensing, there is a dire need to develop sensors that are able to measure multiple antibiotics in a single scan. For this, aptamers have evolved as excellent recognition elements as they impart high affinity toward their targets [171]. Shen et al. [172] developed a multiplexed electrochemical aptasensor for multiplex antibiotics detection in a single experiment. The aptasensor exploited metal ions (Cd(II) and Pb(II))–encoded apoferritin probes for KANA and AMP. The detection limits obtained for KANA and AMP were 18 and 15 fM, respectively, by a single run in SWV, which is enabled only due to double-stirring bars that assisted target recycling. Chen et al. [173] also developed an electrochemical

Table 18 Various available immunosensors for different classes of antibiotics.

Sensor	Characterization techniques	Analytical technique used	Antibiotics detected	LOD	Linear range for detection	Refs.
Ab-MNPs-CS/Au	CV, EIS	DPV	TC	0.07×10^{-9} M	0.18×10^{-9} to 2.25×10^{-9} M	[151]
TC-Py/Py-COOH/MNPs/Au, anti-TC polyclonal sheep Ab	—	EIS	TC	2.7×10^{-12} M	0.22×10^{-12} to 0.22×10^{-7} M	[152]
PVA-co-PE NFM/anti-CAP/SPCE	SEM, FTIR	CA	CAP	0.014×10^{-9} M	0.03×10^{-9} to 0.03×10^{-6} M	[153]
CAP/immobion membrane, anti-CAP monoclonal PAMAM-Au	—	CA	CAP	8.0×10^{-7} M	—	[154]
SA2-BSA/Py/Py-COOH/MNPs/Au, polyclonal antibody Ab155	TEM	DPV	NFX	1.20×10^{-9} M	3.13×10^{-9} to 3.13×10^{-5} M	[155]
	AFM	EIS	Sulfa-pyridine	1.6×10^{-15} M	8.02×10^{-15} to 0.2×10^{-6} M	[156]
Ab2@Ag NPs@SWCNHs/Ab1/Cag/Au NDs/GCE	—	LSV	Sulfa-methazine	0.43×10^{-9} M	1.18×10^{-9} to 0.229×10^{-6} M	[157]
NiFe PBA nanocubes@TB	SEM, EDX, XRD, BET, TEM, FTIR	DPV	Pro-calcitonin	0.0003 ng/mL	0.001–25 ng/mL	[158]
AuNPs-HRP-PEG-Ab2	TEM, UV-Vis	CA	Pro-calcitonin	0.1 pg/mL	0.05–100 ng/mL	[159]
Monensin/BSA/Ab/AuNPs/Zn/Ni-ZIF-8-800@graphene/GCE	SEM, EDS, XRD	DPV	Monensin	0.16×10^{-9} M	0.37×10^{-9} to 149×10^{-9} M	[160]
hemin@MOFs/AuPt-Ab2-HRP	SEM, TEM, XPS	CA	Maduramicin	0.045×10^{-9} M	0.109×10^{-9} to 54.5×10^{-9} M	[161]

Table 19 Various available aptamer-based sensors for detection of antibiotics.

Sensor	Characterization techniques	Analytical technique used	Antibiotics detected	LOD (M)	Linear range for detection (M)	Refs.
Py-M-COF	FTIR, XRD, XPS	EIS	Enrofloxacin (ENR) and AMP	0.016×10^{-12} and 0.11×10^{-15}	0.027×10^{-12} to 5.56×10^{-9}	[162]
Ladder-shaped DNA/Au electrode	CV	DPV	AMP	1×10^{-12}	7×10^{-12} to 100×10^{-9}	[163]
SPCE/4-CP/aptamer	–	CV	TC	0.078×10^{-9}	$(0.1\text{--}45) \times 10^{-9}$	[164]
Apt-CSs	–	DPV	TC	0.45×10^{-9}	$(1.5\text{--}5000) \times 10^{-9}$	[165]
Anti-TET/PGA/GCE	CV, EIS	EIS	TC	3.7×10^{-17}	1×10^{-6} to 1×10^{-16}	[166]
BSA/Apt/AgNPs/GQDs-N-S/AuNPs/GCE	TEM, UV, XRD, FESEM, EDAX	EIS	Streptomycin	5.6×10^{-15}	17.19×10^{-15} to 1.39×10^{-9}	[167]
GR–Fe$_3$O$_4$–AuNPs/PCNR/GCE	SEM, XRD, BET, EDAX	DPV	Streptomycin	0.04×10^{-9}	0.08×10^{-9} to 0.34×10^{-6}	[168]

aptasensor for the detection of multiplex antiobiotics using Y-shaped DNA probes utilizing NMOFs as substrates with DNA polymerization (CSRP) target. The analytes CAP and oxytetracycline (OTC) trigger disassembly of Y-DNA probes resulting in the release of signal tags, which were detected using SWV. The proposed strategy exhibited a high sensitivity with high LOD of 33 and 48 fM toward CAP and OTC, respectively. Chen et al. [174] put forward novel electrochemical biocodes utilizing metal organic frameworks at nanoscale (NMOFs) for multiplex detection of KANA and CAP. Here, amine-derivatized NMOF (UiO-66-NH$_2$) was exploited as a substrate to carry different metal ions Pb(II) or Cd(II) and then labeled with complementary DNA strands of aptamers toward KANA and CAP to obtain two signal tags (Scheme 21). Finally, to obtain biocodes, these tags are followed by conjugation with aptamer-functionalized magnetic beads. In the presence of both analytes for simultaneous determination, aptamers bind to a target specifically, and signal tags go into solution. The resulting response toward Pb(II) and Cd(II) in SWV is directly proportional to the target concentrations.

Wang et al. [175] developed an electrochemical DNA sensor utilizing multiple recycling amplification with signal-on sensing strategy and kanamycin (KANA) as a model compound. The method is divided into two steps; polymerase-catalyzed target recycling

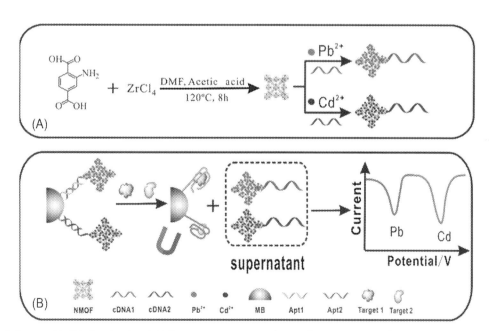

Scheme 21 (A) Synthesis of UiO-66-NH$_2$ nanoparticles and preparation of two signal probes. (B) Schematic of simultaneous electrochemical detection of Kanamycin (KANA) and chloramphenicol (CAP) based on aptamer-nanoscale MOF as biocodes. *(Reproduced from M. Chen, N. Gan, Y. Zhou, T. Li, Q. Xu, Y. Cao, Y. Chen, A novel aptamer-metal ions-nanoscale MOF based electrochemical biocodes for multiple antibiotics detection and signal amplification, Sens. Actuat. B Chem. 242 (2017) 1201–1209.)*

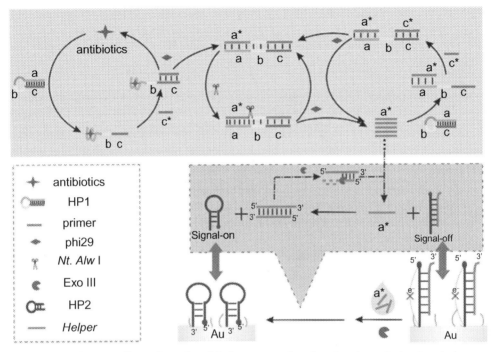

Scheme 22 Schematic illustration of antibiotics assay using the signal-on electrochemical sensor based on target-aptamer binding triggered multiple recycling amplification. *(Reproduced from H. Wang, Y. Wang, S. Liu, J. Yu, Y. Guo, Y. Xu, J. Huang, Signal-on electrochemical detection of antibiotics at zeptomole level based on target-aptamer binding triggered multiple recycling amplification, Biosens. Bioelectron. 80 (2016) 471–476.)*

and Exo-III-assisted secondary target recycling (Scheme 22). First, KANA binds with anti-KANA aptamer (a) of HP1 where conformational alterations open the HP1 stem, which then hybridizes with primer (c*). Now, the associated c* causes the extension reaction of displacement of KANA, and then the displaced KANA combines with another HP1 to start a new recycle. This process is followed by the release of short-stranded ssDNA (a*) that can be hybridized either with another HP1 or with Helper (assistant probe), which initiates the cleavage process of Exo III along with recycling of a* (Table 20).

Apart from this, there are also amino acid functionalized materials that have been utilized for detection of antibiotics. In this context, Chen et al. [177] developed an electrochemical sensor utilizing polyglutamic acid (PGA), which was modified with three-dimensional graphene (3D-GE) for detection of AMX. The electrochemical approach was used to determine the protons and electrons in the detection procedure, and the reaction mechanism was explained by DFT analysis. The LOD was attained as 0.118 μM, and for real-time analysis, an electrode was also used to determine concentration of AMX in urine samples. Benvidi et al. [166] developed an aptasensor utilizing poly (L-glutamic acid)/MWCNTs modified electrode for the determination of TC. The

Table 20 Sensor probes available for multiple recycling amplification modes for multiplex antibiotics detection.

Sensor	Characterization techniques	Analytical technique used	Antibiotics detected	LOD (M)	Linear range for detection (M)	Refs.
Aptamer	–	DPV	AMP	4.0×10^{12}	0.02×10^{9} to 40×10^{-9}	[176]
Aptamer	TEM, SEM, UV, EDX	SWV	KANA, AMP	0.02×10^{-12}, 25×10^{-15}	$(5 \times 10^{-5} - 50) \times 10^{-15}$, $(0.1-1200) \times 10^{-9}$	[171]
Aptamer	–	SWV	AMP, KANA	0.15×10^{-13}, 0.18×10^{13}	0.05×10^{-12}, $50 \times 10^{-9}{}^{12}$, 100×10^{-9}	[172]
Aptamer	TEM, SEM, UV, XRD, FTIR	SWV	CAP, OTC	33×10^{-15}, 48×10^{-15}	$(0.0001-50) \times 10^{-9}$	[173]
UiO-66–NH$_2$, aptamer	SEM, TEM, DLS, PXRD, FTIR, EDX	SWV	KANA, CAP	0.16×10^{-12}, 0.19×10^{-12}	0.002×10^{9} to 50×10^{-9}	[174]
Aptamer	–	DPV	KANA	1.3×10^{-15}	$0.002 \times 10^{9 \text{ to } 100} \times 10^{-9}$	[175]

aminoacid L–glutamic acid offers electrostatic interactions and thus offered covalent bond formations that imparted specificity toward a particular analyte. Different parameters like detection time, concentration of aptamer, and time of immobilization of the aptamer were optimized using CV. The determination of TC was done using EIS with LOD of 3.7×10^{-17} M. The sensor has been successfully utilized for the determination of TC in honey and drug samples. Also, Sun et al. [178] developed an electrochemical sensor based on a graphene/L-cysteine composite-modified electrode for the determination of TC. The analytical performance of the composite was compared with that of simple GO and L-cysteine-modified electrodes where the oxidation currents at GR/L-Cys/GCE were noticeably higher than that of individual electrodes indicating better electrocatalytic performance of the amino acid-functionalized surface of graphene. The sensor exhibited linear response from 8.0 to 140.0 μM with LOD 0.12 μM with good performance toward real-time monitoring.

4. Detection of heavy metal ions

Heavy metal ions are a source of pollution owing to their nonbiodegradable nature and hence, they are considered to be a worldwide problem. The toxic heavy metal ions include lead (Pb(II)), mercury (Hg(II)), cadmium (Cd(II)), arsenic (As(III)), and chromium (Cr(III)), which cause health effects to living organisms even at trace levels. Various cost-effective and reliable electrochemical techniques have been developed for detection of these toxic metal ions; voltammetric techniques like DPV and SWV offer high sensitivity and selectivity for in situ determination of these metal ions. The most common principle involved for stripping analysis of trace level determination follows two steps: (a) the preconcentration of analyte by reducing metals under negative potential and (b) collection of the results by potential sweep in anodic direction to obtain current maxima corresponding to oxidation [179–181]. In this concern, various polymeric and bionanomaterial-based electrochemical sensors for detection of heavy metal ions are discussed in the subsequent chapter.

4.1 Polymeric-based electrochemical sensors for the detection of heavy metal ions

4.1.1 Conducting polymers for the detection of heavy metal ions

Huang et al. [182] synthesized one-dimensional, phytic acid (PA)-doped, PANI nanofibers-based nanocomposites (PA-PANI) by a "doping-dedoping-redoping" method, and various techniques such as EIS, FT-IR, SEM, and EDS have been used for characterizing the properties of the resulting nanocomposites. Here, GCE was modified with Nafion (NA), which provided film stability and anions-resistant permselectivity. NA/PA-PANI/GCE provided a lower charge transfer resistance which reduced the energy barrier for the mass transfer and facilitated the charge transfer rate at the electrode/

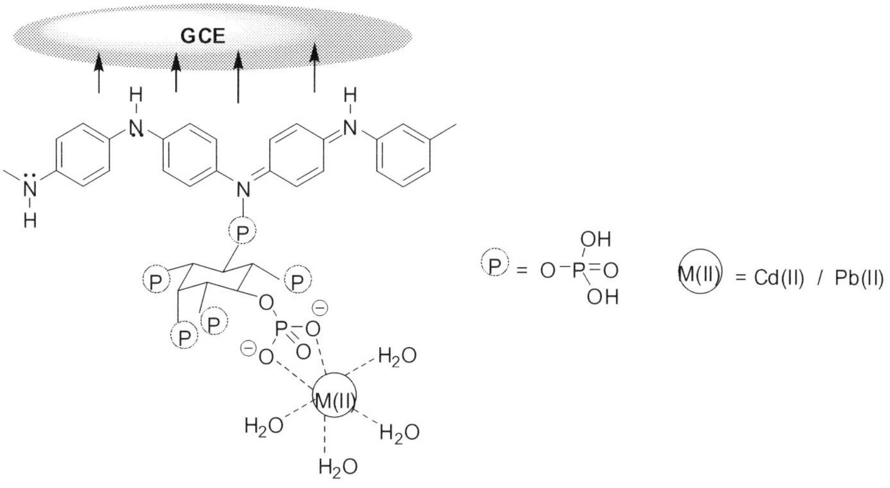

Scheme 23 Schematic representation of interaction between Cd(II)/Pb(II) ions and GCE-modified PA-PANI nanocomposite.

electrolyte interface. Moreover, DPASV curves of PA–PANI-modified electrode displayed better defined and increased stripping peaks for both Cd(II) and Pb(II) ions due to —PO$_3$H$_2$ groups in these modified electrodes that helped in binding with metal ions by which conductivity was enhanced (Scheme 23). Under optimal conditions, the LOD for Cd(II) was calculated to be 0.02 μg/L with good linear relationship in a range of 0.05–60 μg/L, whereas for Pb(II) ions, the LOD was 0.05 μg/L in a range of 0.1–60 μg/L. At higher concentration of Cd(II), the stripping peaks showed little splitting due to the surface inhomogeneity of the modified electrode.

Zhu et al. [183] prepared a MWCNT-emeraldine-based PANI (EBP)-NA composite-modified GCE (MWCNT-EBP-NA/GCE), and in-situ plating of Bi was done on this modified electrode to form Bi/MWCNT-EBP-NA/GCE. The microstructure of MWCNT-EBP-NA/GCE displayed a fibrous structure similar to the hairs on the surface of skin, where NA wrapped around the hydrophobic MWCNT surface and separated them effectively due to its hydrophobic and hydrophilic regions. Moreover, the hair-like structure on the electrode surface improved the anti-interference ability of the modified electrode. The modified Bi/MWCNT-EBP-NA/GCE electrode was further employed for sensitive detection of Cd(II) and Pb(II) ions in the linear range of 1.0–50 μg/L by SWASV. By extending the deposition time to 300 s, the LOD was further improved to 0.04 μg/L for Cd(II) and 0.06 μg for Pb(II). This improvement was attributed to the presence of negatively charged polymeric membrane composed of NA with sulfonate groups that facilitated the preconcentration of Cd(II)/Pb(II) on the electrode surface (Table 21).

Table 21 PANI-based electrochemical sensors for the detection of heavy metal ions.

Sensor	Characterization techniques	Analytical technique used	Metal ions detected	LOD	Linear range for detection	Refs.
PANI-NTs/GCE	FE-SEM, CV, UV–vis–NIR	SWASV	Pb(II), Cd(II)	9.65×10^{-11} mol/L, 2.67×10^{-10} mol/L	0.05–50 µg/L	[183]
EDTA-PANI/SWCNTs/SS	FT-IR, SEM, AFM, EIS	DPV	Cu(II), Pb(II), Hg(II)	0.08 µM, 1.65 µM, 0.68 µM	1.2–2 mM, 2 µM–37 µM, 1 µM–2 mM	[184]
[Fe_3O_4@PANI]/NA/GCE	SEM, HR–TEM, FT–IR, XRD	DPV	Pb(II), Cd(II)	0.03 nmol/L, 0.03 nmol/L	$0.1–10^4$ nmol/L, $1.0–0.9 \times 10^4$ mol/L	[185]
EDTA-PANI/SWCNTs/SSE	FT–IR, CV, AFM	DPV	Cu(II)	1.4 µM	2 mM–4 µM	[186]
SPAN@UIO-66-NH_2@PANI/SPCE	TEM, XRD, FT–IR, TGA, N_2 adsorption/desorption, CV, EIS	SWASV	Cd(II)	0.17 µg/L	0.5–100 µg/L	[187]

Zhuang et al. [189] invoked the phenomenon of interfacial barrier in ZnO/rGO/ PPy heterostructure for electrochemical sensing of Hg(II) ions with excellent anti-interference capacity due to Schottky barrier. The Hg(II) ions were selectively adsorbed to the PPy surface because of –NH functional groups of PPy. Thus, a positively charged layer was formed by the adsorbed Hg(II) that trapped e-s from ZnO through rGO to PPy. Resultantly, a higher decrease in current was observed with adsorption of more Hg(II) ions. Suvina et al. [190] prepared PPy–rGO hydrogel composites by in-situ polymerization of pyrrole in presence of rGO dispersions having very high surface area of $21.48 \ m^2 \ g^{-1}$. As shown in Scheme 24, initially during the preconcentration process, a complex was formed between residual electronegative atoms such as oxygen and nitrogen present on the surface of rGO and different heavy metal ions present in an electrolyte solution. This caused the reduction of corresponding metal ions electrochemically and, during stripping, the metal is re-oxidized and stripped back into the electrolyte.

Table 22 Polypyrrole (PPy)-based electrochemical sensors for the detection of heavy metal ions.

Sensor	Characterization techniques	Analytical technique used	Metal ions detected	LOD	Linear range for detection	Refs.
ZnO/ rGO/ PPy	SEM, TEM, XRD, Raman spectroscopy, EDS	DPV	Hg(II)	1.9 nM	2–10 nM	[189]
PPy–rGO/ GCE	XRD, FESEM, BET	SWASV	Pb(II)	0.3 nM	0.5–450 nM	[190]
pGO/ PPy/Au	SEM, TEM, Raman spectroscopy, FT-IR, CV, EIS	DPSV	Cd(II)	0.05 μg/L	1–100 μg/L	[191]
NH$_2$- MIL-53 (Al)/ PPy/Au	FE-SEM, TEM, XRD, ATR-IR	DPV	Pb(II), Cu(II)	0.315 μg/L, 0.244 μg/L	1–400 μg/L	[192]

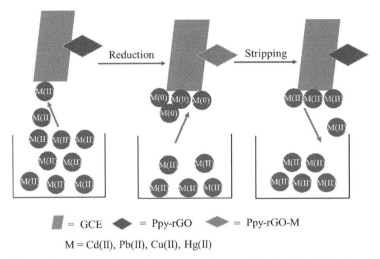

Scheme 24 Schematic representation of heavy metal ion sensing by PPy-rGO hydrogel composite-modified GCE.

Another PA-functionalized PPy/GO-modified GCE detected Cd(II) and Pb(II) ions with a linear working range of 5–150 µg/L. The in situ chemical oxidation polymerization technique was used for synthesis of PPy/GO nanocomposites. The PA molecule with negative charge was adsorbed on the surface of the PPy/GO nanocomposites via electrostatic interactions (Table 22). The DPV responses of the modified electrode were used for detection of Cd(II) and Pb(II) ions, by which LOD was calculated to be 2.13 and 0.41 µg/L for Cd(II) and Pb(II) ions, respectively [188].

Park et al. [193] developed composite using the GO-anchored functionalized polyterthiophene (poly PATT) for detection of various metal ions such as Zn(II), Cd(II), Pb(II), Cu(II), and Hg(II). The performance of developed sensor using chronocoulometry (CC) without predeposition of target ions was further compared with the results of SWASV with predeposition. Wu et al. [194] prepared PEDOT graphitic carbon nitride (PEDOT/g-C$_3$N$_4$) composites by bromine catalyzed polymerization (BCP), solid-sate polymerization (SSP), and metal oxidative polymerization (MOP) methods and compared the electrochemical determination of Cd(II) and Pb(II) ions using the prepared composite-modified GCEs. In the composite, the g-C$_3$N$_4$ ultrathin sheets possessed strong affinity toward heavy metal ions due to the intrinsic —NH and —NH$_2$ groups of g-C$_3$N$_4$ and has been modified with PEDOT to enhance the electron transfer process. Upon comparison, it was found that BCP method displayed higher peak currents than the SSP and MOP methods. Moreover, the inventible mixing of the inorganic and organic phase as well as low p-p interaction between PEDOT and g-C$_3$N$_4$ in the SSP method and iron ion-doping MOP method displayed a negative effect for the electrocatalytic oxidation of the metal ions (Table 23). Resultantly, the composite prepared by BCP method

Table 23 Polythiophenesbased electrochemical sensors for the detection of heavy metal ions.

Sensor	Characterization techniques	Analytical technique used	Metal ions detected	LOD	Linear range for detection	Refs.
NA/PATT/ GO/GCE	FESEM,EDAX, XPS	CC, SWASV	Zn(II), Cd(II), Pb(II), Cu(II), Hg(II), 2G1, Zn(II), Cd(II), Pb(II), Cu(II), Hg(II)	0.05 ppb, 0.08 ppb, 0.20 ppb, 0.09 ppb, 0.10 ppb, 0.30 ppb, 0.15 ppb, 0.08 ppb, 0.10 ppb, 0.09 ppb	1.0 ppb– 10.0 ppm, 1.0 ppb– 1.0 ppm	[193]
PEDOT/g-C_3N_4 /GCE (BCP, SSP, MOP)	FT-IR, UV-Vis, XRD, EDS, SEM, TEM	DPV	Cd(II), Pb(II)	0.0068 μM, 0.0079 μM	0.04–11.4 μM	[194]
Poly(BPE) HNs/GCE Poly(EPE) HNs/GCE	UV-Vis, FT-IR, XRD, TGA, SEM, TEM, EDS	DPV	Cu(II), Pb(II), Cu(II), Pb(II)	0.012 μM, 0.005 μM, 0.02 μM, 0.01 μM	0.016–4.0 μM, 0.012–4.0 μM, 0.04–3.8 μM, 0.02–3.8 μM	[195]
PProDOT (MeSH)$_2$@Si	FT-IR, UV-vis, XRD, SEM, TEM, BET	DPV	Cd(II), Pb(II), Hg(II)	5.75 nM, 2.7 nM, 1.7 nM	0.04–2.8 μM, 0.024–2.8 μM, 0.16–3.2 μM	[196]

(PEDOT/10 wt% g-C_3N_4/GCE) exhibited better electrochemical activity toward Cd(II) and Pb(II) ions (Scheme 25).

Şahutoğlu et al. [197] investigated the optical and electrochemical properties of the rhodamine-based polymer (P(RD-CZ)), which was electropolymerized on an ITO surface (Scheme 26). The selectivity and sensitivity behavior of the P(RD-CZ) film was examined by chronopotentiometric, colorimetric, and electrochemical techniques that displayed excellent selectivity toward Hg(II) ions. Amongst different tested metal ions, a tremendous potential change ($\Delta E = 0.025$ V) was observed with only Hg(II) ions in 50 s. Also, a new band at 560 nm emerged in the absorption spectrum of P(RD-CZ) film with Hg(II) ions, causing the color of the film to change from orange to purple. The Hg(II)–induced optical and electrochemical responses were attributed to the interaction of Hg(II) ions with the stearic groups. This observation was further confirmed by FT-IR analysis, where stretching vibration of C—O of P(RD-CZ) shifted from 1044 to 1079 cm^{-1} after Hg(II) coordination, resulting in potential change of the P(RD-CZ) film sensor. The excellent selectivity of P(RD-CZ) film toward Hg(II) was attributed to the soft acid and heavy atom effect of Hg(II) along with the stearic effect from P(RD-CZ) backbone.

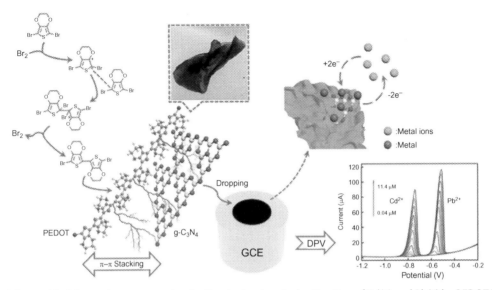

Scheme 25 Schematic representation for the electrochemical estimation of Cd(II) and Pb(II) by PEDOT/ g-C$_3$N$_4$ composites prepared by BCP. *(Reproduced from W. Wu, A. Ali, R. Jamal, M. Abdulla, T. Bakri, T. Abdiryim, A bromine-catalysis-synthesized poly(3,4-ethylenedioxythiophene)/graphitic carbon nitride electrochemical sensor for heavy metal ion determination, RSC Adv. 9 (2019) 34691–34698.)*

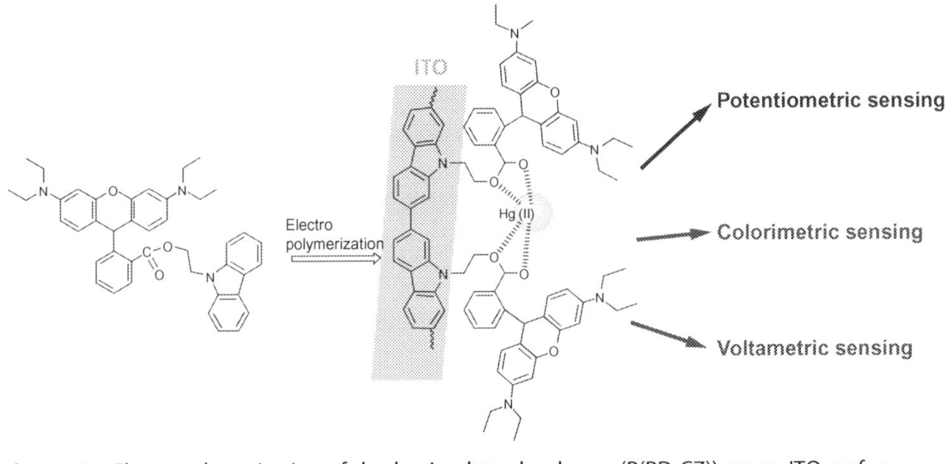

Scheme 26 Electropolymerization of rhodamine-based polymer (P(RD-CZ)) on an ITO surface.

Kaya et al. [197] carried out esterification reaction between Rhodamine B and 9H-carbazole-9-ethanol to form a heterocyclic structure containing nitrogen and high electron density groups. This monomer was further electropolymerized by CV to form Pro-Carb. The polymer displayed selective sensing behavior toward Hg(II) ions. To compare the sensor properties, the polymer was coated both on ITO as well as SPE,

Table 24 Fluorescent conjugated polymers-based electrochemical sensors for the detection of heavy metal ions.

Sensor	Characterization techniques	Analytical technique used	Metal ions detected	LOD (M)	Linear range for detection (M)	Refs.
P(RD-CZ)/ITO	^{1}H NMR, FT-IR, FE-SEM, AFM	UV–Vis, CV, Chrono-potentiometric	Hg(II), Hg(II), Hg(II)	3.1×10^{-8}, 1×10^{-7}, 9.7×10^{-8}	4×10^{-7} to 4×10^{-4}, 1×10^{-4} to 1×10^{-7}, 1×10^{-4} to 1×10^{-7}	[197]
PRoCarb/ITO, PRoCarb/SPE	FT-IR, UV–Vis, NMR, LC–MS, EDX, SEM, CV	Potentiometry	Hg(II), Hg(II)	5.4×10^{-9}, 2.02×10^{-12}	10^{-2}–10^{-9}, 10^{-2}–10^{-11}	[197]

where working linear range of the polymer film coated on SPE was wideer than working linear range of the polymer film coated on ITO (Table 24).

4.1.2 Ion-imprinted polymers for the detection of heavy metal ions

Motlagh et al. [198] prepared a nanostructured Cd(II) ion imprinted polymer (Cd-IIP) via sol–gel process. The resulting Cd-IIP polymer was further employed for modification of CPE, which displayed high sensitivity and response to DPV signals due to high availability of adsorption cavities created during polymerization process. The modified electrode provided stable and selective detection of Cd(II) over the linear range of 0.5–40 µg/L with LOD of 0.15 µg/L, when a paste composition of 10% (*w/w*) of Cd-IIP, 65% (*w/w*) of graphite powder, and 25% (*w/w*) of paraffin oil using PBS buffer (pH 5) with accumulation potential of -1.1 V and accumulation time of 300 s was used. Moreover 100-fold excess of Cr(III), Fe(III), Mn(II), UO_2(II), and Ni(II); 50-fold excess of Co(II), Zn(II), Hg(II), and Ag(I); 30-fold excess of Pb(II); and 10-fold excess of Cu(II) showed negligible changes in the DPV signal of 25 µg/LCd(II).

Hu et al. [199] developed an electrochemical sensor utilizing PPy as IIP for quantification of Cd(II) in aqueous samples. The development of IIP first involves the modification of electrode via rGO followed by imprinting on the same surface via electropolymerization of PPy using Cd(II) as a template. The morphological analysis was done using SEM and TEM with optimizing experimental conditions for attaining good sensitive performance. The sensor imparted good linear range with LOD of 0.26 µg/L with successful application to real water samples. Also, nanosized CN^- sensor was developed by fabricating MIP utilizing MAA and vinyl pyridine (VP) as copolymers

Table 25 Various ion-imprinted sensors for the detection of heavy metal ions.

Sensor	Characterization techniques	Analytical Technique used	Metal ions detected	LOD	Linear range for detection	Refs.
MWCNT-IIP	FTIR, TGA, XRD, EDAX, FESEM, TEM	CV	Cd(II)	0.03 μM	–	[198]
IIP/rGO/GCE	SEM, TEM	SWASV	Cd(II)	0.26 μg/L	1–100 μg/L	[199]
nanoIP-MWCNTs-CP	SEM	Potentiometry	CN^-	7.5×10^{-7} mol/L	10^{-6}–10^{-1} mol/L	[200]
Pb(II) imprinted polymer	SEM, FTIR	DPV	Pb(II)	1.3×10^{-11} mol/L	1.0×10^{-9} to 7.5×10^{-7} mol/L	[201]
RGO-IIP	FESEM, TEM, TGA, FTIR	SWASV	Hg(II)	0.02 μg/L	0.07–80 μg/L	[202]

with CN^- as template (Table 25). The different parameters were optimized and LOD calculated to be 7.5×10^{-7} mol/L [24].

4.2 Bionanomaterials for the detection of heavy metal ions

4.2.1 Amino acids functionalized materials for the detection of heavy metal ions

Since poly(amino acid)s contain a large number of functional groups, they efficiently bind with different analytes [203]. Guo et al. [204] used a rGO-CS/poly-L-lysine (PLL) nanocomposite-modified GCE for simultaneous determination of Cd(II), Pb(II), and Cu(II) ions. Here, positively charged CS interacted strongly with negatively charged rGO and prevented their aggregation. Additionally, CS showed strong interaction toward metal ions because of its number of amino groups (Table 26).

Lu et al. [205] manufactured miniaturized, flexible, and micropatterened poly-L-cysteine (PLC)-modified laser engraved grapheme electrodes (LEGE), transformed from polyimide (PI) films, that were further decorated with IL and IL/PLC/LEGE electrochemical sensors. The designed IL/PLC/LEGE sensor possessed synergistic effect of the large specific surface area of LEGE, specific complexing ability of PLC, and good electron-transfer ability of IL. These properties enabled IL/PLC/LEGE sensor for Pb(II) ions with LOD value of 0.17 μg/L (Scheme 27).

Yi et al. [206] prepared poly(L–glutamic acid) (PGA) and GO composite material and modified GCE with this prepared composite material for detection of heavy metal ions such as Cu(II), Cd(II), and Hg(II) ions. Here, the presence of many carboxyl groups on the side chain of PGA increased its binding ability toward metal ions, whereas GO

Table 26 Various electrochemical sensors for heavy metal ions based on amino acids as modifying materials.

Sensor	Characterization techniques	Analytical technique used	Metal ions detected	LOD	Linear range for detection	Refs.
rGO-CS/PLL/GCE	XRD, FT-IR, XPS, SEM, CV	DPASV	Cd(II), Pb(II), Cu(II)	0.01 μg/L, 0.02 μg/L, 0.02 μg/L	0.05–10 μg/L	[204]
IL/PLC/LEGE	SEM, EDX, XRD, Raman spectroscopy, BET, XPS, CV	SWASV	Pb(II)	0.17 μg/L	1–180 μg/L	[205]
PGA/GO/GCE	SEM, EIS, CV	DPASV	Cu(II), Cd(II), Hg(II)	0.024 μM, 0.015 μM, 0.032 μM	0.25–5.5 μM	[206]
PGA/rGO	SEM, ATR, AFM, CV	SWASV	Pb(II)	0.06 μg/L	0.2–11.5 μg/L	[207]
Gly/rGO/PANI/GCE	XRD, FT-IR, UV-vis, Raman spectroscopy, SEM, BET, CV, EIS	SWASV	Cd(II), Pb(II)	0.07 nM, 0.072 nM	1–0.0001 μM	[208]
T/CA/AuNPs/rGO/GCE	ATR-IR, SEM, CV, EIS	DPV	Hg(II)	1.5 ng/L	10 ng/L–1.0 μg/L	[209]

provided conductivity and chemical stability to the sensor. Hanif et al. [208] developed glycine-functionalized rGO (Gly/rGO)/PANI nanocomposite for trace level detection of Cd(II) and Pb(II). The modified Gly/rGO/PANI/GCE electrode displayed eightfold higher electrochemical response toward Cd(II) and Pb(II) ions than that of bare GCE. This was attributed to the synergistic complexation effect of $-NH_2$ and $-COO^-$ groups of glycine, high surface area of rGO, and high electrical conductivity of PANI along with its ability to bind with metal ions via imine ($=N-$) and amine ($-NH-$) functional groups on its surface. Moreover, Gly/rGO/PANI/GCE-modified electrode exhibited negligible changes in value of anodic peak currents of Cd(II) and Pb(II) ions in the presence of 100 equiv. of Cu(II), Co(II), Cl$^-$, and nitrobenzene (NB). Here, only Cu(II) showed little interference due to its high affinity toward the amine groups ($-NH_2$) of glycine and aniline of modified electrode. Wang et al. [209] deposited AuNPs onto the surface of rGO on which covalently coupled thymine-1-acetic acid (using its

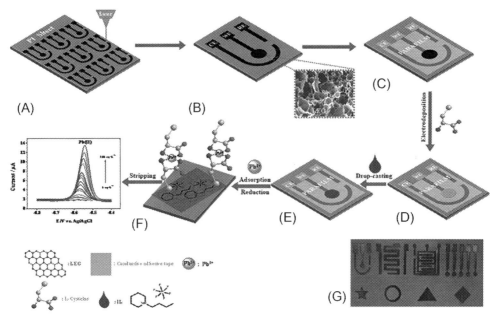

Scheme 27 Schematic representation of fabricated IL/PLC/LEGE sensor for Pb(II) sensing, where (A) laser engraving fabrication of LEGE arrays on PI sheet; (B) three-electrode system of LEGE; (C) selective surface passivation of LEGE areas by parafilm to define the WE area; (D) electropolymerization of L-cysteine to form PLC/LEGE; (E) formation of IL/PLC/LEGE; (F) electrochemical sensing of Pb(II); and (G) optical image of different micro-patterned electrode arrays on PI sheet. *(Reproduced from Z. Lu, X. Lin, J. Zhang, W. Dai, B. Liu, G. Mo, J. Ye, J. Ye, Ionic liquid/poly-L-cysteine composite deposited on flexible and hierarchical porous laser-engraved graphene electrode for high-performance electrochemical analysis of lead ion. Electrochim. Acta 295 (2019) 514–523.)*

—COOH group) and cysteamine (using —NH$_2$ group) self-assembled to produce T/CA/AuNPs/rGO-modified electrode (Scheme 28). The modified electrode displayed strong affinity toward Hg(II) ions because of strong affinity of Hg(II) with thymine bases and electrochemical signal amplification induced by AuNPs/rGO nanostructures. Moreover, the weaker space resistance between the small molecule and Hg(II), compared with T-bases chained in the DNA strand, led to easy coordination.

4.2.2 Enzymatic sensors for the detection of heavy metal ions

Metal ions, being highly toxic, act as enzyme inhibitors and hence decline the activity of an enzyme. This change in activity of biosensing materials offers substitution methodology for determination of heavy metal ions [210]. In this regard, different enzymes such as urease [211], acetylcholine esterase [212], tyrosinase [213], glucose oxidase [214], and horseradish peroxidase [215] have been utilized for construction of biosensors for trace level determination of various heavy metal ions. However, for better biocompatibility to

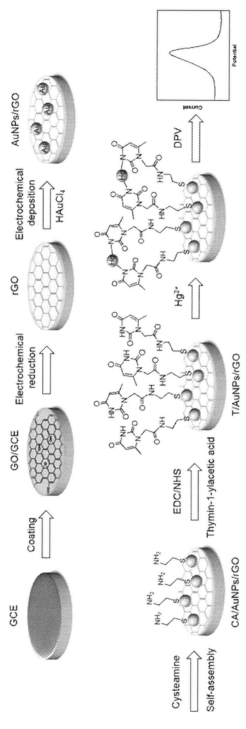

Scheme 28 Schematic representation of T/CA/AuNPs/rGO-modified electrode for Hg(II) detection. (Reproduced from N. Wang, M. Lin, H. Dai, H. Ma, Functionalized gold nanoparticles/reduced graphene oxide nanocomposites for ultrasensitive electrochemical sensing of mercury ions based on thymine-mercury-thymine structure, Biosens. Bioelectron. 79 (2016) 320–326.)

enhance the catalytic response of the enzyme, the requisite of developing advanced electrode materials are in demand. Do et al. [211] developed an electrochemical biosensor by immobilizing urease on the surface of nanostructures (PANi)–Nafion(NSPN)/Au/Al$_2$O$_3$ working on the chronopotentiometry. The wide linear range 0.1–1.0 ppm for determination of Pb(II) has been obtained, and kinetic modeling of proposed biosensor for Hg(II) and Pb(II) has been proposed.

Silva et al. [214] developed an electrochemical sensor based on glucose oxidase (GOx) for trace level monitoring of metal ions (Hg(II), Cd(II), Pb(II), and Cr(VI)). The modified electrode utilized for this purpose was poly(brilliantgreen) films onto MWCNTs (Scheme 29). The surface attributes created by DES in fabricating PBG polymer film had significantly enhanced the electroanalytical response of PBG/MWCNT/GCE sensor thus showing better sensitivity and inhibition of 50%. The sensor offers detection limits in the nanomolar range and has been exploited for determination in milk samples. Pandey et al. [215] developed an electrochemical biosensor for determination of Cr(VI) using HRP enzyme. The electrochemical sensor was modified with ERGO followed by formation of thionine layer via electrodeposition process, and Cr(VI) was used to inhibit

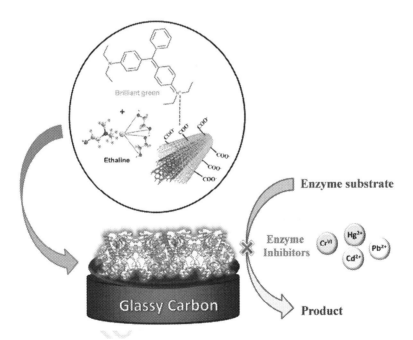

Scheme 29 The fabrication of poly(brilliantgreen)-deep eutectic solvent/CNT-modified electrode. *(Reproduced from W. da Silva, M.E. Ghica, C.M.A. Brett, Biotoxic trace metal ion detection by enzymatic inhibition of a glucose biosensor based on a poly(brilliant green)-deep eutectic solvent/carbon nanotube modified electrode, Talanta 208 (2020) 120427.)*

enzyme activity. The response of the developed sensor is linear from 0.039 to 0.78 mM with LOD 0.02 mM.

4.2.3 Aptasensors for the detection of heavy metal ions

The recent approach for biosensing of heavy metal ions involves the use of aptamers as they impart high sensitivity and specificity to the developed biosensors. The modification of an electrode surface for fabrication of an aptasensor in heavy metal ions sensing is an attractive approach. In this regard, AuNPs offers advantages of enhanced electron transfer catalysis and promote the sensitivity of an electrode. The biocompatibility and electron transfer rate can be enhanced and direct electron transfer can be achieved by sulfhydryls [216]. Taghdisi et al. [217] developed an aptasensor for selective determination of Pb(II) wherein AuNPs and hairpin structure of complementary structure of aptamer and thionine were included. The sensing mechanism of the developed sensor was based on direct electron transfer mechanism between AuNPs and thionine via conjugated aptamer. After introduction of Pb(II), the aptamer intervenes and blocks the electron transfer. Using this, LOD of sensor was 312 pM. Later, they developed another aptamer-based electrochemical sensor with LOD of 149 pM [218]. Ge et al. [219] fabricated Pb(II)-selective biosensor using Pb(II)-dependent DNAzyme strands attachment, where Au/FrGO/Au-PWE was used as an amplified signal sensing platform and S3/H-Mn_2O_3/HRP/GOx as an amplified signal tag. Initially, AuNPs layer was grown on the surfaces of cellulose fibers in the hydrophobic zone of the paper to fabricate a paper-working electrode. Flower-like rGO (FrGO) with large specific surface area and good electrical conductivity was fabricated on the Au-PWE through an electroreduction process to form Fr/GO/Au-PWE electrodes. HRP and GOx molecules were covalently conjugated onto the surface of hierarchical hollow spheres (H-Mn_2O_3) to form H-Mn_2O_3/HRP/GOx bioconjugates, which were further complexed with the signal strand (S3) to serve as molecular tags. In the presence of Pb(II) ions, the DNAzyme (S1) was activated and the substrate strand (S2) was cleaved. Upon immersing Au/FrGO/Au-PWE/S1/BSA/Pb(II)/S3/H-Mn_2O_3/HRP/GOx electrode in 30 mM aniline and 15 mM glucose, a pair of well-defined redox peaks were observed due to redox activity of PANI. The LOD for Pb(II) was found to be 0.002 nM over a wide range of concentration from 0.005 to 2000 nM using developed biosensors. Zhao et al. [220] combined Pb(II)-dependent DNAzymes and toehold-mediated strand displacement reaction (TSDR) for the selective recognition of target Pb(II) ions and the efficient amplification of a response signal. The electrode surface was modified with AuNPs-loaded Fe_3O_4 nanocomposites and a DNA duplex possessing capture probe (NH$_2$-CP) and two specific toehold sequences (AS and PS). Target Pb(II) ions selectively recognized and cleaved Pb(II)-specific DNAzymes to release substrate fragment (rSS). The released rSS activated the TSDR process and resulted in the in situ assembly of complete signal probes labeled with MB-SP on a modified electrode surface (Scheme 30). The resultant spatial proximity of electroactive MB

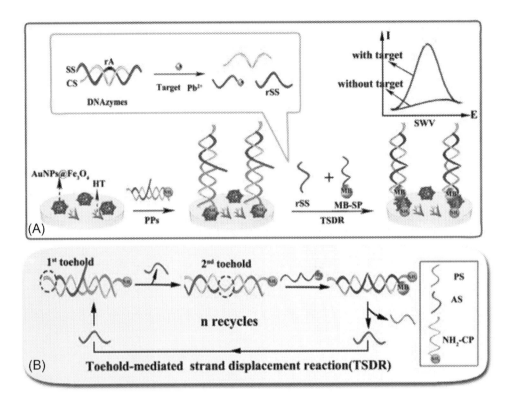

Scheme 30 Schematic representation of the (A) Pb(II)-specific DNAzyme-based electrochemical biosensor for monitoring Pb(II) ions and (B) TSDR reaction. Here, SS means substrate strand; CS means catalytic strand; PS means protection strand; AS means assistant strand; NH$_2$-CP means capture strand; and MB-SP means signal probe. *(Reproduced from J. Zhao, T. Zheng, J. Gao, W. Xu, Toehold-mediated strand displacement reaction triggered by nicked DNAzymes substrate for amplified electrochemical detection of lead ion. Electrochim. Acta 274 (2018) 16–22.)*

and AuNPs@Fe$_3$O$_4$ significantly enhanced the electrochemical response via TSDR and detected Pb(II) ions over linear range of 1 pM to 500 nM with LOD value of 0.32 pM.

Lu et al. [221] fabricated in-situ amplified electrochemical sensing platform based on MB-labeled poly-T(15) oligonucleotide-functionalized GO by coupling with DNase I–assisted target recycling amplification for monitoring of Hg(II) ions (Scheme 31). Target recycling involves a target molecule to interact with multiple nuclei acid–based signaling probes. The DPV measurements showed that peak current decreased linearly with increasing concentration of Hg(II) from 0.5 to 50 nM with LOD value to be 0.12 nM.

Zhang et al. [222] made use of T–Hg(II)–T approach for selective determination of Hg(II). In this approach, DNA1 was attached onto the surface of a modified Au electrode and complementary DNA works as a probe. The sensor offered a wide linear range of 0 to 200 nM and LOD of 0.05 nM. Similar methodology has been adopted by He et al. [223]

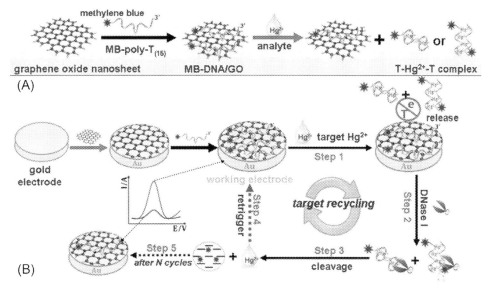

Scheme 31 Schematic representation of (A) reaction process and (B) assay principle of MB-DNA/GO-modified electrode target recycling amplification process for monitoring of Hg(II) ions. *(Reproduced from M. Lu, R. Xiao, X. Zhang, J. Niu, X. Zhang, Y. Wang, Novel electrochemical sensing platform for quantitative monitoring of Hg(II) on DNA-assembled graphene oxide with target recycling, Biosens. Bioelectron. 85 (2016) 267–271.)*

for Hg(II) determination and LOD of 0.2 pM has been achieved. Mei et al. [224] developed electrochemical biosensor for Hg(II) through coupling thymine (T)-Hg(II)-thymine (T) base pairs with surface initiated enzymatic polymerization (SIEP) strategy for signal amplification. Initially, Au electrode was modified with capture DNA labeled with thiols at its $3'$ terminals and MCH to form SH-DNA/MCH self-assembly monolayer. In the presence of Hg(II), the T-T mismatched complementary DNA was hybridized to the capture DNA due to the formation of T-Hg(II)-T complex. Then, use of terminal deoxynucleotidyl transferase (TDT) catalyzed the sequential addition of deoxynucleotides (dNTPs) at the 3-OH group of an oligonucleotide primer and incorporated biotinlated $2'$-deoxyadenosine $5'$-triphosphate (bio-dATP) into the long single-stranded DNA (ssDNA) possessing numerous biotins. A large number of streptavidin- functionalized AgNPs are attached to the electrode surface due to the binding between biotin and streptavidin, thus generating electrochemical stripping signal of Ag to monitor Hg(II) concentration. In the absence of Hg(II), T-T mismatched DNA was not captured and SIEP did not occur, resulting in weak background response due to nonspecific adsorption of AgNPs. However, in the presence of Hg(II) and SIEP amplification, a sharp stripping peak with rapid increase of stripping current was observed. The stripping peak current of

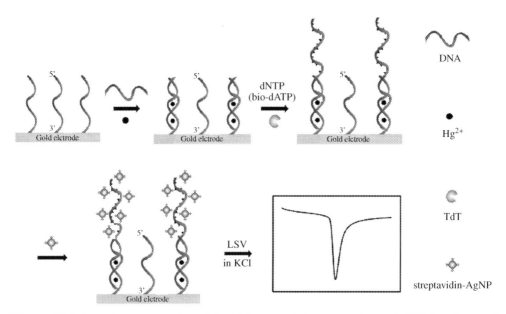

Scheme 32 Schematic representation of the fabrication of biosensor using T-Hg(II)-T chemistry and SIEP signal amplification strategy for detection of Hg(II). *(Reproduced from C. Mei, D. Lin, C. Fan, A. Liu, S. Wang, J. Wang, Highly sensitive and selective electrochemical detection of Hg(2+) through surface-initiated enzymatic polymerization, Biosens. Bioelectron. 80 (2016) 105–110.)*

AgNPs showed gradual increase with increase in Hg(II) concentration over the linear range of 0.05–100 nM and LOD was found to be 24 pM (Scheme 32).

Hu et al. [225] developed a label-free biosensor on Au nanoclusters (GNCs) as a transducer platform with DNAzyme catalytic bacons for the detection of Cu(II) using EIS with LOD value as low as 0.0725 nM. A reversible redox couple, $Fe(CN)_6^{3-/4-}$, was used as a marker using EIS to study the difference in the interfacial e^--transfer resistance. In the presence of Cu(II), the phosphodiester of the DNAzyme is cleaved to form two fragments, causing a decrease of the interfacial charge transfer resistance of the electrodes toward $[Fe(CN)_6]^{3-/4-}$ redox couple. This occurred due to easier electron transfer by GNCs. Thus, Cu(II) detection by prepared biosensor was possible by measuring the changes in the charge transfer resistance of DNA films in its presence and absence. An electrochemical biosensor combined with a 3D origami device based on AuNPs-modified, paper-working electrode (Au-PWE) as a sensor platform and DNA-functionalized, iron-porphyrinic, metal-organic framework ($(Fe-P)_n$-MOF-Au-GR) hybrids as signal probes has been developed by Wang et al. [226] for detection of Pb(II). In the presence of Pb(II), GR was specifically cleaved at the ribonucleotide (rA) site, which resulted in the formation of a short $(Fe-P)_n$-MOF linked oligonucleotide fragment. This fragment further hybridized with hairpin DNA immobilized on the surface of Au-PWE and displayed amplified electrochemical signal, attributed to the mimic peroxidase property of $(Fe-P)_n$-MOF.

Miao et al. [227] prepared DNA-modified Fe_3O_4@AuNPs ad magnetic GCE for electrochemical sensing of Ag(I) and Hg(II) ions. Three DNA probes possessing mismatched base pairs were designed and used. The hybridization of DNA1/DNA2 and DNA1/DNA3 occurred on the basis of stable cytosine-Ag(I)-cytosine and thymine-Hg(II)-thymine complexes, which made simultaneous detection of Ag(I) and Hg(II) possible without significant interference (Table 27).

Another Cu(II)-anchored MOF prepared by Zhang et al. [228] showed catalytic activity to glucose oxidation and offered an alternative mimetic catalyst to amplify the electrochemical signal for selective Hg(II) detection (Scheme 33). sDNA/MOF-Au were used as signal probes, where Cu(I)-anchored MOFs acted as mimetic catalyst and thiolated ssDNA acted as a recognition element. For the preparation of electrochemical sensor, GO@Au was immobilized on GCE surface to prepare high-conductive sensing interface for thiolated cDNA. GO@Au/GCE was passivated with 6-mercapto-hexanol (MCH) to remove nonspecific absorption sites. After incubation with Hg(II) samples and prepared signal probes (sDNA/MOF-Au), selective T-Hg(II)-T complex was formed. Based on the catalytic oxidation of glucose by MOF, the detectable electrochemical signal was generated by which Hg(II) concentration was monitored over the range of 0.10 aM to 100 nM with a LOD of 0.001 aM.

Yuan et al. [229] invoked DNAzyme-assisted target recycling and controllably efficient disassembly of branched DNA polymer induced by strand displacement reaction (SDR) for Pb(II) detection over a wide linear range from 1 pM to 100 nM with a LOD of 0.24 pM. Hybridization chain reaction (HCR) was used to form branched DNA polymer, which was immobilized on an electrode surface and served as an ideal platform for loading abundant AgNPs to generate high current signal output. However, in the presence of Pb(II), cleavage of substrate strand in Pb(II)-specific DNAzyme was activated, which released large quantities of dual triggers (T_1 and T_2). The released triggers dissembled the backbone and side chain of the branched DNA polymer, which resulted in decreased current output for detection of Pb(II).

Carbon-based matrices have also been employed in this category of aptamer-based sensors for heavy metal ions. An electrochemical biosensor possessing 8–17 DNAzyme as the recognition element and the reaction of H_2O_2 catalyzed by MOF Fe-MIL-101 as the signal probe was fabricated by Yu et al. [230] for Pb(II) detection over 10^{-13} to 10^{--7} mol/L with LOD value of 1.74×10^{-14} mol/L. The —COOH of the carboxyl-functionalized graphene (CFGR-COOH) and —NH_2 of DNAzyme formed an amide bond that helped in immobilization of 8–17 DNAzyme on the electrode surface. In the presence of Pb(II), the substrate strand (sDNA) of 8–17 DNAzyme was cleaved and the other sDNA strand remained on the electrode surface hybridized with hDNA. Due to the recognition between biotin from hDNA and the streptavidin (SA) from

Table 27 Aptamer-based bionanomaterials for the detection of heavy metal ions.

Sensor	Characterization techniques	Analytical technique used	Metal ions detected	LOD (M)	Linear range for detection (M)	Refs.
Apt-CS	—	DPV	Pb(II)	312×10^{-12}	$(0.7–100) \times 10^{-9}$	[217]
Aptamer	—	DPV	Pb(II)	149×10^{-12}	$(0.7–300) \times 10^{-9}$	[218]
Au/FrGO/Au-PWE	SE, TEM, EDX	DPV	Pb(II)	0.002×10^{-9}	$(0.005–2000) \times 10^{-9}$	[219]
HT/PPs/AuNPs@Fe$_3$O$_4$/ GCE.	PAGE, RAMAN, XPS	SWV	Pb(II)	0.3×10^{-12}	$(0.001–100) \times 10^{-9}$	[220]
MB-DNA/GO/Au	TEM, CV	DPV	Hg(II)	0.12×10^{-9}	$(0.5–50) \times 10^{-9}$	[221]
AuNPs functionalized DNA 3-DNA 2-DNA 1/Au	—	EIS	Hg(II)	0.05×10^{-9}	$(0 – 200) \times 10^{-9}$	[222]
DFNP/theavidin/AuNps/ NGP/nafion/GCE	TEM, SEM	DPV	Hg(II)	0.21×10^{-12}	$(0.35–3500) \times 10^{-12}$	[223]
Thymine-Hg^{2+}-thymine (T-Hg^{2+}-T)-based aptamer	—	LSV	Hg(II)	0.024×10^{-9}	$(0.05–100) \times 10^{-9}$	[224]
Detection Cu^{2+}/EIS/ DNAzyme	—	EIS	Cu(II)	0.0725×10^{-9}	$(0.1–400) \times 10^{-9}$	[225]
DNA functionalized ((Fe-P)n– MOF-AuGR)	SEM, TEM	DPV	Pb(II)	0.02×10^{-9}	$(0.03–1000) \times 10^{-9}$	[226]
DNA-functionalized Fe$_3$O$_4$@AuNPs	SEM, XRD, UV	SWV	Hg(II)	1.7×10^{-9}	$(10 – 100) \times 10^{-9}$	[227]
sDNA/MOF-Au	SEM	DPV	Hg(II)	1.7×10^{-9}	0.10×10^{-18} to 100×10^{-9}	[228]
T1 + T2/DMs/HT/DI/ depAu/GCE	CV, EIS	LSV	Pb(II)	0.24×10^{-12}	1.0×10^{-12} to 100×10^{-9}	[229]

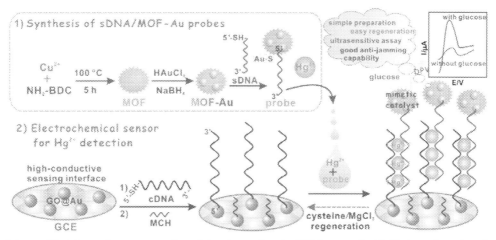

Scheme 33 Schematic representation of the electrochemical DNA sensor for Hg(II) detection using Cu (II)-anchored MOF as mimetic catalyst. *(Reproduced from X. Zhang, Y. Jiang, M. Zhu, Y. Xu, Z. Guo, J. Shi, E. Han, X. Zou, D. Wang, Electrochemical DNA sensor for inorganic mercury(II) ion at attomolar level in dairy product using Cu(II)-anchored metal-organic framework as mimetic catalyst, Chem. Eng. J. 383 (2020) 123182.)*

SA@Fe-MIL-101 composites, MOF is immobilized on the electrode surface. This immobilized Fe-MIL-101 electrocatalyzed the H_2O_2 oxidation at the electrode that served as a detectable electrochemical indicator [230]. Gao et al. [231] developed a label-free electrochemical aptasensor for determination of Pb(II) utilizing graphene-thionine (GR–TH) as an electrode modifier. The sensor works on the general formation of G–quadruplex structure in presence of Pb(II). This triggers the release of GR–TH assembly into solution affecting electrochemical response of sensor. The sensor exhibited linearity and low LOD of 3.2×10^{-14}. Wang et al. [232] developed an electrochemical aptasensor utilizing composites rGO/graphitic carbon nitride (g-C_3N_4) (GCN) with aptamers for quantification of Cd(II). The sensor exhibited high sensitivity and selectivity for Cd(II) determination and LOD is 0.337 nM (Table 28).

5. Concluding remarks

Worldwide globalization and industrialization have caused the release of environmental pollutants like pesticides, antibiotics, and heavy metal ions. This has resulted in various health effects on humans, living organisms, and aquatic biota. For this, electrochemistry has paved a way toward ultratrace level determination of these environmental pollutants as they are cost–effective and offer portability. The recent chapter deals with how polymeric and bionanomaterials have a paved the way for the detection of environmental pollutants using electroanalytical techniques. Conducting polymers, owing to their high

Table 28 Aptamer-based bionanomaterials (carbon structures) for detection of heavy metal ions.

Sensor	Characterization techniques	Analytical technique used	Metal ions detected	LOD (M)	Linear range for detection (M)	Refs.
Fe-MIL 101@SA/ CFGR–COOH	RAMAN, SEM, XRD, FTIR	CV	Pb(II)	3.2×10^{-14}	10^{-13} to 10^{-7}	[230]
TH/GR/ LSA-MCH/ AuE	UV–vis	DPV	Pb(II)	3.2×10^{-14}	1.6×10^{-13} to 1.6×10^{-10}	[231]
TH/GR/ LSA-MCH/ AuE	SEM, TEM, EDX, XRD, FTIR	DPASV	Cd(II)	0.33×10^{-9}	1.0×10^{-9} to 1.0×10^{-6}	[232]

conductivity, have offered a way for ultratrace level detection of toxic pollutants like pesticides, heavy metal ions, and antibiotics. The rise of MIP has also gained considerable attention due to more efficient active sites obtained via imprinting methodology, which also imparts specificity and selectivity toward electrochemical analysis for environmental applications. Besides these, the urge of development of electrochemical sensors for environmental applications made bionanomaterial-based electrochemical sensors an important class. The chapter includes the recent advancements based on either biological elements like enzymes, aptamers, and immunosensors or nanostructured materials for determination of pesticides, antibiotics, or heavy metal ions in the environmental samples. The content of tables summarizes the different materials in different sections included in this chapter along with different analytical techniques deployed for sensing purposes. Also each section summarizes how various amplification strategies like multiple recycling amplification techniques have been employed for ultratrace level or simultaneous detection of environmental pollutants. The mode of action of various electrochemical sensors have also been discussed like direct or indirect detection either from the view point of electroactive nature of analyte or by regarding analytes as substrates or inhibitors. This chapter is beneficial for understanding electrochemical analysis as it provides knowledge of advancements in polymeric and bionanomaterials for environmental applications.

Acknowledgments

Financial grant from Department of Science and Technology-Science and Engineering Research Board (Grant No. SERB/2016/000046) is gratefully acknowledged by the authors. Ranjeet Kaur acknowledges council of scientific and industrial research (CSIR, File no. 09/135(0826)/2018-EMR-I) for junior research fellowship (JRF).

References

[1] R. Kelishadi, Environmental pollution: health effects and operational implications for pollutants removal, J. Environ. Public Heal. 2012 (2012) 341637.

[2] W.A. Suk, H. Ahanchian, K.A. Asante, D.O. Carpenter, F. Diaz-Barriga, E.H. Ha, X. Huo, M. King, M. Ruchirawat, E.R. da Silva, L. Sly, P.D. Sly, R.T. Stein, M. van den Berg, H. Zar, P.J. Landrigan, Environmental pollution: an under-recognized threat to children's health, especially in low- and middle-income countries, Environ. Health Persp. 124 (2016) A41–A45.

[3] L. Rassaei, F. Marken, M. Sillanpää, M. Amiri, C.M. Cirtiu, M. Sillanpää, Nanoparticles in electro-chemical sensors for environmental monitoring, TrAC Trend Anal. Chem. 30 (2011) 1704–1715.

[4] S. Su, S. Chen, C. Fan, Recent advances in two-dimensional nanomaterials-based electrochemical sensors for environmental analysis, Green Energy Environ. 3 (2018) 97–106.

[5] G. Maduraiveeran, W. Jin, Nanomaterials based electrochemical sensor and biosensor platforms for environmental applications, Trends Environ. Anal. Chem. 13 (2017) 10–23.

[6] C.B. Ojeda, F.S. Rojas, Process analytical chemistry: applications of ultraviolet/visible spectrometry in environmental analysis: an overview, Appl. Spectrosc. Rev. 44 (2009) 245–265.

[7] F.F. Donato, M.L. Martins, J.S. Munaretto, O.D. Prestes, M.B. Adaime, R. Zanella, Development of a multiresidue method for pesticide analysis in drinking water by solid phase extraction and determi-nation by gas and liquid chromatography with triple quadrupole tandem mass spectrometry, J. Braz. Chem. Soc. 26 (2015) 2077–2087.

[8] Z. Shi, J. Hu, Q. Li, S. Zhang, Y. Liang, H. Zhang, Graphene based solid phase extraction combined with ultra high performance liquid chromatography-tandem mass spectrometry for carbamate pesti-cides analysis in environmental water samples, J. Chromatogr. A1355 (2014) 219–227.

[9] R.A. Alvarez-Puebla, D.S. dos Santos Jr., R.F. Aroca, SERS detection of environmental pollutants in humic acid-gold nanoparticle composite materials, Analyst 132 (2007) 1210–1214.

[10] M.A. De Jesús, K.S. Giesfeldt, M.J. Sepaniak, Improving the analytical figures of merit of SERS for the analysis of model environmental pollutants, J. Raman Spectrosc. 35 (2004) 895–904.

[11] R. Kaur, S. Rana, R. Singh, V. Kaur, P. Narula, A Schiff base modified graphene oxide film for anodic stripping voltammetric determination of arsenite, Microchim. Acta 186 (2019) 741.

[12] G. Hanrahan, D.G. Patil, J. Wang, Electrochemical sensors for environmental monitoring: design, development and applications, J. Environ. Monitor. 6 (2004) 657–664.

[13] X. Zhao, L. Lv, B. Pan, W. Zhang, S. Zhang, Q. Zhang, Polymer-supported nanocomposites for environmental application: a review, Chem. Eng. 170 (2011) 381–394.

[14] T.R. Sahoo, Polymer nanocomposites for environmental applications, in: D.K. Tripathy, B.P. Sahoo (Eds.), Properties and Applications of Polymer Nanocomposites, Springer-Verlag GmbH, Germany, 2017, pp. 77–106.

[15] T.P. Nguyen, S.H. Yang, Hybrid materials based on polymer nanocomposites for environmental applications, in: M. Jawaid, M.M. Khan (Eds.), Polymer Based Nanocomposites for Energy and Envi-ronmental Applications, Elsevier, 2018, pp. 507–551.

[16] R. Das, A.J. Pattanayak, S.K. Swain, Polymer nanocomposites for sensor devices, in: M. Jawaid, M. M. Khan (Eds.), Polymer Based Nanocomposites for Energy and Environmental Applications, Elsevier, 2018, pp. 205–218.

[17] P. Arunachalam, Polymer-based nanocomposites for energy and environmental applications, in: M. Jawaid, M.M. Khan (Eds.), Polymer Based Nanocomposites for Energy and Environmen-tal Applications, Elsevier, 2018, pp. 185–203.

[18] N.B. Singh, A.B.H. Susan, Polymer nanocomposites for water treatments, in: M. Jawaid, M.M. Khan (Eds.), Polymer Based Nanocomposites for Energy and Environmental Applications, Elsevier, 2018, pp. 569–595.

[19] V. Pichon, F. Chapuis-Hugon, Role of molecularly imprinted polymers for selective determination of environmental pollutants—a review, Anal. Chim. Acta 622 (2008) 48–61.

[20] J. Ashley, M.A. Shahbazi, K. Kant, V.A. Chidambara, A. Wolff, D.D. Bang, Y. Sun, Molecularly imprinted polymers for sample preparation and biosensing in food analysis: progress and perspectives, Biosens. Bioelectron. 91 (2017) 606–615.

[21] J. Fu, L. Chen, J. Li, Z. Zhang, Current status and challenges of ion imprinting, J. Mater. Chem. A3 (2015) 13598–13627.

[22] L. Wang, X. Peng, H. Fu, C. Huang, Y. Li, Z. Liu, Recent advances in the development of electrochemical aptasensors for detection of heavy metals in food, Biosens. Bioelectron. 147 (2020) 111777.

[23] R. Ramachandran, T.-W. Chen, S.-M. Chen, T. Baskar, R. Kannan, P. Elumalai, P. Raja, T. Jeyapragasam, K. Dinakaran, G.P. Gnana Kumar, A review of the advanced developments of electrochemical sensors for the detection of toxic and bioactive molecules, Inorg. Chem. Front. 6 (2019) 3418–3439.

[24] G. Zhao, Y. Guo, X. Sun, X. Wang, A system for pesticide residues detection and agricultural products traceability based on acetylcholinesterase biosensor and Internet of things, Int. J. Electrochem. Sci. 10 (2015) 3387–3399.

[25] S. Weng, W. Zhu, R. Dong, L. Zheng, F. Wang, Rapid detection of pesticide residues in paddy water using surface-enhanced raman spectroscopy, Sensors 19 (2019).

[26] Z. Zhang, S. Zhai, M. Wang, L. He, D. Peng, S. Liu, Y. Yang, S. Fang, H. Zhang, Electrochemical sensor based on a polyaniline-modified SnO_2 nanocomposite for detecting ethephon, Anal. Methods 7 (2015) 4725–4733.

[27] Y. Wang, J. Jin, C. Yuan, F. Zhang, L. Ma, D. Qin, D. Shan, X. Lu, A novel electrochemical sensor based on zirconia/ordered macroporous polyaniline for ultrasensitive detection of pesticides, The Analyst 140 (2015) 560–566.

[28] S. Ebrahim, R. El-Raey, A. Hefnawy, H. Ibrahim, M. Soliman, T.M. Abdel-Fattah, Electrochemical sensor based on polyaniline nanofibers/single wall carbon nanotubes composite for detection of malathion, Synth. Met. 190 (2014) 13–19.

[29] D. Akyüz, A. Koca, An electrochemical sensor for the detection of pesticides based on the hybrid of manganese phthalocyanine and polyaniline, Sensor Actuat. B Chem. 283 (2019) 848–856.

[30] B.M. Hryniewicz, E.S. Orth, M. Vidotti, Enzymeless PEDOT-based electrochemical sensor for the detection of nitrophenols and organophosphates, Sensor Actuat. B Chem. 257 (2018) 570–578.

[31] R. Zamora, Development of poly(3,4- ethylenedioxythiophene)(PEDOT)/carbon nanotube electrodes for electrochemical detection of mancozeb in water, Int. J. Electrochem. Sci. 13 (2018) 1931–1944.

[32] Ş.U. Karabiberoğlu, Ç.C. Koçak, Z. Dursun, Electrochemical determination of dicofol at nickel nanowire modified poly(p-aminophenol) film electrode, Electroanalysis 31 (2019) 1304–1310.

[33] F.L. Migliorini, R.C. Sanfelice, L.A. Mercante, M.H.M. Facure, D.S. Correa, Electrochemical sensor based on polyamide 6/polypyrrole electrospun nanofibers coated with reduced graphene oxide for malathion pesticide detection, Mater. Res. Express 7 (2019) 015601.

[34] L.F. Sgobbi, S.A.S. Machado, Functionalized polyacrylamide as an acetylcholinesterase-inspired biomimetic device for electrochemical sensing of organophosphorus pesticides, Biosens. Bioelectron. 100 (2018) 290–297.

[35] D. Capoferri, F. Della Pelle, M. Del Carlo, D. Compagnone, Affinity sensing strategies for the detection of pesticides in food, Foods 7 (2018) 148.

[36] A. Motaharian, F. Motaharian, K. Abnous, M.R. Hosseini, M. Hassanzadeh-Khayyat, Molecularly imprinted polymer nanoparticles-based electrochemical sensor for determination of diazinon pesticide in well water and apple fruit samples, Anal. Bioanal. Chem. 408 (2016) 6769–6779.

[37] I. Sadriu, S. Bouden, J. Nicolle, F.I. Podvorica, V. Bertagna, C. Berho, L. Amalric, C. Vautrin-Ul, Molecularly imprinted polymer modified glassy carbon electrodes for the electrochemical analysis of isoproturon in water, Talanta 207 (2020) 120222.

[38] M. Khadem, F. Faridbod, P. Norouzi, A. Rahimi Foroushani, M.R. Ganjali, S.J. Shahtaheri, R. Yarahmadi, Modification of carbon paste electrode based on molecularly imprinted polymer for electrochemical determination of diazinon in biological and environmental samples, Electroanalysis 29 (2017) 708–715.

[39] M.E. Khalifa, A.B. Abdallah, Molecular imprinted polymer based sensor for recognition and determination of profenofos organophosphorous insecticide, Biosens. Bioelectron. 2 (2019) 100027.

[40] T. Xie, M. Zhang, P. Chen, H. Zhao, X. Yang, L. Yao, H. Zhang, A. Dong, J. Wang, Z. Wang, A facile molecularly imprinted electrochemical sensor based on graphene: application to the selective determination of thiamethoxam in grain, RSC Adv. 7 (2017) 38884–38894.

[41] M. Zhang, H.T. Zhao, T.J. Xie, X. Yang, A.J. Dong, H. Zhang, J. Wang, Z.Y. Wang, Molecularly imprinted polymer on graphene surface for selective and sensitive electrochemical sensing imidacloprid, Sens. Actuat. B: Chem. 252 (2017) 991–1002.

[42] M. Khadem, F. Faridbod, P. Norouzi, A.R. Foroushani, M.R. Ganjali, S.J. Shahtaheri, Biomimetic electrochemical sensor based on molecularly imprinted polymer for dicloran pesticide determination in biological and environmental samples, J. Iran. Chem. Soc. 13 (2016) 2077–2084.

[43] S. Li, C. Liu, G. Yin, J. Luo, Z. Zhang, Y. Xie, Supramolecular imprinted electrochemical sensor for the neonicotinoid insecticide imidacloprid based on double amplification by Pt-In catalytic nanoparticles and a Bromophenol blue doped molecularly imprinted film, Microchim. Acta 183 (2016) 3101–3109.

[44] F.-R. Wang, G.-J. Lee, N. Haridharan, J.J. Wu, Electrochemical sensor using molecular imprinting polymerization modified electrodes to detect methyl parathion in environmental media, Electrocatalysis 9 (2017) 1–9.

[45] S. Kumar, P. Karfa, S. Patra, R. Madhuri, P.K. Sharma, Molecularly imprinted star polymer-modified superparamagnetic iron oxide nanoparticle for trace level sensing and separation of mancozeb, RSC Adv. 6 (2016) 36751–36760.

[46] Y. Li, L. Zhang, Y. Dang, Z. Chen, R. Zhang, Y. Li, B.C. Ye, A robust electrochemical sensing of molecularly imprinted polymer prepared by using bifunctional monomer and its application in detection of cypermethrin, Biosens. Bioelectron. 127 (2019) 207–214.

[47] Y. Li, J. Liu, Y. Zhang, M. Gu, D. Wang, Y.Y. Dang, B.C. Ye, Y. Li, A robust electrochemical sensing platform using carbon paste electrode modified with molecularly imprinted microsphere and its application on methyl parathion detection, Biosens. Bioelectron. 106 (2018) 71–77.

[48] S. Feng, Y. Li, R. Zhang, Y. Li, A novel electrochemical sensor based on molecularly imprinted polymer modified hollow N, S-Mo2C/C spheres for highly sensitive and selective carbendazim determination, Biosens. Bioelectron. 142 (2019) 111491.

[49] P. Qi, J. Wang, X. Wang, Z. Wang, H. Xu, S. Di, Q. Wang, X. Wang, Se/nsitive and selective detection of the highly toxic pesticide carbofuran in vegetable samples by a molecularly imprinted electrochemical sensor with signal enhancement by AuNPs, RSC Adv. 8 (2018) 25334–25341.

[50] J. Xu, Y. Zhang, K. Wu, L. Zhang, S. Ge, J. Yu, A molecularly imprinted polypyrrole for ultrasensitive voltammetric determination of glyphosate, Microchim. Acta 184 (2017) 1959–1967.

[51] C. Zhang, F. Zhao, Y. She, S. Hong, X. Cao, L. Zheng, S. Wang, T. Li, M. Wang, M. Jin, F. Jin, H. Shao, J. Wang, A disposable molecularly imprinted sensor based on Graphe@AuNPs modified screen-printed electrode for highly selective and sensitive detection of cyhexatin in pear samples, Sens. Actuat. B Chem. 284 (2019) 13–22.

[52] M. Roushani, A. Nezhadali, Z. Jalilian, An electrochemical chlorpyrifos aptasensor based on the use of a glassy carbon electrode modified with an electropolymerized aptamer-imprinted polymer and gold nanorods, Microchim. Acta 185 (2018) 551.

[53] C. Zhang, Y. She, T. Li, F. Zhao, M. Jin, Y. Guo, L. Zheng, S. Wang, F. Jin, H. Shao, H. Liu, J. Wang, A highly selective electrochemical sensor based on molecularly imprinted polypyrrole-modified gold electrode for the determination of glyphosate in cucumber and tap water, Anal. Bioanal. Chem. 409 (2017) 7133–7144.

[54] M. Trozanowicz, Determination of pesticides using electrochemical enzymatic biosensors, Electroanalysis 14 (2002) 19–20.

[55] J. Bao, C. Hou, Q. Dong, X. Ma, J. Chen, D. Huo, M. Yang, K. Galil, W. Chen, Y. Lei, ELP-OPH/BSA/TiO$_2$ nanofibers/c-MWCNTs based biosensor for sensitive and selective determination of p-nitrophenyl substituted organophosphate pesticides in aqueous system, Biosens. Bioelectron. 85 (2016) 935–942.

[56] S. Chen, J. Huang, D. Du, J. Li, H. Tu, D. Liu, A. Zhang, Methyl parathion hydrolase based nanocomposite biosensors for highly sensitive and selective determination of methyl parathion, Biosens. Bioelectron. 26 (2011) 4320–4325.

[57] G. Liu, W. Guo, Z. Yin, Covalent fabrication of methyl parathion hydrolase on gold nanoparticles modified carbon substrates for designing a methyl parathion biosensor, Biosens. Bioelectron. 53 (2014) 440–446.

[58] B. Ma, L.-Z. Cheong, X. Weng, C.-P. Tan, C. Shen, Lipase@ZIF-8 nanoparticles-based biosensor for direct and sensitive detection of methyl parathion, Electrochim. Acta 283 (2018) 509–516.

[59] Z. Wang, B. Ma, C. Shen, L.Z. Cheong, Direct, selective and ultrasensitive electrochemical biosensing of methyl parathion in vegetables using Burkholderia cepacia lipase@MOF nanofibers-based biosensor, Talanta 197 (2019) 356–362.

[60] M. Asal, O. Ozen, M. Sahinler, I. Polatoglu, Recent developments in enzyme, DNA and immuno-based biosensors, Sensors (2018) 18.

[61] D. Du, X. Huang, J. Cai, A. Zhang, Amperometric detection of triazophos pesticide using acetylcholinesterase biosensor based on multiwall carbon nanotube-chitosan matrix, Sens. Actuat. B Chem. 127 (2007) 531–535.

[62] N. Chauhan, C.S. Pundir, An amperometric acetylcholinesterase sensor based on Fe3O4 nanoparticle/multi-walled carbon nanotube-modified ITO-coated glass plate for the detection of pesticides, Electrochim. Acta 67 (2012) 79–86.

[63] S.P. Sharma, L.N.S. Tomar, J. Acharya, A. Chaturvedi, M.V.S. Suryanarayan, R. Jain, Acetylcholinesterase inhibition-based biosensor for amperometric detection of Sarin using single-walled carbon nanotube-modified ferrule graphite electrode, Sens. Actuat. B Chem. 166–167 (2012) 616–623.

[64] H.F. Cui, W.W. Wu, M.M. Li, X. Song, Y. Lv, T.T. Zhang, A highly stable acetylcholinesterase biosensor based on chitosan-TiO2-graphene nanocomposites for detection of organophosphate pesticides, Biosens. Bioelectron. 99 (2018) 223–229.

[65] L. Zhou, X. Zhang, L. Ma, J. Gao, Y. Jiang, Acetylcholinesterase/chitosan-transition metal carbides nanocomposites-based biosensor for the organophosphate pesticides detection, Biochem. Eng. 128 (2017) 243–249.

[66] J. Bao, T. Huang, Z. Wang, H. Yang, X. Geng, G. Xu, M. Samalo, M. Sakinati, D. Huo, C. Hou, 3D graphene/copper oxide nano-flowers based acetylcholinesterase biosensor for sensitive detection of organophosphate pesticides, Sens. Actuat. B Chem. 279 (2019) 95–101.

[67] X. Lu, L. Tao, Y. Li, H. Huang, F. Gao, A highly sensitive electrochemical platform based on the bimetallic Pd@Au nanowires network for organophosphorus pesticides detection, Sens. Actuat. B Chem. 284 (2019) 103–109.

[68] D. Song, Y. Li, X. Lu, M. Sun, H. Liu, G. Yu, F. Gao, Palladium-copper nanowires-based biosensor for the ultrasensitive detection of organophosphate pesticides, Anal. Chim. Acta 982 (2017) 168–175.

[69] X. Lu, L. Tao, D. Song, Y. Li, F. Gao, Bimetallic Pd@Au nanorods based ultrasensitive acetylcholinesterase biosensor for determination of organophosphate pesticides, Sens. Actuat. B Chem. 255 (2018) 2575–2581.

[70] D. Song, Q. Li, X. Lu, Y. Li, Y. Li, Y. Wang, F. Gao, Ultra-thin bimetallic alloy nanowires with porous architecture/monolayer MoS_2 nanosheet as a highly sensitive platform for the electrochemical assay of hazardous omethoate pollutant, J. Hazard. Mater. 357 (2018) 466–474.

[71] N. Nesakumar, S. Sethuraman, U.M. Krishnan, J.B. Rayappan, Electrochemical acetylcholinesterase biosensor based on ZnO nanocuboids modified platinum electrode for the detection of carbosulfan in rice, Biosens. Bioelectron. 77 (2016) 1070–1077.

[72] M. Pumera, A. Ambrosi, A. Bonanni, E.L.K. Chng, H.L. Poh, Graphene for electrochemical sensing and biosensing, Trends Anal. Chem. 29 (2010) 954–965.

[73] E. Mahmoudi, H. Fakhri, A. Hajian, A. Afkhami, H. Bagheri, High-performance electrochemical enzyme sensor for organophosphate pesticide detection using modified metal-organic framework sensing platforms, Bioelectrochemistry 130 (2019) 107348.

[74] P. Zhang, T. Sun, S. Rong, D. Zeng, H. Yu, Z. Zhang, D. Chang, H. Pan, A sensitive amperometric AChE-biosensor for organophosphate pesticides detection based on conjugated polymer and Ag-rGO-NH2 nanocomposite, Bioelectrochemistry 127 (2019) 163–170.

[75] S. Rana, R. Kaur, R. Jain, N. Prabhakar, Ionic liquid assisted growth of poly(3,4-ethylenedioxythiophene)/reduced graphene oxide based electrode: an improved electro-catalytic performance for the detection of organophosphorus pesticides in beverages, Arab. J. Chem. 12 (2019) 1121–1133.

[76] P. Dong, B. Jiang, J. Zheng, A novel acetylcholinesterase biosensor based on gold nanoparticles obtained by electroless plating on three-dimensional graphene for detecting organophosphorus pesticides in water and vegetable samples, Anal. Methods 11 (2019) 2428–2434.

[77] M. Xu, S. Jiang, B. Jiang, J. Zheng, Organophosphorus pesticides detection using acetylcholinesterase biosensor based on gold nanoparticles constructed by electroless plating on vertical nitrogen-doped single-walled carbon nanotubes, Int. J. Environ. Anal. Chem. 99 (2019) 913–927.

[78] B. Jiang, P. Dong, J. Zheng, A novel amperometric biosensor based on covalently attached multilayer assemblies of gold nanoparticles, diazo-resins and acetylcholinesterase for the detection of organophosphorus pesticides, Talanta 183 (2018) 114–121.

[79] Z. Bin, C. Yanhong, X. Jiaojiao, Y. Jing, Acetylcholinesterase biosensor based on functionalized surface of carbon nanotubes for monocrotophos detection, Anal. Biochem. 560 (2018) 12–18.

[80] R. Luo, Z. Feng, G. Shen, Y. Xiu, Y. Zhou, X. Niu, H. Wang, Acetylcholinesterase biosensor based on mesoporous hollow carbon spheres/core-shell magnetic nanoparticles-modified electrode for the detection of organophosphorus pesticides, Sensors 18 (2018).

[81] L. Peng, S. Dong, W. Wei, X. Yuan, T. Huang, Synthesis of reticulated hollow spheres structure NiCo$_2$S$_4$ and its application in organophosphate pesticides biosensor, Biosens. Bioelectron. 92 (2017) 563–569.

[82] N. Kaur, H. Thakur, N. Prabhakar, Conducting polymer and multi-walled carbon nanotubes nanocomposites based amperometric biosensor for detection of organophosphate, J. Electroanal. Chem. 775 (2016) 121–128.

[83] K.M. Soropogui, A.T. Jameel, W.W.A.W. Salim, Enzyme-based biosensors for electrochemical detection of pesticides—a mini review, Indonesian J. Elec. Engg. Informat. 6 (2018) 161–171.

[84] J.H. Lee, Y.D. Han, S.Y. Song, T.D. Kim, H.C. Yoon, Biosensor for organophosphorus pesticides based on the acetylcholine esterase inhibition mediated by choline oxidase bioelectrocatalysis, BioChip J. 4 (2010) 223–229.

[85] Y. Zhang, M.A. Arugula, M. Wales, J. Wild, A.L. Simonian, A novel layer-by-layer assembled multi-enzyme/CNT biosensor for discriminative detection between organophosphorus and non-organophosphrus pesticides, Biosens. Bioelectron. 67 (2015) 287–295.

[86] R.P. Singh, Y.J. Kim, B.-K. Oh, J.-W. Choi, Glutathione-s-transferase based electrochemical biosensor for the detection of captan, Electrochem. Commun. 11 (2009) 181–185.

[87] H. Borah, R.R. Dutta, S. Gogoi, T. Medhi, P. Puzari, Glutathione-S-transferase-catalyzed reaction of glutathione for electrochemical biosensing of temephos, fenobucarb and dimethoate, Anal. Methods 9 (2017) 4044–4051.

[88] A. Samphao, P. Suebsanoh, Y. Wongsa, B. Pekec, J. Jitchareon, K. Kalcher, Alkaline phosphatase inhibition-based amperometric biosensor for the detection of carbofuran, Int. J. Electrochem. Sci. 8 (2013) 3254–3264.

[89] A. Sahin, K. Dooley, D.M. Cropek, A.C. West, S. Banta, A dual enzyme electrochemical assay for the detection of organophosphorus compounds using organophosphorus hydrolase and horseradish peroxidase, Sens. Actuat. B Chem. 158 (2011) 353–360.

[90] J.C. Vidal, L. Bonel, J.R. Castillo, A modulated tyrosinase enzyme-based biosensor for application to the detection of dichlorvos and atrazine pesticides, Electroanalysis 20 (2008) 865–873.

[91] A. Sassolas, B. Prieto-Simón, J.-L. Marty, Biosensors for pesticide detection: new trends, Am. J. Anal. Chem. 03 (2012) 210–232.

[92] K.G. Reddy, G. Madhavi, B.E.K. Swamy, S. Reddy, A.V.B. Reddy, V. Madhavi, Electrochemical investigations of lipase enzyme activity inhibition by methyl parathion pesticide: voltammetric studies, J. Mol. Liq. 180 (2013) 26–30.

[93] A. de Moura Barboza, A. Brunca da Silva, E. Mendonça da Silva, W. Pietro de Souza, M.A. Soares, L. Gomes de Vasconcelos, A.J. Terezo, M. Castilho, A biosensor based on microbial lipase immobilized on lamellar zinc hydroxide-decorated gold nanoparticles for carbendazim determination, Anal. Methods 11 (2019) 5388–5397.

[94] A. Hayat, J.L. Marty, Aptamer based electrochemical sensors for emerging environmental pollutants, Front. Chem. 2 (2014) 41.

[95] R. Rapini, G. Marrazza, Electrochemical aptasensors for contaminants detection in food and environment: recent advances, Bioelectrochemistry 118 (2017) 47–61.

[96] M. Liu, A. Khan, Z. Wang, Y. Liu, G. Yang, Y. Deng, N. He, Aptasensors for pesticide detection, Biosens. Bioelectron. 130 (2019) 174–184.

[97] Y. Jiao, H. Jia, Y. Guo, H. Zhang, Z. Wang, X. Sun, J. Zhao, An ultrasensitive aptasensor for chlorpyrifos based on ordered mesoporous carbon/ferrocene hybrid multiwalled carbon nanotubes, RSC Adv. 6 (2016) 58541–58548.

[98] G. Xu, D. Huo, C. Hou, Y. Zhao, J. Bao, M. Yang, H. Fa, A regenerative and selective electrochemical aptasensor based on copper oxide nanoflowers-single walled carbon nanotubes nanocomposite for chlorpyrifos detection, Talanta 178 (2018) 1046–1052.

[99] Y. Jiao, W. Hou, J. Fu, Y. Guo, X. Sun, X. Wang, J. Zhao, A nanostructured electrochemical apta-sensor for highly sensitive detection of chlorpyrifos, Sens. Actuat. B Chem. 243 (2017) 1164–1170.

[100] N. Prabhakar, H. Thakur, A. Bharti, N. Kaur, Chitosan-iron oxide nanocomposite based electro-chemical aptasensor for determination of malathion, Anal. Chim. Acta 939 (2016) 108–116.

[101] J. Fu, X. An, Y. Yao, Y. Guo, X. Sun, Electrochemical aptasensor based on one step co-electrodeposition of aptamer and GO-CuNPs nanocomposite for organophosphorus pesticide detection, Sens. Actuat. B Chem. 287 (2019) 503–509.

[102] X. Shi, J. Sun, Y. Yao, H. Liu, J. Huang, Y. Guo, X. Sun, Novel electrochemical aptasensor with dual signal amplification strategy for detection of acetamiprid, Sci. Total Environ. 705 (2019) 135905.

[103] C. Hongxia, H. Ji, L. Zaijun, L. Ruiyi, Y. Yongqiang, S. Xiulan, Electrochemical aptasensor for detection of acetamiprid in vegetables with graphene aerogel-glutamic acid functionalized graphene quantum dot/gold nanostars as redox probe with catalyst, Sens. Actuat. B Chem. 298 (2019) 126866.

[104] J. Yi, Z. Liu, J. Liu, H. Liu, F. Xia, D. Tian, C. Zhou, A label-free electrochemical aptasensor based on 3D porous CS/rGO/GCE for acetamiprid residue detection, Biosens. Bioelectron. 148 (2020) 111827.

[105] L. Fan, G. Wang, W. Liang, W. Yan, Y. Guo, S. Shuang, C. Dong, Y. Bi, Label-free and highly selective electrochemical aptasensor for detection of PCBs based on nickel hexacyanoferrate nanopar-ticles/reduced graphene oxides hybrids, Biosens. Bioelectron. 145 (2019) 111728.

[106] C. Zhu, D. Liu, Z. Chen, L. Li, T. You, An ultra-sensitive aptasensor based on carbon nanohorns/gold nanoparticles composites for impedimetric detection of carbendazim at picogram levels, J.Colloid Interf Sci. 546 (2019) 92–100.

[107] S. Eissa, M. Zourob, Selection and characterization of DNA aptamers for electrochemical biosensing of carbendazim, Anal. Chem. 89 (2017) 3138–3145.

[108] S. Hassani, M.R. Akmal, A. Salek-Maghsoudi, S. Rahmani, M.R. Ganjali, P. Norouzi, M. Abdollahi, Novel label-free electrochemical aptasensor for determination of Diazinon using gold nanoparticles-modified screen-printed gold electrode, Biosens. Bioelectron. 120 (2018) 122–128.

[109] R. Rapini, A. Cincinelli, G. Marrazza, Acetamiprid multidetection by disposable electrochemical DNA aptasensor, Talanta 161 (2016) 15–21.

[110] S.M. Taghdisi, N.M. Danesh, M. Ramezani, K. Abnous, Electrochemical aptamer based assay for the neonicotinoid insecticide acetamiprid based on the use of an unmodified gold electrode, Microchim. Acta 184 (2016) 499–505.

[111] F. Bettazzi, A. Romero Natale, E. Torres, I. Palchetti, Glyphosate determination by coupling an immuno-magnetic assay with electrochemical sensors, Sensors 18 (2018) 2965–2977.

[112] J. Mehta, N. Bhardwaj, S.K. Bhardwaj, S.K. Tuteja, P. Vinayak, A.K. Paul, K.-H. Kim, A. Deep, Graphene quantum dot modified screen printed immunosensor for the determination of parathion, Anal. Biochem. 523 (2017) 1–9.

[113] N.V. Chuc, N.H. Binh, C.T. Thanh, N.V. Tu, N.L. Huy, N.T. Dzung, P.N. Minh, V.T. Thu, T. D. Lam, Electrochemical immunosensor for detection of atrazine based on polyaniline/graphene, J. Mater. Sci. Technol. 32 (2016) 539–544.

[114] N. Belkhamssa, C.I. Justino, P.S. Santos, S. Cardoso, I. Lopes, A.C. Duarte, T. Rocha-Santos, M. Ksibi, Label-free disposable immunosensor for detection of atrazine, Talanta 146 (2016) 430–434.

[115] J. Mehta, P. Vinayak, S.K. Tuteja, V.A. Chhabra, N. Bhardwaj, A.K. Paul, K.H. Kim, A. Deep, Gra-phene modified screen printed immunosensor for highly sensitive detection of parathion, Biosens. Bioelectron. 83 (2016) 339–346.

[116] G. Fusco, F. Gallo, C. Tortolini, P. Bollella, F. Ietto, A. De Mico, A. D'Annibale, R. Antiochia, G. Favero, F. Mazzei, AuNPs-functionalized PANABA-MWCNTs nanocomposite-based impedi-metric immunosensor for 2,4-dichlorophenoxy acetic acid detection, Biosens. Bioelectron. 93 (2017) 52–56.

[117] B. Perez-Fernandez, J.V. Mercader, A. Abad-Fuentes, B.I. Checa-Orrego, A. Costa-Garcia, A. Escosura-Muniz, Direct competitive immunosensor for Imidacloprid pesticide detection on gold nanoparticle-modified electrodes, Talanta 209 (2020) 120465.

[118] W.H. Wang, Z.J. Han, P.J. Liang, D.Q. Guo, Y.J. Xiang, M.X. Tian, Z.L. Song, H.R. Zhao, Co_3O_4/PAn magnetic nanoparticle-modified electrochemical immunosensor for chlorpyrifos, Dig. J. Nanomater. Bio. 12 (2017) 1–9.

[119] A. Talan, A. Mishra, S.A. Eremin, J. Narang, A. Kumar, S. Gandhi, Ultrasensitive electrochemical immuno-sensing platform based on gold nanoparticles triggering chlorpyrifos detection in fruits and vegetables, Biosens. Bioelectron. 105 (2018) 14–21.

[120] M.A. Abedalwafa, Y. Li, C. Ni, L. Wang, Colorimetric sensor arrays for the detection and identification of antibiotics, Anal. Methods 11 (2019) 2836–2854.

[121] N.A. Mungroo, S. Neethirajan, Biosensors for the detection of antibiotics in poultry industry—a review, Biosensors 4 (2014) 472–493.

[122] S.A. Kraemer, A. Ramachandran, G.G. Perron, Antibiotic pollution in the environment: from microbial ecology to public policy, Microorganisms 7 (2019) 180–204.

[123] C. Llor, L. Bjerrum, Antimicrobial resistance: risk associated with antibiotic overuse and initiatives to reduce the problem, Ther. Adv. Drug Saf. 5 (2014) 229–241.

[124] X. Liu, J.C. Steele, X.Z. Meng, Usage, residue, and human health risk of antibiotics in Chinese aquaculture: a review, Environ. Pollut. 223 (2017) 161–169.

[125] J. Chen, G.G. Ying, W.J. Deng, Antibiotic residues in food: extraction, analysis, and human health concerns, J. Agr. Food Chem. 67 (2019) 7569–7586.

[126] S. Hannah, E. Addington, D. Alcorn, W. Shu, P.A. Hoskisson, D.K. Corrigan, Rapid antibiotic susceptibility testing using low-cost, commercially available screen-printed electrodes, Biosens. Bioelectron. 145 (2019) 111696.

[127] N. Li, K.W. Ho, G.G. Ying, W.J. Deng, Veterinary antibiotics in food, drinking water, and the urine of preschool children in Hong Kong, Environ. Int. 108 (2017) 246–252.

[128] A. Wong, A.M. Santos, F.H. Cincotto, F.C. Moraes, O. Fatibello-Filho, M. Sotomayor, A new electrochemical platform based on low cost nanomaterials for sensitive detection of the amoxicillin antibiotic in different matrices, Talanta 206 (2020) 120252.

[129] N.R. Jalal, T. Madrakian, A. Afkhami, M. Ghamsari, Polyethylenimine@Fe_3O_4@carbon nanotubes nanocomposite as a modifier in glassy carbon electrode for sensitive determination of ciprofloxacin in biological samples, J. Electroanal. Chem. 833 (2019) 281–289.

[130] M. Tiwari, A. Kumar, U. Shankar, R. Prakash, The nanocrystalline coordination polymer of AMT-Ag for an effective detection of ciprofloxacin hydrochloride in pharmaceutical formulation and biological fluid, Biosens. Bioelectron. 85 (2016) 529–535.

[131] F. Tan, D. Sun, J. Gao, Q. Zhao, X. Wang, F. Teng, X. Quan, J. Chen, Preparation of molecularly imprinted polymer nanoparticles for selective removal of fluoroquinolone antibiotics in aqueous solution, J. Hazard. Mater. 244–245 (2013) 750–757.

[132] Y. Liu, M. Wei, Y. Hu, L. Zhu, J. Du, An electrochemical sensor based on a molecularly imprinted polymer for determination of anticancer drug Mitoxantrone, Sens. Actuat. B Chem. 255 (2018) 544–551.

[133] J. Liu, H. Tang, B. Zhang, X. Deng, F. Zhao, P. Zuo, B.C. Ye, Y. Li, Electrochemical sensor based on molecularly imprinted polymer for sensitive and selective determination of metronidazole via two different approaches, Anal. Bioanal. Chem. 408 (2016) 4287–4295.

[134] Z. Liu, M. Jin, H. Lu, J. Yao, X. Wang, G. Zho, L. Shui, Molecularly imprinted polymer decorated 3D-framework of functionalizedmulti-walled carbon nanotubes for ultrasensitive electrochemical sensing ofNorfloxacin in pharmaceutical formulations and rat plasma, Sens. Actuat. B Chem. 288 (2019) 363–372.

[135] G. Moro, F. Bottari, N. Sleegers, A. Florea, T. Cowen, L.M. Moretto, S. Piletsky, K. De Wael, Conductive imprinted polymers for the direct electrochemical detection of β-lactam antibiotics: the case of cefquinome, Sens. Actuat. B Chem. 297 (2019) 126786.

[136] A. Munawar, M.A. Tahir, A. Shaheen, P.A. Lieberzeit, W.S. Khan, S.Z. Bajwa, Investigating nanohybrid material based on 3D CNTs@Cu nanoparticle composite and imprinted polymer for highly selective detection of chloramphenicol, J. Hazard. Mater. 342 (2018) 96–106.

[137] N. Karimian, M.B. Gholivand, G. Malekzadeh, Cefixime detection by a novel electrochemical sensor based on glassy carbon electrode modified with surface imprinted polymer/multiwall carbon nanotubes, J. Electroanal. Chem. 771 (2016) 64–72.

[138] Z. Xu, X. Jiang, S. Liu, M. Yang, Sensitive and selective molecularly imprinted electrochemical sensor based on multi-walled carbon nanotubes for doxycycline hyclate determination, Chinese Chem. Lett. 31 (2019) 185–188.

[139] L. Devkota, L.T. Nguyen, T.T. Vu, B. Piro, Electrochemical determination of tetracycline using AuNP-coated molecularly imprinted overoxidized polypyrrole sensing interface, Electrochim. Acta 270 (2018) 535–542.

[140] M. Bougrini, A. Florea, C. Cristea, R. Sandulescu, F. Vocanson, A. Errachid, B. Bouchikhi, N. El Bari, N. Jaffrezic-Renault, Development of a novel sensitive molecularly imprinted polymer sensor based on electropolymerization of a microporous-metal-organic framework for tetracycline detection in honey, Food Control 59 (2016) 424–429.

[141] S. Jafari, M. Dehghani, N. Nasirizadeh, M.H. Baghersad, M. Azimzadeh, Label-free electrochemical detection of Cloxacillin antibiotic in milk samples based on molecularly imprinted polymer and graphene oxide-gold nanocomposite, Measurement 145 (2019) 22–29.

[142] M. Dehghani, N. Nasirizadeh, M.E. Yazdanshenas, Determination of cefixime using a novel electrochemical sensor produced with gold nanowires/graphene oxide/electropolymerized molecular imprinted polymer, Mater. Sci. Eng. C 96 (2019) 654–660.

[143] S. Jafari, M. Dehghani, N. Nasirizadeh, M. Azimzadeh, An azithromycin electrochemical sensor based on an aniline MIP film electropolymerized on a gold nano urchins/graphene oxide modified glassy carbon electrode, J. Electroanal. Chem. 829 (2018) 27–34.

[144] D. Dechtrirat, P. Yingyuad, P. Prajongtat, L. Chuenchom, C. Sriprachuabwong, A. Tuantranont, I. M. Tang, A screen-printed carbon electrode modified with gold nanoparticles, poly(3,4-ethylenedioxythiophene), poly(styrene sulfonate) and a molecular imprint for voltammetric determination of nitrofurantoin, Microchim. Acta 185 (2018) 261.

[145] M. Roushani, Z. Rahmati, S.J. Hoseini, R. Hashemi Fath, Impedimetric ultrasensitive detection of chloramphenicol based on aptamer MIP using a glassy carbon electrode modified by 3-ampy-RGO and silver nanoparticle, Colloids Surf. B 183 (2019) 110451.

[146] Z. Li, C. Liu, V. Sarpong, Z. Gu, Multisegment nanowire/nanoparticle hybrid arrays as electrochemical biosensors for simultaneous detection of antibiotics, Biosens. Bioelectron. 126 (2019) 632–639.

[147] L.M. Gonçalves, W.F.A. Callera, M.D.P.T. Sotomayor, P.R. Bueno, Penicillinase-based amperometric biosensor for penicillin G, Electrochem. Commun. 38 (2014) 131–133.

[148] K. Zeng, W. Wei, L. Jiang, F. Zhu, D. Du, Use of carbon nanotubes as a solid support to establish quantitative (centrifugation) and qualitative (filtration) immunoassays to detect gentamicin contamination in commercial milk, J. Agr. Food Chem. 64 (2016) 7874–7881.

[149] A. Pollap, J. Kochana, Electrochemical immunosensors for antibiotic detection, Biosensors 9 (2019) 61–88.

[150] F.S. Felix, L. Angnes, Electrochemical immunosensors—a powerful tool for analytical applications, Biosens. Bioelectron. 102 (2018) 470–478.

[151] X. Liu, S. Zheng, Y. Hu, Z. Li, F. Luo, Z. He, Electrochemical immunosensor based on the chitosan-magnetic nanoparticles for detection of tetracycline, Food Anal. Methods 9 (2016) 2972–2978.

[152] N. El Alami El Hassani, A. Baraket, S. Boudjaoui, E. Taveira Tenorio Neto, J. Bausells, N. El Bari, B. Bouchikhi, A. Elaissari, A. Errachid, N. Zine, Development and application of a novel electrochemical immunosensor for tetracycline screening in honey using a fully integrated electrochemical Bio-MEMS, Biosen. Bioelectron. 130 (2019) 330–337.

[153] A.Y. El-Moghazy, C. Zhao, G. Istamboulie, N. Amaly, Y. Si, T. Noguer, G. Sun, Ultrasensitive label-free electrochemical immunosensor based on PVA-co-PE nanofibrous membrane for the detection of chloramphenicol residues in milk, Biosens. Bioelectron. 117 (2018) 838–844.

[154] M. Tomassetti, R. Angeloni, E. Martini, M. Castrucci, L. Campanella, Enzymatic DMFC device used for direct analysis of chloramphenicol and a comparison with the competitive immunosensor method, Sens. Actuat. B Chem. 255 (2018) 1545–1552.

[155] B. Liu, M. Li, Y. Zhao, M. Pan, Y. Gu, W. Sheng, G. Fang, S. Wang, A sensitive electrochemical immunosensor based on PAMAM dendrimer-encapsulated Au for detection of norfloxacin in animal-derived foods, Sensors 18 (2018) 1946–1957.

[156] N. Hassani, A. Baraket, E.T.T. Neto, M. Lee, J.P. Salvador, M.P. Marco, J. Bausells, N.E. Bari, B. Bouchikhi, A. Elaissari, A. Errachid, N. Zine, Novel strategy for sulfapyridine detection using a fully integrated electrochemical Bio-MEMS: application to honey analysis, Biosens. Bioelectron. 93 (2017) 282–288.

[157] Z. Zhang, M. Yang, X. Wu, S. Dong, N. Zhu, E. Gyimah, K. Wang, Y. Li, A competitive immunosensor for ultrasensitive detection of sulphonamides from environmental waters using silver nanoparticles decorated single-walled carbon nanohorns as labels, Chemosphere 225 (2019) 282–287.

[158] Z. Gao, Y. Li, C. Zhang, S. Zhang, Y. Jia, Y. Dong, An enzyme-free immunosensor for sensitive determination of procalcitonin using NiFe PBA nanocubes@TB as the sensing matrix, Anal. Chim. Acta. 1097 (2020) 169–175.

[159] P. Liu, C. Li, R. Zhang, Q. Tang, J. Wei, Y. Lu, P. Shen, An ultrasensitive electrochemical immunosensor for procalcitonin detection based on the gold nanoparticles-enhanced tyramide signal amplification strategy, Biosens. Bioelectron. 126 (2019) 543–550.

[160] M. Hu, X. Hu, Y. Zhang, M. Teng, R. Deng, G. Xing, J. Tao, G. Xu, J. Chen, Y. Zhang, G. Zhang, Label-free electrochemical immunosensor based on AuNPs/Zn/Ni-ZIF-8-800@graphene composites for sensitive detection of monensin in milk, Sens. Actuat. B Chem. 288 (2019) 571–578.

[161] M. Hu, Y. Wang, J. Yang, Y. Sun, G. Xing, R. Deng, X. Hu, G. Zhang, Competitive electrochemical immunosensor for maduramicin detection by multiple signal amplification strategy via hemin@Fe-MIL-88NH₂/AuPt, Biosens. Bioelecton. 142 (2019) 111554.

[162] M. Wang, M. Hu, J. Liu, C. Guo, D. Peng, Q. Jia, L. He, Z. Zhang, M. Du, Covalent organic framework-based electrochemical aptasensors for the ultrasensitive detection of antibiotics, Biosens. Bioelectron. 132 (2019) 8–16.

[163] S.M. Taghdisi, N.M. Danesh, M.A. Nameghi, M. Ramezani, M. Alibolandi, K. Abnous, An electrochemical sensing platform based on ladder-shaped DNA structure and label-free aptamer for ultrasensitive detection of ampicillin, Biosens. Bioelectron. 133 (2019) 230–235.

[164] A. Alawad, G. Istambouliè, C. Calas-Blanchard, T. Noguer, A reagentless aptasensor based on intrinsic aptamer redox activity for the detection of tetracycline in water, Sens. Actuat. B Chem. 288 (2019) 141–146.

[165] S.M. Taghdisi, N.M. Danesh, M. Ramezani, K. Abnous, A novel M-shape electrochemical aptasensor for ultrasensitive detection of tetracyclines, Biosens. Bioelectron. 85 (2016) 509–514.

[166] A. Benvidi, S. Yazdanparast, M. Rezaeinasab, M.D. Tezerjani, S. Abbasi, Designing and fabrication of a novel sensitive electrochemical aptasensor based on poly (L-glutamic acid)/MWCNTs modified glassy carbon electrode for determination of tetracycline, J. Electroanal. Chem. 808 (2018) 311–320.

[167] M. Roushani, K. Ghanbari, S. Jafar Hoseini, Designing an electrochemical aptasensor based on immobilization of the aptamer onto nanocomposite for detection of the streptomycin antibiotic, Microchem. J. 141 (2018) 96–103.

[168] J. Yin, W. Guo, X. Qin, J. Zhao, M. Pei, F. Ding, A sensitive electrochemical aptasensor for highly specific detection of streptomycin based on the porous carbon nanorods and multifunctional graphene nanocomposites for signal amplification, Sens. Actuat. B Chem. 241 (2017) 151–159.

[169] S. Song, L. Wang, J. Li, C. Fan, J. Zhao, Aptamer-based biosensors, TrAC Trends Anal. Chem. 27 (2008) 108–117.

[170] M. Chen, N. Gan, Y. Zhou, T. Li, Q. Xu, Y. Cao, Y. Chen, An electrochemical aptasensor for multiplex antibiotics detection based on metal ions doped nanoscale MOFs as signal tracers and RecJf exonuclease-assisted targets recycling amplification, Talanta 161 (2016) 867–874.

[171] Z. Yan, N. Gan, T. Li, Y. Cao, Y. Chen, A sensitive electrochemical aptasensor for multiplex antibiotics detection based on high-capacity magnetic hollow porous nanotracers coupling exonuclease-assisted cascade target recycling, Biosens. Bioelectron. 78 (2016) 51–57.

[172] Z. Shen, L. He, Y. Cao, F. Hong, K. Zhang, F. Hu, J. Lin, D. Wu, N. Gan, Multiplexed electrochemical aptasensor for antibiotics detection using metallic-encoded apoferritin probes and double stirring bars-assisted target recycling for signal amplification, Talanta 197 (2019) 491–499.

[173] M. Chen, N. Gan, T. Li, Y. Wang, Q. Xu, Y. Chen, An electrochemical aptasensor for multiplex antibiotics detection using Y-shaped DNA-based metal ions encoded probes with NMOF substrate and CSRP target-triggered amplification strategy, Anal. Chim. Acta 968 (2017) 30–39.

[174] M. Chen, N. Gan, Y. Zhou, T. Li, Q. Xu, Y. Cao, Y. Chen, A novel aptamer- metal ions- nanoscale MOF based electrochemical biocodes for multiple antibiotics detection and signal amplification, Sens. Actuat. B Chem. 242 (2017) 1201–1209.

[175] H. Wang, Y. Wang, S. Liu, J. Yu, Y. Guo, Y. Xu, J. Huang, Signal-on electrochemical detection of antibiotics at zeptomole level based on target-aptamer binding triggered multiple recycling amplification, Biosens. Bioelectron. 80 (2016) 471–476.

[176] X. Wang, S. Dong, P. Gai, R. Duan, F. Li, Highly sensitive homogeneous electrochemical aptasensor for antibiotic residues detection based on dual recycling amplification strategy, Biosens. Bioelectron. 82 (2016) 49–54.

[177] C. Chen, X. Lv, W. Lei, Y. Wu, S. Feng, Y. Ding, J. Lv, Q. Hao, S.M. Chen, Amoxicillin on poly-glutamic acid composite three-dimensional graphene modified electrode: reaction mechanism of amoxicillin insights by computational simulations, Anal. Chim. Acta 1073 (2019) 22–29.

[178] X.-M. Sun, Z. Ji, M.-X. Xiong, W. Chen, The electrochemical sensor for the determination of tet-racycline based on graphene /L-cysteine composite film, J. Electrochem. Soc. 164 (2017) B107–B112.

[179] Y. Lu, X. Liang, C. Niyungeko, J. Zhou, J. Xu, G. Tian, A review of the identification and detection of heavy metal ions in the environment by voltammetry, Talanta 178 (2018) 324–338.

[180] J. Holmes, P. Pathirathna, P. Hashemi, Novel frontiers in voltammetric trace metal analysis: toward real time, on-site, in situ measurements, Trends Anal. Chem. 111 (2019) 206–219.

[181] B. Bansod, T. Kumar, R. Thakur, S. Rana, I. Singh, A review on various electrochemical techniques for heavy metal ions detection with different sensing platforms, Biosens. Bioelectron. 94 (2017) 443–455.

[182] H. Huang, W. Zhu, X. Gao, X. Liu, H. Ma, Synthesis of a novel electrode material containing phytic acid-polyaniline nanofibers for simultaneous determination of cadmium and lead ions, Anal. Chim. Acta 947 (2016) 32–41.

[183] G. Zhu, Y. Ge, Y. Dai, X. Shang, J. Yang, J. Liu, Size-tunable polyaniline nanotube-modified elec-trode for simultaneous determination of Pb(II) and Cd(II), Electrochim. Acta 268 (2018) 202–210.

[184] M.A. Deshmukh, R. Celiesiute, A. Ramanaviciene, M.D. Shirsat, A. Ramanavicius, EDTA_PANI/SWCNTs nanocomposite modified electrode for electrochemical determination of copper (II), lead (II) and mercury (II) ions, Electrochim. Acta 259 (2018) 930–938.

[185] Y. Kong, T. Wu, D. Wu, Y. Zhang, Y. Wang, B. Du, Q. Wei, An electrochemical sensor based on Fe3O4@PANI nanocomposites for sensitive detection of Pb^{2+} and Cd^{2+}, Anal. Methods 10 (2018) 4784–4792.

[186] M.A. Deshmukh, H.K. Patil, G.A. Bodkhe, M. Yasuzawa, P. Koinkar, A. Ramanaviciene, M.D. Shirsat, A. Ramanavicius, EDTA-modified PANI/SWNTs nanocomposite for differential pulse voltammetry based determination of Cu(II) ions, Sens. Actuat. B Chem. 260 (2018) 331–338.

[187] X. Zhao, W. Bai, Y. Yan, Y. Wang, J. Zhang, Core-shell self-doped polyaniline coated metal-organic-framework (SPAN@UIO-66-NH2) screen printed electrochemical sensor for Cd2 + ions, J. Electrochem. Soc. 166 (2019) B873–B880.

[188] H. Dai, N. Wang, D. Wang, H. Ma, M. Lin, An electrochemical sensor based on phytic acid func-tionalized polypyrrole/graphene oxide nanocomposites for simultaneous determination of Cd(II) and Pb(II), Chem. Eng. 299 (2016) 150–155.

[189] Y. Zhuang, M. Zhao, Y. He, F. Cheng, S. Chen, Fabrication of ZnO/rGO/PPy heterostructure for electrochemical detection of mercury ion, J. Electroanal. Chem. 826 (2018) 90–95.

[190] V. Suvina, S.M. Krishna, D.H. Nagaraju, J.S. Melo, R.G. Balakrishna, Polypyrrole-reduced graphene oxide nanocomposite hydrogels: a promising electrode material for the simultaneous detection of multiple heavy metal ions, Mater. Lett. 232 (2018) 209–212.

[191] Y. Song, C. Bian, J. Hu, Y. Li, J. Tong, J. Sun, G. Gao, S. Xia, Porous polypyrrole/graphene oxide functionalized with carboxyl composite for electrochemical sensor of trace cadmium (II), J. Electrochem. Soc. 166 (2019) B95–B102.

[192] N. Wang, W. Zhao, Z. Shen, S. Sun, H. Dai, H. Ma, M. Lin, Sensitive and selective detection of Pb (II) and Cu (II) using a metal-organic framework/polypyrrole nanocomposite functionalized electrode, Sens. Actuat. B Chem. 304 (2020) 127286.

[193] M.-O. Park, H.-B. Noh, D.-S. Park, J.-H. Yoon, Y.-B. Shim, Long-life heavy metal ions sensor based on graphene oxide-anchored conducting polymer, Electroanalysis 29 (2017) 514–520.

[194] W. Wu, A. Ali, R. Jamal, M. Abdulla, T. Bakri, T. Abdiryim, A bromine-catalysis-synthesized poly (3,4-ethylenedioxythiophene)/graphitic carbon nitride electrochemical sensor for heavy metal ion determination, RSC Adv. 9 (2019) 34691–34698.

[195] Z. Zhong, A. Ali, R. Jamal, R. Simayi, L. Xiang, S. Ding, T. Abdiryim, Poly(EDOT-pyridine-EDOT) and poly(EDOT-pyridazine-EDOT) hollow nanosphere materials for the electrochemical detection of Pb 2+ and Cu 2+, J. Electroanal. Chem. 822 (2018) 112–122.

[196] M. Abdulla, A. Ali, R. Jamal, T. Bakri, W. Wu, T. Abdiryim, Electrochemical sensor of double-thiol linked PProDOT@Si composite for simultaneous detection of Cd(II), Pb(II), and Hg(II), Polymers 11 (2019) 815–834.

[197] A.K. Şahutoğlu, İ. Kaya, Synthesis and a new mercury (II) ion sensor application of conductive polymer containing rhodamine B, React. Func. Poly. 141 (2019) 50–57.

[198] M. Ghanei-Motlagh, M.A. Taher, A. Heydari, R. Ghanei-Motlagh, V.K. Gupta, A novel voltammetric sensor for sensitive detection of mercury(II) ions using glassy carbon electrode modified with graphene-based ion imprinted polymer, Mater. Sci. Eng. C 63 (2016) 367–375.

[199] S. Hu, An electrochemical sensor based on ion imprinted PPy/rGO composite for Cd(II) determination in water, Int. J. Electrochem. Sci. 14 (2019) 11714–11730.

[200] T. Alizadeh, R.E. Sabzi, H. Alizadeh, Synthesis of nano-sized cyanide ion-imprinted polymer via non-covalent approach and its use for the fabrication of a CN(-)-selective carbon nanotube impregnated carbon paste electrode, Talanta 147 (2016) 90–97.

[201] X. Luo, W. Huang, Q. Shi, W. Xu, Y. Luan, Y. Yang, H. Wang, W. Yang, Electrochemical sensor based on lead ion-imprinted polymer particles for ultra-trace determination of lead ions in different real samples, RSC Adv. 7 (2017) 16033–16040.

[202] A. Aravind, B. Mathew, Tailoring of nanostructured material as an electrochemical sensor and sorbent for toxic Cd(II) ions from various real samples, Anal. Sci. Technol. 9 (2018) 88–105.

[203] K.M. Lee, K.H. Kim, H. Yoon, H. Kim, Chemical design of functional polymer structures for biosensors: from nanoscale to macroscale, Polymers 10 (2018) 551–585.

[204] Z. Guo, D.D. Li, X.K. Luo, Y.H. Li, Q.N. Zhao, M.M. Li, Y.T. Zhao, T.S. Sun, C. Ma, Simultaneous determination of trace Cd(II), Pb(II) and Cu(II) by differential pulse anodic stripping voltammetry using a reduced graphene oxide-chitosan/poly-l-lysine nanocomposite modified glassy carbon electrode, J. Colloid Interf. Sci. 490 (2017) 11–22.

[205] Z. Lu, X. Lin, J. Zhang, W. Dai, B. Liu, G. Mo, J. Ye, J. Ye, Ionic liquid/poly-l-cysteine composite deposited on flexible and hierarchical porous laser-engraved graphene electrode for high-performance electrochemical analysis of lead ion, Electrochim. Acta 295 (2019) 514–523.

[206] W. Yi, Z. He, J. Fei, X. He, Sensitive electrochemical sensor based on poly(l-glutamic acid)/graphene oxide composite material for simultaneous detection of heavy metal ions, RSC Adv. 9 (2019) 17325–17334.

[207] J. Jency Feminus, P.N. Deepa, Electrochemical sensor based on composite of reduced graphene and poly-glutamic acid for selective and sensitive detection of lead, J. Mater. Sci. 30 (2019) 15553–15562.

[208] F. Hanif, A. Tahir, M. Akhtar, M. Waseem, S. Haider, M.F. Aly Aboud, I. Shakir, M. Imran, M.F. Warsi, Ultra-selective detection of Cd2+ and Pb2+ using glycine functionalized reduced graphene oxide/polyaniline nanocomposite electrode, Synth. Met. 257 (2019) 116185.

[209] N. Wang, M. Lin, H. Dai, H. Ma, Functionalized gold nanoparticles/reduced graphene oxide nanocomposites for ultrasensitive electrochemical sensing of mercury ions based on thymine-mercury-thymine structure, Biosens. Bioelectron. 79 (2016) 320–326.

[210] G. March, T.D. Nguyen, B. Piro, Modified electrodes used for electrochemical detection of metal ions in environmental analysis, Biosensors 5 (2015) 241–275.

[211] J.-S. Do, K.-H. Lin, Kinetics of urease inhibition-based amperometric biosensors for mercury and lead ions detection, J. Taiwan Inst. Chem. Eng. 63 (2016) 25–32.

[212] M. Stoytcheva, V. Sharkova, M. Panayotova, Electrochemical approach in studying the inhibition of acetylcholinesterase by arsenate(III): analytical characterization and application for arsenic determination, Anal. Chim. Acta 364 (1998) 195–201.

[213] O. Domı́nguez Renedo, M.A. Alonso Lomillo, M.J. Arcos Martinez, Optimisation procedure for the inhibitive determination of chromium(III) using an amperometric tyrosinase biosensor, Anal. Chim. Acta 521 (2004) 215–221.

[214] W. da Silva, M.E. Ghica, C.M.A. Brett, Biotoxic trace metal ion detection by enzymatic inhibition of a glucose biosensor based on a poly(brilliant green)-deep eutectic solvent/carbon nanotube modified electrode, Talanta 208 (2020) 120427.

[215] S.K. Pandey, S. Sachan, S.K. Singh, Electrochemically reduced graphene oxide modified with electrodeposited thionine and horseradish peroxidase for hydrogen peroxide sensing and inhibitive measurement of chromium, Mater. Sci. Ener. Technol. 2 (2019) 676–686.

[216] G.K. Mishra, V. Sharma, R.K. Mishra, Electrochemical aptasensors for food and environmental safeguarding: a review, Biosensors 8 (2018) 28–41.

[217] S.M. Taghdisi, N.M. Danesh, P. Lavaee, M. Ramezani, K. Abnous, An electrochemical aptasensor based on gold nanoparticles, thionine and hairpin structure of complementary strand of aptamer for ultrasensitive detection of lead, Sens. Actuat. B Chem. 234 (2016) 462–469.

[218] S.M. Taghdisi, N.M. Danesh, M. Ramezani, M. Alibolandi, K. Abnous, Voltammetric determination of lead(II) by using exonuclease III and gold nanoparticles, and by exploiting the conformational change of the complementary strand of an aptamer, Microchim. Acta 184 (2017) 2783–2790.

[219] S. Ge, K. Wu, Y. Zhang, M. Yan, J. Yu, Paper-based biosensor relying on flower-like reduced graphene guided enzymatically deposition of polyaniline for Pb(2+) detection, Biosens. Bioelectron. 80 (2016) 215–221.

[220] J. Zhao, T. Zheng, J. Gao, W. Xu, Toehold-mediated strand displacement reaction triggered by nicked DNAzymes substrate for amplified electrochemical detection of lead ion, Electrochim. Acta 274 (2018) 16–22.

[221] M. Lu, R. Xiao, X. Zhang, J. Niu, X. Zhang, Y. Wang, Novel electrochemical sensing platform for quantitative monitoring of Hg(II) on DNA-assembled graphene oxide with target recycling, Biosens. Bioelectron. 85 (2016) 267–271.

[222] Y. Zhang, C. Zhang, R. Ma, X. Du, W. Dong, Y. Chen, Q. Chen, An ultra-sensitive Au nanoparticles functionalized DNA biosensor for electrochemical sensing of mercury ions, Mater. Sci. Eng. C 75 (2017) 175–181.

[223] L.-L. He, L. Cheng, Y. Lin, H.-F. Cui, N. Hong, H. Peng, D.-R. Kong, C.-D. Chen, J. Zhang, G.-B. Wei, H. Fan, A sensitive biosensor for mercury ions detection based on hairpin hindrance by thymine-Hg(II)-thymine structure, J. Electroanal. Chem. 814 (2018) 161–167.

[224] C. Mei, D. Lin, C. Fan, A. Liu, S. Wang, J. Wang, Highly sensitive and selective electrochemical detection of Hg(2+) through surface-initiated enzymatic polymerization, Biosens. Bioelectron. 80 (2016) 105–110.

[225] W. Hu, X. Min, X. Li, S. Yang, L. Yi, L. Chai, DNAzyme catalytic beacons-based a label-free biosensor for copper using electrochemical impedance spectroscopy, RSC Adv. 6 (2016) 6679–6685.

[226] X. Wang, C. Yang, S. Zhu, M. Yan, S. Ge, J. Yu, 3D origami electrochemical device for sensitive Pb(2+) testing based on DNA functionalized iron-porphyrinic metal-organic framework, Biosens. Bioelectron. 87 (2017) 108–115.

[227] P. Miao, Y. Tang, L. Wang, DNA modified Fe3O4@Au magnetic nanoparticles as selective probes for simultaneous detection of heavy metal ions, ACS Appl. Mater. Interfaces 9 (2017) 3940–3947.

[228] X. Zhang, Y. Jiang, M. Zhu, Y. Xu, Z. Guo, J. Shi, E. Han, X. Zou, D. Wang, Electrochemical DNA sensor for inorganic mercury(II) ion at attomolar level in dairy product using Cu(II)-anchored metal-organic framework as mimetic catalyst, Chem. Eng. J. 383 (2020) 123182.

[229] X. Xie, Y. Chai, Y. Yuan, R. Yuan, Dual triggers induced disassembly of DNA polymer decorated silver nanoparticle for ultrasensitive electrochemical Pb^{2+} detection, Anal. Chim. Acta 1034 (2018) 56–62.

[230] Z. Yu, N. Li, X. Hu, Y. Dong, Y. Lin, H. Cai, Z. Xie, D. Qu, X. Li, Highly efficient electrochemical detection of lead ion using metal–organic framework and graphene as platform based on DNAzyme, Synth. Met. 254 (2019) 164–171.

[231] F. Gao, C. Gao, S. He, Q. Wang, A. Wu, Label-free electrochemical lead (II) aptasensor using thionine as the signaling molecule and graphene as signal-enhancing platform, Biosens. Bioelectron. 81 (2016) 15–22.

[232] X. Wang, W. Gao, W. Yan, P. Li, H. Zou, Z. Wei, W. Guan, Y. Ma, S. Wu, Y. Yu, K. Ding, A novel aptasensor based on graphene/graphite carbon nitride nanocomposites for cadmium detection with high selectivity and sensitivity, ACS Appl. Nano Mater. 1 (2018) 2341–2346.

CHAPTER 15

Intelligent nanosensors (INS) for environmental applications

Hassan Shokry Hassan[a,b], **Marwa Farouk Elkady**[c,d], and **Nourwanda Mohamed Serour**[c]

[a]Environmental Engineering Department, Egypt-Japan University for Science and Technology, New Borg El-Arab City, Egypt
[b]Electronic Materials Researches Department, Advanced Technology and New Materials Researches Institute, City of Scientific Researches and Technological Applications (SRTA-City), New Borg El-Arab City, Egypt
[c]Chemical and Petrochemical Engineering Department, Egypt-Japan University for Science and Technology, New Borg El-Arab City, New Borg El-Arab City, Egypt
[d]Fabrication Technology Researches Department, Advanced Technology and New Materials and Research Institute, City of Scientific Research and Technological Applications (SRTA-City), Alexandria, Egypt

1. Introduction and historical overview

1.1 Introduction

It is inconceivable how technology has improved and the ways we are ready to witness its impact on our lives. At the same time, we can expect the influence that future expansions will have. The reducing of devices and the modulation of smart-features opened new potentials not only in diagnosis and treatment but also in continued monitoring and health self-assessment [1].

One among the "key" technologies and know-how in sensor systems is the development of "intelligent sensors" that integrate sensing performs with signal extraction, processing, and "understanding." This distinguished target has been pursued with specific effort in application fields such as control and remote sensing, where minimum size and high level multifunction performances as the main accomplishments to be reached wherever they are considered. The term "intelligent sensors" has been constructed to designate sensing structures capable of combining with an "intelligent" method of preprocessing the obtained signal to award direct and selected information [2].

The present chapter presents an outline relating to smart and intelligent devices that take full feature and potential from intelligent nanosensors. Starting with this introduction, to the main connotation of nanosensors, to the identification of current limitations and outlook trends, a few examples of smart and intelligent devices are shown. We shall start with the identification of the main concepts concerning smart devices based on intelligent nanosensors, the meaning of the efficiency, and the size features that define them [3].

Handbook of Nanomaterials for Sensing Applications
https://doi.org/10.1016/B978-0-12-820783-3.00017-8

1.2 Historical overview

The terms *smart* and *intelligent* sensors have been used repeatedly in the sensor industry to depict devices that offer additional practicality than simply providing estimates of the measurand. However, there are varying developing sanders utilized by sensors, which are intelligent or smart, ranging from merely merging an operational amplifier for the output signals to sophisticated data modeling techniques for status monitoring [4].

History has shown that progression in material sciences and engineering have been significant drivers in the expansion of sensor technologies. For example, the temperature sensitivity of electrical resistance in a diversity of materials was celebrated in the early 1800s and was applied in 1860 by Wilhelm von Siemens to develop a temperature sensor based on a copper resistor. The high resonance stabilization of single-crystal quartz, as well as its piezoelectric futures, made possible a substantially wide range of high-performance, reasonable sensors that have played significant roles in everyday life and national defines. Later, a new age in sensor technology was ushered in by the growth of large-scale silicon processing, authorizing the investment of silicon to create new procedures for transducing physical phenomena into electrical output that can be easily processed by a computer. Continuing progress in material technology will permit better control of material characteristics and performance, thus offering possibilities for new sensors with greater fidelity, lower cost, increased reliability, and advanced features [5–8].

During the most recent years, sensors have expanded their abilities more and more thanks to the integration of various functionalities, which has advanced devices such as microcontrollers or computers, enabling them to reach a standard of intelligence that goes beyond a simple measurement and can be appointed for a broader range of applications. Various industrial sectors are taking advantage of these innovative devices, including automotive, production and automation process, robotics, aeronautics, and others where accuracy must fulfill perfect levels, even if human involvement is limited [9–12].

One of the most motivating fields that has the capability of greatly improving in the future regards human healthcare. Indeed, technology is making important contributions to it, considering just prostheses or rehabilitation systems, which try to improve the subject's living conditions as much as possible. Sensors are part of such participation, as they are substantial in sensing vital parameters when the patient is using the biomedical device [13].

Starting with this point of view, and being aware that this is an extremely topical subject, the present chapter aims to provide an overview of the exploitation of smart sensors in such a domain and other environmental applications. In the following sections and subsections, the connotation of intelligent sensors will be first introduced and then different devices will be described, demonstrating why they are the most appropriate for certain applications [14].

2. Sensors

Sensor technology is a rapidly developing field that has considerable potential to better the operation, serviceability, reliability, and utility of many engineering systems. Progresses in material science and engineering have paved the way for the expansion of new and more intelligent sensors [15].

In spite of the comprehensive published literature that addresses the fundamentals of sensor technology, considerable opacity exists in sensor classification and identification, as clarified by a recent buyer's guide for sensors in which two lists of sensor suppliers are provided, one based on features sensed and the other on technologies used. The second list includes both physical phenomena (such as electrochemical, acoustic, infrared sensors, and Hall Effect) and material types (for example, fiberoptic, bimetallic, and thick and thin film) [16].

A sensor can be defined as "*a device/component capable of quantifying a physical quantity.*" However, we can still state that this is a primitive definition of a sensor, without any technical vision. Thus, another potential definition would be to consider a sensor as "*a component capable of detecting a given variation in input energy and converting it to another or the same type of energy as output.*" This definition deals with the transducer definition as a component capable of transforming a form of energy into another form [17].

Notwithstanding, other definitions for transducers may be found. However, for many of us, these definitions are called instrumentations. For the target of the present chapter, the intelligent sensor is considered as a module capable of implementing the sensing task, subsequently enabling the quantification of a given measurand (outputs information). This means that the transducer must be the component capable of converting a given energy form into another, which makes all sensors transducers. Similarly, it also makes all actuators (the opposite of a sensor) transducers as well. In spite of being a sensor, actuator, or transducer, all these components interact with different energy forms, including electrical, thermal, mechanical, optical, acoustic, or electromagnetic, among others [18].

2.1 Micro- and nanosensors

Sensors, like other many technologies, are the future of the development of microfabrication procedures. The capability of miniaturizing sensor devices has opened up the priority for making them implantable in ways that would not have been conceivable a few decades ago. Nevertheless, miniaturization is not closed to the downsizing of the sensor. All of the other devices and components required to work with the sensor need to be miniaturized as well and fabricated into a single device that would enable access to many measurement places and signals. Driven by semiconductor manufacturing, through lithographic and thin–film technology processes, we certified the development from the macroscale into the microscale, and now we are looking at the opening of the nanoscale world [19].

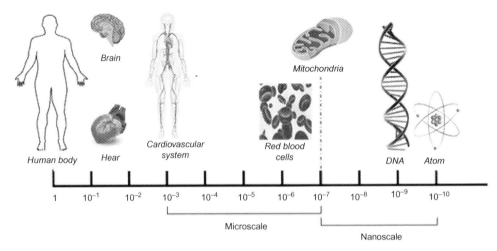

Fig. 1 Dimensions domain for several parts of the human body [20].

Fig. 1 shows the domain of dimensions of several parts and components in the human body. As sensors downsize, it becomes convenient to gain access to the complexities of the smaller elements, enabling the conception of physiological phenomena at a major level. This is what is expected with the development of these devices. It is well known that a microscale device represents feature's size-characteristics within the 1–100 nm range; however, nanoscale is defined by sub-100 nm. Furthermore, the ability to access such structures at this scale has promoted research into sensory systems, which is key in biomimicry and biomimetics [20, 21].

Driven by these defiances, there is no doubt that the capability of downscale devices and components has a massive effect on science due to the technical challenges that downscaling presupposes. At the miniaturization scale we can obtain today, it is no longer a question of simply linearly sizing down components. There is a need to review the manufacturing procedures because, at the micro- and nanoscale, tolerances barely match the normal standards. Further at the microlevel, thermal, mechanical, and optical effects do not work similarly as in the macrosize level. This leads to a review of the manufacturing process to work at such small sizes [22].

As new design and processing technologies became available, they rapidly led to development of new components, which are known nowadays known as microelectro-mechanical systems (MEMS). MEMS sensors may now be found in many electronic devices that we have such as mobile phones, automobiles, and home appliances. However, development has not stopped at MEMS. After breaching the critical barrier of 100 nm, the nanoworld became within our reach, and nanoelectromechanical systems (NEMS) have emerged [23].

At this time, MEMS and NEMS present unique aspects for new work and performance improvements in sensing components. However, as there were challenges at the microscale, so are there other challenges when trying to reduce the scale of MEMS into NEMS because of not all the microfabrication techniques advanced so far are eligible for preparing nanostructures. Subsequently, as the size decreases, we should not neglect the fact that these components/devices need to interface with the real world, which becomes more difficult to do. Examples of such difficulties are low levels of signal, poor signal-to-noise ratio, surrounding parasitics, among others. Furthermore, on a micro- and nanoscale, other parameters have effects, such as friction, adhesion, surface contamination, and lubrication. All these parameters may not form a major problem on the same magnitude with bulk-size material systems; however, at micro- and nanoscale, they may cause considerable problems. Therefore, the overall performance of micro- and nanostructures devices is dependent on the structure's careful design [24, 25].

2.2 Intelligent nanosensors

Starting in the 1970s, the development of intelligent sensor technology was focused and highlighted as a result of the integration between processing capabilities based on readout and signal processing. However, this technology was still far from the complexity needed in advanced IR surveillance and warning systems. This is due to the massive quantities of undesirable noise and signals produced by an operating scenario, particularly in martial applications [26].

The main policies of the operation of intelligent sensors devices is based upon the integration between a simple instrument specified mainly for performing measurements (i.e., a sensor) and some transformation, elaboration (which could be analog and/or digital), and communication modules. Such a combination is systematized in Fig. 2 [12].

The previously described circuits are provided by apposite units existing inside the system or connected by an exterior source. So, this type of architecture offers additional functionalities compared with traditional transducers such as primary data elaboration,

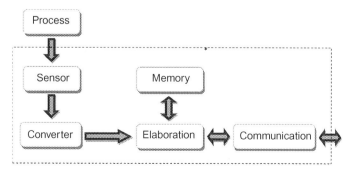

Fig. 2 General conceptual scheme of an intelligent sensor [12].

auto identification (regarding IEEE 1451.4 standard), smart calibration, communication purposes for remote monitoring, and configuration [27].

Initially, for signal processing, only analog elaboration features such as amplification, filtering, and equalization have been introduced. These features have their role not only to correct the sensor's nonideal performances concerning transduction processes like nonlinearity, offsets, or disturbances but also to achieve high accuracy for measurements. These analog elaboration features are followed by implementation in digital segments such as the central processing unit (CPU) and nonvolatile memories. These digital modules may be either analog-to-digital (AD) or digital-to-analog (DA) converters and serial communication ports for significant measurements of operation conditions (especially the critical ones). This development and progress of the intelligent sensor devices enable these devices to acquire various functionalities that were available earlier using microcontrollers only. Accordingly, sensors realize the competences to processing numerous incoming signals at the same time, even if these signals have diverse natures (such as temperature, humidity, and pressure), to derive the corresponding measurements as required. These integrated devices offer several advantages; first, cost dimension of the acquiring subsystems and setup. Second, transferring ability of any developed technologies at other related fields such as electronics will improve and progress the operation of these devices. Also, the opportunity of constructing sensor networks that can interface with a central controller managing the entire system permits realizing novel applications never previously assembled. Finally, due to the continuous growth and development of both sensors and microcontrollers technologies that reduce their expenses, these technologies may be combined together to manufacture highly efficient modules in terms of both functionalities and energy consumption [28, 29].

3. Intelligent sensors identifying needs

Intelligent sensors are vastly used in many different applications. Sensor technology also has become a requisite enabling technology in many applications. The fast growth in the interest in sensors has been supported by numerous applications, such as intelligent fabrication processing in which intelligent sensors can provide great advantages. Also, intelligent sensors are of great significance in safety-related areas with appreciated applications ranging from safety assessment of aircraft to environmental monitoring for hazardous chemicals [30].

Selected examples of intelligent sensor types and applications were chosen to clarify the different driving forces and considerations discussed earlier. These clarifications are not aimed to be inclusive. Each example describes an implementation and discusses the key technical issues related to the sensor's usage and classifies key intelligent sensor material needs. At the same time, each sensor-type model describes the physical phenomena being sensed, classification of the different types of sensors, and sensor materials related to the applications. It is very important to know that these cases are facilitations of reality. For example, sensor needs are presented as if each sensor were an independent entity, although in fact many applications necessitate sensor arrays or a collection of information obtained from different kinds of sensors. Also, there are differences in the level of technical detail for each example according to the scope being discussed [31].

With advanced materials that are produced using developed processing technologies, new kinds of intelligent sensors are becoming possible. To a greater domain than this technology push, market pull is driving expanded activity in intelligent sensors. Economic stimulus for enhanced sensor materials and technology include increasing a product's features, reducing the cost of product manufacturing, and enhancing the quality of the product. These motivations also encourage product competitiveness. For example, many sensors integrated into the operation of modern automobiles have greatly enhanced the safety, quality, and comfort of the vehicle. Likewise, the cost of industrialization and frequency of trouble in vehicles have been dramatically reduced by increased usage of sensors during manufacturing. An equally important economic factor is the expansion and combination of intelligent sensors into products that help prolongation of usable life. For example, intelligent sensors for engine oil that monitor the benignity of motor oil in an engine, alerting a user to change engine oil only when it is necessary because of lubricant degeneration. Another example is intelligent sensors that can detect corrosion or metal erosion in older aircraft in lieu of more expensive external sensors [32].

Intelligent sensors have been substantial in satisfying a massive number of government-mandated regulatory requirements that contain such implementations as monitoring exhaust gases from automobiles and harmful chemical emissions from factories. These sensors can also have an important economic impact and effect on the quality of environment and life [33].

Governmental agencies and associations have many unique wide-ranging sensor requirements. Due to the variety of intelligent sensor technologies and applications, and the resulting diversity of materials needs for intelligent sensors, it is extremely possible to satisfy a given requirement with more than one kind of sensor. This fact has a considerable effect on planning research and development (R&D) of sensor materials, structures, and systems. R&D strategy that preserves a broad applications-driven research base is necessary to accelerate sensor expansion. This matter demands the support and identification of critical core competencies. This chapter will contain many examples and applications of intelligent nanosensor and sensor materials. Sensor technology development depends on a wide diversity of technical disciplines from chemistry to material science and from physics to engineering to process engineering. So, the varied nature of sensor technology enhancement requires interdisciplinary cooperation. Also, it requires a creative applications focus on chosen sensor materials and technologies. Scientific and technological effect, risk, and advancement of a knowledge base for intelligent sensors must be considered to recognize the most promising chance areas that can have a major effect and lead to a great return on investment [34].

4. Sensing principles

According to previous investigation, sensors are defined as devices that have competence to convert one energy type into electrical energy with the assistance of acquisition and processing equipment to display the measurement for the end users.

Sensors can be categorized according to the sensing principles associated to their comprised transducers. The sensing principles are essentially physical principles including electromagnetism, harmonic vibrations, piezoelectricity, etc. that permit the energy conversion from one category to another form. Most of these principles are sequenced by some basic application as will be discussed.

4.1 Direct against indirect sensing

Generally, sensors can be classified according to their conversion of measurements into electrical signals to direct or indirect sensors. In a direct sensor, the measurements have an electrical nature or can directly convert the various energy categories into electricity (Fig. 3A). The regular example of utilized direct sensors is the physiological electrodes that translate bioelectricity into electron-based electricity. However, there are a large number of measurements that cannot be converted directly into electricity. So, for these cases, the indirect conversion is obligatory. In indirect sensors, they have an ability to convert the measurements into another energy category before converting it into electricity (Fig. 3B). So, indirect sensors may have one or more transduction phases according to the requirements of the sensor [35].

The operation concept of an indirect sensing process may be clarified through investigating commercial thermometers for body temperature measurements. These devices are based upon two-step measuring processes. First, at the tip of the thermometer, the

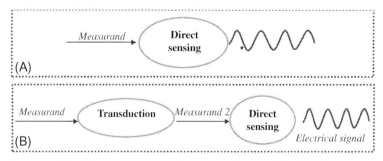

Fig. 3 Classifications of intelligent sensor according to sensing principle. (A) Direct sensor. (B) Indirect sensor [35].

temperature changes the sensor resistivity inside the thermometer. The incorporated electronic circuit in the device senses and measures the induced resistivity then translates its value into an electrical signal. This electronic signal can be handled and displayed on a monitor for the user as the measured temperature. Accordingly, in this sensor type, the temperature variation changes the resistivity, which is then translated into an electrical signal [36].

4.2 Active against passive sensing

On the other hand, sensors can be categorized into active and passive sensors. The passive sensors are devices that don't require external energy for electrical signal acquisition. Photodiodes represent the most famous example of passive sensors, as these devices have electrical current generation ability when exposed to light. However, active sensors need this type of external energy for electrical signal acquisition. A commonly used model for active sensors is the accelerometer utilized in cell phones. These sensors permit the discovery of cell phone orientation in addition to the vibration occurrence. Also, the same sensor type is utilized in biomedical applications for heart rate detection in patents [37].

5. Applications of intelligent nanosensors

We are frequently asked questions from our field employees on a regular basis related to intelligent sensors. They usually ask something like: "What type of digital interface will be required for our intelligent sensors?" or "When are we going to come out with an intelligent sensor?" Our customary answer to these kinds of questions is simply: "When can you tell us what the customer needs in an intelligent sensor?"

Many definitions for intelligent transducers confirm that anything with a digital ability would, by definition, be considered intelligent. From our point of view, we do not think that this is a real requirement of intelligence. We believe that, to define any intelligent components, you should first look at the systems that will integrate these components. We may define the sum of systems as an integration of sensor inputs, actuation outputs, and control. To consider the system as intelligent, it should include some combination of these elements and must always incorporate some portion of the control function. This requirement of integrating functionality appears to be missing in most definitions of intelligent components. The fact that a transducer has an interior digital function, whether it is used for calibration, communication, or some means of information or unit identification such as electronic data sheets, does not make the transducer intelligent. However, if we add a function such as signal analysis or integral fault detection to the unit, then we are of course getting to an intelligent element [38].

Fig. 4 shows a generalized control system. Generally, for any transducer to be considered intelligent, it must contain collections of two or more elements from the system. This would mean that an intelligent transducer could include the transduction function

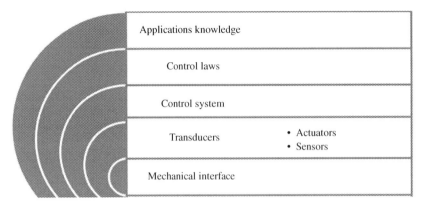

Fig. 4 General control system [39].

and a part of the control function. This is clearly the most unique feature of an intelligent transducer. This means that a transducer that includes sensing and actuation, or multiple sensors in a single package, would not be intelligent unless they also include some form of control or computation action. It is very clear to distinguish between actuators, intelligent sensors, and other subsystems [39].

Now, the boundary of the definition of "intelligent" can be easily described. An intelligent component must include some grouping of the elements of an application system such as computation, control, or decision making. If the component incorporates an actuation function, then it is considered an intelligent actuator. On the other hand, if the component deals with the basic definition but contains only of sensing elements, it is an intelligent sensor. If the component comprises a major part of the application system, including the basic requirements, it can be considered an intelligent subsystem [40].

The last significant consideration for intelligent transducers is that they improve the performance, functionality, or cost of the end application system. This is why most efforts at intelligent components will fail. What appears to be valuable to the engineer or technologist and meets the requirements previously outlined will not be accepted in real applications if it does not demonstrate some advantage to the customer. In the following sections and subsections, some selected types of intelligent nanosensors and their applications will be discussed [39].

5.1 Sensors for environmental applications

5.1.1 Chemical sensors

Response of chemical sensors occurs to the measurands through various chemicals and chemical reactions. These sensors have the capacity to identify and measure liquid or gaseous chemical types. These kinds of sensors are presently used in many chemical analysis techniques such as infrared technology, chromatography, mass spectroscopy, and others due to the miniaturization of sensors. There are large variations in intelligent chemical sensors with various methods of operation. The procedures of operation systems of intelligent chemical sensors can be classified into three classes according to their modes of measurement: (i) electrical and electrochemical properties, (ii) changes in the physical properties, and (iii) optical absorption of the chemical analytes to be measured [41].

Electrochemical sensors are considered the most common approach. These types of intelligent sensors operate by reacting with chemical solutions to produce an electrical signal proportional to the analyte concentration. These kinds of intelligent electrochemical sensors can be distributed into three main categories according to their working mode [42]:

I.   Potentiometric, which depends on the measurement of voltage. To measure the concentration of analytes, it is important for a combination of electric and ionic current to flow in a closed circuit [43].

II. Amperometric, which depends on the measurement of current. These intelligent sensors have been shown to be efficient in a broad range of environmental applications, such as volatile organic compound (VOC) detection in groundwater and soils, detection of mines, and blood analytes [44].

III. Conductometric, which relies on the measurement of either conductivity or resistivity. This kind of intelligent sensor involves a capacitor that changes its capacitance when exposed to the required analyte. These changes in capacitance occur due to a selectively absorbing material such as polymers or other insulators. These absorbing materials operate as a dielectric layer of the capacitor, and then their permittivity changes with exposure to the analyte. These types of intelligent sensors are generally used to detect humidity, carbon dioxide, and VOCs [45].

Previous research in the same areas looked into applying new technologies and features to chemical sensor research and analytical chemistry. At the advanced stages of research, attention is generally concentrated to exploring and demonstrating the basics by which modern technologies can be applied to measure chemical compounds. Open access to specialized facilities and instruments, such as those required for lithographic patterning, can be crucially significant to promote progress and interest as applications to chemical sensing measurements and specific practical analytical start to appear [46, 47].

As mentioned before, the most important material-related chances to develop direct-reading intelligent chemical sensors include the choice of utilized materials to acquire stable selectivity of interaction with the target analyte. Table 1 summarizes different material and compound needs for direct-reading chemical-intelligent sensors. All requirements are presented in terms of material functionality instead of material type (e.g., polymer, semiconductor, ceramic) to avoid unsuitable expectations based on existing solutions [48].

Table 1 Material needs for selective direct-reading chemical sensors [48].

Material	Forms applications	Functional requirements	Possible mechanisms
Membranes	Amperometric, conductimetric, potentiometric Electrochemical sensors	Analyte selectivity Stability	Analyte binding or partitioning Permselectivity Catalytic reactivity Sensing electrode arrays
Coatings/ thin films	Amperometric, conductimetric, potentiometric Electrochemical sensors Optical fibers and waveguides Piezoelectric devices Surface acoustic waves	Analyte selectivity Stability	Analyte binding or partitioning Enzyme or antibody properties Sensing electrode, optical fiber, waveguide arrays Permselectivity Electrocatalytic activity Changes in light propagation or luminescence Viscoelastic changes
Bulk materials	Amperometric and electrochemical sensors	Analyte selectivity Stability	Solid or polymer electrolytes with selective binding sites
Fibers (optical)	Optical fibers and waveguides	Extended operational wavelength range	Improved near- and extended infrared transparency and reflection

5.1.2 Biosensors

A biosensor can be easily defined as an integration between a bioreceptor and a transducer (Fig. 5). A bioreceptor is a biomolecule that recognizes the objective chemical substance, whereas the transducer is a converter tool that converts the signal generated by the biomolecule into an electrical signal. This type of intelligent sensor is derived from the expansion of the chemical sensors. The initial biosensors were capable of measuring the amount of oxygen in the blood. After that, other additional improvements were added to these devices making it able to sense and measure other analytes in the body by adding different types of enzymes to the oxygen electrode. This kind of biosensor is considered the first biosensor and is called "the enzyme electrode." Generally, the main difference between biosensors and chemical sensors is that the first depends on biomolecules for the transducing process and the latter does not [49].

The enzyme electrode or any other primary biosensors are based on the monitoring of the electron-transfer processes, therefore following the amperometric class. In this kind of intelligent biosensor, the resulting current by the device is proportional to the analyte concentration. The electrical signal of these types of sensors is generated by the electron

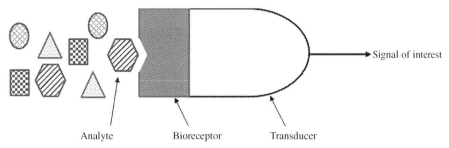

Fig. 5 Biosensors model.

exchange between the electrode and the biological system. The first step occurs when the analyte in the target solution undergoes oxidation–reduction reactions with the bioreceptor. The bioreceptor is a biomolecule that is mainly responsible for selectively catalyzing the target analyte. The oxidation–reduction reaction promoted by the bioreceptor under specific conditions will generate an electron current within the close electrode. For this current to exist, it is important to apply a fixed voltage between the solution and the electrode so as to produce electrons for the oxidation–reduction reactions occurring at the electrode. This operation mode is known as chronoamperometry and performs a measure of current as a function of time whereas applying a constant potential at the electrode [50].

Many other kinds of intelligent biosensors have been advanced using a large variety of totally different approaches. All of these types of intelligent biosensors are based on a biological element before a transducer. Several of the previous transducing principles are often combined with differing types of bioreceptors [46].

5.1.3 Medical sensor

It now becomes motivating to present and discover real medical devices supporting micro- and nanosensors. These models are aimed at supporting the value and focusing on new chances that micro- and nanosensors highlight in the monitoring, diagnosis, and treatment of several diseases.

This section focused on implantable devices based on micro- and nanosensors. These devices were designed, fabricated, and developed to be carefully implanted inside the body. This means that, besides understanding the most convenient sensing principle, the target measurand the intelligent sensor should have conditions in which the device will be implanted. Consequently, implantation technique, compatibility with the surroundings of the implantation place, and also physiological and anatomical characteristics must be considered from the design phase of the device [51].

Nowadays, an aortic aneurysm is a popular case often related to cardiovascular diseases. This status is known as a swelling of the aortic artery section led by the native

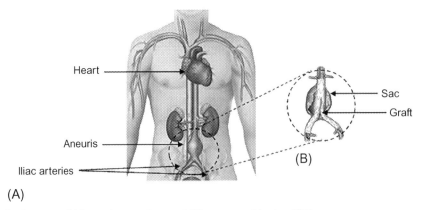

Fig. 6 Clarification of (A) aortic aneurism and (B) stent positioning [52].

weakness of its walls and systemic pressure (Fig. 6). So, we can simply understand the danger of fatality related to a burst on the artery wall, resulting in an enormous loss of blood [52].

There are two possible methods for treatment of this condition: either reconstruction of the aortic artery by open chest surgery or an endovascular aneurism repair procedure. In medical procedures, the present trend is to reduce the open chest surgery and prioritize minimally invasive processes whenever possible. The latter leads to less physiological disturbance and faster recovery times [52].

5.1.4 Food packaging sensor

New food packaging technologies are expanded in response to consumer requests and orientations in conservation of industrial products, flavor stability, product freshness, longer service life, monitoring, and traceability. The mentioned subjects permit quality control and compliance with health needs. In this, packaging is a continuously developing field, achieving a leading role in the conservation and enhancement of the flavor of food [53].

One of the major objectives of food law is the safety of food products. Packaging process protects food from external factors, such as level of oxygen, dust, heat, humidity, light, gaseous emissions, contaminants, enzymes, micro-organisms, insects, relative pressure, odors, and lipid oxidation. To extend the life of food, it is essential to retard biochemical reactions and enzymatic activity using varied strategies, such as humidity management and temperature control by adding chemical components and other additives such as sugar, salt, carbon dioxide, natural acids, antioxidants, antimicrobial, antifungal, and removing oxygen or any other residual compounds of these in packaging [54, 55].

Intelligent nanosensor interacts with internal agents such as food components and/or environmental factors. As a result of this interaction, intelligent nanosensors will produce a response (electric signal or visual signal) that correlates with the status of the food product. The resulting information is not only valuable for communication with consumers, providing them with data on safety and quality of products, but also can be used by producers in their decision support systems to set when and what measures should be taken over the whole production procedure and distribution channels for products [56].

Ongoing scientific research and modern technological advances are being directed toward many fields such as printed electronics, silicon photonics, biotechnology, and nanotechnology. These scientific advances present the potential of developing a new generation of intelligent nanosensors in food packaging applications. Intelligent nanosensors have great capacity to accelerate the rate of detection, quantification, and identification of decaying substances, pathogens, and allergy-causing proteins. Hence, these intelligent nanodevices have the possibility to significantly affect food security (Fig. 7). Generally, intelligent nanosensors are placed on food packaging to monitor external and internal status of the products and in the accurate identification of different contaminants in food [57–59].

5.1.5 Pesticide sensor

Agrochemicals have become one of the most widely used category of components that are commonly found worldwide on Earth surfaces and water resources as a result of their massive applications in agricultural fields and other domestic applications. Regardless of being inherently carcinogenic, mutagenic in nature, and toxic, these compounds have some characteristics such as movement to different areas and persistence in the environment. They have the ability to target the endocrine systems of living organisms, together with humans [60].

An effective way to assess the risks resulting from occupational and environmental exposures is biomonitoring using an intelligent nanobiosensor, which provides an approximation of the total absorbed doses and provides indirect access to locate target

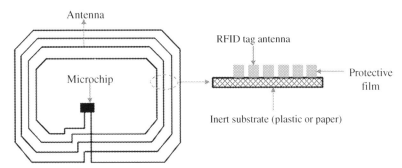

Fig. 7 Schematic exemplification of food packaging nanodevices [57].

site concentrations. To detect the presence of pollutants and to measure their toxicity, intelligent nanobiosensors can be used as biological indicators. Direct and indirect monitoring are two general methods to evaluate pollutants and their toxic effects at different standards, from the types to the society level of any ecosystem. In indirect monitoring, reduction of biodiversity, removal of sensitive species, and degradation of the ecosystem can be detected as adverse results of pollution at the standard of populations, whereas at the level of the individuals, aggregation of toxic compounds in samples, tissues, and organs are indicating pollution in the environment and can be traced. In direct monitoring, the response of modified or artificial populations; particularly functions of organs such as feeding, movement, reproduction, and respiration; the neural regulation; behavioral patterns of specimens; and also cellular and subcellular reactions are studied for the effect of toxic compounds [61].

5.1.6 Sensors for air quality monitoring system

Air pollution is one of the most serious problems present every year. According to the World Health Organization (WHO), air pollution was a main reason in early death of 7 million people globally in 2012. It is especially predominant in large industrial cities in all countries, where most people are exposed to the air pollution without any detailed knowledge about the pollution contents and protective means to preserve their own health. Different kinds of air pollutants can cause significant problems to human health and the environment. So, periodic detection and strict control are very significant. All types of air pollutants have their own standard allowable limits, which is a bit different from one country to another [62].

More cheap, portable, and simpler intelligent sensing devices can be used as an alternative technique to standard air quality monitoring tools. Regardless of whether they do not offer highly dependable data as the standard air quality detecting equipment, they can be utilized for low-power, customized, and mobile gas-sensing applications, which can be adequate enough for early-stage warning systems. Main classifications of compact micro- and nanoelectronic air quality sensors are listed in the following [63]:

- Metal oxide-based conductometric sensor: The main concept of this type of intelligent sensor is that the electrical conductivity of metal oxide particle-based thick or thin films is changing by the exposure to the oxidizing or reducing gases.
- Catalytic reaction sensor (pellistor): This kind of sensor is based on the changes in temperature by the catalytic chemical reaction. The sensor contains ceramic body-based pellets of catalyst and temperature sensor.
- Electrochemical sensor: The idea of this intelligent sensor is based on the electrical chemical reaction of gaseous molecule with electrolyte generating electrical potential and current.

All of these intelligent sensor devices are summarized in Fig. 8 [64].

Micro/nanotechnology are based on the integration of functional elements including electronics, sensors, and actuators in micro/nanoscale devices. Nanotechnology presents

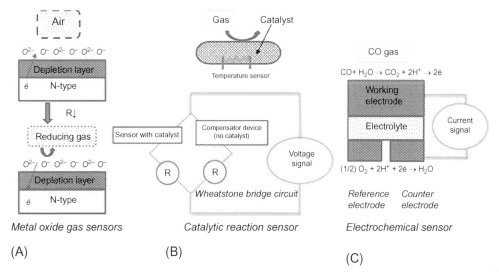

Fig. 8 Basic mechanisms of gas sensors electronic devices: (A) Gas sensor based on semiconducting metal oxide thin film, (B) catalytic reaction sensor, and (C) amperometric electrochemical sensor for CO gas detection [64].

new features of novel nanomaterials and nanostructures with unique chemical and physical characteristics to enhance the devices' performances [65, 66].

5.1.7 Sensors for heavy metal ion detection

Regarding the increment of environmental pollution with heavy metals and harmful chemicals, there is an essential requirement for monitoring these toxic emissions to be removed and controlled according to their levels in the ecosystem. Simultaneously prompt and accurate evaluating devices must be developed to monitor the pollution progress. Accordingly, construction of a pollution prevention system is obligatory for determination of the heavy metals or harmful chemical contamination level at the released field of these pollutants. Lately, there are many molecular recognition technologies that were developed based upon high selective and sensitive detection of these targeted heavy metal ions with low cost. Table 2 investigated the recommended concentrations of heavy metals in drinking water considered by both WHO and the United States Environmental Protection Agency (US-EPA) [62].

Recently, the development of accurate, simple, and inexpensive detection technologies for heavy metals represents a major challenge for scientists. This is owing to the high negative influence of heavy metals as pollutants on human health and the environment even at low concentrations. Amongst the various analytical technologies for heavy metal determination, the electrochemical-based sensors have been widely applied [67].

Voltammetry is an electrochemical analytical method that measures the change in current that correlates to the supplied voltage between a working electrode and a counter

Table 2 Standards limits for heavy metals in drinking water mentioned by WHO and the EPA [62].

Metal	WHO (mg/L)	EPA (mg/L)
Ni	0.07	0.04
Cu	2	1.3
Zn	3	5
Cd	0.003	0.005
Hg	0.001	0.002
Pb	0.010	0.015
As	0.010	0.010

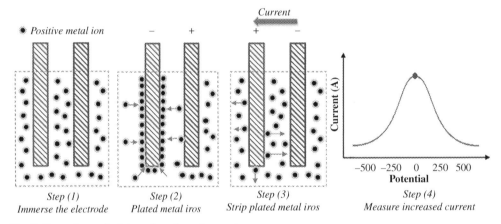

Fig. 9 Operating principles of cyclic voltammetry.

electrode. This technique offers data about the analyzed sample. Voltammetry comprises four processes; immersing electrodes, plating metal ions, stripping plated metal, and increased measuring current as presented in Fig. 9 [64].

This voltammetry technique was handled for measuring the heavy metal concentration where the maximum value of the change in the current defines the concentration of the metal ions. Anodic stripping voltammetry (ASV) is mainly utilized for determination of metal cations, and cathode stripping voltammetry (CSV) is utilized for determination of metal anions. The main advantage of this technique is its ability to measure and determine the type and concentration of different metals at the same time by stripping after plating them with a voltage adjustment as illustrated in Fig. 10 [64].

6. Conclusions and futures

Predicting the future it is often a risky play unless we study our education lessons from the past and observe them in the present; in both cases, it is believed that intelligent sensing

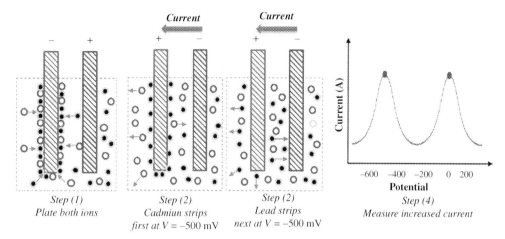

Fig. 10 Detection of multiple heavy metal ions using stripping voltammetry.

will maintain admirable development, and not only because of the abusing the "intelligent" terminology, but also thanks either to the impressive progress in consolidate solid-state Very Large Integration Microcircuits (VLSI) comprising MEMs Technology. The recent market for smart sensors includes about 90 million devices, but it is expected to reach 3 trillion devices by 2025. The market of the smart sensors is expected to grow up to 10% a year and is estimated to reach US $100 billion by the year 2025, thanks to microtechnology and nanotechnology that are responsible for the cost reduction of these smart sensor devices [68].

Additionally, the effective outcomes in both the performance and cost of these smart sensors and the excessive yields in their applications in the fields corresponding to environment, transportation, and energy have disbursed many investments in these sensor applications (e.g., intelligent cities, intelligent water resources, etc.). The intelligent sensors will choose and select the most relevant data rather than just passing on collected information. So, these sensors will screen and estimate the information's quality that performing advanced analysis and explanations that permit territorial planning.

In line to the conventional "intelligent sensors," there will be excessive progress at "intelligent dust," which is known as a "Cloud of Sensors." These sensor types include microsize dimensions that spread over large area and are linked by wireless networks that permit the transmission of data to central computer systems. The expectation with the explosion of intelligent dust thanks to nanotechnology is the corresponding building science of molecule-size electronic devices. However, smart dust networks haven't reached their potential as a technology until now that will revolutionize environmental applications such as security and pollution control. Nevertheless, these wireless tracking sensors have accomplished comparatively low-cost performances, being relatively economic. These novel intelligent sensor technologies, including the system's constituents and signal processing, will be integrated in most applications into so-called intelligent grids/intelligent networks including environmental and energy applications.

Regarding the small size of these novel sensors to be put anywhere, they have the ability to work wirelessly and share data, thus "intelligent dust nanosensors" are capable of being revolutionary. Smart dust is dependent on microelectromechanical systems (MEMs) that are coupled together using tiny microcomputer chips that can evaluate the variation of temperatures, vibrations, or surface pressures generating an automatic action such as governing the building's temperature and controlling the oil flux flow at pipeline [69].

In addition to progressive integrated smart sensing systems that mainly depend upon the integration between the advanced technologies including microelectronics, MEMs, and nanotechnology into so-called smart grids and intelligent networks, it is sensible to forecast the excessive growth of "smart sensors" created by their aimed specific intelligence, which is similar to the first original intelligent sensors (i.e., development of "Task Aimed Intelligent Sensors," which are smart sensors designed and developed for specific applications), even if improved in their performances by using progressive technologies. Generally, through nanotechnology, we may control the matter at the atomic and molecular scales to develop functional materials and devices. This might permit sensing and interactive actions exploiting novel phenomena through considering that most physical, chemical, and biological sensors depend on the presence of interactions at these levels that function on an atomic and molecular scale. So, emerging nanotechnologies have dramatic fluctuations in sensor designs and capabilities [70].

Nanomaterials and nanostructures represent talented application areas that may offer two purposes in the field of sensor applications especially for chemicals and biological substances, which are recognition of concerned molecules and transduction of that recognition event into valuable signals for activating the process. Consequently, nanotechnology hasn't been considered as simply a trend toward the smallest that started with the miniaturization realized by microtechnology and that permits better specificity in less size, smarter, and less costly sensors with the help of utilization ICs, fiber optics, and MEMS technologies. In fact, nanosensors can be categorically smart if designed and realized for certain applications such as detection of chemicals and biologic molecules for environmental pollution control [71, 72]. Although excitement over nanotechnology and its prospective utilization is commonly well originated, the development and integration of nanosensors must take into consideration the realities imposed by physics, chemistry, biology, and engineering. For example, excessive limitations appear during the integration between nanotechnologies and macro-sized systems with the requirements for controlling the flow of matter, energy, and information between the nano- and macroscales. Accordingly, several considerations for nanosensor design are similar to those for microsensors, particularly the interface requirements such as heat dissipation and the requirement for dealing with interference and noise for both electrical and mechanical forces. Each interface in a microsystem may be subjected to unwanted transmission of electrical, mechanical, thermal, and possibly chemical and optical fluxes. Flow control is particularly critical in chemical and biological sensors, into which gaseous or liquid analytes are contacted and then expelled. Additionally, very sensitive and tailored surfaces of these sensors are commonly degraded due to the actions of foreign substances and thermal stresses. Mostly, these limits may be overcome by installing larger nanosensor units thanks to their infinitely small size and controlled cost that permit malfunctioning devices to be ignored in favor of good ones, thus accomplishing a successful system's work with extended useful lifetime [73].

Acknowledgments

This work was supported by the Egyptian Science and Technology Development Fund (STDF) in Egypt (Grant no. 30735).

References

[1] C. Corsi, International NATO Electronic Warfare Conference, 1978.
[2] C. Corsi, DoD/AOC Symposium, Washington, 1979.
[3] T.F. Tao, D. Hilmers, B. Evenor, D.B. Yehoshua, Focal plane processing techniques for background clutter suppression and target detection, in: Proc. SPIE 0178 Smart Sensors, 2 1979, p. 12.
[4] E.T. Powner, F. Yalcinkaya, From basic sensors to intelligent sensors: definitions and examples, Sens. Rev. 15 (4) (1995) 19–22.
[5] J.K. Gimzewski, C. Gerber, E. Meyer, Observations of a chemical reaction using a micromechanical sensor, Chem. Phys. Lett. 217 (5/6) (1994) 589.
[6] W. Göpel, J. Hesse, J.N. Zemel, Sensors, A Comprehensive Survey, vol. 1, VCH. Instrument Society of America Electrical Transduce, New York, 1989.
[7] K.S. Lion, Transducers—problems and prospects, IEEE Trans. Ind. Electron. 16 (1) (1969) 2–5.
[8] S. Middlehoek, D.J.W. Noorlag, Three-dimensional representation of input and output transducers, Sensors Actuators 2 (1) (1982) 29–41.
[9] R. Matsuzaki, A. Todoroki, Passive wireless strain monitoring of actual tire using capacitance-resistance change and multiple spectral features, Sensors Actuators 126 (2006) 277–286.
[10] C. Ferreira, P. Ventura, C. Grinde, R. Morais, A. Valente, C. Neves, M.J.C.S. Reis, A novel monolithic silicon sensor for measuring acceleration, pressure and temperature on a shock absorber, Procedia Chem. 1 (2009) 88–91.
[11] L. Zhining, W. Peng, M. Jianmin, L. Jianwei, T. Fei, Vehicle fault diagnose based on smart sensor, Phys. Procedia 24 (2012) 1060–1067.
[12] X. Desforges, B. Archimède, Multi-agent framework based on smart sensors/actuators for machine tools control and monitoring, Eng. Appl. Artif. Intell. 19 (2006) 641–655.
[13] M.R. Kirchhoff, C. Boese, J. Güttler, M. Feldmann, S. Büttgenbach, Innovative high precision position sensor systems for robotic and automotive applications, Procedia Chem. 1 (2009) 501–504.
[14] K. Moreau, V. Rouet, Presentation of platforms for wireless advanced networks of sensors for aeronautics application, in: Integration Issues of Miniaturized Systems—MOMS, MOEMS, ICS and Electronic Components (SSI), 2nd European Conference & Exhibition, 2008, pp. 1–8.
[15] A.B. Kashyout, H.M.A. Soliman, H. Shokry Hassan, A.M. Abousehly, Fabrication of ZnO and ZnO: Sb nanoparticles for gas sensor applications, J. Nanomater. 2010 (2010) 1–8.
[16] Sensors, Buyer's guide. Sensors, J. Machine Perception 9 (1993) 12.
[17] I.R. Sinclair, Sensors and Transducers, third ed., Elsevier, Amsterdam, 1969.

[18] D. Patranabis, Sensors and Transducers, second ed., Prentice-Hall, New Delhi, 2003.
[19] D. Son, S.Y. Park, B. Kim, J.T. Koh, T.H. Kim, S. An, Nanoneedle transistor- based sensors for the selective detection of intracellular calcium ions, ACS Nano 5 (5) (2011) 3888–3895.
[20] P. Rolfe, Micro- and nano sensors for medical and biological measurement, Sens. Mater. 24 (6) (2012) 275–302.
[21] R. Müller, R. Kuc, Bio sonar-inspired technology: goals, challenges and insights, Bioinspir. Biomim. 2 (4) (2007) S146–S161.
[22] P. Kobel, C. Reymond, Miniaturization challenges and their impact on the micro-factory concept and manipulators, J. Jpn Soc. Precis. Eng. 77 (3) (2011) 263–268.
[23] Z. Cui, C. Gu, Nanofabrication challenges for NEMS, in: 1st IEEE International Conference on Nano/Micro Engineered and Molecular Systems, 2006, pp. 607–610.
[24] A.M. Ionescu, Nano-electro-mechanical-systems: from ideas to reality check, in: New Developments in Micro Electro Mechanical Systems for Radio Frequency and Millimeter Wave Applications, 2009, pp. 89–96.
[25] B. Bhushan, MEMS/NEMS and BioMEMS/BioNEMS: materials, devices, and biomimetics, in: B. - Bhushan (Ed.), Nanotribology and Nanomechanics II, Springer, Berlin, Heidelberg, 2011, pp. 833–945.
[26] H. Shokry Hassan, A.B. Kashyout, H.M.A. Soliman, M.A. Uosif, N. Afify, Effect of reaction time and Sb doping ratios on the architecturing of ZnO nanomaterials for gas sensor applications, Appl. Surf. Sci. 277 (2013) 73–82.
[27] E.Y. Song, K. Lee, Understanding IEEE 1451—networked smart transducer interface standard, IEEE Instrumen. Meas. Mag. (2008) 11–17.
[28] H.S. Hassan, M.F. Elkady, Semiconductor nanomaterials for gas sensor applications, Environ. Nanotechnol. 3 (2020) 305–355.
[29] R. Abdelghani, H. Shokry Hassan, I. Morsi, A.B. Kashyout, Nano-architecture of highly sensitive SnO_2–based gas sensors for acetone and ammonia using molecular imprinting technique, Sensors Actuators B Chem. 297 (2019) 126668.
[30] Defense Base Forecast, Long Ride of Growth Predicted for Sensor Market Worldwide, 73 National Defense, 1993, 4.
[31] H. Shokry Hassan, A.B. Kashyout, I. Morsi, A.A.A. Nasser, H. Abuklill, Development of polypyrrole coated copper nanowires for gas sensor application, Sensing Bio-Sensing Res. 5 (2015) 50–54.
[32] L.M. Shepard, Automotive sensors improve driving performance, Ceram. Bull. 71 (6) (1992) 905–913.
[33] G.F. ElFawal, H.S. Hassan, M.R. El-Aassar, M.F. Elkady, Electrospun polyvinyl alcohol nanofibers containing titanium dioxide for gas sensor applications, Arab. J. Sci. Eng. 44 (2019) 251–257.
[34] G. Zorpette, Sensing climate change, IEEE Spectru. 30 (7) (1993) 20–27.
[35] A.C. Peixoto, A.F. Silva, Smart devices: micro- and nanosensors, in: Bioinspired Materials for Medical Applications, 2017, pp. 297–329.
[36] H. Shokry Hassan, A.B. Kashyout, I. Morsi, A.A.A. Nasser, I. Ali, Synthesis, characterization and fabrication of gas sensor devices using ZnO and ZnO:In nanomaterials, Beni-Suef Univ. J. Basic Appl. Sci. 3 (2014) 216–221.
[37] P.D. Hung, S. Bonnet, R. Guillemaud, E. Castelli, P.T.N. Yen, Estimation of respiratory waveform using an accelerometer, in: 5th IEEE Int Symp Biomed Imaging from Nano to Macro, Proceedings, ISBI, 2008, pp. 1493–1496.
[38] S. Singh, D. Galar, D. Baglee, S. Björling, Self-maintenance techniques: a smart approach towards self-maintenance system, Int. J. Syst. Assur. Eng. Manag. 5 (2014) 75–83.
[39] S.J. Prosser, E.D.D. Schmidt, Smart sensors for industrial applications, Sens. Rev. 17 (1997) 217–222.
[40] G.B. Gebremeske, C. Yi, C. Wang, Z. He, Self-maintenance techniques: a smart approach towards self-maintenance system, Int. J. Syst. Assur. Eng. Manag. 115 (2015) 1151–1178.
[41] J. Fraden, Handbook of Modern Sensors, fourth ed., Springer, New York, NY, 2010.
[42] J. Chou, Electrochemical sensors, in: Hazardous Gas Monitors: A Practical Guide to Selection, Operation and Applications, First ed, McGraw-Hill, New York, NY, 1999.
[43] C.M. Brotherton, D.R. Wheeler, Development of Chemiresponsive Sensors for Detection of Common Homemade Explosives, Sandia National Laboratories, USA, 2012.

[44] M. Guenther, G. Gerlach, T. Wallmersperger, M.N. Avula, S.H. Cho, X. Xie, et al., Smart hydrogel-based biochemical microsensor array for medical diagnostics, Adv. Sci. Technol. 85 (2015) 47–52.

[45] N. Lazarus, G.K. Fedder, Integrated vertical parallel-plate capacitive humidity sensor, J. Micromech. Microeng. 21 (6) (2011) 065028.

[46] R.W. Murray, R.E. Dessy, W.R. Heineman, J. Janata, W.R. Seitz, Chemical sensors and micro instrumentation, in: Proceedings of Symposium. American Chemical Society National Meeting held in Los Angeles, California, ACS Symposium Services, Washington, D.C.: American Chemical Society, 1988, p. 403.

[47] E. Betzig, J.K. Trautman, T.D. Harris, J.S. Weiner, R.L. Kostelak, Breaking the diffraction barrier: optical microscopy on a nanometric scale, Science 251 (1991) 1468–1470.

[48] K. Seiler, D.J. Harrison, A. Manz, Planar glass chips for capillary electrophoresis: repetitive sample injection, quantitation, and separation efficiency, Anal. Chem. 65 (1993) 1481–1488.

[49] M.S. Belluzo, M.É. Ribone, C.M. Lagier, Assembling amperometric biosensors for clinical diagnostics, Sensors 8 (3) (2007) 1366–1399.

[50] J. Wilson, Sensor Technology Handbook, Newnes, Burlington, MA, 2005.

[51] B.T. Katzen, A.A. MacLean, Complications of endovascular repair of abdominal aortic aneurysms: a review, Cardiovasc. Intervent. Radiol. 29 (2006) 935–946.

[52] K.W. Johnston, R.B. Rutherford, M.D. Tilson, D.M. Shah, L. Hollier, J.C. Stanley, Suggested Standards for Reporting on Arterial Aneurysms, Subcommittee on Reporting Standards for Arterial Aneurysms, Ad Hoc Committee on Reporting Standards, Society for Vascular Surgery and North American Chapter, International Society for Cardiovascular, 1991.

[53] L. Tammaro, V. Vittoria, V. Bugatti, Dispersion of modified layered double hydroxides in poly(ethylene terephthalate) by high energy ball milling for food packaging applications, Eur. Polym. J. 52 (1) (2014) 172–180.

[54] J. Gómez-Estaca, C. López-de-Dicastillo, P. Hernández-Muñoz, R. Catalá, R. Gavara, Advances in antioxidant active food packaging, Trends Food Sci. Technol. 35 (1) (2014) 42–51.

[55] D. Restuccia, U.G. Spizzirri, O.I. Parisi, et al., New EU regulation aspects and global market of active and intelligent packaging for food industry applications, Food Control 21 (11) (2010) 1425–1435.

[56] S.D.F. Mihindukulasuriya, L.T. Lim, Nanotechnology development in food packaging: a review, Trends Food Sci. Technol. 40 (2) (2014) 149–167.

[57] K. Ramachandraiah, S.G. Han, K.B. Chin, Nanotechnology in meat processing and packaging: potential applications—a review, Asian Australas. J. Anim. Sci. 28 (2) (2015) 290–302.

[58] M. Vanderroost, P. Ragaert, F. Devlieghere, B. De Meulenaer, Intelligent food packaging: the next generation, Trends Food Sci. Technol. 39 (1) (2014) 47–62.

[59] C. Silvestre, D. Duraccio, S. Cimmino, Food packaging based on polymer nanomaterials, Prog. Polym. Sci. 36 (12) (2011) 1766–1782.

[60] E.D. Armas, R.T.R. Monteiro, P.M. Antunes, M.A.P.F. Santos, P.B. Camargo, Uso de agrotóxicos em cana-de açúcar na bacia do rio Corumbataí e o risco de poluição hídrica, Quim Nova 30 (2007) 1119.

[61] R. Vander Oost, J. Beyer, N.P.E. Vermeulen, Fish bioaccumulation and biomarkers in environmental risk assessment: a review, Environ. Toxicol. Pharmacol. 13 (2003) 57–149.

[62] U.S. Environment Protection Agency, Air and Radiation, http://www.epa.gov/air/criteria.html/, 2019. Accessed 9 November 2019.

[63] H. Shokry Hassan, A.B. Kashyout, I. Morsi, A.A.A. Nasser, A. Raafat, Fabrication and characterization of gas sensor micro-arrays, Sensing Bio-Sensing Res. 1 (2014) 34–40.

[64] C.M. Kyung, Handbook of Smart Sensors for Health and Environment Monitoring, Springer, 2015.

[65] M. Adel Abozeid, H. Shokry Hassan, I. Morsi, A.B. Kashyout, Development of Nano-WO_3 doped with NiO for wireless gas sensors, Arab. J. Sci. Eng. 44 (2019) 647–654.

[66] H.S. Hassan, A.B. Kashyout, I. Morsi, A.A.A. Nasser, A. Raafat, Fabrication and characterization of nano-gas sensor arrays, AIP Conf. Proc. 1653 (2015) 020042.

[67] M.F. Elkady, H.S. Hassan, Invention of hollow zirconium tungsto-vanadate at nanotube morphological structure for radionuclides and heavy metal pollutants decontamination from aqueous solutions, Nanoscale Res. Lett. 10 (2015) 474.

[68] J. Kasperkecic, Demand for Smart Sensors is on the Rise, 2013.

[69] S. Smith, D.J. Nagel, Nanotechnology-enabled sensors: possibilities, realities and applications, in: Sensor Technology Handbook, Elsevier, 2003.

[70] Y. Cui, Q. Wei, H. Park, C.M. Lieber, Nanowire nano sensors for highly sensitive and selective detection of biological and chemical species, Science 293 (5533) (2001) 1289–1292.

[71] J.A. Cassell, DoD grants $3M to study nanoshells for early detection, treatment of breast cancer, Nano-Biotech News 1 (3) (2003).

[72] H. Pei, L. Liang, G. Yao, J. Li, Q. Huang, C. Fan, Reconfigurable three-dimensional DNA nanostructures for the construction of intracellular logic sensors, Angew. Chem. Int. Ed. 51 (36) (2012) 9020–9024.

[73] Smart Sensors, A Global Strategic Business Report, Global Industry Analysts, 2012.

Intelligent nano sensors (INS)—Electronics applications

CHAPTER 16

Phosphorene-based intelligent nanosensor for wearable electronics applications

R. Ramesh[a], Arkaprava Bhattacharyya[a], Adhithan Pon[a], D. Nirmal[b], and J. Ajayan[c]
[a]SASTRA Deemed University, Thanjavur, India
[b]Karunya Institute of Technology and Sciences, Coimbatore, India
[c]SR University, Warangal, India

1. Introduction

Recently, many researchers are attracted to biosensor research as it helps to improve the quality of life. Among many types of biosensors, field-effect transistor (FET)-based biosensors have drawn much attention due to its attractive features such as label-free detection, hypersensitive detection, mass production, and cost-effective manufacturing [1]. The influential importance to identify ions and molecules for laboratory tests have also motivated the exploration for more FET-based sensors. FET-based biosensors work on the principle of change in electrical property [2] (i.e., conductance, current, or threshold voltage) due to the presence of charged molecules between the ionic solution and the gate dielectric. The metal-oxide-semiconductor FET (MOSFET) architecture is modified by introducing a cavity in the dielectric material for the biomolecules to immobilize inside. This alters the effective coupling between the gate and the channel and forms the working principle of a dielectric-modulated FET (DM-FET) [3, 4]. The presence, absence, or properties of a biomolecule changes the electrical parameters of the FET device [5]. This change is used to measure the sensitivity and label-free detection purposes of both neutral (biotin-streptavidin) and charged biomolecules (DNA). To overcome the challenges of conventional complementary metal-oxide-semiconductor (CMOS) biosensor, such as power supply scaling and short channel effects, new device architectures are to be analyzed. In this view, a dielectric-modulated tunnel FET (DM-TFET)-based biosensor exhibits a higher sensitivity. The concept of dielectric modulation (DM) in TFET has already been implemented and demonstrated that it reduces the ambipolar conduction and increases the gate modulation near the source end [6, 7].

Layered materials are called "Materials within Materials." They provide a means to create a new material that has a series of submaterials (or *layers*). The proper definition for *two-dimensional (2D) materials*, also called *single-layered materials*, abide by a single layer of atoms [8]. The first 2D material, Graphene (i.e., a single layer of graphite), was separated from graphite in 2004 [9]. After that, other 2D materials such as MoS_2,

Handbook of Nanomaterials for Sensing Applications
https://doi.org/10.1016/B978-0-12-820783-3.00012-9

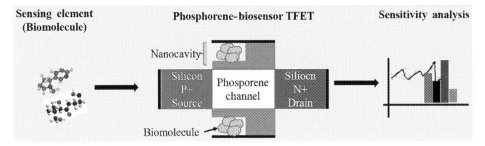

Fig. 1 Graphical abstract for Phosphorene TFET biosensor.

Germanene, Phosphorene, Silicene, etc. were identified. Biosensors made of 2D materials such as Graphene [10] and MoS_2 [11] have been already reported. There has been very less work reported so far using Phosphorene as material for biosensing applications. Chen et al. described a FET–based black phosphorus (BP) nanosheet biosensor [12]. Also, a few-layer Phosphorene-based colorimetric biosensor to detect carcinoembryonic antigen (CEA) in colon and breast cancer has also been reported for the first time [13]. In the recent past, a biochemical sensor made up of few-layer BP and Graphene/TMDCs has been designed and reported. A surface plasmon resonance (SPR) biochemical sensor made with a combination of BP and bilayer WSe_2 works on the refractive index principle and offers high sensitivity [14]. The coating of BP with other 2D materials protects BP from atmospheric oxidation and also improves the sensitivity of the device. This type of biosensor will be able to detect analytes of different refractive indices with 2.4 times more sensitivity than the conventional biochemical sensors [15].

In this chapter, we present the application of Phosphorene-based tunnel field-effect transistor (TFET) as intelligent nanosensors in wearable electronics (Fig. 1).

2. Biosensor

Biosensors are analytical devices that detect the presence and concentration of a biological analyte such as a biomolecule, a biological structure, or a microorganism as an electrical signal. It consists of three elements as shown in Fig. 2 (i) a device that recognizes the analyte and produces a signal, (ii) a signal transducer, and (iii) a reader device. The biosensors are classified as enzyme-based, tissue-based, immunosensors, DNA biosensors, thermal, and piezoelectric biosensors [16]. The biosensors find application in numerous fields. Blood glucose biosensor is the most common commercial biosensor that is used to measure blood glucose levels [17]. In the food industry, it is used to check the quality and safety of food products; in fermentation industry and saccharification process to detect accurate glucose concentrations; and in metabolic engineering to enable in vivo monitoring of cellular metabolism [18]. They are also used to detect analytes such as a cytokine, human papillomavirus, etc. Fluorescent-type biosensors play a vital role in drug discovery

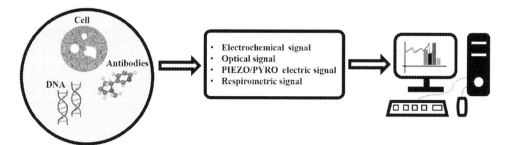

Fig. 2 Signal flow diagram for biosensing.

and in cancer detection [19, 20]. Other applications of biosensors are involved in plant biology [20], defense, the clinical sector, and for marine applications [20–22]. DNA-based biosensors have been used as molecular diagnostic tools for clinical assessment [23, 24]. Additionally, biosensors for the detection of ozone layer depletion and metastasis [25, 26] have also been reported. Biosensors must be highly specific, independent of physical parameters, and should be reusable [27].

Enzyme-based biosensors have been constructed on adsorption of enzymes by Van der Waals forces, ionic bonding, or covalent bonding [28]. The commonly used enzymes are oxidoreductases, polyphenol oxidases, peroxidases, and amino oxidases.

Divies materialized the first microbe-based or cell-based sensor [23]. An inhibitor or substrate was used for the sensor. To determine the amino acid arginine, Rechnitz developed the first tissue-based sensor [29]. Organelle-based sensors were made using membranes, chloroplasts, mitochondria, and microsomes. This type of biosensor offers high stability but with reduced selectivity and extensive detection time. Immunosensors work due to a high affinity of antibodies toward their respective antigens. The DNA biosensors were constructed on the fact that single-strand nucleic acid molecule recognizes and binds to its complementary strand in a sample due to the development of hydrogen bonds between the two nucleic acid strands.

Another type of biosensor detects magnetic micro- and nanoparticles in microfluidic channels using the magnetoresistance effect, which offers a high level of sensitivity and reduced size. Thermometric biosensors use the fundamental property of biological reactions that involves absorption or evolution of heat. This produces a change in the temperature within the reaction medium [30].

Piezoelectric biosensors work on the measurement of changes in the resonance frequency of a piezoelectric crystal due to mass changes on the crystal structure. These type of biosensors record the affinity interactions without the need for any specific reagents [31]. Optical biosensors are the most common types of biosensors. Optical detection is performed by accomplishing the interaction of the optical signal with a biological element. They are characterized by high selectivity, sensitivity, small size, and low cost compared with conventional techniques. Optical biosensing can be broadly classified as label-free and label-based [32]. In a label-free mode, the signal is generated due to the interaction of the material to be analyzed and transducer. Alternately, label-based sensing is disturbed with the use of a label and the optical signal is then developed. The main aim of using an optical biosensor is to generate a signal equivalent to the concentration of the substance to analyzed. This type of sensor has the capability to study various types of biological substances such as enzymes, antibodies, antigens, receptors, nucleic acids, whole cells, and tissues. Optical sensing techniques such as SPR, evanescent wave fluorescence, and optical waveguide interferometry [33] make use of the evanescent field in the near vicinity to the biosensor surface to determine the synergy of the biorecognition element with the analyte.

The electrochemical transducer-based biosensor makes use of variations either in current, potential, conductivity, or capacitance in the testing sample after the interaction between the biological element and analyte. In this category, the biosensors might be amperometric, potentiometric, conductometric [34], or impedimetric biosensors [35]. If the interaction between the biocatalyst and the analyte changes the optical properties, the biosensors are grouped as bioluminescence/chemiluminescence biosensor, fluorescence biosensor, or colorimetric biosensor [36]. Other than this division, they could be piezoelectric or gravimetric biosensors [37], which detect the change in resonating frequency of the piezoelectric material depending upon the adsorption or desorption of molecules from its surface. The change in resonating frequency generates a current production that is measured. The pyroelectric biosensor works on the principle of pyroelectricity, i.e., certain properties of materials produce current with a change in temperature [38]. Baroxymeter is another type of biosensor that produces a signal based on the variation in microbial respiration to determine water toxicity [39].

FET–based biosensors are an exclusive type of biosensor that produces a change in conductance of FET due to biological reaction. The next section discusses the basic concepts of FET (MOSFET and TFET) and its application as a biosensor.

3. FET biosensor

The application of FET as a biosensor, particularly using MOSFET, was reported in 1986 and again discussed in 1996 to study enzymes as a short communication by Bergveld, and again in 1996 as a proof-of-concept study of enzyme FETs [40, 41]. Caras et al. reported the first use of a FET-based biosensor to detect penicillin that involves the study of pH-based enzyme. After that, several works have been reported for FET-based biosensor [5, 41–44]. Due to the excellent electrical characteristics and small size, FET-based biosensors are used in clinical diagnosis. The metal gate in the FET device is replaced by a biofilm layer material such as a receptor, enzyme, antibody, or DNA. This modification in the gate structure modulates the channel conductivity of the device and produces a change in the drain current. During the past three decades, many efforts have been made to develop a better FET architecture with improved performance.

3.1 FET basics

FETs are a recent development, and they operate on a greatly different mechanism to realize signal amplification. FETs are voltage-controlled devices compared with the bipolar transistor (BJT), as they are current-controlled devices. Also, FETs are termed as unipolar devices as the conduction is due to the movement of either type of carrier (electron or holes) [45].

All FETs have three terminals, called the source (S), the drain (D), and the gate (G). There is no physical contact between source and drain, but a current path called the conduction channel is formed between the source and the drain. The gate-to-source voltage (V_{gs}) will turn on/off the device. Because of this mechanism, a FET can function as an on/off switch. Due to the application of gate voltage (V_{gs}), an electric field is produced in the device that influences the conduction of carriers from source to drain. The current flow is determined by the actual motion of the carriers, electrons for the n-type FET, or the holes for the p-type FET. For an n-type FET, the application of positive voltage to the gate creates a channel and allow the electrons to pass through the channel from the source to the drain. In contrast, if a negative gate voltage is applied, the n-type channel will pinch off and there will not be conduction of carriers. For a p-type FET, the opposite occurs, positive (negative) gate voltage will turn off (on) the transistor device.

The amount of change in the electric field that feels vertically downward to the electron motion depends on the applied gate voltage (V_{gs}). When the applied gate voltage (V_{gs}) reaches the threshold voltage (V_{th}), a channel is formed between source and drain; it increases the number of carriers, and conductivity increases with the gate voltage. The threshold voltage is defined as the value of V_{gs} that is needed to turn on the device. When V_{gs} is equal or higher than V_{gs} ($V_{gs} \geq V_{th}$), the n-type FET starts to turn on the device and the drain current (I_{ds}) starts flowing. For $V_{gs} < V_{th}$, there is no channel formed and zero current flows from drain to source. But in the case of p-type FET, V_{gs} should be lower than V_{th} ($V_{gs} < V_{th}$) to create a p-type channel. The drain current is given by Eq. (1):

$$I_D = \mu\, C_{ox} \left(\frac{W}{L} \right) \left[\left(V_{gs} - V_{th} \right) V_{ds} - V_{ds}^2 \right] \tag{1}$$

where μ is the carrier mobility, C_{ox} is the gate capacitance, L is gate length, W is gate width, V_{th} is threshold voltage, V_{gs} is applied gate bias and drain bias (V_{ds}).

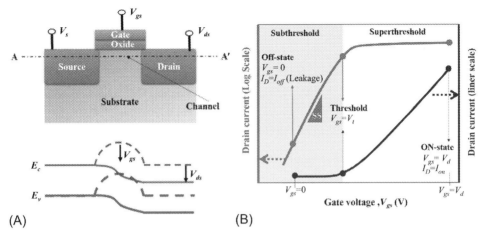

Fig. 3 Schematic diagram of MOSFET and its $I_d - V_g$ characteristics.

3.2 MOSFET

The MOSFET is a primary switch and widely used in CMOS technology, which has both n-type and p-type devices. This MOSFET consists of mainly three terminals; they are source, drain, and gate. The primary working mechanism utilizes an electric field (due to applied gate voltage (V_{gs})) and modulates the drain current (I_d). When the gate voltage is zero ($V_{gs} = 0$ V), the channel between source to drain is not created due to an increase in the barrier height of the channel region (dotted lines) as shown in the band diagram (Fig. 3A). This is called off-state, but a small amount current will flow called the leakage current (I_{off}). When the applied gate voltage (V_{gs}) is above the threshold (V_{th}), the MOSFET is in the conduction state similar to the normal FET. In the band diagram, a reduction in the barrier height is observed (solid lines) when the applied voltage is increased. Fig. 3A clearly mentions the three terminals of a MOSFET with a band diagram and its drain current characteristics (Fig. 3B) labelling the various states of operation.

The performance metrics for MOSFET are the drain current (I_d), subthreshold current (I_{off}), transconductance (g_m), output conductance (g_0), I_{on}/I_{off} ratio, and subthreshold swing (SS), and these formulas are tabulated in Table 1.

MOSFET also finds application as a biosensor and can be used to detect biomolecules, DNA, etc. [42]. Out of many methods used for biomolecule detection using MOSFET, DM can be used [4]. This method offers more sensitivity, low power consumption, and fabrication adaptability. In the device structure, a nanocavity is formed at the oxide region and the biomolecules are placed in the nanocavity. The presence of biomolecules changes the gate capacitance of the device and changes the drain current. The change in drain current values is used to detect the presence and type of biomolecule. It has been proven that the sensing mechanism of the MOSFET biosensor can be improved by changing the device structure and using suitable materials [3, 6, 7, 15, 46].

Table 1 Performance metrics for MOSFET.

S.no	Parameter	Symbol	Formula	Desirable value
1	Drain current	I_{ds}, I_{on}	Maximum current from the characteristics curve	Maximum
2	Leakage current	I_{off}	Current at $V_{gs} = 0$ from the characteristics curve	Minimum
3	Transconductance	g_m	$G_m = \dfrac{d(I_d)}{d(V_{gs})}$	Maximum
4	Output conductance	g_0	$G_0 = \dfrac{d(I_d)}{d(V_{ds})}$	Maximum
5	I_{on}/I_{off} ratio	I_{on}/I_{off}	$\dfrac{I_d}{I_{off}}$	Maximum
6	Subthreshold swing	SS	$SS = d\dfrac{V_{gs}}{d(\text{Log } I_d)}$	Minimum

The sensitivity of biosensor can be maximized by using proper binding material in the nanocavity, which also reduces the biomolecule distribution dependency for improving the sensitivity. The biosensor figure of merits are (i) sensitivity (Eqs. 2 and 3) and (ii) selectivity (Eq. 4), and they are calculated as follows:

(i) Sensitivity

It is the ratio of drain current values in the presence and absence of a biomolecule [3, 47].

$$\text{Sensitivity} = \frac{I_d^{Bio} - I_d^{air}}{I_d^{Bio}} \tag{2}$$

$$\text{Sensitivity} = \frac{I_d^{Bio} - I_d^{air}}{I_d^{air}} \tag{3}$$

where, I_d^{Bio}, drain current with presence of biomolecule and I_d^{air}, drain current without or absence of biomolecule.

(ii) Selectivity

It is the ratio of transconductance and drain current value in the presence of a biomolecule and unit of selectivity is $[V^{-1}]$ [6].

$$\text{Selectivity} = \frac{G_m}{I_d} \tag{4}$$

where, G_m, transconductance and I_d, drain current.

The DM-MOSFET biosensor possesses excellent potential to act as a label-free biosensor. However, it suffers short-channel effect (SCE) and high subthreshold swing (SS) (>60 mV/decade) value. This high SS value of MOSFET is a major hindrance in improving the sensitivity as the response time (time required to detect biomolecules) increases for high SS values. In this context, TFET was proposed as a biosensor

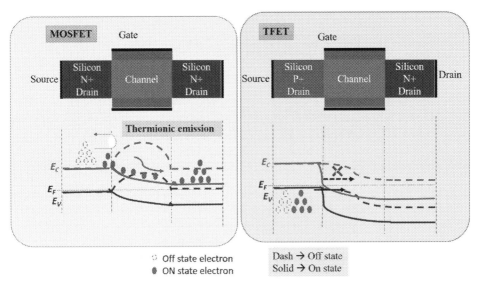

Fig. 4 Principle of operation of dual-gate MOSFET and TFET with schematic energy band profile.

[43, 44, 48], as it offers less SS values compared with MOSFET due to its inherent tunneling process thus increasing the sensitivity with the reduced response time.

3.3 TFET

TFET is also a three-terminal device similar to MOSFET, but the source and drain regions are made up of p-type and n-type semiconductors or vice-versa as shown in Fig. 4. The working mechanism of TFET is based on tunneling mechanism at the source-channel junction [49]. TFET has low SS due to band-to-band tunneling (BTBT) that provides less response time and is also suitable for low-power applications due to its reduced supply voltage (V_{dd}). The energy profile for TFET for both off-state and on-state are shown in Fig. 4.

In MOSFET, the drain current (I_d) is produced due to thermionic emission, and the SS value is set to a fundamental limit of 60 mV/dec. The electrons at the source side retain the same position because it does not have sufficient energy to cross over the energy barrier at the off-state (dash line). After the gate voltage (V_{gs}) is applied (on-state), the energy band is bent toward the downside, so the electron has sufficient energy and flows from source to drain side.

In contrast to MOSFET, TFET is a tunneling-based device, which means charge carriers tunnel from one energy band to another. In off-state, the band bending is less (dotted line), so the carriers do not tunnel from source to channel region. When the gate voltage (V_{gs}) is applied, the band bending is more pronounced (solid line) and the carriers can easily tunnel from source to the channel region. This tunneling phenomenon is reflected in the device characteristics and reduces the SS values and leakage current.

4. TFET biosensor

TFETs have been considered for biosensing applications due to certain advantages such as low SS (< 60 mV/dec), scalability, low power dissipation, and compatibility with existing CMOS technology [3, 4, 6, 7, 18, 43, 44, 47, 48, 50]. Also, its sensing capability is higher in comparison to MOSFET-based biosensors, as it is not affected by short channel effects when the device dimensions are reduced. Narang et al. [18] reported for the first time the concept of dielectric modulated DM–TFET for sensing biomolecule using TCAD simulation software [51]. To mimic the biomolecule property, the dielectric constant and charge of each biomolecule are included in the simulation [5, 52, 53].

A nanocavity is created in the gate oxide region at the top of the source/channel (S/C) region using appropriate binding material (as shown in Fig. 5). In a TFET, the high electric fields exist only at the junction, so this nanocavity is located either at S/C or channel/drain (C/D) region. The nanocavity thickness and the position/distribution of biomolecule play a vital role in biosensing. To mimic the biomolecules, we use different dielectric constant values as suggested by Kim et al. [5], and it is an accurate method to measure the sensing capability of FET architecture. The various types of biomolecules and their corresponding dielectric values are shown in Table 2.

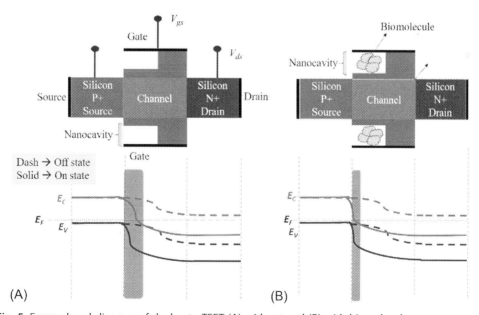

Fig. 5 Energy band diagram of dual gate TFET (A) without and (B) with biomolecule.

Table 2 Biomolecule dielectric values.

S.no	Biomolecule	Dielectric value
1.	Streptavidin [46, 53–55]	2.1
2.	Biotin	2.63
3.	3-Aminopropyltriethoxysilane (APTES)	3.57
4.	Low-hydrated protein powders	5
5.	DNA	8–12 [23, 25]

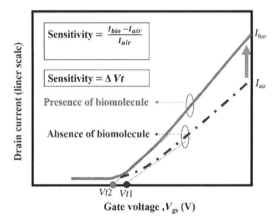

Fig. 6 $I_d - V_{gs}$ Characteristics of TFET biosensor with the presence and absence of biomolecule.

Fig. 5 shows the TFET energy band in the presence and absence of biomolecules. It is found that the tunneling current is increased due to the presence of biomolecule in the cavity as shown in Fig. 6. It creates more band bending (rectangular region; light gray in print version) compared with the absence of biomolecules (rectangular region; light gray in print version). It is found that the tunneling probability is high when the cavity is filled with a biomolecule.

The advantages of TFET such as fast response time and reduced SS values makes it suitable for biosensing applications compared with MOSFET biosensor. However, it has some disadvantages such as ambipolarity [56, 57] and low on–current [58, 59] that reduces the sensitivity of the device. To enhance the sensitivity, many methods such as source engineering, pocket doping, drain engineering, etc. are proposed [60–63]. TFET made up of materials such as germanium, GaAs, InAs, and lower bandgap are also proposed to increase the on–current of the device [64]. TFETs fabricated using 2D materials such as Graphene, MoS_2, and Phosphorene offers higher mobility, lower bandgap, and lighter effective mass, which increases the overall device performance [65]. In the next section, biosensors designed using 2D materials are discussed.

5. 2D material-based FET biosensor

5.1 2D layered materials

The need for new semiconductors material for ultra-scaled down process laid the path for developing 2D material fabrication. They are called layered materials as they have a layer of atoms. The first 2D material, Graphene, a single layer of graphite, was isolated in 2004 [66]. After successful preparation and performance of Graphene, many research groups again rediscovered the material allotropes. After that, many other 2D materials such as Germanene, Silicene, MoS_2, and Hexagonal Boron Nitrate were identified.

There are many methods to fabricate 2D materials, and some of them are shown in Fig. 7. Micromechanical Cleavage (MC) using adhesive tape to cleave thin layers from bulk material [67]. In this method, high-quality 2D materials are extracted, but it requires a skilled person and is difficult to produce a large area of the 2D material. At the same time, effective computational methods were also developed, and electrical band structure was theoretically analyzed to study the characteristics of 2D materials. The computational methods are highly sophisticated and consume less computational time.

For characterization of 2D material, transmission–electron microscopy, atomic force microscopy, Raman scattering spectroscopy, microphotoluminescence, and X-ray photoelectron spectroscopy are used. The properties of widely used 2D materials are listed in Table 3.

5.2 Fabrication steps for 2D material biosensor

The fabrication of 2D material FET biosensor follows the CMOS fabrication flow as shown in Fig. 8. In this method, first, the bulk silicon is oxidized to form SiO_2 on its upper side with required thickness. The SOI wafer thickness is 400 nm, and silicon is heavily doped because it sometimes acts as a back gate. The second step involves the

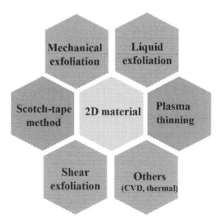

Fig. 7 Fabrication types of 2D material.

Table 3 Widely used 2D materials and 2D material properties.

Properties	Material parameters	Phosphorene	Graphene	MoS$_2$
Electrical	Bandgap (eV)	0.5–1.45	0	1.2–1.8
	Effective mass (electron)	0.19	0.01	0.6
	Affinity (eV)	3.8–4.1	–	4.05
	Dielectric constant	Layer dependent	Layer dependent	Layer dependent
	Thermal conductance (W/m K)	36–110	2000–5000	52
Mechanical	Critical strain	27%–30%	19%–34%	19.5%–36%
	Young's modulus	44 Gpa	1 TPa	170–370 GPa
	Passion's ration	0.4	0.186	0.21–0.27
General	Flexibility	Yes	Yes	Yes
	Stretch ability	Yes	Yes	Yes
	Adaptability	Yes	Yes	Yes
	Preparation	Cleavage with tape Exfoliation Plasma-assisted fabrication	Cleavage with tape Exfoliation Plasma-assisted fabrication CVD Hydrothermal synthesis	Cleavage with tape Exfoliation Plasma-assisted fabrication CVD hydrothermal synthesis

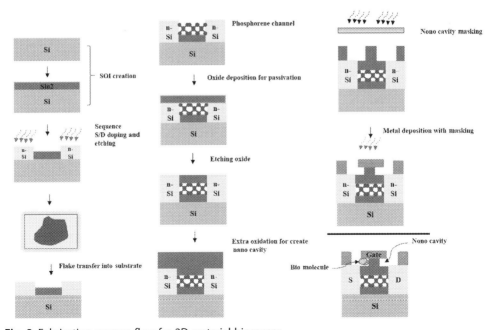

Fig. 8 Fabrication process flow for 2D material biosensor.

creation of source and drain regions using proper masking and ion implantation/diffusion process with a concentration of $1 \times 10^{20}/cm^3$. In the third step, Phosphorene flake is separated using mechanical or liquid exploitation methods. The separated flake is then transferred to the SOI wafer and forms the channel of the device. The fourth step consists of deposition of silicon oxide for passivation and etching. At this stage, oxidation at the top of the entire device is done to create a nanocavity by masking. After masking, metallization is performed for external contacts as shown in Fig. 8.

5.3 Graphene FET biosensor

Graphene is one of the allotropes of Carbon. In this form, 2D hexagonal Carbon atoms are arranged in a single layer, and the lattice parameters are shown in Fig. 9. Graphene possesses good mechanical, optical, and electrical properties. Here we explore the electrical properties of Graphene in terms of mobility, bandgap, and effective mass. Electrons in Graphene act like Dirac fermions, i.e., the electronic energy dispersion is linear near the Brillion zone corner. The velocity of this Dirac fermion is nearly 1/300 times the speed of light. So it can be used in FET, and its performance is better than silicon FET. Graphene FET offers less short channel effects, reduces mobility degradation at higher bias, and has high performance.

For electronic logic application, the desirable values of materials used in FET fabrication are bandgap = 0.4 eV, mobility between 500 and $10,000/cm^2/VS$, and carrier effective mass is $(m_e) < 0.1$. Graphene has all the desirable values but has zero bandgap. Therefore, it is not suitable for digital application, however it can be used for analog/RF application. It is reported that Graphene nanoribbon produces a bandgap inversely proposal to its width. So, Graphene nanoribbon–based, FET-based biosensors can be fabricated with improved performance compared with silicon FET biosensors.

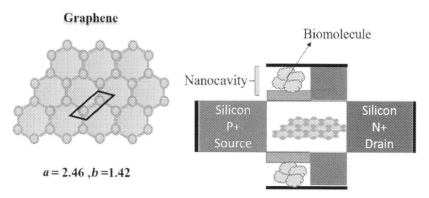

Fig. 9 Graphene layer schematic with lattice parameter and Graphene biosensor schematic.

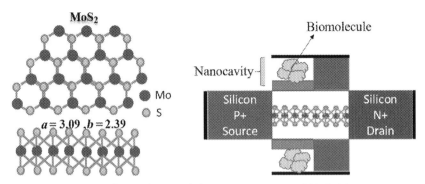

Fig. 10 Schematic diagram of MoS$_2$ layer with lattice parameter and MoS$_2$ biosensor.

5.4 MoS$_2$ FET biosensor

An alternative 2D material is transition metal dichalcogenides (TMDC) that have natural bandgap. The bulk MoS$_2$ is an inorganic compound composed of molybdenum (M) and sulfur (S) atoms are connected with covalent bonds. MoS$_2$ in bulk form is an indirect bandgap semiconductor like silicon and germanium. Layers are exfoliated from bulk MoS$_2$ using methods explained in Section 5.1. It possesses hexagonal lattice, but the lattice atoms are arranged out of the plane (Fig. 10). A single-layer MoS$_2$ consists of one molybdenum and two sulfur atoms. Similar to Graphene, it also possesses good mechanical (lubricant nature, elastic), optical, and electrical properties. So, the monolayer MoS$_2$ can be used as a FET for low-power digital applications. It can be also be used for memory devices, optoelectronic devices due to desirable values of bandgap (1.8 eV), mobility (200–300/cm^2/V/S), and carrier effective mass ($<$0.2 m$_0$) [11].

MoS$_2$-based FET devices show good device performance with 10^8 on/off ratio and SS $\sim$60 mV/dec, more preferable for logic switching application. Due to the previously mentioned advantages, many works are reported that use MoS$_2$ FET as a biosensor [11].

5.5 Phosphorene FET biosensor

The new arrival in the 2D material family is Phosphorene, discovered in 2014 [68–71], which has semiconducting properties [72]. Bulk Black Phosphorous (BP) is a layered material with Van der Waals interactions between the layers of phosphorus atoms that forms a puckered orthorhombic crystal lattice [69]. To synthesize Phosphorene, mechanical and liquid exfoliation methods are used. In mechanical exfoliation method, adhesive tape is used to fabricate single- to few-layered BP. It is then cleaned using different organic solvents. Liquid exfoliation may be accomplished using organic solvents, mixed solvents, and water.

Analogous to Graphene and MoS$_2$, it also possesses good electrical, optical, and mechanical properties. Notably, the electrical properties of Phosphorene have a good trade-off between mobility and bandgap; hence it is more suitable for digital and analog/ RF applications. One of the most attractive properties of multilayered BP is its high carrier mobility (i.e., 600–1000 cm^2/V s at room temperature), which makes Phosphorene a potential candidate for future nanoelectronic applications [65]. This high carrier mobility depends on the thickness of BP layer and increases with decreasing the thickness of multilayered BP [69]. The mono- and multilayer BP possess direct and tunable bandgap (0.3–1.45 eV), making it suitable for applications in photoelectrical conversion [73]. This layer-dependent bandgap property of BP varies the carrier mobility and is suitable for electronic and photonic applications. In Fig. 11A, X and Y direction represents the lightweight mass direction (armchair) and heavyweight mass direction (zigzag), respectively. It is more biodegradable, metal free, and flexible, hence it is a promising candidate for biosensing application.

The band structure of monolayer and few-layer Phosphorene is shown in Fig. 11B. The bandgap is calculated using the first-principle study. The electrical characteristics of monolayer to few layers are calculated using Quantum wise Atomistic Tool kit (ATK) [74] with DFT method utilizing Meta Generalized Gradient Approximation (MGGA) and PBE functional. The obtained bandgap and effective mass values are listed in Table 4.

The bandgap value decreases as the number of layers increases. This produces Phosphorene with a tunable bandgap that ranges from 1.45 eV (Mono) to 0.57 eV (5 layers). The effective mass of armchair is less compared to the zigzag orientation direction. As a result of the properties of Phosphorene, it is used as a channel material that further improves the FET performance. Since TFET is desirable for biosensing applications as mentioned in Section 4, incorporating Phosphorene as a channel material improves its

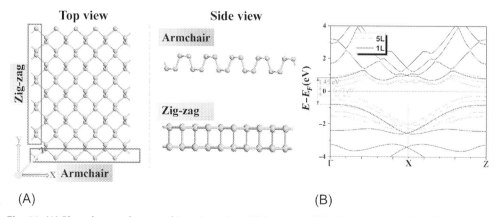

(A)   (B)

Fig. 11 (A) Phosphorene layer and its orientation, (B) bandgap (*E-K* diagram) 1L, and 5L Phosphorene.

Table 4 Electrical parameters of monolayer to few-layer Phosphorene for armchair and zig-zag directions.

Layers	Bandgap, E_g (eV)	m_{ex}/m_o	m_{hx}/m_o	m_{ey}/m_o	m_{hy}/m_o
1	1.45	0.19	0.2	1.2	7.6
2	1.05	0.2	0.26	1.23	1.42
3	0.79	0.19	0.29	1.23	3.57
4	0.65	0.17	0.31	1.22	1.22
5	0.57	0.16	0.32	1.21	3.36

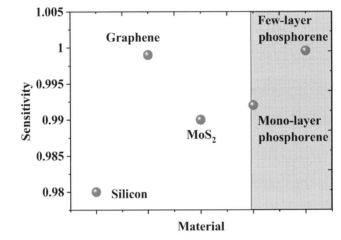

Fig. 12 Sensitivity performance of Phosphorene biosensor compared with other 2D material biosensors.

biosensing capability to a greater extent compared with Graphene/MoS$_2$ biosensors as shown in Fig. 12.

Fig. 13 shows a dual-cavity double-gate Phosphorene TFET label-free biosensor (n-type) [DCDG-phTFET] having a gate length of 40 nm. The source/drain is made up of silicon with doping concentration $1 \times 10^{20}/cm^3$ (p+) at source and (n+) at drain sides, respectively. The thickness of the oxide layer is 1 nm and made of HfO$_2$ material, which enhances the tunneling at the source to channel junction. To study the behavior of DCDG-phTFET, we have simulated its characteristics using Sentaurus TCAD by appropriately including the device working mechanism [51]. In TCAD, the tunneling phenomenon at the source side is included using nonlocal band-to-band tunneling model along with Shockley-Read Hall (SRH) model, field-depended mobility, and Fermi Dirac statistics. The channel is made of Phosphorene with mono/few layers. The Phosphorene layer properties are listed in Table 4. These values are used for device simulation and the biosensor characteristics are obtained.

The nanocavity thickness and the position/distribution of biomolecule plays a very important role in biosensing. Here, we assumed that the biomolecules are uniformly

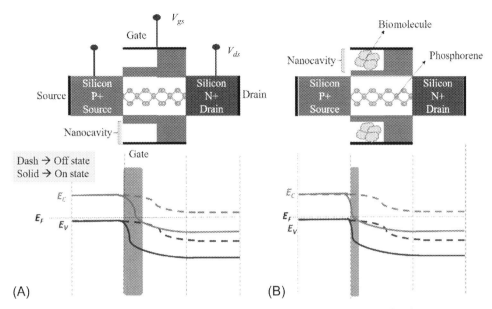

Fig. 13 Energy band diagram of DCDG-phTFET (A) without and (B) with biomolecule.

distributed in the cavity. The cavity thickness varied from 3 to 10 nm based on the size of the molecule and binding probability with the Phosphorene channel. In this device, nanocavity is located above the tunneling junction to accelerate the sensitivity, decrease the location dependency of biomolecule, and reduce the ambipolar conduction.

To mimic the biomolecule's property, we use different dielectric constant values as suggested by Kim et al. [5], and it is an accurate method to measure the sensing capability of TFET architecture.

Fig. 13 shows the DCDG-phTFET biosensor in the presence and absence of biomolecule with their corresponding energy diagram. It is found that tunneling current increases due to the presence of biomolecule in the nanocavity and creates more band bending (rectangular region; light gray in print version). The tunneling width is less compared with the width in the absence of biomolecule (rectangular region; light gray in print version). This increases the tunneling probability that increases the selectivity and sensitivity of the biosensor.

Fig. 14A shows the drain current characteristics of a mono/few-layer Phosphorene DCDG-phTFET in the absence of biomolecule (filled with air, i.e., $K = 1$). In this case, monolayer armchair DCDG-phTFET shows more I_d values, improved I_{on}/I_{off} ratio, and good SS compared with the few-layer device. The sensitivity analysis of the biosensor is obtained from the drain current characteristics of the device. The same device when filled with biomolecules ($K = 12$) (Fig. 14B) in a few-layer armchair DCDG-phTFET shows excellent performance compared with its few-layer zig-zag and monolayer armchair/zig-zag DCDG-phTFET devices due to less bandgap (0.57 eV) and less effective mass.

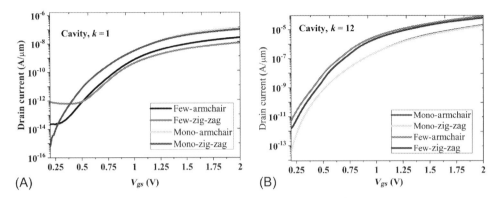

Fig. 14 The drain current variation for mono/few-layer Phosphorene for armchair/zig-zag orientation and (A) with and (B) without biomolecules.

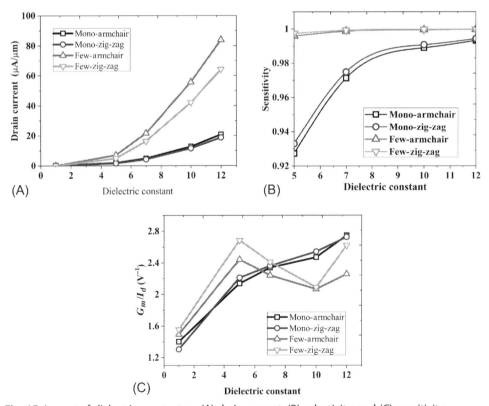

Fig. 15 Impact of dielectric constant on (A) drain current, (B) selectivity, and (C) sensitivity.

Moreover, at the tunneling junction, the modulation of the energy band depends on the dielectric constant values. Since the biomolecule has $K = 15$, the drain current modulates significantly.

The sensitivity of mono/few-layer Phosphorene is calculated using Eq. (2). Fig. 15A shows the variation of drain current as a function of dielectric constant, and it is clearly observed that drain current increases as dielectric constant value increases. It is found that higher dielectric value is more desirable for sensing. Fig. 15B shows the sensitivity variation of DCDG–phTFET biosensor with two configurations. It is found that few–layers show more sensitivity compared with its monolayer counterpart. The selectivity is obtained from g_m/I_{on} ratio, as shown in Fig. 15C. For a biosensor, a higher selectivity is desirable to easily identify a molecule. DCDG–phTFET with $K = 5$ and 12 has higher selectivity compared with other dielectric material biosensors. In particular, for $K = 5$ in few-layer zig-zag and $K = 12$ monolayer, armchair DCDG–phTFET is desirable.

The performance of DCDG Ph-TFET biosensor is analyzed by varying the cavity thickness in the presence of biomolecules in armchair direction. It is established, as the nanocavity thickness is increased from 3 to 10 nm, the drain current decreases (Fig. 16A).

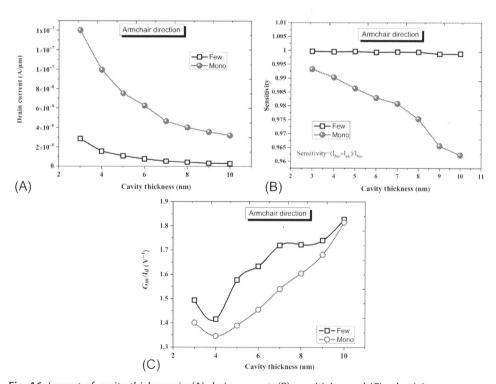

Fig. 16 Impact of cavity thickness in (A) drain current, (B) sensitivity, and (C) selectivity.

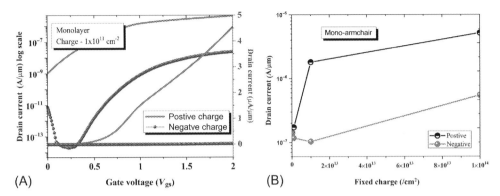

Fig. 17 Impact of fixed charge (A) drain current, (B) function of fixed charge concentration.

This reduction in drain current is rapid in monolayer device than the few-layer, which reduces the sensitivity (Fig. 16B), but it is not reflected in the selectivity of the device (Fig. 16C).

To analyze the effect of a fixed charge, it is categorized into three types: positive, negative and neutral charges. This charge is uniformly placed at the interface (channel/oxide) region with different concentration $(0–1 \times 10^{14}/cm^2)$. Here, 0 concentration represents neutral. If the biomolecule is positively charged, it gives more drain current (Fig. 17A) compared with negatively charged ones. The threshold of a positively charged biomolecule is less compared with its counterpart, hence we clearly classified which type of molecule it is. Fig. 17B shows that drain current is a function of fixed charge concentration. From the plot, it's seen that positive charge gives more drain current, so it has more sensitivity compared with its counterpart.

6. Summary

In this chapter, we reviewed the concepts of various biosensors, working mechanisms of FET devices as a biosensors, characteristics of 2D materials, and their fabrication procedures for biosensing applications. Also, we discussed the influence of 2D materials on the characteristics of FET (MOSFET and TFET) biosensors by studying the various figure of merits. We investigated TFET device as it is one of the steep slope devices along with Phosphorene as channel material for biomolecules detection. We designed a novel DCDG Ph-TFET biosensor and analyzed its performance by varying its physical parameters using DFT studies and TCAD simulation. It is concluded that the proposed device shows better performance in terms of sensitivity and selectivity compared with other 2D material FET biosensors and found that it can be used as an intelligent nanosensor.

References

[1] A.P.F. Turner, Biosensors: sense and sensibility, Chem. Soc. Rev. 42 (8) (2013) 3184–3196.

[2] M.-Y. Shen, B.-R. Li, Y.-K. Li, Silicon nanowire field-effect-transistor based biosensors: from sensitive to ultra-sensitive, Biosens. Bioelectron. 60 (2014) 101–111.

[3] R. Narang, M. Saxena, M. Gupta, Comparative analysis of dielectric-modulated FET and TFET-based biosensor, IEEE Trans. Nanotechnol. 14 (3) (2015) 427–435.

[4] R. Narang, K.V.S. Reddy, M. Saxena, R.S. Gupta, M. Gupta, A dielectric-modulated tunnel-FET-based biosensor for label-free detection: analytical modeling study and sensitivity analysis, IEEE Trans. Electron Devices 59 (10) (2012) 2809–2817.

[5] S. Kim, D. Baek, J.-Y.Y. Kim, S.-J.J. Choi, M.-L.L. Seol, Y.-K.K. Choi, A transistor-based biosensor for the extraction of physical properties from biomolecules, Appl. Phys. Lett. 101 (7) (2012) 73703.

[6] P. Dwivedi, A. Kranti, Applicability of transconductance-to-current ratio (gm/Ids) as a sensing metric for tunnel FET biosensors, IEEE Sensors J. 17 (4) (2017) 1030–1036.

[7] P. Dwivedi, A. Kranti, Dielectric modulated biosensor architecture: tunneling or accumulation based transistor? IEEE Sensors J. 18 (8) (2018) 3228–3235.

[8] Wikipedia contributors, Two-Dimensional Materials - - - {Wikipedia}{,} The Free Encyclopedia, 2019.

[9] K.S. Novoselov, et al., Electric field effect in atomically thin carbon films, Science 306 (5696) (2004) 666–669.

[10] M. Pumera, Graphene in biosensing, Mater. Today 14 (7–8) (2011) 308–315.

[11] S. Barua, H.S. Dutta, S. Gogoi, R. Devi, R. Khan, Nanostructured MoS 2-based advanced biosensors: a review, ACS Appl. Nano Mater. 1 (1) (2018) 2–25.

[12] Y. Chen, R. Ren, H. Pu, J. Chang, S. Mao, J. Chen, Field-effect transistor biosensors with two-dimensional black phosphorus nanosheets, Biosens. Bioelectron. 89 (2017) 505–510.

[13] J. Peng, Y. Lai, Y. Chen, J. Xu, L. Sun, J. Weng, Sensitive detection of carcinoembryonic antigen using stability-limited few-layer black phosphorus as an electron donor and a reservoir, Small 13 (15) (2017) 1603589.

[14] X. Zhao, et al., Sensitivity enhancement in surface plasmon resonance biochemical sensor based on transition metal dichalcogenides/graphene heterostructure, Sensors (Switzerland) 18 (7) (2018).

[15] B. Meshginqalam, J. Barvestani, Performance enhancement of SPR biosensor based on phosphorene and transition metal dichalcogenides for sensing DNA hybridization, IEEE Sensors J. 18 (18) (2018) 7537–7543.

[16] A. Sassolas, B. Prieto-Simón, J.-L. Marty, Biosensors for pesticide detection: new trends, Am. J. Anal. Chem. 03 (03) (2012) 210–232.

[17] J. Wang, Glucose_sensors (40 Years of Advances).pdf, 2001, pp. 983–988.

[18] R. Narang, M. Saxena, R.S. Gupta, M. Gupta, Dielectric modulated tunnel field-effect transistor—a biomolecule sensor, IEEE Electron Device Lett. 33 (2) (2012) 266–268.

[19] C.C. Wang, et al., Array-based multiplexed screening and quantitation of human cytokines and chemokines, J. Proteome Res. 1 (4) (2002) 337–343.

[20] I.E. Tothill, Biosensors for cancer markers diagnosis, Semin. Cell Dev. Biol. 20 (1) (2009) 55–62.

[21] S. Kröger, R.J. Law, Biosensors for marine applications: we all need the sea, but does the sea need biosensors? Biosens. Bioelectron. 20 (10 SPEC. ISS) (2005) 1903–1913.

[22] S. Kröger, S. Piletsky, A.P.F. Turner, Biosensors for marine pollution research, monitoring and control, Mar. Pollut. Bull. 45 (1–12) (2002) 24–34.

[23] J. Zhai, H. Cui, R. Yang, DNA based biosensors, Biotechnol. Adv. 15 (1) (1997) 43–58.

[24] J. Spadavecchia, M.G. Manera, F. Quaranta, P. Siciliano, R. Rella, Surface plamon resonance imaging of DNA based biosensors for potential applications in food analysis, Biosens. Bioelectron. 21 (6) (2005) 894–900.

[25] G. RONTÓ, S. Gaspar, Z. Gugolya, Ultraviolet dosimetry in outdoor measurements based on bacteriophage T7 as a biosensor, Photochem. Photobiol. 59 (2) (1994) 209–214.

[26] S.C. Garrett, et al., A biosensor of S100A4 metastasis factor activation: inhibitor screening and cellular activation dynamics, Biochemistry 47 (3) (2008) 986–996.

[27] L.J. He, M.S. Wu, J.J. Xu, H.Y. Chen, A reusable potassium ion biosensor based on electrochemilu-minescence resonance energy transfer, Chem. Commun. 49 (15) (2013) 1539–1541.

[28] B. Wu, G. Zhang, S. Shuang, M.M.F. Choi, Biosensors for determination of glucose with glucose oxi-dase immobilized on an eggshell membrane, Talanta 64 (2) (2004) 546–553.

[29] P. Mehrotra, Biosensors and their applications—a review, J. Oral Biol. Craniofacial Res. 6 (2) (2016) 153–159.

[30] K. Ramanathan, B. Danielsson, Principles and applications of thermal biosensors, Biosens. Bioelectron. 16 (6) (2001) 417–423.

[31] S. Babacan, P. Pivarnik, S. Letcher, A.G. Rand, Evaluation of antibody immobilization methods for piezoelectric biosensor application, Biosens. Bioelectron. 15 (11–12) (2000) 615–621.

[32] M.A. Cooper, Optical biosensors in drug discovery, Nat. Rev. Drug Discov. 1 (7) (2002) 515–528.

[33] J. Homola, S.S. Yee, G. Gauglitz, Surface plasmon resonance sensors: review, Sens. Actuators B Chem. 54 (1) (1999) 3–15.

[34] Z. Muhammad-Tahir, E.C. Alocilja, A conductometric biosensor for biosecurity, Biosens. Bioelec-tron. 18 (5–6) (2003) 813–819.

[35] J.G. Guan, Y.Q. Miao, Q.J. Zhang, Impedimetric biosensors, J. Biosci. Bioeng. 97 (4) (2004) 219–226.

[36] J. Liu, Y. Lu, Adenosine-dependent assembly of aptazyme-functionalized gold nanoparticles and its application as a colorimetric biosensor, Anal. Chem. 76 (6) (2004) 1627–1632.

[37] M. DeMiguel-Ramos, et al., Gravimetric biosensor based on a 1.3 GHz AlN shear-mode solidly mounted resonator, Sens. Actuators B Chem. 239 (2017) 1282–1288.

[38] R.W. Whatmore, Pyroelectric devices and materials, Rep. Prog. Phys. 49 (12) (1986) 1335–1386.

[39] Y. Lei, W. Chen, A. Mulchandani, Microbial biosensors, Anal. Chim. Acta 568 (1–2) (2006) 200–210.

[40] P. Bergveld, The future of biosensors, Sens. Actuators A Phys. 56 (1–2) (1996) 65–73.

[41] P. Bergveld, The development and application of FET-based biosensors, Biosensors 2 (1) (1986) 15–33.

[42] J. Lee, et al., A highly responsive silicon nanowire/amplifier MOSFET hybrid biosensor, Sci. Rep. 5 (2015) 1–6.

[43] P. Venkatesh, K. Nigam, S. Pandey, D. Sharma, P.N. Kondekar, A dielectrically modulated electrically doped tunnel FET for application of label free biosensor, Superlattice. Microst. 109 (2017) 470–479.

[44] D. Singh, S. Pandey, K. Nigam, D. Sharma, D.S. Yadav, P. Kondekar, A charge-plasma-based dielectric-modulated junctionless TFET for biosensor label-free detection, IEEE Trans. Electron Devices 64 (1) (2017) 271–278.

[45] Y. Tsividis, C. McAndrew, Operation and Modeling of the MOS Transistor, vol. 2, Oxford University Press, Oxford, 1999.

[46] N. Kannan, M.J. Kumar, Dielectric-modulated impact-ionization Mos transistor as a label-free biosen-sor, IEEE Electron Device Lett. 34 (12) (2013) 1575–1577.

[47] D. Soni, D. Sharma, Design of NW TFET biosensor for enhanced sensitivity and sensing speed by using cavity extension and additional source electrode, Micro Nano Lett. 14 (8) (2019) 901–905.

[48] D.B. Abdi, M.J. Kumar, Dielectric modulated overlapping gate-on-drain tunnel-FET as a label-free biosensor, Superlattice. Microst. 86 (2015) 198–202.

[49] S. Saurabh, M.J. Kumar, Fundamentals of Tunnel Field-Effect Transistors, CRC Press, 2016.

[50] A. Gao, N. Lu, Y. Wang, T. Li, Robust ultrasensitive tunneling-FET biosensor for point-of-care diag-nostics, Sci. Rep. 6 (2016) 1–9 November 2015.

[51] S. D. M. TCAD, "Synopsys," Inc., Mt. View, CA, USA, 2012.

[52] P. Li, D. Zhang, C. Jiang, X. Zong, Y. Cao, Ultra-sensitive suspended atomically thin-layered black phosphorus mercury sensors, Biosens. Bioelectron. 98 (June) (2017) 68–75.

[53] S. Kim, J.Y. Kim, J.H. Ahn, T.J. Park, S.Y. Lee, Y.K. Choi, A charge pumping technique to identify biomolecular charge polarity using a nanogap embedded biotransistor, Appl. Phys. Lett. 97 (7) (2010) 2010–2012.

[54] H. Im, X.J. Huang, B. Gu, Y.K. Choi, A dielectric-modulated field-effect transistor for biosensing, Nat. Nanotechnol. 2 (7) (2007) 430–434.

[55] S. Kyu Kim, H. Cho, H.J. Park, D. Kwon, J. Min Lee, B. Hyun Chung, Nanogap biosensors for elec-trical and label-free detection of biomolecular interactions, Nanotechnology 20 (45) (2009).

[56] D.B. Abdi, M.J. Kumar, Controlling ambipolar current in tunneling FETs using overlapping gate-on-drain, IEEE J. Electron Devices Soc. 2 (6) (2014) 187–190.

[57] A. Pon, K.S.V.P. Tulasi, R. Ramesh, Effect of interface trap charges on the performance of asymmetric dielectric modulated dual short gate tunnel FET, AEU Int. J. Electron. Commun. 102 (2019) 1–8.

[58] J. Madan, R. Chaujar, Gate drain-overlapped-asymmetric gate dielectric-GAA-TFET: a solution for suppressed ambipolarity and enhanced ON state behavior, Appl. Phys. A 122 (11) (2016) 973.

[59] A.M. Ionescu, H. Riel, Tunnel field-effect transistors as energy-efficient electronic switches, Nature 479 (7373) (2011) 329–337.

[60] M.G. Upasana, et al., Influence of dielectric pocket on electrical characteristics of tunnel field effect transistor: A study to optimize the device efficiency, in: Proceedings of the 2015 IEEE International Conference on Electron Devices and Solid-State Circuits, EDSSC 2015, 2015, pp. 762–765.

[61] G.T. Sayah, Enhancement of the performance of TFET using asymmetrical oxide spacers and source engineering, Int. J. Comput. Appl. 150 (10) (2016) 14–18.

[62] S.W. Kim, W.Y. Choi, M.C. Sun, H.W. Kim, B.G. Park, Design improvement of L-shaped tunneling field-effect transistors, in: Proc. IEEE Int. SOI Conf, 2012, pp. 7–8.

[63] M.J. Kumar, S. Janardhanan, Doping-less tunnel field effect transistor: design and investigation, IEEE Trans. Electron Devices 60 (10) (2013) 3285–3290.

[64] R. Li, et al., AlGaSb/InAs tunnel field-effect transistor with on-current of 78 μA/μm at 0.5 v, IEEE Electron Device Lett. 33 (3) (2012) 363–365.

[65] F. Schwierz, J. Pezoldt, R. Granzner, Two-dimensional materials and their prospects in transistor electronics, Nanoscale 7 (18) (2015) 8261–8283.

[66] K.S. Novoselov, et al., Electric field in atomically thin carbon films, Science 306 (5696) (2004) 666–669.

[67] C.E. Banks, D.A.C. Brownson, 2D Materials: Characterization, Production and Applications, CRC Press, Taylor & Francis Group, 2018 ISBN : 9781498747394.

[68] F. Xia, H. Wang, Y. Jia, Rediscovering black phosphorus as an anisotropic layered material for optoelectronics and electronics, Nat. Commun. 5 (2014).

[69] H. Liu, A.T. Neal, Phosphorene : an unexplored 2D semiconductor with a high hole mobility, ACS Nano 8 (4) (2014) 4033–4041.

[70] H. Wang, et al., Black phosphorus radio-frequency transistors, Nano Lett. 14 (11) (2014) 6424–6429.

[71] S. Das, M. Demarteau, A. Roelofs, Ambipolar phosphorene field effect transistor, ACS Nano 8 (11) (2014) 11730–11738.

[72] P.W. Bridgman, Two new modifications of phosphorus, J. Am. Chem. Soc. 36 (7) (1914) 1344–1363.

[73] J. Lu, J. Yang, A. Carvalho, H. Liu, Y. Lu, C.H. Sow, Light-matter interactions in phosphorene, Acc. Chem. Res. 49 (9) (2016) 1806–1815.

[74] A. T. K. QuantumWise, "14.2 Reference Manual [Online] http://www.quantumwise.com/documents/manuals/latest," Ref. html.

CHAPTER 17

Fabrication and analysis process of TCR-based carbon nanotube resistive sensor

Soheli Farhana
MIIT, University of Kuala Lumpur (UniKL), Kuala Lumpur, Malaysia

1. Fabrication process

The carbon nanotube (CNT) filaments were set up using pregrowth CNT forest on a Si wafer deposited from a chemical vapor deposition (CVD) process. Multiwalled CNT was examined by Scanning Electron Microscopy (SEM) and Transmission Electron Microscopy (TEM). Manufacturing of a hybrid resistor was used from a combination of CNT fiber winding with a metallic bar. Both terminals of the metallic bar were connected with metallic caps. CNT TCR estimated the electrical properties. Nickel and copper with different nuclear structures were used to produce a composite of a metal resistor. TCR is applied to fabricate the CNT fiber winding resistor that will be built in this work [1–7].

Every lump composed of a brass compound was manufactured by dissolving a planned measure in the softening heater. Every bar was manufactured and toughened for uniform scattering. Furthermore, the focus was balanced by the isolation disposal technique. At last, metal composite wires were created by a hot moving technique. Energy-dispersive X-ray spectroscopy (EDX) was imposed for affirming the synthesis configuration.

The newly designed resistor was manufactured by composing CNT and metallic fiber winding together into a bar. The schematic diagram of the crossbreed resistor is shown in Fig. 1. To begin with, a metal composite was twisted onto a bar made of lead wires, clay pole, and metal tops. Afterward, the fixed two parts of the metallic bar were electrically associated with a spot-fusing technique. At last, sticking glue was used to fix the wires onto the resistor body that was electrically conductive. TCR estimation was used to find electrical properties. TCR estimation was finished in the heater by estimating the resistances.

Handbook of Nanomaterials for Sensing Applications
https://doi.org/10.1016/B978-0-12-820783-3.00003-8

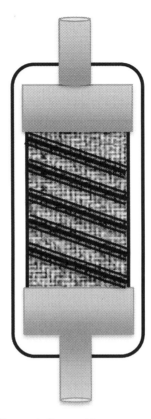

Fig. 1 Proposed diagram of CNT fibers winding resistor.

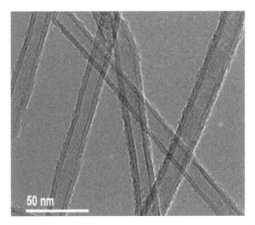

Fig. 2 CNT fiber wire from TEM image.

2. Analysis

A length of nanotube fiber of 75 μm, shown in Fig. 2, was utilized for the development. Multiwalled CNT (MWCNT) fibers are used to prepare numerous strings as shown in Fig. 2.

Fig. 3 shows the estimation of CNT electrical properties using TEM analysis. Nanotube resistance can be expanded by increasing the number of multiwalled tubes. In this design, 10 fiber nanotube strings were used to fabricate four resistors [8–12]. Different temperatures from 25°C to 125°C changed the different resistances. Fig. 3 shows the variation of TCR of nanotube resistances over different temperatures applied on the resistors.

However, the reduction of resistances of nanotube resistors are realized by expanding the number of fibers provided in the TCR of nanotube resistor from −800 to −900 ppm/°C.

Negative TCR of nanotube fibers indicate semiconducting behavior of the newly designed hybrid nanotube resistor. The CVD technique is used to fabricate nanotube fiber-winded resistors. Among the developed resistors prototype, a few percentages of the resistor behave like semiconductors and the rest have metallic behavior.

TCR of nanotube composite resistor forms a balanced and fixed resistor of a combination of Ni and Cu. Resistances of 400–1300 ppm/°C allows measurement of Ni included in the compound, which shifted to modify the metallic combination of the TCR of resistors. By expanding and shortening the length, composite resistors can be set with optimum resistance for better performance. Four nanotube resistors were developed with different TCR from 400 to 1300 ppm/°C, and their resistances ranged from 7.94 to 10.11 Ω.

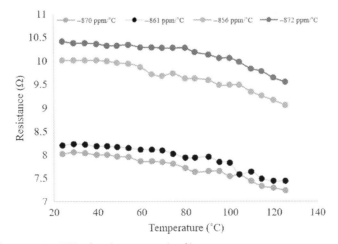

Fig. 3 Electrical properties TCR of carbon nanotube fibers.

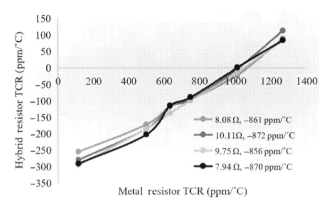

Fig. 4 Population of nanotube resistors over hybrid resistors in the measurement of resistances at different TCRs.

All fixed nanotube and metal composite resistors were made and constructed in parallel. There are four composite resistors fabricated with different TCR associated with several numbers of composite resistors with positive TCR in 400 and 1300 ppm/°C that brought about 24 mixes of half-breed resistors.

Fig. 4 shows estimated resistances over different TCR of metal resistors. The TCR of metal resistors have effects on the variations of hybrid resistors' TCR [13–16]. The hybrid resistor shows its TCR measurement as close to zero. But the nanotube-embedded resistor shows a marvelous opposite TCR reading of −865 ppm/°C, and the metallic resistor shows a positive TCR reading of +865 ppm/°C. The first test was conducted to vary the temperature between 25°C and 125°C, while the nanotube resistance was found to be 7.94 Ω at −870 ppm/°C. In this way, resistance value change ought to be remunerated to zero by interfacing them parallelly. It may happen when the metallic resistor TCR is nearly zero at 1100 ppm/°C, while the hybrid resistors really indicated zero TCR. Accordingly, TCR at 1100 ppm/°C for metal resistor and −865 ppm/°C for hybrid resistor were found. This whole experiment was conducted to find all measurements using bundles of nanotube fibers and metallic resistors as shown in Fig. 4.

Finally, we hypothesized that the resistance of metallic compounds, nanotube winding fibers, and boundary resistors showed a step-by-step admirably at zero TCR, which is shown in Fig. 5 over temperature reading. As shown in Fig. 5, fluctuations of resistances result in positive TCR. However, regarding the TCR stated in Fig. 5, the CNT fiber resistor decreases with temperature, which brings a negative TCR. Accordingly, metallic and hybrid resistors and embedded of CNT fiber resistors show zero viscosity demonstrated at zero TCR.

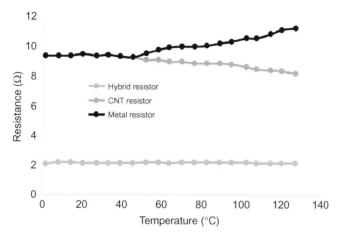

Fig. 5 Demonstration of three different resistors at zero TCR.

3. Conclusion

A complete process of hybrid resistor fabrication steps has been explained in this chapter to reach a near-zero TCR. Nanotube fibers are set up parallelly with the resistor to demonstrate negative TCR. Many CNT filaments have been used to fabricate nanotube resistors by wrapping the filaments. Nickel and copper were used to fabricate metallic resistors in this work. Another two hybrid resistors were also fabricated by placing them parallelly in near-zero TCR. The main goal was to fabricate the resistors in near-zero TCR, which was successfully adjusted by varying the resistances in different temperatures. The whole process of the fabrication for different resistors was performed in TCR of 0 and -2 ppm/$^{\circ}$C.

References

[1] S. Chen, L. Qiu, H.M. Cheng, Carbon-based fibers for advanced electrochemical energy storage devices, Chem. Rev. 120 (5) (2020) 2811–2878.
[2] H.H. Gatzen, V. Saile, J. Leuthold, Micro and Nano Fabrication, Springer-Verlag, Berlin AN, 2016.
[3] E. Menard, J. Rogers, Nanofabrication techniques with high-resolution molded rubber stamps, in: Nanomanufacturing Handbook, CRC Press, United States, 2017 Dec 19, pp. 147–160.
[4] D. Papkov, A. Goponenko, O.C. Compton, Z. An, S.T. Nguyen, Y.A. Dzenis, Controlled nanofabrication of uniform continuous graphene oxide/polyacrylonitrile nanofibers for templated carbonization, J. Micro Nanomanuf. 7 (4) (2019).
[5] J.A. Rogers, E. Menard, Stamping techniques for micro-and nanofabrication, in: Springer Handbook of Nanotechnology, Springer, Berlin, Heidelberg, 2017, pp. 143–161.
[6] D. Oran, S.G. Rodriques, R. Gao, S. Asano, M.A. Skylar-Scott, F. Chen, P.W. Tillberg, A. H. Marblestone, E.S. Boyden, 3D nanofabrication by volumetric deposition and controlled shrinkage of patterned scaffolds, Science 362 (6420) (2018) 1281–1285.
[7] K. Bazaka, O. Baranov, U. Cvelbar, B. Podgornik, Y. Wang, S. Huang, L. Xu, J.W. Lim, I. Levchenko, S. Xu, Oxygen plasmas: a sharp chisel and handy trowel for nanofabrication, Nanoscale 10 (37) (2018) 17494–17511.

[8] Y. Shao, F. Xu, W. Liu, M. Zhou, W. Li, D. Hui, Y. Qiu, Influence of cryogenic treatment on mechanical and interfacial properties of carbon nanotube fiber/bisphenol-F epoxy composite, Compos. Part B 125 (2017) 195–202.

[9] F. Ebrahimi, S. Habibi, Nonlinear eccentric low-velocity impact response of a polymer-carbon nanotube-fiber multiscale nanocomposite plate resting on elastic foundations in hygrothermal environments, Mech. Adv. Mater. Struct. 25 (5) (2018) 425–438.

[10] R.J. Headrick, M.A. Trafford, L.W. Taylor, O.S. Dewey, R.A. Wincheski, M. Pasquali, Electrical and acoustic vibroscopic measurements for determining carbon nanotube fiber linear density, Carbon 144 (2019) 417–422.

[11] K. Koziol, J. Vilatela, A. Moisala, M. Motta, P. Cunniff, M. Sennett, A. Windle, High-performance carbon nanotube fiber, Science 318 (5858) (2007) 1892–1895.

[12] J. Wang, R.P. Deo, P. Poulin, M. Mangey, Carbon nanotube fiber microelectrodes, J. Am. Chem. Soc. 125 (48) (2003) 14706–14707.

[13] McMullan RC, Pushkarakshan BK, Narayan SJ, Sankaran S, Kunz KE, Inventors; Texas Instruments Inc, Assignee. Embedded Tungsten Resistor. United States Patent US 10,461,075 2019 Oct 29.

[14] W.H. Lee, K.C. Chung, Investigation of a copper–nickel alloy resistor using co-electrodeposition, J. Appl. Electrochem. (2020) 1–3.

[15] M. Kolpe, S. Gosavi, G. Phatak, Effect of graphite content on the polymer based resistor paste for an intergated resistor on printed circuit boards (PCB), J. Nanoelectron. Optoelectron. 14 (7) (2019) 1030–1036.

[16] K. Bansal, S. Chander, S. Gupta, M. Gupta, Design and analysis of thin film nichrome resistor for GaN MMICs, in: 2016 3rd International Conference on Devices, Circuits and Systems (ICDCS), IEEE, 2016 Mar 3, pp. 145–148.

Intelligent nano sensors (INS)—Medical/bio applications

CHAPTER 18

Recent progress for nanotechnology-based flexible sensors for biomedical applications

Anindya Nag[a], Samta Sapra[b], and Subhas Chandra Mukhopadhyay[b]

[a]DCI-CNAM Institute, Dongguan University of Technology, Dongguan, People's Republic of China
[b]School of Engineering, Macquarie University, Sydney, NSW, Australia

1. Introduction

With the exponential leap of science and technology and their conglomeration with human beings in their daily usage, research work has greatly increased to improve the quality of electrical and electronic devices. Among the different sectors that have flourished in the last 3–4, sensors have been one of the sectors [1] where researchers have invested a lot of time due to their ubiquitous usage in different forms. Different types of sensors, with their variation in structure, efficiency, dimensions, robustness, and working nature, are being used every day to assist day-to-day life. The popularization in the use of sensors started with the development of microelectromechanical systems (MEMS) around 4 decades ago [2]. Among a range of processing materials available to design and form the MEMS-based sensors, researchers have largely favored the use of silicon [3, 4] and gold [5, 6] for their distant advantages over other materials. Even though these sensors have addressed a lot of applications in industrial [7, 8] and environmental [9, 10] sectors, there are some constraints that deter their dynamicity in the real world. One of the main ones is their utilization for biomedical applications, due to their rigid nature, brittle nature, high cost of fabrication, low life cycles, and limitation in the materials that can be associated with these sensors to increase their sensitivity and selectivity for detecting analyte [11, 12]. This led to the development of flexible sensing prototypes that can deal with the drawbacks of MEMS-based sensors [13]. The flexible sensors are much more advantageous in terms of electrical, mechanical, and thermal characteristics. The fabrication and processing of the flexible materials are done differently, depending on the application of the developed sensors. Some of the common techniques used to fabricate the flexible sensors are photolithography [14, 15]; some printing techniques like microprinting [16, 17], laser printing [18, 19], 3D printing [20, 21], inkjet printing [22, 23], and screen printing [24, 25]; electrodeposition [26, 27]; chemical oxidation [28, 29]; spraying [30, 31]; and vapor deposition polymerization [32, 33]. The raw materials used to process

Handbook of Nanomaterials for Sensing Applications
https://doi.org/10.1016/B978-0-12-820783-3.00009-9

with these mentioned techniques depend on the desired characteristics of the sensing prototypes. For the formation of substrates and electrodes, a range of polymers and conductive materials have been used. Some of the common polymers used to form flexible sensors are polydimethylsiloxane (PDMS) [34, 35], polyethylene terephthalate (PET) [36, 37], polyimide (PI) [38, 39], and poly(3,4-ethylenedioxythiophene) polystyrene sulfonate (PEDOT:PSS) [40, 41]. Similarly, some of the common conductive materials used to form electrodes are carbon nanotubes (CNTs) [42, 43], graphene [44, 45], gold [46, 47], aluminum [48, 49], and iron [50, 51]. The fabrication and implementation of flexible sensors have been done in different sectors in defense [52], automobile industry [53], robotics [54], telecommunication [55], and biomedical [56] applications. Among these areas, the utilization in the biomedical sector appears to be paramount due to the extra advantages these sensors provide compared with other types of sensing prototypes. Fig. 1 [57] shows a market survey of the use of the different types of printed and flexible sensors that was done in 2018. It is seen among different categories showcased here that the budget for printed and flexible sensors is around $3.6 billion USD, which is likely to increase in the future based on their applications mentioned in the figure. An estimation was done by IDTechEx [58] on the use of the different types of printed and flexible sensors that would be used for the next 5 years. Thus, it can be inferred that a lot of research work would be going on in the upcoming years to design, fabricate, characterize, and implement various types of flexible sensors for biomedical applications.

For the development of flexible sensors, the use of nanotechnology has been one of the popular methodologies in the recent era. There has been a significant drift in the last decade [59, 60] to drive the use of nanoparticles and nanomaterials to form sensing

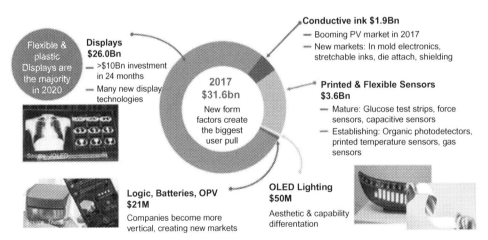

Fig. 1 A market survey of the classification done on the total budget for printed and flexible electronics for the year 2018 [57]. *Image reproduced from https://www.printedelectronicsworld.com/ articles/15712/printed-electronics-key-trends.*

prototypes. This is because the assistance of nanotechnology in the flexible sensors has dramatically affected its sensing range, sensitivity, range of operating temperature, rolling resistance, robustness, and mechanical characteristics. Also, the portability, data storage, and processing capability of the prototypes increases drastically. The size of nanomaterials used for biomedical applications ranges from a few microns to a few nanometers, depending on the nature of the sensing prototype. The type of nanomaterials that are used in the formation also differs, as some of the common ones are nanoparticles [61, 62], nanotubes [63, 64], nanowires [65, 66], nanoballs [67, 68], and engineering organic molecules [69, 70]. These particles serve as potent tools to influence the performance of semiconductors, transparent electrodes, and dielectric and conductive materials for flexible electronics applications. The nanomaterials have also helped to form wearable sensors by making them thin and highly flexible. The high sensitivity obtained by the prototypes is due to the high electrical conductivity as a result of their high charge carrier mobility. They also increase the tensile strength of the formed products, which makes them very susceptible to stress fractures. Another significant aspect imparted on the nanotech-based sensors is their high stability toward changes in temperature, pressure, and corrosive environments [71].

2. Nanotechnology-based flexible sensors

The rise of nanotechnology for their use in the fabrication of flexible sensing prototypes has assisted in dealing with some challenges like flexibility, robustness, and customization of the design of the prototypes [72]. This has also broadened their use as wearable sensors. One of the common types where nanotechnology-based sensors have been used for biomedical applications is as drug delivery systems and therapeutic systems. Fig. 2 [73] shows the schematic diagram of some of the probable uses of nanoparticles for designing and fabricating wearable sensors. This is due to the biocompatibility of the individual materials that are being processed. To employ nanoparticles for sensing purposes, the working nature of the sensors differ based on the application they are being exploited for. Certain types of flexible sensors like capacitive [74, 75] and piezoresistive [76, 77] sensors are very popular due to their multifunctional characteristics. Recently, certain types of sensors like nanoharvesters [78, 79] and nanogenerators [80, 81] are largely favored due to the voltage-inducing capability. Although the range of voltage induced with devices is very low, it is hopeful that they will serve a great purpose with respect to power supply and generation. The signal generated by these nanotechnology-based flexible sensors is being transmitted using a conditioning circuit that is embedded with an individual or array of sensors. The fabrication of flexible printed circuit boards (FPCBs) [82, 83] has greatly assisted in the wearable aspect of the sensors. Signal-conditioning circuits embedded with the sensors are now being developed providing superior mechanical and electrical performances compared with earlier ones. Among the different types of

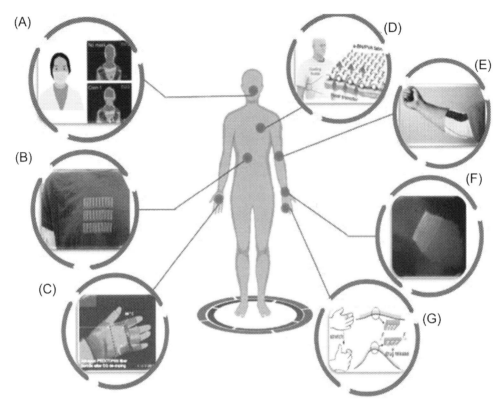

Fig. 2 Schematic diagram of the use of nanomaterials for the fabrication of wearable sensors. These materials have been used for (A) thermal management, (B) personalized clothing, (C) heating gloves, (D) thermal textiles, (E) textile exoskeletons, (F) stretchable heaters, and (G) drug-delivery systems [73]. *Image reproduced from https://www.nanowerk.com/spotlight/spotid=52432.php.*

nanotechnology-based materials used to form the sensors, nanocomposites [84, 85] are a popular choice where the type of nanoparticles mixed as fillers in the polymer matrix is based on the achievable percolation threshold. The inclusion of the type of nanomaterials involved in forming the nanocomposites is 1D nanowires and nanofibers. These nanomaterials are mixed in a polymer matrix using techniques like coating methods of spin [86] and spray [87], casting methods like drop [88] and dip [89], layer-by-layer assembly [90], vacuum filtration [91], and other printing techniques [92]. The use of these prototypes has been stretched on biomedical applications to increase the prospective of wearable sensing systems for real-time data collection and processing when monitoring the human physiological central system.

For some flexible sensors, the choice of nanomaterials depends on a specific application, and others are based on a certain type that can be utilized for multiple applications. The sensors that are designed for monitoring human physiological parameters [93] like heart rate, body temperature, blood pressure, body movement, cardiovascular diseases, and neurological activities are multifunctional and dynamic in nature. These prototypes are primarily designed with a certain type of working nature, and then they are subsequently implemented to all the applications that fit into this operating principle. For a particular type of sensor, the classification of a group of applications is based on the flexibility, sensitivity, and robustness of the prototypes. These parameters, in turn, are a function of the nature of the nanoparticles and fabrication techniques to generate the prototypes. The sensors function as a single sensor or an integration of several sensors for conducting experiments. The applications of these sensors are classified into groups like personal health and environment, intelligent human-machine interfaces, and replicas for human skin in robotics and prosthetics [94, 95]. Another attribute that decides the quality of the flexible sensors is their capability to adjust in different ambiances while maintaining high efficiency in the transmission of collected data. These sensors are fused with other functional components to make them ready to be used for real-world applications. Some of the integrated flexible sensing systems used for biomedicine are transistors [96], amplifiers [97], biosensors [98], photodetector arrays [99], photovoltaics [100], and energy storage elements [101]. The fabrication of the nanomaterials to be used for each flexible sensing prototype can be fabricated using different methods like top-down and bottom-up approaches. Some of the recently developed therapeutic functional nanomaterials [102, 103] are being combined with soft bioelectronics to operate on unconventional diagnostics capabilities. The wearable nature of the sensors increases their sensitivity as they can be intimately attached to the skin or tissue to detect vital physiological and biological markers. The presence of the nanoparticles in terms of nanowires or other nanomaterials induces the selectivity of the sensor toward the biomarker. To increase the nature of selectivity, these nanowires are developed with different materials like ZnO [104], In_2O_3 [105], GeSi [106], GaAs [107], etc. Similar to the range of processed materials, the structures of the sensors were also varied based on the working nature of the electrodes. For example, the printed sensors are based on 2D [108, 109] and 3D [110, 111] types, based on the type of nanoparticles imparted on them. In recent days, the alignment of fabricated nanowires also plays an important role in the response of the sensor via varying the current conducted within the sensor. Also, with vertically aligned nanowires, their crystallographic orientations are maintained and the capability to fabricate electronic and optoelectronic devices is increased [112]. The electrodes and substrates for nanotechnology-based sensors are chosen in such a way that the change in response of the sensors creates a simultaneous change in the resistance between the different contact sites like two electrodes and electrode-electrolyte interfaces. One of the objectives of these sensors is to keep the cost of fabrication low while having a high quality of the operating device. For example, even though the electrical and mechanical characteristics of platinum are better than gold, the cost of fabricating gold is lower than that of platinum, and only a low voltage of 1.5 V is required to drive sensors consisting of gold nanowires electrodes [113]. Among the different sectors mentioned before, where nanotechnology-based flexible sensors are implemented, the biomedical sector is one of the most pivotal ones. The presence of these kinds of sensors has largely affected this sector as many acute and chronic diseases are being monitored and dealt with. Some of these types of sensors are discussed in the succeeding sections to showcase the use of these sensing prototypes for biomedical and related applications.

3. Nanotechnology-based flexible sensors for biomedical applications

This section describes some of the significant works done on the fabrication and implementation of nanotechnology-based flexible sensors for biomedical applications. There are four major types of flexible sensors developed with nanoparticles, namely physical, chemical, biopotential, and bioanalytic sensing prototypes. Each of these types has been subclassified into several types based on their structure, operating principle, and type of detected analyte. Apart from the research works exemplified, there are also other works that have extensively led to an increase in these types of sensors for biomedical applications, as shown in the last section of this chapter.

3.1 Physical sensors

The first type of nanotechnology-based flexible sensors is the physical ones, which differ in terms of their physical structure and operation principle. The structure and dimension of each of these prototypes decide the manner in which they will operate. These sensors have been excellent for biomedical applications due to their exceptional operational capabilities. The physical properties of the biomedical analyte are measured and converted to electrical signals for transmission and analysis purposes. Table 1 shows a comparative study of some of the significant parameters of different types of nanotechnology-based physical sensors. It is seen from the table that the use of nanotechnology has great potential to develop sensors of each type where certain advantages are associated with them. These sensors have been beneficial for the use in real-time monitoring of different strain, electrochemical, and electrical applications.

3.1.1 Mechanical sensors

The first category involves the mechanical sensors, which are based on the change in the shape and size of the prototypes in accordance with the input. These sensors are mostly used for certain types of biomedical sensing like tactile and strain sensing for real-time applications like smart tattoos, artificial skins, and e-skins [124–129]. One of them can be described from the work of Nag et al. [130], where nanocomposite-based sensors were developed by mixing multiwalled carbon nanotubes (MWCNTs) with PDMS at definite proportions. Optimization was done on the amount of MWCNTs to be mixed with PDMS to form the nanocomposites. A trade-off was done between the flexibility of

Table 1 Comparison between the works on bioanalytical sensors, based on their design and corresponding advantages.

Sl. No.	Sensor type	Materials used	Advantages	Application	Reference
1	Mechanical	Reduced Graphene oxide, polyimide	Excellent flexibility, superelasticity, high recovery rate	Strain sensing	[38]
2	Mechanical	Graphene	Good reliability, ultrafast response, high sensitivity	Tactile sensing	[114]
3	Mechanical	Carbon Nanotubes, Silver nanoparticles	High gauge factor, sensitivity, stretchability	Strain sensing	[43]
4	Thermal	Platinum, polymer	High sensitivity, good stability, linear response	Implantable temperature sensing	[115]
5	Thermal	Silver nanowires, thermoplastic polymer	Highly stretchable, highly conductive	Thermal therapy, physiotherapy	[116]
6	Thermal	Nickel, polymer	High reproducibility in response, high sensitivity	Wireless temperature sensing	[117]
7	Electric	Reduced graphene oxide	Ultra-sensitivity, stability, repeatability	Strain sensing	[118]
8	Electric	Graphene, gold, PET	Label-free detection, simplicity in the structure	Bio-electronic nose	[119]
9	Electric	Silicon nanowire,	Low-cost, ultrasensitive, small, rapid response	Point-of-care diagnostics	[120]
10	Optical	Graphene, polypropylene	Roll-to-roll compatible patterning, transparent, highly sensitive	Touch-sensing, light-emitting	[121]

Continued

Table 1 Comparison between the works on bioanalytical sensors, based on their design and corresponding advantages—cont'd

Sl. No.	Sensor type	Materials used	Advantages	Application	Reference
11	Optical	Silver nanowires, PDMS	Low optical transmission loss, suitable impedance for extracellular recordings	Electrophysiological recordings from neurons	[122]
12	Optical	Graphene, PDMS	High sensitivity, high waterproofing performance, high stability	Motion detection	[123]

the developed patches and the electrical conductivity of the electrodes to obtain a wt% of 4% of MWCNTs to be mixed in the polymer matrix. Initially, substrates of PDMS were cast on a PMMA template, followed by adjusting its height to 1000 μm. The samples were cured, followed by casting and curing of the nanocomposite layer. Finally, the laser cutting technique was used to scan the top layer of the samples to form the electrodes. The sensors were then characterized and used for measuring limb movements and respiration. The schematic diagram of fabrication and their application are described in Fig. 3 [130].

Due to the interdigital nature of the electrodes, the change in response of the sensors were measured in terms of capacitance for the flexed and bent positions of the limbs and upper part of the diaphragm. For monitoring of limb movements, the sensors were attached to the joints of the arm and leg to detect the flexed and extended situation via the change in the structure of the sensor and simultaneously the dimensions of the interdigital electrodes. The experiments on monitoring of respiration was done by attaching the sensor patches on the lower part of the diaphragm to detect its outward and inward movement.

Another work related to the use of nanoparticles for fabricating flexible sensors for biomechanical sensing is related to the development of strain sensors using nanoparticle composites consisting of carbon black and silver [131]. The sensors were low cost with excellent static and dynamic stability. The sensors were fabricated for monitoring certain human motions like finger bending, wrist rotation, and elbow flexion. The poly (vinyl-pyrrolidone) grafting process was used to functionalize the carbon nanomaterials via the initial treatment of carbon black with benzoyl peroxide. The mixture was then ultrasonicated, and this process was repeated cumulatively five times to obtain a homogenous

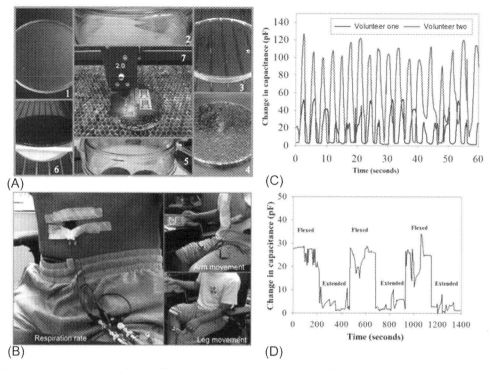

Fig. 3 Representation of the fabrication and implementation of CNT-PDMS based sensors for monitoring of human physiological parameters. (A) Seven individual steps of fabrication. (B) The sensor patches were connected to different body parts using LCR meter for monitoring purposes. (C) Monitoring of respiration of two different individuals. (D) Monitoring of limb movement [130]. *Image reproduced from A. Nag, S.C. Mukhopadhyay, J. Kosel, Flexible carbon nanotube nanocomposite sensor for multiple physiological parameter monitoring. Sens. Actuators A: Phys. 251 (2016) 148–155.*

solution. Finally, anhydrous alcohol was used to dilute the solution, and a part of this PVP-grafted carbon black was dispersed in a solution of deionized (DI) water and silver nanoparticles. The mixture between the carbon black and silver nanowire had a mass ratio of 3:1. The conductive solution was mixed with the thermoplastic polyurethane (TPU) solution to obtain a sensitive film with a thickness of approximately 0.1 mm. Aluminum foil was used as electrodes by putting two pieces on the two sides of the film using conductive silver paste. The patches operated as resistive sensors where the application of strain increased the resultant resistance due to the addition of a variable strain. The value of this variable strain was functional to the magnitude of the applied strain. The change in response was determined based on the change in relative resistance for an oscillatory motion of the different body parts. The electrical conductivity of the sensors was controlled by tuning the level of composite mixed in the TPU. A similar work related to the use of nanowires for developing mechanical sensors for biosensing is related to the fabrication and deployment of pressure sensors formulated by silver nanowires (AgNW) embedded in polyimide (PI) substrates [132]. These nanowires were chosen due to their excellent conductivity, transparency, and flexible endurance. The fabrication was done using selective wet etching techniques that allows for surface modification of AgNW-PI films. The sensor consisted of two face-to-face AgNWs-PI ultrathin layers. The prototypes obtained the highest sensitivity of around 1.3294 kPa^{-1} at a pressure of 600 kPa. Along with the use of PI substrates, polyvinylchloride (PVC) and nickel (Ni) were used as encapsulation layers and electrodes, respectively. The patches had a sensing area of 9 cm^2, containing nanowires embedded on one of the PI surfaces and exposed on the other PI substrate. The increase in pressure led to a simultaneous increase in the conductive pathways between the nanowires, which subsequently decreased the resultant resistance. The fabrication process was carried out using a glass substrate on top of which a seed layer of chromium and copper was layered using a magnetron sputtering technique. The samples were then spin-coated with a positive photoresist, AZ 4620, followed by the process of electroplating to form a thin layer of nickel. The thickness of this metal layer was around 3–5 μm to employ it as electrodes of these prototypes. Finally, the plasma etching was done on the samples along with the spin-coating of AgNW-PI nanocomposites. This was done by using another positive photoresist, AZ 4903, which was spin-coated along with the AgNW-PI mixture. Finally, acetone and sodium bicarbonate were applied to the samples to remove the photoresist and expose the silver nanowires, respectively. The schematic diagram of the explained fabrication process is explained in Fig. 4 [132]. The change in response was analyzed in terms of relative resistance, where a sensitivity of around 1.3294 kPa^{-1} was obtained with pressure extending to 600 kPa. The sensors were tested under static and dynamic conditions, where the changing rate of resistance varied between 0.8 and 1.0. Another similar work on the use of silver nanowires to form nanocomposites for strain sensing was based on their association with PDMS, to form highly stretchable and sensitive prototypes [133]. A sandwich structure was formed using two PDMS layers, with a thin film of silver nanowires being embedded between them. The sensors exhibited a tunable gauge factor ranging from 2 to 14, where the stretchability was as high as 70%. The sensors were piezoresistive in nature, where they were integrated into a glove to detect the motion of fingers. The synthesis of silver nanowires was done using the modified polyol method. These nanowires were drop-cast on a glass slide and patterned with a Kapton tape. The nanowires were dried and deposited on the glass slide via uniform heating using a lamplight. The gradual heating from the lamp assisted in the formation of a uniform and homogeneous mixture of the nanowires. Then, the Kapton tape was removed and the samples were annealed at 200°C to increase their electrical conductivity. Finally, the sensing structures were formed by casting the silver nanowires onto the liquid PDMS having a thickness of 0.5 mm. The samples were cured, peeled off, and attached to copper wires with silver paste for characterization purposes. In the end, another PDMS layer, having the same thickness as the other, was cast and

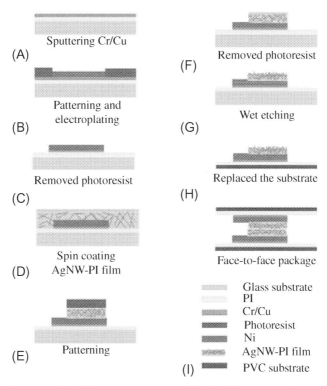

Fig. 4 Schematic diagram of the fabrication process of the flexible sensors developed from AgNW/PI composites. The process included several steps like (A) sputtering, (B) electroplating, (C) removal of photoresist, (D) and (E), spin-coating and patterning of the nanocomposite film and chemical etching [132]. *Image reproduced from H. Li, G. Ding, Z. Yang, A High Sensitive Flexible Pressure Sensor Designed by Silver Nanowires Embedded in Polyimide (AgNW-PI). Micromachines 10(3) (2019) 206.*

embedded onto the samples. The characterization was done based on the change in the relative resistance of the prototypes with respect to the applied strain. The application of pressure on the sensors caused a change in the electrical interconnections between two adjacent nanowires, thus leading to a change in response. The analysis was done to see the movement between 0 and 120 degrees to determine the change in the relative resistance. The human detection was carried out by bending the sensors within angles ranging between 10 and 90 degrees. The index and middle fingers were mostly moved to determine their movements in the virtual environment using a wireless smart glove system. Some similar work in this area [134–138] employs nanowires to develop sensors for strain–sensing purposes. These sensors have been employed on a range of substrates like TPU to depict the change in resistances for experiments conducted over 200 cycles. The nanowires have been mixed with other conductive materials like silver flakes to increase the electrical conductivity of the electrodes of the prototypes. A very high sensitivity of over 33% was obtained for strain conducted of every 1%. Other research works [139] have used certain nanoparticles like tin-doped indium oxide (ITO) to form flexible sensors with high gauge factors. A gauge factor ranging from 18 to 157 was obtained where the sensors contained the alignment of the ITO-based nanoparticles in parallel lines.

Another interesting work related to the use of nanotechnology for developing physical sensors has been the use of graphene platelets to form stretchable prototypes [140]. The sensors were developed to be used as wearable strain sensors, where a composite film of polyurethane and graphene platelets were formed using layer-by-layer laminating techniques. The sensors obtained a high sensitivity of around 1430 ± 50 S/cm along with a gauge factor of 150. The prototypes were employed for determining the real-time monitoring of certain human physiological parameters like pulse movement, finger movement, and check movement. A substrate of polyurethane of around 0.04 mm was obtained along with the thermal expansion of commercial graphite powder that was done to obtain worm-like structure-shaped graphene platelets. The fabrication was done by using a pressure-sensitive adhesive over the ultrathin polyurethane substrates to paste the graphene platelets at the center of the film. Another polyurethane film was placed on top of the graphene platelet film, along with connecting the copper wires on both ends of the graphene platelet film. The thickness of the entire film ranged from 0.12 to 0.2 mm. The sensors had good stability and reproducibility of the results that were obtained by dividing the changes in the response of the sensors into two equal halves and then comparing them. For measuring the pulse movement, a bass–pass filter with a frequency of 0.5–10 Hz was considered. For measuring the strain-induced limb movements, the sensors with a thickness of 0.17 mm were attached to the joints of fingers as well as the top layer of skin. The resistance values of the sensors increased with the increase in temperature when the experiments were conducted between temperatures ranging from 20°C to 160°C.

3.1.2 Thermal sensors

Thermal sensors hold an important aspect in the field of sensing as they are normally conjugated with another type of sensing application. The type of thermal sensors that are being developed using flexible materials depends on certain factors like thermal conductivity, thermal expansion, thermal diffusivity, coefficient of thermal expansion, and others for their uses in the fabrication process. The following are some of the significant types of thermal sensors that are developed fabricated using popular carbon-based allotropes.

One of the works includes the use of CNTs to form flexible sensors that were used as electrothermal film heaters [141]. The sensors were developed using rod-coating and spraying methods where conductive films of CNTs were generated. The sensors had a heating rate of $6.1°C\,s^{-1}$ at 35 V, along with sheet resistance of $94.7\,\Omega\,sq^{-1}$. The highest optical transmittance obtained from these transparent sensors was 72.04% at a wavelength of 550 nm. Initially, a solution of single-walled carbon nanotubes (SWCNTs) with DI water was prepared using sodium dodecylbenzene sulfonate (SDBS) as a surfactant. The mixture was ultrasonicated, centrifuged, and suspended to obtain the SWCNT's solutions. PET was used as substrates to form the heaters due to their high resistance and stiffness toward change in temperature. The sensors were also very lightweight and exhibited exceptional electrical insulating properties, which makes them a popular choice for developing temperature sensors. The PET substrates were initially ultrasonicated using DI water and ethanol to purify them. The substrates were then heated at 105°C, followed by the spraying of carbon nanotubes to obtain proper adherence. The spraying time was controlled to obtain sensors with different sheet resistances. Finally, the samples were post-treated with DI water and then annealed again to enhance the adherence between the SWCNTs and PET substrates. The residual surfactants from SWCNTs were removed by rinsing the samples with DI water, followed by drying them at 80°C. The response time of the sensors was less than 30 s, where real-time measurements were done with a temperature distribution having a variation of around 1.5°C. Another similar work based on the use of CNTs for the same was based on the design and development of the ionization gas temperature sensor [142]. These sensors showed different sensitivities to air and nitrogen based on a fixed temperature of 110°C. The sensors consisted of a three-electrode system where each of the electrodes were developed with silicon, having sizes of 27 mm × 8 mm × 450 µm. The sensors were processed through a certain number of processes like masking, photolithography, dry etching, and finally cleaning to form the electrodes. Two of the slices contained holes having diameters of 4 mm and 3 mm, for the cathode's migration and diffusion of ions, respectively. The collecting electrode of the sensor consisted of a rectangular square blind having a surface area of depth of 48 mm^2 and 200 µm, respectively. This blind was for collecting the positive ions as a result of the migration and diffusion processes. Fig. 5 shows the image of the tri-electrode CNTs-based temperature sensor [142]. Three different metals, namely titanium, nickel, and gold, were sputtered at the thicknesses of 50 nm, 125 nm, and 400 nm, respectively, on the inner side of the cathode and collecting electrodes. The samples were then annealed at 450°C for around 50 s to strengthen the bonds between the substrate and sputtered metals. This was followed by growing of CNTs vertically on the silicon substrates using thermal chemical vapor deposition (CVD) technique. The grown CNTs were finally transferred to the inner side of the cathode by the wetting transfer technique. The electrodes were separated using a polyester film having a thickness of 170 µm, each of which connected to three golden wires for connection purposes. The change in the output of these sensors were recorded in terms of collecting current with respect to the change in temperature. The value of current increased in terms of nA for six testing voltages ranging between 25 and 70 V. The sensors also showed a repeatable response when they were tested with the same

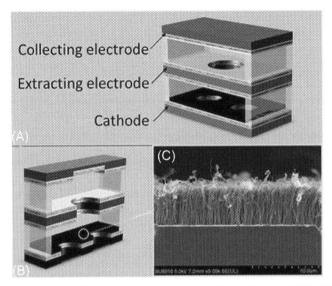

Fig. 5 (A), (B) Representation of the tri-electrode system formed using CNTs. (C) FE-SEM of the vertically formed CNTs film [142]. *Image reproduced from H. Song, Y. Zhang, J. Cao, Sensing mechanism of an ionization gas temperature sensor based on a carbon nanotube film. RSC Adv. 7(84) (2017) 53265–53269.*

experimental conditions having a gap of 1 month between the consecutive tests. The sensitivities of the sensors were 4.74 µA°C^{-1} and 22.72 µA°C^{-1} to air and nitrogen at equal temperature durations, respectively.

The use of nanoparticles to develop temperature sensors can also be varied based on the working principle of the sensors. For example, one of the interesting works involved the combination of SWCNTs, MWCNTs, and silver nanowires in forming stretchable temperature sensors that operate on a Seebeck effect [143]. The sensors had a stretchability of around 40%, along with high linearity and repeatability in their responses. The sensors were formed by printing a couple of nanostructured wires and embedding them inside a soft polymeric film. The lengths and diameters of each type of nanoparticle were different from each other. The lengths varied between 5 and 30 µm for SWCNTs, whereas it was around 3 to 30 µm for MWCNTs. Suspension solutions were made prior to the printing technique by mixing SWCNTs, MWCNTs, and silver nanowires in different solutions. SWCNTs and/or MWCNTs were mixed in methanesulfonic acid, whereas the suspension solution considered for silver nanowires was isopropyl alcohol. The printing process was immediately followed by the removal of the solvent using a vacuum–assisted deposition technique to improve the quality of printing. The residual quantity of the solvent was evaporated spontaneously or was removed by washing the samples with DI water. This was followed by the encapsulation of the printed nanoparticles on the surface of the cured PDMS. The thickness of PDMS was also optimized by varying its quantity from 2 to 8 g. The nanofiller used to embed the nanoparticles on the PDMS surface specifically was taken off by using dichloromethane (DCM) to dissolve it. The used nanowires had a length and diameter of 10 μm and 60 nm, respectively. The prototypes operated as a thermocouple to determine the generated seeback voltage, which was generated as a difference between hot and cold points. One of the major goals of these sensors was the change in their response due to the change in temperature, irrespective of the degree in the change in strain. The sensors were tested for repeatability by varying the temperature between 0°C to 90°C, while the strain was maintained at 0°C. The sensors also had repeatable responses for strain cycles varied between 1 and 1000 cycles, where the strain value was changed from 0% to 40%. A few others [144–148] include certain fabrication techniques like nanoprinting to form temperature sensors used for certain applications like biometric sensing, soft robotics, and cryopreservation of organs. These sensors are developed on flexible Kapton substrates where the electrodes were formed using a composite of copper and nickel. Aerosol jet printing technique was employed where powers of 100 mW and 400 mW were used to form the ultrathin layers of the composites. The sensors should have very high sensitivity toward the change in temperature, along with the porosities ranging between 9% and 24%. The sensors also showed excellent repeatability in their results during the bending cycles performed more than 200 times at three different radii. The use of thermoelectric nanogenerators has come to a very interesting stage where they are employed on different applications as a result of their dynamic behavior. These sensors have also been used as self-powered temperature sensors [149], where nanocomposites were developed using tellurium nanowires and poly (3-hexyl thiophene) (P_3HT) to form thermoelectric-based polymer composites. The sensors had a quick response time of 17 s, where a detection sensitivity of around 0.15 K was obtained for temperature under the ambient atmosphere. Another interesting work related to the wireless operation of the sensors was done to determine the real-time monitoring of the temperature of blood packages [150]. The sensors were developed using silver nanowires conjugated with flexible colorless polyimide (CPI) films that were integrated together in a data transmission circuit. Three-dimensional printed mold-based technique was employed to pattern the silver nanowires embedded with CPI film. The response of the sensors was monitored with the change in resistance, where a considerable change was observed between −20°C and 200°C. The sensors also showed a stable response for 5000 cycles having a bending radii of 5 mm. Rat blood tested with these sensors obtained stable and consistent results. Another work in the use of graphene for the development of temperature sensors is related to the fabrication and implementation of wearable sensors that were used as infrared (IR) photodetectors as well as temperature sensing [151]. The conductive materials considered to form the electrodes were solar exfoliated reduced graphene oxide (SrGO) and graphene flakes, whereas PI was considered as substrates to form the prototypes. The choice of PI was due to assistance in the reduction of mobility and recombination of the photo-generated electrons of graphene, thus causing IR detection. The sensors were capable of measuring body temperature ranging between 35°C and 45°C, with temperature coefficients of -41.30×10^{-4}°C^{-1} and -74.29×10^{-4}°C^{-1} for graphene and SrGO, respectively. The utilization of reduced graphene oxide (GO) has been done quite extensively [151–155] for developing flexible temperature sensors due to their high stability, high electrical conductivity, capability to form, and can be stored without forming clumps. The response of the temperature sensors using reduced graphene oxide (rGO) has shown high linearity ($R^2 = 0.99$), accuracy (0.1°C), and durability for measurements done between 25°C and 45°C. One of them were developed using polyethyleneimine and facile spray-dipping as substrates and fabrication technique, respectively [156]. One of the works involving the use of rGO to form flexible temperature sensors was related to a comparative study in the fabrication and implementation of sensors developed by SWCNTs, MWCNTs, and rGO [157]. Even though all the sensors showed good linearity, sensitivity, and repeatability in their responses, the results obtained from rGO-based temperature sensors were the most optimal ones. These sensors also showed excellent mechanical characteristics, which was essential as flexible prototypes.

The sensors were stable and unresponsive toward the change in other attributes like humidity and gases. The sensors showed good results for them to be considered to be used as an artificial skin for robots. The sensors were fabricated with PET as substrates, with three different types of substrates. Fig. 6 depicts the schematic diagram of the fabrication process of the rGo-based temperature sensors [157] Initially, after purifying the substrates with acetone, alcohol, and DI water, plasma etching on was done to create irregular microstructures on the sensing surface. This was to increase the degree of adherence of the conductive materials on the PET substrates. This was followed by using screen printing technology to print two conductive thin wires attached with silver paste. The samples were then annealed for half an hour at 100°C, followed by the use of air-spray coating method to coat the microstructures on the surface of PET with conductive material. The spraying time was adjusted to 10 s, which was immediately followed by exerting a transparent tape on the conductive material with a force of 500 N. This tape was used as an insulating layer to the prototypes. Experiments were conducted to determine the change in resistance with respect to the change in temperature from 30°C to 100°C, with a step of 5°C. The variation of resistance with temperature was obtained as linear for both rGo and MWCNTs, whereas it was nonlinear for SWCNTs. The response and recovery times of these sensors when tested with a temperature of 45°C was 1.2 s and around 7 s, respectively.

Another work related to the use of CNTs to develop temperature-based sensors was done by Karimov et al. [158], where MWCNTs were mixed with other conductive materials on elastic substrates. The MWCNTs, having a diameter ranging between 10 and 30 nm, were deposited on elastic polymeric tapes having a thickness of 35 μm. The thicknesses of the deposited CNTs were around 300 to 430 μm. The length and

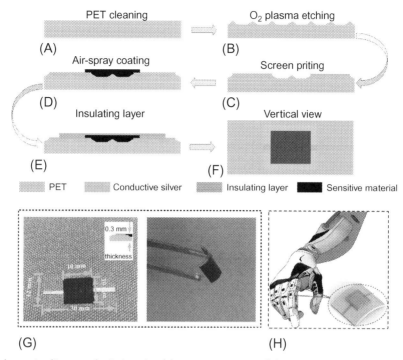

Fig. 6 Schematic diagram depicting the fabrication process of the temperature sensors developed from rGO, MWCNTs, and SWCNTs. Five steps were followed during the fabrication, namely (A) cleaning, (B) O_2 plasma etching, (C) screen printing, (D) air-spray coating, and (E) employing a commercial tape as an insulating layer [157]. *Image reproduced from G. Liu, Q. Tan, H. Kou, L. Zhang, J. Wang, W. Lv, H. Dong, J. Xiong, A flexible temperature sensor based on rGO for robot skin used in internet of things. Sensors 18(5) (2018) 1400.*

width of the developed electrodes had a range between 4 and 6 mm and 3 to 4 mm, respectively. The changes in the responses of the sensors were analyzed in terms of DC with respect to the increase in temperature. The sensors depicted a wide range of sensitivity in terms of its experimental results. The experimental error during the measurements was around $\pm 1°C$. A similar work with the growth and use of CNTs to form nano temperature sensors was explained by Kuo et al. [159], where the conductive material was laterally grown between two electrodes using microwave plasma CVD techniques. The fabrication of the sensors was done using MEMS-based technique, followed by using wire bonding as the ultimate step. The current-voltage relationship was measured using these sensors at different temperatures to obtain a linear relationship between the resultant resistance and input temperature. Initially, thin films of oxide and nickel were deposited on the silicon wafers, followed by patterning of titanium films on them. This was exposed to the side where CNTs were grown in horizontal directions. With the interdigital distance between two consecutive electrodes varying from 0.6 to

2.7 μm, the sensors were sent to electron cyclotron to grow the CNTs. Methane was used as a source in addition to nitrogen to increase the bonding force between the carbon atoms in the nanotubes. The chamber pressure, RF power, and substrate temperature were set to 30 Torr, 800 W, and 400°C, respectively. Finally, the wire bonding was done using a printed circuit board. The entire circuit containing the sensors were kept inside the oven to maintain fixed temperature conditions during the experiments. Two different types of results were obtained due to the metallic and semiconducting nature of the nanotubes. The metallic nature increased the resistance with an increase in temperature, whereas the semiconducting nature decreased it. The slope values for the metallic and semiconducting nature were 1.196 and -0.897, respectively. The maximum and minimum resistance values were obtained at 65°C and room temperature, respectively.

3.1.3 Electric sensors

The use of nanoparticles-based sensors for developing electrical sensors has mostly been in the design and fabrication of field-effect transistors (FETs). Even though FETs have been developed for quite some time [160–165], the rise of flexible FETs has been much more advantageous compared with its counterparts due to their high sensitivity, lower fabrication cost, and higher mechanical flexibility. These devices have also been used for the detection of certain types of analyte, like proteins [166–168] and DNA [160, 169, 170], which are separately explained under the section of bioanalytic sensors. Here, we showcase some of the electrical sensors that have been developed using some of the popularly used carbon-based nanoparticles like CNTs and graphene.

Graphene has been primarily favored to develop flexible FETs due to their high electron mobility, zero bandgap, and very high value of breaking strengths [171]. The use of graphene in devices has formulated the transduction of biomolecular charges into current-voltage characteristics. Researchers are constantly trying to develop graphene-based FET having a low-noise, low-voltage operation; in vivo biocompatibility; and high selectivity [172]. The array formation of graphene-FET (GFET) has also been found to have a high throughput that is advantageous for integration on a chip-scale. One of the works [173] involves the use of graphene to develop FET array, having polymer and ion gel as substrate and gate, respectively. A high on-current was achieved with a low voltage. The mobility shown by the holes and electrons of these devices were 203 ± 57 and 91 ± 50 cm^2/(V.s), respectively, having a drain bias of -1 V. Graphene has also been mixed with other nanowires [174] to form nanowire field-effect transistor (NW-FETs), that have been subsequently used as nanoscale bioelectronics interfaces to detect analyte from tissues and cells. The combination of graphene and nanowires on electrogenic cells have been used to detect the extracellular signals from embryonic chicken cardiomyocytes. The devices had a signal-to-noise ratio of less than 4, where the peak-to-peak widths of the recorded signals increased as a function with the increase in the area of the graphene-based FET devices. One of the intriguing characteristics at work lies in the conjugated use of one-dimensional silicon nanowires, along with the two-dimensional graphene, to form the FET devices. Even though pH sensing is not directly involved with biomedical applications, the prevention of certain gastroesophageal diseases like celiac and ulcerative colitis requires the body to reach a specific pH value. The detection of pH using certain types of devices like solution-gate field-effect transistors (SGFET) [175] has been done using graphene. The principle of polarization was used for graphene and aqueous electrolyte interfaces to perform two separate charges of hydroxyl and hydroxonium on the sensing surfaces. A very high hole and electron mobility of 3600 cm^2/Vs and 2100 cm^2/Vs were obtained due to the ambipolar characteristics of graphene. A supra–Nernstian response of 99 meV/pH was obtained for these devices by performing a shift of the negative gate potential.

One of the interesting works for graphene-based FETs operating at high frequencies was based on their performances at the gigahertz-frequency range [176]. Flexible substrates were used to develop the devices, where graphene was developed using a CVD technique. The sensors generated unity current-gain and unity-power gain for frequencies ranging up to 10.7 GHz and 3.7 GHz, respectively. The channel was fixed at 500 nm for these GFET devices. A power gain was achieved in the GHz regime for strain applied over 0.5%. The devices were fabricated using a bottom-gate device structure having polyethylene naphthalate (PEN) with a thickness of 127 μm as substrates. Electron beam lithography and lift-off processes were employed to develop the bottom gates, formed with titanium and gold-palladium alloy. The dielectric of the gate was formed with a hafnium dioxide where the channel length was fixed at 6 nm. The patterning of the CVD-developed graphene was done using lithography and reactive ion etching techniques. Finally, metals (Ti/Pd/Au) used to form the source and drain electrodes were evaporated to contact the formed graphene. The gate length and source-to-drain spacing were fixed at 500 nm and 900 nm, respectively, having an effective channel of 30 μm. Negligible leakage current was obtained for the devices, having a source-to-gate current below 0.5 pA. The mobility obtained for these devices was around 1500 cm^{2}V^{-1}s^{-1} for a gate to source voltage of −0.25 V. The contact resistance of the devices was around 300 Ω-μm. The measured values of g_m and output resistance obtained with strain ranging between 0% to 1.75% were 5.1 mS and 259 Ω, respectively. A maximum current density obtained from these devices was 0.28 mA/μm, which is at par as per other CVD-based graphene used for GFETs. The devices had a channel length of 190 μm, where electron and hole mobility were obtained as 204 cm^{2}V^{-1}s^{-1} and 476 cm^{2}V^{-1}s^{-1}, respectively. The resulting dielectric constant and gate capacitance values were 6.9 and 2.05 × 10^{-7} – F/cm^2, respectively.

Another interesting work related to the use of nanoparticles to design and develop transistors on flexible substrates was described by Fisichella et al. [177]. The researchers employed a wafer-scale strategy to develop GFETs on PEN substrates. These devices have a wide range of applications, from consumer devices to biomedical applications.

The conductivity of the channel was modulated by a buried gate via an insulating thin film of Al_2O_3. The deposition of this thin film was done using plasma-enhanced atomic layer deposition, where the conditions were optimized to minimize the degradation of the dielectric performances. Initially, PEN was thermally flattened and bonded with a silicon wafer via a mechanical lamination process. The lamination was done using a double-face thermal release tape then cutting them along the edges of the wafer. The insulating layer was coated on the polymer with a thickness of 100 nm, carrying out the process at a temperature lower than 100°C. This was followed by the formation of gate pads made of aluminum, having a thickness of 200 nm. The pads were formed using the sputtering metal technique. Graphene was grown using the CVD process on large-area copper foils. Following their growth, the membranes were transferred to the substrates using PMMA via a wet transfer process. The source and drain of the FET was formed using the lift-off process with nickel and gold, having a thickness of 30 nm and 120 nm, respectively.

Apart from graphene, CNTs have also been used to develop FETs [178–180], where the gate and the conducting channels have been developed using networks of CNTs. Some of them [181–183] have used Parylene as a gate insulator, obtaining device mobility and on/off ratios of $1 \ cm^2 \ V^{-1} \ s^{-1}$, respectively. The value of on/off operated as a function of the insulating layer, formed by the Parylene N. The use of these sensors for detection of biomolecules like chromogranin A has also been done [184] where critical organs of the body, like synaptic terminals of cortical neurons, were monitored to detect these biomolecules. In-situ immobilization was done with an antibody against CgA to obtain prototypes with high selectivity. The sensors also depicted high sensitivity and real-time detection capability. Another study [185] showed the fabrication and implementation of SWCNTs-based devices to study the biocompatible interactions with living cells, where the noncovalent functionalization of the carbon nanotubes with bioactive sugar moieties was studied to determine their sensing capability. The CNTs were surface-functionalized with bioactive sugars, where PC12 cells were adhered and grown. As soon as the solutions with high potassium content were flown through the cells causing an influx of calcium ions, the CNTs secreted catecholamine molecules. The secretion of these molecules caused a change in the response of the SWCNTs in terms of current due to the noncovalent attachment of the aromatic rings of these molecules with the sidewalls of the nanotubes. The operating principle is shown in Fig. 7 [185].

3.1.4 Optical sensors

The use of nanotechnology to develop optical fibers has been popularized for the last 2 decades [186, 187]. These sensors have been largely preferred over their counterparts due to some of their distinct advantages like large bandwidth, low power loss, lighter weight, lower cost, higher flexibility, higher security in the transmitted data, lesser attenuation, and lesser interference with noise during the transmission of data. Some of the optical

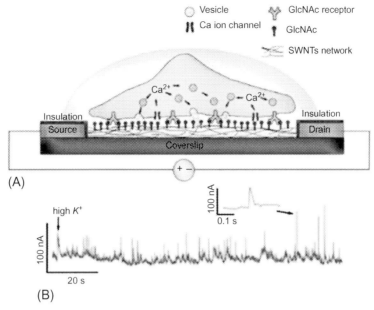

Fig. 7 Schematic diagram of the use of SWCNTs-based FET for (A) triggered exocytosis. (B) Representation of the response of the CNTs for exocytosis that were triggered by high potassium stimulation, causing a change in the influx of calcium ions. The drain-to-source voltage set for these experiments was 0.4 V [185]. *Image reproduced from S. Liu, X. Guo, Carbon nanomaterials field-effect-transistor-based biosensors. NPG Asia Mater. 4(8) (2012) e23.*

sensors like surface plasmon resonance (SPR) has been used as the most advanced and powerful label-free detector of biomolecules. Among biomedical applications, these sensors have a vast range like drug screening, medical diagnostics, biotechnology, food safety, and others due to their analytical capability. Similar to the electrical sensors, CNTs had been preferred to develop flexible optical sensors, but with the advent of graphene, researchers today largely prefer graphene. SPR has also been used for other biological applications like the determination of diseases among livestock [188]. SPR executes either of the methods, namely physical absorption or covalent attachment for the detection of the analyte using biorecognition elements. Three different detection techniques, namely direct, sandwich, and competitive inhibition assay, have been used for measurements by SPR biosensors. Some of the metallic nanoparticles, i.e., gold [189], palladium, [190] and platinum [191], have been used to increase the sensitivity of these devices for the detection of biomolecules. One of the major advantages of these sensors is their rapid and label-free detection of biomolecules. One of the works [192] showcasing the use of optical fibers has been done for the detection of in-situ DNA hybridization using reflective microfiber Bragg grating. Temperature compensation of the refractive index was done during the measurement of the wavelength separation between two resonances. The sensing surface was functionalized using a monolayer of poly-L-lysine (PLL) containing synthetic DNA sequences. These sequences assisted in the label-free biorecognition via achieving high specificity for the target molecule. Real-time monitoring was done with the optical fibers with a detection limit of 0.5 μM via testing different concentrations of DNA solutions. The advantages of these sensors were their rapid response, high sensitivity, and compact and portable nature of the sensing platform. The sensors were formed using a 193 nm excimer laser and a phase mask method. The phase mask contained a period of 1070.49 nm, having a repetition rate and energy set to 200 Hz and 3 mJ/pulse, respectively. The microfiber in the sensors had an energy density of 120 mJ/cm^2 that took around 5 mins for grating inscription. The diameters of the microfibers ranged from 3 to 10 μm, while the reflection spectrum had a diameter of 3.9 μm. The bandwidth set at 3 dB had two modes, namely fundamental and high-order, having their wavelengths set at 2.1 nm and 0.27 nm, respectively. When the sensors were immersed in the ssDNA solution having a concentration of 20 μM, the refractive index of the surface of the fiber gradually increased with time due to the binding of the ssDNA. The responses of the sensors were repeatable in nature with minimal problems of non-negligible memory effect. Similar to DNA, the detection of proteins was also done [193] using these types of sensors. The sensors were developed with GO and employed for the monitoring of C-reactive protein (CRP) using optical fiber Bragg gratings (FBGs). The highest sensitivity of 10 μg/L was achieved at low concentrations along with the linear response of the sensors. The linearity was maintained for concentrations ranging between 0.01 mg/L to 100 mg/L, which had a detection limit of 0.01 mg/L. The FBGs were etched and coated with an anti-CRP antibody GO complex to increase their sensitivity. The sensors were also tested in the presence of other interfering factors like urea, creatinine, and glucose to validate the specificity of the coating. The affinity constant extracted from the experiments was around 1.1×10^{10} M^{-1}. rGO has also been used for the detection of biomolecules using these optical sensors. Sridevi et al. [194] have explained the detection of glucose and glycated hemoglobin using etched FBGs. The sensors were coated with minophenylboronic acid (APBA) and functionalized with rGO. The adsorption of glucose or glycated hemoglobin in the sensors caused an immediate shift in the Bragg wavelength due to the sensitivity caused by the APBA–rGO complex. The immediate response of the sensors was due to the formation of a five-membered cyclic ester bond formation complex created between the APBA and glucose molecules. Fig. 8 [194] shows the operating principle of this explained phenomenon. The detection of the sensors for D-glucose was 1 nM, having a linear response ranging between 1 and 10 nM. The linear response of the sensors for glycated hemoglobin was little bit higher than that of glucose, specifically from 86 to 023 nM. A change in the concentration to 10 nM created a simultaneous change in the 4 pm in the Bragg wavelength.

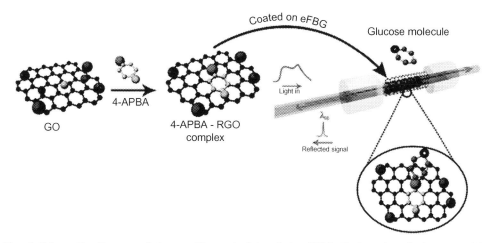

Fig. 8 Schematic diagram of the working principle of the APBA-rGo-based optical sensors. The prototypes were used for the detection of D-glucose and glycated hemoglobin using the selectivity coated on their sensing surface [194]. *Image reproduced from S. Sridevi, K.S. Vasu, S. Sampath, S. Asokan, A.K. Sood, Optical detection of glucose and glycated hemoglobin using etched FBGs coated with functionalized rGO. J. Biophoton. 9(7) (2016) 760–769.*

CNTs have also been considered to develop optical sensors where they have been employed for different biomedical applications as wearable sensing prototypes for strain sensing of human motion, movement, and breathing. CNTs have been a popular choice for developing optical fibers due to unique optical properties [195]. These materials are deposited via different methods like Langmuir-Blodgett (LB) [196, 197], optical deposition [198, 199], spray coating [200, 201], drop-casting [202, 203], and dip coating [204, 205] with controlled deposition parameters. Among these techniques, LB would be preferred for the experiments where high precision control is required over the thickness of the monolayers deposited on the CNTs. The sensors developed by Choi et al. [206] consisted of multiple electrodes operating with multipoint detection. PDMS was used as the substrate of the sensors, whereas a single polyvinylidene fluoride (PVDF) was used as the receiver. The sensors were able to detect the multipoint strain sensing using a single electrode up to a maximum strain of 30%. Three-point CNTs and six-point CNTs were also conducted to determine S-curve contained with expansion and compression components for finger bending and flexible phantom bending, respectively. Due to the emission of high acoustic pressure, CNTs were used as the optical absorption material, whereas PDMS and PVDF were used as the stretchable material and receiver of the system, respectively. The signals generated by the CNTs travelled through the PDMS tube to reach PVDF, which was operating on a large bandwidth without an operation voltage. The sensors were fabricated using a mold-based technique via an acrylic structure with holes. The length, width, and thickness of the molds were 150 mm, 5 mm, and

3 mm, respectively. The spacing between the CNT's points was 30 mm. A composite of CNT–PDMS was formed and poured parallel and perpendicular to the acrylic fixture and axial directions of the tube, respectively. The core diameter of the optical fiber bundle was 600 μm with a wire length of 1 m. The signals received by PDVF was amplified with a preamplifier performing an average of eight acquisitions. The experimental results showed a maximum light power irradiation of 0.65 mJ/cm^2, along with an acoustic pressure of 60.5 kPa. The optical and acoustic loss of this system were calculated to be 0.192 dB/mm and 0.301 dB/m, respectively, along with a standard deviation of 0.48%. For the glove-bending experiments, real-time experiments were performed where fingers were bent at angles of 0, 45, and 90 degrees. Another work explaining the use of CNTs and PDMS for developing optical fibers was related to the application based on ultrasound imaging [207]. MWCNTs–PDMS nanocomposites were formed and used to coat the distal ends of the optical fibers. Two types of organogel were used as solvents for mixing with MWCNTs, which were then subsequently used to overcoat with PDMS. These devices had a broad frequency bandwidth of 39.8 GHz that was used for all-optical, pulse-echo, ultrasound imaging applications. Initially, MWCNTs were dispersed in xylene using oleylamine-functionalized pyrene as a ligand. This was mixed with PDMS, followed by exerting dip-coating technique to deposit three separate composites on the optical fibers. The distal ends of the optical fibers were situated at a distance of 3 mm from the sensing surface. Mechanical stripping of buffer layers was done to the multimodal optical fibers and made perpendicular to their corresponding axes. The mixture of PDMS and Xylene was done at a ratio of 1:8, followed by curing at room temperature for 24 h. During the generation of optical ultrasound, the efficiencies achieved by the MWCNTs–PDMS, MWCNTs–Xylene/PDMS, and MWCNTs–gel/PDMS coatings were 0.37, 0.42, and 0.65 MPa mJ^{-1} cm^{-2}, respectively. The bandwidths for the MWCNTs–gel/PDMS and MWCNTs–PDMS coatings were 29.2 and 26.2 MHz, respectively, whereas the peak frequencies were obtained at 19.4 and 20.9 MHz, respectively. The experimental process of ultrasound imaging was carried out using a miniature probe comprised of two fibers for transmission and reception purposes. One of the attributes of this work was the variation in the results of the all–optical ultrasound transmitters, corresponding to the variation caused to the homogeneity and thickness of MWCNT's layer in the composite coatings.

3.2 Electrochemical sensors

The fabrication and execution of electrochemical sensors using nanotechnology have been done for the last 2 decades [208–211]. Although the range of electrochemical applications is pretty wide, the explanation of research work in the following text shows the electrochemical sensing related to biomedical applications. The electrochemical sensors largely depend on the reactions taking place on the sensing surface, subjective to the change in ionic current with respect to the concentration of the target material. Different kinds of nanomaterials have been associated with the electrochemical sensors to increase their selectivity and specificity toward the target molecule. The ionic reactions taking place in the electrochemical cell large depends on the type of conductive material opted to develop the sensors. Table 2 presents a comparative study between some of the major works done in this area in recent years.

One of the works [219] explains the use of copper oxide nanowires to develop flexible sensors for nonenzymatic glucose detection purposes. The nanowires were fabricated using a facile two-step process with wet-chemistry synthesis and direct calcination methods. The glucose sensors exhibited high sensitivity of 648.2 $\mu A\ cm^{-2}\ mM^{-1}$ when they were used with 50 nM of sodium hydroxide solutions. The prototypes had a fast response time of less than 5 s when the experiments were conducted using cyclic

Table 2 Comparison between the works on electrochemical sensors based on their design and corresponding advantages.

Sl. No.	Materials used	Advantages	Application	Reference
1	Silicon nanowires, gold, polyimide	Scalable and transferable fabrication technique, high sensing performance	Detection of Avidin	[212]
2	Graphene, platinum, gold, Manganese oxide	High sensitivity, wide linear range, low detection limit, satisfactory selectivity, excellent reproducibility	Glucose sensing	[213]
3	Zinc nanowires, gold, polyester	High sensitivity, low Michaelis-Menten constant, fast response time	Glucose sensing	[104]
4	Zinc oxide, gold, chromium	Low detection, high sensitivity	Ethyl glucuronide in human sweat	[214]
5	Copper nanowires	High sensitivity, stability in response	Glucose sensing	[215]
6	Graphene, Molybdenum disulphide	High reproducibility, high flexibility, good selectivity, fast response	Glucose and lactate sensing	[216]
7	Carbon nanotubes, graphene, platinum	High stability in response, high sensitivity, high selectivity, and reproducibility	Secretion of hydrogen peroxide from live cells macrophages	[217]
8	Carbon nanotubes, polyaniline	Simple preparation technique, fast response, wide linearity in response, high stability	pH sensing	[218]

voltammetry (CV) method. Further results were obtained regarding the sensing of electroactive compounds like ascorbic acid, uric acid, acetaminophen, and fructose using the copper nanowires based nonenzymatic glucose sensors. Some of the attributes of these sensors include high accuracy in their measurements, high precision in the quantification of the tested concentrations in real-time samples obtained from human serum, and high sensitivity. Similar oxide-based nanostructured sensors were also presented by Shetti et al. [220], where zinc oxide nanostructures were used to develop electrodes for electrochemical sensing in biomedical applications. Zinc-based nanostructures had been largely favored by the researchers for biomedical applications due to their availability in various forms like nanolayers, nanoparticles, and nanowires. They assist in increasing the sensitivity of the prototypes via enhancing the binding ability of the sensing surface with the targeted biomolecules. These nanostructures have also been considered for forming sensors for certain applications like detection of infectious diseases and pharmaceutical compounds. In one of the works related to zinc oxide nanowires, Ali et al. [221] used the selective potentiometric process for the detection of uric acid. Immobilization of uricase was done onto the zinc oxide nanowires to increase the sensitivity of the sensing surface. The nanowires have a length and diameter of 1.5 µm and 900 nm, respectively, where grown on gold-coated flexible plastic substrates. The immobilization of uricase led to an increase in selectivity, stability, and reproducibility of the results of the developed sensors. The results had a linear range of 1 to 650 µM, which was extended to 1000 µM by using Nafion membrane on the sensors. The sensors had a high durability with the presence of a membrane. Their responses were not affected from the interference of other molecules like ascorbic acid, glucose, and urea. These prototypes had a response time between 6.25 and 9 s.

Along with the use of nanowires, the substrates have also been varied to optimize the cost of fabrication and mechanical attributes of the formed sensors. One of the works [222] explained the use of paper as substrates for low-cost, point-of-care prototypes for medical diagnostics. Template-assisted electrodeposition and adhesive-based tape techniques were used to fabricate high-density nanowires directly on the paper. Three different types of conductive metals, namely platinum, nickel, and copper, were fabricated to determine the ones with the optimal performances on paper substrates. The work was done to determine the performance of dry paper-based electrodes toward the change in impedance when subjected to electrocardiogram signals without the assistance of wet-gel adhesives. The sensors were also employed as cathodes in batteries, targeted for energy- harvesting applications. The batteries had dimensions of 4 cm^2 and were capable of generating power up to 6 mW. The use of nanocomposites was also encouraged [223] to develop wearable electrochemical sensors for detection of glucose at low concentrations. rGO was used as one of the conductive materials that was synthesized and micropatterned using MEMS-based technique on flexible substrates. PI was used as a substrate due to its biocompatible nature. Then, electrochemical deposition technique was employed to coat the rGO layers with an alloy of gold and platinum nanoparticles. Finally, composites developed from chitosan and glucose were integrated on the sensing surface to make them selective toward the target molecule. The testing was done using real samples where human sweat mixed glucose was tested to validate the capability of the sensors to detect low concentrations. The sensors exhibited a high sensitivity of 45 μA/ mMcm^2 with a detection limit of 5 μM. The detection range, response time, and linearity of these sensors were 0–2.4 mM, 20 s, and 0.99, respectively. Another paper-based work as explained by Dong et al. [224] was related to fabrication and implementation of flexible sensors using manganese oxide and graphene. The sensors were used for monitoring the secretion of hydrogen peroxide in live cells. Afterward, the prototypes were fabricated using a single-step of electrochemical reduction of GO and loading of manganese oxide nanowires. The operation was carried out based on electrocatalytic activity for nonenzymatic detection of hydrogen peroxide. The output of the sensors was stable and reproducible, achieving a highly selective nature. The linear range limits of detection of the sensors were 0.1–45.5 mM and 10 μM respectively. The sensitivity attained by these prototypes was 59 μA cm^{-2}. mM^{-1} during their experimentation on real-time tracking of the secretion of hydrogen peroxide from live cells macrophages. Other advantages of these sensors were found in their flexible nature, tailored shapes, and customized properties to obtain high quality performances during the point-of-care testing.

The modification of electrodes with conductive polymer-based nanocomposites has been done in another way [225] to form the electrodes of the sensors for the detection of neurotransmitter acetylcholine. The nanocomposites were formed using iron oxide nanoparticles and PEDOT at definite proportions. The fabricated prototypes were used to test acetylcholine taken from Alzheimer's patients. Different types of enzymes like acetylcholinesterase (AChE) and choline oxidase (ChO) were immobilized on the nanocomposite-based surface that had been modified using fluorine-doped tin oxide (FTO). Electrochemical impedance spectroscopy and CV were used to determine the change in the output of the sensors. The sensors showed high sensitivity and stability in their responses toward the tested samples. The selectivity of the sensing assisted the prototypes to obtain a linear range between 4.0 nM and 800 μM. The response time and detection limit of the sensors for the chosen concentrations were 4 s and 4 nM, respectively. The testing of glucose using nanowires has been shown in one of the interesting works by Fan et al. [226] where flexible, disposable, and transparent electrodes were formed for testing purposes. The electrodes modified with graphene and copper nanowires were developed as a substitute for glassy carbon electrodes (GCE). Four different steps were followed to form the electrodes, where spin-coating technique was used to deposit copper nanowires on the graphene-modified PET substrates. The characterization of the sensors was done using scanning electron microscope (SEM) and X-ray power diffraction to study the morphology and phase structures, respectively. Experimental setup consisted of four-point probe system and CV to determine the changes in the resistance and electrochemical properties of the sensing prototypes. The sensors depicted a high sensitivity of 1100 μA/mM cm^2 along with a detection of 1.6 μM. The linear range of the sensors was from 0.005 to 6 mM, while it showed excellent anti-interference ability. One of the interesting uses of nanowires in connection with wireless use of the sensors was done by Ali et al. [227] to develop a sensing system for detection of glucose. The sensing surface contained with zinc nanowires was immobilized with glucose oxidase enzyme along with a coating of Nafion membrane. A communication protocol employing Short Message Service was used to transmit the data wirelessly to the monitoring unit. The sugar levels of a patient were monitored and transmitted to a centralized system for analysis purposes. The fabrication of the zinc oxide nanowires was done on silver wire having a length and diameter of 3 cm and 250 μm, respectively. The silver wire was washed and dipped into a seed solution twice that was made of 0.025 M zinc nitrate and 0.025 M HMT. The wire was then washed with DI water and dried at ambient temperature. The immobilization was done by rinsing the nanowire-based wire with glucose oxidase, followed by rinsing it with PBS to achieve a hydrophilic surface. The linking between the glucose oxidase and zinc nanowires were strengthened via adding 2 μL aqueous solution containing 2.5% glutaraldehyde and 0.5% Nafion. The developed sensors were tested with glucose solutions of 100 μM to determine their electrochemical responses. The signals collected from the sensors were passed through an amplifier that filtered the noise generated during the transmission.

Printing techniques have also been used to develop electrochemical sensors for biomedical applications, where a range of substrates were used to form the biosensors. For example, Knieling et al. [228] described the fabrication of wearable lactate sensors using PET foils as the substrates. Redox cycling was done to amplify the signals obtained from the measurements done on the biomolecules' concentration in fluids. The voltages obtained for oxidation and reduction during the redox reactions were 300 mV and −250 mV, respectively. Evaporating, wet etching, and screen printing technique were used to develop the gold electrodes that acted as counterelectrodes during the redox reactions. The working and reference electrodes were also developed using screen printing technique. Another application of the printing technique to develop electrochemical sensors for biomedical applications was done by Kuretake et al. [229] to form prototypes for the detection of ethanol vapor. The sensors were applicable as enzymatic biosensors for breath analysis. Chromatography paper was used as the enzyme-supporting layer, along with a liquid phase layer on top of the electrodes. The electrodes were formed by placing chromatography paper on top of the electrodes and performing screen printing technique. Two types of enzymes, namely alcohol oxidase (AOD) and hydrogen peroxidase (HRP), were used along with an electron mediator for the ethanol reactions to take place. Screen printing was done to develop the carbon-based electrodes. The sensors were able to measure ethanol vapors at a concentration ranging between 50 and

500 ppm. The ethanol was oxidized to acetaldehyde via using oxygen as the electron acceptor to generate hydrogen peroxide as the by–product. The response time of sensors was around 40 s for the currents to increase due to the reaction of the enzymes. Another interesting work [230] on the use of flexible printed sensors for electrochemical applications was done to develop screen-printed thick film reference electrodes on PET substrates. A composite made of glass and potassium chloride was formed, on which the screen printing was done to develop thick film Ag/AgCl/KCl electrodes. The characterization and experiments were performed using the formed sensors at a fixed voltage for three different bending radii of 3, 5, and 7 mm. The sensors were used for pH sensing and supercapacitor applications. Some of the attributes of these prototypes were their flexibility, high endurance, and small size. Sensitivities of 9 μA/pH and 75 μF/cm^2 were achieved for pH measurements and supercapacitive sensing, respectively. Initially, potassium chloride (KCl) powder was mixed with isopropyl alcohol and annealed at 80°C for 2 h to remove the alcohol content. This KCl powder was then mixed with free glass powder composite at a ratio of 1:1. Then, ball milling was done on this mixture to form a homogeneous blend. This mixture was again cured at 80°C for 2 h and then subsequently mixed in an agate motor with poly (methyl methacrylate) and poly (butyl methacrylate) as the polymers and butyl carbitol as the solvent. This blend was again cured at 120°C to form a paste for screen printing purposes. Fig. 9 [230] shows the schematic diagram of the fabrication steps of the screen printing process. After the silver paste was printed on the PET substrates, the samples were cured at 120°C for an hour. The two ends of the silver paste were dip coated separately on chloride solution and sodium

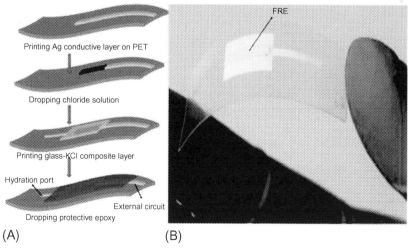

Fig. 9 (A) Schematic diagram of the steps of fabrication of the screen-printed electrodes. (B) Final product [230]. *Image reproduced from L. Manjakkal, D. Shakthivel, R. Dahiya, Flexible printed reference electrodes for electrochemical applications. Adv. Mater. Technol. 3(12) (2018) 1800252.*

hypochlorite for 1 min. The screen printing of KCl-glass composite was done the silver chloride layer, followed by curing it again at 120°C for 2 h. A layer of polyurethane resin was used as an epoxy layer to delay the decaying time of the composite. The coating of this resin was done on the surface by leaving a small portion of the electrode called hydration port, to allow the penetration of the tested solutions to the electrodes. CV was performed with a sweep rate of 100 mV/s to determine the changes in the current of the sensors for pH experiments conducted between 4 and 10. The impedance change of the sensors was determined to calculate the change in capacitance for the supercapacitors. The sensors showed high change in the response toward both the applications, where pH and supercapacitor measurements can be related to the gastroesophageal diseases and energy harvesting of the wearable devices, respectively. The reader should refer to some other works on nanotechnology-based sensors for electrochemical sensing of biomedical applications [211, 231–233].

3.3 Bioanalytical sensors

One of the significant applications of nanotechnology has been the development of sensors used to detect the proteins and enzymes secreted from the body at different concentrations. The high selectivity and sensitivity of the nanoparticles have increased their usage in the flexible sensors for bioanalytical detection purposes. Some of the significant works in this area describing the detection of proteins, DNA hybridization, and other enzymes have been explained in the succeeding sections. Table 3 gives a comparative study of some of important parameters of the nanotech-based biosensors designed and fabricated in recent years. All of these prototypes have been used for the detection of proteins and enzymes.

3.3.1 Protein-based sensors

The measurement of proteins and enzymes has been done in different ways [240–242] when flexible sensors have been used. The type of sensor used for protein detection depends on their sensitivity, selectivity, cost, and in some cases, wearability [243, 244]. Some researchers have used graphene sensors [245] for a comparative study between the dry and wet sensing of the proteins. The performance of the sensors to determine the monitoring of bovine serum albumin (BSA) was studied using a novel dry sensing process and was later compared with the wet sensing procedure. Even though both techniques obtained a high sensitivity, the signals obtained from wet sensing process were not stable or consistent. Also, the response time and detection were broader for wet sensing process compared with its counterpart. The drifting of the obtained signals is much less in the case of dry sensing process. The detection limit of this novel method is around 1 min. Nanowires-based sensors have also been considered that revolutionized the healthcare industry via detection of biomolecules such as nucleic acids and proteins [246]. The nanowires have been used in different forms like suspension arrays and FETs

Table 3 Comparison between the works on bioanalytical sensors, based on their design and corresponding advantages.

Sl. No.	Materials used	Advantages	Application	Reference
1	Graphite	Excellent electrocatalytic activity, high sensitivity and good selectivity	Protein detection	[234]
2	Silicon nanowires	High sensitivity, label-free, real-time detection	Protein detection	[235]
3	Gold, PET, Ferricynaide	High selectivity, high sensitivity, high stability in the response	Protein detection	[236]
4	Graphene	Label-free, high specificity	Detection of DNA hybridization	[237]
5	Silicon nanowires	Low signal-to-noise ratio, high sensitivity	DNA sensing	[238]
6	Multiwalled Carbon Nanotubes, PDMS	Simple and quick fabrication process, low detection limit	DNA sensing	[239]

by integrating them with microchip technology to form point-of-care, low power, electronic biosensors. For example, the types of nanowires have also varied to form different segments via altering their sequences [247] or changing their diameters [248]. The detection of certain proteins like nucleic acids were carried out using nanowires having lengths and cross-sectional diameter of 6 μm and 500 nm, respectively. Very low concentrations of 5 ng of proteins like ovalbumin were detected to achieve a sensitivity of 1×10^5 using multiplexed detecting sensors [249]. Proteins have also been used as adducts [250] with conductive materials like CNTs to detect some of the analytes like blood, glucose, cholesterol, triglyceride, and Hb1AC. These sensors increased the sensitivity by 5% for the current 15% available on the available standard devices. The sensors served perfectly for the point-of-care devices for clinical and emergency uses as well as for ubiquitous monitoring of patients. The conjugation of CNTs and proteins have been interestingly used [251] for the label-free detection of single proteins via developed of near-IR florescent devices. Florescent SWCNTs were used to form microarrays that were subsequently employed for protein recognition. Hexahistidine was able to capture proteins via SWCNTs/chitosan microarrays that were bound to nickel ions. These ions were further bounded to $N\alpha$, $N\alpha$-bis(carboxymethyl)-L-lysine in a grafted manner to chitosan that surrounded the CNTs. The analyte proteins were captured by the nickel ions, which acted as a proximity quencher. The detection limit, ensemble average, and observation time of the sensors toward the captured proteins were 100 nM, 10 pM, and 600 s,

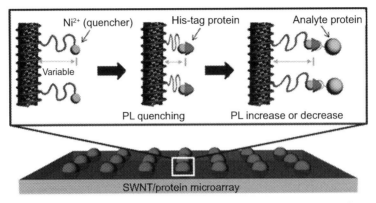

Fig. 10 Schematic diagram of the use of CNTs and protein to form mircoarrays for single protein detection [251]. *Image reproduced from J.H. Ahn, J.H. Kim, N.F. Reuel, P.W. Barone, A.A. Boghossian, J. Zhang, H. Yoon, A.C. Chang, A.J. Hilmer, M.S. Strano, Label-free, single protein detection on a near-IR fluorescent single-walled carbon nanotube/protein microarray fabricated by cell-free synthesis. Nano Lett. 11(7) (2011) 2743–2752.*

respectively. These sensors were also used to analyze the environment of staurosporine-induced apoptosis of SH–SY5Y cells to study the network of 1156 protein-protein interactions with the formation of SWCNTs arrays. Fig. 10 [251] shows the schematic diagram of the previously explained phenomenon. CNTs have also been used [241] to develop transistors for the label-free detection of proteins from physiological solutions.

Apart from CNTs, graphene has also been used to form prototypes for real-time detection of protein molecules [252, 253]. Some of the sensors were based on 3D hierarchical biocomposite developed from a graphene coated layer on hollow, natural pollen microcapsules. Ultrathin PET substrates were considered to form the modular assembly of the graphene-coated microcapsules. Highly flexible sensing prototypes were developed that contained tunable selectivity using the antigen–antibodies covalent immobilization. The sensors showed a sensitivity of 1.7×10^{-15} M for the detection of prostate specific antigen (PSA). One of the significant attributes of these prototypes was in their real-time feedback of the generated output that helped them achieve superior performance compared with the conventional 2D graphene-coated sensors. The sensors were also able to perform under different bending conditions due to the ultrahigh flexible nature of the prototypes. Similar to graphene, nanowires have also been considered [246] to develop sensors for multiplexed detection of biomolecules. The nanowires have been used to form suspension arrays and semiconductor nanowire-based FETs to detect biomolecules like nucleic acids and proteins. Some of the advantages of this approach is in their low input power, portability of the systems, and point-of-care devices. Similar to nanowires, nanofibers have also been employed [254] to form sensing prototypes and biodegradable and stretchable protein-based sensors. Silk fibroin (SF) was considered as a substrate to these sensors due to its natural biodegradable biocompatible nature. Silver nanofibers were used alongside SF using a water-free process. The developed sensors had a sheet resistance of 10.5 Ω/sq. along with 90% transmission. They also displayed excellent stability in their responses for consecutive bending cycles over 2200. The sensors had a high stretchability of 70%, which was exploited to use them for tactile sensing for a pressure ranging between 35 and 700 kPa.

Another sector used for the design and development of sensors for protein detection are the silk–conductive polymer-based biocomposites [255]. The attributes of these sensors are their flexibility, organic and biodegradable nature, and label-free impedimetric responses. The sensors were fabricated using a photolithography technique to pattern a conductive ink on a photoreactive silk sericin that was coupled with a conductive polymer. The design of the conductive electrodes was customized to form free-standing or conforming to soft surfaces by controlling their thickness on the flexible fibroin substrates. Experiments were conducted to determine the capability of the sensors to detect vascular endothelial growth factor (VEGF) via attaching them to the antibodies and immobilizing them within the conducting matrix. The sensors were tested with real-time samples of human serum to examine their selectivity and sensitivity.

One of the interesting works related to the use of nanoparticles to detect proteins was done by Shafiee et al. [256] to showcase the fabrication of sensitive, robust, portable, and point-of-care biosensors. Paper-based sensors were formed by using the integrated form of cellulose paper and polyester films as substrates. The sensors were used for HIV detection through viral lysate impedance spectroscopic technique. The detection was done via spiking of the samples with HIV-1, which was isolated using magnetic beads that were conjugated with antibodies. The sampling frequency and voltage during the testing were 1 kHz and 1 V, respectively, where the electrical conductivity was monitored to determine its changes subsequent to the experiments. An average change of around 20% to 30% was observed in the impedance when the sensors were tested with control samples. The magnitude shift for the blood and plasma samples were $13 \pm 2\%$ and $16 \pm 5\%$, respectively. The repeatability of the responses for these chosen concentrations was between 88% and 99%. The sensors were further used to detect E. coli and S. aureus by modified gold nanoparticles-based electrodes to recognition sites for the specific analyte. Capillary effect was used to distribute the modified gold nanoparticles solution after they were shifted to a disposable paper for uniform distribution. Certain coupling agents like 1-Mercaptoundeconoic acid (MUA), N-Ethyl-N′-(3-dimethylaminopropyl) carbodiimide hydrochloride (EDC)/N-hydroxysuccinimide (NHS) coupling agents and proteins like Lipopolysaccharide Binding Protein (LBP) were used for the modification of gold nanoparticles to determine E. coli. The experiments were conducted considering only a surface area of around 0.25 cm^2 of the cellulose paper to generate the changes in the middle of the nanoparticles. Another similar work utilizing flexible substrates and nanoparticles to develop sensors for protein detection was done by Visser et al. [257], where the mobility of a particle was sensed using a coupling technique with flexible molecular tether. The sensors were sensitive toward proteins present in the buffers and blood plasma in very low concentrations. Other attributes of this detection process was the reversibility in their reactions, single-molecular resolution, and direct and self-contained working principle. The molecules that were targeted for sensing purposes are specific nucleic acids and protein thrombin. Surface functionalization of a DNA system and a thrombin system was done prior to the experiments; later, a specific amount was injected into the measurement chamber for motion and recording purposes. The particle motion was recorded in the chamber at a frequency of 30 Hz. A lifetime of 1 s was achieved for the experiments conducted with ssDNA and thrombin, when the experimental duration was around 5 min. Nanomolar and picomolar concentrations of the analyte were detected using the developed sensing prototypes. One of the interesting works related to the detection of proteins using metallic nanoparticles was done by Etezadi et al. [258], where nanoplasmonic mid-IR biosensors were designed and fabricated for in-vitro detection purposes. Some of the attributes of this method was its noninvasive and label-free nature to develop mid-IR nanoantennas. An arrays of sensors was formed to monitor nanometer-thin protein layers in different aqueous solutions. The sensors were sensitive toward the measurements done to differentiate between the secondary-structural variations in native β-sheet proteins monolayers from cross-β sheets. The electrodes were formed with multiple arrays of gold nanoantennas that were in the form of rods. Calcium fluoride was used as substrates due to their capability of plasmonic resonant excitations in the mid-IR frequency range. The substrates were spin-coated with PMMA and methyl methacrylate (MMA) using electron-beam exposure technique. A lift-off process was used to form the nanoantenna arrays on multiple calcium fluoride substrates, via evaporation of thin layers of chromium and gold of thickness of 100 nm. The gold surface of the nanoantennas was functionalized with COOH-functional alkane thiol self-assembled monolayer (SAM), MHDA (16-mercaptohexadecanoic acid, from Sigma-Aldrich), and synthesized amino-(PEG)2-Maleimide as a linker. The sensors were used to monitor samples developed with fibrillated protein samples formed with a mixture of 10% A140C and 90% WT α-Syn monomers. The structural dependency of the protein decided the range of frequencies where the multiple absorption components took place in the amide I band. The β-sheets varied the wavelengths with ranges between 1615 and 1640 cm^{-1} or 1680–1700 cm^{-1}.

3.3.2 Ligands, cells, and DNA

Lastly, one of the major significant parts of bioanalytical sensing done by the nanoparticle-based flexible sensors was done to determine the signals and/or information obtained from the DNA and ligands present between the cells. Due to the large amount of information obtained from molecules, the detection of DNA and ligands have been done simultaneously with the development of flexible sensing prototypes. Some of the works [259–261] have developed DNA bionsensors that were based on the electrochemical detection of labelling with nanoparticles. These nanoparticles attached on the platforms detect the targeted DNA on the biorecognition reporting probe. Different techniques like pulse voltammetry (PV) and stripping voltammetry (SV) were used to analyze the formation of DNA bioassay and subsequent changes in the oxidation of the ions. The nanotags for the voltammetry operation were carried out using specific nanoparticles like gold, silver, cadmium, and lead. These nanoparticles were attached to three-electrode systems having a range of polymer and paper-based flexible substrates. Other than nanoparticles, graphene is one of the common materials that has been considered [262] for detection of DNA and other biomolecules. The high thermal conductivity, excellent material strength, and biocompatibility of graphene make it an excellent choice for DNA sensing. These sensors, along with the diagnosis, have also been used for therapeutics. The conductance model of this material has been developed to detect the DNA hybridization by defining the concentration of DNA as a function of gate voltage [263]. Certain forms of graphene like nanopores, nanoribbons, oxides, reduced oxides, and nanoparticle composites have been designed and fabricated to operate as DNA detecting sensing prototypes. These sensors have unique characteristics with improved performances. Apart from these works, the use of oxides in other forms such as platinum has been used [264] to form flexible sensors for label-free DNA sensing. The sensors were formed using thin films of platinum oxides, having a thickness of 100 nm. Reactive ion sputtering technique was utilized to form the sensors, which opted for a p-type semiconducting behavior. The bandgap, resistivity, and activity of the sensors were 1.5 eV, 0.16 Q-m, and 0.22 eV, respectively. The prototypes were formed using two terminal sensors on transparent, flexible, and acetate substrates, having an active layer of surface area of 8 mm $\times$ 60 μm. The change in response of the sensors was detected with a corresponding change in conductance. This change occurred when DNA was attached to the device, and simultaneously the charge carrier density changed. The measurements were further verified using Fourier transform IR spectroscopy and fluorescence techniques. The developed prototypes as shown in Fig. 11 [265], showed a proper linear response along with a sensitivity of 0.5 nM. The results were also reproducible in nature, along with a standard of less than 10%. Label-free DNA sensing was done using integrated electronic forms like the conjugation of organic transistors operating on electrochemical sensing and flexible microfluidic systems [265]. The devices were formed using the immobilization technique on single-stranded DNA probes with gate electrodes. Pulsed-enhanced hybridization technique was used to monitor the changes in the complementary DNA targets at low concentrations. High flexibility and high sensitivity are some of the attributes of these sensors. GO-based FETs have also been considered [160] to form ultrasensitive sensors for label-free detection of PNA-DNA hybridization.

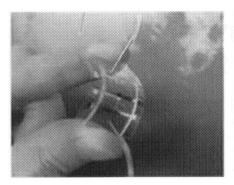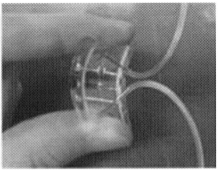

Fig. 11 Representation of the integrated sensing prototypes developed from flexible microfluidic systems and organic electrochemical transistors [265]. *Image reproduced from P. Lin, X. Luo, I.M. Hsing, F. Yan, Organic electrochemical transistors integrated in flexible microfluidic systems and used for label-free DNA sensing. Adv. Mater. 23(35) (2011) 4035–4040.*

The rGO was developed using hydrazine, followed by using drop-casting technique to suspend it on the sensing surface of the FET sensors. The sensors had a very low detection limit of 100 fM, along with the proper distinction of complementary DNA from the mismatched and noncomplementary DNA. The sensors, having a high specificity, were a popular choice for being a point-of-care tool for diagnosis of diseases.

One of the interesting works related to the use of both nanowires and graphene for DNA hybridization was shown by Kim et al. [266] where graphene and silicon nanowires were employed to form diode-type biosensors. Large-scale fabrication was done using CVD technique to form graphene-surface modified vertical silicon nanowire array junctions. The primary advantage of these sensors over single silicon-based nanowire-based DNA sensors was the reduction in noise with the increase in the magnitude of source-drain current. Another advantage of having high–density and thin diameter of silicon nanowires was the generation of large charge carriers on their surfaces. The probe on the surfaces of the sensing arrays were decorated with oligonucleotide formed with a fragment of human DENND2D promoter sequence and subsequently with the complementary of this oligonucleotide. This increased the current from 19% to 120% as the hybridization led to a doping effect on the surface of the nanowires. The sensors were tested with concentrations ranging between 0.1 and 500 nM, having a deviation of around 1% to 10%. The silicon nanowires were formed using metal-assisted chemical etching (MaCE) technique, having gold film with the hole arrays and anodic aluminum oxide as the catalyst and template, respectively. To produce silicon nanowires, the gold films were sputtered on the silicon substrates, which were mixed with hydrofluoric acid and hydrogen peroxide at a ratio of 1:1 for 3 min. After the nanowires were formed, the samples were washed with DI water and moved to aqua regia to selectively remove the gold films. The nanowires were then again washed with DI water. Graphene was formed inside a graphite–heater–based CVD quartz tube furnace and placed in copper foils having a thickness of 70 μm. The temperature and gas present inside the tube furnace were 1000°C and methane, respectively. The formed graphene was then spin-coated with PMMA and treated with nickel for etching of the attached copper. This was followed by the transfer of PMMA-coated CVD-formed graphene on the tips of the surface-modified silicon nanowires. The samples were dried and subsequently backed at 75°C for 4 h. Some regions of the nanowires were covered, whereas other portions were left open for dropping DNA solutions on them. The fabrication process was completed with the coating of gold film having a thickness of 50 nm, on graphene to form the top electrodes. Fig. 12 shows the described fabrication process [266].

Samples of formed oligonucleotides were then dropped on the sensing layer to perform the experiments. The responses of the sensors were measured in terms of current and voltage graphs, with response to changing concentration. The current increased with the increase in concentration from 10 nM to 500 nM, after which it saturated.

Other than graphene, CNTs have also been used [267] to develop flexible sensors for DNA sensing. Nanotubes were used to form sensors having different radii of curvatures and to determine the changes in the responses in terms of resistance with the simultaneous experimentation with biomolecules. With the variation of the substrates, one of the works presented by Ferrara et al. [268] was related to the fabrication of DNA-based biosensors using flexible nylon substrates. Dip-pen lithography technique was used to form sensors, which were then used for topoisomerase detection. The prototypes consisted of

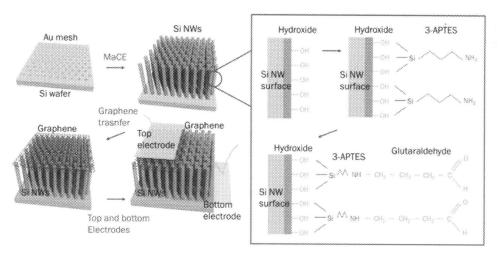

Fig. 12 Steps of fabrication of the graphene/silicon nanowire-based DNA sensors [266]. *Image reproduced from J. Kim, S.Y. Park, S. Kim, D.H. Lee, J.H. Kim, J.M. Kim, H. Kang, J.S. Han, J.W. Park, H. Lee, S.H. Choi, Precise and selective sensing of DNA-DNA hybridization by graphene/Si-nanowires diode-type biosensors. Sci. Rep. 6 (2016) 31984.*

an array of DNA substrates that were employed as a drug screening platform for anticancer molecules. A novel protocol was formulated that assisted in the immobilization of minute quantities of DNA oligonucleotides on the surface of the flexible substrates. SWCNTs have also been used [269] to detection DNA sensors that operated on generating random sequence of 15 mer and 30 mer oligonucleotides. The hybridization of DNA was done on gold, instead of SWCNTs sidewalls, to detect the change in electrical conductance. The change in response occurred as a result of the modulation of energy level alignment due to the contact between the gold and SWCNTs. Other forms of flexible conductive materials used to develop the DNA sensors are the nanowires [270] that have been used as novel biosensing tools. These materials have been processed to form multifunctional nanowires for obtaining high sensitivity toward the detected analyte. Certain semiconducting nanowires like indium oxide [271, 272], polymer [273, 274], silicon [275, 276], and gold [277, 278] have been engineered to optimize their electronic and optical properties for DNA detection.

4. Conclusion and future work

This chapter presented some of the significant works done on the use of nanotechnology to form flexible sensing prototypes for biomedical applications. Different types of flexible sensors have been showcased, which varied in terms of their working principle, processed materials, and performance. Comparison between similar type of sensors has been done in the tables to analyze the differences between the prototypes, as well as highlighting their advantages. With the current era of using nanotechnology in almost every sector of science, it is expected that their exploitation would be done more in the recent future. This range is expected to further increase in the next decade, with the rapid growth of the use of nanotechnology. With the miniaturization of electronic goods, the nanomaterials serve a great purpose due to their enhanced electrical, mechanical, and thermal attributes. Efficient utilization of the nanoparticles-based flexible sensing prototypes would be highly beneficial to improve the quality of life.

References

[1] S.M. Sze, Semiconductor Sensors, Wiley, New York, 1994.
[2] History of MEMS, Available:https://nanohub.org/resources/26535/download/App_Intro_PK10_PG.pdfs, 2020. Accessed 23 July 2020.
[3] M. Van Putten, M. Van Putten, A. Van Putten, J. Pompe, H. Bruining, A silicon bidirectional flow sensor for measuring respiratory flow, IEEE Trans. Biomed. Eng. 44 (2) (1997) 205–208.
[4] A. Errachid, A. Ivorra, J. Aguilo, R. Villa, N. Zine, J. Bausells, New technology for multi-sensor silicon needles for biomedical applications, Sensors Actuators B Chem. 78 (1–3) (2001) 279–284.
[5] J.G. Noel, Review of the properties of gold material for MEMS membrane applications, IET Circuits Devices Syst. 10 (2) (2016) 156–161.
[6] N. Mahalik, Introduction to Microelectromechanical Systems (MEMS), Tata McGraw-Hill, New Delhi, India, 2007.

[7] A. Nag, A.I. Zia, X. Li, S.C. Mukhopadhyay, J. Kosel, Novel sensing approach for LPG leakage detection: part I—operating mechanism and preliminary results, IEEE Sensors J. 16 (4) (2016) 996–1003.

[8] K.S. Szajda, C.G. Sodini, H.F. Bowman, A low noise, high resolution silicon temperature sensor, IEEE J. Solid State Circuits 31 (9) (1996) 1308–1313.

[9] M.J. Schöning, et al., A highly long-term stable silicon-based pH sensor fabricated by pulsed laser deposition technique, Sensors Actuators B Chem. 35 (1–3) (1996) 228–233.

[10] L. Pancheri, C. Oton, Z. Gaburro, G. Soncini, L. Pavesi, Very sensitive porous silicon NO2 sensor, Sensors Actuators B Chem. 89 (3) (2003) 237–239.

[11] Disadvantages of Silicon Sensors, Available:https://www.printedelectronicsworld.com/articles/52/problems-with-silicon-chips, 2020. Accessed 23 July 2020.

[12] Advantages and Disadvantages of Silicon, Available:http://www.rfwireless-world.com/Terminology/Advantages-and-Disadvantages-of-Silicon.html, 2020. Accessed 23 July 2020.

[13] J.C. Costa, F. Spina, P. Lugoda, L. Garcia-Garcia, D. Roggen, N. Münzenrieder, Flexible sensors—from materials to applications, Technologies 7 (2) (2019) 35.

[14] M. Acuautla, S. Bernardini, L. Gallais, T. Fiorido, L. Patout, M. Bendahan, Ozone flexible sensors fabricated by photolithography and laser ablation processes based on ZnO nanoparticles, Sensors Actuators B Chem. 203 (2014) 602–611.

[15] R.K. Pal, A.A. Farghaly, M.M. Collinson, S.C. Kundu, V.K. Yadavalli, Photolithographic micropatterning of conducting polymers on flexible silk matrices, Adv. Mater. 28 (7) (2016) 1406–1412.

[16] S. Miller, Z. Bao, Fabrication of flexible pressure sensors with microstructured polydimethylsiloxane dielectrics using the breath figures method, J. Mater. Res. 30 (23) (2015) 3584–3594.

[17] S.-J. Woo, J.-H. Kong, D.-G. Kim, J.-M. Kim, A thin all-elastomeric capacitive pressure sensor array based on micro-contact printed elastic conductors, J. Mater. Chem. C 2 (22) (2014) 4415–4422.

[18] A. Nag, S.C. Mukhopadhyay, J. Kosel, Tactile sensing from laser-ablated metallized PET films, IEEE Sensors J. 17 (1) (2016) 7–13.

[19] A. Nag, S.C. Mukhopadhyay, J. Kosel, Sensing system for salinity testing using laser-induced graphene sensors, Sensors Actuators A Phys. 264 (2017) 107–116.

[20] A. Nag, S. Feng, S. Mukhopadhyay, J. Kosel, D. Inglis, 3D printed mould-based graphite/PDMS sensor for low-force applications, Sensors Actuators A Phys. 280 (2018) 525–534.

[21] A. Kapoor, et al., Soft, flexible 3D printed fibers for capacitive tactile sensing, in: SENSORS, 2016 IEEE, IEEE, 2016, , pp. 1–3.

[22] C.-T. Wang, K.-Y. Huang, D.T. Lin, W.-C. Liao, H.-W. Lin, Y.-C. Hu, A flexible proximity sensor fully fabricated by inkjet printing, Sensors 10 (5) (2010) 5054–5062.

[23] L. Xiang, Z. Wang, Z. Liu, S.E. Weigum, Q. Yu, M.Y. Chen, Inkjet-printed flexible biosensor based on graphene field effect transistor, IEEE Sensors J. 16 (23) (2016) 8359–8364.

[24] W.-Y. Chang, T.-H. Fang, H.-J. Lin, Y.-T. Shen, Y.-C. Lin, A large area flexible array sensors using screen printing technology, J. Disp. Technol. 5 (6) (2009) 178–183.

[25] S. Khan, L. Lorenzelli, R.S. Dahiya, Bendable piezoresistive sensors by screen printing MWCNT/PDMS composites on flexible substrates, in: 2014 10th Conference on Ph.D. Research in Microelectronics and Electronics (PRIME), IEEE, 2014, , pp. 1–4.

[26] Y. Sun, H.H. Wang, Electrodeposition of Pd nanoparticles on single-walled carbon nanotubes for flexible hydrogen sensors, Appl. Phys. Lett. 90 (21) (2007) 213107.

[27] Y.K. Kim, S.-H. Hwang, S. Kim, H. Park, S.K. Lim, ZnO nanostructure electrodeposited on flexible conductive fabric: a flexible photo-sensor, Sensors Actuators B Chem. 240 (2017) 1106–1113.

[28] T. Xue, et al., Patterned flexible graphene sensor via printing and interface assembly, J. Mater. Chem. C 7 (2019) 6317–6322.

[29] Y.H. Kwak, et al., Flexible glucose sensor using CVD-grown graphene-based field effect transistor, Biosens. Bioelectron. 37 (1) (2012) 82–87.

[30] A. Abdelhalim, A. Falco, F. Loghin, P. Lugli, J.F. Salmerón, A. Rivadeneyra, Flexible NH3 sensor based on spray deposition and inkjet printing, in: 2016 IEEE SENSORS, IEEE, 2016, , pp. 1–3.

[31] S. Bernardini, M. Bendahan, M. Acuautla, E. Pietri, Flexible Gas Sensors Fabricated by Ultrasonic Spray Deposition, The First International Conference on Advances in Sensors, Actuators, Metering and Sensing, Venice, Italy, 2016.

[32] H. Choi, H.Y. Jeong, D.-S. Lee, C.-G. Choi, S.-Y. Choi, Flexible NO 2 gas sensor using multilayer graphene films by chemical vapor deposition, Carbon Lett. 14 (3) (2013) 186–189.

[33] L. Gomez De Arco, Y. Zhang, C.W. Schlenker, K. Ryu, M.E. Thompson, C. Zhou, Continuous, highly flexible, and transparent graphene films by chemical vapor deposition for organic photovoltaics, ACS Nano 4 (5) (2010) 2865–2873.

[34] S. Khan, S. Tinku, L. Lorenzelli, R.S. Dahiya, Flexible tactile sensors using screen-printed P (VDF-TrFE) and MWCNT/PDMS composites, IEEE Sensors J. 15 (6) (2014) 3146–3155.

[35] C.-Y. Chen, C.-L. Chang, T.-F. Chien, C.-H. Luo, Flexible PDMS electrode for one-point wearable wireless bio-potential acquisition, Sensors Actuators A Phys. 203 (2013) 20–28.

[36] U. Yaqoob, D.-T. Phan, A.I. Uddin, G.-S. Chung, Highly flexible room temperature NO 2 sensor based on MWCNTs-WO 3 nanoparticles hybrid on a PET substrate, Sensors Actuators B Chem. 221 (2015) 760–768.

[37] M.-x. Jing, C. Han, M. Li, X.-q. Shen, High performance of carbon nanotubes/silver nanowires-PET hybrid flexible transparent conductive films via facile pressing-transfer technique, Nanoscale Res. Lett. 9 (1) (2014) 1–7.

[38] Y. Qin, et al., Lightweight, superelastic, and mechanically flexible graphene/polyimide nanocomposite foam for strain sensor application, ACS Nano 9 (9) (2015) 8933–8941.

[39] J.A. Dobrzynska, M.A. Gijs, Flexible polyimide-based force sensor, Sensors Actuators A Phys. 173 (1) (2012) 127–135.

[40] Y. Seekaew, S. Lokavee, D. Phokharatkul, A. Wisitsoraat, T. Kerdcharoen, C. Wongchoosuk, Low-cost and flexible printed graphene–PEDOT: PSS gas sensor for ammonia detection, Org. Electron. 15 (11) (2014) 2971–2981.

[41] G. Latessa, F. Brunetti, A. Reale, G. Saggio, A. Di Carlo, Piezoresistive behaviour of flexible PEDOT: PSS based sensors, Sensors Actuators B Chem. 139 (2) (2009) 304–309.

[42] S.-J. Young, Z.-D. Lin, Ethanol gas sensors composed of carbon nanotubes with au nanoparticles adsorbed onto a flexible PI substrate, ECS J. Solid State Sci. Technol. 6 (10) (2017) M130–M132.

[43] S. Zhang, et al., Highly stretchable, sensitive, and flexible strain sensors based on silver nanoparticles/carbon nanotubes composites, J. Alloys Compd. 652 (2015) 48–54.

[44] E. Singh, M. Meyyappan, H.S. Nalwa, Flexible graphene-based wearable gas and chemical sensors, ACS Appl. Mater. Interfaces 9 (40) (2017) 34544–34586.

[45] C. Lou, et al., Flexible graphene electrodes for prolonged dynamic ECG monitoring, Sensors 16 (11) (2016) 1833.

[46] B. Putz, O. Glushko, M.J. Cordill, Electromigration in gold films on flexible polyimide substrates as a self-healing mechanism, Mater. Res. Lett. 4 (1) (2016) 43–47.

[47] K. Lee, et al., Highly sensitive, transparent, and flexible gas sensors based on gold nanoparticle decorated carbon nanotubes, Sensors Actuators B Chem. 188 (2013) 571–575.

[48] M. Akiyama, N. Ueno, K. Nonaka, H. Tateyama, Flexible pulse-wave sensors from oriented aluminum nitride nanocolumns, Appl. Phys. Lett. 82 (12) (2003) 1977–1979.

[49] L. Natta, et al., Soft and flexible piezoelectric smart patch for vascular graft monitoring based on Aluminum Nitride thin film, Sci. Rep. 9 (1) (2019) 8392.

[50] K.K. Sadasivuni, D. Ponnamma, H.-U. Ko, H.C. Kim, L. Zhai, J. Kim, Flexible NO2 sensors from renewable cellulose nanocrystals/iron oxide composites, Sensors Actuators B Chem. 233 (2016) 633–638.

[51] D. Bandgar, S. Navale, M. Naushad, R. Mane, F. Stadler, V. Patil, Ultra-sensitive polyaniline–iron oxide nanocomposite room temperature flexible ammonia sensor, RSC Adv. 5 (84) (2015) 68964–68971.

[52] Modern Sensors for Defence and Military Applications, Available:https://electronicsforu.com/market-verticals/aerospace-defence/modern-sensors-defence-military-applications, 2020. Accessed 23 July 2020.

[53] Flexible and Printed Electronics and the Automotive Market, Available:https://www.printedelectronicsnow.com/contents/view_online-exclusives/2019-02-12/flexible-and-printed-electronics-and-the-automotive-market/, 2020. Accessed 23 July 2020.

[54] A. Nag, B. Menzies, S.C. Mukhopadhyay, Performance analysis of flexible printed sensors for robotic arm applications, Sensors Actuators A Phys. 276 (2018).

[55] K. Xu, Y. Lu, K. Takei, Multifunctional skin-inspired flexible sensor systems for wearable electronics, Adv. Mater. Technol. (2019) 1800628.

[56] D. Vilela, A. Romeo, S. Sánchez, Flexible sensors for biomedical technology, Lab Chip 16 (3) (2016) 402–408.

[57] Printed Electronics: Key Trends, Available:https://www.printedelectronicsworld.com/articles/15712/printed-electronics-key-trends, 2020. Accessed 23 July 2020.

[58] Printed And Flexible Sensors, Available:https://www.idtechex.com/ja/research-article/printed-and-flexible-sensors-there-will-be-many-winners/5853, 2020. Accessed 23 July 2020.

[59] V. Subramanian, T. Lee, Nanotechnology-based flexible electronics, Nanotechnology 23 (34) (2012) 340201.

[60] A.R. Jha, MEMS and Nanotechnology-based Sensors and Devices for Communications, Medical and Aerospace Applications, CRC Press, 2008.

[61] Q. Nie, et al., Sensitivity enhanced, stability improved ethanol gas sensor based on multi-wall carbon nanotubes functionalized with Pt-Pd nanoparticles, Sensors Actuators B Chem. 270 (2018) 140–148.

[62] G. Niarchos, et al., Humidity sensing properties of paper substrates and their passivation with ZnO nanoparticles for sensor applications, Sensors 17 (3) (2017) 516.

[63] T. Han, A. Nag, S.C. Mukhopadhyay, Y. Xu, Carbon nanotubes and its gas-sensing applications: a review, Sensors Actuators A Phys. 291 (2019) 107–143.

[64] S. Young, Z. Lin, Ammonia gas sensors with Au-decorated carbon nanotubes, Microsyst. Technol. 24 (10) (2018) 4207–4210.

[65] H. Zhu, Semiconductor nanowire MOSFETs and applications, in: Nanowires: New Insights, 2017p. 101.

[66] T. Lee, W. Lee, S.W. Kim, J.J. Kim, B.S. Kim, Flexible textile strain wireless sensor functionalized with hybrid carbon nanomaterials supported ZnO nanowires with controlled aspect ratio, Adv. Funct. Mater. 26 (34) (2016) 6206–6214.

[67] X.-S. Zhang, B. Meng, F.-Y. Zhu, W. Tang, H.-X. Zhang, Switchable wetting and flexible SiC thin film with nanostructures for microfluidic surface-enhanced Raman scattering sensors, Sensors Actuators A Phys. 208 (2014) 166–173.

[68] T. Huang, et al., Porous fibers composed of polymer nanoball decorated graphene for wearable and highly sensitive strain sensors, Adv. Funct. Mater. 29 (2019).

[69] A.N. Sokolov, B.C. Tee, C.J. Bettinger, J.B.-H. Tok, Z. Bao, Chemical and engineering approaches to enable organic field-effect transistors for electronic skin applications, Acc. Chem. Res. 45 (3) (2011) 361–371.

[70] Y. Zang, D. Huang, C.A. Di, D. Zhu, Device engineered organic transistors for flexible sensing applications, Adv. Mater. 28 (22) (2016) 4549–4555.

[71] Advantages of Nanotechnology in Flexible Electronics, Available:https://www.mouser.com/blog/how-nanotechnology-has-helped-to-realize-flexible-and-wearable-electronics, 2020. Accessed 23 July 2020.

[72] W.A.D.M. Jayathilaka, et al., Significance of nanomaterials in wearables: a review on wearable actuators and sensors, Adv. Mater. 31 (7) (2019) 1805921.

[73] How Nanotechnology Enables Wearable Electronics, Available:https://www.nanowerk.com/spotlight/spotid=52432.php, 2020. Accessed 23 July 2020.

[74] A. Nag, S.C. Mukhopadhyay, Fabrication and implementation of printed sensors for taste sensing applications, Sensors Actuators A Phys. 269 (2018) 53–61.

[75] T. Li, et al., Flexible capacitive tactile sensor based on micropatterned dielectric layer, Small 12 (36) (2016) 5042–5048.

[76] A. Chhetry, P.S. Das, H. Yoon, J.Y. Park, A sandpaper-inspired flexible and stretchable resistive sensor for pressure and strain measurement, Org. Electron. 62 (2018) 581–590.

[77] M.O.F. Emon, J.-W. Choi, Flexible piezoresistive sensors embedded in 3D printed tires, Sensors 17 (3) (2017) 656.

[78] C.K. Jeong, C. Baek, A.I. Kingon, K.I. Park, S.H. Kim, Lead-free perovskite nanowire-employed piezopolymer for highly efficient flexible nanocomposite energy harvester, Small 14 (19) (2018) 1704022.

[79] A. Chandrasekhar, N.R. Alluri, M. Sudhakaran, Y.S. Mok, S.-J. Kim, A smart mobile pouch as a biomechanical energy harvester toward self-powered smart wireless power transfer applications, Nanoscale 9 (28) (2017) 9818–9824.

[80] M.S. Rasel, et al., An impedance tunable and highly efficient triboelectric nanogenerator for large-scale, ultra-sensitive pressure sensing applications, Nano Energy 49 (2018) 603–613.

[81] W. Seung, et al., Nanopatterned textile-based wearable triboelectric nanogenerator, ACS Nano 9 (4) (2015) 3501–3509.

[82] M. Lindeberg, K. Hjort, Interconnected nanowire clusters in polyimide for flexible circuits and magnetic sensing applications, Sensors Actuators A Phys. 105 (2) (2003) 150–161.

[83] A. Eshkeiti, et al., Screen printing of multilayered hybrid printed circuit boards on different substrates, IEEE Trans. Compon. Packag. Manuf. Technol. 5 (3) (2015) 415–421.

[84] Z. Zou, C. Zhu, Y. Li, X. Lei, W. Zhang, J. Xiao, Rehealable, fully recyclable, and malleable electronic skin enabled by dynamic covalent thermoset nanocomposite, Sci. Adv. 4 (2) (2018) eaaq0508.

[85] Y. Hu, J. Ding, Effects of morphologies of carbon nanofillers on the interfacial and deformation behavior of polymer nanocomposites—a molecular dynamics study, Carbon 107 (2016) 510–524.

[86] K.-Y. Chun, et al., Free-standing nanocomposites with high conductivity and extensibility, Nanotechnology 24 (16) (2013) 165401.

[87] C. Robert, J.F. Feller, M. Castro, Sensing skin for strain monitoring made of PC–CNT conductive polymer nanocomposite sprayed layer by layer, ACS Appl. Mater. Interfaces 4 (7) (2012) 3508–3516.

[88] B. Paczosa-Bator, Ion-selective electrodes with superhydrophobic polymer/carbon nanocomposites as solid contact, Carbon 95 (2015) 879–887.

[89] F. Mirri, et al., High-performance carbon nanotube transparent conductive films by scalable dip coating, ACS Nano 6 (11) (2012) 9737–9744.

[90] D. Zhang, J. Tong, B. Xia, Humidity-sensing properties of chemically reduced graphene oxide/polymer nanocomposite film sensor based on layer-by-layer nano self-assembly, Sensors Actuators B Chem. 197 (2014) 66–72.

[91] J. Huang, S.-C. Her, X. Yang, M. Zhi, Synthesis and characterization of multi-walled carbon nanotube/graphene nanoplatelet hybrid film for flexible strain sensors, Nanomaterials 8 (10) (2018) 786.

[92] J.F. Christ, N. Aliheidari, A. Ameli, P. Pötschke, 3D printed highly elastic strain sensors of multiwalled carbon nanotube/thermoplastic polyurethane nanocomposites, Mater. Des. 131 (2017).

[93] S. Yao, P. Swetha, Y. Zhu, Nanomaterial-enabled wearable sensors for healthcare, Adv. Healthc. Mater. 7 (1) (2018) 1700889.

[94] S. Yao, et al., Nanomaterial-enabled flexible and stretchable sensing systems: processing, integration, and applications, Adv. Mater. 32 (2019).

[95] S. Choi, H. Lee, R. Ghaffari, T. Hyeon, D.H. Kim, Recent advances in flexible and stretchable bio-electronic devices integrated with nanomaterials, Adv. Mater. 28 (22) (2016) 4203–4218.

[96] H. Fuketa, et al., 30.3 Organic-transistor-based 2kV ESD-tolerant flexible wet sensor sheet for biomedical applications with wireless power and data transmission using 13.56 MHz magnetic resonance, in: 2014 IEEE International Solid-State Circuits Conference Digest of Technical Papers (ISSCC), IEEE, 2014, , pp. 490–491.

[97] T. Yokota, et al., Sheet-type flexible organic active matrix amplifier system using pseudo-CMOS circuits with floating-gate structure, IEEE Trans. Electron Devices 59 (12) (2012) 3434–3441.

[98] M.D. Angione, et al., Carbon based materials for electronic bio-sensing, Mater. Today 14 (9) (2011) 424–433.

[99] H. Deng, et al., Flexible and semitransparent organolead triiodide perovskite network photodetector arrays with high stability, Nano Lett. 15 (12) (2015) 7963–7969.

[100] Z. Yin, et al., Organic photovoltaic devices using highly flexible reduced graphene oxide films as transparent electrodes, ACS Nano 4 (9) (2010) 5263–5268.

[101] G.T. Hwang, et al., Self-powered wireless sensor node enabled by an aerosol-deposited PZT flexible energy harvester, Adv. Energy Mater. 6 (13) (2016) 1600237.

[102] T. Ji, Y. Zhao, Y. Ding, G. Nie, Using functional nanomaterials to target and regulate the tumor microenvironment: diagnostic and therapeutic applications, Adv. Mater. 25 (26) (2013) 3508–3525.

[103] S. Gai, et al., Recent advances in functional nanomaterials for light–triggered cancer therapy, Nano Today 19 (2018) 146–187.

[104] D. Pradhan, F. Niroui, K. Leung, High-performance, flexible enzymatic glucose biosensor based on ZnO nanowires supported on a gold-coated polyester substrate, ACS Appl. Mater. Interfaces 2 (8) (2010) 2409–2412.

[105] C. Li, et al., Complementary detection of prostate-specific antigen using In2O3 nanowires and carbon nanotubes, J. Am. Chem. Soc. 127 (36) (2005) 12484–12485.

[106] P.D. Bhuyan, A. Kumar, Y. Sonvane, P. Gajjar, R. Magri, S.K. Gupta, Si and Ge based metallic core/shell nanowires for nano-electronic device applications, Sci. Rep. 8 (1) (2018) 16885.

[107] Y. Sun, H.S. Kim, E. Menard, S. Kim, I. Adesida, J.A. Rogers, Printed arrays of aligned GaAs wires for flexible transistors, diodes, and circuits on plastic substrates, Small 2 (11) (2006) 1330–1334.

[108] F. Xiao, J. Song, H. Gao, X. Zan, R. Xu, H. Duan, Coating graphene paper with 2D-assembly of electrocatalytic nanoparticles: a modular approach toward high-performance flexible electrodes, ACS Nano 6 (1) (2011) 100–110.

[109] X. Zan, H. Bai, C. Wang, F. Zhao, H. Duan, Graphene paper decorated with a 2D array of dendritic platinum nanoparticles for ultrasensitive electrochemical detection of dopamine secreted by live cells, Chem. Eur. J. 22 (15) (2016) 5204–5210.

[110] X. Xu, et al., Self-sensing, ultralight, and conductive 3D graphene/iron oxide aerogel elastomer deformable in a magnetic field, ACS Nano 9 (4) (2015) 3969–3977.

[111] Z. Li, et al., Ag nanoparticle-grafted PAN-nanohump array films with 3D high-density hot spots as flexible and reliable SERS substrates, Small 11 (40) (2015) 5452–5459.

[112] W.-C. Hou, T.-H. Wu, W.-C. Tang, F.C.-N. Hong, Nucleation control for the growth of vertically aligned GaN nanowires, Nanoscale Res. Lett. 7 (1) (2012) 373.

[113] S. Gong, et al., A wearable and highly sensitive pressure sensor with ultrathin gold nanowires, Nat. Commun. 5 (2014).

[114] B. Zhu, et al., Microstructured graphene arrays for highly sensitive flexible tactile sensors, Small 10 (18) (2014) 3625–3631.

[115] Z. Yang, Y. Zhang, T. Itoh, A flexible implantable micro temperature sensor on polymer capillary for biomedical applications, in: 2013 IEEE 26th International Conference on Micro Electro Mechanical Systems (MEMS), IEEE, 2013, , pp. 889–892.

[116] S. Choi, et al., Stretchable heater using ligand-exchanged silver nanowire nanocomposite for wearable articular thermotherapy, ACS Nano 9 (6) (2015) 6626–6633.

[117] J. Jeon, H.B.R. Lee, Z. Bao, Flexible wireless temperature sensors based on Ni microparticle-filled binary polymer composites, Adv. Mater. 25 (6) (2013) 850–855.

[118] T.Q. Trung, N.T. Tien, D. Kim, M. Jang, O.J. Yoon, N.E. Lee, A flexible reduced graphene oxide field-effect transistor for ultrasensitive strain sensing, Adv. Funct. Mater. 24 (1) (2014) 117–124.

[119] S.J. Park, O.S. Kwon, S.H. Lee, H.S. Song, T.H. Park, J. Jang, Ultrasensitive flexible graphene based field-effect transistor (FET)-type bioelectronic nose, Nano Lett. 12 (10) (2012) 5082–5090.

[120] A. Gao, N. Lu, Y. Wang, T. Li, Robust ultrasensitive tunneling-FET biosensor for point-of-care diagnostics, Sci. Rep. 6 (2016) 22554.

[121] E.T. Alonso, et al., Graphene electronic fibres with touch-sensing and light-emitting functionalities for smart textiles, NPJ Flex. Electron. 2 (1) (2018) 25.

[122] C. Lu, et al., Flexible and stretchable nanowire-coated fibers for optoelectronic probing of spinal cord circuits, Sci. Adv. 3 (3) (2017) e1600955.

[123] D. Wang, B. Sheng, L. Peng, Y. Huang, Z. Ni, Flexible and optical fiber sensors composited by graphene and PDMS for motion detection, Polymers 11 (9) (2019) 1433.

[124] H. Yousef, J.-P. Nikolovski, E. Martincic, Flexible 3D force tactile sensor for artificial skin for anthropomorphic robotic hand, Procedia Eng. 25 (2011) 128–131.

[125] K. Takei, et al., Nanowire active-matrix circuitry for low-voltage macroscale artificial skin, Nat. Mater. 9 (10) (2010) 821.

[126] R. Park, et al., One-step laser patterned highly uniform reduced graphene oxide thin films for circuit-enabled tattoo and flexible humidity sensor application, Sensors (Basel, Switz.) 18 (6) (2018).

[127] A.J. Bandodkar, W. Jia, J. Wang, Tattoo-based wearable electrochemical devices: a review, Electroanalysis 27 (3) (2015) 562–572.

[128] B. Dyatkin, E-skin sensor self-heals and can be recycled, MRS Bull. 43 (5) (2018) 319.

[129] S.-W. Kim, et al., A triple-mode flexible E-skin sensor interface for multi-purpose wearable applications, Sensors 18 (1) (2017) 78.

[130] A. Nag, S.C. Mukhopadhyay, J. Kosel, Flexible carbon nanotube nanocomposite sensor for multiple physiological parameter monitoring, Sensors Actuators A Phys. 251 (2016) 148–155.

[131] W. Zhang, Q. Liu, P. Chen, Flexible strain sensor based on carbon black/silver nanoparticles composite for human motion detection, Materials 11 (10) (2018) 1836.

[132] H. Li, G. Ding, Z. Yang, A high sensitive flexible pressure sensor designed by silver nanowires embedded in polyimide (AgNW-PI), Micromachines 10 (3) (2019) 206.

[133] M. Amjadi, A. Pichitpajongkit, S. Lee, S. Ryu, I. Park, Highly stretchable and sensitive strain sensor based on silver nanowire–elastomer nanocomposite, ACS Nano 8 (5) (2014) 5154–5163.

[134] F. Meng, W. Lu, Q. Li, J.H. Byun, Y. Oh, T.W. Chou, Graphene-based fibers: a review, Adv. Mater. 27 (35) (2015) 5113–5131.

[135] Z. Jing, Z. Guang-Yu, S. Dong-Xia, Review of graphene-based strain sensors, Chin. Phys. B 22 (5) (2013) 057701.

[136] S. Shengbo, et al., Highly sensitive wearable strain sensor based on silver nanowires and nanoparticles, Nanotechnology 29 (25) (2018) 255202.

[137] M.M. Ali, et al., Printed strain sensor based on silver nanowire/silver flake composite on flexible and stretchable TPU substrate, Sensors Actuators A Phys. 274 (2018) 109–115.

[138] Y. Zang, F. Zhang, C.-a. Di, D. Zhu, Advances of flexible pressure sensors toward artificial intelligence and health care applications, Mater. Horiz. 2 (2) (2015) 140–156.

[139] D.H. Lee, et al., Highly sensitive and flexible strain sensors based on patterned ITO nanoparticle channels, Nanotechnology 28 (49) (2017) 495501.

[140] Q. Meng, et al., A facile approach to fabricate highly sensitive, flexible strain sensor based on elastomeric/graphene platelet composite film, J. Mater. Sci. 54 (15) (2019) 10856–10870.

[141] S.-L. Jia, et al., Carbon nanotube-based flexible electrothermal film heaters with a high heating rate, R. Soc. Open Sci. 5 (6) (2018) 172072.

[142] H. Song, Y. Zhang, J. Cao, Sensing mechanism of an ionization gas temperature sensor based on a carbon nanotube film, RSC Adv. 7 (84) (2017) 53265–53269.

[143] Y. Xin, J. Zhou, G. Lubineau, A highly stretchable strain-insensitive temperature sensor exploits the Seebeck effect in nanoparticle-based printed circuits, J. Mater. Chem. A 7 (42) (2019) 24493–24501.

[144] M.T. Rahman, et al., High performance flexible temperature sensors via nanoparticle printing, ACS Appl. Nano Mater. 2 (5) (2019) 3280–3291.

[145] B.F. Monea, E.I. Ionete, S.I. Spiridon, D. Ion-Ebrasu, E. Petre, Carbon nanotubes and carbon nanotube structures used for temperature measurement, Sensors 19 (11) (2019) 2464.

[146] M. De Volder, D. Reynaerts, C. Van Hoof, S. Tawfick, A.J. Hart, A temperature sensor from a self-assembled carbon nanotube microbridge, in: SENSORS, 2010 IEEE, IEEE, 2010, , pp. 2369–2372.

[147] A. Saraiya, D. Porwal, A. Bajpai, N. Tripathi, K. Ram, Investigation of carbon nanotubes as low temperature sensors, Synth. React. Inorg., Met.-Org., Nano-Met. Chem. 36 (2) (2006) 163–164.

[148] S. Sarma, J. Lee, Developing efficient thin film temperature sensors utilizing layered carbon nanotube films, Sensors 18 (10) (2018) 3182.

[149] Y. Yang, Z.-H. Lin, T. Hou, F. Zhang, Z.L. Wang, Nanowire-composite based flexible thermoelectric nanogenerators and self-powered temperature sensors, Nano Res. 5 (12) (2012) 888–895.

[150] D.-Y. Youn, et al., Wireless real-time temperature monitoring of blood packages: silver nanowire-embedded flexible temperature sensors, ACS Appl. Mater. Interfaces 10 (51) (2018) 44678–44685.

[151] P. Sahatiya, S.K. Puttapati, V.V. Srikanth, S. Badhulika, Graphene-based wearable temperature sensor and infrared photodetector on a flexible polyimide substrate, Flexible Printed Electron. 1 (2) (2016) 025006.

[152] G. Khurana, S. Sahoo, S. Barik, N. Kumar, G. Sharma, Reduced graphene oxide as an excellent temperature sensor, J. Nanosci. Nanotechnol. Appl. 2 (2018) *101 Abstract Keywords: Graphene Oxide.*

[153] P. Sehrawat, S. Islam, P. Mishra, Reduced graphene oxide based temperature sensor: extraordinary performance governed by lattice dynamics assisted carrier transport, Sensors Actuators B Chem. 258 (2018) 424–435.

[154] Y. Zeng, T. Li, Y. Yao, T. Li, L. Hu, A. Marconnet, Thermally conductive reduced graphene oxide thin films for extreme temperature sensors, Adv. Funct. Mater. (2019) 1901388.

[155] M. Segev-Bar, H. Haick, Flexible sensors based on nanoparticles, ACS Nano 7 (10) (2013) 8366–8378.

[156] Q. Liu, H. Tai, Z. Yuan, Y. Zhou, Y. Su, Y. Jiang, A high-performances flexible temperature sensor composed of polyethyleneimine/reduced graphene oxide bilayer for real-time monitoring, Adv. Mater. Technol. 4 (3) (2019) 1800594.

[157] G. Liu, et al., A flexible temperature sensor based on reduced graphene oxide for robot skin used in internet of things, Sensors 18 (5) (2018) 1400.

[158] K.S. Karimov, et al., Carbon nanotubes based flexible temperature sensors, Optoelectron. Adv. Mater. Rapid Commun. 6 (2012) 194–196.

[159] C.Y. Kuo, C.L. Chan, C. Gau, C.-W. Liu, S.H. Shiau, J.-H. Ting, Nano temperature sensor using selective lateral growth of carbon nanotube between electrodes, IEEE Trans. Nanotechnol. 6 (1) (2007) 63–69.

[160] B. Cai, S. Wang, L. Huang, Y. Ning, Z. Zhang, G.-J. Zhang, Ultrasensitive label-free detection of PNA–DNA hybridization by reduced graphene oxide field-effect transistor biosensor, ACS Nano 8 (3) (2014) 2632–2638.

[161] S. Mao, K. Yu, J. Chang, D.A. Steeber, L.E. Ocola, J. Chen, Direct growth of vertically-oriented graphene for field-effect transistor biosensor, Sci. Rep. 3 (2013) 1696.

[162] D. Sarkar, W. Liu, X. Xie, A.C. Anselmo, S. Mitragotri, K. Banerjee, MoS2 field-effect transistor for next-generation label-free biosensors, ACS Nano 8 (4) (2014) 3992–4003.

[163] B. Cai, L. Huang, H. Zhang, Z. Sun, Z. Zhang, G.-J. Zhang, Gold nanoparticles-decorated graphene field-effect transistor biosensor for femtomolar MicroRNA detection, Biosens. Bioelectron. 74 (2015) 329–334.

[164] K.-I. Chen, B.-R. Li, Y.-T. Chen, Silicon nanowire field-effect transistor-based biosensors for biomedical diagnosis and cellular recording investigation, Nano Today 6 (2) (2011) 131–154.

[165] N. Liu, R. Chen, Q. Wan, Recent advances in electric-double-layer transistors for bio-chemical sensing applications, Sensors 19 (15) (2019) 3425.

[166] Y. Ohno, K. Maehashi, Y. Yamashiro, K. Matsumoto, Electrolyte-gated graphene field-effect transistors for detecting pH and protein adsorption, Nano Lett. 9 (9) (2009) 3318–3322.

[167] D.-J. Kim, I.Y. Sohn, J.-H. Jung, O.J. Yoon, N.-E. Lee, J.-S. Park, Reduced graphene oxide field-effect transistor for label-free femtomolar protein detection, Biosens. Bioelectron. 41 (2013) 621–626.

[168] K. Maehashi, T. Katsura, K. Kerman, Y. Takamura, K. Matsumoto, E. Tamiya, Label-free protein biosensor based on aptamer-modified carbon nanotube field-effect transistors, Anal. Chem. 79 (2) (2007) 782–787.

[169] S. Sorgenfrei, et al., Label-free single-molecule detection of DNA-hybridization kinetics with a carbon nanotube field-effect transistor, Nat. Nanotechnol. 6 (2) (2011) 126.

[170] T. Sakata, H. Otsuka, Y. Miyahara, Potentiometric detection of DNA molecules hybridization using gene field effect transistor and intercalator, MRS Online Proc. Libr. Arch. 782 (2003).

[171] K. Bhatt, et al., A comparative study of graphene and graphite-based field effect transistor on flexible substrate, Pramana 90 (6) (2018) 75.

[172] M. Donnelly, D. Mao, J. Park, G. Xu, Graphene field-effect transistors: the road to bioelectronics, J. Phys. D. Appl. Phys. 51 (49) (2018) 493001.

[173] B.J. Kim, H. Jang, S.-K. Lee, B.H. Hong, J.-H. Ahn, J.H. Cho, High-performance flexible graphene field effect transistors with ion gel gate dielectrics, Nano Lett. 10 (9) (2010) 3464–3466.

[174] T. Cohen-Karni, Q. Qing, Q. Li, Y. Fang, C.M. Lieber, Graphene and nanowire transistors for cellular interfaces and electrical recording, Nano Lett. 10 (3) (2010) 1098–1102.

[175] P.K. Ang, W. Chen, A.T.S. Wee, K.P. Loh, Solution-gated epitaxial graphene as pH sensor, J. Am. Chem. Soc. 130 (44) (2008) 14392–14393.

[176] N. Petrone, I. Meric, J. Hone, K.L. Shepard, Graphene field-effect transistors with gigahertz-frequency power gain on flexible substrates, Nano Lett. 13 (1) (2012) 121–125.

[177] G. Fisichella, et al., Advances in the fabrication of graphene transistors on flexible substrates, Beilstein J. Nanotechnol. 8 (1) (2017) 467–474.

[178] F. Xu, et al., Highly stretchable carbon nanotube transistors with ion gel gate dielectrics, Nano Lett. 14 (2) (2014) 682–686.

[179] S. Aikawa, E. Einarsson, T. Thurakitseree, S. Chiashi, E. Nishikawa, S. Maruyama, Deformable transparent all-carbon-nanotube transistors, Appl. Phys. Lett. 100 (6) (2012) 063502.

[180] C. Wang, K. Takei, T. Takahashi, A. Javey, Carbon nanotube electronics–moving forward, Chem. Soc. Rev. 42 (7) (2013) 2592–2609.

[181] E. Artukovic, M. Kaempgen, D. Hecht, S. Roth, G. Grüner, Transparent and flexible carbon nanotube transistors, Nano Lett. 5 (4) (2005) 757–760.

[182] S. Selvarasah, X. Li, A. Busnaina, M.R. Dokmeci, Parylene-C passivated carbon nanotube flexible transistors, Appl. Phys. Lett. 97 (15) (2010) 153120.

[183] B. Lee, Y. Chen, A. Cook, A. Zakhidov, V. Podzorov, Stable doping of carbon nanotubes via molecular self assembly, J. Appl. Phys. 116 (14) (2014) 144503.

[184] C.W. Wang, et al., In situ detection of chromogranin a released from living neurons with a single-walled carbon-nanotube field-effect transistor, Small 3 (8) (2007) 1350–1355.

[185] S. Liu, X. Guo, Carbon nanomaterials field-effect-transistor-based biosensors, NPG Asia Mater. 4 (8) (2012) e23.

[186] X. Guo, Surface plasmon resonance based biosensor technique: a review, J. Biophotonics 5 (7) (2012) 483–501.

[187] M. Hernaez, C.R. Zamarreño, S. Melendi-Espina, L.R. Bird, A.G. Mayes, F.J. Arregui, Optical fibre sensors using graphene-based materials: a review, Sensors 17 (1) (2017) 155.

[188] P.R. Sahoo, P. Swain, S.M. Nayak, S. Bag, S.R. Mishra, Surface plasmon resonance based biosensor: a new platform for rapid diagnosis of livestock diseases, Vet. World 9 (12) (2016) 1338.

[189] M. Riskin, R. Tel-Vered, O. Lioubashevski, I. Willner, Ultrasensitive surface plasmon resonance detection of trinitrotoluene by a bis-aniline-cross-linked au nanoparticles composite, J. Am. Chem. Soc. 131 (21) (2009) 7368–7378.

[190] K. Lin, Y. Lu, J. Chen, R. Zheng, P. Wang, H. Ming, Surface plasmon resonance hydrogen sensor based on metallic grating with high sensitivity, Opt. Express 16 (23) (2008) 18599–18604.

[191] D. Beccati, et al., SPR studies of carbohydrate–protein interactions: signal enhancement of low-molecular-mass analytes by organoplatinum (II)-labeling, Chembiochem 6 (7) (2005) 1196–1203.

[192] D. Sun, T. Guo, Y. Ran, Y. Huang, B.-O. Guan, In-situ DNA hybridization detection with a reflective microfiber grating biosensor, Biosens. Bioelectron. 61 (2014) 541–546.

[193] S. Sridevi, K. Vasu, S. Asokan, A. Sood, Sensitive detection of C-reactive protein using optical fiber Bragg gratings, Biosens. Bioelectron. 65 (2015) 251–256.

[194] S. Sridevi, K. Vasu, S. Sampath, S. Asokan, A. Sood, Optical detection of glucose and glycated hemoglobin using etched fiber Bragg gratings coated with functionalized reduced graphene oxide, J. Biophotonics 9 (7) (2016) 760–769.

[195] Y.C. Tan, Chemical sensing applications of carbon nanotube-deposited optical fibre sensors, Chemosensors 6 (4) (2018) 55.

[196] Y. Kim, N. Minami, W. Zhu, S. Kazaoui, R. Azumi, M. Matsumoto, Langmuir–Blodgett films of single-wall carbon nanotubes: layer-by-layer deposition and in-plane orientation of tubes, Jpn. J. Appl. Phys. 42 (12R) (2003) 7629.

[197] X. Li, et al., Langmuir – Blodgett assembly of densely aligned single-walled carbon nanotubes from bulk materials, J. Am. Chem. Soc. 129 (16) (2007) 4890–4891.

[198] J. Nicholson, R. Windeler, D. DiGiovanni, Optically driven deposition of single-walled carbon-nanotube saturable absorbers on optical fiber end-faces, Opt. Express 15 (15) (2007) 9176–9183.

[199] K. Kashiwagi, S. Yamashita, S.Y. Set, Optically manipulated deposition of carbon nanotubes onto optical fiber end, Jpn. J. Appl. Phys. 46 (10L) (2007) L988.

[200] S.Y. Set, H. Yaguchi, Y. Tanaka, M. Jablonski, Laser mode locking using a saturable absorber incorporating carbon nanotubes, J. Lightwave Technol. 22 (1) (2004) 51.

[201] Y.-W. Song, K. Morimune, S.Y. Set, S. Yamashita, Polarization insensitive all-fiber mode-lockers functioned by carbon nanotubes deposited onto tapered fibers, Appl. Phys. Lett. 90 (2) (2007) 021101.

[202] T. Sreekumar, T. Liu, S. Kumar, L.M. Ericson, R.H. Hauge, R.E. Smalley, Single-wall carbon nanotube films, Chem. Mater. 15 (1) (2003) 175–178.

[203] A. Shabaneh, et al., Dynamic response of tapered optical multimode fiber coated with carbon nanotubes for ethanol sensing application, Sensors 15 (5) (2015) 10452–10464.

[204] M.E. Spotnitz, D. Ryan, H.A. Stone, Dip coating for the alignment of carbon nanotubes on curved surfaces, J. Mater. Chem. 14 (8) (2004) 1299–1302.

[205] E.Y. Jang, T.J. Kang, H.W. Im, D.W. Kim, Y.H. Kim, Single-walled carbon-nanotube networks on large-area glass substrate by the dip-coating method, Small 4 (12) (2008) 2255–2261.

[206] W.Y. Choi, H.G. Jo, S.W. Kwon, Y.H. Kim, J.Y. Pyun, K.K. Park, Multipoint-detection strain sensor with a single electrode using optical ultrasound generated by carbon nanotubes, Sensors 19 (18) (2019) 3877.

[207] S. Noimark, et al., Carbon-nanotube–PDMS composite coatings on optical fibers for all-optical ultrasound imaging, Adv. Funct. Mater. 26 (46) (2016) 8390–8396.

[208] F. Xiao, Y. Li, X. Zan, K. Liao, R. Xu, H. Duan, Growth of metal–metal oxide nanostructures on freestanding graphene paper for flexible biosensors, Adv. Funct. Mater. 22 (12) (2012) 2487–2494.

[209] M. Zhang, A. Halder, C. Hou, J. Ulstrup, Q. Chi, Free-standing and flexible graphene papers as disposable non-enzymatic electrochemical sensors, Bioelectrochemistry 109 (2016) 87–94.

[210] W. Cai, J. Lai, T. Lai, H. Xie, J. Ye, Controlled functionalization of flexible graphene fibers for the simultaneous determination of ascorbic acid, dopamine and uric acid, Sensors Actuators B Chem. 224 (2016) 225–232.

[211] H. Huang, et al., Graphene-based sensors for human health monitoring, Front. Chem. 7 (2019).

[212] L. Maiolo, D. Polese, A. Pecora, G. Fortunato, Y. Shacham-Diamand, A. Convertino, Highly disordered array of silicon nanowires: an effective and scalable approach for performing and flexible electrochemical biosensors, Adv. Healthc. Mater. 5 (5) (2016) 575–583.

[213] F. Xiao, Y. Li, H. Gao, S. Ge, H. Duan, Growth of coral-like PtAu–MnO2 binary nanocomposites on free-standing graphene paper for flexible nonenzymatic glucose sensors, Biosens. Bioelectron. 41 (2013) 417–423.

[214] A.P. Selvam, S. Muthukumar, V. Kamakoti, S. Prasad, A wearable biochemical sensor for monitoring alcohol consumption lifestyle through ethyl glucuronide (EtG) detection in human sweat, Sci. Rep. 6 (2016) 23111.

[215] J. Huang, Y. Zhu, X. Yang, W. Chen, Y. Zhou, C. Li, Flexible 3D porous CuO nanowire arrays for enzymeless glucose sensing: in situ engineered versus ex situ piled, Nanoscale 7 (2) (2015) 559–569.

[216] Z. Wang, et al., Graphene paper supported MoS2 nanocrystals monolayer with cu submicron-buds: high-performance flexible platform for sensing in sweat, Anal. Biochem. 543 (2018) 82–89.

[217] Y. Sun, K. He, Z. Zhang, A. Zhou, H. Duan, Real-time electrochemical detection of hydrogen peroxide secretion in live cells by Pt nanoparticles decorated graphene–carbon nanotube hybrid paper electrode, Biosens. Bioelectron. 68 (2015) 358–364.

[218] M. Kaempgen, S. Roth, Transparent and flexible carbon nanotube/polyaniline pH sensors, J. Electroanal. Chem. 586 (1) (2006) 72–76.

[219] Y. Zhang, et al., CuO nanowires based sensitive and selective non-enzymatic glucose detection, Sensors Actuators B Chem. 191 (2014) 86–93.

[220] N.P. Shetti, S.D. Bukkitgar, R.R. Kakarla, C. Reddy, T.M. Aminabhavi, ZnO-based nanostructured electrodes for electrochemical sensors and biosensors in biomedical applications, Biosens. Bioelectron. (2019) 111417.

[221] S.M.U. Ali, N. Alvi, Z. Ibupoto, O. Nur, M. Willander, B. Danielsson, Selective potentiometric determination of uric acid with uricase immobilized on ZnO nanowires, Sensors Actuators B Chem. 152 (2) (2011) 241–247.

[222] P. Mostafalu, S. Sonkusale, A high-density nanowire electrode on paper for biomedical applications, RSC Adv. 5 (12) (2015) 8680–8687.

[223] X. Xuan, H.S. Yoon, J.Y. Park, A wearable electrochemical glucose sensor based on simple and low-cost fabrication supported micro-patterned reduced graphene oxide nanocomposite electrode on flexible substrate, Biosens. Bioelectron. 109 (2018) 75–82.

[224] S. Dong, et al., High loading MnO2 nanowires on graphene paper: facile electrochemical synthesis and use as flexible electrode for tracking hydrogen peroxide secretion in live cells, Anal. Chim. Acta 853 (2015) 200–206.

[225] N. Chauhan, S. Chawla, C. Pundir, U. Jain, An electrochemical sensor for detection of neurotransmitter-acetylcholine using metal nanoparticles, 2D material and conducting polymer modified electrode, Biosens. Bioelectron. 89 (2017) 377–383.

[226] Z. Fan, et al., A flexible and disposable hybrid electrode based on cu nanowires modified graphene transparent electrode for non-enzymatic glucose sensor, Electrochim. Acta 109 (2013) 602–608.

[227] S.M.U. Ali, T. Aijazi, K. Axelsson, O. Nur, M. Willander, Wireless remote monitoring of glucose using a functionalized ZnO nanowire arrays based sensor, Sensors 11 (9) (2011) 8485–8496.

[228] T. Knieling, E. Nebling, L. Blohm, C. Beale, M. Fahland, Printed and flexible electrochemical lactate sensors for wearable applications, Multidiscip. Digit. Publ. Inst. Proc. 1 (8) (2017) 828.

[229] T. Kuretake, S. Kawahara, M. Motooka, S. Uno, An electrochemical gas biosensor based on enzymes immobilized on chromatography paper for ethanol vapor detection, Sensors 17 (2) (2017) 281.

[230] L. Manjakkal, D. Shakthivel, R. Dahiya, Flexible printed reference electrodes for electrochemical applications, Adv. Mater. Technol. 3 (12) (2018) 1800252.

[231] N. Nasiri, C. Clarke, Nanostructured gas sensors for medical and health applications: low to high dimensional materials, Biosensors 9 (1) (2019) 43.

[232] N. Nasiri, C. Clarke, Nanostructured chemiresistive gas sensors for medical applications, Sensors 19 (3) (2019) 462.

[233] C.M. Hung, D.T.T. Le, N. Van Hieu, On-chip growth of semiconductor metal oxide nanowires for gas sensors: a review, J. Sci. Adv. Mater. Devices 2 (3) (2017) 263–285.

[234] W. Cai, T. Lai, H. Du, J. Ye, Electrochemical determination of ascorbic acid, dopamine and uric acid based on an exfoliated graphite paper electrode: a high performance flexible sensor, Sensors Actuators B Chem. 193 (2014) 492–500.

[235] G. Zheng, C.M. Lieber, Nanowire biosensors for label-free, real-time, ultrasensitive protein detection, in: Nanoproteomics, Springer, 2011, , pp. 223–237.

[236] Y.-L. Chen, C.-Y. Lee, H.-T. Chiu, Growth of gold nanowires on flexible substrate for highly sensitive biosensing: detection of thrombin as an example, J. Mater. Chem. B 1 (2) (2013) 186–193.

[237] X. Dong, Y. Shi, W. Huang, P. Chen, L.J. Li, Electrical detection of DNA hybridization with single-base specificity using transistors based on CVD-grown graphene sheets, Adv. Mater. 22 (14) (2010) 1649–1653.

[238] Z. Li, Y. Chen, X. Li, T. Kamins, K. Nauka, R.S. Williams, Sequence-specific label-free DNA sensors based on silicon nanowires, Nano Lett. 4 (2) (2004) 245–247.

[239] J. Li, E.-C. Lee, Carbon nanotube/polymer composite electrodes for flexible, attachable electrochemical DNA sensors, Biosens. Bioelectron. 71 (2015) 414–419.

[240] M.J. Whitcombe, et al., The rational development of molecularly imprinted polymer-based sensors for protein detection, Chem. Soc. Rev. 40 (3) (2011) 1547–1571.

[241] M.S. Filipiak, et al., Highly sensitive, selective and label-free protein detection in physiological solutions using carbon nanotube transistors with nanobody receptors, Sensors Actuators B Chem. 255 (2018) 1507–1516.

[242] P. Salvo, et al., Sensors and biosensors for C-reactive protein, temperature and pH, and their applications for monitoring wound healing: a review, Sensors 17 (12) (2017) 2952.

[243] D. Brennan, P. Galvin, Flexible substrate sensors for multiplex biomarker monitoring, MRS Commun. 8 (3) (2018) 627–641.

[244] Y. Yu, H.Y.Y. Nyein, W. Gao, A. Javey, Flexible electrochemical bioelectronics: the rise of in situ bioanalysis, Adv. Mater. (2019).

[245] Y. Zhang, A. Black, N. Wu, Y. Cui, Comparison of "dry sensing" and "wet sensing" of a protein with a graphene sensor, IEEE Sens. Lett. 2 (4) (2018) 1–4.

[246] B. He, T.J. Morrow, C.D. Keating, Nanowire sensors for multiplexed detection of biomolecules, Curr. Opin. Chem. Biol. 12 (5) (2008) 522–528.

[247] C.D. Keating, M.J. Natan, Striped metal nanowires as building blocks and optical tags, Adv. Mater. 15 (5) (2003) 451–454.

[248] S. Matthias, J. Schilling, K. Nielsch, F. MüLLER, R.B. Wehrspohn, U. Gösele, Monodisperse diameter-modulated gold microwires, Adv. Mater. 14 (22) (2002) 1618–1621.

[249] J.B.H. Tok, et al., Metallic striped nanowires as multiplexed immunoassay platforms for pathogen detection, Angew. Chem. Int. Ed. 45 (41) (2006) 6900–6904.

[250] S. Viswanathan, P. Li, W. Choi, S. Filipek, T. Balasubramaniam, V. Renugopalakrishnan, Protein–carbon nanotube sensors: single platform integrated micro clinical lab for monitoring blood analytes, in: Methods in Enzymology, vol. 509, Elsevier, 2012, , pp. 165–194.

[251] J.-H. Ahn, et al., Label-free, single protein detection on a near-infrared fluorescent single-walled carbon nanotube/protein microarray fabricated by cell-free synthesis, Nano Lett. 11 (7) (2011) 2743–2752.

[252] L. Wang, J.A. Jackman, W.B. Ng, N.J. Cho, Flexible, graphene-coated biocomposite for highly sensitive, real-time molecular detection, Adv. Funct. Mater. 26 (47) (2016) 8623–8630.

[253] S. Viswanathan, et al., Graphene–protein field effect biosensors: glucose sensing, Mater. Today 18 (9) (2015) 513–522.

[254] C. Hou, et al., A biodegradable and stretchable protein-based sensor as artificial electronic skin for human motion detection, Small 15 (11) (2019) 1805084.

[255] M. Xu, V.K. Yadavalli, Flexible biosensors for the impedimetric detection of protein targets using silk-conductive polymer biocomposites, ACS Sens. 4 (4) (2019) 1040–1047.

[256] H. Shafiee, et al., Paper and flexible substrates as materials for biosensing platforms to detect multiple biotargets, Sci. Rep. 5 (2015) 8719.

[257] E.W. Visser, J. Yan, L.J. Van IJzendoorn, M.W. Prins, Continuous biomarker monitoring by particle mobility sensing with single molecule resolution, Nat. Commun. 9 (1) (2018) 2541.

[258] D. Etezadi, J.B. Warner IV, F.S. Ruggeri, G. Dietler, H.A. Lashuel, H. Altug, Nanoplasmonic mid-infrared biosensor for in vitro protein secondary structure detection, Light Sci. Appl. 6 (8) (2017) e17029.

[259] C. Kokkinos, Electrochemical DNA biosensors based on labeling with nanoparticles, Nanomaterials 9 (10) (2019) 1361.

[260] J. Wang, Nanoparticle-based electrochemical DNA detection, Anal. Chim. Acta 500 (1–2) (2003) 247–257.

[261] H.-I. Peng, B.L. Miller, Recent advancements in optical DNA biosensors: exploiting the plasmonic effects of metal nanoparticles, Analyst 136 (3) (2011) 436–447.

[262] X. Wu, F. Mu, Y. Wang, H. Zhao, Graphene and graphene-based nanomaterials for DNA detection: a review, Molecules 23 (8) (2018) 2050.

[263] F. Karimi, M. Ahmadi, M. Rahmani, E. Akbari, M.J. Kiani, M. Khalid, Analytical modeling of graphene-based DNA sensor, Sci. Adv. Mater. 4 (11) (2012) 1142–1147.

[264] N. Basu, A.K. Konduri, P.K. Basu, S. Keshavan, M.M. Varma, N. Bhat, Flexible, label-free DNA sensor using platinum oxide as the sensing element, IEEE Sensors J. 17 (19) (2017) 6140–6147.

[265] P. Lin, X. Luo, I.M. Hsing, F. Yan, Organic electrochemical transistors integrated in flexible microfluidic systems and used for label-free DNA sensing, Adv. Mater. 23 (35) (2011) 4035–4040.

[266] J. Kim, et al., Precise and selective sensing of DNA-DNA hybridization by graphene/Si-nanowires diode-type biosensors, Sci. Rep. 6 (2016) 31984.

[267] K. Akhmadishina, et al., Flexible biological sensors based on carbon nanotube films, Nanotechnol. Russ. 8 (11–12) (2013) 721–726.

[268] V. Ferrara, et al., DNA-based biosensor on flexible nylon substrate by dip-pen lithography for topoisomerase detection, in: Convegno Nazionale Sensori, Springer, 2018, , pp. 309–316.

[269] X. Tang, S. Bansaruntip, N. Nakayama, E. Yenilmez, Y.-l. Chang, Q. Wang, Carbon nanotube DNA sensor and sensing mechanism, Nano Lett. 6 (8) (2006) 1632–1636.

[270] P. Arora, A. Sindhu, N. Dilbaghi, A. Chaudhury, Engineered multifunctional nanowires as novel biosensing tools for highly sensitive detection, Appl. Nanosci. 3 (5) (2013) 363–372.

[271] K. Kalyanikutty, G. Gundiah, C. Edem, A. Govindaraj, C. Rao, Doped and undoped ITO nanowires, Chem. Phys. Lett. 408 (4–6) (2005) 389–394.

[272] P. Nguyen, et al., Epitaxial directional growth of indium-doped tin oxide nanowire arrays, Nano Lett. 3 (7) (2003) 925–928.

[273] S.J. Hurst, E.K. Payne, L. Qin, C.A. Mirkin, Multisegmented one-dimensional nanorods prepared by hard-template synthetic methods, Angew. Chem. Int. Ed. 45 (17) (2006) 2672–2692.

[274] T. Dürkop, S. Getty, E. Cobas, M. Fuhrer, Extraordinary mobility in semiconducting carbon nanotubes, Nano Lett. 4 (1) (2004) 35–39.

[275] R.S. Kane, A.D. Stroock, Nanobiotechnology: protein-nanomaterial interactions, Biotechnol. Prog. 23 (2) (2007) 316–319.

[276] J. Li, C. Lu, B. Maynor, S. Huang, J. Liu, Controlled growth of long GaN nanowires from catalyst patterns fabricated by "dip-pen" nanolithographic techniques, Chem. Mater. 16 (9) (2004) 1633–1636.

[277] M.A. Lapierre-Devlin, C.L. Asher, B.J. Taft, R. Gasparac, M.A. Roberts, S.O. Kelley, Amplified electrocatalysis at DNA-modified nanowires, Nano Lett. 5 (6) (2005) 1051–1055.

[278] M. Basu, et al., Nano-biosensor development for bacterial detection during human kidney infection: use of glycoconjugate-specific antibody-bound gold nanowire arrays (GNWA), Glycoconj. J. 21 (8–9) (2004) 487–496.

CHAPTER 19

Fiber optic biosensors with enhanced performance assisted by two-dimensional (2D) materials

Anuj Kumar Sharma and Ankit Kumar Pandey
Department of Applied Sciences (Physics Division), National Institute of Technology Delhi, New Delhi, India

1. Introduction

A single layer material of a few nanometers thickness is commonly referred to as a two-dimensional (2D) material. For example, graphene, a monolayer counterpart of graphite, has a monolayer thickness of ~ 0.34 nm. The last decade has witnessed a growth in the applications of 2D materials in various optoelectronics devices [1–3]. The 2D materials such as graphene, blue/black phosphorene, transition metal dichalcogenides (TMDCs), hexagonal boron nitride (BN), graphene-like GaN (g-GaN) and their heterostructures, etc. are focused upon in a number of photonic applications [1,4,5 owing to their intriguing physical, chemical, and optical properties [6]. The high-quality 2D inorganic and organic materials are introduced due to significant developments in various micromechanical exfoliation techniques [7]. 2D materials possess a high surface area, which provides a large number of reactive sites making these materials suitable for catalysis, energy storage technologies, and efficient sensing [8]. The 2D structure provides advantages such as: (i) high surface-to-volume ratio and (ii) high electron transport sensitivity through 2D material to the adsorption of gas molecules, which are best suited for gas–sensing applications. Apart from gas sensing, 2D materials are being utilized in various biosensing applications [4, 9–11]. The optical sensors based on 2D materials can be fabricated in several ways, e.g., transferring of 2D materials grown by chemical vapor deposition (CVD) onto optical sensors, CVD growth of 2D materials directly onto sensor structure, and drop-casting 2D materials solution onto optical sensors. In the context of biosensors, fiber optic (FO)-based optical sensors are prevalent in a number of biological applications [12–14]. FO-based optical sensors possess several advantages as follows:

I. Compact, low-cost, and provide label-free sensing
II. High stability, sensitivity, and immunity toward electromagnetic interferences
III. The sensor's performance can be tuned by modifying the core geometry in microstructured optical fibers (MOF)

Handbook of Nanomaterials for Sensing Applications
https://dx.doi.org/10.1016/B978-0-12-820783-3.00006-3
429

Apart from the previously mentioned advantages, flexible and simple design, miniaturized sensor system, and capability of remote sensing are added advantages. There are two prominent FO-based sensing schemes. The first category is evanescent wave absorption-based FO sensors (EWFOS). EWFOS have also been very useful in various sensing and measurement applications [15, 16]. These sensors are based on the absorption of an evanescent wave, which is an exponentially decaying wave generated at an interface where attenuated total reflection (ATR) occurs by the surrounding medium (i.e., sensing layer). The second category is surface plasmon resonance (SPR)-based FO sensor. SPR FO sensors have obtained extensive popularity because of their accuracy, fast response, and ease of fabrication [17–19]. P-polarized waves that exist at metal-dielectric interface (MDI) are referred as surface plasmon polaritons (SPPs), and their resonant oscillation under the influence of an external field is known as SPR. In view of this discussion of two prominent FO-based sensing schemes, this chapter focuses on the application of 2D materials (graphene and TMDCs) in FO sensors emphasizing evanescent absorption and SPR-based sensing schemes.

2. Evanescent wave absorption-based fiber optic sensor

EWFOS possesses relatively easier design in which a bare fiber or the one with an absorbing clad (such as a core coated with dyes) can be used to sense an analyte, which can absorb light. This light absorption will lead toward a change in output power emanating from the fiber end. Furthermore, EWFOS possesses benefits of independent measurements from bulk solution as the penetration depth of evanescent wave is generally 10 to several hundred nanometers [20]. In addition, multiple reflections occurring in a short sensing region makes it more sensitive than single point ATR sensors [21, 22]. Furthermore, the measurement system of EWFOS is portable, economical, smart structured, and possesses good compatibility [23]. The stability and performance of EWFOS can be improved by using either 2D materials (e.g., graphene, MoS_2), additional absorbing layers (e.g., silicon), or a combination of both. In this section, the sensing performance of EWFOS is demonstrated utilizing a combination of Si and 2D material (graphene or MoS_2) coated over fiber.

2.1 Calculation of power transmitted from fiber sensor probe

The basic sensing principle of a EWFOS is the absorption of a launched light signal due to total internal reflection (TIR), taking place at the core/cladding interface of the fiber [24]. When the fiber core diameter, D, is significantly greater than the wavelength of incident light, ray analysis is chosen for power calculations. Considering this, each propagating mode can be thought as a ray propagating through the fiber. The angular (or modal) power distribution for a collimated (i.e., light is assumed to be well-focused on fiber axis) source can be expressed as [25]:

$$p(\theta) \propto \frac{n_{co}^2 \sin\theta \cos\theta}{\left(1 - n_{co}^2 \cos^2\theta\right)^2} \tag{1}$$

Here, θ is the angle with which light ray propagates through the fiber. Each ray suffers a number of reflections (N_{ref}) at the interface during its propagation within the fiber of length L. So, the normalized power at the output for all guided rays with θ varying between critical angle ($\theta_{cr} = \sin^{-1}\{n_{cl}/n_{co}\}$) and maximum angle ($\pi/2$) can be estimated as [26]:

$$P = \frac{\int_{\theta_{cr}}^{\pi/2} R^{N_{ref}(\theta)} p(\theta) d\theta}{\int_{\theta_{cr}}^{\pi/2} p(\theta) d\theta} \tag{2}$$

The reflection coefficient (R) is calculated using transfer matrix method (TMM) for multilayer system [27] and N_{ref} is:

$$N_{ref}(\theta) = \frac{L \cot\theta}{D} \tag{3}$$

For simulation of results, the MATLAB® platform is the preferred one considering all of the previously mentioned necessary parameters.

2.2 Performance parameters: Sensitivity and resolution

The evaluation of the sensor's performance is generally defined in terms of sensitivity and resolution. Sensitivity (S_r and S_i) is defined as the change in transmitted power at output (δP) corresponding to a change in real and imaginary part of the analyte RI (n_r and n_i, respectively) [26, 28, 29]:

$$S_{r,i} = \left| \frac{\delta P}{\delta n_{r,i}} \right| \tag{4}$$

Resolution is defined as the smallest change in analyte RI detectable by the sensor. Numerically, resolution (R_r and R_i) corresponding to real and imaginary RI of analyte can be expressed as:

$$R_{r,i} = \Delta P_R \times \left| \frac{\delta n_{r,i}}{\delta P} \right| = \frac{\Delta P_R}{S_{r,i}} \tag{5}$$

Here, ΔP_R is the resolution of the power detector. The present light power detectors can possess the magnitude of ΔP_R as small as 100 pW.

2.3 Fluoride fiber-based EWFOS utilizing 2D material and amorphous silicon layers

A schematic outline of five-layered EWFOS is shown in Fig. 1 [30]. Initially, a light source is used to launch light inside the fiber core. The intensity variation (after modulation by sensor probe) is measured by a detector at the output end. Different core (fluoride glasses) and clad (fluoride-based and silica-based glasses) materials are used considering the requirement of TIR (i.e., core RI, n_{co} > clad RI, n_{cl}) at core-clad interface and greater numerical aperture (NA) to accommodate a large number of modes (angles) [31].

The performance comparison (sensitivity and resolution) indicated that ZBLA ($ZrF4_{57}BaF2_{34}LaF3_5AlF3_4$) fluoride glass core and GeO_2 (19.3%)–doped silica clad is found to be the most preferred combination. Dispersion relations in terms of Sellmeier expressions for RIs of ZBLA and GeO_2 (19.3%)–doped silica can be considered as:

$$n_{co}(\lambda) = A_1\lambda^{-4} + A_2\lambda^{-2} + A_3 + A_4\lambda^2 + A_5\lambda^4 \tag{6}$$

$$n_{cl}(\lambda) = \sqrt{1 + \frac{a_1\lambda^2}{\lambda^2 - b_1} + \frac{a_2\lambda^2}{\lambda^2 - b_2} + \frac{a_3\lambda^2}{\lambda^2 - b_3}} \tag{7}$$

In Eqs. (6) and (7), λ is the wavelength (in μm) and A_1–A_5, a_1–a_3, and b_1–b_3 are the corresponding Sellmeier coefficients [31]. The ultrathin (~50 nm) cladding is considered so that a sufficiently large extent of interaction between evanescent field and analyte medium (discussed further) can take place.

A thin layer of α-Si is considered to be deposited over ultrathin cladding in the five-layer scheme. Wavelength-dependent complex RI values of amorphous silicon are adapted [32]. Monolayers of graphene (thickness: 0.34 nm) and MoS_2 (thickness: 0.72 nm) are considered over the cladding (one at a time) seeking larger absorption of light propagating through the fluoride fiber core. The complex RI values of graphene

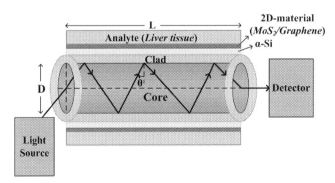

Fig. 1 Schematic representation of five-layer EW absorption-based fiber optic sensor. The analyte is liver tissue.

(up to $\lambda = 1$ μm) and MoS_2 (up to $\lambda = 1.236$ μm) in an IR region are considered from the previously reported works [33, 34]. For the analysis purpose, two liver tissues, i.e., normal (N) and cancerous/malignant (HCC) are considered as analyte media due to different absorbing and scattering properties. Extent of liver tissue malignancy causes the changes in complex RI. The complex RI values of N and HCC liver tissues are taken from an earlier reported work in the spectral range of 0.45–1.551 μm [35]. The sensing is based on the concept of measurement in RI variation of "HCC" with "N" as reference.

2.3.1 Effect of different core-clad combinations

In this section, the study emphasized the performance of EWFOS with different core-clad combinations of fiber. Several glasses have been considered with different spectral variations of their corresponding RI values. As a first result, Fig. 2 shows the spectral variation of output power (P) for (A) N and (B) HCC tissue. This fiber sensor belongs to a three-layer sensor design (i.e., without 2D material and α-Si layers). Clearly, the power spectrum takes different forms for different core-clad combinations for any given analyte due to their diverse dispersion relations. The transmitted power varies differently for two analytes (N and HCC tissue) in the case of given core/clad combination, which is primarily because of different dispersions of their complex RI values [22].

2.3.2 Effect of 2D materials

The simulated variation of sensitivity with wavelength for four-layer sensor design (ZBLA core—19.3% GeO_2-doped silica clad—2D material—analyte) is shown in Fig. 3. The resolution variation is shown in the inset. For both 2D materials (here, MoS_2 and graphene), the trend of sensitivity increment with wavelength remains the same. However, MoS_2 monolayer-based sensor design has larger S_r and S_i values than the one with a graphene monolayer. In addition, the difference between their corresponding sensitivities increases with wavelength. On calculation, it is found that, for a four-layer sensor with a MoS_2 monolayer, S_r increases from 80.0 mW/RIU (at

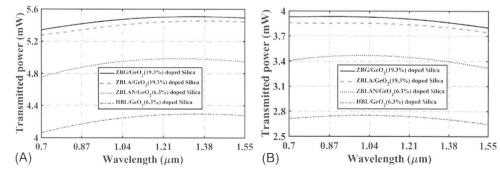

Fig. 2 Simulated variations of transmitted power with wavelength considering (A) N tissue and (B) HCC tissue as analyte media for different core-clad combinations.

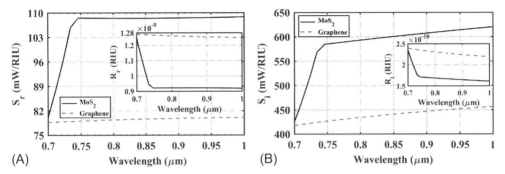

Fig. 3 Sensitivity and resolution (in inset) variation with the wavelength corresponding to (A) real and (B) imaginary RIs of HCC tissue for MoS$_2$ and graphene-coated four-layer sensor structure.

$\lambda = 0.7$ μm) to 108.8 mW/RIU (at $\lambda = 1$ μm) and S_i varies from 425.4 mW/RIU (at $\lambda = 0.7$ μm) to 620.5 mW/RIU (at $\lambda = 1$ μm). In this view, the maximum values of S_r and S_i are obtained as 108.8 mW/RIU and 620.5 mW/RIU, respectively. Furthermore, the MoS$_2$-based sensor design is likely to provide better sensitivities compared with three-layer conventional sensor design (without 2D material) [30]. The same is the case with resolution R_r and R_i, which achieve the best values of 9.19×10^{-10} RIU and 1.61×10^{-10} RIU, respectively, at $\lambda = 1$ μm. So, it is clear that the performance enhancement can be achieved by the introduction of 2D material (MoS$_2$, in particular) owing to its prominent absorption characteristics.

2.3.3 Effect of α-Si layer

The variation of sensitivities with wavelength for five-layer sensor configuration (ZBLA core—19.3% GeO$_2$-doped silica clad—MoS$_2$ monolayer—α-Si layer—analyte), is shown in Fig. 4 [30]. It should be noted that the simulation is performed for two different thicknesses (10 nm and 60 nm) of the α-Si layer. In addition, the same plot also shows the

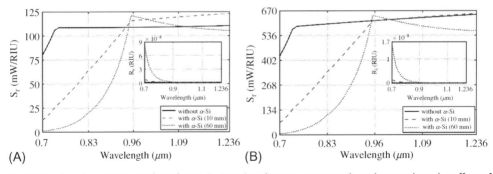

Fig. 4 Simulated sensitivity and resolution (in inset) with respect to wavelength to analyze the effect of α-Si layer corresponding to (A) real and (B) imaginary HCC tissue RIs. Here, the 2D material is MoS$_2$.

Table 1 Resolution comparison among some of the fiber optic sensors.

S. no.	Fiber sensor configuration	Resolution	References
1.	Multicore flat fiber-based SPR sensor	10^{-5}–10^{-6} RIU	[27]
2.	D-shaped plastic optical fiber-based plasmonic sensor	1.2×10^{-3} RIU	[28]
3.	Optical refractometer based on LMR	3.28×10^{-9} RIU	[36]
4.	Silicon Fabry-Pérot cavity-based fiber optic sensor	6×10^{-4} °C	[37]
5.	D-shaped plastic optical fiber sensor	6.5×10^{-3} RIU	[38]
6.	Modeling of stripped clad multimode fiber sensor	2.2×10^{-5} RIU	[15]
7.	Chalcogenide core and silica clad with 2D material (graphene and MoS$_2$) coating	10^{-9}–10^{-10} RIU	[29]
8.	ZBLA fiber core—19.3% GeO$_2$-doped silica clad—MoS$_2$ monolayer—α-Si layer—analyte (H$_2$O as reference)	10^{-10}–10^{-11} RIU	[30]

sensitivity variation for four-layer sensor (i.e., without α-Si). It is evident that there are coarsely two zones of λ with different patterns of sensitivity variation among three curves. The four-layer sensor is able to provide better sensitivity (both S_r and S_i) than the two variants of five-layer sensor for $λ < 0.93$ μm. However, for $λ > 0.93$ μm, the five-layer sensor shows better values of S_r and S_i with the variant corresponding to 10 nm thickness of α-Si layer with maximum S_r and S_i values at $λ = 1.236$ μm. Numerically, for five-layer sensor design with 10 nm-thick α-Si layer, (S_r, S_i) and (R_r, R_i) become (123.0 mw/RIU, 653.9 mw/RIU) and $(8.13 \times 10^{-10}$ RIU, 1.53×10^{-10} RIU), respectively, at $λ = 1.236$ μm. It is noteworthy that, for four-layer sensor (i.e., without α-Si layer), the corresponding S_r and S_i are 110.3 mw/RIU and 647.8 mw/RIU, respectively, at $λ = 1.236$ μm. Thus, a 11.5% enhancement in S_r (while S_i values are almost identical) is observed for five-layer sensor configuration (with 10 nm α-Si layer).

Furthermore, to summarize the performance of some of the previously reported FO sensors, Table 1 shows a comparative analysis in terms of resolution as a performance parameter.

2.3.4 Effect of temperature and graphene's chemical potential on the performance parameters

Apart from the incorporation of 2D materials for performance enhancement in EWFOS, this section provides further performance enhancement in near-infrared (NIR) region while simultaneously tuning the temperature (T) and graphene's chemical potential

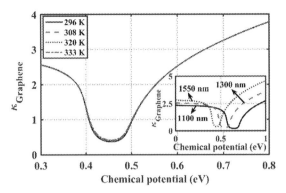

Fig. 5 Simulated variation of extinction coefficient ($\kappa_{Graphene}$) of graphene monolayer with μ for different temperatures at $\lambda = 1550$ nm. The inset shows the corresponding $\kappa_{Graphene}$ vs. μ variations for different wavelengths at $T = 296$ K.

(μ) at three different NIR wavelengths (1.10 μm, 1.30 μm, and 1.55 μm). The primary objective of this study is to find the set of parameters (i.e., T, μ, and λ) leading to an improved sensing performance (in terms of sensitivity and resolution) in NIR. The variation of graphene's extinction coefficient ($\kappa_{Graphene}$) with μ for four different temperatures at $\lambda = 1.55$ μm is shown in Fig. 5. A dip is observed that has a significant dependence on λ (inset of Fig. 5). It is also evident from Fig. 5 that the μ value corresponding to this dip in $\kappa_{Graphene}$ is nearly unaffected by the variation in T (in the range 296–333 K).

Conceptually, the minimum absorption by a graphene monolayer is demonstrated by a dip in $\kappa_{Graphene}$, which leads to minimum variation in P_{out}. This variation causes a decrease in S_n as apparent from Fig. 6. Thus, it can be inferred that longer λ with greater values μ are the preferred choice. However, taking a practical scenario into consideration, achieving μ of the order of 0.8 eV is not an easy task as there will be a requirement of dynamic electrical gating or very high levels of doping, etc. Notably, if μ is set at 0.6 eV, chemical doping can incessantly adjust graphene layers for robust device applications. In this case, the isolation of a graphene layer from the environment can be considered [39]. Table 2 shows the maximum S_n and minimum R_n with their corresponding tunable parameters (μ, T) at three NIR wavelengths.

Hence, the value of μ should not exceed 0.6 eV from a practical viewpoint, which is also beneficial from the perspective that no dip is observed in S_n at $\mu = 0.6$ eV (for $\lambda = 1.55$ μm). Revisiting Fig. 6 based on this approach at $\mu = 0.6$ eV (for 1.55 μm, $T = 311.1$ K), it is found that the best values of S_n and R_n are 112.211 mW/RIU and 8.91×10^{-10} RIU, respectively. Remarkably, the optimized temperature value (~311.1 K) is very close to the human body temperature (~310 K), which makes this sensor probe an ideal candidate for malignancy detection at body temperature only.

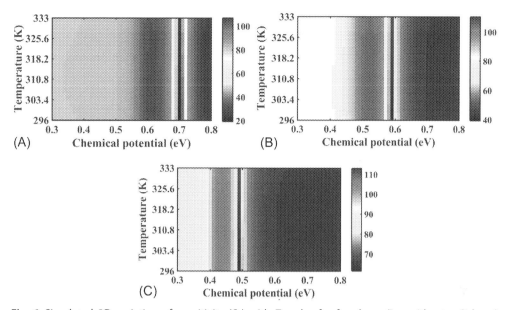

Fig. 6 Simulated 2D variation of sensitivity (S_n) with T and μ for four-layer (i.e., without α-Si layer) EWFOS at (A) $\lambda = 1.10$ μm, (B) $\lambda = 1.30$ μm, and (C) $\lambda = 1.55$ μm.

Table 2 Maximum S_n and minimum R_n with their corresponding tunable parameters (μ,T) at three NIR wavelengths.

λ (μm)	Performance parameters		Corresponding tunables	
	Max. S_n (mW/RIU)	Min. R_n (RIU)	μ (eV)	T (K)
1.10	106.933	9.35×10^{-10}	0.8	312.2
1.30	110.373	9.06×10^{-10}	0.8	309.8
1.55	113.036	8.85×10^{-10}	0.8	311.6
1.55	112.211	8.91×10^{-10}	0.6	311.1

Moreover, (S_n, R_n) is almost identical to the values achieved at $\mu = 0.8$ eV, $\lambda = 1.55$ μm, and $T = 311.6$ K, i.e., (113.036 mW/RIU, 8.85×10^{-10} RIU).

3. SPR-based fiber optic sensor utilizing ORD

FO SPR sensors are preferred owing to miniaturized and flexible design suitable for remote sensing applications with easy integrations [12, 40]. As a feasible solution, in recent years, several experimental and theoretical research works have been reported on fiber SPR sensors with inclusion of 2D material coating [41, 42]. Discussing SPR sensors and 2D materials together, it can be observed that multilayer graphene has been

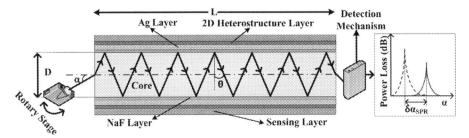

Fig. 7 Schematic diagram of multilayered FOSPR sensor with output power loss spectra. The sensing length of fiber is *L* and *D* is the fiber core diameter. The 2D heterostructure layer acts as analyte interacting layer.

employed in SPR-based sensors for performance enhancement due to increased surface adsorption [10, 13, 41]. In this continuation, Mishra et al. demonstrated the utilization of MoS_2-coated graphene in FOSPR-based sensor. A high sensitivity of 6.2 μm/RIU was reported with fiber coated with Al/graphene/MoS_2. Recently, some of the research works have been reported based on application of 2D material (graphene and MoS_2) in FOSPR sensors for figure of merit (FOM) enhancement by tuning the radiation damping (RD) [43, 44]. In addition, the 2D material-based heterostructures have also shown their application in performance enhancement of FO SPR sensor [45]. A FO SPR sensor utilizing 2D heterostructure material is shown in Fig. 7. The output power loss (in dB) variation with α of light beam is the operating principle of the FO-based multilayered sensor utilizing ORD. The modulation effects are responsible for power loss variation of plasmonic structure coated on small length "L" of the fiber (Fig. 7).

The source of monochromatic light is a laser diode mounted on a rotary stage. According to Snell's law, the incident angle "α" and "θ" (inside the fiber) variations are interrelated. The resonance will be satisfied at a particular value of incident angle (i.e., $\alpha = \alpha_{SPR}$). The N-layer TMM is employed for calculation of normalized reflection coefficient (*R*) of transverse magnetic (TM) incident light [46]. The normalized output power after plasmonic modulation and taking Snell's law into account can be calculated as:

$$P(\alpha) = R(\theta)^{N_{ref}(\theta)} \tag{8}$$

In this equation, $N_{ref}(\theta) = L/(D \tan \theta)$ represents the number of reflections corresponding to a light ray propagating at an angle "θ" inside the fiber core. Here, *D* is the fiber core diameter. Finally, the power loss (PL) can be calculated as:

$$PL \; (in \; dB) = 10 \log_{10} \left(\frac{P_{ref}}{P_{out}} \right) \tag{9}$$

Here, P_{ref} is the normalized reference power and P_{out} $(= P(\alpha))$ is the normalized output power calculated from Eq. (1).

3.1 Performance parameters

FOM is the overall performance evaluation parameter and is represented as the product of sensitivity $(\frac{\delta\alpha_{SPR}}{\delta n_s})$ and detection accuracy (D.A. = 1/FWHM) as:

$$FOM = \frac{\delta\alpha_{SPR}}{\delta n_s} \times \frac{1}{FWHM} \tag{10}$$

Here, $\delta\alpha_{SPR}$ is the angular shift in the power loss peaks due to small variation in RI of analyte (δn_s). Full width at half maximum (FWHM) is the angular width of the power loss spectrum corresponding to the reference analyte. Moreover, the power loss ratio (PLR) and Rayleigh scattering factor (RSF, i.e., λ^{-4}) and are also important factors for a complete evaluation of sensor's performance from a practical viewpoint. A combined figure of merit (C-FOM in μm^4/RIU) of the sensor that incorporates the important factors, maximum FOM (M-FOM), RSF, and corresponding PLR, can be presented as:

$$C-FOM = MFOM \times \frac{PLR}{RSF} \tag{11}$$

3.2 Comprehensive analysis of fluoride fiber SPR sensor utilizing multilayer variants of 2D materials (graphene and MoS₂) under optimum radiation damping in NIR

In this section, the performance of a fluoride fiber optic SPR sensor using 2D material (graphene, MoS_2) (number of monolayers "N" varying from 1 to 2) is analyzed in the NIR region for the detection of alcohols. The evaluation of sensing performance is presented as a 2D function of Ag layer thickness and NIR wavelength for variable "N."

ZBLAN ($ZrF4_{55.8}BaF2_{14.4}LaF2_{5.2}Al- F3_{3.8}NaF_{20.2}$) fluoride glass is considered as the fiber core material owing to its favorable optical properties in NIR such as broad transmission window, low mean dispersion, Rayleigh scattering, and optical nonlinearity. The λ-dependent RI of ZBLAN glass core (n_{core}) is taken into account [47]. Furthermore, the light acceptance angle (α_{acc}) of the fiber is defined in relation to numerical aperture (NA) as:

$$\alpha_{acc} = \sin^{-1}(NA), \tag{12}$$

$$NA = \sqrt{n_{core}^2 - n_{clad}^2} \tag{13}$$

Here, n_{clad} is the RI of clad material. So, from Eq. (11), we can see that, to increase the acceptance angle, the value of n_{clad} should be considerably smaller than n_{core}. For this purpose, a thin NaF film (say, 5 nm) is selected as clad material for the present study. The RI of NaF film is taken from work reported earlier [48]. In this study, Ag (with thickness d_{Ag}) is considered as a SPR–active metal deposited over NaF. The fourth layer is 2D material (graphene or MoS₂; one at a time). H_2O is considered as the reference sample and ethanol as an analyte at room temperature for this study. The RI values of both the analytes are adapted from the work of Kedenburg et al. [49]. Based on the previously discussed methodologies, Fig. 8 depicts the 2D (d_{Ag}-λ) variation of FOM for the fiber optic SPR sensor with monolayer (A) graphene and (B) MoS$_2$. It is clear that multiple (d_{Ag}, λ) combinations are obtained for which the FOM attains significantly large magnitudes for both the cases. In Fig. 8A, two most prominent (d_{Ag}, λ) combinations are magnified. A M-FOM value of 12,623.25 RIU^{-1} is achieved for $d_{Ag} = 47.1$ nm and $\lambda = 841.6$ nm, which can be attributed to ORD for the corresponding SPR sensor structure. Similarly, in the case of monolayer MoS$_2$ (Fig. 8B) a M-FOM value of 26,184.36 RIU^{-1} corresponding to $d_{Ag} = 48.4$ nm and $\lambda = 925.3$ nm is obtained. Remarkably, the M-FOM in case of sensor structure with monolayer MoS$_2$ is more than double the M-FOM corresponding to graphene monolayer (i.e., 12,623.25 RIU^{-1}).

To understand the simulated sensor model as close to practical realization as possible, a more comprehensive analysis is required. In this regard, a large operating wavelength should be maintained that will provide a smaller RSF (i.e., λ^{-4}) as well as a smaller photodamage of the analyte (biosamples, in particular) in NIR [50]. Also, the PLR, which is the ratio of peak PL (ethanol) to peak PL (H$_2$O) at corresponding ORD conditions, should be as large and RSF should be as small as possible (both within the practical limits). Table 3 presents the values of parameters required for C-FOM. Table 3 clearly indicates that the overall superior performance is achieved in the case of monolayer MoS$_2$-based sensor design. This is mainly due to longer optimum wavelength (leading to smaller RSF) and greater PLR. Graphene-based sensor design may not be opted due to reasonably smaller PLR and significantly larger RSF. In a comparative manner, monolayer MoS$_2$ based sensor design ($N = 1$) has nearly 33% greater M-FOM, but its C-FOM is merely 10% greater than bilayer MoS$_2$-based design ($N = 2$). The reason behind this is nearly identical PLR and almost 24% greater RSF. Hence, MoS$_2$ bilayer-based sensor design may be preferred for specific sensing applications requiring as small signal scattering and photodamage as possible (such as biochemical detection).

The role of temperature in performance of a multilayered (ZBLAN/NaF/Ag/WS$_2$/analyte) FO SPR sensor has been reported recently [51]. It is worthwhile to envisage the role of temperature on ORD condition in view of thermo-optic properties of different media (e.g., dielectric, metallic, and TMDC) involved with the fiber SPR sensor design. Above room temperature, it has been found that the optical properties of metal and dielectric materials are affected considerably owing to different effects such as thermal expansion, phonon-electron scattering, electron-electron scattering, and polarizability variation [13, 14]. Generally, the temperature (T)-dependent RI of a material may be calculated as:

$$n(T) = n(T_0) + (T - T_0)\frac{dn}{dT} \tag{14}$$

In Eq. (14), $n(T_0)$ is the RI of the material at reference (room) temperature (T_0) and dn/dT is the thermo-optic coefficient of the respective material [15].

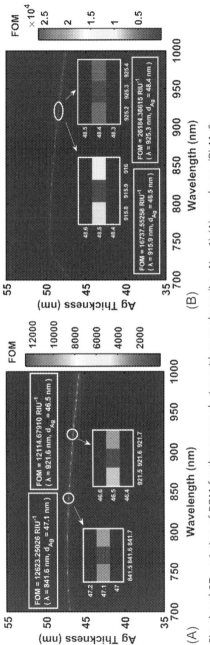

Fig. 8 Simulated 2D variation of FOM for the sensor design with monolayer (i.e., $N = 1$) (A) graphene, (B) MoS$_2$.

Table 3 Combined performance parameter (C-FOM) for four different sensor designs.

N	M-FOM	λ (μm)	RSF (μm^{-4})	PLR	C-FOM
Graphene-based sensor design					
1	12,623.25	0.8416	1.993	2.501	15,840.82
2	13,526.02	0.7556	3.067	2.193	9671.52
MoS$_2$-based sensor design					
1	26,814.36	0.9253	1.364	2.929	57,580.10
2	20,112.28	0.9762	1.101	2.860	52,244.43

With this backdrop, this simulation study reports on tuning and enhancement of fiber optic SPR sensor's performance using monolayer and multilayer structures (number of monolayers N varying from 2 to 5) of WS$_2$. The SPR sensor structure is simulated at NIR wavelength of 785 nm for detection of alcohols for a temperature range (20°C–63°C). Furthermore, the simulation of the sensor scheme is based on experimentally verified wavelength-dependent RI and thermo-optic coefficient values/phenomena of constituent material layers. The corresponding simulation result is depicted in Fig. 9.

Here, it is necessary to mention that d_m varies between 35 nm and 55 nm and temperature varies from 20°C (room temperature) to 63°C. However, the maximum FOM values are observed for d_m range 54.5–56 nm only, which is visible in Fig. 2. At $\lambda = 785$ nm, the maximum FOM of 22,489.95 RIU^{-1} is attained for combination "$d_m = 55.5$ nm and $T = 37.4$°C" for WS$_2$ monolayer-based sensor design. Notably, there is another discrete combination "$d_m = 55.4$ nm and $T = 41.2$°C" leading to next highest

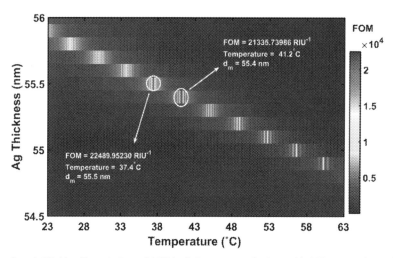

Fig. 9 Simulated 2D (d_m, T) variation of FOM of the sensor design with WS$_2$ monolayer ($N = 1$) at $\lambda = 785$ nm.

FOM value of 21,335.74 RIU^{-1}. There are other such combinations, but their corresponding FOM values keep decreasing, therefore, we chose not to mark (in Fig. 2) or discuss them any further. The previously two values of FOM indicate that, at fixed λ (i.e., 785 nm), the extremely high values of FOM achieved are largely dependent on T rather than d_m. More specifically, d_m remains nearly constant (55.4–55.5 nm), but T undergoes a considerable variation from 37.4 nm (for FOM = 22,489.95 RIU^{-1}) to 41.2 nm (for FOM = 21,335.74 RIU^{-1}).

It can be interpreted as a case of ORD under the influence of thermo-optic effect of multilayered plasmonic structure (ZBLAN-NaF-Ag-WS$_2$-analyte). There are established details of dynamic RD as a function of metal thickness and wavelength [1]. Smaller photodamage of biosamples and low Rayleigh scattering at longer wavelengths should also be analyzed alongside RD in plasmonic sensors. The Rayleigh scattering coefficient (μ_{scat}) of optical signals in optical fiber is defined as [27]:

$$\mu_{scat} = \frac{8\pi^3}{\lambda^4} n_p^8 p^2 \kappa T \beta \tag{15}$$

In this expression, n_p and p, respectively, are the RI and photoelastic coefficient of the fiber core material, κ is the Boltzmann constant, T is temperature, and β is the isothermal compressibility. At constant T, μ_{scat} is primarily dependent on λ^{-4} factor as other factors remain constant at any defined T. With variation in T, these factors show their cumulative effect in the variation of μ_{scat} as the previous study shows nearly 0.4% variation in μ_{scat} with per-degree change in T [28]. Correspondingly, the RD should also be affected by the variation in temperature due to variation in Rayleigh scattering. Certainly, it is a very compound correlation among λ, d_m, and T and its effect on FOM of fiber SPR sensor owing to heterogeneous nature of the media involved with the multilayered structure. Nevertheless, the outcome is very crucial as it led to excessively large FOM values.

The potential experimental setup for the discussed FO SPR sensor scheme may consist of the following steps:

(a) ZBLAN fluoride fiber is available commercially [52].

(b) Thermal or CVD method can be used for deposition of a multilayered structure with desired thickness of NaF, Ag, and graphene or MoS$_2$ layers on fluoride fiber core.

(c) Fiber-coupled laser source is available commercially for application in NIR spectral region corresponding to ORD condition. The angular shift can be measured by position-sensitive detector (PSD) with an angular precision of 0.001 degree or smaller [50].

(d) A temperature-controlled environment is suggested for practical implementation. There are commercially available temperature controllers, which can provide controlled temperature variation of as small as 0.001°C [53].

(e) Furthermore, power loss can be measured by most of the power meters connected at the fiber's output end. The loss in transmitted power will be maximum at SPR

Table 4 Maximum FOM comparison among some of the recently reported works based on ORD.

Ref	SPR modalities (ORD-based)	FOM (RIU^{-1})
Sharma and Kaur [43]	Samerium-doped fiber coated with 2D material	6904.012 (graphene) 5897.082 (MoS$_2$)
Sharma and Kaur [54]	Chalcogenide fiber sensor with polymer and 2D layer	1647 ($\lambda = 1200$ nm)
Sharma and Gupta [51]	ZBLAN fiber coated with NaF/Ag/WS$_2$	27,563.575 ($\lambda = 785$ nm)
Sharma and Gupta [55]	ZBLAN fiber coated with NaF/Ag/MoS$_2$	26,814.36
Sharma et al. [56]	Fluoride fiber coated with metal and (BlueP/TMD) heterostructure	12,409.30 (BlueP/WS$_2$) 15,650.75 (BlueP/MoS$_2$)

condition (i.e., $\alpha = \alpha_{SPR}$). The final outcome has to be processed with the help of a computing system to measure the sensor's FOM by analyzing the corresponding power loss spectra.

As a final note, a comparative analysis in terms of maximum achievable FOM (at ORD condition) is presented in Table 4 based on recently reported works on FO SPR sensor utilizing 2D materials.

4. Conclusion and future scope

This chapter summarizes the application of 2D materials in fiber optic sensors. Furthermore, introduction of thin α-Si layer in five-layer system, i.e., ZBLA core—19.3% GeO$_2$-doped silica clad—MoS$_2$ monolayer—α-Si layer (10 nm)—analyte provides superior performance in longer wavelength region. To continue, the simultaneous effect of temperature and graphene's chemical potential on the performance of EWFOS is presented for application in the NIR region. Another important type of fiber-based sensing scheme i.e., FO sensor based on SPR is demonstrated. The demonstration is carried out based on the condition of ORD. Furthermore, the effects of monolayer graphene and MoS$_2$ on overall performance are presented. There is a lot of scope for the application of 2D materials in FO sensors. For example, very little work has been reported based on photonic crystal fiber sensor utilizing 2D materials.

References

[1] S. Balendhran, S. Walia, H. Nili, S. Sriram, M. Bhaskaran, Elemental analogues of graphene: silicene, germanene, stanene, and phosphorene, Small 11 (6) (2015) 640–652.

[2] G. Rubio-Bollinger, et al., Enhanced visibility of MoS2, MoSe2, WSe2 and black-phosphorus: making optical identification of 2D semiconductors easier, Electronics 4 (4) (2015) 847–856.

[3] J.S. Ponraj, Z. Xu, S. Chander, F. Wang, Z. Wang, 2D materials advances : from large scale synthesis and controlled heterostructures to improved characterization techniques, defects and applications, 2D Mater. 3 (2016) 042001.

[4] L. Wu, et al., Sensitivity enhancement by using few-layer black phosphorus-graphene/TMDCs heterostructure in surface plasmon resonance biochemical sensor, Sensors Actuators B Chem. 249 (2017) 542–548.

[5] K.S. Novoselov, A. Mishchenko, A. Carvalho, A.H.C. Neto, 2D materials and van der Waals heterostructures, Research 353 (6298) (2016) aac9439.

[6] J. Guo, Z. Zhou, H. Li, H. Wang, C. Liu, Tuning electronic properties of blue phosphorene/graphene-like GaN van der Waals heterostructures by vertical external electric field, Nanoscale Res. Lett. 14 (2019) 174–176.

[7] R. Mas-Ballesté, C. Gómez-Navarro, J. Gómez-Herrero, F. Zamora, 2D materials: to graphene and beyond, Nanoscale 3 (1) (2011) 20–30.

[8] S. Varghese, S. Varghese, S. Swaminathan, K. Singh, V. Mittal, Two-dimensional materials for sensing: graphene and beyond, Electronics 4 (3) (2015) 651–687.

[9] B. Kaur, A.K. Sharma, Plasmonic biosensor in NIR with chalcogenide glass material: on the role of probe geometry, wavelength, and 2D material, Sens. Imaging 19 (1) (2018) 1–9.

[10] K.N. Shushama, M.M. Rana, R. Inum, M.B. Hossain, Graphene coated fiber optic surface plasmon resonance biosensor for the DNA hybridization detection: simulation analysis, Opt. Commun. 383 (2017) 186–190.

[11] N. Huo, Y. Yang, J. Li, Optoelectronics based on 2D TMDs and heterostructures, J. Semicond. 38 (3) (2017) 031002.

[12] A.K. Sharma, A.K. Pandey, B. Kaur, A review of advancements (2007–2017) in plasmonics-based optical fiber sensors, Opt. Fiber Technol. 43 (2018) 20–34.

[13] Y. Zhao, X. Li, X. Zhou, Y. Zhang, Review on the graphene based optical fiber chemical and biological sensors, Sensors Actuators B Chem. 231 (2016) 324–340.

[14] J. Kim, et al., Label-free quantitative immunoassay of fibrinogen in Alzheimer disease patient plasma using fiber optical surface plasmon resonance, J. Electron. Mater. 45 (5) (2016) 2354–2360.

[15] H. Apriyanto, et al., Comprehensive modeling of multimode fiber sensors for refractive index measurement and experimental validation, Sci. Rep. 8 (1) (2018) 1–13.

[16] R. Taheri Ghahrizjani, H. Sadeghi, A. Mazaheri, A novel method for online monitoring engine oil quality based on tapered optical fiber sensor, IEEE Sensors J. 16 (10) (2016) 3551–3555.

[17] Y. Yuan, et al., Investigation for terminal reflection optical fiber SPR glucose sensor and glucose sensitive membrane with immobilized GODs, Opt. Express 25 (4) (2017) 3884.

[18] K. Gasior, T. Martynkien, M. Napiorkowski, K. Zolnacz, P. Mergo, W. Urbanczyk, A surface plasmon resonance sensor based on a single mode D-shape polymer optical fiber, J. Opt. 19 (2017) 025001.

[19] N.M.Y. Zhang, et al., Design and analysis of surface plasmon resonance sensor based on high-birefringent microstructured optical fiber, J. Opt. 18 (6) (2016) 065005.

[20] N. Zhong, et al., A high-sensitivity fiber-optic evanescent wave sensor with a three-layer structure composed of Canada balsam doped with GeO2, Biosens. Bioelectron. 85 (2016) 876–882.

[21] X.D. Wang, O.S. Wolfbeis, Fiber-optic chemical sensors and biosensors (2013-2015), Anal. Chem. 88 (1) (2016) 203–227.

[22] X. Xin, N. Zhong, Q. Liao, Y. Cen, R. Wu, Z. Wang, High-sensitivity four-layer polymer fiber-optic evanescent wave sensor, Biosens. Bioelectron. 91 (January) (2017) 623–628.

[23] N. Zhong, M. Chen, H. Chang, T. Zhang, Z. Wang, X. Xin, Optic fiber with Er3 +:YAlO3/SiO2/TiO2 coating and polymer membrane for selective detection of phenol in water, Sensors Actuators B Chem. 273 (July) (2018) 1744–1753.

[24] V.V.R. Sai, T. Kundu, C. Deshmukh, S. Titus, P. Kumar, S. Mukherji, Label-free fiber optic biosensor based on evanescent wave absorbance at 280 nm, Sensors Actuators B Chem. 143 (2) (2010) 724–730.

[25] B.D. Gupta, C.D. Singh, Fiber-optic evanescent field absorption sensor: a theoretical evaluation, Fiber Integr. Opt. 13 (4) (1994) 433–443.

[26] V. Ruddy, An effective attenuation coefficient for evanescent wave spectroscopy using multimode fiber, Fiber Integr. Opt. 9 (2) (1990) 143–151.

[27] E. Hecht, Optics, fifth ed., Pearson, 2017.

[28] A.K. Sharma, J. Gupta, R. Basu, Simulation and performance evaluation of fiber optic sensor for detection of hepatic malignancies in human liver tissues, Opt. Laser Technol. 98 (2018) 291–297.

[29] A.K. Sharma, J. Gupta, Fiber optic sensor's performance enhancement by tuning NIR wavelength, polarization, and 2D material, IEEE Photon. Technol. Lett. 30 (12) (2018) 1087–1090.

[30] I. Sharma, A.K. Sharma, Multilayered evanescent wave absorption based fluoride fiber sensor with 2D material and amorphous silicon layers for enhanced sensitivity and resolution in near infrared, Opt. Fiber Technol. 50 (April) (2019) 277–283.

[31] A. Ghatak, K. Thyagarajan, An Introduction to Fiber Optics, Cambridge University Press, Cambridge, 1998.

[32] D.T. Pierce, W.E. Spicer, Electronic structure of amorphous Si from photoemission and optical studies, Phys. Rev. B 5 (8) (1972) 3017–3029.

[33] J.W. Weber, V.E. Calado, M.C.M. van de Sanden, Optical constants of graphene measured by spectroscopic ellipsometry, Appl. Phys. Lett. 97 (9) (2010) 130–132.

[34] H.P. Hughes, A.R. Beal, Kramers-Kronig analysis of the reflectivity spectra of 2H-MoS_2, 2H-$MoSe_2$ and 2H-$MoTe_2$, Solid State Phys. 881 (12) (1979) 881.

[35] P. Giannios, et al., Visible to near-infrared refractive properties of freshly-excised human-liver tissues: marking hepatic malignancies, Sci. Rep. 6 (February) (2016) 1–10.

[36] F.J. Arregui, I. Del Villar, C.R. Zamarreño, P. Zubiate, I.R. Matias, Giant sensitivity of optical fiber sensors by means of lossy mode resonance, Sensors Actuators B Chem. 232 (2016) 660–665.

[37] G. Liu, M. Han, W. Hou, High-resolution and fast-response fiber-optic temperature sensor using silicon Fabry-Pérot cavity, Opt. Express 23 (6) (2015) 7237–7247.

[38] F. Sequeira, et al., Refractive index sensing with D-shaped plastic optical fibers for chemical and biochemical applications, Sensors 16 (12) (2016) 2119 1–11.

[39] M. Aliofkhazraei, N. Ali, W.I. Milne, C.S. Ozkan, S. Mitura, J.L. Gervasoni, Graphene Science Handbook, CRC Press, Taylor & Francis Group, Boca Raton, FL, 2016.

[40] G. Liang, Z. Luo, K. Liu, Y. Wang, J. Dai, Y. Duan, Fiber optic surface plasmon resonance–based biosensor technique: fabrication, advancement, and application, Crit. Rev. Anal. Chem. 46 (3) (2016) 213–223.

[41] H. Fu, S. Zhang, H. Chen, J. Weng, Graphene enhances the sensitivity of fiber-optic surface plasmon resonance biosensor, IEEE Sensors J. 15 (10) (2015) 5478–5482.

[42] A.K. Mishra, S.K. Mishra, R.K. Verma, Graphene and beyond graphene MoS2 : a new window in surface-plasmon-resonance-based fiber optic sensing, J. Phys. Chem. C 120 (5) (2016) 2893–2900.

[43] A.K. Sharma, B. Kaur, Fiber optic SPR sensing enhancement in NIR via optimum radiation damping catalyzed by 2D materials, IEEE Photon. Technol. Lett. 30 (23) (2018) 1.

[44] A.K. Sharma, B. Kaur, Simulation of multilayered heterojunction-based chalcogenide fiber SPR sensor with ultrahigh figure of merit in near infrared, IEEE Sensors J. 19 (11) (2019) 4074–4078.

[45] A.K. Sharma, A.K. Pandey, B. Kaur, Fluoride fiber-based plasmonic biosensor with two–dimensional material heterostructures: enhancement of overall figure-of-merit via optimization of radiation damping in near infrared region, Materials (Basel) 12 (9) (2019) 1542.

[46] A.K. Sharma, Plasmonic biosensor for detection of hemoglobin concentration in human blood: design considerations, J. Appl. Phys. 114 (4) (2013) 044701.

[47] L. Zhang, F. Gan, P. Wang, Evaluation of refractive-index and material dispersion in fluoride glasses, Appl. Opt. 33 (1) (1994) 50.

[48] H.H. Li, Refractive index of alkali halides and its wavelength and temperature derivatives, J. Phys. Chem. Ref. Data Monogr. 5 (2) (1976) 329–528.

[49] S. Kedenburg, M. Vieweg, T. Gissibl, H. Giessen, Linear refractive index and absorption measurements of nonlinear optical liquids in the visible and near-infrared spectral region, Opt. Mater. Express 2 (11) (2012) 1588–1611.

[50] R. Ziblat, V. Lirtsman, D. Davidov, B. Aroeti, Infrared surface plasmon resonance: a novel tool for real time sensing of variations in living cells, Biophys. J. 90 (7) (2006) 2592–2599.

[51] A.K. Sharma, J. Gupta, Fluoride fiber plasmonic sensor with multilayer variants of tungsten disulfide (WS2): seeking enhanced figure-of-merit via thermo-optic tuning of radiation damping, Opt. Fiber Technol. 53 (September) (2019) 102037.

[52] ZBLAN Fluoride Glass Fibers & Cables, [Online]. Available https://www.fiberlabs.com/fiber_index/, 2019 [Accessed: 22-Nov-2019].

[53] High Accuracy Temperature Controllers, [Online]. Available https://www.tc.co.uk/temperature_control/F9000.html, 2019 [Accessed: 22-Nov-2019].

[54] A.K. Sharma, B. Kaur, Optical fiber technology chalcogenide fiber-optic SPR chemical sensor with MoS2 monolayer polymer clad and polythiophene layer in NIR using selective ray launching, Opt. Fiber Technol. 43 (March) (2018) 163–168.

[55] A.K. Sharma, J. Gupta, Simulation and comprehensive analysis of fluoride fiber SPR sensor with multilayer variants of 2D materials (graphene and MoS2) under optimum radiation damping in NIR, IEEE Sensors J. 19 (19) (2019) 8775–8780.

[56] A.K. Sharma, A.K. Pandey, B. Kaur, Simulation study on comprehensive sensing enhancement of BlueP/MoS2—and BlueP/WS2—based fluoride fiber surface plasmon resonance sensors: analysis founded on damping, field, and optical power, Appl. Opt. 58 (16) (2019) 4518.

CHAPTER 20

Soft sensors for screening and detection of pancreatic tumor using nanoimaging and deep learning neural networks

K. Sujatha[a], R. Krishnakumar[b], B. Deepalakshmi[c], N.P.G. Bhavani[d], and V. Srividhya[d]

[a]Department of EEE, Dr. MGR Educational & Research Institute, Chennai, India
[b]Department of EEE, Vels Institute of Science Technology and Advanced Studies, Chennai, India
[c]Department of ECE, Ramco Institute of Technology, Chennai, India
[d]Department of EEE, Meenakshi College of Engineering, Chennai, India

1. Introduction

The leading cause of death is cancer and, nowadays, it has become the largest public health issue worldwide. The early detection of nanosized pancreatic tumors is a challenging task, and it has been the most common disease among many people around the world. An occurrence of nearly a million of new cases has been estimated by the World Health Organization (WHO) [1]. The pancreatic tumor stands as a major cause for death around the world [2]. About 80% of survival rates are recorded for pancreatic tumors in economically elevated countries to below 40% in economically lowered countries [3]. The nonavailability of early detection schemes for detecting nanosized tumors has led to low survival rate in certain countries. These schemes have a chief impact on the success of pancreatic cancer treatment, because the treatment becomes complicated in the chronic stages [4]. However, visual analysis of MRI is a difficult task, even for radiologists. Great variability in analysis using imaging methods makes diagnosis a complex task [5]. The diagnosis made during clinical practice may not match the traditional images and hypothetical descriptions [6]. This is the reason for Computer Aided Diagnosis (CAD) to be an effective tool helping radiologists improve the accuracy of diagnosis. A number of studies are using conventional image processing tools for diagnosis in the medical field. Therefore, the combination of professional's specialized knowledge and pattern recognition computational tools may improve diagnosis accuracy [7–27]. An Intelligent scheme can help radiologists in making unique decisions, thus improving the efficiency in identifying the anatomical abnormalities of the pancreas [5, 8–28]. In this chapter, DWNN, a deep learning tool based on multiple levels of wavelet decomposition, is used. For example, this method is applied to detect and classify the nanosized tumors present in the MRI

Handbook of Nanomaterials for Sensing Applications
https://doi.org/10.1016/B978-0-12-820783-3.00002-6

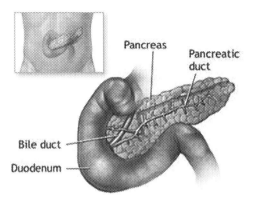

Fig. 1 Location of pancreas in the thoracic cavity.

of the pancreas. The results are obtained using Haralick descriptors, Wavelet transform and intelligent classifiers are compared with other approaches that are in current use.

There are two major categories of pancreatic cancer. The first one is called an exocrine tumor where the exocrine cells are affected during the onset of cancer. Exocrine cells produce enzymes, and nearly 95% of pancreatic cancers belong to this type. Adenocarcinoma is a common type where the duct of the pancreas is affected and nearly 80% belong to this type. The second type is called as an endocrine tumor that originates from the cells that produce hormones. Less than 5% of pancreatic cancer belongs to this type.

The depth of the abdominal cavity contains the pancreas, which is a flattened, lengthened organ. This gland secrets insulin hormone to reduce blood sugar levels and forms a part of the human digestive system [7, 8]. The in-depth location of the pancreas is a major hurdle in the detection of nanosized tumors. Such tumors are left unidentified by radiologists during external diagnosis of the abdomen and manual visualization of MRI [9, 10]. Many times, pancreatic tumors are diagnosed only in the chronic stage, which produces notable symptoms like malfunctioning of the neighboring organs like gall bladder, liver, small intestine, large intestine, and lungs as shown in Fig. 1.

The use of image processing technology to detect pancreatic tumors in the range of 10^{-9} level is called nanoimaging. Nanotechnology differs from traditional methods of diagnosing and treating pancreatic tumors and provides some advantages over them. Nanoimaging is a branch of the recently budding field of nanotechnology.

2. Related works

The various medical images are classified using convolutional neural network (CNN) approaches, including the detection of the nanosized tumors. For example, it is evident that the multilayer perceptron (MLP) is used to classify normal and abnormal MRI images containing tumors [28–30]. Yet, there are not too many investigations carried out with deep learning applied to MRI images of the pancreas. Above and beyond, in general, all these investigations are restricted to categorize only normal and abnormal images. It is evident that researchers have used infrared (IR) images to diagnose tumors in such complicatedly situated organs [31]. Segmentation of IR images is done to remove regions like gall bladder, lungs, duodenum, and intestine. Finally, the Deep-Neural Network (DNN) with three layers is used for classification purposes in detecting pancreatic tumors. In some cases, the support vector machine (SVM) receives the features as a resulting matrix of features that are used as inputs to classify the images accordingly with the patient having the possibility of cancer. The efficiency of this system is 78% for healthy images with 94% accuracy. One approach that is considered in this scheme, further than classifying images into normal or abnormal, is an initiative that has been taken to classify abnormal images into categories, such as benign, malignant, and cysts when it is nanosized [27].

This section creates awareness regarding the various methods used to diagnose pancreatic tumors. Cancer in the pancreas can be detected at an early stage by preprocessing the CT images using minimum distance classifier with 60% accuracy as proposed by Jeenal Shah et al. (2015) [1,2,32]. Various filters like median and Gaussian filters are used along with image segmentation and Artificial Neural Network (ANN) classifiers [3, 4]. K. Jayaprakash et al. [33] have discussed that, out of 800 CT samples of the pancreas for diagnosis of pancreatic cancer, nearly 50 images were selected to extract the Gray level co-occurrence matrix (GLCM), which is used to extract features like energy, entropy, homogeneity, contrast, and correlation and then classified using minimum distance approach [5, 6]. The deviation in pancreatic cancer diagnosis by this method has a deviation of 0.00573.

3. Challenges in detecting pancreatic cancers

The challenge in pancreatic cancer detection is that the cancerous cells would have grown in size and spread to the adjacent organs and blood vessels located near the pancreas. Normally, the pancreatic cancers are detected only during metastasis. This may prove to be life-threatening for the patients. Usually, pancreatic cancers are detected only in the T1 stage where the size is smaller than 2 cm.

4. Research highlights

The motivation behind this work includes the following:
• Diagnosis of nanosized pancreatic tumors using image processing and deep learning algorithms

- Extraction of the wavelet coefficients for benign and malignant nanosized pancreatic tumors
- Deep Wavelet Neural Networks (DWNN) for classification of the nanosized pancreatic tumors
- Increased sensitivity for the proposed method of diagnosis

5. Materials and methods

This section describes the various algorithms used. Section 5.1, includes the general concept of DWNN, and in Section 5.2, the algorithm for DWNN is discussed. Section 5.3 throws light on the performance analysis.

5.1 Understanding of DWNN

A deep learning method for feature extraction based on multilevel wavelet decomposition (DWNN) is used to investigate the nanosized pancreatic tumors present in MRI images [34]. For wavelet-based feature extraction, filter bank consisting of a set of low-pass and high-pass filters are convolved with the MRI image of the pancreas bearing nanosized tumors. The filtered output images resulting from low-pass and high-pass filters are called approximations and details [35]. In the approximations, the smoothness of the original pancreatic image is depicted, whereas in the details its boundaries are depicted. This pattern recognition strategy is used in detecting nanosized pancreatic tumors by image analysis in both spatial and frequency domains [36]. In the DWNN approach, a neuron is created by combining a given filter with pancreatic image size reduction, called decimation.

The set of low-pass filters on the input side and high-pass filters on the output side is used in DWNN to create a filter bank. For example, the filter bank has "n" low-pass and high-pass filters. Thus, an input image of the pancreas is presented to "n" processing elements, which form the first intermediate layer of the neural network. In the second layer, the processed images, which are the output from the first layer, are presented to the same filter bank and decimated as processed in the first layer. The process repeats itself for the third and successive intermediate layers to form the wavelet coefficients. The synthesis block is available, which is responsible for extracting information from the resultant images. The wavelet coefficients of the final intermediate layer are given as the input to the neurons of the output layer of the MLP model. The neurons of the output layer have sigmoid activation function. This combination of wavelet-based filter bank with the MLP structure has generated the novel DWNN model with deep learning concept. This approach is sketched in Fig. 2.

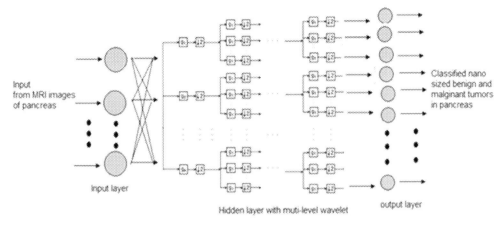

Fig. 2 Architecture for DWNN.

5.2 Algorithm for DWNN

The algorithm for DWNN is split into two phases. The first phase is the forward propagation phase, and the second phase is the reverse propagation phase. The algorithm for BPA is explained as follows:

Steps 1–5 denotes the forward propagation.

- Apply the inputs to the network and obtain the output. The initial outputs are random numbers.
- Calculate the wavelet coefficients

$$x[m] = x(t_0 + m\Delta t), (m = 0, 1, \ldots, N-1) \tag{1}$$

$$x[m] = 1/\sqrt{N}\left\{\sum W_\emptyset[j_0, k]\emptyset_{j0,k}[m]\right\} + 1/\sqrt{N}\left\{\sum\sum W_\emptyset[j, k]\emptyset_{j,k}[m]\right\},$$

$$(m = 0, 1, \ldots, N-1) \tag{2}$$

- Calculate the error for neuron "B" present in the hidden layer:

$$\text{Error}_B = \text{Output}_B (1 - \text{Output}_B)(\text{Target}_B - \text{Output}_B) \tag{3}$$

Output $(1 - \text{Output})$ is necessary in the equation because of the Sigmoid Function. (Target—Output) is needed if threshold activation function is used.

- Change the weight. Let "W_{AB}^+" be the newly trained weights and 'W_{AB}' the initial weights. Note that it is the output of the connecting neuron A, the neurons of the input layer. Update all the weights in the output layer:

$$W^+_{AB} = W^+_{AB} + (Error_B \times Output_A) \qquad (4)$$

- Calculate the errors for the hidden layer neurons. Unlike the output layer, it is not possible to calculate these directly (because there is no target), so back propagate them from the output layer (hence the name of the algorithm). This is done by taking the errors from the output neurons and running them back through the weights to get the hidden layer errors. For example, if neuron "A" is connected as shown to B and C, then take the errors from "B" and "C" (the neurons of the output layer) to generate an error for "A."

$$Error_A = Output_A (1 - Output_A) (Error_B W_{AB} + Error_C W_{AC}) \qquad (5)$$

Again, the factor "Output $(1 - \text{Output})$" is present because of the sigmoid squashing function.

- Having obtained the error for the hidden layer neurons, now proceed as in Stage 3 to change the hidden layer weights. By repeating this method, a network can be trained for any number of layers.

Eqs. (1)–(5) denote the calculation of the outputs in the forward propagation phase. The procedure for the reverse propagation is as follows:

- Calculate errors of the output neurons:

$$\delta_\alpha = out_\alpha (1 - out_\alpha) (Target_\alpha - out_\alpha) \qquad (6)$$

$$\delta_\beta = out_\beta (1 - out_\beta) \left(Target_\beta - out_\beta\right) \qquad (7)$$

- Update the weights between the output layer and hidden layer:

$$W^+_{A\alpha} = W_{A\alpha} + \eta \delta_\alpha out_A \qquad (8)$$

$$W^+_{B\alpha} = W_{B\alpha} + \eta \delta_\alpha out_B \qquad (9)$$

$$W^+_{C\alpha} = W_{C\alpha} + \eta \delta_\alpha out_C \qquad (10)$$

$$W^+_{A\beta} = W_{A\beta} + \eta \delta_\beta out_A \qquad (11)$$

$$W^+_{B\beta} = W_{B\beta} + \eta \delta_\beta out_B \qquad (12)$$

$$W^+_{C\beta} = W_{C\beta} + \eta \delta_\beta out_C \qquad (13)$$

- Calculate the (back propagate) hidden layer errors:

$$\delta_A = out_A \left(1 - out_A\right) \left(\delta_\alpha W_{A\alpha} + \delta_B W_{A\beta}\right) \tag{14}$$

$$\delta_B = out_B \left(1 - out_B\right) \left(\delta_\alpha W_{B\alpha} + \delta_B W_{B\beta}\right) \tag{15}$$

$$\delta_C = out_C \left(1 - out_C\right) \left(\delta_\alpha W_{C\alpha} + \delta_B W_{C\beta}\right) \tag{16}$$

- Update the weights between hidden layer and input layer:

$$W^+_{\lambda A} = W_{\lambda A} + \eta \delta_A in_\lambda \tag{17}$$

$$W^+_{\lambda B} = W_{\lambda B} + \eta \delta_B in_\lambda \tag{18}$$

$$W^+_{\lambda C} = W_{\lambda C} + \eta \delta_C in_\lambda \tag{19}$$

$$W^+_{\Omega A} = W^+_{\Omega A} + \eta \delta_A in_\Omega \tag{20}$$

$$W^+_{\Omega B} = W^+_{\Omega B} + \eta \delta_B in_\Omega \tag{21}$$

$$W^+_{\Omega C} = W^+_{\Omega C} + \eta \delta_C in_\Omega \tag{22}$$

Eqs. (6)–(22) denote the weight updating process in the reverse propagation stage. The constant η (called the learning rate, and nominally equal to one) is put in to speed up or slow down the learning if required.

5.3 Performance analysis using sensitivity

This study evaluates a new test that screens people for benign and malignant pancreatic tumors. Each person taking the test may have a benign or malignant pancreatic tumor or even the pancreas may be normal and free from disease. The test outcome can be positive (categorizing the patient as having the benign or malignant pancreatic tumor) or negative (categorizing the patient as not having pancreatic tumor). The test results for pancreatic tumor may or may not match the actual status of the pancreas. With this background:

- True positive: Patients with benign or malignant pancreatic tumor correctly identified as benign or malignant
- False positive: Healthy pancreas incorrectly identified as benign or malignant pancreatic tumor
- True negative: Healthy pancreas correctly identified as healthy
- False negative: Patients with benign or malignant pancreatic tumor incorrectly identified as healthy

Generally, Positive is denoted as "identified" and negative is denoted as "rejected." Therefore the following inference can be made:

- True positive = correctly identified as benign or malignant pancreatic tumor
- False positive = incorrectly identified as benign or malignant pancreatic tumor

- True negative = correctly rejected as benign or malignant pancreatic tumor
- False negative = incorrectly rejected as benign or malignant pancreatic tumor

Sensitivity refers to the test's ability to correctly detect ill patients who have benign or malignant pancreatic tumor. It is also called the Detection rate.

- Sensitivity = Number of true positives/(Number of true positives + Number of false negatives) or Number of true positives/total number of patients affected by benign or malignant tumors

6. Methodology for image-based diagnosis of nanosized pancreatic tumor using DWNN

The first and foremost step is the database formation related to pancreatic tumor. Normally, the tumor cells multiply at a rapid rate and, due to this uncontrolled growth, there arises symptoms of constipation, high and continual temperature, loss of appetite, and extreme weight loss leading to fluid accumulation inhibiting the respiratory tract causing malicious tumors in pancreas. Screening of pancreatic tumors is presently done using CT imaging or MRI scanning techniques that are not sophisticated enough to detect nanosized pancreatic tumors. Radiologists may not be able to confirm the presence of nanosized pancreatic tumors whether it is benign or malignant. Detection of nanosized tumors in a pancreas may be fruitful for patients because variation in diagnosis leads to adverse effect on the patients. The block diagram of the proposed work is given in Fig. 3. Hence to improve the accuracy of this automatic nanosized tumor detection using machine

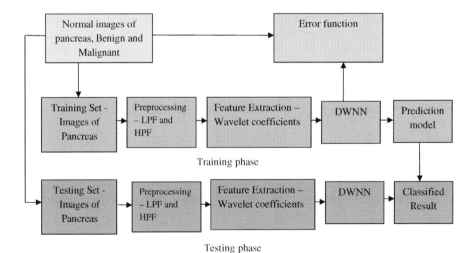

Fig. 3 Classification using DWNN in diagnosis of pancreatic tumor.

vision algorithms, DWNN is used. The MRI image consists of benign and malignant images. The important steps in detection of nanosized, benign, and malignant tumors in a pancreas include denoising using filter high- and low-pass filter banks. Then, the wavelet coefficients are extracted. It contains nearly 200 images, out of which 100 images have benign pancreatic tumors, and the remaining 100 images have malignant pancreatic tumors. All these tumors are nanosized. The classification is done using DWNN. The entire analysis is done using MATLAB software.

7. Results and discussion

The images of the pancreas with benign and malignant tumor are gathered from the website https://pubs.rsna.org/doi/full/10.1148/radiographics, which is an open source site. The images of the normal pancreas and those with nanosized benign and malignant tumors are depicted in Figs. 4–6, respectively.

7.1 Denoising for quality enhancement in detection of nanosized pancreatic

The denoised output for noise removal by the filter banks using high-pass and low-pass filters are illustrated in Fig. 7 for images of the pancreas containing nanosized benign and malignant tumors. The Peak Signal-to-Noise Ratio (PSNR) is 44 dB. Then, the wavelet coefficients, approximations, and details are extracted and recorded in Table 1 for a few MRI samples of the pancreas pertaining to both categories. The approximations represent the coefficients of low-pass filter when the pancreatic images are denoised and the details are the wavelet coefficients corresponding to the high-pass filter output when the pancreatic images are denoised.

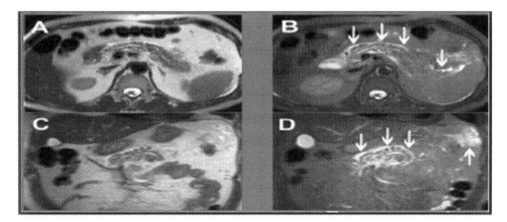

Fig. 4 Normal MRI images of pancreas.

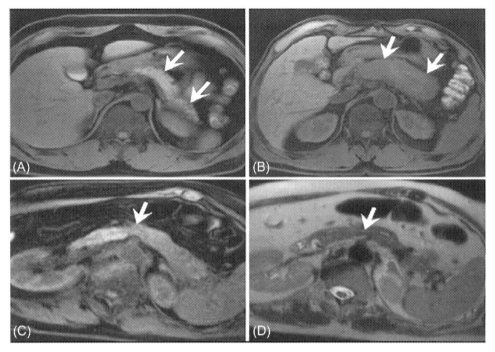

Fig. 5 MRI images of pancreas with nanosized benign tumor.

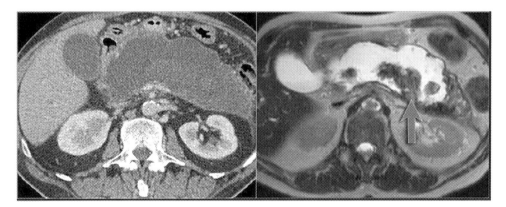

Fig. 6 MRI images of pancreas with nanosized malignant tumor.

7.2 Wavelet-based feature extraction

The wavelet coefficients consist of approximations and details that are used as input vectors for the DWNN. The distinct variation in approximations and details for the pancreas with nanosized tumors is inferred from Table 1.

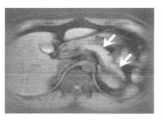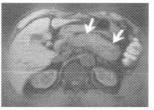

De-noised output for normal pancreatic images using filter banks

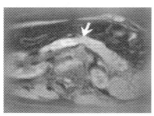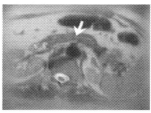

De-noised output for pancreatic images with nano sized benign tumour
using filter banks

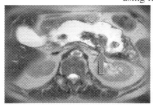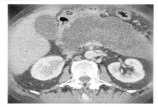

De-noised output for pancreatic images with nano sized malignant
tumour using filter banks

Fig. 7 Denoising by Filter bank.

7.3 Classification by DWNN

This is a type of supervised training where both inputs (wavelet coefficients) and outputs (detection of nanosized tumors) are provided to the DWNN. The inputs are processed by the network using wavelet transform, and the result is compared with the desired outputs. The computed error is propagated back, so the weights are adjusted continually and the network is optimized. The optimization function is the objective function in which the means squared error is optimized. In the data set, nearly 50%–70% are used for training the network, which is called a "training set." The connection strengths are refined many times during the training and the same set of data is processed many times so that the objective function is minimized. Presently, the tools available allow training of the ANN and have the ability to converge and predict the exact output. Finally, if the system is correctly trained and no further learning is needed, the desired set of weights is considered to be the optimal weights. The MSE value for the benign category is in the range of 0.0001–0.00125, and for the malignant category, the MSE value is in the range of 0.00126–0.0018 as shown in Fig. 8.

Table 1 Samples of extracted features for MRI images of pancreas with nanosized benign and malignant tumors.

	Wavelet coefficients for benign tumors (inputs)		Output	Wavelet coefficients for malignant tumors (inputs)		Output
S. no.	Approximations $\times 10^{-3}$	Details	Range for nanosized tumors (nm)	Approximations $\times 10^{-2}$	Details	Range for nanosized tumors (nm)
1.	0.4243	0.6444	50	0.7702	0.9398	100
2.	0.7847	0.3181	52	0.0358	0.8217	102
3.	0.7218	0.1389	54	0.1527	0.8013	104
4.	0.6074	0.4423	56	0.7384	0.9428	106
5.	0.9174	0.9899	58	0.7655	0.8329	108
6.	0.2875	0.5828	60	0.5762	0.9329	110
7.	0.5466	0.4609	62	0.6476	0.9792	112
8.	0.4243	0.6444	64	0.6963	0.9938	114
9.	0.7847	0.3181	66	0.6393	0.8217	116
10.	0.7218	0.1389	68	0.6073	0.8013	118
11.	0.6074	0.4423	70	0.5861	0.8790	120
12.	0.9174	0.9899	72	0.5860	0.8329	122
13.	0.2875	0.5828	74	0.6260	0.9329	124
14.	0.5466	0.6444	76	0.6476	0.9398	126
15.	0.4243	0.4609	78	0.6963	0.0938	128
16.	0.7847	0.4714	80	0.6393	0.8217	130
17.	0.7218	0.4735	82	0.6073	0.8013	132
18.	0.6074	0.1917	84	0.5861	0.5747	134
19.	0.9174	0.2691	86	0.5860	0.8329	136
20.	0.2875	0.0911	88	0.6260	0.9329	138
21.	0.5466	0.3309	90	0.6476	0.8843	140
22.	0.4243	0.4609	92	0.6963	0.9398	142
23.	0.7847	0.4714	94	0.6393	0.8790	144
24.	0.7218	0.4735	96	0.6073	0.8843	146
25.	0.6074	0.1917	98	0.5861	0.8790	148

7.4 Mathematical evaluation of sensitivity for nanosized pancreatic tumor identification

Numerous side effects occur in traditional cancer treatments, and seldom have the patients faced a lot of pain. Nowadays, nanotechnology uses targeted drug delivery systems that take over the entire scenario. The identification of nanosized tumors helps to control their growth rate and probably provides total relief to the patients from the disease. The number of true positives and total number of patients affected by benign and malignant pancreatic tumors are illustrated in Table 2.

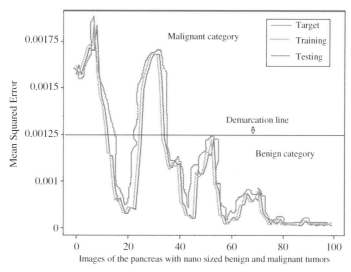

Fig. 8 Result for training and testing of DWNN.

Table 2 Calculation of sensitivity for identification of nanosized tumors.

S. no.	Number of true positives for benign tumor	Total number of patients affected by benign tumors	% Sensitivity for benign category	Number of true positives for malignant tumor	Total number of patients affected by malignant tumor	% Sensitivity for malignant category
1.	99	100	99	99	100	99

7.5 Comparative analysis of machine learning algorithms for detection of nanosized pancreatic tumors

The performance of the proposed technique is compared with other algorithms like Minimum Distance Classifier, Texture features with minimum distance approach, and texture features with Bacterial Foraging Algorithm (BFA). It is inferred from Table 3 that the

Table 3 Performance analysis.

S. no.	Algorithm used for classification	% Sensitivity	
		Benign	Malignant
1.	Minimum distance classifier	62	60
2.	Texture features with minimum distance approach	74	72
3.	Texture features with BFA	89	89
4.	Classification by DWNN	99	99

DWNN shows outstanding performance in efficiently classifying the nanosized benign and malignant pancreatic tumors.

8. Conclusion

In this work, the preprocessing of the images is done by filter banks composed of a set of low-pass and high-pass filters. The Peak Signal-to-Noise Ratio (PSNR) is 44 dB, and then the feature extraction is done to extract the wavelet coefficients, which include the approximations and details. The DWNN classifier is used to train, test, and classify the benign and malignant nanosized tumors in the pancreas, and the performance is compared with the conventional classifiers like minimum distance classifier and texture feature-based minimum distance approach. The proposed work can be enhanced by considering many images of the pancreas along with additional features extracted so there would be paradigm shift in the sensitivity of the classifier.

References

[1] P. Keshvan, R. Anandan, R. Balakrishna, Segmentation of pancreatic tumor using region based active contour, J. Adv. Res. Dyn. Control Syst. 7 (4S2) (2017).
[2] What is Cancer? Defining Cancer, National Cancer Institute, National Institutes of Health, 7 March 2014 Retrieved 5, December 2014.
[3] C.L. Wolfgang, J.M. Herman, D.A. Laheru, A.P. Klein, M.A. Erdek, E.K. Fishman, R.H. Hruban, Recent progress in pancreatic cancer, CA Cancer J. Clin. 63 (5) (2013) 318–348.
[4] A. Vincent, J. Herman, R. Schulick, R.H. Hruban, M. Goggins, Pancreatic cancer, Lancet 378 (9791) (2011) 607–620.
[5] Can pancreatic cancer be prevented?, American Cancer Society, 11 June 2014 Retrieved 13 November 2014.
[6] M. Bardou, I. Le Ray, Treatment of pancreatic cancer: a narrative review of cost-effectiveness studies, Best Pract. Res. Clin. Gastroenterol. 27 (6) (2013) 881–892.
[7] B. Agarwal, A.M. Correa, L. Ho, Survival in pancreatic carcinoma based on tumor size, Pancreas 36 (2008) e15–e20.
[8] C. Doulaverakis, et al., Panacea, a semantic-enabled drug recommendations discovery framework, J. Biomed. Semantics 5 (2014) 13.
[9] M. Dumontier, N. Villanueva-Rosales, Towards pharmacogenomics knowledge discovery with the semantic web, Brief. Bioinform. 10 (2) (2009) 153–163.
[10] A. Ben Abacha, P. Zweigenbaum, Medical question answering: translating medical questions into sparql queries, in: Proceedings of the 2nd ACM SIGHIT, ACM, 2012, , pp. 41–50.
[11] A. Jemal, R. Siegel, E. Ward, Y. Hao, J. Xu, M.J. Thun, Cancer statistics, 2009. CA Cancer J. Clin. 59 (4) (2009) 225–249, https://doi.org/10.3322/caac.20006.
[12] Y. Wang, P. Zhao, Advances in early diagnosis of pancreatic cancer, Oncol. Prog. 4 (4) (2006) 327–332.
[13] F. Fraioli, G. Serra, R. Passariello, CAD (computed-aided detection) and CADx (computer aided diagnosis) systems in identifying and characterising lung nodules on chest CT: overview of research, developments and new prospects. Radiol. Med. 115 (3) (2010) 385–402, https://doi.org/10.1007/s11547-010-0507-2.
[14] H. Jiang, D. Zhao, T. Feng, S. Liao, Y. Chen, Construction of classifier based on MPCA and QSA and its application on classification of pancreatic diseases, Comput. Math. Methods Med. 2013 (2013).

[15] H. Lu, K.N. Plataniotis, A.N. Venetsanopoulos, MPCA: multilinear principal component analysis of tensor objects. IEEE Trans. Neural Netw. 19 (1) (2008) 18–39, https://doi.org/10.1109/tnn.2007.

[16] B. Kovalerchuk, E. Triantaphyllou, J.F. Ruiz, J. Clayton, Fuzzy logic in computer-aided breast cancer diagnosis: analysis of lobulation, Artif. Intell. Med. 11 (1) (1997) 75–85.

[17] P.C. Pendharkar, J.A. Rodger, G.J. Yaverbaum, N. Herman, M. Benner, Association, statistical, mathematical and neural approaches for mining breast cancer patterns, Expert Syst. Appl. 17 (3) (1999) 223–232.

[18] M. Antonie, O. Zaiane, A. Coman, Application of data mining techniques for medical image classification, in: Knowledge Discovery and Data Mining, 2001, pp. 94–101.

[19] J. Zhang, Y. Wang, Y. Dong, Y. Wang, Ultrasonographic feature selection and pattern classification for cervical lymph nodes using support vector machines, Comput. Methods Prog. Biomed. 88 (1) (2007) 75–84.

[20] J. Ramírez, R. Chaves, J. Górriz, et al., Bioinspired applications in artificial and natural computation, in: Third International Work-Conference on the Interplay Between Natural and Artificial Computation, IWINAC, 2009.

[21] D. Tsai, K. Kojma, Enhancement of CT pancreatic features by a simple cascading filter, in: Proceedings of the Nuclear Science Symposium and Medical Imaging Conference, 1993 IEEE Conference Record, 1993.

[22] T. Takada, H. Yasuda, K. Uchiyama, H. Hasegawa, T. Iwagaki, Y. Yamakawa, A proposed new pancreatic classification system according to segments: operative procedure for a medial pancreatic segmentectomy, J. Hepato-Biliary-Pancreat. Surg. 1 (3) (1994) 322–325.

[23] S. He, H.J. Cooper, D.G. Ward, X. Yao, J.K. Heath, Analysis of premalignant pancreatic cancer mass spectrometry data for biomarker selection using a group search optimizer, Trans. Inst. Meas. Control. 34 (6) (2012) 668–676.

[24] V.N. Vapnik, The Nature of Statistical Learning Theory, Springer, New York, NY, 1995.

[25] W.-T. Pan, A new fruit fly optimization algorithm: taking the financial distress model as an example, Knowl.-Based Syst. 26 (2012) 69–74.

[26] T. Wang, X. Ye, L. Wang, H. Li, Grid search optimized SVM method for dish-like underwater robot attitude prediction, in: Proceedings of the 5th International Joint Conference on Computational Sciences and Optimization (CSO '12), June 2012, pp. 839–843.

[27] M. Zhao, C. Fu, L. Ji, K. Tang, M. Zhou, Feature selection and parameter optimization for support vector machines: a new approach based on genetic algorithm with feature chromosomes, Expert Syst. Appl. 38 (5) (2011) 5197–5204.

[28] F.-H. Yu, H.-B. Liu, Structural damage identification by support vector machine and particle swarm algorithm, J. Jilin Univ. Eng. Technol. Ed. 38 (2) (2008) 434–438.

[29] M. Dorigo, V. Maniezzo, A. Colorni, Ant system: optimization by a colony of cooperating agents, IEEE Trans. Syst. Man Cybern. B Cybern. 26 (1) (1996) 29–41.

[30] X. Xu, J. Jiang, J. Jie, H. Wang, W. Wang, An improved real coded quantum genetic algorithm and its applications, in: Proceedings of the International Conference on Computational Aspects of Social Networks (CASoN '10), September 2010.

[31] P.K. Tiwari, D.P. Vidyarthi, A variant of quantum genetic algorithm and its possible applications, Adv. Intell. Soft Comput. 130 (1) (2012) 797–811.

[32] J. Shah, V. Turkar, S. Surve, Pancreatic tumor detection using image processing, Procedia Comput. Sci. 49 (2015) 11–16.

[33] R. Jayaprakash, K. Rajamohanan, P. Anil, Determinants of symptom profile and severity of conduct disorder in a tertiary level pediatric care set up: a pilot study, Indian J. Pyschiatry 56 (4) (2014) 330–336.

[34] S. Xu, H. Zhao, Y. Xie, Grey SVM with simulated annealing algorithms in patent application filings forecasting, in: Proceedings of the International Conference on Computational Intelligence and Security Workshops (CIS '07), December 2007.

[35] F. Huber, H.P. Lang, J. Zhang, D. Rimoldi, C. Gerber, Nanosensors for cancer detection, Swiss Med. Wkly. 145 (2015) w14092.

[36] J. Bosch, K. Heister, T. Hofmann, R.U. Meckenstock, Nanosized iron oxide colloids strongly enhance microbial iron reduction, Appl. Environ. Microbiol. 76 (2010) 184–189.

CHAPTER 21

Modern applications of quantum dots: Environmentally hazardous metal ion sensing and medical imaging

Pooja[a], Meenakshi Rana[b], and Papia Chowdhury[a]
[a]Department of Physics and Materials Science and Engineering, Jaypee Institute of Information Technology, Noida, Uttar Pradesh, India
[b]Department of Physics, School of Sciences, Uttarakhand Open University, Haldwani, Uttarakhand, India

1. Introduction

The United Nations Children's Fund (UNICEF) has studied groundwater samples collected from some open defecation-free (ODF) and non-ODF villages of different parts of the world and found that most of the villages are 11.25 times more likely to have contaminated groundwater sources, 1.13 times more likely to have contaminated soil, 1.48 times more likely to have contaminated food, and 2.68 times more likely to have contaminated household drinking water [1]. So contamination is the main reason for bad human health conditions in many parts of the world especially in underdeveloped and developing countries [2]. According to the UNICEF study, it was found that pollution of water, soil, and food is the biggest problem of society facing nowadays where pollution is arising mainly due to contamination [3]. Water and soil pollution are the biggest concerned areas of civic society, which we are dealing with right now. Water contamination usually appears from different sanitation and nonsanitation sources [4]. Under sanitation-related sources, households may dispose both black water (untreated/partially treated sewage) and grey water (domestic wastewater other than sewage) into open drains, which contaminate local water bodies and that further contaminate the groundwater. In the case of nonsanitation-related sources, these contaminants include fertilizers and pesticides (chemical pollution), industrial wastages from leather tanning, electroplating, paint pigment, electronics chemical industries, etc. that include many hazardous metal ions like zinc (Zn^{2+}), lead (Pb^{2+}), chromium (Cr^{3+}), mercury (Hg^{3+}), copper (Cu^{3+}), cobalt (Co^{2+}), and their complexes, etc., and animal waste disposal (biological pollution) released to the groundwater, rivers, ponds, etc. by the process of leaking or direct mixing in periurban towns. Discharge of these unwanted hazardous heavy metal ions to groundwater, rivers, soil, and food lead to the contamination of the environment, which further causes different diseases. Many of these metal ions (e.g., Cr^{3+}, Zn^{2+}, Fe^{3+}, Cd^{2+}, Cu^{2+}) are also essential nutrients for our body as they bind with amine, imidazole, carboxyl, and

Handbook of Nanomaterials for Sensing Applications
https://doi.org/10.1016/B978-0-12-820783-3.00025-7

enzyme groups of a variety of metabolites and are associated with major metabolic pathways such as uptake of oxygen to tissues [5]. These ions are also involved in metabolism of carbohydrates, nucleic acids, and proteins, and thus they are necessary for health in moderate ingestion. Insufficiency of Cr^{3+} and Cu^{2+} minerals can cause some diseases like muscle weakness, anemia, inhibition of protein synthesis, heart disease, diabetes, and cardiovascular disorders [6]. However, their excess presence greatly affects many cellular activity like cell damage, memory loss, anemia, lung cancer, low white blood cell count, and neurological problems. Mainly, Cr^{3+}, Pb^{2+}, and Cu^{2+} are the mostly detected toxic metallic pollutants, which mainly are found dissolved in water and mixed with various pesticides [7]. The concentration of metals as minerals required in groundwater are approximately: 0.0005 mg/L for Cr^{3+} [8], 0.005 mg/L for Pb^{2+} [9], 1 mg/L for Cu^{2+} [10], etc. The permissible limit for health hazards of dissolved metal ions are between 0.01 mg/L for Cr^{3+}, 0.01 mg/L for Pb^{2+}, and 1.3 mg/L for Cu^{2+} [11]. In India (Delhi/NCR and U.P. region), due to the presence of many industries around Ganga, Yamuna, and Hindon rivers, the concentration of Cr^{3+}, Pb^{2+}, and Cu^{2+} are found to be 0.390, 0.10 and 1.5 mg/L [11], respectively, which are definitely harmful for a living body and the environment. Directly or indirectly, a large amount of human/animal population used to consume the river water for drinking and bathing purposes. So, deficiency and excess of these nutrients are associated with many risk factors [6, 12]. These facts indicate the need for the designing of efficient and economical sensors for metal ions so that the risk associated with deficiency or excess presence can be identified and necessary steps can be taken. So whether a metal ion is biologically essential, technoeconomically important, or environmentally hazardous, its sensitive and selective detection is one of the most important issues for environment and disease prevention. For agricultural soil, pesticides are widely used in agriculture for controlling weeds and pests to improve food production and to meet the demands of an ever-increasing population [13]. Because there is no available regulatory guideline, limitless pesticides are being continuously introduced into the market everyday [14]. Continuous exposure of these pesticide residues can pose serious public health risks like neurotoxicity, endocrine disruption, genotoxicity, mutagenicity, and carcinogenesis [13]. For control measures to be enforced, highly sensitive, selective, and affordable analytical sensing methods are required to monitor these compounds. Therefore, detection of different toxic materials, even in low concentration levels, is very important to devise plans to control environmental (water and soil) pollution [15].

In today's medical industry, one of the major concerned areas is selection of efficient tools for live cell imaging [16]. The point of interest in imaging for researchers is the possibility to study the various biochemical, physical, and kinetic processes in the cells or the whole body by means of some markers. In the biomedical sciences, for labeling, imaging, tracking, detection, and therapy, fluorescence is used as a powerful tool [17]. The fluorescent labeling of biological molecules use various organic fluorophores as a detector. Most all traditional organic-based fluorophores show narrow absorption and broad emission spectra with low luminescence quantum yield and rapid photobleaching. These characteristics limit their vast applications in most biomedical applications. Limitations also arise due to the small size of these traditional fluorescent dyes for the labeling of other small molecules, drugs, transporters, and probes to cell-surface receptors. Conjugates of these dyes often lack sensitivity or specificity in the detection of the desired targets [18]. In addition, most of the industrially used classical fluorophores emit light in the infrared and near-infrared regions, but the fact is that the absorption of tissues is minimal in this region. For medical imaging, various research reports have suggested the demand for a replacement of traditional organic fluorophores markers [19], which should have a constant selection of a suitable source of excitement.

Technically, sensitive, selective, and intelligent detection is one of the most important issues for environment protection and disease prevention [20, 21]. In industry, available discrimination and detection techniques like Raman, atomic absorption [22], mass [23], infrared [24], coupled plasma [25], voltammetry [26], chromatography [27], etc. are sophisticated, time-consuming, and need high maintenance and skilled manpower for their operation and interpretation [28]. The mentioned techniques are also not suitable for in situ detection due to destruction of tissues. So an alternative to these tools is required. Fluorescent sensing emerges as an excellent alternative due to its high sensitivity, low-cost, simple operation and ability to perform real-time imaging [29]. Different available fluorescent molecular and material-based probes like organic fluorophores [30] and metal nanoparticles [31] are used as attractive fluorescent detectors. However, organic fluorophores are associated with multistep preparation, narrow excitation, and poor photostability [32]. In some specific sensing purposes, gold and silver nanoparticles are used, but they are not cost-effective materials for probing metal ions and living tissues [33]. Therefore, it is highly desirable to search for materials that are cost-effective, easy to synthesize, and have better function in aqueous medium and appreciable optical properties.

Very effective and intelligent nanosized Quantum dots (QDs) can act as alternative intelligent optical-based nanosensors due to their tunable optical properties including low cost, high biocompatibility, and better stability [34]. QDs based on metals and carbonaceous nanomaterials can have most of the previously mentioned desirable characteristics. Tunable fluorescence property, broad excitation, and high photostability of QDs make them occupy a special space in the today's sensor industry [35]. In the coming sections, we will discuss details about different categories of sensors, QDs as intelligent nanosensors, their synthesis techniques, characteristics, and their applications including their advantages and disadvantages in the area of today's sensing industry including biology and medicine.

2. Sensor

A device that detects and responds differently to different types of inputs is known as a sensor (Fig. 1). The input could be physical or chemical in nature. The physical inputs may be categorized by various environmental phenomena such as light, heat, motion, moisture, pressure, humidity, movement, etc. Similarly, the chemical inputs may be classified as concentration, polarity, acidity or basicity, etc. After receiving the input signal, some further processing should be carried out inside the sensory device (detector, transducer), which converts the input signal to an output signal. The output signal is then converted to human readable and usable form and measured by a receiver or display recorder. There are many types of sensors available in industry. We will discuss some of them.

2.1 Types of some available sensors

2.1.1 Chemical sensor

A chemical sensor is used to measure the physical and chemical property of an analyte into a measurable signal. Normally, the magnitude of the measurable signal is proportional to the concentration of analyte. There are two major detection mechanisms in chemical sensors. They are photochemical and photometric, and are used to find the concentration or changes in the chemical reactions with most accuracy [36]. The chemical sensors are highly sensitive in nature. They can even detect a very small amount of analyte from its vapour as well. Chemical sensors are also efficient in detecting single molecules. Examples include a breathalyzer, carbon monoxide sensor, and electrochemical gas sensor.

2.1.2 Biosensor

Biosensors are the one of the most intelligent and advanced sensors that can be fitted internally into the human body for the investigation of changes like protein, enzymes, DNA, etc. A biosensor is an independent integrated device capable of providing specific quantitative or semiquantitative analytical information using a biological recognition element, which is in direct spatial contact with a transducer element[37]. This sensor can also observe and determine the level of protein and enzymes. It is used for

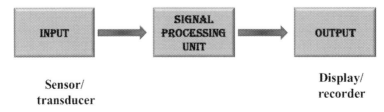

Fig. 1 Block diagram of a sensor as a device.

the detection of an analyte that integrates the biological component with a physico-chemical detector, for example, a glucometer.

2.1.3 Image sensor

The mechanism of an image sensor is based on an array of photo elements (pixels). The sensor is comprised of a device for randomly addressing individual pixels and a device for selectively varying the number of pixels, which can be read out on any one reading cycle [38]. There are two types of image sensors available in the market: complementary metal oxide semiconductor (CMOS) and charge coupled device (CCD). Both CMOS and CCD image sensors are based on metal oxide semiconductor (MOS) technology [39].

2.1.4 Biochemical sensor

A biochemical sensor is a biosensor in nature that is commonly used to convert chemical and biological signals into an electrical signal. The sensor contains one or more than one receptor that is particular for variety of analytes of interest attached to a sensing platform. Basically, the sensing platform is a semiconductor-based transistor that produces an electrical signal in response to analyte binding. The sensor is used for the detection of analytes like toxins, disease markers, nucleic acids, proteins, or other molecules or complexes in a sample or in a human/living organism. The sensor is compatible with a wide spectrum of targets and allows for the simultaneous detection of multiple analytes. The detection of analytes in a sample has applications in many fields like medical, food, and environmental. There are many methods to detect this, but the well-known and advantageous type of detection is optical. Among optical sensors, an important class is integrated optical (IO) sensors that comprise a chip with a waveguide and sensing biolayers [40]. Examples include electrolyte-insulator-semiconductor (EIS) Sensor and ion-sensitive field-effect transistors (ISFET).

2.1.5 Nanosensor

Nanosensors work on the "Nano" scale. "Nano" is a unit of measurement around 10^{-9} m. A nanosensor is a device capable of carrying data and information about the behavior and characteristics of particles at the nanoscale level to the macroscopic level. Nanosensors can be used to detect chemical or mechanical information such as the presence of chemical species and nanoparticles or monitor physical parameters such as temperature on the nanoscale. Nanosensors can be classified based on their structure and application. On the basis of structure, nanosensor are of two types; one is an optical nanosensor and another is an electrochemical nanosensor. Based on application, the nanosensor can be categorized as a chemical nanosensor, biosensor, electrometers, and deployable nanosensor [41]. The nanosensor has wide applications in chemical, optical, biomedical, food, and electronics industries. Specifically nanosensors have their eventful applications in medical diagnostic applications, food and water quality sensing, and other chemicals.

Optical signal detection is one of the most promising application areas of many nanosensors. Depending on their optical selectivity, many classifications of nanosensors are possible [42]. They are carbon nanotube-based nanosensors, QD-based nanosensors, photoacoustic-based nanosensors, etc.

2.1.6 Qunatum dot (QD) sensor

Fluorescence is a very important and beneficial tool for detection and imaging. The signal-to-noise (S/N) ratio of fluorescence is superior than the other optical techniques, and its multiple photophysical measurement approaches to its numerous applications, including sensing. Photon energy (i.e., wavelength/color), photoluminescence (PL) intensity, and PL lifetime can all be modified and thus used as sensor outputs [43]. So many fluorophores, including organic/inorganic dyes and lanthanide-based emitters, have been used to produce fluorescent sensors [44]. Among all of these, QD is used for fluorescence sensors because it has high Quantum yield other than organic dyes and lanthanide-based emitters [18]. So QDs is used as a fluorescence-based sensor. QDs are those nanosized semiconductor nanocrystals having fluorescent properties due to their quantum confinement. QDs are one most applicable newly invented nanoparticles that, due to their unique physical (optical) and chemical properties, are subjected as a promising sensor material in the field of biology and medicine.

3. Quantum dot

Semiconductor QDs are nanoparticles having unique photophysical and optical properties. They have numerous applications in many fields such as imaging, biosensing, and solar cells. QDs are zero-dimensional semiconductors that exhibit well discrete and separated quantized energy levels (Fig. 2) [45]. Discretization of the electronic energy levels are the result of quantum confinement in QDs [46]. Therefore, QDs are a class of quantum-confined semiconductor nanocrystals whose radii are smaller than bulk exciton Bohr diameter (1–10 nm) [47]. Usually, QDs are the fluorescent nanocrystals consisting of atomic elements II-IV or III-V groups of the Mendeleev periodic table. There are a large number of QDs that are available commercially, for example, CdSe, CdS, ZnS, ZnSe, ZnSe/ZnS, CdSe/ZnSe, CdHgTe, CdSeS. Usually, QDs are inorganic and toxic in nature [48, 49]. To improve the quantum yield, stability, and to overcome the toxicity, the concept of core/shell, alloyed QDs were introduced with the introduction of capping the QDs with higher bandgap semiconductor materials [50] or with organic-based capping reagents [51].

3.1 Characteristics of QDs

In a quantum-confined regime, the control of particle sizes allows the bandgap to be "tuned" to give the desired electronic and optical properties such as broad excitation

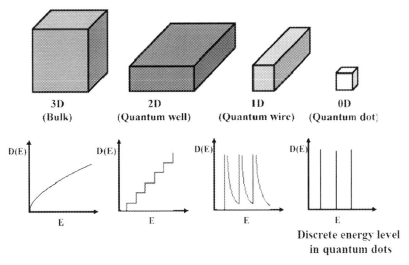

Fig. 2 Discrete energy levels in a quantum dot.

spectra and tunable and symmetric emission spectra with a narrow bandwidth [51, 52]. QDs are also exciting new raw materials for photonics and quantum information processing because their properties can be tailored to a wide extent. However, the main attraction of QDs lies in their high optical extinction coefficients [53], tunable sharp emission profile [54], carrier multiplication ability [55, 56], and variable nonlinear properties [57, 58], etc., which can be tuned by adjusting size and shape. QDs have narrow and tunable emission spectra with intense brightness, which is size-dependent [35]. High quantum yield, photostability, and narrow and tunable emission spectra from different size-dependent QDs can be applicable over a specific spectral range [34]. Due to their numerous fascinating properties, QDs have attracted considerable attention in many applications, such as novel luminescent sensors [59–61], quantum computing [62], solar cell [63], nanoscale display devices [64], QD lasers [65], white light-emitting diode (LEDs) [66], etc. Besides this, QDs are highly resistant to chemical degradation and more resistant to photobleaching compared with the organic fluorophores [67]. All these properties make QDs available as an alternative potential tool compared with the organic fluorescent dyes used in bioimaging and other biological applications [17]. Nowadays, QDs have displayed a vast potential in the biological, biochemical, and biomedical areas [68]. Besides the applications as simple sensors [69–71], the main function of the QDs based on their exceptional fluorescent properties in the biochemical and biomedical research area is their use as unique fluorescent labels [72–76].

3.2 History

Ever since Andre Geim and Konstantin Novoselov were awarded the Nobel Prize in Physics in 2010 for ground-breaking experiments regarding the two-dimensional material graphene, developments and research works in this nanodomain has become rapid, with promising applications in a plethora of fields including sensing, imaging, medicine, energy storage/production, electronics, computing, etc. [77]. The initial discovery of QDs in the early 1980s has been credited to Brus, Efros, and Ekimov by the Optical Society of America [78–80]. The extraordinary optical and electronic properties of nanocrystalline semiconductors were discovered in 1981 by Alexey Ekimov of the Vasilov State Optical Institute in Russia, who first synthesized nanocrystals embedded in a glass matrix (Fig. 3) [79]. He took the help of theoretician Alexander Efros to explain apparent color changes of what appeared to be tiny particles of variable sizes [80]. At the same time, Louis Brus in the United States [81] and Arnim Henglein in Germany [82] discovered striking color changes of II–VI semiconductor nanoparticles grown as aqueous colloidal suspensions, giving rise to the simple description of the bandgap dependence on size that is now familiar to the world over as the "Brus formula" (Fig. 3) [83]. These publications are widely considered to be the inception of what is now the flourishing field of QDs. About 10 years later, procedures for synthesis of high-quality CdSe QDs dispersed in organic solvents were developed by Murray et al. (Fig. 3) [84]. However, it was not until 1998 that QDs entered into their new role as fluorescent probes when two groups simultaneously reported the procedures for making QDs water-soluble and conjugating them to biomolecules [85]. Philippe Guyot-Sionnest from University of Chicago synthesized the first QDs with the core surrounded by a shell in 1996 to observe the multiphoton properties of these materials [86]. Their work made it possible to stabilize the properties of the synthesized particles with their surface chemistry. Following these reports, extensive research has been directed toward developing QDs for use in sensing, information processing, solar cell, and bioimaging [87–91]. The most studied and valued property of QDs is their fluorescence. Although QD research gained momentum in the early 1990s, it was arguably the adoption of QD-bioconjugates as fluorescent labels for biological

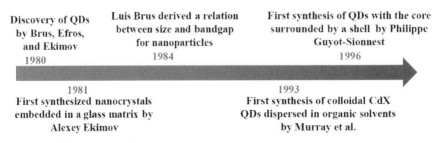

Fig. 3 History of the quantum dot.

imaging [92, 93] that catalyzed significant interest across the bioanalytical, biophysical, and biomedical research communities at the turn of the century.

3.3 Types of QDs

QD semiconductor nanocrystals have sizes almost 1–20 nm and are composed of elements from the II-VI, III-V, or IV-VI group of the periodic table [94]. The quantum confinement effect and the large surface-to-volume ratio are the key factors for the existence of these nanomaterials [80, 95]. Quantum confinement results in discreteness in the electron and hole energy states [81], and therefore the bandgap comes out to be a function of the QD's diameter. Normally, the bandgap of a semiconductor is inversely proportional to its diameter (Fig. 4).

Presently, a number of QDs are available commercially, for example, ZnS, ZnSe, CdSe, CdS, CdSeS, CdSe/ZnSe, CdHgTe, ZnSe/ZnS. QDs are mainly categorized in three parts: the first is core-type QDs (ZnS, ZnSe, CdSe and CdS, etc.), second is core/shell-type QDs (CdSe/ZnSe, ZnSe/ZnS, etc.), and third is alloyed-type QDs (CdSeS, CdHgTe, etc.) (Fig. 5). In 1993, Murray et al. synthesized the first cadmium-based core-type QDs: CdX (X = S, Se, Te) using the fast injection process to improve quantum yields to ~10%. Because Cd is toxic in nature, core-type CdX QDs are also reported to be very toxic [84].

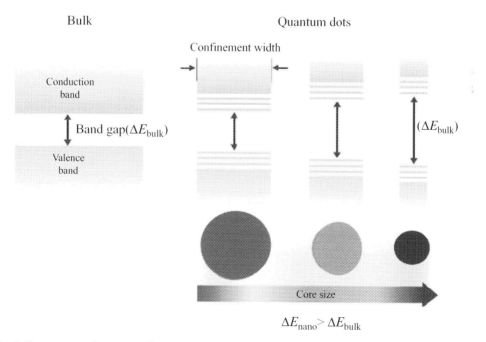

Fig. 4 Quantum confinement effect in quantum dots.

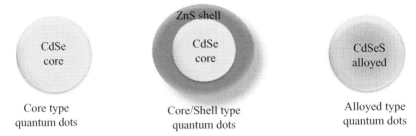

Fig. 5 Schematic diagram of core-, core/shell-, and alloyed-type quantum dots.

QDs are inorganic in nature so they are normally toxic in nature. To improve the quantum yield, stability, and to overcome the toxicity, the concept of core/shell QDs were introduced in 1996 with the introduction of capping the core QDs with higher bandgap semiconductor materials [96]. First reported core/shell QD was CdSe/ZnS with improved quantum yield to ∼50% by Bawendi et al. [84]. There are a number of research studies on core and core/shell-type QDs [82,97]. Recently in 2007, the first alloyed-type QD was synthesized to provide high quantum yields 61% [85]. However, until now, very few works have been reported on the alloyed-type QDs [98].

3.4 Synthesis of QDs

To synthesize QDs, several routes have been used. The commonly used techniques for QD synthesis are bottom–up and top–down approach. In the bottom–up approach, a material is built up atom-by-atom or molecule-by-molecule. In this approach, atom-by-atom deposition leads to formation of self-assembly of atoms/molecules and clusters. In the top–down approach, there is a slicing or successive cutting of a bulk material to get nanosized particles. QDs can be formed on a surface with help of molecular-beam epitaxy, ion implantation, e-beam lithography, and X-ray lithography methods. These QD structures are normally used in optical converters, LED sources, and photovoltaic cells [99]. For practical use and other biomedical studies, only colloidal types of QDs are used [100]. QDs are stabilized in solutions due to the ligands casing them. Depending on ligand structure, they form either organic or aqueous colloidal solutions [101]. Because all reactions proceed in the hydrophilic environment, in live systems QDs in aqueous colloidal solutions is required. Second main important point is that QDs must get rid of toxic organic chemicals. A core/shell/shell complex composition has high fluorescence yield than core-type or core/shell-type QDs. The synthesis of QD's "core/shell" increases the quantum yield because:

1. Shell provides passivation of uncompensated chemical bonds on a surface of the nanocrystal
2. Shell of QD provides safety of a core from oxidation
3. It blocks the nonradiative recombination process because of the development by the cover of a potential barrier for an exciton in core [99]

Trioctyl phosphine oxide (TOPO) is usually used in QDs synthesis and provides them with their hydrophobic nature. QDs surface is covered with a number of hydrophilic covering of mercaptoacids, polyethyleneglycol (PEG), and bovine serum albumin (BSA) [102]. By means of various exchange reactions, these hydrophilic coverings convert QDs to aqueous colloidal solution. Ligand replacement is performed covering QDs with amphiphilic polymers or making micellar encapsulation [103]. However, the conversion of QDs in an aqueous phase results in a decrement in luminescence brightness [104]. Along with this ligand replacement also results in the increase in QDs diameter. One can also synthesize the QDs directly in the aqueous phase. But in this case quantum yield and stability are lower. Along with this, it is also difficult to obtain QDs of different diameters in aqueous media [101], but it is possible in the case of organic media. The chemical and optical properties of QDs, morphology, average size, spectral band width, PL, and quantum yield are influenced by the used precursors, solvents, reaction temperature, and injection parameters [105]. For semiconductor nanocrystal synthesis, DNA, RNA, and nucleic acids can serve as a matrix. The specific groups attached with DNA, RNA, and nucleic acids can control synthesis and manipulate the emission properties. Nowadays, widely used conjugated QDs are with biomolecules. The QDs are normally conjugated with nucleic acids, peptides, and proteins, and are functionalized materials achieved by covalent bonding. All features of QD interaction with living systems should be considered at synthesis of nanoparticles for medical and biological use.

3.5 Solubilization, toxicity, and functionalization of QDs

Different forms of QDs (solid or liquid) determine their applicability. Because it's a semiconductor in nature, most all QDs are insoluble in a hydroxylic medium [106]. QDs are potentially harmful to the environment and biological system. So research works are concentrating on core/shell-type heterostructures. High quantum yield and less toxicity make core/shell QD a good alternative to organic fluorophores for many applications [48,107]. Core/shell-type QDs also have been shown to induce cytotoxic, genotoxic, antimicrobial activities, and oxidative stress responding in different cellular models. The inherent toxicity of heavy metals (Cd, Zn, etc.) can be reduced by encapsulation of QDs with various shells, surfactants, or organic functional probe as capping agents [108] (Fig. 6).

Different functional groups can be easily modified onto the surfaces of QDs through synthesis. In addition, the surface states and modifications of QDs have an effect not only on the luminescence properties, but also on enabling the interaction with other substances [109]. The interaction also caused the fluorescence intensity of QDs to be enhanced or quenched. With these attractive advantages of QDs, many applications have been developed for chem/biosensing and bioimaging platforms based on QDs.

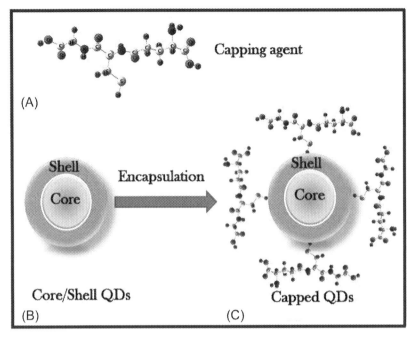

Fig. 6 (A) Capping agent, (B) core/shell quantum dot, and (C) quantum dot with capping agent.

Combinations of enzymes and QDs have shown great promise for health monitoring, environmental applications, and agricultural practices because its unique structural and photophysical properties and excellent specificity can be effectively integrated [110].

3.6 Application areas in current industry

With new synthetic development techniques, the emerging nontoxic QD materials with near-infrared emitting capabilities, and flourishing photophysics, the applicability of QDs is showing a strong and determined push toward real commercial applications. Now, QDs providing unprecedented color purity have already entered the market, commercialized as components in HDTV displays. Photovoltaic applications of QDs are thriving, with new record-certified conversion efficiencies reported regularly, and studies of water-splitting systems utilizing QDs as components of hybrid materials are on the rise. QDs are of interest in a wide variety of applications because of their tunable properties. They are commercially used in areas such as optoelectronics, optics, renewable energy, communications, biology, computing, and security markets. In recent years, companies like LG, Sony, and Samsung have all launched QD TVs, with the latter moving away from OLED TV development in favor of QD LCDs [111]. Amazon has used QD technology in its Kindle Fire HDX 7 and HDX 8.9 tablets [112].

In the healthcare and life sciences sectors, the application of QDs shows promise in sensing different analytes, biomedical aging, cell tracking, and drug delivery applications. The global market for QD-based products is projected to reach US$3.5 billion by 2020. Although for a few years, high-quality research works on QDs have been started all over the world, in view of their applications for today's industry, there exist many major concerns like the toxic nature of QDs. QDs are toxic and hydrophobic in nature and hence biologically not compatible with living cells [32]. Therefore, the main challenge is to design specific hydrophilic QDs having biological compatibility [113]. Another important parameter is to develop techniques for targeted delivery by nontoxic QDs to the diseased cells without harming/affecting the normal cells [114]. However, with the recent advances in the development of surface modifications of QDs, the potential toxicity of QDs is no longer a problem for in-vitro and in-vivo imaging studies as the cellular uptake of QDs can be modulated by their size, shape, and surface functionalization. In fact, the use of multicolor QDs for stem cell imaging is probably the most important and clinically relevant application for regenerative medicine in the immediate future. Despite the enormous potential of QDs in therapeutics, the fundamental information on the interaction between QDs and therapeutic cells is relatively limited.

4. Nanosensors in biology and medicine

Nanomaterials are applied in many fields of physics, chemistry, biology, and engineering. In 1998, the potential of nanoparticles for applications involving biological labeling was first reported [115]. In the field of biology and medicine, the applicable areas of nanomaterials are drug and gene delivery [116,117], fluorescent biological markers [85,118], probing of DNA and protein structure [119,120], tissue engineering [121,122], separation and purification of biological cells [123], diseased cell detection [124] and damaging them [124], etc. In all of the previously mentioned application areas, size is one of most important characteristics of nanoparticles that itself is rarely sufficient, which is to be used as biological tags. Nanoparticles can easily enter cells very efficiently due to their equivalent dimensions. The influx occurs by endocytosis. The particles can be inserted and diffused through the lipid bilayer of the cell membrane. They can also enter into the cells even after linkage to proteins such as antibodies. Nanoparticles have been shown to make a conjugate with antibodies against exclusive cancer cell surface receptors, which have been used to specifically bind with cancerous cells. The functionalized nanoparticles have also been used for targeted entry into cells [125]. In parallel with these for the applications in bioimaging and bioanalysis, QD nanomaterials have evolved as a most interesting detection unit, which provides greater flexibility and capability than the other available nanomaterials. QDs in the size range of 2–10 nm are of considerable interest in the field of biological applications due to their dimensional similarities with some biological macromolecules, for example, nucleic acids and proteins. After the first successful application of QDs to medical diagnostics, QDs are placed at the leading edge of the rapidly developing field of nanotechnology due to their unique size-dependent properties, which make these materials superior and indispensable in many areas of biological-based applications [126]. Understanding different biological processes on the nanoscale level is a strong driving force behind development of nanotechnology. Living organisms are built up of cells that are typically 10 μm in dimension with the cell parts in the submicron domain, whereas the proteins are of typical size of just 5 nm, which is comparable with the dimensions of the smallest manmade QD nanoparticles. This simple size comparison gives an idea of using QD nanoparticles as very small probes that would allow us to spy all levels of cellular machinery without introducing too much interference on their activities. Due to their size-dependent variable physical properties like optical and magnetic effects, they are mostly used in various biological and sensing applications [35]. Size-tunable absorption and emission property of QDs is an extremely valuable property for biological imaging as they can be tuned all the way from the UV to the near-infrared of the spectrum. But before discussing the applications in detail, we should keep in mind that QDs should not be viewed as total replacements for all industrially applied fluorescent dyes, sensors, drugs, etc. QDs can be considered as advantageous in many applications but similarly QD's inherent toxicity emerges as disadvantageous in many other many applied areas in biomedical applications. In the coming sections, we will discuss some working QDs in the areas of biology and medicine. The aim of the next sections is to give a detailed review of the historic prospective of QD nanomaterials for the application to biology, medicine, and everyday life as a sensor and imaging agent with the description of the most recent developments in these fields, and finally to discuss the road to commercialization.

5. QDs for hazardous metal ion sensing

QDs are covered with organic-based, nontoxic, polar-capping agents that decrease the toxicity and increase the solubility in a polar environment [127]. Along this, capping agents also increase the reactivity and stability of QDs and prevent aggregation. Kuno et al. reported the first observation on the modification of properties of QDs by surface capping of QDs [128]. Later, various capping agents have been studied for detoxification, increased solubilization, and surface modification, etc.

Nowadays, water pollution in urban and periurban areas is a life-threatening problem caused by the industrial wastages. These industrial wastages mostly include water-dissolvable metal ions such as Cu^{2+}, Zn^{2+}, Co^{2+}, Pb^2, and Cr^{3+}. Optical sensing is a useful techniques for the detection of many environmentally hazardous metal ions in a water environment. There are a number of core/shell QDs having convenient size and high luminescent strength, and narrow emission spectra have been used as fluorescent metal sensors in detecting ions such as CdTe/ZnS for Cu^{2+} detection [51], CdSe/ZnS for Hg^{2+} detection [129], CdSe/CdS for Cu^{2+} detection, etc. [130]. To minimize the toxicity and enhance the stability of QDs as fluorescent sensors, different capping agents have been used, such as 3-mercaptopropionic acid (MPA), L–cysteine, thioglycolic acid (TGA), cysteamine, mercaptosuccinic acid (MSA), 2-mercaptoethanol, etc. Recently, L–glutathione (fluorescent sensor L-GSH) is enormously being used as a capping agent for the sensing industry [131]. Glycine, glutamate, and cysteine are the main three components of antioxidant protein GSH, which is found in most organisms. It reduces its disulfide bonds and converts to a reduced form of glutathione disulfide (GSSG), which is also known as L–GSH. It can be easily conjugated with various QDs, metal ions, and biomolecules due to the presence of carboxylate or amine groups in it [132–134]. Through polymerization, it can form GSH–capped core/shell QDs, which can be used to detoxify heavy metal ions by binding with heavy metal ions [135]. Let us discuss the working of some L-GSH–capped QDs (CdSeS/ZnS) as metal ion sensors. The nonpolar nature of a CdSeS/ZnS QD makes it insoluble in a water environment. L-GSH capping makes CdSeS/ZnS water-soluble. Capped CdSeS/ZnS QDs are found to be more biocompatible than other available water-soluble QDs [136]. Also due to high interaction capability of L-GSH, the capped Qd can easily interact with heavy metal ions and form complex structure [136]. The surface alteration with ligand makes nonpolar CdSeS/ZnS suitable for interaction with the targeted metal ions (Fig. 7).

5.1 SEM and EDX analysis

SEM image of uncapped CdSeS/ZnS QDs are shown in Fig. 8A. The shape of uncapped QD structures is spherical, and size measured as less than 20 nm. At lower capping concentration of L-GSH, QDs are irregular in shape due to agglomeration. The increased

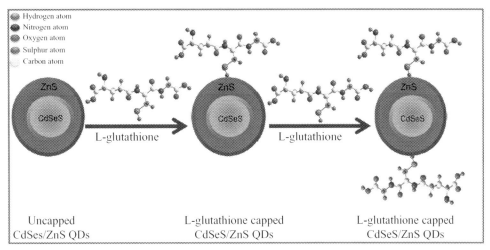

Fig. 7 Schematic illustration of the core/shell uncapped CdSeS/ZnS QDs and L-GSH-CdSeS/ZnS core/shell QDs.

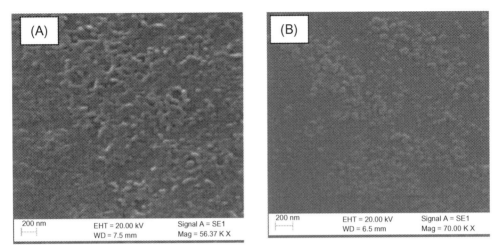

Fig. 8 SEM characterization of (A) uncapped and (B) L-GSH capped CdSeS/ZnS QDs.

concentration of capping ligand causes a slight increment in the particle size, and we observed a string-like structure that covers the QDs (Fig. 8B). This string-like structure appears due to the formation of covalent disulfide bonding between the S atom of thiol group of L-GSH and S atom of ZnS surface of QDs. EDX analysis of L-GSH-CdSeS/ZnS QD provides detail of various elements present in the structure. L-GSH-CdSeS/ZnS QD consists of mostly Kα and Lα peaks having different values of energy (in keV). The observed different elements present in EDX with corresponding energy are C (Kα line, 0.2 keV), Zn (Kα line, 1 keV), O (Kα line, 0.5 keV), Se (Lα line, 1.4 keV), S (Kα line, 2.3 keV), and Cd (Lα line, 3.1 keV) (Fig. 9). The observed elements confirm the existence of C, Zn, O, Se, S, and Cd in capped QD.

5.2 UV-Vis absorption

The sizes and optical properties of nanomaterials can be monitored by UV-Vis absorption spectroscopy. Functionalized CdSeS/ZnS QDs show UV-Vis absorption spectrum having a narrow absorption band at ~268 nm, shoulders at ~261 nm, and ~350 nm in hydroxylic medium (Fig. 10). With increasing concentration of L-GSH (0, 0.05, 0.1, and 0.2 M), we observed the increment in the overall absorbance of L-GSH-CdSeS/ZnS QDs with a small red shift in the lower energy absorption peak [137]. With increasing concentration of L-GSH (0, 0.05, 0.1, and 0.2 M), shifted absorption peak positions were observed at 513, 534, and 550 nm. This red shift in peak position suggested that, at a higher capping agent concentration, bigger size QDs are formed (Fig. 10). The average particle size (2R) of the capped L-GSH-CdSeS/ZnS QDs can be determined by using the absorption edge of the spectrum using the formula [138]:

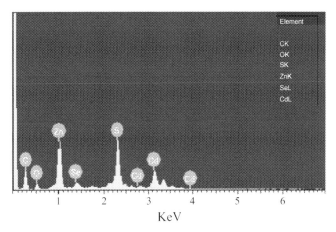

Fig. 9 EDX analysis of L-GSH-CdSe/ZnS QDs.

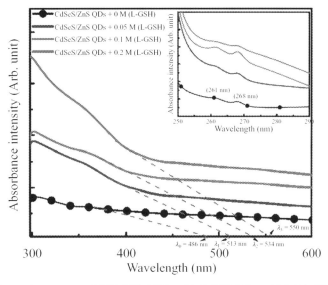

Fig. 10 UV-Vis absorption spectra of CdSeS/ZnS QDs with 0, 0.05, 0.1, and 0.2 M L-GSH concentrations in hydroxylic medium.

$$2R_{(CdSeS/ZnS\ \ or\ \ L-GSH-CdSeS/ZnS)} = \frac{0.1}{(0.138 - 0.0002345\lambda_C)} \text{nm} \qquad (1)$$

Where 2R is the diameter of the particle and λ_C is the absorption edge. The average particle size of QDs has been estimated as ~4.16 nm (λ_C at 486 nm). The particle size is estimated as ~6 nm in nonpolar toluene medium. The decrement in the particle size in aqueous medium is due to the high polarity of the medium. The average particle size with different concentration of L-GSH (0.05, 0.1, and 0.2 M) in capped QDs are estimated as 5.64, 7.82, and 10 nm (λ_C at 513, 534, and 550 nm).

5.3 FTIR absorption

FTIR spectra of capped CdSeS/ZnS QDs and L-GSH have been analyzed in hydroxylic environment (Fig. 11). FTIR spectra of CdSeS/ZnS does not show any absorption in a water environment (Fig. 11). A wide and very strong multiple split IR absorption peaks for L-GSH was observed between 1900 and 2750 cm^{-1} due to symmetric stretching of thiol (ν_{SH}) group. Along this, a strong IR absorbance observed at ~1690 cm^{-1} for ν_{COO} stretching mode in L-GSH. A weak and wide absorbance between 2900 and 3440 cm^{-1} is observed for symmetric ν_{NH} stretching mode. In presence of 0.1 M concentration of L-GSH, due to strong intermolecular interaction between CdSeS/ZnS and L-GSH, a weak complex-like structure L-GSH-CdSeS/ZnS is formed that shows a change in IR signal. In the capped CdSeS/ZnS, peaks between 2900 and 3440 cm^{-1} range indicate the presence of ν_{NH} stretching, and peak between 1600 and 1700 cm^{-1} assigned to ν_{COO} mode, which confirms the presence of carboxylic ($\geq$COO) and amine (>N-H) groups on the surface of QDs. We observed no ν_{SH} mode in IR spectrum of capped CdSeS/ZnS. Disappearance of ν_{SH} mode in capped CdSeS/ZnS is certainly due to the deprotonation of the thiol group and binding of the thiol group to ZnS shell [139] due to the formation of new covalent bonding. In addition, the blue shift in ν_{COO} mode may appear due to the disappearance of ν_{SH} mode in capped QD structure.

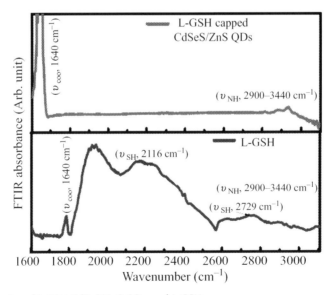

Fig. 11 FTIR spectra of L-GSH-CdSeS/ZnS QDs and L-GSH.

5.4 UV-Vis fluorescence

Pure CdSeS/ZnS QDs show a wide and low intense emission band between 400 and 450 nm (Fig. 12). The presence of 0.05 M concentration of L-GSH causes high enhancement in the emission. However, there is no specific change in position of emission wavelength reported due to this concentration of capping (Fig. 12). Uniformity of shape of both capped and uncapped QDs is confirmed by the symmetric shape of emission profile. The higher concentration (0.1 M, 0.2 M) of L-GSH present in QD causes agglomeration, which a decrement in the emission intensity. The reason of the decrement in the intensity may be due to the hydrogen bonding between the adsorbed thiols. As we observed, the highest intensity is in the concentration 0.05 M concentration of L-GSH, so for better sensing applications, this is the suitable concentration for capped L-GSH-CdSeS/ZnS QDs.

5.5 Effect of pH on L-GSH-CdSeS/ZnS QDs

Studies on the effect of pH on fluorescence intensity of functionalized L-GSH-CdSeS/ZnS QDs are very necessary to obtain an optimum condition to develop a sensitive fluorescence sensor [140]. The fluorescence intensity of L-GSH-CdSeS/ZnS nanosensor is checked with variable pH (Fig. 13). We observed an enhancement in the fluorescence intensity by increasing the pH of the environment. The reason for this is when pH increases, deprotonation in the thiol group of capping agent L-GSH occurs. The covalent interaction between CdSeS/ZnS and L-GSH is strengthened by the deprotonation in the thiol group (Fig. 13) [141]. In lower pH value means (acidic environment), the observed

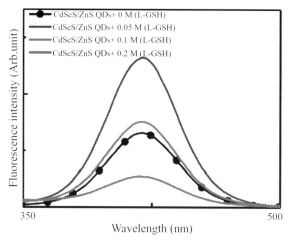

Fig. 12 Emission spectra of CdSeS/ZnS QDs with 0, 0.05, 0.1, and 0.2 M L-GSH concentrations in hydroxylic medium.

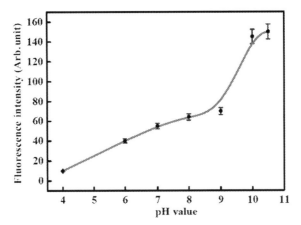

Fig. 13 Effect of pH on the fluorescence intensity of L-GSH-CdSeS/ZnS QDs.

minimum fluorescence intensity appear caused by the presence of a large number of protons (H^+).

In higher pH value of the medium (>7), the solubility of L-GSH-CdSeS/ZnS decreased. The reason of the lesser solubility at higher pH medium may be the dissociation of L-GSH-CdSeS/ZnS binding [142].

5.6 Optical sensing for hazardous metal ions

Fluorescence response of L-GSH-CdSeS/ZnS QDs in the presence hazardous heavy metal ions (Cr^{3+}, Pb^{2+}, Cu^{2+}, Zn^{2+}, and Co^{2+}) were observed in a water phase. Fluorescence response of L-GSH-CdSeS/ZnS QDs in a water medium show strong fluorescence intensity at 400–450 nm. The concentrations of hazardous heavy metal ions Pb^{2+}, Zn^{2+}, Cr^{3+}, Co^{2of+}, and Cu^{2+} have been used in the range between 0.12×10^{-6} M to 1.0×10^{-6} M (Figs. 14–16).

A spectacular enhancement in the emission profile of L-GSH-CdSeS/ZnS QD is observed in presence of Pb^{2+} ions (Fig. 14), while the presence of Cr^{3+} ion causes a decrement in the intensity of L-GSH-CdSeS/ZnS (Fig. 15). Yet Zn^{2+} ion causes an enhancement in the fluorescence intensity (Fig. 16) [70]. For heavy metal ions Co^{2+} and Cu^{2+} ions, no response is observed.

To check the stoichiometry of complexation with hazardous metal ions evaluated by binding constant (K) of complexation, Benesi-Hildebrand Eq. (2) has been used to find the K [143]:

$$\frac{1}{I - I_0} = \frac{1}{K(I_1 - I_0)[X^{n+}]} + \frac{1}{I_1 - I_0} \qquad (2)$$

where I_0 and I_1 are the emission intensity of capped QDs in absence and in presence of quencher Q (concentrations of different metal ion (X^{n+}) of maximum concentrations. The fluorescence intensity at given X^{n+} concentration is denoted by I. The slope

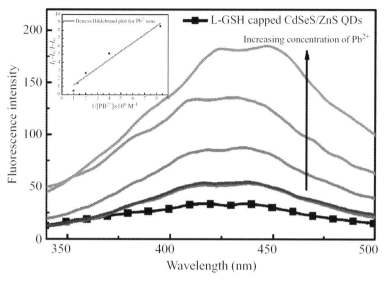

Fig. 14 Emission spectra of L-GSH-CdSeS/ZnS QDs in water and in the presence of Pb^{2+} ion.

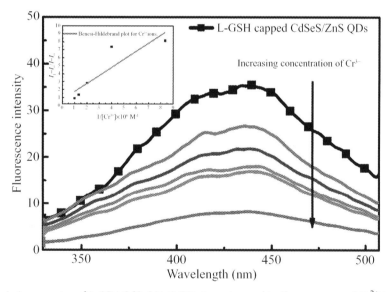

Fig. 15 Emission spectra of L-GSH-CdSeS/ZnS QDs in water and in the presence of Cr^{3+} ion.

obtained from the plot of $\frac{I_1-I_0}{I-I_0}$ versus $\frac{1}{X^{n+}}$(Figs. 14–16) provides the value of K. A linear Benesi–Hildebrand plot for Pb^{2+}, Cr^{3+}, and Zn^{2+} ions show the characteristic of 1:1 (X^{n+}: L-GSH-CdSeS/ZnS) complexation behavior [143]. The calculation of K represents strong complexation for L-GSH-CdSeS/ZnS with Pb^{2+} ($K_{Pb} = 1.05 \times 10^6$ M^{-1}) and Zn^{2+} ($K_{Zn} = 1.25 \times 10^6$ M^{-1}) (Table 1).

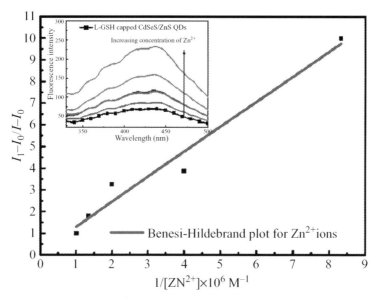

Fig. 16 Benesi-Hildebrand plot for Zn^{2+} ions. Inset of the figure: Emission spectra of L-GSH-CdSeS/ZnS QDs in water and in the presence of Zn^{2+} ion.

Table 1 Binding constants for X^{n+}: L-GSH-CdSeS/ZnS complexation, where $X^{n+} = Zn^{2+}$, Pb^{2+}, Cu^{2+} and Cr^{3+}.

	Ion binding constant	
S. No.	Metal ion	Binding constant (K) in M^{-1}
1	Zn^{2+}	1.25×10^6
2	Pb^{2+}	1.05×10^6
3	Cu^{2+}	0.78×10^6
4	Cr^{3+}	0.90×10^6

Job plot applies "Job's continuous variation method for the determination of the stoichiometry of inclusion complex." The observed value of mole fraction of 0.5 for metal ions: Pb^{2+}, Cr^{3+} and Zn^{2+} validates the existence of strong 1:1 complexation (X^{n+}: L-GSH-CdSeS/ZnS) between guest (COOH-L-GSH-CdSeS/ZnS) and host (Pb^{2+}, Cr^{3+} and Zn^{2+}) [30] (Figs. 17 and 18A and B).

Furthermore, to check the applicability of L-GSH-CdSeS/ZnS QDs as an active optical metal ion sensor for metal ions, the concept of relative fluorescence has been used. The ratio of change in fluorescence intensity and initial intensity of probe sensor is known as relative fluorescence intensity ($F - F_0/F_0$), where F_0 and F are the initial fluorescence of sensor and fluorescence of complex, respectively. The relative fluorescence intensity of

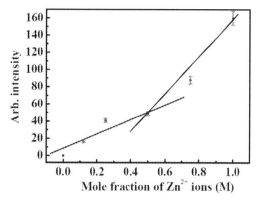

Fig. 17 Job plot between L-GSH-CdSeS/ZnS QDs and Zn^{2+} ions.

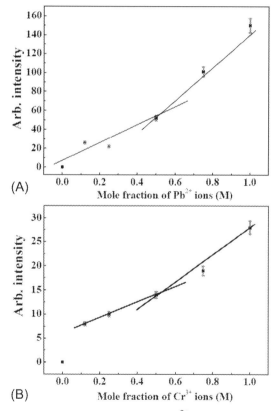

Fig. 18 (A) Job plot between L-GSH-CdSeS/ZnS QDs and Pb^{2+} ions. (B) Job plot between L-GSH-CdSeS/ZnS QDs and Cr^{3+} ions.

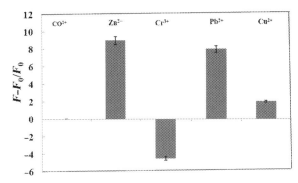

Fig. 19 Relative fluorescence intensity ($F - F_0/F_0$) of L-GSH-CdSeS/ZnS QDs to metal ions: Co^{2+}, Zn^{2+}, Cr^{3+}, Pb^{2+}, and Cu^{2+}.

X^{n+}: L-GSH–CdSeS/ZnS complex in the presence of different concentrations of metal ions were observed in terms of negative response values: 9, 8, and 5 for Zn^{2+}, Pb^{2+}, and Cr^{3+} metal ions, respectively (Fig. 19). Appearance of higher value of relative fluorescence intensity validate L-GSH–CdSeS/ZnS as a better sensitive sensor for Zn^{2+}, Pb^{2+}, and Cr^{3+} ions.

5.7 Detection limit for proposed sensor

The lowest concentration of a dissolved metal ion that can be detected at a specified level of 99% confidence is known as limit of detection (LOD). LOD has been calculated using the equation given here [144]:

$$LOD = \frac{3\sigma}{k} \tag{3}$$

where K is the stern volmer constant and σ is the standard deviation of the blank measurements. The LOD of different heavy metal ions are obtained as 0.775×10^{-6} M (Pb^{2+}), 1.42×10^{-6} M (Cr^{3+}), 4.78×10^{-6} M (Cu^{2+}), and 0.536×10^{-6} M (Zn^{2+}).

5.8 Analysis with real samples

The practical applicability of caped CdSeS/ZnS QDs metal ion sensors has been studied with different real-life samples. To check the concentration of different dissolved metal ions, we have collected water samples from different areas of our society, mainly urban river water (near to leather and chemical industry), rainwater in industrial areas, and normal paint water (Fig. 20). A slight red shift in the fluorescence peak of L-GSH–CdSeS/ZnS QD shows the presence of Pb^{2+} ions, however no shift observed in the fluorescence peak shows the presence of Zn^{2+} ions (Fig. 20). An enhancement in the fluorescence intensity with a wide fluorescence band in the presence of river water (Fig. 20) confirms

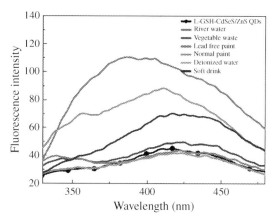

Fig. 20 Emission spectra of L-GSH-CdSeS/ZnS QDs in water and in the presence of real samples.

the presence of Zn^{2+} and Pb^{2+}, both ions having a concentration of 0.25×10^{-6} M and 0.5×10^{-6} M. A slight increment in the fluorescence intensity in the vegetable waste confirms the presence of Pb^{2+} ion with a concentration of 0.12×10^{-6} M. A red shift in the fluorescence peak is also found in the case of normal paint and soft drink, which verify the presence of Pb^{2+} ion having concentrations of 0.25×10^{-6} M and 0.12×10^{-6} M. We didn't observe any presence of Pb^{2+}, Zn^{2+}, or Cr^{3+} ions in lead–free paint and deionized water. This shows that QD can work very selectively as it can detect a very small concentration of water-solvable hazardous metal. Thus, L–GSH–CdSeS/ZnS QDs can perform as a very sensitive detecting tool for a majority of water–dissolvable hazardous metal ions.

6. Fluorescent markers in biological imaging

Scientists relied on fluorescent dyes to make biological molecules visible. Proteins, nucleic acids, lipids, or small molecules when labeled with an extrinsic fluorophore (organic or inorganic) are called fluorescent dyes. Fluorescent dye can be a small/big molecule, a protein, or a QD. Fluorescent labeling and staining, when combined with an appropriate imaging instrument, is a sensitive and quantitative method that is widely used in today's biomedical, optoelectronics [145], and chemical [146] industries for a variety of experimental, analytical, industrial, and quality control applications. Fluorescent detection offers a number of advantages such as sensitivity, multilevel possibility, stability, low hazard, easier commercial availability, and lost cost over other available detection methods (potentiometric, calorimetric, etc.) [147].

For peeking inside cells, staining organelles, and other imaging experiments, fluorescence is used as a powerful tool for labeling, imaging, tracking, detection, and therapy where chemistry was behaving as one and only method by the presence of many synthetic-based fluorescent dyes. Some important markets available for synthetic-based dyes are rhodamine 6G, rhodamine B, uranine A, croceine orange G, cyanine, etc. [148]. Rhodamines are bright and cell-permeable and so they can easily slip into cells to make them glow. There are many types of rhodamine dyes having different color output that are available in industry, though making new rhodamines was not easy [149]. Although for more than 100 years, chemists have worked with rhodamines, despite this chemists had created only a few dozen colors. Same problem exists for most of the other available synthetic dyes also. Therefore, the "king" was dethroned by a glowing green organic jellyfish protein dye called green fluorescent protein (GFP). GFPs are some proteins or small molecules in cells and are naturally fluorescent in nature. GFP is widely used as a reporter molecule for the study of protein localization, protein binding events, and gene expression [150]. There are many types of fluorescent proteins of different colors currently in use as fluorescent tags for bioimaging [151]. To choose a fluorescent protein to use for imaging, one must decide on which color variant to use. But one of the major problems with these GFPs is that they may lose their fluorescence during tissue fixation or subsequent processing [152]. Numerous GFP variants have therefore been engineered to overcome these limitations [153]. The most available GEPs in this category are ECFP, EBFP, DsRed, GFP-S65T, etc. [154,155]. Still with the engineered versions, also the expression of a fluorescent protein in a cell has shown the indication of cytotoxic effects [156].

The next player in the imaging game is organic-based dyes having excellent advantages. There are many organics-based fluorescent sensors available in the market now like coumarine, xanthene, bodipy, cyanine, etc. and their derivatives [157]. For "Coumarins," the skeleton is 1-benzopyran-2-one, which comes from a Caribbean word "*coumarou*" for the tonka bean. Because of their sweet scent, coumarins have been used in perfumes since 1882. Coumarin itself is nonfluorescent, but its derivatives with electron-donating groups (EDG) at the C-7 position develop the intense fluorescence. Coumarins are well known as the largest class of laser dyes ("blue-green" region). Fluorescence of coumarin derivatives changes drastically with electronic characteristics and regiochemical positions of their substituent. However, there are a number of limitations with all these existing organic dyes. The absorption spectra of organic dyes have discrete bands, and one can hardly found the fluorescence from different dyes for the similar excitation. So switching of wavelength is required. In addition, the emission spectra of these dyes are broad and asymmetric in nature, and they cause self-absorption because of their small Stokes shifts. All organic dyes suffer from photobleaching. Also they are less valuable for long-term cell tracking strategies and have a very short lifetime.

To overcome all these problems, QDs with unique optical properties make a significant entry in the field of biomedical imaging. QDs are brighter and more photostable than all other available organic/inorganic dyes for labeling macromolecules and can therefore be used to complement the existing fluorophore and fluorescent protein-fusion techniques [118]. The main attraction of QDs lie in their high optical extinction coefficients, tunable sharp emission profile, carrier multiplication ability, and variable nonlinear properties, etc., which can be tuned by adjusting size and shape. Another important advantage of QDs over classical fluorophore is that they emit light in the infrared and near-infrared regions, as the absorption of tissues is minimal in this region. There are many reports of using QDs for imaging purposes in lymph nodes, tumor-specific receptors, and as malignant tumor detectors. Study shows that, since 2000 onward, active exploration of QD-based biomedical applications have shown more than 300% increase according to relevant peer-reviewed publications like PubMed and Nature.com. These research works provide a detail of the key achievements in nanoscience that have initiated the work on QDs for biomedical applications and validated by the recent developments that have converted QDs into clinically relevant tools.

Development of QDs of high stability with their high solubility and low toxicity to healthy cells initiates many researchers nowadays to apply the QDs for cell imaging. In 1998, Alivisatos et al. [118] reported the first water-soluble QDs for Hela cells and Nie's et al. [85] for 3-day transfer, inoculum 3×10^5 (3T3) cells labeling. Goldman et al. demonstrated the simultaneous detection of four toxins: cholera, ricin, shiga-like toxin 1, and staphylococcal enterotoxin B—in a sandwich immune assay with QD labels. They reported the corresponding reporter antibodies were conjugated with QD_{510}, QD_{555}, QD_{590}, and QD_{610}, and offered LODs in the range of 3–300 ng/mL. QDs are also well suited to multiplexed staining of tissue biopsy specimens. Five-color molecular profiling of human prostate cancer cells by using QD-IHC has been described by Xing et al. In addition to labeling fixed cells and tissues, epifluorescence microscopy and QDs have been widely applied for live cell labeling and imaging. In the next section, we will discuss the working of CdSeS/ZnS QDs as a medical imaging tool.

6.1 Working of QD as a medical imaging tool

CdSeS/ZnS QD was tested as a medical imaging tool with GSH as a surface-coating agent, whose carboxylate or amine groups on GSH can easily be conjugated with QDs. L-GSH-coated core/shell CdSeS/ZnS QDs will be tested as an imaging tool to view different living cells including some cancerous cell. MTT cell viability assay is studied here for percent cell viability and cellular toxicity of cancerous and noncancerous/healthy cells detection in presence of bare and L-GSH-coated CdSeS/ZnS QDs. The fluorescence microscopy is used to study the cellular uptake of the QDs and their subcellular localization. MCF-7 is the first hormone-responsive breast cancer cell line that is being used in the area of breast cancer research in the past 45 years. H9C2 cardiomyoblast cells are used as a healthy cell line, as reports suggest that it displays comparable expressions of biomarkers that are actually present in live organisms and serves as a significant in-vitro experimental model for cardiac-related studies.

6.1.1 QD induction and 4′,6-diamidino-2-phenylindole (DAPI) fluorescent nuclear staining to visualize the effects of QDs

MCF-7 cells are seeded on sterilized cover-slips in a six-well culture dish and kept overnight at 37°C. For concentrations of 1, 5, and 50 pg/μL, media was replaced and then QD induction was given. The cells were incubated for 6, 24, and 48 h time period. With 50 ng/mL concentration of DAPI dye, cells are counterstained. Similar steps were followed for H9C2 cells. For H9C2 cells, the previously mentioned concentrations were studied for 24 h and, at different magnifications, images were captured.

6.1.2 Effect of L-GSH-capped CdSeS/ZnS QDs on cell viability

The effect of different doses of capped QDs on cell viability has been studied. By using MTT cell viability assay, an initial dose-dependent study was made to analyze cancer cell viability upon increasing concentration of capped QDs. MTT assay is performed for calculating the number of viable cells leading any treatment of cells. MTT dye is yellow in color caused by oxidoreductase enzymes present in the mitochondria of live cells; it was converted into purple color substance called formazon crystals. These crystals converted in purple color by the live cells. The prepared cells were solubilized by adding DMSO. Furthermore, this was quantified at 570 nm by spectrophotometer. In the present study, MTT assay was done for increasing QDs concentration (0.1–100 pg/μL) ranging for 24 h on MCF-7 cancer cells. The calculated percent cell viability shows that up to the concentration of 5 pg/μL QD, no significant cell death was observed. Further increments in the QD concentrations causes a decrease in cell viability where up to 50% cell death at the dose of 50 pg/μL QD, concentration was recorded; 50% cell death was recorded at 100 pg/μL concentration was further reduced to 25% (Fig. 21); and 50 pg/μL concentration was selected as the inhibitory concentration (IC_{50}).

For any cytotoxicity effect of L-GSH–CdSeS/ZnS core/shell QDs on healthy cells, the effect of the previously mentioned QD concentrations was also studied on noncancerous cells H9C2 cardiomyoblasts. These cells represent normal cardiac cells and were used as healthy cells. MTT assay was performed for the same concentration range of QDs as for MCF-7 cells. MTT assay suggested that, at the dose concentration of 5 pg/μL, cell viability is up to 95% and until the dose of 20 pg/μL, QDs had no significant cytotoxicity of these concentrations on the healthy cells (Fig. 22). So we can use capped CdSeS/ZnS QDs up to 20 pg/μL concentrations for cancer cells imaging as it will display minimal cytotoxicity in any other healthy cells/tissues. For additional experiments, 50 pg/μL was used as an IC_{50} dose, and 1 and 5 pg/μL concentrations were used as safe doses.

At much lower concentrations of 1 pg/μL, a significant decrease in cell viability of healthy cells was observed. Also, in healthy cardiac cells, this decrease in cell viability was further exaggerated.

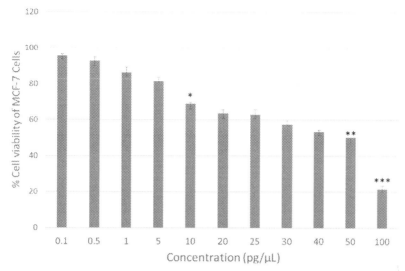

Fig. 21 MTT assay of MCF-7 breast cancer cell line for concentrations L-GSH-CdSeS/ZnS QDs ranging from to 0.1 pg/µL to 100 pg/µL (*$P < .05$).

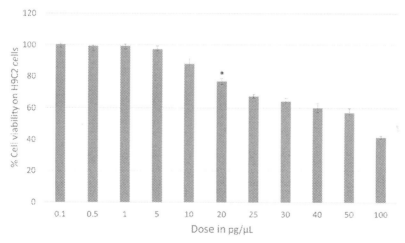

Fig. 22 MTT assay of H9C2 cardiomyocytes for L-GSH-CdSeS/ZnS QD concentrations ranging from 0.1 pg/µL to 100 pg/µL (*$P < .05$).

6.1.3 Effect of L-GSH-capped CdSeS/ZnS QDs on morphological alterations in MCF-7 cells

Cells morphological change indicates the undergoing stress. Therefore preliminary microscopic analysis of MCF-7 cells was done. To study the effects of selected concentrations of QD on the cancer cells, cells were made with the three selected doses and any morphological changes are observed after 24 h. At the dose concentration of 1 pg/μL QD, no significant change is observed in the morphology of QDs and cells look like the control untreated cells. At slightly higher concentration of 5 pg/μL, less significant growth inhibition was observed with few dead cells. On the other hand, as the concentration of the dose increased to 50 pg/μL, a significant growth inhibition was observed as well as cell death. This validates that, at this concentration, 50% cell death observed in MTT assay (Fig. 23).

6.1.4 Fluorescent imaging and cellular uptake of L-GSH-capped CdSeS/ZnS QDs in cancer and healthy cells

MCF-7 cancer cell's cellular uptake and localization of QDs were studied by fluorescent microscopy with variation of concentration and time. L-GSH-CdSeS/ZnS QDs can easily be visualized under the FIITC filter of fluorescent microscope. For the time period of 6, 24, and 48 h, three selected doses of QDs were analyzed. Nucleus-specific DAPI fluorescent stain was used to visualize the cells under the fluorescent microscope. DAPI binds with DNA and produces a strong blue fluorescence. As shown in Fig. 24, the DAPI-stained MCF-7 cells show some faded fluorescence in FIITC filter when excited in

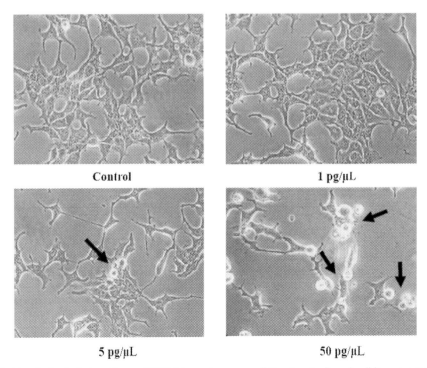

Control   1 pg/μL

5 pg/μL   50 pg/μL

Fig. 23 Morphological alterations of MCF-7 breast cancer cell line (control) and cell lines treated with various concentrations (1, 5, 50 pg/μL) of L-GSH-CdSeS/ZnS QDs.

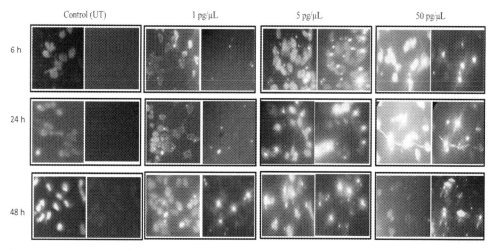

Fig. 24 Concentration and time-dependent cellular uptake and localization of MCF-7 cancer cells with L-GSH-CdSeS/ZnS QD by fluorescent microscopy at 100× magnification.

the dark field and bright blue fluorescence under DAPI filter. In addition, L-GSH-CdSeS/ZnS QDs were also displaying fluorescence in both DAPI and FIITC filters. For untreated (UT) control cells, no fluorescence is observed in FIITC filter. Fluorescence microscopic analysis shows that QDs are not only distributed within the cytoplasmic compartment of MCF-7 cells, however, the nuclear compartment as evident from the microscopic pictures displaying QD florescence immediately next to the nuclear wall of DAPI stained cells. It suggests that the capped QDs are too small to enter the cell nucleus. At 50 pg/µL concentration, maximum QD fluorescence is observed. However, this dose cannot be used in biological systems, because this dose results in a decrement in the cell viability of noncancerous cells. The concentration of 5 pg/µL was treated as the safe dose to be used for cancer detection in-vitro because, at this concentration, cancer cells can easily be detected.

Time-dependent cellular uptake study shows that, from 6 to 24 h, the signal intensity increases, while above this time signal intensity remains almost constant. Therefore, for detecting cancer cells, 24-h time period was derived. This time was also validated on noncancerous healthy cells. It is important to study the effect of the time period and QD concentration to be used for cancer cell detection on normal healthy cells. At 24 h time-point for cellular uptake, different QD concentrations were studied by H9C2 healthy cells. Up to the concentration of 5 pg/µL in the microscopic pictures, no significant cellular uptake of QD was observed (Fig. 25). Increased concentration of QDs (up to 50 pg/µL) significantly increased the cellular uptake of QDs further validating the cytotoxicity of this dose as observed by MTT assay and morphological

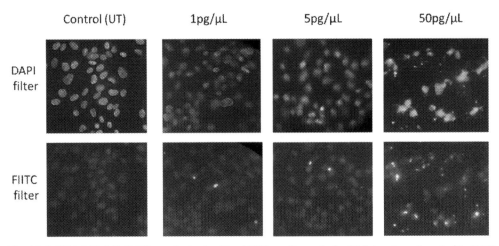

Fig. 25 L-GSH-CdSeS/ZnS QD uptake by healthy H9C2 cardiac cells for 24 h treatment period in DAPI and FIITC filters.

analysis. In healthy cells, the membrane receptors are properly working, which does not permit the entry for any foreign unidentified entity within the cells.

L-GSH-CdSeS/ZnS QDs are stable in cytoplasm and suitable for in-vitro, cell tracking, cancer cell labeling, and bioimaging applications. L-GSH-CdSeS/ZnS QDs can be easily visualized under the FIITC filter of a fluorescent microscope because of the presence of strong absorption and emission spectra between the range 400–450 nm. The concentration of L-GSH-CdSeS/ZnS QDs 5 pg/μL can be concluded as the safe dose to be used for cancer detection in vitro. The probe-capped QDs show less reactivity with healthy living cells. Due to its less reactivity with healthy living cells, the probe QDs may serve as a favorable tool for in-vivo imaging for the current medical industry.

7. Conclusion

Until now, we observed that semiconductor base-capped QDs can be applicable for different biomedical applications. Still, biocompatibility, good solubility, and low cytotoxicity of the bioimaging agent in the cell environment is one of the most essential requirements of different QDs for biomedical applications [158]. Because QDs are composed of heavy metals, they are potentially toxic during in-vitro and in-vivo imaging. The toxicity to cells and animals is time- and dose-dependent. Therefore, when imaging is applicable with a low concentration of QDs within a certain time, they can be ideal probes. Although many cases have been reported, there are still many work scopes of organic-capped QDs for the application in medical industry as imaging tools are still in demand. Similarly, other organic-based, small-dimensional, carbon-based nanomaterials such as CQD, nanodiamonds, carbon nanotubes, and graphene QDs are now appearing in biomedical applications, and they are regarded as appropriate alternative candidates of a fore-mentioned semiconducting II-VI QD because of their good chemical stability, biocompatibility, high water-solubility, sufficient fluorescence quantum yield, and low toxicity against living cell. There are already few reports available in the literature regarding application of various organic QDs as cell imaging agents in biomedical research. Work scopes of organic-capped and organic-based QDs for the applications in biomedical industry as imaging tools and sensors are in high demand. New QD-based applications need to be explored.

References

[1] https://www.downtoearth.org.in/news/water/new-study-points-out-how-odf-villages-are-better-off-64947.
[2] R.P. Schwarzenbach, T. Egli, T.B. Hofstetter, U.V.O. Gunten, B. Wehrli, Global water pollution and human health, Annu. Rev. Env. Resour. 35 (2010) 109–136.
[3] M.T. Amin, A.A. Alazba, U. Manzoor, A review of removal of pollutants from water/wastewater using different types of nanomaterials, Adv. Mater. Sci. Eng. 2014 (2014) 1–24.
[4] A. Peal, B. Evans, C.V.D. Voorden, Hygiene and Sanitation Software: An Overview of Approaches, Water Supply and Sanitation Collaborative Council, 2010.
[5] M. Jaishankar, T. Tseten, N. Anbalagan, B.B. Mathew, K.N. Beeregowda, Toxicity, mechanism and health effects of some heavy metals, Interdiscip. Toxicol. 7 (2014) 60–72.
[6] F.S.A. Fartusie, S.N. Mohssan, Essential trace elements and their vital roles in human body, Indian J. Adv. Chem. Sci. 5 (2017) 127–136.
[7] N. Singh, V.K. Gupta, A. Kumar, B. Sharma, Synergistic effects of heavy metals and pesticides in living systems, Front. Chem. 5 (2017) 70.
[8] Chromium in Drinking-water, Background Document for Development of WHO Guidelines for Drinking-Water Quality, World Health Organization, Geneva, Switzerland, 2003.
[9] Lead in Drinking-water, Background Document for Development of WHO Guidelines for Drinking-water Quality, World Health Organization, Geneva, 2011.
[10] Copper in Drinking-water, Background Document for Development of WHO Guidelines for Drinking-Water Quality, World Health Organization, Geneva, 2004.
[11] D. Paul, Research on heavy metal pollution of river Ganga: a review, Ann. Agrarian Sci. 15 (2017) 1–9.
[12] R.A. Wuana, F.E. Okieimen, Heavy metals in contaminated soils: a review of sources, chemistry, risks and best available strategies for remediation, ISRN Ecol. 2011 (2011) 1–20.
[13] I. Mahmood, S.R. Imadi, K. Shazadi, A. Gul, K.R. Hakeem, Effects of pesticides on environment, in: Plant, Soil and Microbes, Springer, Cham, 2016, , pp. 253–269.
[14] FAO/WHO, Guidelines on Pesticide Legislation, International Code of Conduct on Pesticide Management, Food and Agriculture Organization of the United Nations, 2015.
[15] T. Gong, J. Liu, X. Liu, J. Liu, J. Xiang, Y. Wu, A sensitive and selective sensing platform based on CdTe QDs in the presence of L-cysteine for detection of silver, mercury and copper ions in water and various drinks, Food Chem. 213 (2016) 306–312.
[16] M. Rana, A. Jain, V. Rani, P. Chowdhury, Glutathione capped core/shell CdSeS/ZnS quantum dots as a medical imaging tool for cancer cells, Inorg. Chem. Commun. 112 (2020) 107723.
[17] J. Liab, J.J. Zhu, Quantum dots for fluorescent biosensing and bio-imaging applications, Analyst 138 (2013) 2506.
[18] J. Tang, R.A. Marcus, Determination of energetics and kinetics from single-particle intermittency and ensemble-averaged fluorescence intensity decay of quantum dots, J. Chem. Phys. 125 (2006) 044703.

[19] I. Martinic, S.V. Eliseeva, S. Petoud, Near-infrared emitting probes for biological imaging: organic fluorophores, quantum dots, fluorescent proteins, lanthanide(III) complexes and nanomaterials, JOL 189 (2017) 19–43.

[20] D. Huang, C. Niu, X. Wang, X. Lv, G. Zeng, "Turn-On" fluorescent sensor for Hg^{2+} based on single-stranded DNA functionalized Mn: CdS/ZnS quantum dots and gold nanoparticles by time-gated mode, Anal. Chem. 85 (2013) 1164–1170.

[21] L.N. Neupane, E.T. Oh, H.J. Park, K.H. Lee, Selective and sensitive detection of heavy metal ions in 100% aqueous solution and cells with a fluorescence chemosensor based on peptide using aggregation-induced emission, Anal. Chem. 88 (2016) 3333–3340.

[22] A.S.N. Trindade, A.F. Dantas, D.C. Lima, S.L.C. Ferreira, L.S.G. Teixeira, Multivariate optimization of ultrasound-assisted extraction for determination of Cu, Fe, Ni and Zn in vegetable oils by high-resolution continuum source atomic absorption spectrometry, Food Chem. 185 (2015) 145–150.

[23] S.M. Bak, E. Hu, Y. Zhou, X. Yu, S.D. Senanayake, S.J. Cho, K.B. Kim, K.Y. Chung, X.Q. Yang, K.W. Nam, Structural changes and thermal stability of charged $LiNixMnyCozO_2$ cathode materials studied by combined in situ time-resolved XRD and mass spectroscopy, ACS Appl. Mater. Interfaces 24 (2014) 22594–22601.

[24] H. Huang, M. Zou, X. Xu, X. Wang, F. Liu, N. Li, Near-infrared fluorescence spectroscopy of single-walled carbon nanotubes and its applications, Trends Anal. Chem. 30 (2011) 1109–1119.

[25] N.M. Taufek, D. Cartwright, M. Davies, A.K. Hewavitharana, P. Koorts, P.N. Shaw, et al., The simultaneous analysis of eight essential trace elements in human milk by ICP-MS, Food Anal. Methods 9 (2016) 2068–2075.

[26] S.A. Tukur, N.A. Yusof, R. Hajian, Linear sweep anodic stripping voltammetry: determination of chromium (VI) using synthesized gold nanoparticles modified screen-printed electrode, J. Chem. Sci. 127 (2015) 1075–1081.

[27] V. Arancibia, M. Valderrama, K. Silva, T. Tapia, Determination of chromium in urine samples by complexation-supercritical fluid extraction and liquid or gas chromatography, J. Chromatogr. B 785 (2003) 303–309.

[28] E. Kopysc, K. Pyrzynska, S. Garbos, E. Bulska, Determination of mercury by cold vapor atomic absorption spectrometry with pre concentration on a gold-trap, Anal. Sci. 16 (2000) 1309–1312.

[29] A. Kumar, A. Ray Chowdhuri, D. Laha, T.K. Mahto, P. Karmakar, S.K. Sahu, Green synthesis of carbon dots from *Ocimum sanctum* for effective fluorescent sensing of Pb^{2+} ions and live cell imaging, Sens. Actuators B 242 (2017) 679–686.

[30] Y. Zhang, X. Guo, W. Si, L. Jia, X. Qian, Ratiometric and water-soluble fluorescent zinc sensor of carboxamidoquinoline with an alkoxyethylamino chain as receptor, Org. Lett. 10 (2008) 473–476.

[31] W. Fritzsche, A. Taton, Metal nanoparticles as labels for heterogeneous, chip-based DNA detection, Nanotechnology 14 (2003) R63–R73.

[32] J.K. Jaiswal, S.M. Simon, Potentials and pitfalls of fluorescent quantum dots for biological imaging, Trends Cell Biol. 14 (2004) 497–504.

[33] I. Khan, K. Saeed, I. Khan, Nanoparticles: properties, applications and toxicities, Arab. J. Chem. 12 (2019) 908–931.

[34] U.R. Genger, M. Grabolle, S.C. Jaricot, R. Nitschke, T. Nann, Quantum dots versus organic dyes as fluorescent labels, Nat. Methods 5 (2008) 763.

[35] T. Jamieson, R. Bakhshi, D. Petrova, R. Pocock, M. Imani, A.M. Seifalian, Biological applications of quantum dots, Biomaterials 28 (2007) 4717–4732.

[36] R. Kumar, R. Singh, Prospect of graphene for use as sensors in miniaturized and biomedical sensing devices, in: Reference Module in Materials Science and Materials Engineering, Elsevier, Oxford, 2018, , pp. 1–13.

[37] M. Farre, L. Kantiani, D. Barcelo, Microfluidic Devices: Biosensors, Chemical Analysis of Food: Techniques and Applications, Elsevier, 2012, pp. 177–217.

[38] J. Wilder, W.F. Kosonocky, Multiple Resolution Image Sensor, (1993).

[39] E.R. Fossum, CMOS image sensors: electronic camera-on-a-chip, IEEE Trans. Electron Devices 44 (1997) 1689–1698.

[40] A.P. Selvam, S. Muthukumar, V. Kamakoti, S. Prasad, A wearable biochemical sensor for monitoring alcohol consumption lifestyle through ethyl glucuronide (EtG) detection in human sweat, Sci. Rep. 6 (2016) 23111.

[41] S. Agrawal, R. Prajapati, Nanosensors and their pharmaceutical applications: a review, Int. J. Pharm. Sci. Nanotechnol. 4 (2012) 1528–1535.

[42] G. Rong, E.E. Tuttle, A.N. Reilly, H.A. Clark, Recent developments in nanosensors for imaging applications in biological systems, Annu. Rev. Anal. Chem. 12 (2019) 109–128.

[43] M. Chern, J.C. Kays, S. Bhuckory, A.M. Dennis, Sensing with photoluminescent semiconductor quantum dots, Methods Appl. Fluoresc. 7 (2019) 012005.

[44] X. Qian, Y. Xiao, Y. Xu, X. Guo, J. Qiana, W. Zhu, "Alive" dyes as fluorescent sensors: fluorophore, mechanism, receptor and images in living cells, Chem. Commun. 46 (2010) 6418.

[45] A.D. Yoffe, Semiconductor quantum dots and related systems: electronic, optical, luminescence and related properties of low dimensional systems, Adv. Phys. 50 (2001) 1–208.

[46] D.M. Willard, Nanoparticles in bioanalytics, Anal. Bioanal. Chem. 376 (2003) 84–286.

[47] C. Yadav, K. Surana, P.K. Singh, B. Bhattacharya, An ultra-simple method for synthesis of violet cds quantum dots at sub-room temperature, J. Nanosci. Nanotechnol. 20 (2020) 3935–3938.

[48] M. Ishikawa, V. Biju, Luminescent quantum dots, making invisibles visible in bioimaging, Prog. Mol. Biol. Transl. Sci. 104 (2010) 53–99.

[49] S.M. Liu Guo, Z.H. Zhang, R. Li, W. Chen, Z.G. Wang, Characterization of CdSe and CdSe/CdS core/shell nanoclusters synthesized in aqueous solution, Phys. E. 8 (2000) 174–178.

[50] M. Rana, P. Chowdhury, Studies on size dependent structures and optical properties of CdSeS clusters, J. Clust. Sci. 31 (2020) 11–21.

[51] W. Bian, F. Wang, H. Zhang, L. Zhang, L. Wang, S. Shuang, Fluorescent probe for detection of Cu2+ using core/shell CdTe/ZnS quantum dots, Luminescence 30 (2015) 1064–1070.

[52] A.P. Alivisatos, Perspectives on the physical chemistry of semiconductor nanocrystals, J. Phys. Chem. 100 (1996) 13226–13239.

[53] W.W. Yu, L. Qu, W. Guo, X. Peng, Experimental determination of the extinction coefficient of CdTe, CdSe, and CdS nanocrystals, Chem. Mater. 15 (2003) 854–2860.

[54] M.A. Walling, J.A. Novak, J.R. Shepard, Quantum dots for live cell and in-vivo imaging, Int. J. Mol. Sci. 10 (2009) 441–491.

[55] R.D. Schaller, V.I. Klimov, High efficiency carrier multiplication in PbSe nanocrystals: implications for solar energy conversion, Phys. Rev. Lett. 92 (2004) 186601.

[56] Y. Kobayashi, T. Udagawa, N. Tamai, Carrier multiplication in CdTe quantum dots by single-photon timing spectroscopy, Chem. Lett. 38 (2009) 830–831.

[57] S.H. Park, M.P. Casey, J. Falk, Nonlinear optical properties of CdSe quantum dots, J. Appl. Physiol. 73 (1993) 8041–8045.

[58] X. Sun, L. Xie, W. Zhou, F. Pang, T. Wang, A.R. Kost, Z. An, Optical fiber amplifiers based on PbS/CdS QD modified by polymers, Opt. Express 21 (2013) 8214–8219.

[59] H. Kurta, M. Yuceb, B. Hussaina, H. Budak, Dual-excitation upconverting nanoparticle and quantum dot aptasensor for multiplexed food pathogen detection, Biosens. Bioelectron. 81 (2016) 280–286.

[60] S. Zhu, X. Lin, P. Ran, Q. Xia, C. Yang, J. Ma, Y. Fu, A novel luminescence-functionalized metal-organic framework nanoflowers electrochemiluminesence sensor via "on-off" system, Biosens. Bioelectron. 91 (2017) 436–440.

[61] Y.Y. Cao, X.F. Guo, H. Wang, High sensitive luminescence metal-organic framework sensor for hydrogen sulfide in aqueous solution: a trial of novel turn-on mechanism, Sens. Actuators B 243 (2017) 8–13.

[62] Y. Wang, N. Wang, X. Ni, Q. Jiang, W. Yang, W. Huang, W. Xu, A core-shell CdTe quantum dots molecularly imprinted polymer for recognizing and detecting p-nitrophenol based on computer simulation, RSC Adv. 5 (2015) 73424–73433.

[63] A. Badawi, N. Al-Hosiny, S. Abdallah, S. Negm, H. Talaat, Tuning photocurrent response through size control of CdTe quantum dots sensitized solar cells, Sol. Energy 88 (2013) 137–143.

[64] S.S. Gandhi, L.C. Chien, High transmittance optical films based on quantum dot doped nanoscale polymer dispersed liquid crystals, Opt. Mater. 54 (2016) 300–305.

[65] S. Chen, M. Liao, M. Tang, J. Wu, M. Martin, T. Baron, A. Seeds, H. Liu, Electrically pumped continuous-wave 1.3 μm InAs/GaAs quantum dot lasers monolithically grown on on-axis Si (001) substrates, Opt. Express 25 (2017) 4632.

[66] X. Dai, Y. Deng, X. Peng, Y. Jin, Quantum-dot light-emitting diodes for large-area displays: towards the dawn of commercialization, Adv. Mater. 29 (2017) 1607022.

[67] P. Sharma, S. Brown, G. Walter, S. Santra, B. Moudgil, Nanoparticles for bioimaging, Adv. Colloid Interface Sci. 123 (2006) 471–485.

[68] N.T. Thanh, L.A. Green, Functionalisation of nanoparticles for biomedical applications, Nano Today 5 (2010) 213–230.

[69] Q. Mu, H. Xu, Y. Li, S. Ma, X. Zhong, Adenosine capped QD based fluorescent sensor for detection of dopamine with high selectivity and sensitivity, Analyst 139 (2014) 93–98.

[70] M. Rana, P. Chowdhury, L-glutathione capped CdSeS/ZnS quantum dots as an environmentally hazardous chemical sensor, J. Basic Appl. Eng. Res. 2 (2015) 1319–1321.

[71] J. Feng, Y. Tao, X. Shen, H. Jin, T. Zhou, Y. Zhou, Highly sensitive and selective fluorescent sensor for tetrabromobisphenol-A in electronic waste samples using molecularly imprinted polymer coated quantum dots, Microchem. J. 144 (2018) 93–101.

[72] S. Jain, S.B. Park, S.R. Pillai, P.L. Ryan, S.T. Willard, J.M. Feugang, Applications of fluorescent quantum dots for reproductive medicine and disease detection, in: Unraveling the Safety Profile of Nanoscale Particles and Materials-From Biomedical to Environmental Applications, InTech, 2018.

[73] S.T. Yang, X. Wang, H. Wang, F. Lu, P.G. Luo, L. Cao, Y. Huang, Carbon dots as nontoxic and high-performance fluorescence imaging agents, J. Phys. Chem. C 113 (2009) 18110–18114.

[74] K. Sapsford, T. Pons, I. Medintz, H. Mattoussi, Biosensing with luminescent semiconductor quantum dots, Sensors 6 (2006) 925–953.

[75] X. Zhang, F. Chen, X. Song, P. He, S. Zhang, Proximity ligation detection of lectin Concanavalin A and fluorescence imaging cancer cells using carbohydrate functionalized DNA-silver nanocluster probes, Biosens. Bioelectron. 104 (2018) 27–31.

[76] Y. Zhang, J. Xiao, Y. Sun, L. Wang, X. Dong, J. Ren, W. He, F. Xiao, Flexible nanohybrid microelectrode based on carbon fiber wrapped by gold nanoparticles decorated nitrogen doped carbon nanotube arrays: in situ electrochemical detection in live cancer cells, Biosens. Bioelectron. 100 (2018) 453–461.

[77] https://www.nobelprize.org/prizes/physics/2010/press-release/.

[78] OSA, Twenty Attain 2006 Top Honors from the Optical Society of America, http://www.osa.org/about_osa/newsroom/news_releases/releases/09.2006/awards.aspx, 2006 (Accessed November 30, 2011).

[79] A.I. Ekimov, A.A. Onushchenko, Quantum size effect in three-dimensional microscopic semiconductor crystals, J. Exp. Theor. Phys. 34 (1981) 363–366.

[80] A.L. Efros, A.L. Efros, Interband absorption of light in a semiconductor sphere, Sov. Phys. Semicond. 16 (1982) 772–775.

[81] R. Rossetti, S. Nakahara, E.L. Brus, Quantum size effects in the redox potentials, resonance raman spectra, and electronic spectra of cadmium sulfide crystallites in aqueous solution, J. Chem. Phys. 79 (1983) 1086–1088.

[82] H. Weller, U. Koch, M. Gutierrez, A. Henglein, Photochemistry of colloidal metal sulfides. 7. Absorption and fluorescence of extremely small zinc sulfide particles (The world of the neglected dimensions), Ber. Bunsenges. Phys. Chem. 88 (1984) 649–656.

[83] L. Brus, Electronic wave functions in semiconductor clusters: experiment and theory, J. Phys. Chem. 90 (1986) 2555–2560.

[84] C.B. Murray, D.J. Norris, M.G. Bawendi, Synthesis and characterization of nearly monodisperse CdE (E = S, Se, Te) semiconductor nanocrystallites, J. Am. Chem. Soc. 115 (1993) 8706–8715.

[85] W.C.W. Chan, S. Nie, Quantum dot bioconjugates for ultrasensitive nonisotopic detection, Science 281 (1998) 2016–2018.

[86] M.A. Hines, P.G. Sionnest, Synthesis and characterization of strongly luminescing ZnS-Capped CdSe nanocrystals, J. Phys. Chem. 100 (1996) 468–471.

[87] A. Rogach, Quantum dots still shining strong 30 years on, ACS Nano 8 (2014) 6511–6512.

[88] H. Fan, et al., Differentiation of heavy metal ions by fluorescent quantum dot sensor array in complicated samples, Sens. Actuators B 295 (2019) 110–116.

[89] Y. Chen, H. Liang, Applications of quantum dots with upconverting luminescence in bioimaging, J. Photochem. Photobiol. B Biol. 135 (2014) 23–32.

[90] A. Small, et al., Quantum information processing using quantum dot spins and cavity QED, Phys. Rev. Lett. 83 (1999) 4204–4207.

[91] H.C. Kuo, et al., A critical review on two-dimensional quantum dots (2D QDs): from synthesis toward applications in energy and optoelectronics, Prog. Quantum Electron. 68 (2019) 100226.

[92] Y.W. Bao, X.W. Hua, Y.H. Li, H.R. Jia, F.G. Wu, Hyperthemia-promoted cytosolic and nuclear delivery of copper/carbon quantum dot-crosslinked nanosheets: multimodal imaging-guided photothermal cancer therapy, ACS Appl. Mater. Interfaces 10 (2018) 1544–1555.

[93] B.A. Kairdolf, X. Qian, S. Nie, Bioconjugated nanoparticles for biosensing, in-vivo imaging, and medical diagnostics, Anal. Chem. 89 (2017) 1015–1031.

[94] A.K. Geim, K.S. Novoselov, The rise of graphene, in: Nanoscience and Technology: A Collection of Reviews from Nature Journals, 2010, pp. 11–19.

[95] A.I. Ekimov, A.A. Onushchenko, Quantum-size effects in optical spectra of semiconductor microcrystals, Sov. Phys. Semicond. 16 (1982) 775–778.

[96] R.S. Pawar, P.J. Upadhaya, V.B. Patravale, Chapter 34—quantum dots: novel realm in biomedical and pharmaceutical industry, in: Micro and Nano Technologies, Elsevier, 2018, , pp. 621–637.

[97] L.E. Brus, Electron-electron and electron-hole interactions in small semiconductor crystallites; the size dependence of the lowest excited electronic state, J. Chem. Phys. 80 (1984) 4403–4409.

[98] B.O. Dabbousi, J.R. Viejo, F.V. Mikulec, J.R. Heine, H. Mattoussi, R. Ober, K.F. Jensen, M. G. Bawendi, (CdSe)ZnS core-shell quantum dots: synthesis and characterization of a size series of highly luminescent nanocrystallites, J. Phys. Chem. B 101 (1997) 9463–9475.

[99] Z. Pan, K. Zhao, J. Wang, H. Zhang, Y. Feng, X. Zhong, Near infrared absorption of $CdSe_xTe_{1-x}$ alloyed quantum dot sensitized solar cells with more than 6% efficiency and high stability, ACS Nano 7 (2013) 5215–5222.

[100] P. Samokhvalov, M. Artemyev, I. Nabiev, Basic principles and current trends in colloidal synthesis of highly luminescent semiconductor nanocrystals, Chem. Eur. J. 19 (2013) 1534–1546.

[101] E.S. Speranskaya, I.Y. Goryacheva, Fluorescent quantum dots: synthesis, modification, and application in immunoassays, Nanotechnol. Russ. 8 (2013) 685–699.

[102] V.A. Oleinikov, A.V. Sukhanova, I.R. Nabiev, Fluorescent Semiconductor Nanocrystals in Biology and Medicine, 2 Ross. Nanotekhnol., 2007, pp. 160–173

[103] A.J. Sutherland, Quantum dots as luminescent probes in biological systems, Curr. Opinion Solid State Mater. Sci. 6 (2002) 365–370.

[104] E. Casals, T. Pfaller, A. Duschl, G.J. Oostingh, V.F. Puntes, Hardening of the nanoparticle–protein corona in metal (Au, Ag) and oxide (Fe_3O_4, CoO, and CeO_2) nanoparticles, Small 7 (2011) 3479–3486.

[105] H. Li, C. Wang, Z. Peng, X. Fu, A review on the synthesis methods of CdSeS-based nanostructures, J. Nanomater. 2015 (2015) 1–16.

[106] E. Oh, R. Liu, A. Nel, K.B. Gemill, M. Bilal, Y. Cohen, I.L. Medintz, Meta-analysis of cellular toxicity for cadmium containing quantum dots, Nat. Nanotechnol. 11 (2016) 49–486.

[107] C. Kirchner, T. Liedl, S. Kudera, T. Pellegrino, A. Muñoz Javier, H.E. Gaub, W.J. Parak, Cytotoxicity of colloidal CdSe and CdSe/ZnS nanoparticles, Nano Lett. 5 (2005) 331–338.

[108] M. Rana, P. Chowdhury, L-glutathione capped CdSeS/ZnS quantum dot sensor for the detection of environmentally hazardous metal ions, JOL 206 (2019) 105–112.

[109] H.R. Rajabi, M. Fars, Study of capping agent effect on the structural, optical and photocatalytic properties of zinc sulfide quantum dots, Mater. Sci. Semicond. Process. 48 (2016) 14–22.

[110] O.Z. Saglam, Y. Dilgin, Fabrication of photoelectrochemical glucose biosensor in flow injection analysis system using ZnS/CdS-carbon nanotube nanocomposite electrode, Electroanaylsis 29 (2017) 1368–1376.

[111] J.S. Steckel, J. Ho, C. Hamilton, J. Xi, C. Breen, W. Liu, P. Allen, S.C. Sullivan, Quantum dots: the ultimate down-conversion material for LCD displays, J. Soc. Inf. Disp. 23 (2015) 294–305.

[112] T.D. Costa, A.L. Nunes, Far Touch: Integrating Visual and Haptic Perceptual Processing on Wearables, ProQuest Dissertations Publishing, 2015.

[113] L. Chen, S. Xu, J. Li, Recent advances in molecular imprinting technology: current status, challenges and highlighted applications, Chem. Soc. Rev. 40 (2011) 2922–2942.

[114] B.D. Chithrani, A.A. Ghazani, W.C.W. Chan, Determining the size and shape dependence of gold nanoparticle uptake into mammalian cells, Nano Lett. 6 (2006) 662–668.

[115] B. Djezzar (Ed.), Ionizing Radiation Effects and Applications, BoD–Books on Demand, 2018.

[116] C. Mah, I. Zolotukhin, T.J. Fraites, J. Dobson, C. Batich, B.J. Byrne, Microsphere-mediated delivery of recombinant AAV vectors *in-vitro* and *in-vivo*, Mol. Ther. 1 (2000) S239.

[117] D. Panatarotto, C.D. Prtidos, J. Hoebeke, F. Brown, E. Kramer, J.P. Briand, S. Muller, M. Prato, A. Bianc, Immunization with peptide-functionalized carbon nanotubes enhances virus-specific neutralizing antibody responses, Chem. Biol. 10 (2003) 961–966.

[118] M. Bruchez, M. Moronne, P. Gin, S. Weiss, A.P. Alivisatos, Semiconductor nanocrystals as fluorescent biological labels, Science 281 (2) (1998) 2013–2016.

[119] M.J. Nam, C.C. Thaxton, C.A. Mirkin, Nanoparticles-based bio-bar codes for the ultrasensitive detection of proteins, Science 301 (2003) 1884–1886.

[120] R. Mahtab, J.P. Rogers, C.J. Murphy, Protein-sized quantum dot luminescence can distinguish between "straight", "bent", and "kinked" oligonucleotides, J. Am. Chem. Soc. 117 (1995) 9099–9100.

[121] J. Ma, H. Wong, B.L. Kong, K.W. Peng, Biomimetic processing of nanocrystallite bioactive apatite coating on titanium, Nanotechnology 14 (2003) 619–623.

[122] A. Isla, W. Brostow, B. Bujard, M. Estevez, J.R. Rodriguez, S. Vargas, V.M. Castano, Nanohybrid scratch resistant coating for teeth and bone viscoelasticity manifested in tribology, Mater. Res. Innov. 7 (2003) 110–114.

[123] R.S. Molday, D. MacKenzie, Immunospecific ferromagnetic iron dextran reagents for the labeling and magnetic separation of cells, J. Immunol. Methods 52 (1982) 353–367.

[124] J. Yoshida, T. Kobayashi, Intracellular hyperthermia for cancer using magnetite cationic liposomes, J Magn Magn Mater. 194 (1999) 176–184.

[125] X. Cai, Y. Luo, W. Zhang, pH-Sensitive ZnO quantum dots-doxorubicin nanoparticles for lung cancer targeted drug delivery, ACS Appl. Mater. Interfaces 8 (2016) 22442–22450.

[126] Pooja, M. Rana, P. Chowdhury, Influence of size and shape on optical and electronic properties of CdTe quantum dots in aqueous environment, AIP Conf. Proc. 2136 (2019) 040006.

[127] M.S. Liu, Q.H. Guo, H.Z. Zhang, R. Li, W. Chen, G.Z. Wang, Z. G, Characterization of CdSe and CdSe/CdS core/shell nanoclusters synthesized in aqueous solution, Physica E 8 (2000) 174–178.

[128] M. Kuno, K.J. Lee, O.B. Dabbousi, V.F. Mikulec, G.M. Bawendi, The band edge luminescence of surface modified CdSe nanocrystallites: probing the luminescing state, J. Chem. Phys. 106 (1997) 9869–9882.

[129] C. Yuan, K. Zhang, Z. Zhang, S. Wang, Highly selective and sensitive detection of mercuric ion based on a visual fluorescence method, Anal. Chem. 84 (2012) 9792–9801.

[130] Y.H. Zhang, H.S. Zhang, X.F. Guo, H. Wang, L-Cysteine-coated CdSe/CdS core-shell quantum dots as selective fluorescence probe for copper (II) determination, J. Microchem. 89 (2008) 142–147.

[131] K.K. Jain, Nanotechnology in clinical laboratory diagnostics, Clin. Chim. Acta 358 (2005) 37–54.

[132] Y. Shan, L. Wang, Y. Shi, H. Zhang, H. Li, H. Liu, B. Yang, T. Li, X. Fang, W. Li, NHS-mediated QDs-peptide/protein conjugation and its application for cell labeling, Talanta 75 (2008) 1008–1014.

[133] Z. Li, P. Huang, R. He, J. Lin, S. Yang, X. Zhang, Q. Ren, D. Cui, Aptamer-conjugated dendrimer-modified quantum dots for cancer cell targeting and imaging, Mater. Lett. 64 (2010) 375–378.

[134] M. Adeli, M. Kalantari, M. Parsamanesh, E. Sadeghi, M. Mahmoudi, Synthesis of new hybrid nanomaterials: promising systems for cancer therapy, Nanomed.: Nanotechnol., Biol. Med. 7 (2011) 806–817.

[135] P. Reiss, J. Bleuse, A. Pron, Highly luminescent CdSe/ZnSe core/shell nanocrystals of low size dispersion, Nano Lett. 2 (2002) 781–784.

[136] L. Ding, P.J. Zhou, H.J. Zhan, C. Chen, W. Hu, T.F. Zhou, C.W. Lin, Microwave-assisted synthesis of L-glutathione capped ZnSe QDs and its interaction with BSA by spectroscopy, JOL 142 (2013) 167–172.

[137] M. Pal, N.R. Mathews, P. Santiago, X. Mathew, A facile one-pot synthesis of highly luminescent CdS nanoparticles using thioglycerol as capping agent, J. Nanopart. Res. 14 (2012) 916.

[138] L. Spanhel, M. Haase, H. Weller, A. Henglein, Photochemistry of colloidal semiconductors surface modification and stability of strong luminescing CdS particles, J. Am. Chem. Soc. 109 (1987) 5649–5655.

[139] A. Khan, CdS nanoparticles with a thermo responsive polymer: synthesis and properties, J. Nanomater. 2012 (2012) 66.

[140] N. Murase, M. Gao, Preparation and photoluminescence of water-dispersible ZnSe nanocrystals, Mater. Lett. 58 (2004) 3898–3902.

[141] M. Gao, S. Kirstein, H. Möhwald, A.L. Rogach, A. Kornowski, A. Eychmüller, H. Weller, Strongly photo luminescent CdTe nanocrystals by proper surface modification, J. Phys. Chem. B 102 (1998) 8360–8363.

[142] C. Landes, C. Burda, M. Braun, M.A. El-Sayed, Photoluminescence of CdSe nanoparticles in the presence of a hole acceptor: n–butylamine, J. Phys. Chem. B 105 (2001) 2981–2986.

[143] H.A. Benesi, J.H. Hildebrand, A spectrophotometric investigation of the interaction of iodine with aromatic hydrocarbons, J. Am. Chem. Soc. 71 (1949) 2703–2707.

[144] Y. Ding, S.Z. Shen, H. Sun, K. Sun, F. Liu, Synthesis of L-glutathione-capped-ZnSe quantum dots for the sensitive and selective determination of copper ion in aqueous solutions, Sens. Actuators B 203 (2014) 35–43.

[145] E.R. Goldman, E.D. Balighian, H. Mattoussi, M.K. Kuno, J.M. Mauro, P.T. Tran, G.P. Anderson, Avidin: a natural bridge for quantum dot-antibody conjugates, J. Am. Chem. Soc. 124 (2002) 6378–6382.

[146] Y. Chen, Z. Rosenzweig, Luminescent CdS quantum dots as selective ion probes, Anal. Chem. 74 (2002) 5132–5138.

[147] P. Vadgama, W.P. Crump, Biosensors: recent trends. A review, Analyst 117 (1992) 1657–1670.

[148] M. Beija, A.C. Afonso, M.J. Martinho, Synthesis and applications of Rhodamine derivatives as fluorescent probes, Chem. Soc. Rev. 38 (2009) 2410–2433.

[149] M. Aldén, E.P. Bengtsson, H. Edner, Rotational CARS generation through a multiple four-color interaction, Appl. Optics 25 (1986) 4493–4500.

[150] T. Misteli, L.D. Spector, Applications of the green fluorescent protein in cell biology and biotechnology, Nat. Biotechnol. 15 (1997) 961–964.

[151] A. Reisch, S.A. Klymchenko, Fluorescent polymer nanoparticles based on dyes: seeking brighter tools for bioimaging, Small 12 (2016) 1968–1992.

[152] K.J. Guo, C.E. Cheng, L. Wang, S.E. Swenson, A.T. Ardito, M. Kashgarian, S.D. Krause, The commonly used β-actin-GFP transgenic mouse strain develops a distinct type of glomerulosclerosis, Transgenic Res. 16 (2007) 829–834.

[153] T.T. Yang, P. Sinai, G. Green, A.P. Kitts, T.Y. Chen, L. Lybarger, R.S. Kain, Improved fluorescence and dual color detection with enhanced blue and green variants of the green fluorescent protein, J. Biol. Chem. 273 (1998) 8212–8216.

[154] A.F. Fradkov, Y. Chen, L. Ding, E.V. Barsova, M.V. Matz, S. Lukyanov, Novel fluorescent protein from Discosoma coral and its mutants possesses a unique far-red fluorescence, FEBS Lett. 479 (2000) 127–130.

[155] D.D. Deheyn, K. Kubokawa, J.K.M. Carthy, A. Murakami, M. Porrachia, G.W. Rouse, Endogenous green fluorescent protein (GFP) in amphioxus, Biol. Bull. 213 (2007) 95–100.

[156] R. Stripecke, M.C. Villacres, C.D. Skelton, N. Satake, S. Halene, B.D. Kohn, Immune response to green fluorescent protein: implications for gene therapy, Gene Ther. 6 (1999) 1305–1312.

[157] M. Gsänger, D. Bialas, L. Huang, M. Stolte, F. Würthner, Organic semiconductors based on dyes and color pigments, Adv. Mater. 28 (2016) 3615–3645.

[158] S.J. Rosenthal, J.C. Chang, O. Kovtun, J.R. Mcbride, L.D. Tomlinson, Biocompatible quantum dots for biological applications, Chem. Biol. 18 (2011) 10–24.

CHAPTER 22

Potent aptamer-based nanosensors for early detection of lung cancer

Neelam Verma[a,b], Ashish Kumar Singh[b], Rajni Sharma[b,c], and Mohsen Asadnia[c]
[a]Division of Research and Development, Lovely Professional University, Phagwara, India
[b]Biosensor Technology Laboratory, Department of Biotechnology, Punjabi University, Patiala, India
[c]School of Engineering, Macquarie University, Sydney, NSW, Australia

1. Introduction

Cancer is the general term used for a large family of disease that involves the rapid increase of cells with unregulated growth and ability to spread to other parts of body. It is one of the leading global health problems and has affected around 2.09 million people just in 2018, and moreover, the number of cases is estimated to increase with population expansion [1]. Among 200 different types of cancers, lung cancer is the most lethal type accounting for one-fourth of mortality of all cancer-related deaths. It has higher possibilities of metastasis as the lungs are engaged in transportation of oxygen to all body cells [2]. Smoking remains the predominant risk factor for lung cancer (80%–90%), whereas exposure to radon gas (9%–15%) is the leading cause of lung cancer in nonsmokers and the second leading cause in chain smokers after smoking. Besides this, genetic/epigenetic variations and contact to environmental pollutants (1%–2%) are major risk factors for lung cancer [3].

Lung cancer is incurable due to tumor cells' dissemination into other body parts like the central nervous system in an advanced metastasis stage. Most of the time, cancer metastasizes all over the body before being detected, which lessens the survival rate of patients. To cure the deadly disease, an accurate diagnosis at an early stage is as significant as the development of new treatment therapies and anticancer drugs [4]. Detection at the initial stage can improve the 5-year survival rate from 20% to >60% [5]. However, the early detection of lung cancer is very complicated due to the appearance of nonspecific symptoms in the pr-metastatic stage such as mild cough, breathing shortness, and weight loss. The conventional diagnostic methods lack the high sensitivity to detect premature tumor cells. Though the technical advancement in last few decades has brought about easy diagnostic techniques, there is no improvement in lung cancer survival rates. Therefore, a highly sensitive, rapid, and miniaturized detection tool is critically required for the screening and early diagnosis of cancer for a large population [6]. In the current scenario, the aptamers came into the limelight for cancer imaging and detection in biological fluid

Handbook of Nanomaterials for Sensing Applications
https://doi.org/10.1016/B978-0-12-820783-3.00004-X
505

due to their distinctive properties to selectively bind to almost any kind of biomarker or metabolite of cancer cells in addition to intact tumor cells even at a very minute concentration [7]. Aptamers are single-stranded oligonucleotides, ssDNA or RNA, which are proficient to identify the specific targets due to their exclusive three-dimensional conformations.

In this chapter, we briefly described the variants of lung cancer, challenges of conventional techniques, and emergence of a new era of ultrasensitive detection methods with aptamer-based nanosensors. Furthermore, the aptamer's synthesis, lung cancer biomarkers, as well as the recent progress in the fabrication of aptamer-based nanosensors for the screening and early detection of lung cancer have been discussed. The aptamer-based nanosensors offer high potential for large-scale application of aptamer proteomic technology for future cancer diagnosis.

2. Histological variants of lung cancer

Histologically, lung cancer acquires different variants depending upon the site of their origin such as alveoli, bronchioles, bronchial epithelium, or bronchial mucous glands [7]. Lung cancer is primarily classified and studied under two major categories: Small Cell Lung Carcinoma (SCLC) and Non-small Cell Lung Carcinoma (NSCLC) (Fig. 1) of which SCLC represents 13% of cases of lung cancer, which usually presents lower survival rate due to a meager response to chemical as well as radiation therapy. It is characterized by rapid tumor growth, lack of differentiation, and excretion of specific hormones and biomarkers. On the other hand, the diagnosis of NSCLC in the early stage could help to increase the survival rate from 10% to >70%. NSCLC accounts for 85% cases of lung cancer, which is histologically classified into subtypes: adenocarcinoma, squamous cell

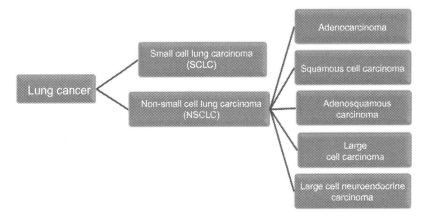

Fig. 1 Histological classification of lung cancer.

carcinoma, adenosquamous carcinoma, large cell neuroendocrine carcinoma, and large cell carcinoma [5, 8].

The most common type of NSCLC is adenocarcinoma, which encompasses around 40% of all lung cancer. It arises from glandular cells of bronchial mucosa. As compared to other types of lung cancer, adenocarcinoma tends to grow slower and has a higher possibility of early diagnosis before becoming malignant [9, 10]. Squamous cell carcinoma encompasses 25%–30% cases of lung cancer and appears from primitive squamous cells in the areolar epithelial cells present in the bronchial tubes at the center position of the lungs. This subtype of NSCLC is strongly correlated with cigarette smoking persons [11]. Adenosquamous carcinoma is a variant with two varieties of cells: slender and plane squamous cells lining some organs and gland-like cells. Large cell neuroendocrine carcinoma originates from neuroendocrine cells of the respiratory path layer or smooth muscle cells of its wall and encompasses large undifferentiated polygonal cells [12]. Large cell carcinoma comprising 5%–10% of lung cancers are undifferentiated cells that do not show squamous or glandular maturation and diagnoses by default excluding other possibilities. It usually arises in the inner part of the lungs, occasionally in either chest wall and proximal lymph nodes or distant organs [13]. The information of histological classification of lung cancer is significant with regard to the study of involved risk factors, response to treatment, invasiveness, and severity of pain [14].

Depending upon the stage and subtype, the lung cancerous cells show differences in molecular level and morphologically, which are apparently distinct from healthy lung cells. The difference in cell morphology, differentiation, and proliferation allocates the overexpression of particular proteins that assists to record a signal and information from the cells about the progression of disease [15].

3. Challenges to conventional techniques: Need of aptasensors

The conventional cancer diagnostic techniques involve computerized tomography (CT), low-dose helical/spiral CT, computerized axial tomography (CAT), magnetic resonance imaging (MRI), positron emission tomography (PET), bone scans, optical coherence, sputum cytology, chest radiography, and virtual bronchoscopy [16–20]. CT and chest radiography are the most popularly used methods for lung cancer diagnosis.

These biopsy methods inflict a considerable challenge for early diagnosis of cancer as these methods lack the sensitivity to detect premature stage of lung cancer cells [5]. In addition, other screening strategies like biomarker-based techniques including radio-immune assay (RIA) and enzyme-linked immune-sorbent assay (ELISA) have been established as competent to diagnose lung cancer at an early stage; nevertheless, no improvement in mortality rate was observed [19, 21].

Henceforth, ultrasensitive and economical diagnostic tools are required for routine check-up and assessment of patients even at a premature stage to prevent the proliferation of lung cancer as compared to laborious, expensive, risk-prone conventional methods requiring expertise personnel and sophisticated instruments.

The tumor-specific biomarkers are considerable targets for the diagnosis and treatment of cancer. Over the past decade, aptamers have become distinctively ideal tools for the specific recognition of biomarkers due to their extreme specificity toward proteins overexpressed in tumor cells [22]. Consequently, the adaptation of aptamer-based technologies as diagnostic tools in clinics could provide a great opportunity for advances in cancer detection.

In 1996, the idea of aptasensors as advanced analytical tools came from the amalgam of two basic techniques, biosensor and aptamers, when Drolet and his team [23] used the aptamer as a biorecognition unit as compared to antibodies in modified ELISA termed as "Enzyme-linked Oligonucleotide Assay" (ELONA). Biosensor is an analytical tool consisting of an immobilized biorecognition element such as enzyme, protein, antibody, or aptamer in close proximity to a physicochemical transducer to provide either qualitative or quantitative signal in readable format [24–27]. Fig. 2 shows the generic schematic diagram for biosensor assembly. Nowadays, various biosensors have been developed for the specific detection of lung cancer cells [2, 28].

4. Aptamers: A biorecognition element

The term "aptamer" was drawn from the Latin *aptus* meaning "to fit" and Greek *meros* meaning "region/part" [29]. These molecules change their conformations and fold into unique three-dimensional conformations (Fig. 3) such as pseudoknots, hairpin-loops, and G-quadruplexes on recognition of a specific target to fit appropriately [30, 31]. Aptamers are single-stranded DNA or RNA molecules containing 10 to 100 nucleotides prepared from combinatorial oligonucleotide libraries by a in-vitro selection process known as systemic evolution of ligands by exponential enrichment (SELEX) and proficient to perform as a catalytic unit as well as a receptor. The synthetic oligonucleotides can recognize the specific target molecules with strong affinity analogous to antigen–antibody interactions and hence illustrious as "artificial antibodies". In addition to features of antibodies, these molecules possess some unique characteristics supporting their huge contribution in clinical diagnosis and cancer research [32].

Similar to antibodies, several aptamers are commercially available in the markets that are selective against a wide range of targets and can be customized by linking to a variety of diagnostic agents via physical or chemical bonding to further enhance the sensitivity and detection range in diagnosis or biosensor assays [27, 33–35]. Moreover, the synthesis of aptamers is easy and inexpensive at a large scale as compared to antibodies' production in living cells. In addition, aptamers have small size, low molecular weight, easy to label, longer shelf-life at ambient temperature, amendable sequences, and programmable structure. Furthermore, the aptamers show long-lasting storage and thermal stability, easy

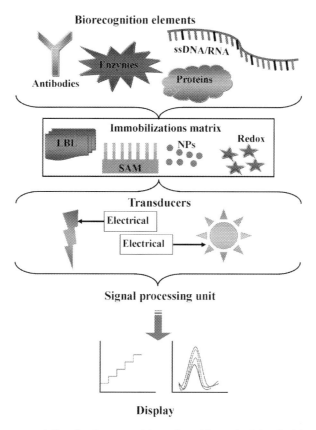

Fig. 2 Schematic representation for the assembly and working principle of a biosensor.

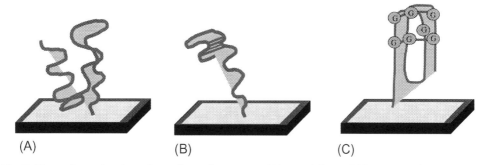

Fig. 3 Three-dimensional conformations of aptamers: (A) pseudoknot, (B) hairpin, and (C) G-quadruplex.

transportation, and reversible denaturation as compared to proteinaceous antibodies. Some of the world's major companies manufacturing customized aptamers for clinical diagnostics and research purpose are listed in Table 1.

Currently, several aptamers have been synthesized to be selective against small organics, peptides, proteins, and intact whole cells. The synthesized aptamers have been used as appropriate replacement of antibodies in various analytical techniques such as affinity chromatography, capillary electochromatography, ELISA, flow cytometry, and biosensor assays [23, 36–40]. Nowadays, aptamers came into view as influential recognition probes for the specific detection of tumor cells. Depending upon the stages, the cancer cells show variation at molecular and morphological levels, which seem apparently different from healthy normal cells. These differences in cell morphology, differentiation, and proliferation assist to record a signal and information from the cells about the progression of disease. A number of aptamers have been selected against tumor-specific biomarkers for cancer cell lines including BNL CL.2,CCRF-CEM, NCI-H69, Ramos, RERF-LC-MA, and A549 [41]. Aptamers that bind to specific biomarkers, for example, overexpressed mucin-1 glycoprotein on lung cancer cells are potential candidates for the fabrication of aptasensors for cancer detection at its early stage.

4.1 Synthesis of aptamers

At present, a wide range of aptamers have been selected against surface proteins present on tumor cells using in–vitro selection process, e.g., systematic evaluation of ligands by exponential enrichment (SELEX). Fig. 4 represents the sequential process of synthesis of aptamer via SELEX against the cancer cells. This technique was first employed in 1990 by two research groups independently [29, 42].

The SELEX process commences with the incubation of ligands that may be specific protein(s) or target cell(s) with combinatorial oligonucleotide library of nucleic acids (DNA/RNA) consisting of $\sim 10^{12}$ to 10^{14} unique sequences with 10–100 nucleotides. In the next step, specific aptamers (oligonucleotides) bound to ligands are eluted following washing and then amplified to prepare a library for next round of selection. The steps are repeated so on, and eventually the desired target-specific aptamers are obtained subsequent to multiple selection rounds. An enriched pool of high affinity aptamers are generated using PCR amplification. Furthermore, a negative selection step can be carried out for the elimination of nonspecific target binding and for the improvement of specificity of the aptamer selection process. The sequences of a highly enriched panel of aptamers can be assessed using next generation sequencing (NGS) for characterization or chemical synthesis of individual aptamers. The advanced sequencing techniques and bioinformatics tools assist in the earlier selection of an appropriate panel of aptamers alleviating the multiple steps of selection [43].

Table 1 World's leading companies manufacturing aptamers for clinical applications.

S. no.	Firm's name	Firm's address	Web address
1.	IBA Lifesciences	IBA GmbH, Rudolf-Wissell-Str. 28, 37,079 Göttingen Germany	https://www.iba-lifesciences.com
2.	Aptagen, LLC	Aptagen, LLC, 250 North Main Street, Jacobus, PA 17407 USA	https://www.aptagen.com/
3.	Bio-Synthesis Inc	Bio-Synthesis Inc., 612 East Main Street Lewisville, TX 75057 USA	https://www.biosyn.com/index.aspx
4.	Cambio	Cambio Ltd., 1 The Irwin Centre, Scotland Road, Dry Drayton Cambridge, CB23 8AR, U.K	https://www.cambio.co.uk/
5.	AM Biotech	Aptamer Synthesis Lab, 12,521 Gulf Fwy Houston, Texas 77,034, USA	https://www.am-biotech.com/
6.	Oak BioSciences	754 N Pastoria Ave, Sunnyvale, CA 94085, USA	https://www.oakbiosciences.com
7.	Taros Chemicals	Taros Chemicals GmbH & Co. KG Emil-Figge-Str. 76a 44,227 Dortmund, Germany	https://www.tarosdiscovery.com/en/
8.	VeZerf lab.	VeZerf Laboratory Syntheses GmbH, Langenfelder Str. 12 D-55743 Idar-Oberstei, Germany	https://www.vezerf.de/
9.	AMS Biotechnology (Europe) Ltd	AMS Biotechnology (Europe) Limited, 184 Park Drive, Milton Park, Abingdon OX14 4SE, UK	http://www.amsbio.com/home.aspx
10.	TAGCyx Biotechnologies Inc.	TAGCyx Biotechnologies Inc., Tokyo, Japan	www.tagcyx.com
11.	IDT Inc.	Integrated DNA Technologies, Inc. 1710 Commercial Park Coralville, Iowa 5224, USA	https://eu.idtdna.com/
12.	2bind	Am Biopark 11, 93,053 Regensburg, Germany	https://2bind.com/
13.	Centauri Therapeutics	Centauri Therapeutics, Discovery Park, Kent. U.K	http://www.centauritherapeutics.com/index.php
14.	AptaMatrix,	AptaMatrix, Inc. Suite 2400A, 841 East Fayette St. Syracuse, NY 13210	http://www.aptamatrix.com/
15.	Aptamer Group	Suite 2.78–2.91, Bio Centre, Innovation Way Heslington, York YO10 5NY	https://www.aptamergroup.co.uk/
16.	Aptamer Sciences Inc.	407, Dolma-ro, Bundang-gu, Seongnam, Gyeonggi-do, Korea	http://aptsci.com/

Continued

Table 1 World's leading companies manufacturing aptamers for clinical applications—cont'd

S. no.	Firm's name	Firm's address	Web address
17.	AptaTargets	Avda. Cardenal Herrera Oria, 29,828,035 Madrid – Spain	http://www.aptatargets.com/
18.	Apta Biosciences Ltd	c/o Anglo Scientific, The Royal Institution of Great Britain 21 Albemarle Street, London U.K.	https://www.aptabiosciences.com/
19.	Apterna		http://apterna.com/
20.	Aptitude Medical Systems Inc	Aptitude Medical Systems Inc., 2219 Bath St Santa Barbara, CA 93105	http://www.aptitudemedical.com/index.html
21.	Aptus Biotech	Avda. Cardenal Herrera Oria, 298, 28,035 Madrid, Spain	http://en.aptusbiotech.com/
22.	Novaptech		
23.	ATDBio Ltd.	ATDBio Ltd., Southampton, UK	https://www.atdbio.com
24.	AuramerBio	AuramerBio, New Zealand	www.auramerbio.com
25.	Base Pair Biotechnologies, Inc.	Base Pair Biotechnologies, Inc., USA	https://www.basepairbio.com/
26.	BBI Group	BBI Group, Cardiff, UK	https://www.the-bbigroup.com/
27.	DSM Biotechnology	The Netherlands	https://www.dsm.com
28.	Firefly Bioworks	Firefly Bioworks, MA, USA	https://www.fireflybio.com/
29.	Izon Science	Izon Science Oxford, UK	https://izon.com/
30.	LC Sciences	LC Sciences, TX, USA	https://www.lcsciences.com/
31.	LFB Biotechnologies	LFB Biotechnologies, France	https://www.groupe-lfb.com/en/
32.	Nal von Minden	Nal von Minden Germany	https://www.nal-vonminden.com/en/
33.	Neoventures	Ontario, Canada	https://neoventures.ca/
34.	Noxxon Pharma	Noxxon Pharma, AG Germany	https://www.noxxon.com/
35.	Ophthotech	Ophthotech NY, USA	https://ivericbio.com/
36.	OTC Biotech	OTC Biotech TX, USA	http://www.otcorp.com/

Table 1 World's leading companies manufacturing aptamers for clinical applications—cont'd

S. no.	Firm's name	Firm's address	Web address
37.	Piculet Biosciences	Piculet Biosciences, Kernhemseweg 26718ZB Ede, Netherlands	https://www.cmocro.com/index.php
38.	Pure Biologics	Pure Biologics, Duńska 11, 54-427 Wrocław, Poland	http://purebiologics.pl/
39.	Ribomic	Shirokanedai Usui Bldg. 6F, 3-16-13 Shirokanedai, Minato-ku Tokyo 108-0071, Japan	https://www.ribomic.com/eng/
40.	SomaLogic	SomaLogic, 2945 Wilderness Pl. Boulder, CO 80301 USA	https://somalogic.com/
41.	Allergan	Clonshaugh Business and Technology Park, Coolock, Dublin, D17 E400, Ireland	https://www.allergan.com/
42.	Bio-Techne Ltd.	19 Barton Lane, Abingdon Science Park, Abingdon OX14 3NB United Kingdom	https://www.bio-techne.com
43.	TriLink Biotechnologies	10,770 Wateridge Circle, Suite 200 San Diego, 9955 Mesa Rim Road San Diego, CA 92121	https://www.trilinkbiotech.com/
44.	Veraptus	Veraptus, China	http://www.veraptus.com/

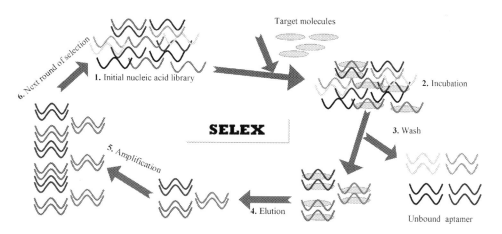

Fig. 4 SELEX (Systematic evolution of ligands by exponential enrichment) process for the generation of ligand specific single-stranded DNA or RNA aptamer. SELEX starts with the incubation of a nucleic acid library with the target protein. Unbound aptamers are washed off and protein-bound aptamers are eluted and amplified to create the library for the next round of selection. After iterative selection rounds, potential target-specific sequences are enriched and dominate the population of aptamers.

The conventional SELEX process is neither applicable for insoluble/unknown protein(s) nor multimeric protein complex. Therefore, several upgraded variants of SELEX are proposed to enhance the selection efficiency and analytical execution of aptamers. The SELEX variants include negative SELEX (subtractive SELEX, counter SELEX), capillary electrophoresis (CE)-SELEX, cell-SELEX, HTS-SELEX, microfluidic-SELEX, primer-free SELEX, genomic SELEX, and qPCR-SELEX [44]. Among these SELEX variant techniques, cell-SELEX has acquired more interest due to its competency to recognize cell surface molecules in native confirmation as well as for unknown membrane receptors. The plus points of cell-SELEX include (a) effectual recognition of cell surface receptors by target aptamers under natural growth conditions, (b) simultaneous screening of multiple aptamers against various cell surface receptors, and (c) identification of binding sites to use as specific biomarkers [45]. Therefore, cell-based SELEX offers a new window to recognize tumor-specific biomarkers. Over the past two decades, several significant biomarkers, for example, platelet–derived growth factor (PDGF), vascular endothelial cell growth factor (VEGF), nuclear factor kappa B (NFKB), human epidermal growth factor 3 (HER3), and tenascin-C have been reported for the detection of cancer at an early stage using cell-SELEX [46, 47].

The cell-SELEX method begins with a negative step, the incubation of the nucleic acid library with normal cells, which do not express the target protein. The aptamers bound to normal cells are eliminated for the next round of positive selection: incubation, elution, and amplification. The selection efficiency of aptamers can be increased by repeating the round until satisfying results are obtained. Furthermore, the changed cell conditions in every round of selection increased the number of noncompetent cells and washing steps support to improve the aptamer affinity toward target cells and consistency of slots [48, 49].

Prior to initiating the selection process, a decision has to be made regarding the type of oligonucleotide library: either ssDNA or RNA depending upon the final application of aptamers. DNA and RNA aptamers obviously execute similar functions, even though these molecules have their own pros and cons. DNA aptamers are comparatively easy and inexpensive to synthesize and exhibit more stability due to inheritance property. On the other hand, RNA aptamers represent higher binding affinity and specificity on the basis of more diverse three-dimensional conformations due to the presence of $2'$-hydoxyl group and stronger RNA-RNA interactions. However, RNA aptamers have to be first transcribed reversibly into cDNA to carry out further amplification with PCR and again transcribed in-vitro for the next round of selection. Therefore, RNA aptamers necessitate the involvement of more steps in the selection process, which make the process expensive and laborious [50, 51].

Furthermore, nonspecific uptake can be avoided by taking care of important facts during aptamer selection. First, the elimination of dead cells favor the identification of specific aptamers since cell death leads to altered protein expression. Second, it is important to choose the proper cell line because cell surface proteins are very complex in some cancers. Because of the complexity and heterogenicity of some cancer cells, it is critical to perform additional selection rounds against nontarget cells to improve aptamer specificity. However, this selection process can be quite complex and may have a negative effect in terms of time and resources spent by the researcher [52].

Nowadays, some novel selection techniques have emerged on the basis of whole cell-based SELEX to improve aptamer screening, such as automated SELEX [53], target expressed on cell surface-SELEX (TECS-SELEX) [54], in-vivo SELEX [55], fluorescence-activated cell sorting-SELEX (FACS-SELEX) [56], cell-internalization SELEX [57], 3D SELEX [58], and click-and SELEX [59].

4.2 Matrix for immobilization of aptamers

The sensitivity and specificity of aptamer-based sensors can be improved by using an immobilized aptamer on a solid surface as a biorecognition element rather than free aptamers in solution. Moreover, it facilitates the usual aptamer recovery for manifold applications. Immobilization of aptamers can be achieved by physical adsorption, covalent binding/cross-linking with functional groups [60] or coupled to self-assembled monolayer (SAM) and layer-by-layer (LBL) approach [61–64].

Among the various immobilization methods for conformation-oriented aptamers to develop biosensors for cancer, SAM is the most frequent adopted technique due to its great stability, simple handling, and higher loading capacity. The formation of SAM can be accomplished either by direct grouping onto the gold electrode surface or via conjugation with gold nanoparticles. SAM with a high degree of stability and orientation along with molecular order packing can be prepared directly on various surfaces using either sulfur-containing compounds such as thiol/dithiol [65] or using various silanes as shown in Fig. 5A.

The second most frequently used method for aptamer immobilization on solid surface is a layer-by-layer (LBL) approach (Fig. 5B). The technique provides more effective probes to enhance amplified signal to improve the detection sensitivity. Aptamers are immobilized on the outermost layer of an electrode through electrostatic interactions, which reduces cost and saves time for developing the aptasensor platform. Liu et al. [66] explained the hindrance of electron transfer on the cytosensor surface due to insulation of the cell membranes when the cancer cells were bonded with aptamer.

Different aptamer immobilization strategies for aptasensor applications have their own advantages and limitations (Table 2).

5. Biomarkers for lung cancer

A number of tumor-specific protein biomarkers are available for lung cancer, even though the sensitivity achieved so far is not 100%. According to Tessitore et al. [69], only

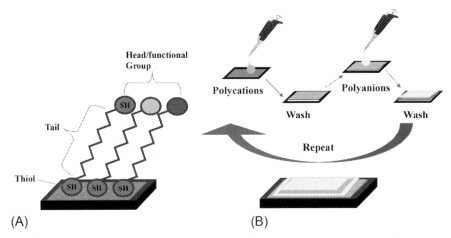

Fig. 5 (A) Schematic of SAM formation on gold and oxide surfaces, respectively; (B) Schematic of the LBL deposition process using drop and cast method.

Table 2 Different types of aptamer immobilization methods for aptasensor applications.

Immobilization methods	Bioconjugation process	Advantages	Limitations	References
LBL	Consecutive adsorption of anions and cations	Versatility, ease of preparation, low cost, and potential for scale-up	Multiple steps and nonspecific attachment	[67]
SAM	Thiols and alkyne disulfides on gold, Alcohols on glass	Stability, oriented identification	More suitability with silicon and gold surfaces	[39]
Covalent	EDC/NHS ester chemistry	Broad range of functional groups Flexibility	Multiple conjugation steps, Nonspecific attachment	[68]
Cross-linking	Gluteraldehyde	Simple and stable	Involvement of toxic chemicals	[38]
Physical adsorption	Direct connection on metal surfaces and surfaces covered with hydrophobic polymers	Simple and fast	Unsystematic orientation of aptamers, weak binding	[64]

LBL, Layer-by-layer; *SAM*, self-assembled monolayer; *EDC*, 1-ethyl-3-[3-(dimethylamino)propyl]-carbodiimide; *NHS*, N-hydroxysuccinimide.

79% sensitivity for first and second stage lung cancer was attained with the proteomic study of tumor-specific biomarkers. The failure to achieve 100% success might be due to (a) loss of trace amount of proteins in blood or tissues before they are noticed, (b) lesser reproducibility of laboratory experiments, (c) tumors' genetic heterogeneity, (d) inadequate research designing, and most significantly due to (e) inefficient techniques for biomarker discovery [22, 70]. Hence, there is the need of ultrasensitive techniques for the discovery of more reliable biomarkers that can differentiate protein isomers and post-translational modified proteins [71].

Mass spectrometry is most usually employed for the discovery of new biomarkers; however, aptamer-based biomarker discovery (AptaBID) is gaining interest nowadays. Various protein biomarkers have been detected using AptaBID including MUC1 (mucin-1), NCL (nucleolin), ANXA2 (annexin A2), ANXA5 (annexin A5), TUB (tubulin), H2B (H2B histone family member M), LMN (lamin), CTSD (cathepsin D) [72], ACT (actin), DEF(defensin), and VIM (vimentin) [7]. The biomarkers discovered by AptaBID are highly useful in the detection of lung cancer at an early stage.

Mucin 1 protein (MUC1), a glycoprotein, is associated with membranes of most of the epithelial cells. However, the unusual overexpression of MUC1 on the apical surface of epithelial cells is generally related to carcinomas, for example, in breast cancers (70%), pancreatic, prostate and epithelial ovarian cancers (90%), and in invasive lung cancers (96.7%) as well [72–74]. Henceforth, it has been widely used in the fabrication of apta-sensors for the early diagnosis of lung cancer. Furthermore, Ostroff and his team [75] discovered a panel of 12 protein biomarkers (CD30 ligand, cadherin-1, endostatin, HSP90a, LRIG3, MIP-4, PRKCI, pleiotrophin, RGM-C, sL-selectin, SCF-sR, and YES) for early detection of lung cancer by a case-control study on serum samples collected from 1326 subjects of four independent NSCLC studies. The outcomes present the large-scale application of aptamer proteomic technology.

The surface nucleolin (NCL) is another overexpressing protein in cancerous cells. The aptamers selective against nucleolin are highly useful for the detection of tumor cells. Li et al. [76] suggested the use of Cu_{64}-labeled aptamers AS1411 selective against nucleolin as a diagnostic imaging agent in positron-emitting tomography (PET) for early assessment of lung cancer. In addition, Vascular Endothelial Growth Factor (VEGF165) is another biomarker for detection of lung cancer, which is either secreted by cancerous cells or produced by normal cells in response to the cancerous cells in the body [3]. Jung et al. [5] developed aptamer-based protein biomarkers to detect NSCLC cells on the basis of a study conducted in 200 clinical samples collected from 100 normal persons and 100 lung cancer patients in Korea. The assessment of the method exhibited 91% specificity and 75% sensitivity with lung nodules. This technology was commercialized by Aptamer Sciences Inc., Korea. A number of aptamer-based nanosensors have been developed for detection of lung cancer at a premetastatic stage.

6. Aptamer-based nanosensors for lung cancer

The aptamer-based nanosensors known as "aptasensors" or "apta–nanosensors" are suitable alternates for antibodies or enzyme-based biosensors for cancer diagnosis. The conjugation of aptamers with novel nanomaterials, in-depth understanding of nucleic–acid aptamers in terms of their conformational, and ligand-binding properties has significantly produced interest in bioassay methods and led to the development of aptamer-based nanosensors.

The early stage detection of lung cancer is tricky but highly needed to improve the survival rate. An aptasensor allows the detection of intact tumor cells by specific biomarker recognition [6]. The diagnosis via whole cells instead of proteins or nucleic acids obtained by disrupting cells can extensively reduce the sample preparation steps, diagnosis cost, as well as chances of false results that would smooth the progress of practical clinical application. Based on the strong aptamer selectivity and sensitivity against tumor-associated biomarkers, a number of aptasensors have been developed. Some electrochemical and optical aptasensors for lung cancer are described in the following sections.

6.1 Electrochemical aptasensors

Electrochemical biosensors are highly recommended to use for cancer detection due to its sensitivity, rapid response, miniaturization, and reproducibility [25, 28, 63, 77]. In electrochemical-based biosensors, the aptamer immobilization on the electrode surface assures the selective capturing and detection of target tumor cells. Fig. 6 presents the working principle of an electrochemical-based aptasensor for lung cancer detection.

An electrochemical aptasensor for the detection of lung adenocarcinoma was described by Sharma and her team [78] on the basis of a MUC-1 biomarker in which a DNA aptamer was immobilized onto silane SAM on indium–tin–oxide (ITO) surface using cross-linker glutaraldehyde, and the response was evaluated with cyclic voltammetry (CV) and differential pulse voltammetry (DPV). The aptasensor was competent to detect A549 lung carcinoma (NSCLC) cells in the range of 10^3 to 10^7 cells/mL with a detection limit of 10^3 cells/mL within 1 min. Later, a label-free, highly sensitive electrochemical aptasensor was fabricated for the detection of VEGF165 in serum of lung cancer patients [3]. During sample analysis, VEGF165 tumor markers were supposed to interact with the aptamers (AntiVEGF165) immobilized on the surface of screen-printed electrode (SPE) with ordered mesoporous carbon–gold nanocomposite (OMC-Aunano). The positive interaction enforced the signal due to interfacial change in impedance, which was recorded by electrochemical impedance spectroscopy (EIS). EIS offers the examination of change in a biological system between the electrode by evaluating the interfacial capacitance and resistance. The aptasensor exhibited exceptionally high sensitivity with detection limit of 1.0 pg/mL and detection range of 10 to 300 pg/mL in spite of being highly selective and reproducible.

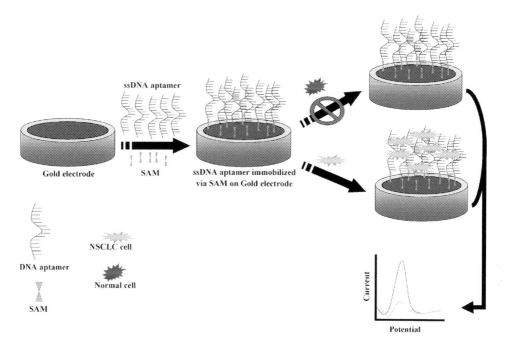

Fig. 6 Schematic representation of the working principle of an aptamer-based electrochemical biosensor.

Another amperometric nanosensor was fabricated for the early detection of lung cancer cells (A549) using MUC1 aptamer with the sequence 5′-GCAGTTGAT-CCTTTGGATACCCTGG-3′ [68]. The recognition probe is fabricated by covalent immobilization of a specific aptamer on a conducting polymer nanocomposite formed through the self-assembly of 4-([2,2′:5′,2″-terthiophen]-3′-yl) benzoic acid (TTBA) on gold nanoparticles. Furthermore, a bioconjugate consisting of hydrazine and an aptamer attached on gold nanoparticles was used to selectively amplify the signal. The signals were measured using chronoamperometry and validated with microscopic and DPV methods based on silver staining experiments. The linear relationship was observed with the cell concentration range of 15 to 10^6 cells/mL, and as few as eight cells could be detected using the developed nanosensor. The aptasensor was highly selective for A549 lung cancer cells as compared to other cells like liver tumors (HepG2) cells, human prostate cancer cells (PC3), as well as for MUC1-negative normal lung cells (MRC-5).

In 2016, Kara et al. [79] fabricated a highly sensitive and selective electrochemical aptasensor for the early detection of lung cancer cells with a detection limit of 164 cells/mL. 5′ amino linked aptamer was immobilized onto the SPEs, and response was measured using EIS. The aptasensor was highly selective for adenocarcinoma (A549) cells as compared to other tumor cells such as human liver hepatocellular carcinoma (HepG2) and human cervical cancer (HeLA) cells.

Additionally, Zamay and his team [80] described an electrochemical aptasensor based on protein biomarker offering the noninvasive and cost-effective approach for routine diagnosis of lung cancer. DNA–aptamer LC-18 selective against lung cancer tissue and circulating cancerous cells was immobilized on gold microelectrode, and analysis of crude blood plasma was performed using square wave voltammetry (SWV), the most superior voltammetric technique. The sensitivity of the aptasensor was boosted by inserting hydrophobic silica-coated iron oxide magnetic beads. The detection limit of the developed aptasensor was as low as 0.023 ng/mL.

Recently, Nguyen et al. [39] reported simple and inexpensive DNA aptasensors on microfluidic chip for detecting human lung cancer cells using EIS. The capacitance change develops on electrode immobilized target cells/proteins by binding to specific receptors. During analysis, the formation of SAM due to the binding of target A549 cells with amine terminated aptamers on gold microelectrode resulting in capacitance change. The responses were assessed with EIS, and a linear relationship between change in capacitance and concentration of target cells was observed in the range of 1×10^5 to 5×10^5 cells/mL at frequency 5 kHz. The designed aptasensor was found highly specific for A549 lung carcinoma cells as compared to Hela cells, Caco-2 cells, MKN45 cells, and RBCs.

6.2 Optical aptasensors

Several aptasensors for lung cancer detection were fabricated based on optical transducers. Fig. 7 represents an example of molecular beacon aptamer-based optical biosensor for cancerous cell detection with fluorophore and quencher at the 5′- and 3′-ends of the aptamer, respectively. In the absence of a target, the hairpin structure of the MB-Apt

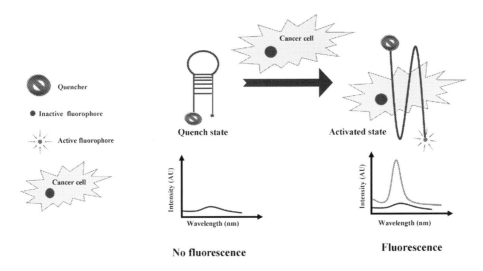

Fig. 7 Working principle of optical aptasensor for cancer detection.

holds the quencher molecule close to the fluorophore, resulting in no fluorescence emission. When the MB-Apt is bound to membrane receptors of the cancerous cell, its conformation is altered, thus resulting in an activated fluorescence signal due to the presence of free fluorophore.

Chen et al. [21] described the recognition of Small-Cell Lung Cancer (SCLC) NCI-H69 and NCI-H128 cells on the basis of molecular biomarkers present on their surface using fluorescein isothiocyanate (FITC)-labeled aptamers (HCC03 and HCH01). In this approach, they can detect only $\sim 10^{-5}$ cells. In 2009, Zhao and his team [81] described the DNA aptamer probe for the first time that can differentiate the subtypes of NSCLC cancer cells in the clinical samples. The detection of NSCLC A549 cells with aptamers was assessed with flow cytometry and fluorescence imaging. The aptamers were extremely specific (K_d 28.2–66.9 nM) to detect lung adenocarcinoma A549 cells (NSCLC) as compared to other lung cancer cells, e.g., large cell carcinoma NCI-H299, squamous carcinoma NCI-H520, Small Cell Lung Carcinoma NCI-H446, as well as for other types of cancer cells including Hela, MCF7, and HepG2 cell lines. The aptamer-based nanosensor was highly useful for the diagnosis and treatment of NSCLC lung cancer accounting for 85% cases of lung cancer.

A robust aptasensor for VEGF165 detection was designed based on plasmon-enhanced fluorescence (PEF) for early cancer diagnosis [82]. The interaction of aptamer with biomarker VEGF165 directed the inactivation of fluorescence probe previously attached to the aptamers and provided the signal corresponding to the concentration of biomarker. The developed aptasensor provided high potential for clinical sample analysis due to wide detection range (1.25 pM to 1.25 µM), lost cost (<10¢ per test), sample miniaturization, and good thermal and biological stability over and above high specificity for VEGF165. Furthermore, the aptasensor performance was validated with standard ELISA in biological fluids.

In 2013, aptamer-based fluorescence imaging probes were developed by Shi et al. [83] for in vitro and ex vivo diagnosis of tumor cells. The Cy5-labeled aptamer S6 was found highly specific for target A549 lung carcinoma as compared to Tca8113 tongue carcinoma. In addition to lung cancer cells, liver cancer cells were also detected with high specificity in flow cytometry assays using labeled ZY8 and LS2 aptamers, which were selected against SMMC-7721 and Bel-7404 liver cancer cells, respectively.

A remarkable strategy of specific detection and therapeutics for lung cancer was advocated using surface-enhanced Raman scattering (SERS) active nanostructures with aptamer-silver-gold shell-core [84]. The aptamer-based nanostructures recognized the target lung adenocarcinoma A549 cells with high specificity as compared to other types of lung cancerous cells including cell lines NCI-H157, NCI-H520, NCI-H1299, and NCI-H446 as well as for different types of cancerous cells such as HeLa and MCF-7 cells. The competency of nanostructure to absorb near-infrared (NIR) facilitated the photothermal therapy of targeted tumor cells at a very low irradiation power density of

0.20 W cm^2 without affecting the surrounding healthy cells. SERS signal produced from the interaction of Rhodamine 6G-labeled nanostructures with target cells assisted in monitoring the progress of therapy in addition to attaining a low detection limit (10 cells/mL). Furthermore, Wu and his team [41] proposed the simple and sensitive assay based on "recognition before labeling" strategy for the diagnosis of lung cancer using target–specific aptamer S11e as a recognition unit and quantum dots (QDs) as fluorescent signal producing unit. The cytosensing was endowed with simple streptavidin-biotin reaction in two steps with detection limit of 6×10^3 cells/mL. First, streptavidin-associated aptamers were used to recognize target A549 cells, which were subsequently tagged with biotinylated QDs for cellular imaging and detection. The use of label-free aptamers for recognition provides the advantage of high specificity without steric hindrance. In addition, QDs being nonaggregates also represent high signal–to–noise ratio, high photostability, and strong fluorescent signal.

Besides the cytosensing of A549 cells to detect lung cancer, adenosine is another important biomarker for investigating cancer progress. Adenosine can be examined in the urine samples of lung cancer patients. A fluorescence resonance energy transfer (FRET)–based aptasensor was fabricated for the detection of adenosine in urine. FRET phenomenon was observed between CdS QDs as a donor/fluorophore and polypyrrole (Ppy) as an acceptor/quencher attached to the recognition element, an antiadenosine aptamer. The quenching of QDs by Ppy was restored by the replacement of Ppy with adenosine, and the fluorescence intensity corresponded to the concentration of adenosine in the range of 23–146 nM. The aptasensor was highly sensitive with a detection limit of 9.3 nM [85].

In 2018, an ultrasensitive and highly specific aptamer-based detection of human A549 lung carcinoma was depicted by Wu and his team [86]. The sensing attained aa combined advantage of reformation property of supramolecule cy-M (3,3′-di(3-sulfopropyl)-4,5,4′,5′-dibenzo-9-methylthiacarbocyanine triethylammonium salt) to aptamer S6 G-quadruplex structure and second, a high affinity of aptamer S6 to lung cancer cells. The unique assembly raised the fluorescence 10^4-fold and made it possible to diagnose the tumor cells in complex samples like human serum and pleural effusion with a wide detection range of 0 to 3×10^5 cells/mL and a detection limit as low as 8 cells/mL. Furthermore, Liu and his workers [38] designed several fluorescent aptasensors for the detection of cancerous cells using specific bifunctional aptamers and catalytic hairpin assembly (CHA). In the presence of lung cancer A549 cells, the quencher was released from the fluorophores, which led to the binding of Cy5 and BHQ2-labeled aptamer AS1411 to the nucleolin of target cancer cells, and an amplified fluorescent signal was produced with triggered CHA reaction. Besides aptamer AS1411 for lung cancer A549 cells, other aptamers were customized for recognition sites of multiple cancer cells as well as for the different recognition sites of a single type of cancer cell. One of the recognition domains of bifunctional aptamer was specific to the receptors on the target cell whereas the other domain was efficient to induce the CHA reaction. These simple and homogeneous aptasensors were efficient to provide a quick response in different body fluids like serum, urine, hydrothorax, and cerebrospinal fluid with a detection limit of 10 cells/mL. The presence of cancerous cells in body fluids substantiates the metastasis and directs treatment. These designed cost-effective aptasensors are highly useful for the diagnosis and clinical management of cancer.

Aptamer-based electrochemical biosensors for lung cancer detection found better performance as compared with the other transducer systems like optical biosensor techniques. It is clear from the table that electrochemical techniques have been the most exploited for lung cancer biomarker detection. Furthermore, it is clear that, among electrochemical techniques, DPV and CV techniques allow various strategies to prepare electrodes and result in excellent linearity, sensitivity, detection limits, and other parameters for various lung cancer biomarkers. Moreover, these techniques allow detection of multiple analytes simultaneously with high sensitivity and stability.

7. Aptasensors for other cancers

Besides lung cancer, aptasensors were designed for other cancer cell lines as well as including breast cancer (MCF-7), hepatocellular carcinoma (HepG2), human chronic myelogenous leukemia (K562), ovarian cancer (A2780), lymphoblastic leukemia (CCRF-CEM), colorectal cancer (CT26), colon cancer (HCT 116), cervix carcinoma (HeLa), and prostate cancer [32, 87–90]. Several aptamers were selected against the biomarkers present on the cell lines (Table 3). The utilization of the validated biomarkers in routine practice for early cancer detection may illuminate the path of alleviation of deadly disease.

8. Conclusion and future perspectives

The chapter emphasizes the significant need of early cancer detection and histological variants of lung cancer. In addition, the recent development of an aptamer as a biorecognition element and aptasensors as advanced analytical tools for early diagnosis and clinical

Table 3 Biomarkers for cancer cell lines.

Cancer cell lines	Biomarkers	References
Breast cancer	HER2/neu, CEA, VEGF, uPA	[32, 91, 92]
Prostate cancer	PSA, VEGF	[93, 94]
Ovarian cancer	CA-125, uPA	[95, 96]
Colorectal cancer	K-Ras mutation, CEA, VEGF	[92, 97]
Pancreatic adenocarcinoma	VEGF	[98]
Burkitt Lymphoma	IGHM	[99]

HER2/neu, Human epidermal growth factor receptor 2; *CEA*, carcinoembryonic antigen; *VEGF*, vascular endothelial cell growth factor; *uPA*, urokinase plasminogen activator; *PSA*, prostate-specific antigen; *CA-125*, carbohydrate antigen 125 (CA-125); *IGHM*, immunoglobulin heavy constant mu.

management of lung cancer based on tumor–specific biomarkers has been precisely elaborated. The potential use of aptamers in clinical diagnostics and therapeutics has supported by the sharp increase in related U.S. patents and peer-review publications in the last two decades. However, in spite of vast available literature, the commercialized diagnostic kits for cancer detection are not yet available, and only a single aptamer, Pegaptanib (anti-VEGF), has been approved by the U.S. Food and Drug Administration (FDA) thus far [100], even though the aptamers possess great potential for clinical applications in the near future, due to their unique characteristics (ease of synthesis, high affinity binding, chemical modification, and low immunogenicity) as compared to antibodies. Various aptamers are under clinical trial, and a few companies including Neo Ventures Biotechnology Inc. (Ontario, Canada), Aptamer Sciences Inc. (AptSci, South Korea), Base Pair (Houston, TX, USA), and SomaLogic, Inc. (Boulder, CO, USA) are taking the initiative to successfully prepare commercial aptamer-based diagnostic products [59, 101]. The commercialization of designed inexpensive and sensitive aptamer-based nanosensors for the early diagnosis of lung cancer may be highly beneficial for the screening of a large population and for improving the survival rate of cancer patients.

References

[1] WHO, WHO Report 2018, https://www.who.int/news-room/fact-sheets/detail/cancer, 2018. Accessed 3 November 2019.
[2] S.K. Arya, S. Bhansali, Lung cancer and its early detection using biomarker-based biosensors, Chem. Rev. 111 (2011) 6783–6809.
[3] M.A. Tabrizi, M. Shamsipur, L. Farzin, A high sensitive electrochemical aptasensor for the determination of VEGF165 in serum of lung cancer patient. Biosens. Bioelectron. (2015), https://doi.org/10.1016/j.bios.2015.07.032.
[4] T. Hussain, Q.T. Nguyen, Molecular imaging for cancer diagnosis and surgery, Adv. Drug Deliv. Rev. 66 (2014) 90–100.
[5] Y.J. Jung, E. Katilius, R.M. Ostroff, Y. Kim, M. Seok, S. Lee, S. Jang, W.S. Kim, C.-M. Choi, Development of a protein biomarker panel to detect non-small cell lung cancer in Korea. Clin. Lung Cancer (2016), https://doi.org/10.1016/j.cllc.2016.09.012.
[6] A. Raouafi, A. Sánchez, N. Raouafi, R. Villalonga, Electrochemical aptamer-based bioplatform for ultrasensitive detection of prostate specific antigen, Sensors Actuators B Chem. 297 (2019) 126762.
[7] T.N. Zamay, G.S. Zamay, O.S. Kolovskaya, R.A. Zukav, M.M. Petrova, A. Gargaun, M.V. Berezovski, A.S. Kichkailo, DNA current and prospective protein biomarkers of lung cancer. Cancer 9 (2017) 155, https://doi.org/10.3390/cancers9110155.
[8] P. Schnabel, K. Junker, Pulmonary neuroendocrine tumors in the new WHO 2015 classification. Start of breaking new grounds? Pathologe 36 (2015) 283–292.
[9] S. Couraud, G. Zalcman, B. Milleron, F. Morin, P.J. Souquet, Lung cancer in never smokers—a review, Eur. J. Cancer 48 (9) (2012) 1299–1311.
[10] L.B. Travis, R.E. Curtis, B. Glimelius, E.J. Holowaty, F.E. Van Leeuwen, C.F. Lynch, A. Hagenbeek, M. Stovall, P.M. Banks, J. Adami, M.K. Gospodarowicz, Bladder and kidney cancer following cyclophosphamide therapy for non-Hodgkin's lymphoma, J. Natl. Cancer Inst. 87 (7) (1995) 524–531.
[11] S.A. Kenfield, E.K. Wei, M.J. Stampfer, B.A. Rosner, G.A. Colditz, Comparison of aspects of smoking among the four histological types of lung cancer, Tob. Control. 17 (3) (2008) 198–204.

[12] W. Travis, Update on small cell carcinoma and its differentiation from squamous cell carcinoma and other non-small cell carcinomas, Mod. Pathol. 25 (2012) S18–S30.

[13] C. Brambilla, F. Fievet, M. Jeanmart, F. Fraipont, S. Lantuejoul, V. Frappat, G. Ferretti, P.Y. Brichon, D. Moro-Sibilot, Early detection of lung cancer: role of biomarkers, Eur. Respir. J. 21 (39) (2003) 36–44.

[14] J. Polański, M. Chabowski, B. Jankowska-Polańska, D. Janczak, J. Rosińczuk, Histological subtype of lung cancer affects acceptance of illness, severity of pain, and quality of life, J. Pain Res. 11 (2018) 727.

[15] H. Abiri, M. Abdolahad, M. Gharooni, S.A. Hosseini, M. Janmaleki, S. Azimi, M. Hosseini, S. Mohajerzadeh, Monitoring the spreading stage of lung cells by silicon nanowire electrical cell impedance sensor for cancer detection purposes, Biosens. Bioelectron. 15 (68) (2015) 577–585.

[16] J.M. Hoffman, S.S. Gambhir, Molecular imaging: the vision and opportunity for radiology in the future, Radiology 244 (2007) 39–47.

[17] A.K. Ganti, J.L. Mulshine, Lung cancer screening, Oncologist 11 (5) (2006) 481–487.

[18] M.F. Kircher, J.K. Willmann, Molecular body imaging: MR imaging, CT, and US. Part II. Applications, Radiology 264 (2012) 349–368.

[19] I.E. Tothill, Biosensors for cancer markers diagnosis, Semin. Cell Dev. Biol. 20 (1) (2009) 55–62 Academic Press.

[20] S. Wu, L. Zhang, J. Zhong, Z. Zhang, Dual contrast magnetic resonance imaging tracking of iron-labeled cells in vivo, Cytotherapy 12 (2010) 859–869.

[21] H.W. Chen, C.D. Medley, K. Sefah, D. Shangguan, Z. Tang, L. Meng, J.E. Smith, W. Tan, Molecular recognition of small-cell lung cancer cells using aptamers, ChemMedChem 3 (2008) 991–1001.

[22] S.C. Chiang, C.L. Han, K.H. Yu, Y.J. Chen, K.P. Wu, Prioritization of cancer marker candidates based on the immunohistochemistry staining images deposited in the human protein atlas, PLoS One 8 (2013) e81079.

[23] D.W. Drolet, L. Moon-McDermott, T.S. Romig, An enzyme-linked oligonucleotide assay, Nat. Biotechnol. 14 (8) (1996) 1021.

[24] S. Kumar, N. Verma, A.K. Singh, Development of cadmium specific recombinant biosensor and its application in milk samples, Sensors Actuators B Chem. 240 (2017) 248–254.

[25] A.K. Singh, M. Singh, N. Verma, Electrochemical preparation of Fe3O4/MWCNT-polyaniline nanocomposite film for development of urea biosensor and its application in milk sample, J. Food Meas. Charact. 13 (3) (2019) 1–13.

[26] N. Verma, A.K. Singh, P. Kaur, Biosensor based on ion selective electrode for detection of L-arginine in fruit juices, J. Anal. Chem. 70 (9) (2015) 1111–1115.

[27] N. Verma, R. Sharma, S. Kumar, Advancement towards microfluidic approach to develop economical disposable optical biosensor for lead detection, Austin J. Biosens. Bioelectron. 2 (2) (2016) 1021.

[28] R.S. Singh, T. Singh, A.K. Singh, Enzymes as diagnostic tools, in: R.S. Singh, R.R. Singhania, A. Pandey, C. Larroche (Eds.), Advances in Enzyme Technology, Elsevier, USA, 2019, pp. 225–271 Chapter 9.

[29] A.D. Ellington, J.W. Szostak, In vitro selection of RNA molecules that bind specific ligands, Nature 346 (1990) 818–822.

[30] M. Jarczewska, L. Gorski, E. Malinowska, Electrochemical aptamer-based biosensors as potential tools for clinical diagnostics, Anal. Methods 8 (2016) 3861.

[31] R. Stoltenburg, C. Reinemann, B. Strehlitz, SELEX—A (r)evolutionary method to generate high-affinity nucleic acid ligands, Biomol. Eng. 24 (2007) 381–403.

[32] M.O. Caglayan, Electrochemical aptasensors for early cancer diagnosis: a review, Curr. Anal. Chem. 13 (2007) 18–30.

[33] D.R. Ciancio, M.R. Vargas, W.H. Thiel, M.A. Bruno, P.H. Giangrande, M.B. Mestre, Aptamers as diagnostic tools in cancer. Pharmaceuticals 11 (2018) 86, https://doi.org/10.3390/ph11030086.

[34] A.R. Rhouati, G. Catanante, G. Nunes, A. Hayat, J.-L. Marty, Label-free aptasensors for the detection of mycotoxins. Sensors 16 (2016) 2178, https://doi.org/10.3390/s16122178.

[35] N. Verma, N. Kaur, A.K. Singh, Study of rs699 SNP of hypertensive patients with gold surface immobilized molecular beacon biosensor, Int. J. Rec. Sci. Res. 7 (4) (2016) 10276–10281.

[36] K.A. Davis, B. Abrams, Y. Lin, S.D. Jayasena, Staining of cell surface human CD4 with $2'$-F-pyrimidine-containing RNA aptamers for flow cytometry, Nucleic Acids Res. 26 (17) (1998) 3915–3924.

[37] Q. Deng, I. German, D. Buchanan, R.T. Kennedy, Retention and separation of adenosine and analogues by affinity chromatography with an aptamer stationary phase, Anal. Chem. 73 (22) (2001) 5415–5421.

[38] J. Liu, Y. Zhang, Q. Zhao, B. Situ, J. Zhao, S. Luo, B. Li, X. Yan, P. Vadgama, L. Su, W. Ma, W. Wang, L. Zheng, Bifunctional aptamer-mediated catalytic hairpin assembly for the sensitive and homogenous detection of rare cancer cells. Anal. Chim. Acta (2018), https://doi.org/10.1016/j.aca.2018.04.068.

[39] N.-V. Nguyen, C.-H. Yang, C.-J. Liu, C.-H. Kuo, D.-C. Wu, C.-P. Jen, An aptamer-based capacitive sensing platform for specific detection of lung carcinoma cells in the microfluidic chip. Biosensors 8 (2018) 98, https://doi.org/10.3390/bios8040098.

[40] M.A. Rehder, L.B. McGown, Open-tubular capillary electrochromatography of bovine β-lactoglobulin variants A and B using an aptamer stationary phase, Electrophoresis 22 (17) (2001) 3759–3764.

[41] C. Wu, J. Liu, P. Zhang, J. Li, H. Ji, X. Yang, K. Wang, Recognition-before-labeling strategy for sensitive detection of lung cancer cells with quantum dots-aptamer complex. Analyst (2015), https://doi.org/10.1039/C5AN01145K.

[42] C. Tuerk, L. Gold, Systematic evolution of ligands by exponential enrichment: RNA ligands to bacteriophage T4 DNA polymerase, Science 249 (1990) 505–510.

[43] W.H. Thiel, P.H. Giangrande, Analyzing HT-SELEX data with the galaxy project tools—a web based bioinformatics platform for biomedical research, Methods 97 (2014) 3–10.

[44] Y. Dong, Z. Wang, S. Wang, Y. Wu, Y. Ma, J. Liu, Introduction of SELEX and important SELEX variants. in: Aptamers for Analytical Applications, 2018, pp. 1–25, https://doi.org/10.1002/9783527806799.ch1.

[45] S. Ohuchi, Cell-SELEX technology, Biores. Open Access 1 (2012) 265–272.

[46] H. Chen, C.H. Yuan, Y.F. Yang, C.Q. Yin, Q. Guan, F.B. Wang, J.C. Tu, Subtractive cell-SELEX selection of DNA aptamers binding specifically and selectively to hepatocellular carcinoma cells with high metastatic potential, Biomed. Res. Int. 2016 (2016) 5735869.

[47] M. Haghighi, H. Khanahmad, A. Palizban, Selection and characterization of single-stranded DNA aptamers binding human B-cell surface protein CD20 by cell-SELEX, Molecules 23 (2018) 715.

[48] W.H. Thiel, T. Bair, A.S. Peek, X. Liu, J. Dassie, K.R. Stockdale, M.A. Behlke, F.J. Miller, P.H. Giangrande, Rapid identification of cell-specific, internalizing RNA aptamers with bioinformatics analyses of a cell-based aptamer selection, PLoS One 7 (2012) e43836.

[49] L. Zhao, W. Tan, X. Fang, Introduction to aptamer and cell-SELEX, in: X. Fang, W. Tan (Eds.), Aptamers Selected by Cell-SELEX for Theranostics, 2016 Springer-Verlag, Heidelberg/Berlin, Germany, 2016, pp. 1–11.

[50] K.W. Thiel, P.H. Giangrande, Therapeutic applications of DNA and RNA Aptamers, Oligonucleotides 19 (2009) 209–222.

[51] H. Ulrich, C.A. Trujillo, A.A. Nery, J.M. Alves, P. Majumder, R.R. Resende, A.H. Martins, DNA and RNA aptamers: from tools for basic research towards therapeutic applications, Comb. Chem. High Throughput Screen. 9 (2006) 619–632.

[52] A.C. Yan, M. Levy, Aptamer-mediated delivery and cell-targeting aptamers: room for improvement, Nucleic Acid Ther. 28 (2018) 194–199.

[53] J.C. Cox, A.D. Ellington, Automated selection of anti-protein aptamers, Bioorg. Med. Chem. 9 (2001) 2525–2531.

[54] S.P. Ohuchi, T. Ohtsu, Y. Nakamura, Selection of RNA aptamers against recombinant transforming growth factor-type III receptor displayed on cell surface, Biochimie 88 (2006) 897–904.

[55] J. Mi, Y. Liu, Z.N. Rabbani, Z. Yang, J.H. Urban, A. Sullenger, B.M. Clary, In vivo selection of tumor-targeting RNA motifs, Nat. Chem. Biol. 6 (2010) 22–24.

[56] G. Mayer, M.S.L. Ahmed, A. Dolf, E. Endl, P.A. Knolle, M. Famulok, Fluorescence-activated cell sorting for aptamer SELEX with cell mixtures, Nat. Protoc. 5 (2010) 1993–2004.

[57] W.H. Thiel, K.W. Thiel, K.S. Flenker, T. Bair, A.J. Dupuy, J.O.I.I. McNamara, F.J. Miller, P.H. Giangrande, Cell-internalization SELEX: method for identifying cell-internalizing RNA Aptamers for delivering siRNAs to target cells, Methods Mol. Biol. 1218 (2015) 187–199.

[58] A.G. Souza, K. Marangoni, P.T. Fujimura, P.T. Alves, M.J. Silva, V.A.F. Bastos, L.R. Goulart, V. A. Goulart, 3D cell-SELEX: development of RNA aptamers as molecular probes for PC-3 tumor cell line, Exp. Cell Res. 341 (2016) 147–156.

[59] F. Pfeiffer, F. Tolle, M. Rosenthal, G.M. Brändle, J. Ewers, G. Mayer, Identification and characterization of nucleobase-modified aptamers by click-SELEX, Nat. Protoc. 13 (2018) 1153–1180.

[60] A.K. Singh, M. Singh, N. Verma, Extraction, purification, kinetic characterization and immobilization of urease from *Bacillus sphaericus* MTCC 5100, Biocat. Agric. Biotechnol. 12 (2017) 341–347.

[61] S. Balamurugan, A. Obubuafo, S.A. Soper, D.A. Spivak, Surface immobilization methods for aptamer diagnostic applications, Anal. Bioanal. Chem. 390 (2008) 1009–1021.

[62] D. Samanta, A. Sarkar, Immobilization of bio-macromolecules on self-assembled monolayers: methods and sensor applications, Chem. Soc. Rev. 40 (2011) 2567–2592.

[63] N. Verma, A.K. Singh, M. Singh, L-arginine biosensor: a comprehensive review, Biochem. Biophys. Rep. 12 (2017) 128–139.

[64] X. Zhang, V.K. Yadavalli, Surface immobilization of DNA aptamers for biosensing and protein interaction analysis, Biosens. Bioelectron. 26 (2011) 3142–3147.

[65] A.K. Singh, N. Verma, Quartz crystal microbalance based approach for food quality, Curr. Biotechnol. 3 (2) (2014) 127–132.

[66] J. Liu, Y. Qin, D. Li, T. Wang, Y. Liu, J. Wang, E. Wang, Highly sensitive and selective detection of cancer cell with a label-free electrochemical cytosensor, Biosens. Bioelectron. 41 (2013) 436–441.

[67] G. Decher, Fuzzy nanoassemblies: toward layered polymeric multicomposites, Science 277 (5330) (1997) 1232–1237.

[68] T.A. Mir, J.H. Yoon, N.G. Gurudatt, M.S. Won, Y.B. Shim, Ultrasensitive cytosensing based on an aptamer modified nanobiosensor with a bioconjugate: detection of human non-small-cell lung cancer cells, Biosens. Bioelectron. 74 (2015) 594–600.

[69] A. Tessitore, A. Gaggiano, G. Cicciarelli, D. Verzella, D. Capece, M. Fischietti, F. Zazzeroni, E. Alesse, Serum biomarkers identication mass spectrometry in high-mortality tumors, Int. J. Proteomics 2013 (2013) 1–12.

[70] G. Sozzi, D. Conte, M. Leon, R. Cirincione, L. Roz, C. Ratcliffe, E. Roz, N. Cirenei, M. Bellomi, G. Pelosi, et al., Quantification of free circulating DNA as a diagnostic marker in lung cancer, J. Clin. Oncol. 21 (2003) 3902–3908.

[71] L. Gold, D. Ayers, J. Bertino, et al., Aptamer-based multiplexed proteomic technology for biomarker discovery, PLoS One 5 (2010) e15004.

[72] L.D. Roy, M. Sahraei, D.B. Subramani, D. Besmer, S. Nath, T.L. Tinder, E. Bajaj, K. Shanmugam, Y.Y. Lee, S.I. Hwang, Oncogene 30 (12) (2011) 1449.

[73] I. Lakshmanan, M.P. Ponnusamy, M.A. Macha, D. Haridas, P.D. Majhi, S. Kaur, M. Jain, S.K. Batra, A.K. Ganti, J. Thorac. Oncol. 10 (1) (2015) 19–27.

[74] J.J. Rahn, L. Dabbagh, M. Pasdar, J.C. Hugh, The importance of MUC1 cellular localization in patients with breast carcinoma: an immunohistologic study of 71 patients and review of the literature, Cancer 91 (11) (2001) 1973–1982.

[75] R.M. Ostroff, W.L. Bigbee, W. Franklin, et al., Unlocking biomarker discovery: large scale application of aptamer proteomic technology for early detection of lung cancer, PLoS One 5 (2010) e15003.

[76] J. Li, H. Zheng, P.J. Bates, T. Malik, X.-F. Li, J.O. Trent, C.K. Ng, Aptamer imaging with Cu-64 labeled AS1411: preliminary assessment in lung cancer, Nucl. Med. Biol. 41 (2014) 179–185.

[77] I. Willner, M. Zayats, Electronic aptamer-based sensors, Angew. Chem. Int. Ed. 46 (34) (2007) 6408–6418.

[78] R. Sharma, V.V. Agrawal, P. Sharma, R. Varshney, R.K. Sinha, B.D. Malhotra, Aptamer-based electrochemical sensor for detection of human lung adenocarcinoma A549 cells, J. Phys. Conf. Ser. 358 (2012).

[79] P. Kara, Y. Erzurumlu, P. Ballar Kirmizibayrak, M. Ozsoz, Electrochemical aptasensor design for label free cytosensing of human non-small cell lung cancer. J. Electroanal. Chem. (2016), https://doi.org/10.1016/j.jelechem.2016.06.008.

[80] G.S. Zamay, T.N. Zamay, V.A. Kolovskii, A.V. Shabanov, Y.E. Glazyrin, D.V. Veprintsev, A. V. Krat, S.S. Zamay, O.S. Kolovskaya, A. Gargaun, et al., Electrochemical aptasensor for lung cancer-related protein detection in crude blood plasma samples, Sci. Rep. 6 (2016) 343–350.

[81] Z. Zhao, L. Xu, X. Shi, W. Tan, X. Fang, D. Shangguan, Recognition of subtype non-small cell lung cancer by DNA aptamers selected from living cells, Analyst 134 (2009) 1808–1814.

[82] H. Cho, E.-C. Yeh, R. Sinha, T.A. Laurence, J.P. Bearinger, L.P. Lee, Single-step nanoplasmonic VEGF165 aptasensor for early cancer diagnosis, ACS Nano 6 (2012) 7607–7614.

[83] H. Shi, W. Cui, X. He, Q. Guo, K. Wang, X. Ye, J. Tang, Whole cell-SELEX aptamers for highly specific fluorescence molecular imaging of carcinomas in vivo, PLoS One 8 (8) (2013) e70476.

[84] P. Wu, Y. Gao, Y. Lu, H. Zhang, C. Cai, High specific detection and near-infrared photothermal therapy of lung cancer cells with high SERS active aptamer–silver–gold shell–core nanostructures, Analyst 138 (2013) 6501.

[85] Z. Hashemian, T. Khayamian, M. Saraji, M.P. Shirani, Aptasensor based on fluorescence resonance energy transfer for theanalysis of adenosine in urine samples of lung cancer patients, Biosens. Bioelectron. 79 (2016) 334–340.

[86] Y. Wu, H. Zhanga, J. Xianga, Z. Maoc, G. Shena, F. Yanga, Y. Liua, W. Wangd, N. Dud, J. Zhangd, Y. Tang, Ultrasensitive and high specific detection of non-small-cell lung cancer cells in human serum and clinical pleural effusion by aptamer-based fluorescence spectroscopy, Talanta 179 (2018) 501–506.

[87] Y. Cao, E. Guangqi, E. Wang, K. Pal, S.K. Dutta, D. Bar-Sagi, D. Mukhopadhyay, VEGF exerts an angiogenesis-independent function in cancer cells to promote their malignant progression, Cancer Res. 72 (16) (2012) 3912–3918.

[88] L.K. Kheyrabadi, M.A. Mehrgardi, E. Wiechec, A.P.F. Turner, A. Tiwari, Ultrasensitive detection of human liver hepatocellular carcinoma cells using a label-free aptasensor, Anal. Chem. 86 (2014) 4956–4960.

[89] F. Li, Q. Wang, H. Zhang, T. Deng, P. Feng, B. Hu, Y. Jiang, L. Cao, Characterization of a DNA aptamer for ovarian cancer clinical tissue recognition and in vivo imaging. Cell. Physiol. Biochem. 1 (6) (2018) 2564–2574, https://doi.org/10.1159/000495925.

[90] C. Zhang, X. Ji, Y. Zhang, G. Zhou, X. Ke, H. Wang, P. Tinnefeld, Z. He, One-pot synthesized aptamer-functionalized CdTe: Zn^{2+} quantum dots for tumor-targeted fluorescence imaging in vitro and in vivo, Anal. Chem. 85 (12) (2013) 5843–5849.

[91] Z. Mitri, T. Constantine, R. O'Regan, The HER2 receptor in breast cancer: pathophysiology, clinical use, and new advances in therapy, Chemother. Res. Pract. 2012 (2012) 743193.

[92] H.F. Qin, L.L. Qu, H. Liu, S.S. Wang, H.J. Gao, Serum CEA level change and its significance before and after Gefitinib therapy on patients with advanced non-small cell lung cancer, Asian Pac. J. Cancer Prev. 14 (2013) 4205–4208.

[93] P.B. Crulhas, A.E. Karpik, F.K. Delella, G.R. Castro, V.A. Pedrosa, Electrochemical aptamer-based biosensor developed to monitor PSA and VEGF released by prostate cancer cells, Anal. Bioanal. Chem. 409 (2017) 6771–6780.

[94] L.G. Gomella, X.S. Liu, E.J. Trabulsi, K.W. Kevin, R. Myers, T. Showalter, A. Dicker, R. Wender, Screening for prostate cancer: the current evidence and guidelines controversy, Can. J. Urol. 18 (5) (2011) 5875.

[95] R.C. Bast, D. Badgwell, Z. Lu, R. Marquez, D. Rosen, J. Liu, K.A. Baggerly, E.N. Atkinson, S. Skates, Z. Zhang, A. Lokshin, New tumor markers: CA125 and beyond, Int. J. Gynecol. Cancer 15 (3) (2005) 274–281.

[96] C. Borgfeldt, S.R. Hansson, B. Gustavsson, A. Masback, B. Casslen, Redifferentiation of serous ovarian cancer from cystic to solid tumors is associated with increased expression of Mrna for Urokinase plasminogen activator (Upa), its receptor (Upar) and its inhibitor (Pai-1), Int. J. Cancer 92 (2001) 497–502.

[97] P.J. Roberts, T.E. Stinchcombe, C.J. Der, M.A. Socinski, Personalized medicine in non-small-cell lung cancer: is KRAS a useful marker in selecting patients for epidermal growth factor receptor–targeted therapy? J. Clin. Oncol. 28 (31) (2010) 4769–4777.

[98] C. Huang, R. Huang, W. Chang, T. Jiang, K. Huang, J. Cao, X. Sun, Z. Qiu, The expression and clinical significance of pSTAT3, VEGF and VEGF-C in pancreatic adenocarcinoma, Neoplasma 59 (1) (2012) 52–61.

[99] S. Bamrungsap, T. Chen, M.I. Shukoor, Z. Chen, K. Sefah, Y. Chen, W. Tan, Pattern recognition of cancer cells using aptamer–conjugated magnetic nanoparticles, ACS Nano 6 (5) (2012) 3974–3981.
[100] A.D. Keefe, S. Pai, A. Ellington, Aptamers as therapeutics, Nat. Rev. Drug Discov. 9 (2010) 537–550.
[101] A. Ruscito, M.C. DeRosa, Small-molecule binding aptamers: selection strategies, characterization, and applications, Front. Chem. 4 (2016) 14.

CHAPTER 23

Micro/nanodeposition techniques for enhanced optical fiber sensors

Aitor Urrutia[a,b], Pedro J. Rivero[c,d], Javier Goicoechea[a,b], and Francisco J. Arregui[a,b]

[a]Nanostructured Optical Devices Laboratory, Department of Electrical, Electronic and Communication Engineering, Public University of Navarra, Pamplona, Spain
[b]Institute of Smart Cities (ISC), Public University of Navarra, Pamplona, Spain
[c]Materials Engineering Laboratory, Department of Engineering, Public University of Navarra, Pamplona, Spain
[d]Institute of Advanced Materials (INAMAT), Public University of Navarra, Pamplona, Spain

1. Introduction

Optical fibers are high-performance optical waveguides for information transmission that are becoming ubiquitous these days, even at domestic communications. With this purpose, optical fibers are specially designed for preserving the integrity of the propagated optical signals shielding the waveguide from external influence. Depending on their application, they can be different sizes, shapes, and materials. The most relevant advantages of optical fiber technology compared with conventional electronic communications systems are the possibility of highly-dense signal multiplexing, robust and low-cost long-distance communications, and the electromagnetic immunity of the signals sent through them, which make this technology an ideal solution for high-speed backbone networks. Nevertheless, since the beginning of the development of this technology, there has been great interest in the application of optical fiber as a guide for optical signals, and at the same time, a sensor probe. For sensing applications, it is necessary to expose the propagated light through the optical fiber waveguide and induce an interaction with the external medium to detect a measurable change in an optical parameter due to the presence of the desired analyte. Traditionally, there have been several approaches used, such as optical fiber cleaved ends, tapered optical fibers, side-polished optical fibers, gratings such as fiber Bragg gratings (FBG) and long-period gratings (LPG), photonic crystal fibers (PCF), multicore fibers, etc.

Since 1980, there have been more than 26,000 research papers published in international journals with the words "Optical Fiber Sensor" in their title, abstract, or keywords, according to the database *Scopus*. Besides this enormous interest of the scientific community, there are also some unsolved challenges for optical sensing technology, such as the configuration of simple optical fiber sensor network, and the relative high cost of optical fiber sensing systems compared with other conventional alternatives. The successful

Handbook of Nanomaterials for Sensing Applications
https://doi.org/10.1016/B978-0-12-820783-3.00018-X

application of optical fiber sensing technology will depend on the development of optical fiber sensors with enhanced properties that help to overcome these barriers.

The first optical fiber sensor approaches were based on physical changes of the fiber or the external medium, but there was a real boost in optical fiber sensing technology when researchers combined those previously cited fibers and structures with sensitive optical coatings. The use of micro- and nanotechnology to create organized optical coatings that transduce the interaction with certain analytes into an optical signal has been the key to the proliferation of different promising devices. In these new devices, new composite materials can be created incorporating polymers, nanoparticles, microfibers, patterns, etc. The enhancement of the properties of this generation of optical fiber sensors is possible thanks to the use of different phenomena (high surface-to-volume ratio, surface biological reactions, optical nanoparticle-quantum confinement, plasmonic interactions, enhanced-Raman scattering, lossy mode resonances [LMRs], etc.), only accessible at the micro- and nanoscale. In this chapter, it is possible to find a review of the most recent relevant contributions in which the use of micro- and nanotechnology to create sensitive coatings has led to enhanced optical fiber sensors, with a special emphasis in the fabrication techniques.

2. Vapor deposition techniques

2.1 Chemical vapor deposition

Chemical vapor deposition (CVD) is a deposition process of solid materials at a high temperature as a result of a chemical reaction directly over the substrate. In other words, CVD is a generic name for a group of processes whereby a thin solid film is deposited onto a substrate through chemical reactions of the gaseous species [1]. This deposition enables the possibility of creating ordered crystalline coatings grown directly on the substrate, starting from vapor precursors, and usually takes place at temperatures around 1000°C.

CVD process can be sequenced in seven well-differenced steps as shown in Fig. 1 [2]. First, reactant gases (1) usually enter a chamber or reactor by forced flow (2). Then, gases diffuse through the boundary layer, and gases arrive at the surface of the substrate (3B). However, homogeneous gas phase reaction (3A) can also occur where the intermediate species undergo subsequent decomposition and/or chemical reaction, forming powders and volatile by-products in the gas phase. From (3B), gaseous reactants are absorbed onto the heated substrate, and the heterogeneous reaction occurs at the gas–solid interface (4). The deposited material is diffused along the heated substrate surface forming the crystallization center (5) and growth of the film. Finally, gaseous by-products of the chemical reaction are diffused away from the surface, through the boundary layer (6), and flowed out of the chamber (7).

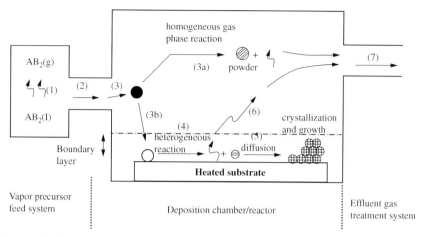

Fig. 1 Steps of CVD reaction process. AB₂: mixture of gaseous reactants and a carrier gas. *Reprinted with permission from K. L. Choy, Chemical vapour deposition of coatings, Prog. Mater. Sci. 48(2) (2003) 57–170.*

CVD offers the deposition of highly pure materials with accurate control at the nanometric range. Moreover, it can use a large list of chemical precursors, including organometallic compounds, halides, and hydrides, among others. Thus, the CVD technique enables the deposition of a wide type of materials such as metals, nitrides, carbides, oxides, polymers, and other nonmetallic elements. The versatility of CVD had led to rapid growth, and it has become one of the main processing methods for the deposition of thin films and coatings for a wide range of applications, including semiconductors; dielectrics; refractory ceramic materials, metallic films; and fiber production or coatings [2].

Nevertheless, CVD method has several disadvantages such as the risk of using toxic, corrosive, flammable, and/or explosive precursor gases. Another concern is the different vaporization rates and pressures for different precursor materials, which hinder the deposition of multicomponent materials and increase the fabrication costs. To overcome these drawbacks, some alternative techniques have emerged based on CVD, such as plasma-enhanced CVD (PECVD), flame-assisted CVD, and aerosol-assisted CVD.

CVD technique can also be employed for the development of sensitive coatings onto optical waveguides. Thus, design and fabrication of optical fiber sensors based on CVD coatings have become an interesting alternative to other conventional approaches. For example, Chauhan et al. [3] deposited a silicon nitride thin film with a CVD process over an Au nanoparticle-based coated region of an optical fiber probe. As a result, they demonstrated an improvement (1.24 times) in refractive index (RI) sensitivity of localized surface plasmon resonance (LSPR) fiber-optic probe by coating it with an optimum thickness of a SiNx film. Fig. 2 shows the RI sensitivity differences in function of the number of SiNx layers.

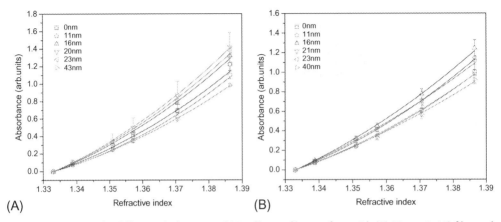

Fig. 2 RI sensitivity for different thicknesses of SiNx film on fiber surface with (A) 18-nm AuNP film and (B) 39-nm AuNP film. *Reprinted with permission from S. K. Chauhan, N. Punjabi, D. K. Sharma, S. Mukherji, A silicon nitride coated LSPR based fiber-optic probe for possible continuous monitoring of sucrose content in fruit juices, Sensors Actuators B Chem. 222 (2016) 1240–1250.*

Other work presented by Luan et al. proposed a remarkable enhancement in the tensile strength of an optical fiber [4], where a smooth and uniform boron nitride thin film was deposited on sapphire optical fibers. The fabricated devices can be used as strain sensors at high temperatures with good light transmission properties.

A relevant application of CVD to optical fiber sensing was proposed by Smietana et al. [5], where diamond-like carbon nano-overlays were deposited onto long period fiber gratings (LPGs) using radiofrequency PECVD. The results demonstrated that DLC-coated LPGs can be as much as 15 times more sensitive than uncoated LPGs to variations within a specified range of the RI of the surrounding medium. The same research group have also developed other enhanced sensors onto LPGs for RI [6] and temperature [7] monitoring using diamond-like carbon and silicon nitride layers, respectively.

Another new CVD variant is atomic layer deposition (ALD) [8], which offers atomic level precision in determining film thickness, extreme surface conformity, and excellent process repeatability and scalability. In the optical fiber sensing field, some first works have been reported recently [9, 10]. Dominik et al. presented LPG sensors based on TiOx thin films with RI sensitivity values around 3400 nm/RIU using this technique [9]. In another article, Mandia et al. studied the effect of ALD-grown Al_2O_3 on the RI sensitivity of previous fabricated CVD-gold coated optical fiber gratings [10]. The results concluded that sensitivity of such sensors can be greatly enhanced by controlling the Al_2O_3 layer thickness.

Other reports with optical fiber physical sensors supported by CVD-based coatings can be found in Table 1.

Table 1 Summary of the different optical fiber sensors to detect physical parameters with their corresponding CVD technique, sensitive coating, optical structure, sensing mechanism, and parameter of study.

Technique	Sensitive coating	Optical structure	Sensing mechanism	Parameter detection	Reference
Hot filament CVD	Silicon nitride + Au nanoparticles	Cladding removed MMF	LSPR	Refractive index	[3]
CVD	Boron nitride	Sapphire optical fiber	–	Strain and temperature	[4]
PEPVD	Diamond like carbon	Long period grating	Wavelength variation	Refractive index	[5], [6]
PECVD	Diamond like carbon	Cladding removed MMF	Wavelength variation	Refractive index	[11]
PECVD	Silicon nitride	Long period grating	Wavelength variation	Temperature	[7]
ALD	TiOx	Long period grating	Wavelength variation	Refractive index	[9]
ALD + CVD	Al2O3 + Au	Tilted fiber Bragg grating	Wavelength variation	Refractive index	[10]

2.2 Physical vapor deposition

Physical vapor deposition (PVD) comprises a group of processes in which atoms or molecules of material are vaporized from a solid or liquid source, transported in the form of a vapor state through a vacuum or low-pressure-gaseous environment, and finally condensed on a substrate [12]. PVD allows the deposition of almost any inorganic material as well as some polymeric materials, generating single materials, layers with a graded composition, multilayer coatings, or very thick deposits ranging between a few nanometers to some microns. Within PVD, there are several deposition techniques [13]: thermal evaporation, sputtering, e beam, etc.

Within the thermal evaporation category, vacuum deposition is the most common process. In this method, the atoms or molecules from a thermal vaporization source reach the substrate without collisions with residual gas molecules in the deposition chamber. Vacuum deposition normally requires a vacuum of better than 10^{-4} Torr to have a long mean-free path between collisions. Other important parameters are the vaporization rate, the temperature, the equilibrium vapor pressure of each material to be deposited, and the geometry and position of the substrate.

Sputtering or sputter deposition is a nonthermal vaporization process where surface atoms are physically ejected by momentum transfer from an atomic-sized energetic

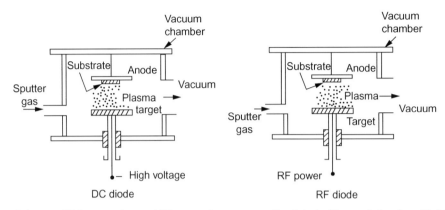

Fig. 3 Scheme of DC-sputtering and RF-sputtering systems. *Reprinted with permission from H. Adachi, K. Wasa, Thin films and nanomaterials, in Handbook of Sputtering Technology, no. Dlc, Elsevier, 2012, pp. 3–39, ©Elsevier 2012.*

bombarding particle, which is usually a gaseous ion accelerated from a plasma or ion gun [12]. To produce the "sputtering" phenomenon, several different systems have been developed, including DC diode, RF diode, magnetron, or ion beam sputtering. Fig. 3 shows both schemes of DC and RF sputtering systems, respectively. It can be appreciated that the only difference between both systems is the way of generating energy necessary to produce the acceleration of ions and further bombarding on the target.

A variant of sputtering is pulsed laser deposition (PLD), which was proposed by Singh and Narayan [14]. PLD is widely used for deposition of alloys and/or compounds with a controlled chemical composition. PLD uses a high-power ultraviolet pulsed laser to increase the energy density on the target source. Thus, atoms are ablated or evaporated from the target, and then collected on the substrates, forming thin films.

Among the wide range of applications of the sputtering deposition process, the fabrication of sensitive coatings for the development of optical fiber sensors has provoked a considerable effect in the field. The resulting coatings have been applied onto diverse optical structures such as cladding removed multimode [15–18] or single-mode [19] fibers, side-polished fiber [20, 21], tapered fibers [22], end facet of fibers [23–25], or gratings [26–29], among others.

As mentioned previously, sputtering allows the deposition of almost any material. This advantage recently opened the door to the generation, monitoring, and analysis of LMRs [30, 31]. Metal oxides such as tin oxide (SNO_2), indium tin oxide (ITO), and indium oxide (In_2O_3) as well as polymers are suitable for the development of LMR-based sensors. Thus, the deposition of the mentioned metal oxides onto optical fiber structures has been studied by Arregui's research group [31] in different works for the design and fabrication of several physical sensors such as refractometers [16, 18] and

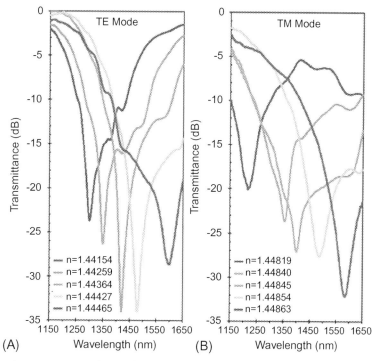

Fig. 4 Transmittance spectra obtained when the sensitive region was submerged in solutions of different refractive index. (A) Spectra corresponding to the TE mode when the fiber is submerged in solutions of 1.4415–1.4447 SMRI. (B) Spectra corresponding to the TM mode when the fiber is submerged in solutions of 1.4481–1.4487 SMRI [21]. *Copyright © 2017, Springer Nature (http://creativecommons.org/licenses/by/4.0/).*

humidity sensors [19, 22], among other applications. LMR phenomena can be tuned depending on the thickness and the effective RI of the coatings. In addition, LMR bands can be split into their transversal electric (TE) and transversal magnetic (TM) modes by polarizing the transmitted light. This signal decomposition allows the monitoring of both TE and TM modes, obtaining a demonstrable improvement in sensitivity ratios against surrounding medium refractive index (SMRI) variations. Arregui et al. [20] obtained an optical LMR-based refractometer with a record sensitivity of 304,360 nm per RI unit (nm/RIU) thanks to SNO_2-sputtered films onto D-shape single-mode fibers. In a further report, Ozcariz et al. [21] presented another optimized refractometer with an unprecedented sensitivity of more than 1 million nm/RIU, which means a sensitivity below 10^{-9} RIU with a pm resolution detector. The results of this device can be appreciated in Figs. 4 and 5, where the transmitted spectrum changed when the sensitive region was immersed in different RI solutions. Fig. 4 represents the wavelength variations of TE and TM modes against the mentioned RI changes.

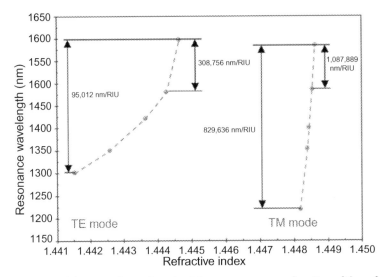

Fig. 5 Representation of the central wavelength of the resonances as a function of the refractive index of the surrounding media (SMRI) [21]. *Copyright © 2017, Springer Nature (http://creativecommons.org/licenses/by/4.0/).*

Gold-sputtered films applied onto optical fibers have been also reported to design SPR-based refractometers [25, 32] or temperature Fabry-Perot based sensors [33, 34]. Other recent works have explored other type of materials such as transition-metal di-chalcogenides (MoS_2) [26] or magnetostrictive compounds (TbDyFe) [27] for the fabrication of thin films onto FBGs, thus developing high-sensitivity temperature sensors.

A summary of different physical sensors in which sputtering has been employed to deposit sensitive regions can be found in Table 2.

3. Wet deposition techniques

3.1 Self-assembled monolayers

Another nanodeposition technique used for the fabrication of thin films onto optical fibers is self-assembled monolayer (SAM). SAMs are ordered molecular assemblies that are formed spontaneously by the adsorption of a surfactant with a specific affinity of its head group to a substrate [42]. The molecules used for the SAM formation have three well-differenced parts: (1) head groups that are usually connected to (2) alkyl chains with different lengths, ending in (3) tail-functionalized groups. The head groups bind onto the substrate, and the molecule chains are highly ordered and can be incorporated with various functional groups, thus providing surfaces with specific sensing functions by fine chemical control [43]. Thus, SAMs are created onto a substrate from either liquid or vapor phase followed by a two-dimensional organization of functionalized tail groups.

Table 2 Summary of the different optical fiber sensors to detect physical parameters using Sputtering techniques.

Sputtering system	Sensitive coating	Optical structure	Sensing mechanism	Parameter detection	Reference
DC	Indium tin oxide and tin oxide	Cladding-removed MMF	LMR	Refractive index	[15]
DC	Indium tin oxide	Cladding-removed MMF	LMR	Refractive index	[16], [17]
DC	Tin oxide layer	Cladding-removed MMF	LMR	Refractive index	[18]
DC	Tin oxide films	D-shape SMF	LMR	Refractive index	[20], [21]
DC	Indium tin oxide	Tapered fibers	LMR	Relative humidity	[22]
DC	Tin oxide layer + LbL thin films	Cladding-removed MMF	LMR	Relative humidity	[35]
DC	Tin oxide film	Low finesse Fabry-Perot	Fabry-Perot interference	Relative humidity	[36]
DC	Gold films	Fabry-Perot cavity	Fabry-Perot interferometry	Temperature	[33]
DC	Gold films	Fabry-Perot	Fabry-Perot interferometry	Temperature	[34]
–	Gold films	SMF with an open microcavity	Fabry-Perot interferometry	Refractive index	[37]
–	Gold film	Tilted Long Period Grating	Wavelength variation	Refractive index	[38]
Pulsed DC	Low/High RI Tin oxide layers	End facet of MMF	Grating interference	Refractive index	[23]
Pulsed DC	Low/High RI Indium Oxide layers	End facet of MMF	Grating interference	Relative Humidity	[24]
Pulsed DC	Tin Oxide	Cladding etched SMF	LMR	Relative Humidity	[19]
RF Magnetron	Gold film	Cladding-removed and end-facet MMF	SPR	Refractive index	[25]

Table 2 Summary of the different optical fiber sensors to detect physical parameters using Sputtering techniques—cont'd

Sputtering system	Sensitive coating	Optical structure	Sensing mechanism	Parameter detection	Reference
Magnetron	Au layer	D-shape macrobend plastic fiber	SPR	Refractive index	[32]
DC magnetron	Molybdenum-di-sulfide multilayers	Etched fiber Bragg grating	Wavelength variation	Temperature	[26]
Magnetron	TbDyFe film	Fiber Bragg grating with groove microstructure	Wavelength variation	Temperature	[27]
Magnetron	Titanium and silver films	Fiber Bragg grating	Wavelength variation	Strain	[28]
Magnetron	Pd/Ag composite film	Photonic crystal fiber-based Sagnac loop	Sagnac interferometry	H_2	[39]
Magnetron	Palladium oxides	Long period fiber grating	Wavelength variation	H_2	[29]
–	Platinum	Sapphire tip	Frequency/phase variation	Temperature	[40], [41]

Moreover, the thickness of a SAM can be fine-controlled by changing the length of the alkyl chain and, depending on the nature of the deposited SAMs onto optical fiber structures, it is possible to develop a wide variety of sensors with different purposes, especially chemical and biological ones.

However, there are few works about the study and development of physical sensors using SAM films. Abdelghani et al. [44] proposed an optical fiber refractometer based on a SAM of silver. In a further report [45], the same authors presented other devices to detect diverse gases and vapors such as trichloroethylene, carbon tetrachloride, chloroform, and methylene chloride with low detection limits and response times lower than 2 min.

3.2 Langmuir-Blodgett films

Langmuir-Blodgett (LB) technique allows the design and generation of well-defined layered structures with molecular-level precision onto a substrate [46]. This technique is based on the fabrication of organic monolayer films, which are first oriented on a subphase and subsequently transferred with a molecule-specific surface pressure onto a solid surface, in a layer-by-layer (LbL) manner [47]. The materials used for the LB deposition

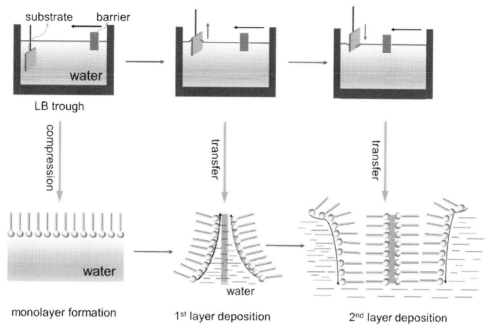

Fig. 6 Scheme of the LB fabrication films. (1) Monolayer formation; (2) Transferring the monolayer onto the substrate; (3) Second monolayer formation with opposite orientation. *Reprinted with permission from M. Jie Yin, B. Gu, Q. F. An, C. Yang, Y. L. Guan, K. T. Yong, Recent development of fiber-optic chemical sensors and biosensors: mechanisms, materials, micro/nano-fabrications and applications, Coord. Chem. Rev. 376 (2018) 348–392, © 2018 Elsevier.*

are commonly amphiphilic molecules that have hydrophobic and hydrophilic tail or head groups that have been previously prepared in solvents. Fig. 6 represents the LB film fabrication process [48]. First, a monolayer of the material is formed by compression with a constant pressure on the surface of an aqueous subphase. The molecules are oriented with the hydrophilic part in the water and the hydrophobic part upward forming the monolayer. Then, the monolayer is transferred onto the substrate using a barrier that allows the compression of the monolayer. This step is analogous to the LbL method, controlling a constant and slow speed of the substrate withdrawal and, in this case, applying a constant surface pressure during the transfer process. Next, it is possible to deposit a second layer with an opposite orientation of the hydrophobic and hydrophilic tails as shown in Fig. 6.

In the LB technique, it is possible to control the thickness with a resolution of ca. 1 nm per layer. Thus, LB became in an interesting fabrication method for waveguide applications and planar substrates [49, 50]. The fabrication of nanostructured LB films for the development of optical fiber sensors had an important relevance in the 1990s with a large list of publications [51–57]. For example, Charter, Ashwell, Tatam et al. [52] first studied the evanescent field provoked by LB films onto polished optical fibers and then developed diverse chemical fiber sensors [53, 57]. They also reported other approaches with LB films onto fiber gratings [58] and Fabry-Perot interferometers [59]. Among other interesting works, ethanol sensors [55] or gas sensors [56] can be found in the literature. During the last decade, volatile organic compound (VOC) sensors using LB films have been also reported [60, 61].

3.3 Electrospinning and electrodeposition techniques

Electrospinning is a technique used to produce synthetic micro- and nanofibers starting, most of the time, from polymeric solutions. The fabrication process of the so-called electrospun nanofibers (NFs) is based on the application of a strong electric field between a precursor solution and the substrate to be coated. The solution is held at the end of a capillary tube, which is subjected to the mentioned electric field that induces charge on the liquid surface. As the intensity of the electrical field is increased, the hemispherical surface of the solution at the tip of the capillary tube elongates to form a conical shape known as the Taylor cone [62, 63]. Eventually, a part of the solution is ejected from the tip of the Taylor cone and the jet is accelerated toward the metallic collector or a different substrate. The jet thins fast because of the elongation and evaporation of the solvent until polymeric micro- or nanofibers are deposited onto the substrate. An illustrative scheme of a simple approach of the electrospinning process is represented in Fig. 7, placing an optical fiber as a substrate in front of the collector.

Electrospinning presents numerous advantages such as simplicity, fast production, effectiveness, and low-cost, which enable the implementation of NFs in a wide range

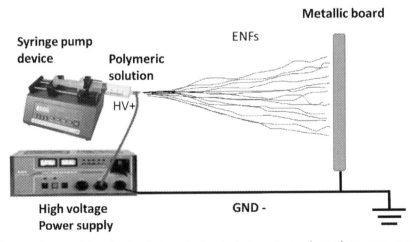

Fig. 7 Setup scheme of the simple electrospinning technique to produce electrospun nanofibers. *Reprinted with permission from P. J. Rivero, A. Urrutia, J. Goicoechea, F. J. Arregui, Nanomaterials for functional textiles and fibers, Nanoscale Res. Lett. 10(1) (2015) 501. © 2015 (http://creativecommons. org/licenses/by/4.0/).*

of research fields [63, 64]. Furthermore, the NFs-based structures present other strengths such as high surface area, small diameter, and specific physicochemical properties [65].

Electrospun natural NFs of hyaluronic acid, chitosan, chitin, silk, cellulose, or collagen, among others, can be found in the literature [66]. On the other hand, synthetic NFs have been produced from a wide list of polymeric compounds by electrospinning: Poly acrylic acid (PAA), poly(lactic acid) (PLA), polyamide (PA), poly(ethylene glycol) (PEG), poly(vinyl alcohol) (PVA), poly(ethylene oxide) (PEO), poly(ethylene-co-vinyl acetate) (PEVA), poly(vinyl acetate) (PVAc), poly(acrylonitril) (PAN), and more [67].

In this deposition technique, there exists a wide number of key fabrication parameters for the electrospun NFs production. Regarding the precursor polymeric solution, viscosity, surface tension, conductivity, and intrinsic solvent characteristics must be considered. Apart from the solution properties, other factors can be determinant in the process, for example, electric voltage applied, distance between the tip and the collector, ambient conditions such as temperature and humidity, or the shape, size, and location of the substrate [68]. The management of these parameters allow the deposition of NFs with the desired properties in terms of fiber diameter, compositions, fiber alignment, robustness, and other functionalities. These characteristics are useful in relevant applications, such antibacterial purposes, water treatment, wound dressing, tissue engineering, catalyst, drug delivery, flame retardants, microwave absorption, sensing, and other biomedical issues [69]. In the sensing field, the fabrication of sensitive coatings suitable to be deposited on optical fibers has become an alternative to design novel optical fiber devices [70]. Thus, a first generation of optical fiber sensors based on electrospun NFs have been recently published, monitoring physical parameters as relative humidity (RH) [71–73], or volatile compounds [74, 75].

Urrutia et al. [71] reported a study about the application of PAA electrospun NF mats on the surface of optical fibers to fabricate humidity sensors. NF mats with thicknesses and fiber diameters were deposited on short segments of cladding-removed multimode optical fibers and the optical response against humidity variations was analyzed. This study demonstrated how the swelling property of the PAA NFs affected the evanescent field of the optical fiber, thus provoking changes in the surrounding RI and eventually in the optical signal. Furthermore, thanks to the high surface area of the PAA NFs, the developed sensors stood out for their fast response times, even allowing real-time human breathing monitoring tests. Other humidity sensors with PVDF nanowebs deposited onto the surface of a segment of a hollow core fiber was presented by Corres et al. [73]. The coated segment was coupled to two multimode fibers at both extremes, fabricating a novel evanescent field fiber device. This design offered an average sensitivity and hysteresis of 0.05 dB/%HR and $\pm 0.5\%$, respectively, as well as a response time below 100 ms.

More recently, electrospun NFs have also been applied to volatile compounds detection. Thus, Bagchi et al. [74] have reported an optical fiber sensor probe coated with

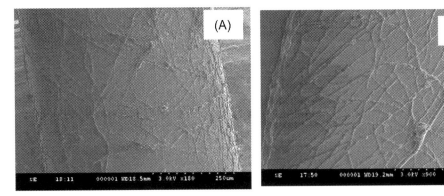

Fig. 8 (A) SEM image of PPO composite fibrous mat on optical fiber and (B) a magnified section of the coated fiber. *Reprinted with permission from S. Bagchi, R. Achla, S. K. Mondal, Electrospun polypyrrole-polyethylene oxide coated optical fiber sensor probe for detection of volatile compounds, Sensors Actuators B Chem. 250 (2017) 52–60. © Elsevier 2017.*

electrospun polypyrrole-polyethylene oxide fibers to monitor ammonia, trimethylamine, ethanol, methanol, and acetone. A 600-μm core fiber was etched by hydrofluoric acid to assure the presence of stronger evanescent wave on the surface of the optical fiber and enhance the interaction with the surrounding media composed of the NFs and the different gases. Fig. 8 shows a SEM image with the detail of the electrospun NFs deposited on the optical fiber. The sensor probe showed different detection limit such as 0.1 ppm, 2 ppm, 2 ppm, and 1 ppm and recovery time of 10 min, 54.4 s, 21.06 s, and 11.08 s for ammonia, ethanol, methanol, and triethylamine vapors, respectively. These results demonstrate that electrospinning can help to design new alternative optical fiber-based physical sensors.

Electrodeposition (ED) is another technique to fabricate coatings from a precursor ion solution onto an electrically conducting surface by applying an electric field between them. Its main advantages are rapidity, low cost, industrial applicability, higher deposition rates, easy to control alloy composition, and production of coatings on widely differing substrates. ED and electrospinning setups are quite similar. Whereas NFs are created by electrospinning, here the charged ions are attracted to the substrate, forming a nanocomposite-based coating, which depends on the ion's nature. To date, this technique is not very common in the development of physical optical fiber sensors.

A variant of the ED called electrostatic deposition pyrolysis (EDP) was used by Cusano et al. [75] to fabricate ammonia fiber sensors. They fabricated SnO_2-sensitive layers on the optical fiber end. Using a tin chloride solution during the process, when the droplets of the solution reach the heated substrate, whose temperature was typically 300–450°C, the chemical reaction of tin chloride with water vapor of the solution, stimulated by the temperature, takes place with formation of the tin oxide film. Then, the fiber was heated at 500°C for 1 h to ensure the eventual transform to present amorphous

SnO_x in SnO_2. This work shows the importance of the layer morphology on the further sensitivity of the fabricated devices. The proposed ammonia sensor presented a sensitivity of 7.7×10^{-3} ppm^{-1} at the 0–1 ppm range, with an estimated resolution of ca. 80 ppb and response times of a few minutes. Other interesting works have developed alkane vapor sensors thanks to an electrospray coating composed of polyisoprene and brilliant blue dye [76], and ammonia sensors by Mohamed et al. by just spraying polyaniline mixed with graphene on etched tapered single-mode fibers [77, 78].

3.4 Spin coating

Spin coating (SC) technique consists of the application of a small amount of a volatile solvent-based solution onto the substrate, which is then rotated at high speed to spread the fluid by centrifugal force. Rotation of the substrate is continued while the fluid spins off the edges of the substrate, until the desired thickness of the film is achieved and the solution evaporates [79]. This method enables the fabrication of uniform thin films with layers under 1 μm of photoresists or other materials and is commonly used in photolithography and microfabrication. The angular speed of spinning is the key parameter to control the desired final film thickness and generally, the higher the speed, the thinner the film. Apart from the spin speed, other factors such as the solute diffusivity, the type of solvent, or the kinematic viscosity affect the final properties of the deposited coatings [80]. Thus, SC allows the fabrication of flat and homogeneous films at a nanoscale level using diverse materials suitable in sensing applications.

Nevertheless, SC process implies some limitations to coat the inherent cylindrical and nonflat shape of optical fibers. To overcome this issue, optical fibers are usually pretreated to prepare a flat surface. Hence, side-polished fibers [81] or D-shape fibers [82] can be successfully coated. An interesting work with a side-polished single-mode fiber spun coated with an appropriate polymer was reported by Nagaraju et al. [81] to design and fabricate an evanescent field-based temperature sensor with a sensitivity of 5.3 nm/°C at the 26–70°C range. Other relevant study using D-shape PCFs deposited by SC have been reported by Kim et al. [82]. They first polished a PCF to produce a D-shape fiber. Then, a photoresist was spin-coated on the flat surface of the D-shape fiber, thus obtaining a photoresist grating. The optical response of the resultant device was measured as a refractometer, obtaining a sensitivity to the external ambient index change ca. 585.3 nm/RIU. Furthermore, the proposed device presented a temperature sensitivity of −0.3 nm/°C in the range from 30°C to 100°C.

Another alternative to polish a segment of optical fiber is to deposit the material on the end facet of such fiber. Thus, Wang et al. [83] developed a repetitive and simple humidity sensor by just SC a multimode optical fiber tip with a $CoCl_2$-PVA/SiO_2 mixture precursor solution Table 3.

Table 3 Summary of the different optical fiber physical sensors fabricated with sensitive coatings using electrospinning, spin coating, and other related techniques.

Technique	Optical structure	Precursor solution	Target	Reference
ES	Cladding-removed MMF	PAA in water	Relative humidity	[71]
ES	Hollow core fiber	PVDF in DMF	Relative humidity	[73]
ES	Etched 600-µm MMF	PVP-FeCl$_3$-PEO	Ammonia, ethanol, triethylamine, methanol, acetone	[74]
EDP	End face of fiber	Tin chloride	Ammonia	[75]
ESP	Plastic-clad fiber	Polyisoprene/brilliant blue	Alkane vapor	[76]
SP	Etched tapered SMF	PANI/graphene fibers	Ammonia	[77], [78]
SC	End face of MMF	CoCl$_2$-doped PVA-SnO$_2$	Relative humidity	[83]
SC	Side-polished SMF	Polymeric mixture	Temperature	[81]
SC	D-shape PCF	NOA81© photoresist	Relative humidity Temperature	[82]

ES, electrospinning; *EDP*, electrostatic deposition pyrolysis; *ESP*, electrospray pyrolysis; *SC*, spin coating.

3.5 Sol-gel technique

Sol-gel deposition is one of the most versatile and simplest techniques for the formation of thin films. This deposition technique is based on the method called by the same name. The sol-gel method is an outstanding technique for synthesizing porous, glass–like materials and ceramics, nanoparticles, or nanocomposites [84]. This process has been extensively investigated by the scientific community because sol-gel reactions can produce a high variety of inorganic networks that are prepared from metal alkoxide solutions [85]. All these reactions consist of the preparation of a sol, successive gelation, and solvent removal to produce a gel, a substance that contains a continuous solid skeleton enclosing a continuous liquid phase [86–88]. The liquid precursors are the mentioned alkoxides. After the whole process, different materials can be obtained depending on the type of precursor, the temperature, the water:alkoxide molar ratio, the pH, the nature and concentration of the catalyst, or the addition of other compounds, among other factors. The sol-gel-derived materials have important applications in fields as diverse as chemistry, optics, electronics, biology, biomedicine, or materials.

The sol-gel process can be employed to coat surfaces or other type of substrates thus obtaining the deposition of sol-gel coatings. Specifically, its application to the optical fiber structures is generally performed by dip-coating. This way, the optical fibers are immersed in a sol-gel solution that contains the precursor alkoxide, water, cosolvent, and catalyst. Once the sol-gel precursor is prepared and aged for a specific period of time, the desired optical fiber substrate, previously cleaned and treated, is immersed into it. After a determined period, the substrate is pulled out from the gel. The changes observed in the solution during the process and the type of catalysts used (acid or basic) have a great influence on the final structure of the gel and, hence, in the resultant coating properties. Another critical parameter is the withdrawal speed of the fiber structures, which determines the film thickness. Afterward, the substrates are dried during a fixed time, and the process can be repeated. In some approaches, a post-treatment is also necessary to stabilize or improve the film properties [89, 90].

One of the most remarkable features of the sol-gel process is the possibility of adding nanoparticles, dyes, or other substances to the precursor solution to fabricate sol-gel films with sensitive properties. The additional molecules are mixed and uniformly distributed in the solution and, after gelation, they remain entrapped in the final sol-gel matrix structure. Thanks to this easy way of entrapping doping agents into thin films, numerous sensing applications have emerged during the two last decades. In the field of optical fiber sensors, several new chemical and physical novel devices have been also reported.

An interesting study of different indicators for the development of temperature fiber optic sensors was reported by Garcia Moreda et al. [91]. In such study, some fluorescent dyes were entrapped in a sol-gel matrix coating, thus obtaining several fluorescence-based temperature sensors. Sol-gel solutions were composed of a determined dye (fluorescein, quinacrine dihydrochloride [QD], among others) and a metal alkoxide colloidal solution containing 7% silica and 93% of organic solvents. Then, optical fibers were immersed in the solutions by a controlled manner to obtain uniform and homogeneous thin films. After a curing process, optical fibers were characterized and measured as temperature sensors. Light was sent from the LED source through a coupler and excited the fluorescence of the sensor that stood inside a furnace at an established temperature. Both fluorescence and reflected excitation signal travelled back crossing the coupler again, reaching a spectrometer where the variations of fluorescence intensity with temperature were monitored. Results of the developed sensors with fluorescein and QD as fluorescent agents are shown in Figs. 9 and 10, respectively.

Regarding RH-sensing approaches, a relevant work was published by Estella et al. [92] using a porous silica xerogel film as a sensing element. The film was synthesized by the sol-gel process and fixed onto the end of the optical fiber by immersing it into a TEOS:Ethanol:water:HCl solution with an specific stoichiometric ratio, pH level, temperature, and mix aging. The resultant coated optical fiber was dried for a week to create the desired xerogel film. Then, the sensor was measured using the experimental

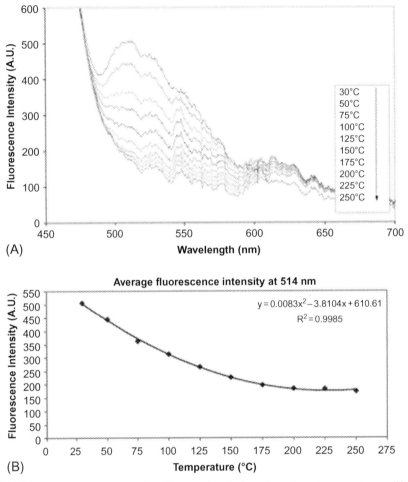

Fig. 9 (A) Fluorescence spectrum of a fluorescein device at various temperatures; (B) average fluorescence intensity calculated over a 10-nm interval around fluorescein emission wavelength (514 nm) at various temperatures. The excitation light was a 440-nm LED. *Reprinted with permission from F. J. García Moreda, F. J. Arregui, M. Achaerandio, I. R. Matias, Study of indicators for the development of fluorescence based optical fiber temperature sensors, Sensors Actuators B Chem. 118 (1–2) (2006) 425–432. © 2006 Elsevier.*

setup shown in Fig. 11. Light is emitted by a white light source and travels through a coupler and reaches the end of the xerogel-coated fiber. Subsequently, the coating acts as an optical nanocavity that provokes a Fabry-Perot interferometer. Thus, the reflected light can be received by a spectrometer. The optical spectrum is analyzed and monitored in a function of the controlled humidity variations from the measuring cell. This sensor was capable of sensing in a very high humidity range, from 4% to 100%, and had response

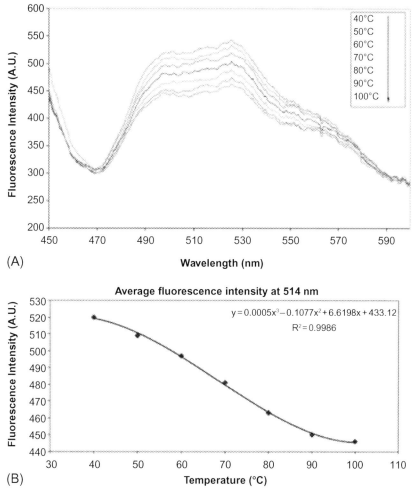

Fig. 10 (A) Fluorescence spectrum of a QD device at various temperatures; (B) average fluorescence intensity calculated over a 10-nm interval around QD emission wavelength (496 nm) at various temperatures. The excitation light was a 405-nm LED. *Reprinted with permission from F. J. García Moreda, F. J. Arregui, M. Achaerandio, I. R. Matias, Study of indicators for the development of fluorescence based optical fiber temperature sensors, Sensors Actuators B Chem. 118(1–2) (2006) 425–432. © 2006 Elsevier.*

times between 10 s and 2 min, depending on the RH percentage and the measuring procedure.

Moreover, the porosity and the high surface area ratio of sol-gel-based coatings are especially convenient for gas detection. Thus, there are a wide range of applications to detect specific gases such as oxygen [93, 94], carbon dioxide [95–97], hydrogen [98–101], ammonia [102, 103], nitrogen dioxide [104], or H_2S [105]. For oxygen concentration

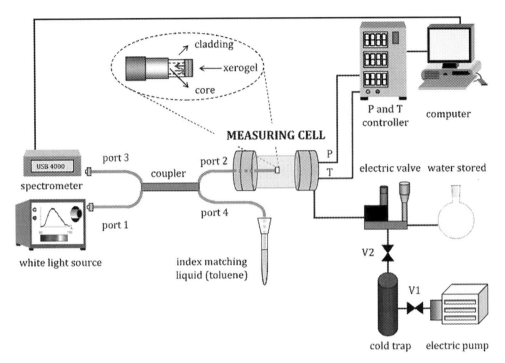

Fig. 11 Experimental setup scheme for the characterization and analysis of the fiber optic humidity sensor reported by Estella et al. [92]. *Reprinted with permission from J. Estella, P. de Vicente, J. C. Echeverría, J. J. Garrido, A fibre-optic humidity sensor based on a porous silica xerogel film as the sensing element, Sensors Actuators B Chem. 149(1) (2010) 122–128. © 2010 Elsevier.*

detection, Cao et al. [93] used methylene blue as an entrapping dye in the silica gel film. Another strategy was reported by Chu et al. [94] including platinum complexes into fluorinated xerogels. Regarding hydrogen sensing, the development of fiber devices are mainly based on metal-oxide films such as titanium oxide [99], tin oxide [98], or WO_3 [101] coatings doped with other metal nanoparticles such as palladium [99] or platinum [100, 101]. Apart from these gases, the monitoring of VOCs have been also studied thanks to the development of optical fiber sensors based on sol-gel coatings. Hence, supporting ZnO films doped with aluminum particles [106, 107] provide a precise behavior for ethanol, methanol, and acetone concentration measurements. Other developed devices based on other organometallic compounds [108] or doping agents such as gadolinium [109] have been reported to detect ethanol and methanol in the vapor phase. Also, other VOCs as n-hexane, 2-propanol, toluene, 1-butylamine, acetone, or cyclohexane have been also studied and detected using optical fiber structures coated by sol-gel films [110, 111].

Nevertheless, sol-gel technique has also some disadvantages or critical points to consider. The most important issue is a lack of precision in the thickness control at the nanoscale range. Thus, in some applications where the required thickness is below 100 nm, sol-gel coatings are usually considered as thick. Another limiting aspect can be the effective RI of the resultant coatings. Generally, sol-gel materials provide an RI up to 1.76. Other possible concerns are microstructural stability, response time, or repeatability. However, sol-gel technique is a simple, versatile, and low-cost processing method for the development of optical fiber sensing devices as mentioned. This fact can be appreciated in the works presented in Table 4.

3.6 Layer-by-layer nanoassembly technique

One of the nanodeposition techniques most employed for the fabrication of nanostructured thin films with an excellent control over the resultant chemical structure is the LbL nanoassembly technique. This deposition process is based on the sequential adsorption of molecular-thickness layers onto a desired substrate, being the opposite of the electrostatic charge of the coating materials of the main adsorption force for the fabrication of a multilayer assembly structure. In addition, this technique is considered as a "wet-chemistry" technology because the materials used for the fabrication of the thin films are found in the form of aqueous solutions (denoted as polyelectrolytes), which show ionizable functional groups to form charged polyions such as polycations (positive charge) or polyanions (negative charge), respectively. The great versatility combined with the simplicity and easily scalability of this nanodeposition technique makes possible a high degree of control over the thickness, showing a well-organized structure with a high uniformity as a function of critical parameters such as pH of charged polyelectrolytes, number of the bilayers (resultant thickness), and the ionization degree of the selected dipping solutions [126]. In the literature, a growth of numerous of scientific articles based on the implementation of LbL nanocoatings can be found for optical fiber sensing applications. In addition, a wide variety of sensitive coatings combined with a specific optical structure and sensing mechanism have been successfully employed for the detection of the desired parameter (mostly RH, refractive index, temperature, or even strain). One of the most common and used polyelectrolytic pairs for the fabrication of LbL nanocoatings are poly(allylamine hydrochloride) (PAH) as polycation and poly(acrylic acid) (PAA) as polyanion. Both polyelectrolytes can be used as an effective multilayer structure for sensing applications. A representative example can be found [127] where PAH/PAA structure has been deposited onto an optical fiber core, being this polymeric coats the resultant sensitive region for RH changes. During the fabrication process of this PAH/PAA multilayer structure, multiple resonances denoted as LMR can be clearly appreciated, showing a different spectral shift as a function of the coating thickness. Another interesting work that consists of a humidity sensor based on hetero-core optical fiber was presented Akita et al. [128]. In this work, the polymeric structure used as a sensitive coating is based on highly hygroscopic polymers such as poly-glutamic acid sodium (poly-Glu) (polycation) and

Table 4 Summary of the different optical fiber sensors based on sol-gel coatings to detect physical parameters with their corresponding sensitive coating, optical structure, sensing mechanism, and parameter of study.

Sensitive coating	Optical structure	Sensing mechanism	Parameter detection	Reference
Al-doped ZnO nanostructures	Tapered plastic optical fiber	Evanescent wave field	Relative humidity	[112]
Organic-silica hybrid material (di-ureasil)	Fiber Bragg grating	Wavelength variation	Relative humidity	[113]
Silica film doped with Methylene Blue	U-bend fiber	Evanescent wave field	Relative humidity	[114]
Silica nanostructured matrix doped with ZnO NPs	End facet of a cladding-removed multimode fiber	Evanescent wave field	Relative humidity	[115]
Xerogel (from TEOS, ethanol and water)	End facet of a multimode fiber	Fabry-Perot interference	Relative humidity	[92]
Porous silica film (from TMOS, water and HCl)	Bent cladding-removed multimode fiber	Evanescent-wave field	Relative humidity	[116]
Magnesium oxide films	U-shape glass rod	Intensity	Relative humidity	[117]
Xerogels (from TEOS, ethanol, water and HCl)	Polished cladding-removed optical fiber	Evanescent-wave field	Refractive index	[118]
Tin dioxide derived gels (from SnO2 in ethanol)	Long Period Gratings	Wavelength variation	Refractive index and gases	[119]
Zirconium acetate gel film onto silver layer	Cladding-removed optical fiber	SPR	Refractive index	[120]
Indium tin oxide films	Etched cladding-removed optical fiber	SPR	Temperature and gases	[121]
Fluorinated xerogels doped with PtTFPP and 5(6)-carboxyfluorescein	End facet of 1000-μm plastic fiber	Relative luminescence intensity	Temperature and oxygen	[122]
GPTMS and APTMS doped with CdSe/ZnS quantum dots	Tip of an optical fiber	Fluorescence intensity	Temperature	[123]
Gels (from Phtes, Mtes, TEOS, ethanol, water, HCl)	Fused tapered coupler	Wavelength variation	Temperature	[124]

Table 4 Summary of the different optical fiber sensors based on sol-gel coatings to detect physical parameters with their corresponding sensitive coating, optical structure, sensing mechanism, and parameter of study—cont'd

Sensitive coating	Optical structure	Sensing mechanism	Parameter detection	Reference
Silica gel doped with Europium (III)	Silica optical fibers	Fluorescence intensity	Temperature	[125]
Silica gels dopes with fluorescent dyes	Perpendicular, irregular, and tapered ends of fibers	Fluorescence intensity	Temperature	[91]
TEOS and TMOS based gels doped with thymol blue	Etched cladding removed optical fiber	Intensity	Carbon dioxide	[95]
Mtes-based gels doped with thymol blue, phenol red, methyl red	Cladding-removed optical fiber	Intensity	Carbon dioxide	[96]
Silica gel doped with methylene blue	Bent U-shape cladding-removed fiber	Evanescent wave field	Oxygen	[93]
Fluorinated xerogels doped with Platinum (II) complexes	End facet of multimode optical fibers	Fluorescence intensity	Oxygen	[94]

TEOS, tetraorthosilicate; *TMOS*, tetramethylorthosilicate; *PtTFPP*, platinum tetrakis pentrafluoropheny porphine; *GPTMS*, 3-glycidoxypropyl trimethoxysilane; *APTMS*, 3-aminopropyl trimethoxysilane; *Phtes*, phenyltriethoxysilane; *Mtes*, methyltriethoxysilane.

poly-lysine hydrobromide (poly-Lys) (polyanion). The experimental results indicate that light intensity is changed in the range of 26.4%–92.9% RH, showing a fast response time.

However, one important advantage of the LbL assembly is that it can be performed on a wide variety of different materials (not only polyelectrolyte nature), making possible the fabrication of advanced LbL coatings [129]. In this sense, several works related to optical fiber sensors based on the incorporation of organic dyes (Polymeric Dye R–478, Prussian Blue), metallic nanoparticles (silver, gold), or metal oxide nanoparticles (Al_2O_3, TiO_2) can be found in the bibliography for the fabrication of high-performance optical fiber sensors [130]. A representative example was presented Corres et al. [131] where the response of a tapered optical fiber humidity sensor is optimized, attending to thickness of the sensitive coating, the dimensions of the taper, the light source, and the utilization of sensitive materials with different refractive indexes by using poly(diallylmethylammonium chloride) (PDDA) and polymeric dye R–478 (Poly-R) as a sensitive coating. In this work, it is demonstrated that sensitive coatings with high refractive indexes show a highly oscillating behavior (up to 35 dB), which allows the use of these nanostructures to obtain very high sensitive humidity sensors.

Another example was presented in which an optical fiber humidity sensor based on LMRs was used as a sensitive coating a multilayer porous structure composed of titanium dioxide (TiO_2) and poly(sodium 4-styrenesulfonate) (PSS) [132]. In this work, the presence of two differentiated resonances within the spectral range (400–1700 nm) makes it possible to perform dual-reference measurements with a high sensitivity and large dynamic range when the surrounding medium relative humidity varies from 20% to 90%.

Other examples of optical fiber humidity sensors based on the immobilization of metallic nanoparticles can be found in the literature [133], [134] [135], [136]. In these works, silver nanoparticles (AgNPs) with spherical shape (PAA-AgNPs) have been successfully synthesized, showing an intense LSPR band in the visible range (around 440 nm). Rivero et al. [135] presented the fabrication and characterization of an optical fiber humidity sensor based on both LMR and LSPR, respectively, for the first time. In addition, a great difference in their sensitivities has been clearly appreciated because the LSPR inherent to the AgNPs only showed a slight wavelength variation, whereas the LMR bands showed a strong wavelength response to Relative Humidity (RH) changes from 20% to 70% RH at 25°C, as can be observed in Fig. 12.

In Rivero et al. [136], they presented an optical fiber sensor based on the immobilization of the silver metallic nanoparticles into LbL films, although the incorporation of the AgNPs is performed by using a chemical route denoted as in situ sythesis (ISS) process of the AgNPs inside a polymeric matrix of PAH/PAA previously deposited onto an optical fiber core. The presence of these AgNPs directly affects the RI of the overlay promoting the observation of a LMR band in the infrared region, which is used as a sensing signal to monitor human breathing, showing good sensitivity (0.455 nm per RH%) and fast response time (692 ms and 839 ms for rise/fall), as seen in Fig. 13.

Another interesting work based on the LMR phenomenon for humidity detection (from 20% RH up to 90% RH) was presented by Hernaez et al. [35]. A thin film made of alternating polyethylenimine (PEI) and graphene oxide (GO) was used as an effective sensitive coating, which is deposited on a SnO_2-sputtered fiber core. The experimental results corroborate that the response of the optical fiber sensor is very sharp at high RH values, which means this device is more sensitive at greater wetness values, and it can be also used to monitor human breathing for consecutive and repetitive inhalation/exhalation cycles. Finally, the sensor exhibits very fast response time (160 ms) and, especially, extremely rapid recovery time (262 ms), clearly improving on those of some ultrafast optical fiber sensors found in the literature.

Another aspect to remark upon is that the sensitive coatings obtained by LbL nanoassembly technique can be also implemented for the detection of RI and, as a result, optical

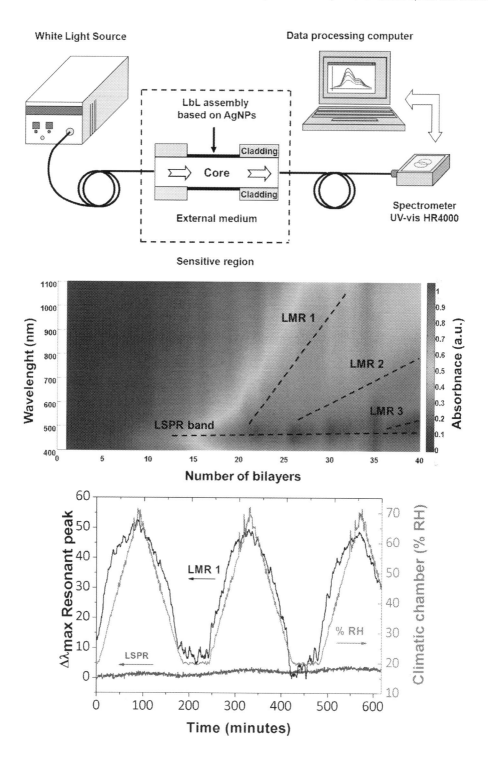

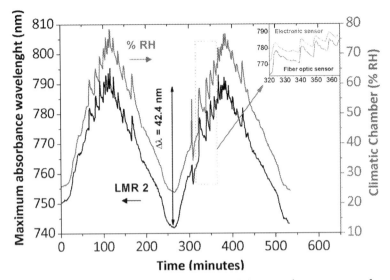

Fig. 12 Schematic representation of the measuring setup (A); spectral response as a function of the number of bilayers added during the LbL deposition technique (B); dynamic response of 25 bilayer device (C) and 40 bilayers device (D), respectively. In all the cases of study, the wavelength shift has been monitored to RH cycles from 20% to 70% RH at 25°C. *Reprinted with permission of P. J. Rivero, A. Urrutia, J. Goicoechea, F. J. Arregui, Optical fiber humidity sensors based on localized surface plasmon resonance (LSPR) and Lossy-mode resonance (LMR) in overlays loaded with silver nanoparticles, Sensors Actuators B Chem. 173 (2012) 244–249. © 2012 Elsevier.*

fiber refractometers have been successfully developed. In another study, an optical fiber tip sensor was shown for RI measurements by using poly(diallylmethylammonium chloride) (PDDA) and poly(sodium 4-styrenesulfonate) (PSS) as multilayer sensitive coating [137]. Another interesting work about optical fiber refractometer was in which LbL method was applied to fabricate a coating of TiO_2 NPs onto the multimode optical fiber core [138]. The resultant device shows an absorption band in the infrared region that is shifted to higher wavelengths when the RI is gradually increased, reaching a dynamic range of 157 nm and a sensitivity of 1987 nm/RIU, as seen in Fig. 14.

Other types of different nanoparticles such as AuNPs can be also incorporated into LbL films [139], [140], [141]. Optical fiber refractometers based on the successive incorporation of AuNPs with a spherical shape make possible the design of a dual LSPR–LMR device where the LSPR band is used as a reference signal and the LMR band is used as a sensing signal to variations of the external refractive index [140]. Other interesting work

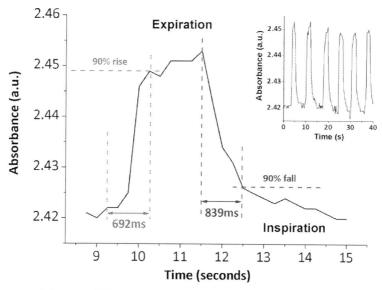

Fig. 13 Response of the optical fiber sensor to several consecutive human breathing cycles (*right*), and adetailed response time of one inhalation-exhalation cycle. *Reprinted with permission of P. J. Rivero, A. Urrutia, J. Goicoechea, I. R. Matias, F. J. Arregui, A Lossy mode resonance optical sensor using silver nanoparticles-loaded films for monitoring human breathing, Sensors Actuators B Chem. 187 (2013) 40–44. © 2013 Elsevier.*

was also presented in which optical fiber sensors based on gold nanorods (GNRs) were embedded in polymeric thin films [141]. These GNRs present two distinct LSPR peaks at different wavelengths as seen in Fig. 15. The first one is located at 530 nm, which is inherent to the transverse plasmon resonance (LSPR-T), whereas the second one is located at 710 nm and is inherent to the longitudinal plasmon resonance (LSPR-L). In addition, the experimental results indicate that LSPR-T shows a better sensitivity (75.69 dB/RIU) to the RI changes compared with the LSPR-L band (50.46 dB/RIU) in the spectral range from 1.33 up to 1.41, respectively.

The great versatility of the LbL nanoassembly technique enables the incorporation of quantum dots (QD) for temperature monitoring [142], [143]. Two different types of CdTe QD with diameter of 4 and 5 nm, which correspond to the green and red emission bands, were discussed in another study [142]. In both cases, the emission peak shifts to higher wavelengths and the luminescence intensity decreases when the temperature varies from 30°C to 100°C. In addition, the wavelength of the emission peak for both green and red QD-based devices is around 0.2 nm/°C, as seen in Fig. 16. Larrió et al. [143] also

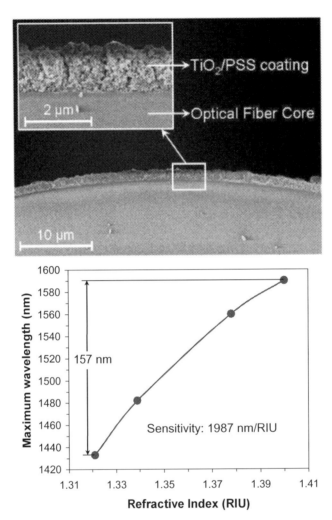

Fig. 14 SEM image and detail of the optical fiber core composed of TiO2/PSS coating (A) and variation of the maximum absorbance wavelength as long as the refractive index of the external medium changes (B). *Reprinted with permission from M. Hernaez, C. R. Zamarreño, I. Del Villar, I. R. Matias, F. J. Arregui, Lossy mode resonances supported by TiO2-coated optical fibers, In Procedia Engineering, vol. 5, 2010, pp. 1099–1102. © 2010 Elsevier.*

presented a temperature sensor by the deposition of QD films inside the holes of a PCF by means of the LbL technique. The experimental results clearly indicate that the optical absorption, the emission intensity, and the emission peak wavelength show a clear exponential behavior in the range from −40°C up to +70°C.

Other works are focused on the multifunctionality of the optical fiber sensors used for the multiparameter monitoring of relative humidity and temperature by using optical

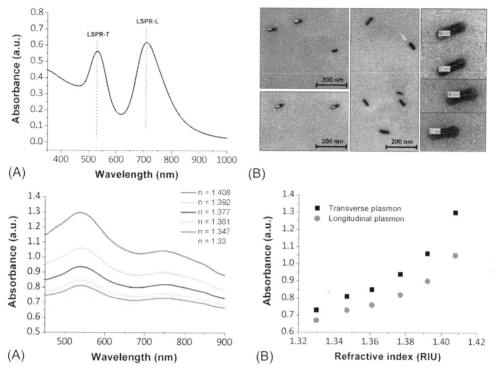

Fig. 15 (A) UV-vis absorption spectrum of the GNRs@PSS solution with two distinct absorption peaks at 530 and 710 nm, respectively; (B) TEM micrographs and some zoom details of the synthesized GNRs; (C) spectral response of the LSPR bands when the sensitive region is immersed in different RI solutions; (D) growth in intensity of the LSPR-T and LSPR-L maxima to external RI variations. *Reprinted with permission of A. Urrutia, J. Goicoechea, P. J. Rivero, A. Pildain, F. J. Arregui, Optical fiber sensors based on gold nanorods embedded in polymeric thin films, Sensors Actuators B Chem. 255 (2018) 2105–2112. © 2018 Elsevier.*

fiber long period gratings (LPGs) [144], [145]. According to this, a humidity-sensitive nanostructured polymeric coating is first deposited onto LPG, after being exposed to RH changes from 20% to 80% RH and temperature tests from 25°C up to 85°C, respectively [144]. And second, half of the LPG coating was chemically removed, and the semi-coated LPG was also exposed to the previous experimental conditions. The results indicate that the dual-wavelength-based measurement showed different sensitivity ratios, that of 62.33 pm/RH% and 410.66 pm/°C for the attenuation band corresponding to the coated contribution, whereas the sensitivity ratios for semi-half-coated LPG showed 55.22 pm/RH% and 405.09 pm/°C, respectively. Finally, the LPG array consists of two gratings, where one was kept bare to monitor the temperature change, whereas the second was modified with a mesoporous film of silica nanoparticles to measure the relative

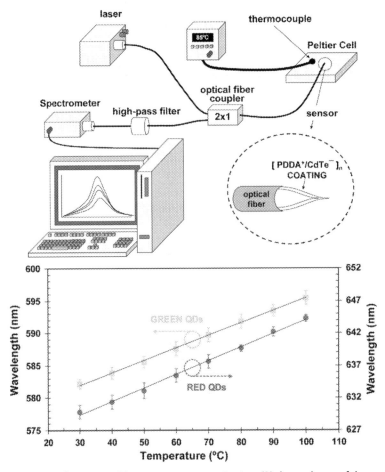

Fig. 16 (A) Experimental setup used for temperature monitoring; (B) dependence of the emission peak wavelength with respect to the temperature where the sensors were submitted to three complete cycles of heating and cooling, respectively. *Reprinted with permission of G. De Bastida, F. J. Arregui, J. Goicoechea, I. R. Matias, Quantum dots-based optical fiber temperature sensors fabricated by layer-by-layer, IEEE Sensors J. 6(6) (2006) 1378–1379. © IEEE 2006.*

humidity, showing a sensitivity of 0.46 ± 0.01 nm/°C and 0.53 nm/RH%, respectively [145] Table 5.

4. Other micro- and nanopatterning techniques

In the previous sections, the main experimental techniques that have proven to be good tools to create micro- and nanostructured coatings onto optical fiber substrates for sensing applications have been summarized. There is recent research interest in creating complex structures over the optical fiber tip, enabling the possibility of using phenomena such as

Table 5 Summary of the different optical fiber sensors to detect physical parameters with their corresponding sensitive coating, optical structure, sensing mechanism, and parameter of study.

Sensitive coating	Optical structure	Sensing mechanism	Parameter detection	Reference
[PAH/PAA]	Cladding-removed MMF	LMR	Relative humidity	[127]
[poly-Glu/poly-Lys]	Hetero-core MMF-SMF-MMF	Intensity	Relative humidity	[128]
[PDDA]/Poly-R]	Taper	Taper	Relative Humidity	[131]
[TiO$_2$/PSS]	Cladding-removed MMF	LMR	Relative humidity	[132]
[PAH/PAA-AgNPs]	Cladding-removed MMF	LSPR	Relative humidity	[134]
[PAH/PAA-AgNPs]	Cladding-removed MMF	LSPR + LMR	Relative humidity	[135]
[PAH/PAA] + ISS of AgNPs	Cladding-removed MMF	LMR	Relative humidity	[136]
[PEI/GO]	Cladding-removed MMF	LMR	Relative humidity	[35]
[PDDA/PSS]	End face of fiber	Intensity	Refractive index	[137]
[TiO$_2$/PSS]	Cladding-removed MMF	LMR	Refractive index	[138]
[PAH/PAA-AuNPs]	Cladding-removed MMF	LSPR + LMR	Refractive index	[140]
[PAH/GNRs@PSS]	Cladding-removed MMF	LSPR-L LSPR-T	Refractive index	[141]
[PDDA/CdTeQD]	Taper	Wavelength variation	Temperature	[142]
[PDDA/CdSeQD]	Hollow Photonic Crystal Fiber	Intensity variation	Temperature	[143]
[PAH/PAA]	Long Period Grating	Wavelength variation	Temperature and Relative Humidity	[144]
[PAH/SiO2]	Long Period Grating	Wavelength variation	Temperature and Relative Humidity	[145]

nanoplasmonics or ultra–high resolution near field optics to create new enhanced optical fiber sensors. These new approaches have attracted the interest of many scientists and researchers [146], and nowadays, it is possible to find the terms "Lab-on-Fiber" and "Lab-on-Tip" in the scientific bibliography [147–149]. Some approaches are based on the application of traditional lithographic techniques to create pattered structures using the cleaved tip of an optical fiber as substrate. For example, interference lithography has been used to create patterned nanopillar structures [150] using UV-curing photoresists for molecular identification by means of highly sensitive surface-enhanced Raman scattering (SERS). In other works, the creation of ordered nanoimprinted pillar-like structures using the natural microstructured tissue of the wing of an insect as a contact mask [151] has been reported. Such micropatterned structures can be metallized with gold or silver to create SERS probes for molecule identification. Other researchers have used different maskless photolithography approaches to reproduce the microstructured patterns, such as the projection of UV-curing images using a Digital Light-Processor chip. For example, Yin et al. proposed a simple and cost-effective 3D patterning approach to create Poly(acrylic acid) hydrogel patterns over a tapered optical fiber for miniature pH sensing applications [152]. In Fig. 17, it is possible to see the optical setup for the direct photoinscription of the hydrogel microstructures and confocal microcopies of different patterns over a tapered optical fiber.

An alternative technique for the creation of micro- and nanostructures is focus ion-beam milling (FIB). The FIB equipment resembles a scanning electron microscope, but instead of having an accelerated focused electron beam image the sample, in FIB an accelerated focused ion beam, typically gallium, is used. The projected gallium ions impact the sample with enough energy to sputter away the first nanometers of the surface, resulting in a micromilling tool. The typical minimum milling spot is around 15 nm; consequently, this technique is very useful for the selective removal of thin coating over the optical fibers with a remarkable submicrometric precision. For example, FIB can be used for etch away the surface of an optical fiber to create Fresnel lenses directly on the endface of an optical fiber [153], as shown in Fig. 18. Such structures control the light distribution in the surroundings of the fiber endface, giving the possibility of detect and manipulating (optical tweezers) small particles, such as PMMA microbeads and single living cells.

Other researchers have used FIB to create grating structures on tapered fibers [154] or in a standard fiber tip [155] that show a wavelength response with the temperature. There are other works in which a metal-coated fiber is micromachined using FIB to metallic gratings or metallic nanopillar arrays onto the perpendicularly cleaved optical fiber endface. For example, Yan et al. have created Pd gratings with a line width of 30 nm and a pitch of 450 nm that show how the Pd grating reflectance (with a peak around 700 nm) can red shift due to the absorption of hydrogen and the change in the geometrical proportions of the grating structure [156] Fig. 19.

A different approach to create microstructured optical fiber sensors is using laser ablation. High-power ultrashort laser pulses (from nanoseconds to femtoseconds) are able to vaporize the coatings or even the optical fiber itself. Such fs laser pulses can be focused in a controlled way to create piercings or grooves in metallic coatings. For example, Rizzo et al. have reported different microperforated metallic coatings over tapered optical fiber tips for dynamically addressable light delivery for optogenetic control of neural activity in the mammalian brain [157].

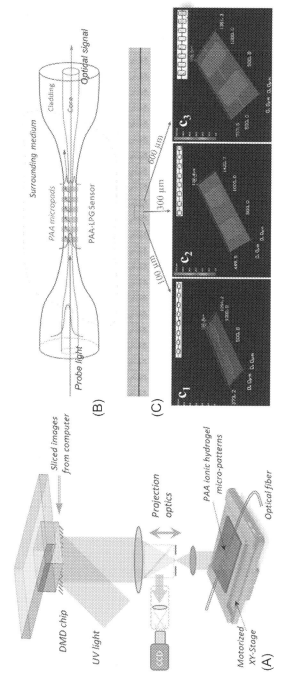

Fig. 17 (A) Schematic diagram of the optical maskless stereolithography system: UV light illuminates the digital-mirror device (DMD) chip, and the generated optical pattern is projected on the photosensitive polymer to fabricate microstructures. (B) Schematic design of a microfiber pH sensor based on micropatterned PAA ionic hydrogel. (C) Optical microscopic image of a tapered optical fiber with a diameter of 30 μm, with three kinds of PAA micropads of different sizes. *Reproduced with permission from M.-J. Yin, M. Yao, S. Gao, A. P. Zhang, H.-Y. Tam, P.-K. A. Wai, Rapid 3D patterning of poly(acrylic acid) ionic hydrogel for miniature pH sensors, Adv. Mater. 28(7) (2016) 1394–1399.*

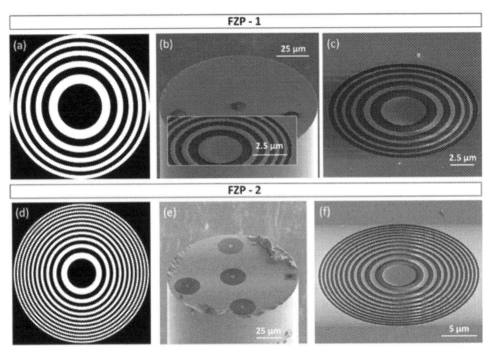

Fig. 18 Concentric circular microstructures created using Focused Ion Beam Milling (FIBM) on the cleaved tip of an optical fiber. Such structures can generate Fresnel zones that can be used to focus the light propagated though the optical fiber to act as an optical tweezer, or even to detect the presence of a single biological cell in the optical fiber tip surrounding [153]. Fresnel zone plates (FZP): (A) amplitude mask used to fabricate FZP-1; (B) image of the optical fibre with FZP-1, (inset) zoom of a FZP fabricated with a metallic film of Au/Pd; (C) zoom of FZP-1; (D) amplitude mask used to fabricate FZP-2; (E) image of the optical fibre with FZP-2; (F) zoom of FZP-2. Published in 2017 by Springer Nature Ltd. under Creative Commons 4.0.

Besides this new family of micromachined optical fiber probes, there are other already available technologies more focused in planar optics devices and the development of microfluidics [158] and Photonic Lab–on–Chip systems (PhLOC) [159]. For example, although some research works show optical fibers incorporated into microfluidic Poly(dimethylsiloxane) (PDMS) chips [160], [161], other researchers have used a different approach creating in–line microholes in the optical fiber with a second harmonic 400-nm femtosecond laser pulse [162]. The width of such holes is approximately 50 µm, acting as microfluidic in-line channels that maximize the interaction with a liquid sample, while minimizing the sample volume. The in-fiber microfluidic cells were coated with gold nanoparticles, and their LSPR absorption band was monitored for further sensing applications.

5. Conclusions and future outlook

In this chapter, there is a brief overview of the main micro- and nanofabrication techniques and tools for the development of enhanced optical fiber sensors. The optical fiber

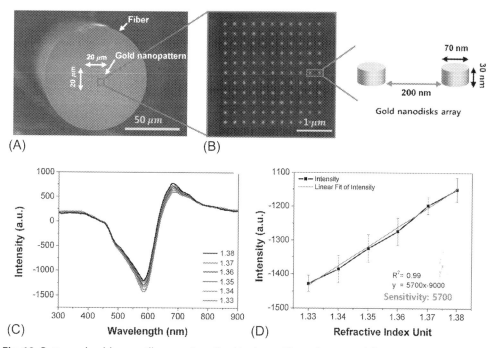

Fig. 19 Patterned gold nanopillar array inscribed in the endface of an optical fiber showing a Localized Surface Plasmon Resonance effect that can be used for biosensing applications. (A) FE-SEM image of the FO LSPR sensor surface after the lift-off process. (B) Enlarged FE-SEM image of the fabricated nanopattern and schematic view of the nanopattern structure. (C)The LSPR spectra are measured according to changes in the refractive index from 1.34 to 1.38. (D) Resonance intensity for various refractive indices. *Reproduced with permission from H.-M. Kim, M. Uh, D. H. Jeong, H.-Y. Lee, J.-H. Park, S.-K. Lee, Localized surface plasmon resonance biosensor using nanopatterned gold particles on the surface of an optical fiber, Sensors Actuators B Chem. 280 (2019) 183–191.*

has been demonstrated to be a powerful platform for sensing applications. The technological development in the last decades have enabled the possibility of creating optical coatings and structures that take advantage of optical phenomena such as high surface-to-volume ratio, optical nanoparticle–quantum confinement, plasmonic interactions, enhanced-Raman scattering, LMRs, among others to create a new generation of optical fiber sensors.

The main advantages of vapor deposition techniques (PVD and CVD) and wet deposition techniques (SAM, LB, electrospinning, SC, sol–gel, and LbL) have been reviewed. PVD and CVD processes provide a rapid growth of uniform structures with accurate control at the nanometric scale using a wide range of materials to fabricate sensitive coatings. Wet-chemistry techniques such as sol–gel and SC are versatile and simple deposition processes that make possible the fabrication of robust nanoporous sensitive thin films. SAM, LB, and LbL present well-defined layered structures with molecular level precision enabling the design of optical fiber sensors with an important improvement in sensitivity terms. Electrospinning and ED offer the development of other types of coatings with high surface-to-area ratios due to the NF nature morphology.

Finally, patterned structures enable additional features. The possibility of having ordered optical microdevices such as gratings, micropillar arrays, etc. make possible the use of phenomena such as plasmonic interactions, SERS, near-field light imaging for molecule identification, and ultrasensitive optical fiber devices.

The current trends seem to pursue further development of the previously described techniques and their combination to create more complex sensor structures and their integration with other technologies such as microfluidics using PDMS chips or even microfluidic channels fabricated in the optical fiber itself. These new approaches have the goal of achieving multiparametric sensing systems, close to the concept of Photonic Lab-on-Chip (PhLOC) and Lab-on-Fiber.

Acknowledgments

This work was supported in part by the Spanish Science, Innovation and Universities Ministry—FEDER Proyecto Retos de la Sociedad RTI2018-096262-B-C41, and by the Public University of Navarre (PJUPNA1936 project).

References

[1] Y. Xu, X.-T. Yan, Chemical Vapour Deposition, Springer, London, 2010.

[2] K.L. Choy, Chemical vapour deposition of coatings, Prog. Mater. Sci. 48 (2) (2003) 57–170.

[3] S.K. Chauhan, N. Punjabi, D.K. Sharma, S. Mukherji, A silicon nitride coated LSPR based fiber-optic probe for possible continuous monitoring of sucrose content in fruit juices, Sensors Actuators B Chem. 222 (2016) 1240–1250.

[4] X. Luan, R. Yu, Q. Zhang, S. Zhang, L. Cheng, Boron nitride coating of sapphire optical fiber for high temperature sensing applications, Surf. Coat. Technol. 363 (February) (2019) 203–209.

[5] M. Smietana, J. Szmidt, M.L. Korwin-Pawlowski, W.J. Bock, J. Grabarczyk, Application of diamond-like carbon films in optical fibre sensors based on long-period gratings, Diam. Relat. Mater. 16 (4–7 SPEC. ISS) (2007) 1374–1377.

[6] M. Śmietana, M. Koba, P. Mikulic, W.J. Bock, Combined plasma-based fiber etching and diamond-like carbon nanooverlay deposition for enhancing sensitivity of long-period gratings, J. Lightwave Technol. 34 (19) (2016) 4615–4619.

[7] M. Smietana, W.J. Bock, P. Mikulic, Temperature sensitivity of silicon nitride nanocoated long-period gratings working in various surrounding media, Meas. Sci. Technol. 22 (11) (2011).

[8] S.M. George, Atomic layer deposition: an overview, Chem. Rev. 110 (1) (2010) 111–131.

[9] M. Dominik, et al., Titanium oxide thin films obtained with physical and chemical vapour deposition methods for optical biosensing purposes, Biosens. Bioelectron. 93 (September 2016) (2017) 102–109.

[10] D.J. Mandia, et al., The effect of ALD-grown Al2 O3 on the refractive index sensitivity of CVD gold-coated optical fiber sensors, Nanotechnology 26 (43) (2015).

[11] M. Śmietana, M. Dudek, M. Koba, B. Michalak, Influence of diamond-like carbon overlay properties on refractive index sensitivity of nano-coated optical fibres, Phys. Status Solidi Appl. Mater. Sci. 210 (10) (2013) 2100–2105.

[12] D.M. Mattox, Physical vapor deposition processes, Prod. Finish. 65 (12) (2001) 72.

[13] H. Adachi, K. Wasa, Thin films and nanomaterials, in: Handbook of Sputtering Technology, Elsevier, 2012, pp. 3–39 no. Dlc.

[14] R.K. Singh, J. Narayan, Pulsed-laser evaporation technique for deposition of thin films: physics and theoretical model, Phys. Rev. B 41 (13) (1990) 8843–8859.

[15] P. Sánchez, C.R. Zamarreño, F.J. Arregui, I.R. Matías, LMR-based optical fiber refractometers for oil degradation sensing applications in synthetic lubricant oils, J. Lightwave Technol. 34 (19) (2016) 4537–4542.

[16] C.R. Zamarreño, S. Lopez, M. Hernaez, I. Del Villar, I.R. Matias, F.J. Arregui, Resonance-based refractometric response of cladding-removed optical fibers with sputtered indium tin oxide coatings, Sensors Actuators B Chem. 175 (2012) 106–110.

[17] S. Lopez, I. Del Villar, C. Ruiz Zamarreño, M. Hernaez, F.J. Arregui, I.R. Matias, Optical fiber refractometers based on indium tin oxide coatings fabricated by sputtering, Opt. Lett. 37 (1) (2012) 28–30.

[18] P. Sanchez, C.R. Zamarreño, M. Hernaez, I.R. Matias, F.J. Arregui, Optical fiber refractometers based on lossy mode resonances by means of SnO2 sputtered coatings, Sensors Actuators B Chem. 202 (2014) 154–159.

[19] J. Ascorbe, J.M. Corres, I.R. Matias, F.J. Arregui, High sensitivity humidity sensor based on cladding-etched optical fiber and lossy mode resonances, Sensors Actuators B Chem. 233 (2016) 7–16.

[20] F.J. Arregui, I. Del Villar, C.R. Zamarreño, P. Zubiate, I.R. Matias, Giant sensitivity of optical fiber sensors by means of lossy mode resonance, Sensors Actuators B Chem. 232 (2016) 660–665.

[21] A. Ozcariz, C.R. Zamarreño, P. Zubiate, F.J. Arregui, Is there a frontier in sensitivity with Lossy mode resonance (LMR) based refractometers? Sci. Rep. 7 (1) (2017) 10280.

[22] J. Ascorbe, J.M. Corres, F.J. Arregui, I.R. Matias, Optical fiber humidity sensor based on a tapered fiber asymmetrically coated with indium tin oxide, in: Proc. IEEE Sensors, vol. 2014, 2014, pp. 1916–1919.

[23] J. Ascorbe, J.M. Corres, F.J. Arregui, I.R. Matias, Refractive index sensing performance of a Bragg grating built up on the tip of an optical fiber by reactive sputtering, in: Proc. IEEE Sensors, vol. 2017, 2017, pp. 1–3.

[24] J. Ascorbe, J.M. Corres, F.J. Arregui, I.R. Matias, Humidity sensor based on bragg gratings developed on the end facet of an optical fiber by sputtering of one single material, Sensors 17 (5) (2017) 991.

[25] P. Hlubina, M. Kadulova, D. Ciprian, J. Sobota, Reflection-based fibre-optic refractive index sensor using surface plasmon resonance, J. Eur. Opt. Soc. 9 (2014).

[26] S. Sridhar, S. Sebastian, S. Asokan, Temperature sensor based on multi-layer MoS 2 coated etched fiber Bragg grating, Appl. Opt. 58 (3) (2019) 535.

[27] T. Li, Y.-T. Dai, Q.-C. Zhao, A new type of high sensitivity optical fiber temperature sensor with microstructure, Guangdianzi Jiguang/J. Optoelectron. Laser 25 (4) (2014) 625–630.

[28] Y. Tu, L. Ye, S.P. Zhou, S.T. Tu, An improved metal-packaged strain sensor based on a regenerated fiber bragg grating in hydrogen-loaded boron–germanium co-doped photosensitive fiber for high-temperature applications, Sensors 17 (3) (2017) 431.

[29] H. Xia, et al., Temperature-dependent fiber optic hydrogen gas sensor response characteristics, in: Photorefractive Fiber and Crystal Devices: Materials, Optical Properties, and Applications XII, vol. 6314, 2006p. 631411 no. 518.

[30] F.J. Arregui, et al., Fiber-optic lossy mode resonance sensors, in: Procedia Engineering, vol. 87, 2014, pp. 3–8.

[31] I. Del Villar, et al., Optical sensors based on lossy-mode resonances, Sensors Actuators B Chem. 240 (2017) 174–185.

[32] N. Jing, J. Zhou, K. Li, Z. Wang, J. Zheng, P. Xue, Refractive index sensing based on a side-polished macrobend plastic optical Fiber combining surface plasmon resonance and macrobending loss, IEEE Sensors J. 19 (14) (2019) 5665–5669.

[33] X. Li, S. Lin, J. Liang, Y. Zhang, H. Oigawa, T. Ueda, Fiber-optic temperature sensor based on difference of thermal expansion coefficient between fused silica and metallic materials, IEEE Photon. J. 4 (1) (2012) 155–162.

[34] C.J. Easley, L.A. Legendre, M.G. Roper, T.A. Wavering, J.P. Ferrance, J.P. Landers, Extrinsic Fabry-Perot interferometry for noncontact temperature control of nanoliter-volume enzymatic reactions in glass microchips, Anal. Chem. 77 (4) (2005) 1038–1045.

[35] M. Hernaez, B. Acevedo, A.G. Mayes, S. Melendi-Espina, High-performance optical fiber humidity sensor based on lossy mode resonance using a nanostructured polyethylenimine and graphene oxide coating, Sensors Actuators B Chem. 286 (2019) 408–414.

[36] A. Lopez Aldaba, et al., SnO2-MOF-Fabry-Perot optical sensor for relative humidity measurements, Sensors Actuators B Chem. 257 (2018) 189–199.

[37] X. Li, Y. Shao, Y. Yu, Y. Zhang, S. Wei, A highly sensitive fiber-optic fabry–perot interferometer based on internal reflection mirrors for refractive index measurement, Sensors 16 (6) (2016) 794.

[38] Y. Zhao, Y. Liu, Q. Guo, F. Zou, T. Wang, Refractive index sensitivity of the tilt long-period fiber gratings coated with nano-scale golden film, in: Proceeding—2013 12th International Conference on Optical Communications and Networks, ICOCN 2013, 2013, pp. 1–4.

[39] Y. Yang, F. Yang, H. Wang, W. Yang, W. Jin, Temperature-insensitive hydrogen sensor with polarization-maintaining photonic crystal fiber-based Sagnac interferometer, J. Lightwave Technol. 33 (12) (2015) 2566–2571.

[40] Q. Zheng, K. Torii, Response of optical fiber thermometer with blackbody cavity sensor (Aiming to measure the gas temperature inside internal combustion engine), JSME Int. J. Ser. B 37 (3) (1994) 588–595.

[41] T. Takahashi, M. Katsuki, Y. Mizutani, Measurement of flame temperature by optical fiber thermometer, Exp. Fluids 6 (8) (1988) 514–520.

[42] F. Schreiber, Structure and growth of self-assembling monolayers, Prog. Surf. Sci. 65 (5–8) (2000) 151–257.

[43] A. Ulman, Formation and structure of self-assembled monolayers, Chem. Rev. 96 (4) (1996) 1533–1554.

[44] A. Abdelghani, et al., Study of self-assembled monolayers of n-alkanethiol on a surface plasmon resonance fibre optic sensor, Thin Solid Films 284–285 (1996) 157–161.

[45] A. Abdelghani, J.M. Chovelon, N. Jaffrezic-Renault, C. Ronot-Trioli, C. Veillas, H. Gagnaire, Surface plasmon resonance fibre-optic sensor for gas detection, Sensors Actuators B Chem. 39 (1–3) (1997) 407–410.

[46] R.H. Tredgold, The physics of Langmuir-Blodgett films, Rep. Prog. Phys. 50 (12) (1987) 1609–1656.

[47] K. Ariga, Y. Yamauchi, T. Mori, J.P. Hill, 25th anniversary article: what can be done with the Langmuir-Blodgett method? Recent developments and its critical role in materials science, Adv. Mater. 25 (45) (2013) 6477–6512.

[48] M.J. Yin, B. Gu, Q.F. An, C. Yang, Y.L. Guan, K.T. Yong, Recent development of fiber-optic chemical sensors and biosensors: mechanisms, materials, micro/nano-fabrications and applications, Coord. Chem. Rev. 376 (2018) 348–392.

[49] I.R. Peterson, Langmuir-Blodgett films, J. Phys. D. Appl. Phys. 23 (4) (1990) 379–395.

[50] X. Chen, S. Lenhert, M. Hirtz, N. Lu, H. Fuchs, L. Chi, Langmuir-Blodgett patterning: a bottom-up way to build mesostructures over large areas, Acc. Chem. Res. 40 (6) (2007) 393–401.

[51] I.V. Turko, I.S. Yurkevich, V.L. Chashchin, Antigen binding properties of Langmuir-Blodgett films of immunoglobulin G deposited onto the optical fiber core, in: Proceedings of SPIE—The International Society for Optical Engineering, vol. 1587, 1992, pp. 346–349.

[52] R.B. Charters, A.P. Kuczynski, S.E. Staines, R.P. Tatam, G.J. Ashwell, Passive and active in-line fiber components using Langmuir-Blodgett films on monomode optical fiber, in: Proceedings of SPIE—The International Society for Optical Engineering, vol. 2290, 1994, pp. 304–314.

[53] D. Flannery, S.W. James, R.P. Tatam, G.J. Ashwell, pH sensor using Langmuir-Blodgett overlays on polished optical fibers, Opt. Lett. 22 (8) (1997) 567–569.

[54] M. Mar, R. Jorgenson, S. Letellier, S. Yee, In-situ characterization of multilayered Langmuir-Blodgett films using a surface plasmon resonance fiber optic sensor, in: Proceedings of the Annual Conference on Engineering in Medicine and Biology, vol. 15, 1993, pp. 1551–1552 no. pt 3.

[55] J.-W. Choi, J.Y. Bae, J. Min, K.S. Cho, W.H. Lee, Fiber-optic ethanol sensor using alcohol dehydrogenase-immobilized langmuir-Blodgett film, Sensors Mater. 8 (8) (1996) 493–504.

[56] C. Fushen, L. Yunqi, X. Yu, L. Qu, Effects of Langmuir-Blodgett-film gas sensors with integrated optical interferometers, Opt. Lett. 21 (20) (1996) 1700–1702.

[57] D. Flannery, S.W. James, R.P. Tatam, G.J. Ashwell, Fiber-optic chemical sensing with Langmuir—Blodgett overlay waveguides, Appl. Opt. 38 (36) (1999) 7370–7374.

[58] N.D. Rees, S.W. James, R.P. Tatam, G.J. Ashwell, Optical fiber long-period gratings with Langmuir-Blodgett thin-film overlays, Opt. Lett. 27 (9) (2002) 686–688.

[59] N.D. Rees, S.W. James, S.E. Staines, R.P. Tatam, G.J. Ashwell, Submicrometer fiber-optic Fabry-Perot interferometer formed by use of the Langmuir-Blodgett technique, Opt. Lett. 26 (23) (2001) 1840–1842.

[60] S.M. Topliss, S.W. James, F. Davis, S.P.J. Higson, R.P. Tatam, Optical fibre long period grating based selective vapour sensing of volatile organic compounds, Sensors Actuators B Chem. 143 (2) (2010) 629–634.

[61] M. Consales, et al., SWCNT nano-composite optical sensors for VOC and gas trace detection, Sensors Actuators B Chem. 138 (1) (2009) 351–361.

[62] G. Taylor, Electrically driven jets, Proc. R. Soc. Lond. A Math. Phys. Eng. Sci. 313 (1515) (1969) 453–475.

[63] J. Doshi, D.H. Reneker, Electrospinning process and applications of electrospun fibers, J. Electrost. 35 (2–3) (1995) 151–160.

[64] D.H. Reneker, A.L. Yarin, Electrospinning jets and polymer nanofibers, Polymer 49 (2008) 2387–2425.

[65] D. Li, Y. Xia, Direct fabrication of composite and ceramic hollow nanofibers by electrospinning, Nano Lett. 4 (5) (2004) 933–938.

[66] J. Venugopal, S. Ramakrishna, Applications of polymer nanofibers in biomedicine and biotechnology, Appl. Biochem. Biotechnol. 125 (3) (2005) 147–158.

[67] Z.-M. Huang, Y.-Z. Zhang, M. Kotaki, S. Ramakrishna, A review on polymer nanofibers by electrospinning and their applications in nanocomposites, Compos. Sci. Technol. 63 (15) (2003) 2223–2253.

[68] J. Deitzel, J. Kleinmeyer, D. Harris, N.B. Tan, The effect of processing variables on the morphology of electrospun nanofibers and textiles, Polymer 42 (1) (2001) 261–272.

[69] P.J. Rivero, A. Urrutia, J. Goicoechea, F.J. Arregui, Nanomaterials for functional textiles and fibers, Nanoscale Res. Lett. 10 (1) (2015) 501.

[70] B. Ding, M. Wang, X. Wang, J. Yu, G. Sun, Electrospun nanomaterials for ultrasensitive sensors, Mater. Today 13 (11) (2010) 16–27.

[71] A. Urrutia, J. Goicoechea, P.J. Rivero, I.R. Matías, F.J. Arregui, Electrospun nanofiber mats for evanescent optical fiber sensors, Sensors Actuators B Chem. 176 (2013) 569–576.

[72] A. Urrutia, P.J. Rivero, J. Goicoechea, F.J. Arregui, I.R. Matias, Optical sensor based on polymer electrospun nanofibers for sensing humidity, in: 2011 Fifth International Conference on Sensing Technology, 2011, pp. 380–383.

[73] J.M. Corres, Y.R. Garcia, F.J. Arregui, I.R. Matias, Optical fiber humidity sensors using PVdF electrospun nanowebs, IEEE Sensors J. 11 (10) (2011) 2383–2387.

[74] S. Bagchi, R. Achla, S.K. Mondal, Electrospun polypyrrole-polyethylene oxide coated optical fiber sensor probe for detection of volatile compounds, Sensors Actuators B Chem. 250 (2017) 52–60.

[75] A. Cusano, et al., Optochemical sensor for water monitoring based on SnO2 particle layer deposited onto optical fibers by the electrospray pyrolysis method, Appl. Phys. Lett. 89 (11) (2006).

[76] A. Nakamura, Y. Suzuki, M. Morisawa, Swelling clad-type plastic optical fiber alkane sensor with multi-layer cladding using electrospray deposition method, in: 25th International Conference on Optical Fiber Sensors, vol. 10323, 2017 p. 1032329.

[77] H.A. Mohammed, S.A. Rashid, M.H. Abu Bakar, S.B. Ahmad Anas, M.A. Mahdi, M.H. Yaacob, Fabrication and characterizations of a novel etched-tapered single mode optical fiber ammonia sensors integrating PANI/GNF nanocomposite, Sensors Actuators B Chem. 287 (January) (2019) 71–77.

[78] H.A. Mohammed, et al., Sensing performance of modified single mode optical fiber coated with nanomaterials-based ammonia sensors operated in the C-band, IEEE Access 7 (2019) 5467–5476.

[79] D.B. Hall, P. Underhill, J.M. Torkelson, Spin coating of thin and ultrathin polymer films, Polym. Eng. Sci. 38 (12) (1998) 2039–2045.

[80] C.J. Lawrence, The mechanics of spin coating of polymer films, Phys. Fluids 31 (10) (1988) 2786–2795.

[81] B. Nagaraju, R.K. Varshney, B.P. Pal, A. Singh, G. Monnom, B. Dussardier, Design and realization of a side-polished single-mode fiber optic high-sensitive temperature sensor, in: Proc. SPIE 7138, Photonics, Devices, and Systems IV, 2008p. 71381H.

[82] H.-J. Kim, O.-J. Kown, S.B. Lee, Y.-G. Han, Measurement of temperature and refractive index based on surface long-period gratings deposited onto a D-shaped photonic crystal fiber, Appl. Phys. B Lasers Opt. 102 (1) (2011) 81–85.

[83] B. Wang, F. Zhang, F. Pang, T. Wang, An optical fiber humidity sensor based on optical absorption, in: Optical InfoBase Conference Paper, vol. 8311, 2011, pp. 1–6.

[84] F. Branda, The sol-gel route to nanocomposites, in: Advances in Nanocomposites—Synthesis, Characterization and Industrial Applications, InTech, 2011.

[85] C.J. Brinker, A.J. Hurd, P.R. Schunk, G.C. Frye, C.S. Ashley, Review of sol-gel thin film formation, J. Non-Cryst. Solids 147–148 (C) (1992) 424–436.

[86] C.J. Brinker, G.W. Scherer, Sol → gel → glass: I. Gelation and gel structure, J. Non-Cryst. Solids 70 (3) (1985) 301–322.

[87] C.J. Brinker, G.W. Scherer, E.P. Roth, Sol → gel → glass: II. Physical and structural evolution during constant heating rate experiments, J. Non-Cryst. Solids 72 (2–3) (1985) 345–368.

[88] G.W. Scherer, C.J. Brinker, E.P. Roth, Sol → gel → glass: III. Viscous sintering, J. Non-Cryst. Solids 72 (2–3) (1985) 369–389.

[89] V. Musat, B. Teixeira, E. Fortunato, R.C. Monteiro, P. Vilarinho, Al-doped ZnO thin films by sol–gel method, Surf. Coat. Technol. 180–181 (2004) 659–662.

[90] W. Tang, D.C. Cameron, Aluminum-doped zinc oxide transparent conductors deposited by the sol-gel process, Thin Solid Films 238 (1) (1994) 83–87.

[91] F.J. García Moreda, F.J. Arregui, M. Achaerandio, I.R. Matias, Study of indicators for the development of fluorescence based optical fiber temperature sensors, Sensors Actuators B Chem. 118 (1–2) (2006) 425–432.

[92] J. Estella, P. de Vicente, J.C. Echeverría, J.J. Garrido, A fibre-optic humidity sensor based on a porous silica xerogel film as the sensing element, Sensors Actuators B Chem. 149 (1) (2010) 122–128.

[93] W. Cao, Y. Duan, Optical fiber evanescent wave sensor for oxygen deficiency detection, Sensors Actuators B Chem. 119 (2) (2006) 363–369.

[94] C.-S. Chu, Y.-L. Lo, High-performance fiber-optic oxygen sensors based on fluorinated xerogels doped with Pt(II) complexes, Sensors Actuators B Chem. 124 (2) (2007) 376–382.

[95] H. Segawa, E. Ohnishi, Y. Arai, K. Yoshida, Sensitivity of fiber-optic carbon dioxide sensors utilizing indicator dye, Sensors Actuators B Chem. 94 (3) (2003) 276–281.

[96] K. Wysokiński, M. Napierała, T. Stańczyk, S. Lipiński, T. Nasiłowski, Study on the sensing coating of the optical fibre CO_2 sensor, Sensors 15 (12) (2015) 31888–31903.

[97] C.-S. Chu, Y.-L. Lo, Highly sensitive and linear optical fiber carbon dioxide sensor based on sol-gel matrix doped with silica particles and HPTS, Sensors Actuators B Chem. 143 (1) (2009) 205–210.

[98] Q. Yan, S. Tao, H. Toghiani, Optical fiber evanescent wave absorption spectrometry of nanocrystalline tin oxide thin films for selective hydrogen sensing in high temperature gas samples, Talanta 77 (3) (2009) 953–961.

[99] A. Yan, R. Chen, M. Zaghloul, Z.L. Poole, P. Ohodnicki, K.P. Chen, Sapphire fiber optical hydrogen sensors for high-temperature environments, IEEE Photon. Technol. Lett. 28 (1) (2015) 47–50.

[100] S. Masuzawa, S. Okazaki, Y. Maru, T. Mizutani, Catalyst-type-an optical fiber sensor for hydrogen leakage based on fiber Bragg gratings, Sensors Actuators B Chem. 217 (2015) 151–157.

[101] Y. Wang, et al., Fiber optic hydrogen sensor based on Fabry-Perot interferometer coated with sol-gel Pt/WO3 coating, J. Lightwave Technol. 33 (12) (2015) 2530–2534.

[102] W. Cao, Y. Duan, Optical fiber-based evanescent ammonia sensor, Sensors Actuators B Chem. 110 (2) (2005) 252–259.

[103] S. Tac, T.V.S. Sarma, A Fiber-optic sensor for monitoring trace ammonia in high-temperature gas samples with a CuCl2 doped porous silica optical fiber as a transducer, IEEE Sensors J. 8 (12) (2008) 2000–2007.

[104] M. Debliquy, et al., Optical fibre NO_2 sensor based on lutetium bisphthalocyanine in a mesoporous silica matrix, Sensors 18 (3) (2018) 740.

[105] T.V.S. Sarma, S. Tao, An active core fiber optic sensor for detecting trace H_2S at high temperature using a cadmium oxide doped porous silica optical fiber as a transducer, Sensors Actuators B Chem. 127 (2) (2007) 471–479.

[106] S. Shaari, A.R.A. Rashid, P.S. Menon, N. Arshad, Al-doped ZnO coated fiber optic for ultraviolet and acetone sensing, Adv. Sci. Lett. 19 (5) (2013) 1306–1309.

[107] A.R.A. Rashid, P.S. Menon, S. Shaari, Optical VOC sensing applications of Al-doped ZnO coated optical fiber, Optoelectron. Adv. Mater. Rapid Commun. 7 (11–12) (2013) 835–839.

[108] M. Bezunartea, et al., Optical fibre sensing element based on xerogel-supported [Au2Ag2(C6F5)4 (C14H10)]n for the detection of methanol and ethanol in the vapour phase, Sensors Actuators B Chem. 134 (2) (2008) 966–973.

[109] J.L. Noel, R. Udayabhaskar, B. Renganathan, S. Muthu Mariappan, D. Sastikumar, B. Karthikeyan, Spectroscopic and fiber optic ethanol sensing properties Gd doped ZnO nanoparticles, Spectrochim. Acta A Mol. Biomol. Spectrosc. 132 (2014) 634–638.

[110] J.C. Echeverría, P. De Vicente, J. Estella, J.J. Garrido, A fiber-optic sensor to detect volatile organic compounds based on a porous silica xerogel film, Talanta 99 (2012) 433–440.

[111] J.C. Echeverría, I. Calleja, P. Moriones, J.J. Garrido, Fiber optic sensors based on hybrid phenyl-silica xerogel films to detect n-hexane: determination of the isosteric enthalpy of adsorption, Beilstein J. Nanotechnol. 8 (1) (2017) 475–484.

[112] Z. Harith, M. Batumalay, N. Irawati, S.W. Harun, H. Arof, H. Ahmad, Relative humidity sensor employing tapered plastic optical fiber coated with seeded Al-doped ZnO, Optik 144 (2017) 257–262.

[113] S.F.H. Correia, et al., Optical fiber relative humidity sensor based on a FBG with a di-ureasil coating, Sensors 12 (7) (2012) 8847–8860.

[114] Z. Zhao, Y. Duan, A low cost fiber-optic humidity sensor based on silica sol–gel film, Sensors Actuators B Chem. 160 (1) (2011) 1340–1345.

[115] R. Aneesh, S.K. Khijwania, Zinc oxide nanoparticle based optical fiber humidity sensor having linear response throughout a large dynamic range, Appl. Opt. 50 (27) (2011) 5310–5314.

[116] L. Xu, J.C. Fanguy, K. Soni, S. Tao, Optical fiber humidity sensor based on evanescent-wave scattering, Opt. Lett. 29 (11) (2004) 1191–1193.

[117] S.K. Shukla, et al., Nano-like magnesium oxide films and its significance in optical fiber humidity sensor, Sensors Actuators B Chem. 98 (1) (2004) 5–11.

[118] K. Cherif, S. Hleli, A. Abdelghani, N. Jaffrezic-Renault, V. Matejec, Chemical detection in liquid media with a refractometric sensor based on a multimode optical fibre, Sensors 2 (6) (2002) 195–204.

[119] Z. Gu, Y. Xu, K. Gao, Optical fiber long-period grating with solgel coating for gas sensor, Opt. Lett. 31 (16) (2006) 2405–2407.

[120] W.B. Lin, M. Lacroix, J.M. Chovelon, N. Jaffrezic-Renault, H. Gagnaire, Development of a fiber-optic sensor based on surface plasmon resonance on silver film for monitoring aqueous media, Sensors Actuators B Chem. 75 (3) (2001) 203–209.

[121] Y. Jee, et al., Plasmonic conducting metal oxide-based optical Fiber sensors for chemical and intermediate temperature-sensing applications, ACS Appl. Mater. Interfaces 10 (49) (2018) 42552–42563.

[122] C.-S. Chu, C.-A. Lin, Optical fiber sensor for dual sensing of temperature and oxygen based on PtTFPP/CF embedded in sol-gel matrix, Sensors Actuators B Chem. 195 (2014) 259–265.

[123] H.D. Duong, J. Il Rhee, Enhancement of the sensitivity of a quantum dot-based fiber optic temperature sensor using the sol–gel technique, Sensors Actuators B Chem. 134 (2) (2008) 423–426.

[124] H. Guo, F. Pang, X. Zeng, N. Chen, Z. Chen, T. Wang, Temperature sensor using an optical fiber coupler with a thin film, Appl. Opt. 47 (19) (2008) 3530–3534.

[125] H. Guo, S. Tao, An active core fiber-optic temperature sensor using an Eu(III)-doped sol-gel silica fiber as a temperature indicator, IEEE Sensors J. 7 (6) (2007) 953–954.

[126] P.J. Rivero, J. Goicoechea, F.J. Arregui, Layer-by-layer nano-assembly: a powerful tool for optical fiber sensing applications, Sensors 19 (3) (2019) 683.

[127] C.R. Zamarreño, M. Hernaez, I. Del Villar, I.R. Matias, F.J. Arregui, Lossy mode resonance-based optical fiber humidity sensor, in: Proceedings of IEEE Sensors, 2011, pp. 234–237.

[128] S. Akita, H. Sasaki, K. Watanabe, A. Seki, A humidity sensor based on a hetero-core optical fiber, Sensors Actuators B Chem. 147 (2) (2010) 385–391.

[129] A. Urrutia, et al., An antibacterial surface coating composed of PAH/SiO$_2$ nanostructurated films by layer by layer, Phys. Status Solidi Curr. Top. Solid State Phys. 7 (11–12) (2010) 2774–2777.

[130] P.J. Rivero, J. Goicoechea, F.J. Arregui, Optical fiber sensors based on polymeric sensitive coatings, Polymers 10 (3) (2018) 280.

[131] J.M. Corres, F.J. Arregui, I.R. Matias, Design of humidity sensors based on tapered optical fibers, J. Lightwave Technol. 24 (11) (2006) 4329–4336.

[132] C.R. Zamarreño, M. Hernaez, P. Sanchez, I. Del Villar, I.R. Matias, F.J. Arregui, Optical fiber humidity sensor based on lossy mode resonances supported by TiO2/PSS coatings, in: Procedia Engineering, vol. 25, 2011, pp. 1385–1388.

[133] P.J. Rivero, A. Urrutia, J. Goicoechea, F.J. Arregui, I.R. Matias, Humidity sensor based on silver nanoparticles embedded in a polymeric coating, in: Proceedings of the International Conference on Sensing Technology, ICST, 2011.

[134] P.J. Rivero, A. Urrutia, J. Goicoechea, F.J. Arregui, I.R. Matías, Humidity sensor based on silver nanoparticles embedded in a polymeric coating, Int. J. Smart Sens. Intell. Syst. 5 (1) (2012) 71–83.

[135] P.J. Rivero, A. Urrutia, J. Goicoechea, F.J. Arregui, Optical fiber humidity sensors based on localized surface plasmon resonance (LSPR) and Lossy-mode resonance (LMR) in overlays loaded with silver nanoparticles, Sensors Actuators B Chem. 173 (2012) 244–249.

[136] P.J. Rivero, A. Urrutia, J. Goicoechea, I.R. Matias, F.J. Arregui, A Lossy mode resonance optical sensor using silver nanoparticles-loaded films for monitoring human breathing, Sensors Actuators B Chem. 187 (2013) 40–44.

[137] M. Jiang, et al., Optical response of fiber-optic Fabry-Perot refractive-index tip sensor coated with polyelectrolyte multilayer ultra-thin films, J. Lightwave Technol. 31 (14) (2013) 2321–2326.

[138] M. Hernaez, C.R. Zamarreño, I. Del Villar, I.R. Matias, F.J. Arregui, Lossy mode resonances supported by TiO2-coated optical fibers, in: Procedia Engineering, vol. 5, 2010, pp. 1099–1102.

[139] P.J. Rivero, M. Hernaez, J. Goicoechea, I.R. Matias, F.J. Arregui, Optical fiber refractometers based on localized surface plasmon resonance (LSPR) and lossy mode resonance (LMR), in: Proceedings of SPIE—The International Society for Optical Engineering, vol. 9157, 2014.

[140] P.J. Rivero, M. Hernaez, J. Goicoechea, I.R. Matías, F.J. Arregui, A comparative study in the sensitivity of optical fiber refractometers based on the incorporation of gold nanoparticles into layer-by-layer films, Int. J. Smart Sens. Intell. Syst. 8 (2) (2015) 822–841.

[141] A. Urrutia, J. Goicoechea, P.J. Rivero, A. Pildain, F.J. Arregui, Optical fiber sensors based on gold nanorods embedded in polymeric thin films, Sensors Actuators B Chem. 255 (2018) 2105–2112.

[142] G. De Bastida, F.J. Arregui, J. Goicoechea, I.R. Matias, Quantum dots-based optical fiber temperature sensors fabricated by layer-by-layer, IEEE Sensors J. 6 (6) (2006) 1378–1379.

[143] B. Larrión, M. Hernáez, F.J. Arregui, J. Goicoechea, J. Bravo, I.R. Matías, Photonic crystal fiber temperature sensor based on quantum dot nanocoatings, J. Sensors 2009 (2009).

[144] A. Urrutia, J. Goicoechea, A.L. Ricchiuti, D. Barrera, S. Sales, F.J. Arregui, Simultaneous measurement of humidity and temperature based on a partially coated optical fiber long period grating, Sensors Actuators B Chem. 227 (2016) 135–141.

[145] J. Hromadka, et al., Simultaneous in situ temperature and relative humidity monitoring in mechanical ventilators using an array of functionalised optical fibre long period grating sensors, Sensors Actuators B Chem. 286 (2019) 306–314.

[146] G. Kostovski, P.R. Stoddart, A. Mitchell, The optical fiber tip: an inherently light-coupled microscopic platform for micro- and nanotechnologies, Adv. Mater. 26 (23) (2014) 3798–3820.

[147] A. Ricciardi, et al., Lab-on-fiber technology: a new vision for chemical and biological sensing, Analyst 140 (24) (2015) 8068–8079.

[148] P. Vaiano, et al., Lab on Fiber technology for biological sensing applications, Laser Photonics Rev. 10 (6) (2016) 922–961.

[149] L.-X. Kong, et al., Lab-on-tip: protruding-shaped all-fiber plasmonic microtip probe toward in-situ chem-bio detection, Sensors Actuators B Chem. 301 (2019).

[150] X. Yang, et al., Nanopillar array on a fiber facet for highly sensitive surface-enhanced Raman scattering, Opt. Express 20 (22) (2012) 24819–24826.

[151] G. Kostovski, D.J. White, A. Mitchell, M.W. Austin, P.R. Stoddart, Nanoimprinted optical fibres: biotemplated nanostructures for SERS sensing, Bioscns. Bioelectron. 24 (5) (2009) 1531–1535.

[152] M.-J. Yin, M. Yao, S. Gao, A.P. Zhang, H.-Y. Tam, P.-K.A. Wai, Rapid 3D patterning of poly(acrylic acid) ionic hydrogel for miniature pH sensors, Adv. Mater. 28 (7) (2016) 1394–1399.

[153] R.S. Rodrigues Ribeiro, P. Dahal, A. Guerreiro, P.A.S. Jorge, J. Viegas, Fabrication of Fresnel plates on optical fibres by FIB milling for optical trapping, manipulation and detection of single cells, Sci. Rep. 7 (1) (2017) 4485.

[154] J.-L. Kou, S.-J. Qiu, F. Xu, Y.-Q. Lu, Demonstration of a compact temperature sensor based on first-order Bragg grating in a tapered fiber probe, Opt. Express 19 (19) (2011) 18452–18457.

[155] J. Feng, M. Ding, J.-L. Kou, F. Xu, Y.-Q. Lu, An optical fiber tip micrograting thermometer, IEEE Photon. J. 3 (5) (2011) 810–814.

[156] H. Yan, et al., A fast response hydrogen sensor with Pd metallic grating onto a fiber's end-face, Opt. Commun. 359 (2016) 157–161.

[157] A. Rizzo, et al., Laser micromachining of tapered optical fibers for spatially selective control of neural activity, Microelectron. Eng. 192 (2018) 88–95.

[158] J. Dietvorst, J. Goyvaerts, T.N. Ackermann, E. Alvarez, X. Muñoz-Berbel, A. Llobera, Microfluidic-controlled optical router for lab on a chip, Lab Chip 19 (12) (2019) 2081–2088.

[159] A. Fernández-Gavela, et al., Full integration of photonic nanoimmunosensors in portable platforms for on-line monitoring of ocean pollutants, Sensors Actuators B Chem. 279 (2019) 126758.

[160] M.-H. Wu, J. Wang, T. Taha, Z. Cui, J.P.G. Urban, Z. Cui, Study of on-line monitoring of lactate based on optical fibre sensor and in-channel mixing mechanism, Biomed. Microdevices 9 (2) (2007) 167–174.

[161] V. Moradi, M. Akbari, P. Wild, A fluorescence-based pH sensor with microfluidic mixing and fiber optic detection for wide range pH measurements, Sens. Actuators A 297 (2019) 111507.

[162] M. Shiraishi, K. Goya, A. Seki, K. Watanabe, An LSPR fiber optic sensor based on in-line micro-holes fabricated by a second harmonic 400nm femtosecond laser, in: Proceedings of SPIE—The International Society for Optical Engineering, vol. 9750, 2016.

CHAPTER 24

The brain-machine interface, nanosensor technology, and artificial intelligence: Their convergence with a novel frontier

Gauri Kalnoor

School of Engineering, Central University of Karnataka Kalaburagi, Kadaganchi, Karnataka, India

1. Introduction

Technological capabilities are a confluence that creates opportunities for Artificial Intelligence (AI) and Machine Learning to enable *Brain-machine Interfaces (BMI)*, a nanoengineered and smart technology. The main aim is to support contextual learning and thus allow communication with the brain such that it can adapt to changes in functional requirements. Invasive technologies are applied that aim to restore the function of neurology. In the case of neuroprosthesis, the signals enabled in noninvasive technologies are called "Electroencephalograph (EEG)." The computational advancement, algorithms, and hardware that learn contextually dependency and adaptation will make it possible to influence nanoengineering capabilities offering design and BMI functionality. Thus, eventually, learning and its adaptation are carried out by these technologies in near real-time. They require an external shift from physiology and the environment. Ultimately, the aim of BMI is to create individual experiences of the user as a personalized experience considering a few applications like gaming. It also allows the device to learn the changing requirements in diseases and accordingly adapt in clinical scenarios. The enabling capabilities that the devices exhibit are explored, as well as the reasons why the technologies and its state are needed to build them. In this chapter, we discuss a number of various open challenges and problems based on technical issues that must be solved to achieve these goals.

2. Brain-computer interfaces (BCI) and the brain-machine: The opportunity

The interfaces used for brain-machine and brain-computer are designed such that they communicate with the central nervous system representing the technology. The three

Handbook of Nanomaterials for Sensing Applications
https://doi.org/10.1016/B978-0-12-820783-3.00013-0

parts they represent include the spinal cord, the brain, and the neural sensory retina. Based on the clinical design and intent of the technology, the goal is to record and then interpret neural signals, such that the neural commands, that are intended through an external device; neural simulation can be achieved, which often restores neural function that follows trauma, disease, or both. The feedback is used by some of the devices such that an attempt can be made to optimize performance, whether through specific intent of patient instructions or through physiological aspects. Also, a growing list of noninvasive technologies of BMI are not meant for clinical applications and are primarily determined by the companies at start-up. The user experience and interface control are technologies that are intended to augment them, mainly for augmented reality (AR) and gaming applications with virtual reality (VR). Although it differs from most of the technologies that are developed aiming at restoring and treating clinical function and a patient's quality of life, it is a market that cannot be ignored. Thus, it can provide resources that could benefit clinically related research. For example, advances in our understanding of the relevant neurophysiology, cognitive neuroscience, mathematical, and engineering aspects of signal processing and hardware can significantly affect both the gaming industry as well as clinical devices and neural prosthesis. The BMI market is projected to reach $1.46B by 2020, with a compound annual growth rate (CAGR) of 11.5% between 2014 and 2020 by one estimate [1], and a comparable $1.72B in 2022, with a predicted CAGR of 11.5% between 2012 and 2022 by another estimate [2]. Much of this projected growth will be due to noninvasive technologies, with the gaming industry as a market driver roughly on par with healthcare applications. As significant as these numbers are, these projections primarily reflect enabling technologies for interfacing between neural control and sensory experiences with machines. They do not reflect opportunities that go beyond what is currently possible with the existing state of the art. BMIs that can learn and adapt reflect the cutting edge of what is technologically possible due to a confluence of BMI technologies, in particular nanotechnologies, machine learning, and AI, alongside a continued increase in understanding of the relevant neuroscience. AI can provide opportunities to create "smart" BMI that contextually learn and adapt to changing functional requirements and demands. This has the potential to produce personalized individual experiences in gaming and AR/VR, and allow for changing requirements associated with patient-specific disease progression and evolution in clinical applications. This latter point cannot be overestimated, because not only would it accommodate the differing clinical demands of different neurological disorders, it would allow for patient-specific adaptation of BMI functionality to the needs of different patients. It would also allow the technology to continue to adapt as disease progression evolves in individuals over time. One of the significant limitations of the current state of the art BMI and neural prosthesis is the assumption of "one size fits all," in other words, the assumption that a technology operating under a specific set or range of functionality will properly treat all patients. Although we are not (yet) aware of a device or technology that reflects the actualized integration of machine learning and nanotechnology applied to BMI, we argue that the potential and effect of doing so make the subject worth exploring. Each on their own, machine learning and nanotechnology are already being used in the design and function of BMI and neural prostheses in a number of ways that align with the vision we propose here.

3. Integration of BMI with machine learning

What advantages does machine learning and AI offer BMI? What exactly are machine learning algorithms learning, and how can they use that information to adapt in a meaningful way? What these algorithms can learn is information provided by feedback and telemetry from the hardware. This could be information about the current state of the output settings of the device or any kind of external information measured by sensors in the BMI, for example, physiological measurements in response to stimulation, feedback from other algorithms external to the BMI-machine learning system, such as haptic or computer vision feedback, or the internal parameter settings of current stimulation or recording protocols to the device. In the case of internal parameters, the algorithms have constant access to variables such as pulse durations and amplitudes, stimulation frequencies, energy consumption by the device, stimulation or recording densities, electrical properties of the neural tissues it is interfacing with (resistances, impedances), and continuous or near-continuous levels of biochemical factors such as neurotransmitters or other metabolites. Of course, none of these are mutually exclusive, with multiple types and streams of information possibly being provided to the algorithms in parallel, albeit likely at different sampling resolutions. With this information, machine learning algorithms could then identify subtle and nontrivial patterns and phenomena in the data, ideally in (near) real-time, to produce desired functional outcomes from the BMI that change dynamically as external (e.g., clinical or functional) requirements demand them. This would necessitate the development and training of machine learning models and algorithms offline as part of the design of the BMI system. The algorithms would need to learn a wide enough range of the parameter spaces to appropriately identify patterns in the data they encounter when online. Subsequent algorithms can then autonomously make decisions about how to use that data. This step does not necessarily have to be part of the BMI system itself and could be executed with algorithms computing in the cloud, if sufficient bandwidth was available, or even offline following periodic data downloads, for example. Clearly though, on-board decision algorithms that operate in real-time with the machine learning algorithms that are identifying patterns in the data would be ideal. This would alleviate issues of data transfer delays and bandwidth insufficiency. This could also allow for the need to store less data on the device, which could be limited due to physical constrains. Data would only have to be stored long enough for the system to make an autonomous decision, essentially as a moving window that matches the processing capabilities of the algorithms. Of course, it may still be valuable or necessary to store some data or types of data for offline analyses even though they may not be needed for the BMI system to make a decision, for example, to understand offline after the fact why the algorithms made the decisions they made and the clinical outcomes of those decisions. With this process complete, we close the loop: information is provided to the machine learning algorithms, followed by learning, pattern identification, and subsequent executable autonomous decisions that in turn dynamically change the output of the BMI and how it interacts with the external environment it is interfacing with. This could be the brain itself, in the case of a neural prosthesis intended to restore clinical function, or software in a noninvasive BMI that is interfacing as part of an AR or VR system.

While still in the early days, a number of research groups have recognized this potential and are beginning to explore how machine learning could inform and integrate neural stimulation and feedback. Nurse et al. [3] have developed a generalized approach that takes advantage of a stochastic machine learning method to classify motor-related signals specifically for BMI applications. Importantly, their classifier does not need to rely on the use of extensive a priori data to train the BMI. Their algorithms outperformed other methods on the Berlin BMI IV 2008 dataset and demonstrated high levels of classification accuracy when tested on datasets derived from EEG signals. In another recent study, The authors have explored different data preprocessing strategies and convolution neural network architectures for classification tasks derived from EEG signals. Interestingly, they found that a rather straightforward network architecture, when combined with a preprocessing step that analyzed spectral power preserving features of the electrode arrangements, was sufficient to handle the analysis of the data. Their network consisted of a single convolution layer, one connection layer, and single linear regression classifier layer. Their approach allowed them to carry out co-adaptive training on the data to achieve online classification. A different study had explored a similar approach. Lawhern et al. [4] have also explored a similar approach.

Early work relied on feedback from external sensory references to compute an error between the output of the system and desired supervised target. These included visual and auditory signals [5], mechanotransduction [6], and direct cortical sensory stimulation [7]. But these approaches are severely limited due to their need for continuous information from an external reference target to adjust the mapping to the output of the BMI. More recent work has addressed some of these limitations by adapting output parameters to unsupervised learning methods such as Bayesian statistical methods and reinforcement learning that do not rely on an external reference [8–10], although in most cases they still require significant training periods. More recent studies have begun to investigate the use of endogenous neural signals directly as the training source in iterative closed feedback loops with BMI that can respond and adapt in a much more direct way [11]. For example, Prassad and colleagues are developing an approach they call Actor-Critic reinforcement learning that does not need to rely on a supervised error signal.

In general, the BMI field and neural prosthesis field in particular are still exploring machine learning. One of the challenges is that key state of the art methods, such as deep learning, that have had huge successes in other applications may not be the best approach for the constraints imposed by the needs of BMI. In a recent paper, an excellent job summarizes the current state and challenges of neural engineering aimed at restoring neural function, including proposing a number of similar requirements discussed in the current paper.

4. Application-based methodology

The human brain is the central organ of the human framework. The brain sensor measures the brain signals from the temple. This project focuses on utilizing BMI technology for wireless communication of a human brain in controlling an elect mechanical device. The biosensor band measures the concentration and peace level of mind and digitizes brainwave signals to operate the electronic devices. The principle behind this interface is bioelectrical action of nerves and muscles. The brain is composed of millions of neurons; when a neuron fires, there is a voltage sent over the cells that create signals on the surface of the brain. EEG is an estimation of electrical action created by the brain as recorded from electrodes set on the forehead.

Thus real-time environment brain impulses are acquired and interfaced with a bank of these input signals, processing these signals with predefined mechanism with the aid of specific firmware in a software that triggers the automation of an electronic device.

EEG signal acquisition: The human brain constitutes the various potential values corresponding to certain brain activity.

Translation into command: The neural impulses are realized to a task functioning through a specialized firmware code.

Preprocessing: The EEG signal that has electric potential in the range of millivolts needs to be extracted the desired signal amplified by removing noise (Fig. 1).

Feature extraction: The brain impulses are amplified and available at this stage; the features of the brain's parameters need to be defined to classify its function.

Classification: This section classifies the categories of signals to translate to the code.

Feedback: This section takes constant feedback signals and determines the functions in an application.

4.1 Robot control with BMI

An EEG sensor receives human attention levels from the forehead, and the firmware code to run a robot is compiled and dumped onto Arduino. The code analyses the attention parameters and decides to control the robot's movement. Robot movement is a basic requirement in most of the automation. In this module, the traditional robot performance is enhanced with an interface through a HC-05 Bluetooth module to the brain sensor EEG. Through the thinking ability of a person, the robot's directions are specified (Fig. 2).

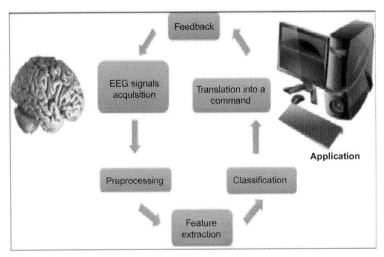

Fig. 1 Block diagram of BMI application. The preprocessing stage does this to transmit brain signal to further stages.

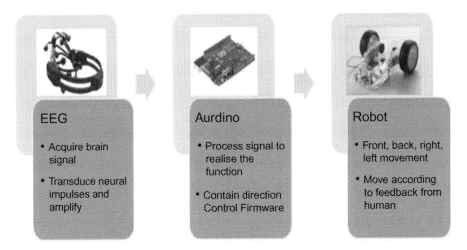

Fig. 2 Brain-controlled robot.

4.2 Switching home appliances with BMI

An EEG sensor acquires a signal from a human and, as per the attention level as it attains the programmed threshold value, the home appliances are turned on/off. In the home, we generally operate many electronic appliances such as a fan, light, cooler, air conditioner, and so on through switches located at one remote place. To avoid the activity of going to that location and operating device, the new technology of remote control through IR sensor signals is used. To avoid even the action of operating a remote or switch, this BMI technology creates a smart home appliance control with human thinking power. Conventional manual switching method is overcome with this BMI technology that switches the appliance on and off without any physical interaction.

Brain signal feature extraction in control functions: Parameters to control: The way our brain works is very much similar to that of computers. We receive data from stimulus; these data impulses are processed and finally computed into thoughts and visualize the information and perform the action.

Human CONCENTRATION: A mental process focusing on a particular task is concentration. For example, to read a book a person needs to focus his concentration on that book reading only in spite of many thoughts running in his/her mind.

Neural functions in concentration: Any action we perform is taken in our brain, which constitutes billions of neurons. These neurons are interconnected throughout the body that gather and transmit electrochemical signals that evoke our thoughts and motor functions. Cerebral cortex is the part of brain associated with our memory, thoughts, attention, awareness, and consciousness.

Human ATTENTION: The object-based centering is what happens in certain areas of concentration.

Neural work in attention: The brain accomplishes this centered attention on faces or other objects in a portion of prefrontal cortex called the inferior frontal junction (IFJ), which is the control center for most cognitive functions. It takes charge of the brain's consideration and controls important parts of the visual cortex, which receives sensory input that controls visual processing regions tuned to recognize a particular category of objects.

5. EEG signal detection and training

Human thinking is correlated with a specific neural activity. A Neurosky headset consists of two electrodes with Thinkgear AM (TGAM) sensor technology that acquires an EEG signal in the range of 3–100 Hz without use of conducting gel or saline.

Training samples are taken in two stages: one is "Rest" (10 s), and other is active state, i.e., "EEG" (10 s) with a time sampling gap of 10–20 s for about a minute to record response and action. These sampled data are split in a length of "L" window. Discrete Fourier transform (DFT) is performed at each of these windows to decompose frequency components. These frequency components are used as attributes and assigned class labels as rest or EEG state. Training classification is done using attributes as training samples. Machine learning is carried out with aid of Support vector machine (SVM), Linear discriminant algorithm (LDA), and Bayesian algorithm.

Let us consider the simple case of predicting a single device output, and then this process can be repeated for each device output independently. Let the system be given a

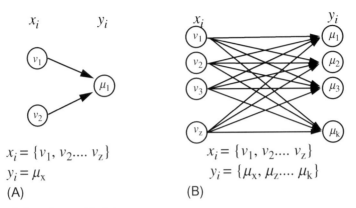

Fig. 3 EEG signal detection and training.

training set D consisting of a set of input data X = {x1,x2, …,xn} and a set of actions Y = {y1,y2, …,yn}; our goal is to predict the value of a new device action yn + 1 given a new test data xn + 1. We can then define yi = sign (wTφ(xi)) for some parameter vector w and a basis function φ(xi) that allows the classification boundary to be nonlinear.

Finally, we use the expectation propagation method to perform Bayesian inference. Expectation propagation chooses each approximation such that the posterior uses the term exactly, which is approximately close in KL-divergence. Expectation propagation turns out to be good method to perform Bayesian inference faster and more accurately than other methods (Fig. 3).

5.1 Asynchronous brain-machine interface

In this system, the EEG continuously monitors various brain states and, from previous actions taken by user in response to different environmental conditions, the system learns user preference and automatically controls devices even without the need to focus. So to accomplish this, we expand the essential Baye's Point Machine structure to adapt to various sources of inputs-outputs and train it utilizing expectation propagation (Table 1) [1].

Environment conditions can be classified as illumination (dark, light), temperature (cold, mild, hot), and humidity (regular, high). These states will trigger action on

Table 1 Different brain signals and their ranges.

Frequency band	Range	Brain state
Delta	0.5–4 Hz	Deep sleep
Theta	4–8 Hz	Intuitive, dreaming
Alpha	8–13 Hz	Meditative, sleepy
Beta	>13 Hz	Awake, active, aware

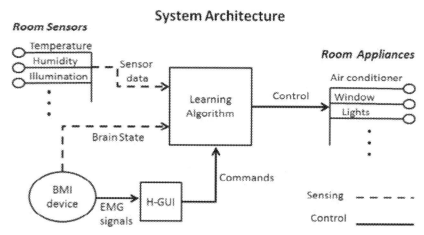

Fig. 4 Home appliances system architecture.

appliances such as light (on/off), fan (on/off), and windows (open/close) in accordance with the brain state of the user (Fig. 4).

6. Experimental analysis

BMI utilizes mind signals, such as Attention Level, through a brainwave headset to control the robot and home automation. We checked for a couple of edges with regard to working the peripherals. It's exceptionally easy to use the headband over the head electrodes, and there's no damage in doing this. You reasonably must be put it over the scalp and check that the affiliation status of Fitting appears in the Mind Wave Mobile center. In this case, the headset is associated accurately, and at that point it's great to go. Attention data is changed between 1 and 100 number values to control machines at home using EEG signals. Likewise proposed and evaluated is the prospect of giving learning capacities to the system so as to unravel the issue of mental shortcoming or stress brought about by constantly controlling gadgets through BMI.

Attention Sense: This unassigned one-byte data reports with the current Sense Attention meter of the client, which shows the quality of a user's dimension of mental "center" or "attention", for example, that which occurs during solid concentration and coordinated (yet unfaltering) mental activity. Its esteem ranges from 0 to 100. Preoccupations, wandering thoughts, absence of focus, or uneasiness may bring down the Attention meter levels.

Meditation Sense: This unassigned one-byte data reports the current sense Meditation meter of the client, which demonstrates the dimension of a client's psychological "calmness" or "unwinding". It ranges from 0 to 100. Note that Meditation might be a level of an individual's psychological dimensions, not physical dimensions, so essentially loosening up every one of the muscles of the body may not suddenly result in an expanded Meditation level. For most people in average condition, loosening up the body every now and again has an effect on the brain to relax also. Meditation is identified with diminished development by the dynamic mental procedures in the cerebrum, and it has for quite some time been a watched effect that closing one's eyes turns the psychological exercises, which process pictures from the eyes, so shutting the eyes is much of the time a viable system for extending the Meditation meter level. Diversions, meandering thoughts, uneasiness, disturbance, and sensory stimulus may bring down the Meditation meter levels. The prepared attention levels are caught and given to the Bluetooth HC-05 module.

The attention levels are categorized into four different levels for the interfacing of four different relays in home automation and for the movement of a robot in four different directions.

For home automation: The attention level from 0 to 40 is set for turning ON the first relay. The attention level from 40 to 60 is set for turning ON the second relay. The attention level from 60 to 80 is set for turning ON the third relay. The attention level from 80 to 100 is set for turning ON the fourth relay (Fig. 5).

For Robot car: The attention level from 0 to 40 is set for FORWARD movement. The attention level from 40 to 60 is set for REVERSE movement. The attention level from 60 to 80 is set for LEFT movement. The attention level from 80 to 100 is set for RIGHT movement.

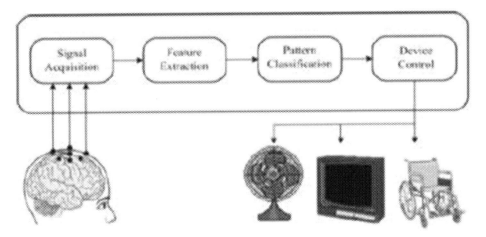

Fig. 5 Home automation applications.

7. Conclusion

Our overall aim is to run the robot car (wheelchair) and home-controlled system utilizing Meditation and Attention values of extracted and acquisitioned brain signals. This plan is executed to control a robot car was expanded to a wheelchair application and home automation from brain signals for physically disabled and paralyzed clients. With the number of disabled individuals, as of now they use diverse innovations that physically provide those who are impaired the capacity to move around. But there are still numbers of individuals who are completely disabled and paralyzed, but their intellect still works properly. So our task for those individuals who are physically disabled is to assist them so that they can move around and control any robotized gadgets utilizing their mind's thinking power. For that we outlined, one can control a robot car and home machine's control system, which are completely robotized, by utilizing human brain's attention and meditation levels using brainwave sensors, which identify brain signals and also use Arduino to control the robot.

The integration of machine learning and AI with nanoengineered brain-machine and brain-computer interfaces offers the potential for significant advances in neurotechnology. BMIs that have the ability to learn and adapt from the environment and situational demands of external requirements offer tremendous possibilities to radically change the treatment and quality of life of patients. It also offers opportunities for noninvasive interactions and collaborations between humans and machines that, at the moment, are still in the realm of science fiction. It is conceivable that we are approaching an era of personalized individual experiences that will affect both clinical and nonclinical applications. Of course, as with any truly disruptive and paradigm-changing progress, there remain many technical challenges that must be overcome, many in no way trivial or easy, and serious ethical questions that have to be thoughtfully considered and navigated. But it is hard not to be excited about the prospects, what it could mean for how we interact with and use technology and computers for everyone, and the life-changing effects it could have on the quality of life and well-being of patients who stand to benefit the most.

We end with one last parting consideration. We have argued the position that the machine learning and AI algorithms that will be required to arrive at "smart" nanoengineered BMI systems may include the use of existing state of the art algorithms but also possibly new neural derived algorithms and machine learning architectures that more directly model computational and systems neuroscience. What we have not argued for, and what is in no way obvious, is a need for artificial general intelligence (AGI) as necessary to achieve this. Advanced applications such as smart adaptive BMI will almost certainly benefit from advanced algorithms that depend on new mathematical models and theory grounded in empirical neurobiological data. But such algorithms in isolation and out of context do not constitute AGI (although they could conceivably contribute to it).

These algorithms need to be able to execute very sophisticated data analyses, pattern recognition, learning, and decision making but only within the context and embodiment of the neurotechnologies they support. The concept of a self-aware or conscious machine is not required and should not be confused with the technical considerations that actually are needed, i.e., the discussion in this paper. This distinction is important, because the serious societal and ethical concerns and on-going conversations surrounding AGI are very different than the societal and ethical questions that we need to discuss involving neurotechnologies.

8. Challenges and open problems

In this final section, we briefly introduce some of the challenges and open problems associated with actually executing the vision discussed earlier. We do not elaborate in this paper but leave them open for further discussion and dialog. First, most (all) of the recent efforts in the development of neurotechnologies aimed at high density recording or stimulation have focused on the physics, chemistry, and engineering of the core nanotechnologies themselves. This is understandable because the fundamental technologies necessary to enable stimulation or recordings at the actual interface with the brain have to come first. They need to precede any methods or technologies intended to modify or make use of data and information such technologies provide. Beyond the actual interface itself, mechanical and operational stability and long-term reliability of the devices is critical to ensure accurate recordings or stimulation. For example, if the electrodes move or there is excessive reactive gliosis, it will severely affect the efficacy and accuracy of the devices, rendering any control or adaptation by machine learning algorithms irrelevant. These reflect fundamental engineering challenges that have attracted significant amounts of work. And although significant progress has been made, it very much remains highly active areas of research. We do not discuss these issues further in this paper (see, for example, [12–14]).

Furthermore, this framework can be created by including highlights like changing power of light and controlling velocity of a fan utilizing EEG signal level. Additionally, FM Channels and TV slots can be chosen by utilizing human attention level. The preparation time frame in learning the user inclination can be limited by adjusting man-made reasoning.

References

[1] Allied Market Research, Available at: https://www.alliedmarketresearch.com/brain-computer-interfaces-market, 2015.
[2] Grand View Research, Grand View Research, Available at: https://www.grandviewresearch.com/industry-analysis/brain-computer-interfaces-market, 2018.

[3] J.A. Unar, W.C. Seng, A. Abbasi, A review of biometric technology along with trends and prospects, Pattern Recognit 47 (8) (2014) 2673–2688.

[4] V.J. Lawhern, A.J. Solon, N.R. Waytowich, S.M. Gordon, C.P. Huang, B.J. Lance, EEGNet: a compact convolution network for EEG-based brain-computer interfaces. J. Neural Eng. 15 (2016) 56013, https://doi.org/10.1088/1741-2552/aace8c.

[5] M. Lebedev, J.M. Carmena, J.E. O'Deherty, M. Zacksenhouse, C.S. Henriquez, J.C. Principe, et al., Cortical ensemble adaption to represent velocity of an artificial actuator controlled by a brain-machine interface. J. Neurosci. 25 (2005) 4681–4693, https://doi.org/10.1523/JNEUROSCI.4088-04.2005.

[6] M.A. Nicolelis, J.K. Chapin, Controlling robits with the mind. Sci. Am. 287 (2002) 46–55, https://doi.org/10.1038/scientificamerican1002-46.

[7] P. Bach-y-Rita, S.W. Kercel, Sensory substitution and the human-machine interface. Trends Cogn. Sci. 7 (2003) 541–546, https://doi.org/10.1016/j.tics.2003.10.013.

[8] R. Bauer, A. Gharabaghi, Reinforcement learning for adaptive threshold control of restorative brain-computer interfaces: a Bayesian simulation. Front. Neurosci. 9 (2015) 36, https://doi.org/10.3389/fnins.2015.00036.

[9] M.J. Bryan, S.A. Martin, W. Cheung, R.P.N. Rao, Probabilistic co-adaptive brain–computer interfacing. J. Neural Eng. 10 (2013), https://doi.org/10.1088/1741-2560/10/6/066008.

[10] Y. Huang, R.P.N. Rao, Reward optimization in the primate brain: a probabilistic model of decision making under uncertainty. PLoS One 30 (2013) 16777–16787, https://doi.org/10.1371/journal.pone.0053344.

[11] J.M. Carmena, Advances in neuroprosthetic learning and control. PLoS Biol. 11 (2013) e1001561https://doi.org/10.1371/journal.pbio.1001561.

[12] V. Gilja, C.A. Chestek, I. Diester, J.M. Henderson, K. Deisseroth, K.V. Shenoy, Challenges and opportunities for next-generation intracortically based neural prostheses. IEEE Trans. Biomed. Eng. 58 (2011) 1891–1899, https://doi.org/10.1109/TBME.2011.2107553.

[13] B.C. Lega, M.D. Serruya, K.A. Zaghloul, Brain-machine interfaces: electrophysiological challenges and limitations. Crit. Rev. Biomed. Eng. 39 (2011) 5–28, https://doi.org/10.1615/CritRevBiomedEng.v39.i1.20.

[14] C.W. Lu, P.G. Patil, C.A. Chestek, Current challenges to the clinical translation of brain machine interface technology. Int. Rev. Neurobiol. 107 (2012) 137–160, https://doi.org/10.1016/B978-0-12-404,706-8.00008-5.

CHAPTER 25

Metal nanoparticles for electrochemical sensing applications

Shambhulinga Aralekallu and Lokesh Koodlur Sannegowda
Department of Studies in Chemistry, Vijayanagara Sri Krishnadevaraya University, Ballari, Karnataka, India

1. Introduction

The 21st century is facing everlasting problems due to many reasons, and that environment is a prime focus. The limitless growth of world population; intensification of industrial activities; contamination of air, soil, and aquatic ecosystems; global climate change; and other environmental factors are becoming major issues of political and scientific attention. It is essential to make efforts in understanding the influence of human activities on the environment and to develop new technologies to mitigate associated environmental implications. Among the various strategies to address these environmental challenges, recent developments in the field of nanotechnology have triggered a lot of interest because of the unique properties of nanomaterials that find numerous environmental applications [1, 2].

The new field of nanoscience and nanotechnology effectively utilizes the metal nanoparticles (MNPs) that have attracted enormous interest and play an important role in various areas such as catalysis, nanosensors, biomedicine, biological labeling, surface enhanced Raman scattering, and microelectronics due to their specific properties like optical, electrical, thermal, and catalytic activities. Specifically, the unique characteristics such as large surface-to-volume ratio, high electrical conductivity, biocompatibility, excellent catalytic ability, and surface reaction activity provide a huge space for MNPs in improving the sensor performance. Furthermore, the structure of matter at the nanosize decides the basis of natural processes and technological applications. For example, ion transportation via lipid membrane and between the electrodes is highly essential for living cells and batteries. Similarly, sense of smell and detection of hazardous materials are based on the identification of tiny molecules. Therefore, the interaction and process optimization at this scale should ideally be able to utilize the matter at the nanometer level. Indeed, in living cells, exactly tuned nanoenvironments are made available by ion channels and receptors that are formed by self-assembly of molecular building blocks. An artificial device, more particularly man-made architectures, typically depend on the synthetic methodology adopted to prepare the MNPs at the desired size for a particular application [3, 4].

Handbook of Nanomaterials for Sensing Applications
https://doi.org/10.1016/B978-0-12-820783-3.00001-4

Since the last decade, numerous research works have been undertaken on the utilization of MNPs for sensing applications. Also, a bundle of review papers concerning MNP-based electrochemical sensors have been published.

MNPs are nanosized materials formed by either physical or chemical treatment of the bulk metal. Over the last two decades, MNPs have grown as promising materials for a wide range of potential applications for society and industry. The chemistry of these MNPs is investigated in terms of their high surface area, intrinsic size reactivity, and response to physical and chemical stimuli. In addition, the synthetic strategies designed enable these characteristics at the nanoscale for desired applications. Hence, MNPs as a high surface area material platform opens a new door for fine tuning of various structure-property relationships [5, 6].

MNPs are typically obtained in a nanometer scale through different synthetic procedures. Therefore, MNP research initially targeted on testing and optimizing the physical and chemical properties in the context of bulk applications, electrochemical applications, and engineering operations (sensing and detection) [7]. This field is constructed based on the existing research work and implementation of related materials, e.g., gold and silver nanoparticles. More recently, the use of the gold nanoparticles are promising for the construction of electrochemical-sensing devices. In general, gold is a poor catalyst in bulk form, but nanometer-sized gold nanoparticles show excellent catalytic performance because of their relative high surface area-to-volume ratio and their interface-dominated properties, which make the nanoparticles significantly different from their bulk counterparts [8–10]. For example, gold nanoparticles (<10 nm) supported on oxides display high catalytic activity for the chemical and electrochemical oxidation of carbon monoxide (CO) and methanol. High catalytic activity of gold nanoparticles in catalyzing CO oxidation is related to the bandgap of metallic-insulator transition for particles that is in the range of a few nanometers [11–13].

The properties of the prepared MNPs should be in line with the desired properties for electrocatalysis. The MNPs are often coated on the surface of an electrode either by self-assembly leading to chemisorption, physically (drop-casting), or electrochemically (electrodeposition). The electrodes with close-packing and uniform orientation of MNPs can be achieved via self-assembly [14]. However, the dense film formation of MNPs on electrode from aqueous solution dispersion is difficult [15]. The literature reports infer that the surface coverage of gold nanoparticles adsorbed from aqueous solution is less than 30% [16]. The increase in the gold nanoparticle loading on the surface can be achieved by using additional linker molecules that bind extra nanoparticles to the surface [16]. However, the covalent binding of molecules adversely affects the activity of MNPs. Therefore, the electrochemical method of coating is superior to physical adsorption [17].

This book chapter highlights the application of MNPs as an integral part of electrochemical-sensing devices. The designing and working of electrochemical-sensing devices have been explained with suitable examples and mechanisms. Various aspects are considered in this chapter to understand the fundamental physical properties of MNPs and sensing device fabrication. Importantly, some examples of MNPs integration in sensing devices are discussed for different analytes; progress and perspectives have also been discussed.

2. Nanoparticles in electrochemical sensing

Electrochemical sensing using portable systems are in demand and growing rapidly after the development of nanomaterials and nanotechnology. Nanomaterial-enabled technologies have the capability to produce electrochemical detection systems that are inexpensive and facilitate on-site analysis. These are user-friendly devices and function as alternatives to conventional analytical methods. MNP-based electrochemical sensors have demonstrated selective and sensitive electrochemical detection of a wide variety of electroactive species. Metal nanomaterials play a vital role in the development of electrochemical sensors as they possess desirable properties necessary to improve the overall electrochemical redox reactions, increase the sensitivity, and mitigate the drawbacks associated with commercialization issues. Fig. 1 shows a schematic representation of different types of nanomaterials that have been used for the electrochemical detection.

2.1 Carbon based nanoparticles

Carbon nanoparticles (CNPs) offer very distinct advantages such as high surface area, electrical conductivity, chemical stability, biocompatibility, and robust mechanical

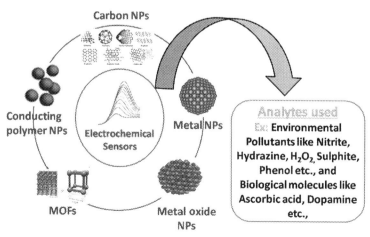

Fig. 1 Schematic illustration of diverse types of nanomaterials employed for the electrochemical monitoring of different analytes.

strength [18, 19]. Therefore, these important properties have been exploited in sensing elements. CNP-based electrochemical sensors usually have higher sensitivities and lower detection limits than their conventional counterparts. The morphology of CNPs provides an additional scope that enables their activity and stable performance in the design of efficient electrochemical sensors [20]. Single-walled carbon nanotubes (SWCNTs) have been used to fabricate stable and sensitive sensors for biological analytes [21]. In addition, doping of heteroatoms to carbon nanostructures has gathered much attraction in recent days. Especially, nitrogen–doped carbon nanotubes (CNTs) have higher potential applications in the development of electrochemical sensors for the sensitive and selective determination of analytes [22]. Graphene is a true two–dimensional material and has received surprising attraction due to its unique physicochemical properties. It has also gained considerable scientific and technological interest. In the following section, it has been elaborately discussed about the applications of carbon-based nanomaterials in the development of electrochemical sensors for different analytes [23].

2.1.1 Single-walled carbon nanotubes

SWCNTs are simple carbon molecules consisting of single graphene that is wrapped into cylindrical tubes. Generally, the diameters of these cylinders vary between 0.4 and 2.5 nm and offer excellent physicochemical properties for a wide range of applications. High electrical conductivity in connection with the extremely small size makes them suitable as individual nanoelectrodes. Furthermore, many researchers have proven that SWCNTs have the ability to promote electron-transfer reactions. Therefore, the incorporation of carbon-based materials has replaced expensive materials used in devices and likely will improve the signal-to-noise (S/N) ratio, which leads to ultrasensitive abilities of the electrochemical sensors. Furthermore, the functionalized CNTs show unique properties that may facilitate a variety of applications including electrochemical sensing [24].

The SWCNT-based carbon paste electrode displayed enhanced electron transfer reaction of hydrogen peroxide (H_2O_2), which resulted in rapid detection of the analyte at lower potential (-0.10 V) [24]. The modified SWCNT-mineral-oil paste electrode was employed for the amperometric monitoring of polyphenolic compounds [24]. High sensitivity value has been reported for the SWCNT-induced electrocatalytic detection of enzymatically generated catechol-quinone [25]. Continuous efforts to develop high performance sensor for dopamine, which is an important neurotransmitter within the central nervous system, has resulted in the fabrication of an efficient sensor based on SWCNT for simultaneous detection of biological molecules such as ascorbic acid, uric acid, and dopamine, which typically coexist in real samples [25].

2.1.2 Multiwalled carbon nanotubes

Multiwalled carbon nanotubes (MWCNTs) are carbonaceous materials that comprise multiple nested graphene sheets and have diameters up to 100 nm. The lengths of these

MWCNTs can range from a few nanometers to several micrometers. The surfaces of these MWCNTs can be chemically modified to improve the surface area and physical properties, therefore MWCNTs are potential candidates for electrochemical applications. They have been efficiently incorporated in the designing of the cyclic voltammetric DNA sensors [26].

The epinephrine biomolecule, an important neurotransmitter, is involved in the signal transmission in the central nervous system. In biological fluids such as blood and urine, epinephrine coexists along with ascorbic acid and uric acid [27]. The ascorbic acid and uric acid interfere with the electrochemical sensing of epinephrine at a pristine electrode. MWCNT-modified electrodes have successfully demonstrated the determination of epinephrine without interference from ascorbic acid and uric acid. In addition, the simultaneous determination of ascorbic acid, epinephrine, and uric acid has been reported at MWCNTs modified with platinum and gold nanoparticle electrode [28].

The ideal cholesterol sensor should work in a concentration range of 2.5–10 mM, because total blood cholesterol of less than 5 mM is considered risk-free, whereas high cholesterol levels of greater than 6 mM are dangerous. A layer-by-layer assembly technique has been used to immobilize cholesterol oxidase and MWCNT on gold electrodes to design a cholesterol-dedicated biosensor that showed linear range of 0.2–6 mM. A screen-printed carbon-paste electrode modified with MWCNT and cholesterol oxidase has the ability to detect cholesterol in blood. The CNTs promote electron transfer and nearly double the sensitivity and improve the linearity range of the electrode for cholesterol [29].

H_2O_2 is one of the important biological analytes produced by the metabolic activities. It is an indicator of oxidative stress and a product of various enzyme-catalyzed reactions. Therefore, the detection of H_2O_2 with high sensitivity and low LOD is highly desirable in clinical diagnostics. Furthermore, a lot of interest has been shown in the study of oxidative stress as it is related to human physiology. Hence, efforts to fabricate an efficient and sensitive sensor based on designing of N-4 macrocycle-embedded CNP sensor exhibited linear concentration range in a nanomolar and micromolar level. The practical application and biological relevance of the fabricated sensor has been evaluated by the determination of H_2O_2 in a human urine sample. The urine samples collected from coffee drinkers showed higher H_2O_2 concentration level with increased oxidative stress [30]. Furthermore, the indirect determination of cholesterol, glucose, and triglyceride may be obtained by the quantification of H_2O_2 produced in enzymatic or enzymatic-like reactions [30].

2.2 Metal nanoparticles

Metallic nanoparticles are nanosized metals with a size range of 10–100 nm. Their properties vary from their bulk counterparts. For example, gold solution does have a golden yellow color, whereas a solution of 20-nm gold nanospheres has a red ruby color, and 200-nm gold nanospheres has a bluish color. MNPs have received profound attention as they find applications in several fields because of their uniform size and sharp size distribution in the nanometer scale. MNPs have unlocked many new pathways in recent years. Their properties can be tuned for particular applications with appropriate functional groups. MNPs can be synthesized and functionalized, which allow them to be efficiently utilized for the desired applications [31]. The noble metals, especially silver and gold, have gained much attention in various branches of science and technology namely catalysis, photography, and medical fields such as anticancer and antimicrobial agents. Faraday (1908) first recognized the existence of metallic nanoparticles in solution, and Mie gave the quantitative explanation on their color [32].

Initially, MNPs were used as decorative materials on the surface due to unique optical properties of noble MNPs. The most important aspect of MNPs is their surface area-to-volume ratio, where it easily allows them to interact with other particles. The high surface area-to-volume ratio makes their diffusion faster and also feasible at lower temperatures. Therefore, even in the medicinal field, MNPs play an interesting role because, without disturbing and poisoning of healthy cells, nanoparticles are able to deliver or treat affected cells and tissues.

2.2.1 Special characteristics of metal nanoparticles

The characteristic properties of MNPs that make them very attractive compared with their bulk are:

(a) Large surface energies.
(b) Large surface area–to-volume ratio.
(c) Quantum confinement.
(d) Plasmon excitation.
(e) Increased number of kinks.

2.2.2 Metal nanoparticles as catalyst

It is well known that MNPs function as a catalyst for various kinds of chemical reactions as catalytic sites are located on a metal surface. Hence, metallic nanoparticles of 1–10 nm size can function as effective catalysts. In MNPs, the ratio of surface atoms increases with decreasing particle size. MNPs are stable under the catalytic condition, otherwise, if it easily coagulates in solution, then it forms aggregates, which are less effective as catalysts. Advantages of using MNPs as catalysts are:

(a) The temperature applied to the metallic nanoparticle catalyst dispersed in solution is below the boiling point of the solvent.
(b) Metallic nanoparticles dispersed in solution can be used as photocatalysts as they are transparent to light.
(c) During preparation, metallic nanoparticles' size and shape can be easily controlled.

(d) Metallic nanoparticles immobilized on solid support acts as a catalyst even for reactions in a gaseous phase.

(e) Bimetallic and trimetallic nanoparticles can be prepared by modifying the condition and composition.

2.2.3 Stabilization of metallic nanoparticles

Because of large surface energy, MNPs coalesce to each other to give thermodynamically favored bulk particles. Coagulation occurs between two MNPs in the absence of repulsive forces. Therefore, stabilization of MNPs is necessary for spatial confinement of the particles in a nanometer range. This stabilization can be achieved either by steric exclusion or electrostatic stabilization by using a capping agent such as surfactant, polymer, solid support, or ligand with suitable functional groups.

Electrostatic stabilization

According to DLVO theory (named after Boris Derjaguin, Lev Landau, Evert Verwey and Theodoor Overbeek), the total interaction between two electrostatically stabilized particles is the combination of Van der Waals attraction and electrostatic repulsion.

Some assumptions considered in the DLVO theory are: (1) uniform surface charge density; (2) no change in concentration profiles of both counterions and surface charge determining ions, i.e., the electric potential remains unchanged; and (3) infinite flat solid surface. However, in spite of the assumptions, the DLVO theory satisfactorily explains the interaction between two approaching charged particles and thus is widely accepted in the research community of colloidal science.

Some limitations of electrostatic stabilization are:

(a) It is a kinetic stabilization method.

(b) Only applicable to dilute systems.

(c) Not applicable to electrolyte-sensitive systems.

(d) Difficult to apply for multiple phase systems, because in a given condition, different solids develop different surface charge and electric potential.

Steric stabilization

Also called polymeric stabilization, this method is widely used in stabilization of colloidal dispersions. However, it has several advantages over electrostatic stabilization:

(i) It is not electrolyte-sensitive.

(ii) It is suitable for multiple phase system.

(iii) It is a thermodynamic method, so particles are always redispersible.

Steric stabilization is achieved by binding of polymer or macromolecules with long alkyl chains to the particle surface.

2.2.4 Synthesis of metal nanoparticles

Generally, two methods are employed for the synthesis of MNPs; top-down and bottom-up methods. *Bottom-up approach* includes the attenuation of material components with further self-assembly process leading to the formation of nanostructures. During self-assembly, the physical forces operating at the nanoscale are used to combine units into large stable structures. Typical examples include Quantum dot and formation of nanoparticles from colloidal dispersion.

Top-down approach includes macroscopic structures that can be externally controlled in the processing of nanostructures. Typical examples are ball milling and application of severe plastic deformation [33, 34].

Top-down method vs. *bottom-up method:* Top-down method starts with a pattern generated on a large scale, which is then reduced to nanoscale. It is quick to manufacture but slow and not suitable for large-scale production. Attrition/milling is a top-down type of method. Bottom-up approach begins with atoms or molecules and builds up to nanostructures. The fabrication process is less expensive, and the bottom-up method involves the production of colloidal dispersion [35].

Chemical reduction method

Noble metal nanoparticles Noble MNPs including platinum, gold, and silver are particularly interesting due to their unique size and shape-dependent optoelectronic properties. In addition, these nanomaterials exhibit various properties such as electrical, optical, magnetic, and chemical and have been intensively studied not only for their fundamental scientific understanding but also for their several technological applications. Furthermore, these noble MNPs can be easily synthesized, characterized, and functionalization on the surface. MNP-based technology is a flourished field with high potential for real-world applications. To understand this potential, especially for electrochemical applications, it is necessary to design and engineer the nanoparticles that may be specifically targeted for various applications. Thus, these engineered nanomaterials serve as unique multidimensional materials whose properties can vary significantly from their bulk material counterparts. Noble MNPs, like platinum, gold, and silver nanoparticles, are the most intensively studied nanomaterials and have led to the development of numerous techniques and methods for many electroanalytical applications [36]. Many of their unique physicochemical properties at the nanoscale have been explored in the development of new electrochemical sensors. In this regard, it is focused on the unique physicochemical properties of noble MNPs that have led to the development of reliable and highly sensitive electrochemical sensors.

2.2.5 Characterization of metal nanoparticles

(a) *Absorption spectroscopy*: Electronic spectroscopy is useful to characterize MNPs as most of the metal NPs possess a bright color that is visible by naked eye. This technique provides qualitative information about the nanoparticle and, by applying Beer's law, absorbance can be measured.

(b) *Infrared spectroscopy*: This method yields information on organic layers surrounding metallic nanoparticles. It also gives valuable information to understand the surface structure of the MNPs.

(c) *Transmission electron microscope*: TEM is widely used to characterize nanomaterials to gain information on morphological features like particle size, shape, crystallinity, and interparticle interaction. TEM is a high spatial resolution structural and chemical characterization microscopic tool and has the capability to directly image atoms in crystalline specimens at resolutions close to 1 to 0.1 nm, smaller than interatomic distance. An electron beam is focused on a particle of diameter smaller than ~0.3 nm, which yields quantitative chemical analysis for the single nanocrystal.

(d) *Scanning electron microscopy*: SEM is a powerful technique for imaging any material surface with a resolution down to 10–50 nm. The interaction of an incident electron beam with the specimen produces secondary electrons with energies smaller than 50 eV. SEM provides information about the morphology, topology, and purity of a nanoparticle sample.

(e) *Atomic force microscopy*: AFM is a better choice for the morphological studies of non-conductive nanomaterials. Typically, it has vertical resolution of less than 0.1 nm and lateral resolution of around 1 nm. It gives detailed information on the atomic scale, which is important for understanding electronic structure and chemical bonding of atoms and molecules.

(f) *X-ray diffraction*: XRD is a useful and widely used technique for determining the crystal structure of crystalline materials. Diffraction line widths are closely related to the size, their distribution, and strain in nanocrystal. The line width is broadened as the size of the nanocrystal decreases due to the loss of long-range order of the bulk material. XRD profile and peaks can be used to determine the particle size by Debye-Scherrer method.

$$D = 0.9\,\lambda/b\cos\theta$$

Where D = nanocrystal diameter, λ = incident X-ray wavelength, b = full width at half maximum of the peak (radians), and θ = Bragg angle.

(g) *Extended X-ray absorption fine structure*: EXAFS is one of the most reliable and powerful characterization techniques to evaluate the structure of metallic nanoparticles, in particular for bimetallic nanoparticles. The sample of metallic nanoparticle should be homogeneous, and this method provides information on the number of atoms surrounding the X-ray absorbing atom and their interatomic distances in the shells.

(h) *X-ray photoelectron spectroscopy*: XPS is used to derive the oxidation state of metal. The oxidation state of the metal atoms on the surface may often get oxidized by air. So, by using this method, valency of surface metal atoms may be confirmed.

2.2.6 General applications of metal nanoparticles
Optical applications
Optical properties of NPs determines potential applications in the domains of imaging, sensor, display, solar cell, photocatalysis, biomedicine, optical detector, lasers, etc. Optical properties and the applications mainly depend on factors such as shape, size, surface area, doping, and interaction with the surrounding molecules. For example, the optical properties of CdSe semiconductor nanoparticles can change with size. Gold nanospheres exhibit changes in the optical properties with variation in the size of metallic nanoparticles. Surface absorption plasmon of AuNPs and AgNPs changes by changing the particle size, form, and shape of the particle and condensation rate.

Thermal properties
The thermal property of NPs varies from that of bulk materials. When the diameter of a nanoparticle is less than 10 nm, the melting point will be lower than a bulk metal. The low melting point of NPs is exploited in the making of electronic wiring.

Mechanical properties
Polymers filled with nanotubes improve the mechanical properties. The improvement is purely dependent on the filler type and the process of filing. The larger particle size of the filler results in poorer properties. Polymer matrix with defoliated phyllosilicates yield excellent mechanical properties. Mechanical properties of metallic nanoparticles can be improved by mixing the nanoparticles with metals or ceramics.

Magnetic properties
Pt, Co, and gold nanoparticles exhibit magnetic property at the nanosized level, but they are nonmagnetic in bulk. By capping, the nanoparticle surface can be improvised due to interaction with other chemical species. The capping with an appropriate molecule provides an opportunity to modify the physical property of nanoparticles.

Catalysis
Catalysts based on metallic nanoparticles are selective, highly active, and exhibit a long lifetime for several types of reactions. There are two types of catalysts; i) Heterogeneous catalysts, which are immobilized on inorganic support and find potential applications in oxidation reactions, synthesis of H_2O_2, water gas shift reaction, and hydrogenation, and ii) homogenous catalysts, in which metallic nanoparticles are surrounded with stabilizers and find applications in nitrile hydrogenation, olefin hydrogenation, etc.

Fuel cell catalysts
A fuel cell is a device that directly converts chemical potential energy into electrical energy. A proton exchange membrane (PEM) cell uses hydrogen gas (H_2) as fuel and

oxygen gas (O_2) as an oxidant, which undergoes oxidation and reduction in presence of suitable catalysts. The precious platinum metal is used as the electrocatalyst for the fuel cell reactions. Also, some nonprecious metal oxides, metal-oxide frameworks (MOFs), organic hybrid molecules, and nanoparticles are being explored to replace the benchmark catalyst platinum. MNPs can be loaded on carbon support to achieve extremely high electrochemically active surface areas. Novel synthetic strategies have been developed to enhance the oxygen reduction reaction (ORR) activity of PtM (M = Co, Ni, Fe, etc.) alloy catalysts by controlling the shapes and composition of nanoparticles. Selective exposure of the highly active surface, (111) and (100) facets, for the ORR was possible by producing PtM nanoparticles of varying shapes, such as octahedra or cubes [37].

Materials science

Nickel nanoparticles are used as electrical conductive pastes, battery materials, etc. The PtRu nanoparticles are used for the direct methanol fuel cells [38]. Noble MNP and nanotubes are used for fuel cell applications toward the enhancement of catalytic activity [39]. Various Pt-based nanoarchitectures and catalysts are being used for fuel cell applications [40].

Medical treatment

Healthy cells can be distinguished from cancer cells by injecting antibodies coupled to the Au nanoparticles. Gold nanoparticles are extensively used for drug delivery and imaging purposes.

Paints

Nano TiO_2 is used in paints to exploit two outstanding properties; photocatalytic activity and UV protection. The addition of nanosilicon dioxide to paint improves the macro- and microhardness, abrasion, and scratch resistance.

Removal of pollutants

Metallic nanoparticles are highly active in terms of physical, chemical, and mechanical properties. They are used as catalysts to prevent environmental pollution arising from coal and gasoline burning as they react with toxic gases such as CO and nitrogen oxide and convert them to nontoxic gases.

Sunscreen lotion

Nanomaterials are very useful as sunscreen lotions as they block UV radiation effectively for a prolonged period of time. The long-time exposure to UV radiation causes skin burns. Sunscreen lotions containing nano-TiO_2 provides sun protection factor (SPF).

Therapeutic applications

As anti-infective agents The AgNPs are proven to be effective for antiviral properties. Metallic nanoparticles like AgNPs have been described as HIV preventative therapeutics and, in addition, silver acts as a viricidal agent and directly binds to the virus through glycoprotein gp120 [62]. This binding in turn prevents the CD4-dependent virion binding, which effectively decreases HIV-1's infectivity. Moreover, the metallic nanoparticles have been found to be effective antiviral agents against herpes simplex virus, influenza, and respiratory syncytial viruses [41].

Antiangiogenic

Angiogenesis is the development of new blood vessels during normal development and in some diseased conditions. It acts as a promoter for a number of diseases such as cancer and rheumatoid arthritis. In normal conditions, angiogenesis is tightly regulated between various proangiogenic growth factors (VEGF, PDGF, and TGF-B) and antiangiogenic factors (platelet factor 4, TSP-1). Under diseased conditions, angiogenesis is turned on and these agents have serious toxicities such as fatal hemorrhage, thrombosis, and hypertension. To overcome these issues, nanoparticles have been tested as efficacious antiangiogenic agents [42].

Radiotherapy

AuNPs are used in radiotherapy of cancer cells. The tumor cells are loaded with gold NPs, which absorb more X-rays as gold is an excellent absorber of X-rays. The deposition of more X-ray beam energy results in a local dose that specifically targets tumor cells.

Drug delivery

Most of the injected chemotherapeutic agents distribute to all parts of the body and results in toxicity with poor compliance in the treatment of patients. Hence, the targeted delivery of therapeutic agents to specific places or organs, for example, tumor cells is a challenge. Imaging of tumor cells is done using metallic nanoparticles by active and passive targeting. Both at surface and inside cells, metallic nanoparticles interact with biomolecules because of their small size, which yields better targeting for therapeutic purpose. The 10–100-nm-sized gold, nickel, silver, and iron [63] metallic nanoparticles with different shapes have been examined as diagnostics and drug delivery systems. Gold nanoparticles' utility in cancer cells' imaging and xenograft tumor mouse model has been experimented upon and, in addition, nontoxic PEG gold nanoparticles have been applied for tumor targeting (in vivo) because of their biocompatibility [43]. Even though the AuNPs are inert and biocompatible, use of metallic nanoparticles for drug delivery is a concern because some fraction of metallic particles may remain in the body after drug administration, which may have side effects at later stages.

These metallic nanoparticles easily conjugate with various agents such as peptides, antibodies, and DNA/RNA to specifically target different cells and, with biocompatible polymers (polyethylene glycol), prolong their circulation in vivo for drug and gene delivery applications. They also transform light into heat, thus enabling thermal ablation of targeted cancer cells [64]. Au nanoparticles have been used as vehicles for the delivery of anticancer drugs such as paclitaxel or cisplatin and oxiplatin (platinum-based drugs) to the target sites or cells. Au nanoparticles of 2 nm covalently bind with the chemotherapeutic drug paclitaxel. The gold-gold sulfide nanoshells produced acts as a photothermal modulated drug delivery system. These nanoshells are covered by a hydrogel matrix that is thermosensitive. These nanoshells are basically designed to strongly absorb NIR light and to release multiple bursts of any soluble material held within the hydrogel matrix in response to repeated NIR irradiation. The 50-nm hollow Au-nanocubes with eight lopped-off porous corners are covered with a thermosensitive polymer containing preloaded effectors that is used to release the effector in a controllable manner using an NIR laser. Protein modified with 10-, 20-, and 40-nm Au nanoparticles and 20-, 50-, and 100-nm Ag nanoparticles obtained from fetal bovine serum have potential effects on radiation-induced killing of glioma cells [65].

3. Sensor electrochemistry

Electroanalytical techniques combined with lab-on-a-chip (LOC) systems are promising detection techniques as they offer on-site analysis, less expensive, high sensitivity, low power requirement, and a distinctive compatibility with other analytical techniques [44, 66, 67]. The working electrode is the heart of state-of-the-art of the electrochemical-sensing system. The sensing reaction will occur on the surface of a working electrode. Redox behavior of the target analyte and the selectivity and sensitivity parameters decide the basis for the selection of an electrode catalyst. Carbon, gold, and platinum are the most common working electrodes used in the sensing devices. Carbon electrodes include glassy carbon, carbon paste, and carbon ink electrodes as working electrodes due to their electrochemical stability, wide potential range, and fast electron transfer [45, 46, 68, 69]. Fig. 2 outlines the general scheme for the fabrication of electrochemical sensors.

Electrodes in microfluidic systems can be produced with thick- or thin-film techniques. Generally, a screen-printing method is employed for fabricating the thick-film electrode and is significantly cheaper than the thin-film method of noble metals; in addition, screen-printing method is also highly compatible with the fabrication of carbon electrodes. Screen-printing of carbon electrodes can be performed on various substrates, e.g., alumina, ceramic plate, polydimethylsiloxane (PDMS), and thermoplastics, and can successfully be integrated into microfluidic systems. [47, 70, 71].

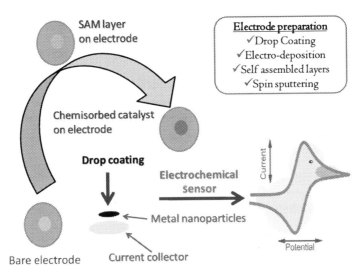

Fig. 2 General schematics for the fabrication of electrochemical sensor.

In addition, inkjet printing serves as an efficient technique to prepare electrodes on flexible substrates. For example, inkjet printing of silver on PDMS has been reported by Wu et al. [72]. Inkjet-printed gold electrodes on paper are reported as well [73]. Common electrode materials for the working and counterelectrode are carbon, copper, gold, or aluminum ink, where the reference electrode (RE) is usually fabricated with Ag/AgCl inks [48, 74, 75].

Thin-film method needs sophisticated instruments to prepare film on an electrode (10–100 nm film thickness), especially for noble metals such as gold and platinum on rigid substrates like glass and silicon wafers.

Sputtering method is widely used to deposit thin film of gold on electrode surface [76]. In addition, Hilmi et al. have reported electroless deposition of thin film of gold onto a glass substrate, which is simple and can be routinely performed in any wet chemistry laboratory [77]. Thin-film fabrication techniques typically require added adhesive layer, e.g., Ti or Cr to support the bonding between metal film and substrates [49]. Patterning of electrodes can be performed by depositing thin film through a mask, using laser ablation, or standard lithography followed by a lift-off process with higher resolution [78]. Furthermore, electrodeposition technique has been employed to form thin film of bismuth on patterned gold electrodes [50].

The REs can be integrated using Ag/AgCl pastes, screen-printing, Ag/AgCl wire, or sputtered Ag. The integration of Ag/AgCl reference into microfluidic system is often challenging. One major problem is the difficulty in transferring an internal electrolyte solution to miniaturized systems to construct the Ag/AgCl RE configuration similar to the stable macro-RE. Thus, most miniaturized Ag/AgCl REs simply rely on "quasi-REs" or "pseudo-REs" where Ag/AgCl is in direct contact with the test solution. Although the RE system has been successfully implemented to LOC-based voltammetric sensors, degradation of AgCl limits the longevity of pseudo-Ag/AgCl [51–53, 79]. Moreover, the unstable ionic strength restricts the application of pseudo-REs in potentiometric measurements. As a result, varieties of strategies are proposed to overcome these problems [54]. A widely used strategy is the generation of a salt bridge between the electrode and the channel [55]. A similar approach without the need of a solid salt bridge is reported by Wongkaew et al. in which Ag RE is implemented into a side channel containing KCl solution that can be pushed uninterrupted into the main channel [80]. The strategy is highly suitable for application in miniaturized potentiometric sensors in which a stable potential regardless of ionic strength of test solution is required.

Alternatively, pseudo-REs made of other metallic materials are integrated into microfluidic systems. For example, Odijk et al. used sputtered palladium as quasi-RE [56], whereas Noh et al. used a RE of nanoporous platinum [57]. Schumacher et al. reported the use of a sputtered iridium/iridium oxide RE on a silicon wafer [58].

4. Noble metal nanoparticles preparation and applications in catalysis

4.1 Platinum nanoparticles

Platinum nanoparticles are attractive for various electrochemical applications due to their size and shape-dependent chemical and electronic properties. The use of these nanoparticles considers the advantage of both the dimensional and functional characteristics of their surface and inorganic domains, which transforms to enhance the specific physical properties. Furthermore, the size of these nanomaterials provide large surface area-to-volume ratio and, as the particle size shrinks, the population of surface atoms increases significantly [59]. By the exploitation of these extraordinary properties, nanoparticles offer feasible alternative platforms to address a wide variety of applications in the electroanalytical field.

During the synthesis of platinum nanoparticles, the platinum metal precursor is taken either in an ionic or a molecular state. The addition of reducing agent leads to chemical changes that convert the precursor to platinum metal atoms. These metal atoms then combine with stabilizer or supported materials to form nanoparticles. For example, the chemical reduction of H_2PtCl_6 by $NaBH_4$ or Zn produces platinum nanoparticles.

$$H_2PtCl_6 + NaBH_4 = Pt^0 + \text{other reaction product}$$

H_2PtCl_6 is the common precursor used in the synthesis of platinum nanoparticles. H_2PtCl_6 is normally dissolved in organic liquid phase or aqueous phase. Either by decomposition or displacement reaction with reducing agent or electrochemical reaction, the dissolved metal precursor is converted into the solid metal. The platinum nanoparticles have also been prepared by physical techniques like irradiation and laser ablation. In one of the methods, irradiation combined with ultrasonication has resulted in monodispersed Pt nanoparticles. The $H_2PtCl_6 \cdot 6H_2O$ was added to polypyrole and sodium dodecyl sulfate (SDS) and particle size was controlled by varying the length and time of ultrasonication and light irradiation [60, 61].

The design of reliable, efficient, and inexpensive tools for the determination of biological molecules is essential in the prevention and monitoring of their deficiencies and, in turn, the resulting diseases. These platforms have broad potential applications in detection of biowarfare and bioterrorism agents as well as forensic and genetic identification. To exploit these opportunities, a variety of assays for the detection of DNA have been developed. Among the conventional DNA detection techniques, electrochemical sensors present a potential approach for the rapid and sensitive determination of genetic disorders. Specifically, nanoparticles are opening up new horizons as efficient electrocatalysts in electrochemical DNA sensors. The detection of low concentration of DNA has been achieved by employing a platinum nanoparticle on gold ultramicroelectrodes combined with hydrazine oxidation reaction [81, 82]. The presence of a target oligonucleotide hybridized with both capture and detection probes facilitated the transit of a platinum nanoparticle onto the electrode surface, where the ensuing electrochemical oxidation of hydrazine resulted in a current response. The demonstrated detection method involved single platinum nanoparticle collision in a sandwich-type DNA sensor. The use of typical ensemble properties of nanoparticles as a recognition signal resulted in the low detection limit for the target DNA of about 100 pM. The ability of nanoparticles to detect individual biomolecules can be enhanced by using high surface area carbon materials. Platinum nanoparticles were used in combination with MWCNTs in the fabrication of a highly sensitive electrochemical DNA sensor [83, 84]. MWCNTs and platinum nanoparticles were dispersed in Nafion and decorated the electrode with the composite, which displayed a superior amplification in the response. The detection limit for target DNA using the Pt nanoparticle-modified MWCNTs was as low as 1.0×10^{-11} M [85].

The development of nonenzymatic catalysts for the electro-oxidation of glucose is of great interest toward the development of sensors for diabetic patients [86]. Mechanistic studies of the electrochemical oxidation of α-glucose and β-glucose on Pt surface in a neutral phosphate buffer solution indicate that glucono-d lactone is the product formed due to the two-electron oxidation of glucose via dehydrogenation at the C_1 carbon, which is hydrolyzed to form the final product, gluconic acid.

4.2 Gold nanoparticles

Since the last decade, enormous growth has been found with the applications of gold nanoparticles in biology and nanomedicine, particularly in therapeutic and imaging applications. From an electroanalytical view, Au nanoparticles have gained significant reputation due to their excellent conductivity, biological compatibility, and high surface-to-volume ratio. Furthermore, owing to their superior stability and complete recovery in biochemical redox processes, gold nanoparticles have been applied as catalysts in numerous electrochemical biomedical applications. In addition, the gold nanoparticles are redox active and open the possibility for the miniaturization of sensing devices to the nanoscale, offering excellent prospects for electrochemical and biological sensing. These noble MNPs are widely utilized as ideal support for the fabrication of electrochemical sensors. These MNPs have significant applications due to: (i) their high chemical stability, which makes them less hazardous; (ii) simple and straightforward synthesis and fabrication process; and (iii) a genuine biocompatibility and noninterference with other labeled biomaterials [87].

From many years, gold nanoparticles have been extensively studied and reported in the literature. Furthermore, in contrast to other noble MNPs, gold nanoparticles can be prepared densely and used for selective oxidation or, in some cases, reduction [88].

Chemical reduction is the most common and widely used method for the preparation of gold nanoparticles. This method includes the reduction of gold salt in the presence of a reducing agent [9]. In 1857, Michael Faraday first studied the gold colloid synthesis in solution phase, where in an aqueous gold chloride was reduced with phosphorus [10]. In 1951, citrate reduction method was reported for the preparation of gold nanoparticles [11]. AuNPs were prepared based on single-phase reduction of gold tetrachloroauric acid by sodium citrate in an aqueous medium yielding particles of about 20 nm [12]. The most prevalent method for the synthesis of monodispersed spherical gold nanoparticles was pioneered by Turkevich et al. in 1951 and later refined by Frens et al. in 1973. This method uses the chemical reduction of gold salts such as hydrogen tetrachloroaurate ($HAuCl_4$) using citrate as the reducing agent. This method produced monodispersed spherical gold nanoparticles in the range of 10–20 nm in diameter, and Frens method showed the following characteristics:

 (i) It allows better control of size and size distribution of gold nanoparticles.
 (ii) It produces particles of higher monodispersity.
(iii) It yields higher concentration.

The major contribution in the synthesis of AuNPs was published in 1994 now popularly known as the Burst–Schifrin method [13]. This process involves two phases in which thiol ligand strongly binds to gold due to the soft character of S and Au. First, with the help of a phase transfer agent such as tetraoctylammonium bromide, gold salt is transferred into an organic solvent, then organic thiol is added. Finally, a strong reducing agent such as sodium borohydride (in excess) is added, which forms thiolate protected AuNPs. The major benefits of synthesizing AuNPs by this process are ease of preparation, size control, thermally stable nanoparticles, and reduced dispersity [3].

Natan investigated the seeded growth of gold nanoparticles by the modification of the Frens synthesis [89]. AuNPs of narrow size distribution were prepared by kinetically controlled seeded growth strategy from HAuCl$_4$ reduction using sodium citrate, which yielded uniform quasispherical-shaped particles with dimensions of 200 nm. During homogeneous growth, the secondary nucleation was controlled by adjusting temperature, pH, and seed particle concentration. The AuNPs can be further functionalized with a wide variety of molecules. Hence, this method seems to be promising in biomedicine, electronics, and photonic fields.

The AuNPs are also formed by the photochemical reduction of gold salt [90]. This method utilizes continuous UV light irradiation (250–400 nm) with PVP as a capping agent and ethylene glycol as the reducing agent. Glycol concentration and viscosity of the solvent mixture are the two factors that affect the AuNP's formation. Furthermore, AuNP's preparation process has been improved by the addition of Ag$^+$ to the Au solution, which led to an increase in the production of Au nanoparticles [91].

Vast improvements in the synthesis of Au nanoparticles have been demonstrated, but researchers still face challenges such as size control, morphology, and monodispersity. The inclusion of gold nanoparticles certainly revitalized the area of research on electrochemical sensors. This strategy could also be utilized for the label-free detection of biomolecules, environmental pollutants, and other molecules.

Chitosan-capped gold nanoparticles coupled with ampicillin have displayed two-fold enhancement in antibacterial activity [8]. Salaün et al. developed an electrochemical sensor using a gold microwire electrode for the simultaneous sensitive electrochemical detection of copper and mercury in seawater by anodic stripping voltammetry in acidic media [92]. Similarly, vibrating gold microwire electrode combined with stripping voltammetry was used to develop a rapid, reliable, and low-cost sensor by Soares and co-workers for the simultaneous electrochemical determination of arsenic, copper, mercury, and lead ions in fresh water [93]. Nanoporous gold prepared by the alloying/dealloying method significantly increased the electrochemical active surface area (EASA), thus enhancing the electrochemical-sensing capability [94]. Ding and co-workers prepared a nanoporous gold leaf by dealloying the Au/Ag alloy in concentrated nitric acid medium. The resulting 3D nanostructured material consisted of ~3 layers of permeable pores and Au ligaments that facilitated the movement of small molecules within the structural framework. The electrocatalytic performance of the nanoporous gold leaf electrode exhibited an excellent activity for nitrite oxidation. In addition, the peak current obtained with the nanoporous gold leaf electrode for the electro-oxidation of nitrite was significantly higher compared with GCE and a planar gold electrode. The higher activity toward nitrite oxidation is attributed to the increased surface area, high electron conductivity, and high specific activity of the 3D nanoporous material. Furthermore, the presence of several exposed Au sites on the radial curvature of the nanostructured material promoted the higher specific activity of the nanoporous gold leaf [95]. Jia et al. developed a nanoporous gold electrode surface through multiple cyclic electrochemical alloying/dealloying process in a "green" electrolyte of ZnCl$_2$ and benzyl alcohol. The resulting

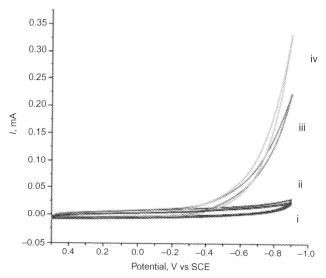

Fig. 3 Electrochemical sensing of H_2O_2 with porphyrin macrocycle-stabilized gold and silver nanoparticles.

nanoporous gold film electrode possessed an ultrahigh surface area and electrochemical activity coupled with excellent selectivity. This study provided a simplistic green route for the preparation of nanoporous metal film electrodes [96]. Our group has successfully developed the H_2O_2 sensor by exploiting the gold and silver nanoparticles stabilized with metal porphyrin macrocycle as shown in Fig. 3 [97, 98]. Gold nanoparticles have also been developed and tested for fuel cell applications.

4.3 Silver nanoparticles

Silver nanoparticles, though not as widely studied as gold nanoparticles, still have made a high impact in the world of nanoscience. Considerable effort is being exerted for developing new applications and protocols for this promising and interesting noble metallic nanoparticle. Silver nanoparticles have proven to be one of the most important groups of nanomaterials for biosensing applications, as well as in biomedical therapeutic applications. Highly sensitive and specific electrochemical sensors based on noble silver nanoparticles have opened up the feasibility of creating new diagnostic platforms for disease markers and detection of biomolecules and environmental pollutants.

AgNPs are one of the most attractive inorganic materials and have been used several applications in various fields like photography, diagnostics, catalysis, biosensor, antimicrobial, etc. [99]. AgNPs can be prepared by one of the following methods:

4.3.1 Reduction by citrate anion

From the pioneering studies, it is now well known that citrate acts as both stabilizer and reducing agent for the metal ion. The citrate molecule plays a major role in stabilizing and controlling the size and shape of AgNPs. [100] Boiling of silver salt at different citrate concentration yielded AgNPs with plasmon maximum absorbance at 420 nm. The increase in the concentration of sodium citrate to silver cation i.e., [citrate]/[Ag+] by one to five times resulted in a decrease in the time required for AgNPs formation from 40 to 20 min, which indicates that a fraction of the Ag^+ was not reduced under equimolar conditions.

4.3.2 Reduction by gallic acid

At room temperature, reduction of Ag^+ in aqueous medium can be achieved by using gallic acid (GA) whose oxidation potential is 0.5 V [101]. The hydroxyl group at determined position in the benzoic acid structure of GA plays an important role in the reduction and stabilization of MNPs. When hydroxyl groups are located at meta-position, nanoparticle synthesis was not successful, but hydroxyl groups at ortho- and parapositions facilitated the nanoparticle formation. Here, carboxylic group acts as a stabilizer and hydroxyl as the reactive part. To obtain silver colloid, NaOH addition is important as it promoted the formation of silver intermediate species Ag_2O, which has been reported as a good precursor for AgNPs preparation by thermal decomposition.

A novel electrochemical sensor based on silver nanoparticles was fabricated by electrodeposition of Ag nanoparticles on poly(ferrocenylsilane) DNA network modified glassy carbon electrode for the detection of H_2O_2 [102]. The developed sensor exhibited good catalytic activity for the reduction of H_2O_2 in the linear range of 2.0 to 353 mM with a detection limit of 0.6 mM. This highly selective and sensitive sensor was employed for practical applications in the detection of H_2O_2. Raoof et al. demonstrated the electrochemical synthesis of silver nanoparticles on a glassy carbon electrode that was modified with p-isopropyl calixarene. The electrochemical experiments revealed that the synthesized silver nanoparticles exhibited superior electrocatalytic activity for H_2O_2 reduction under optimum conditions in the linear range 5.0×10^{-5} to 6.5×10^{-3} M with a detection limit of 2.7×10^{-5} M by an amperometric method [103].

A promising electrochemical-sensing platform has been reported for sensitive nitrite detection using silver nanoparticles decorated with copolymer methyl methacrylate (MMA) and 2-acrylamido-2-methylpropane sulfonic acid (AMPS). The oxidation current of NO_2^- is linear in the concentration range of 1.0–100,000.0 μM and the detection limit was 0.2 μM with a sensitivity 104.6 μA mM^{-1} cm^{-2}. The proposed sensor was also applied for the selective determination of NO_2^- present in water samples, and the results are quite promising [104].

Clenbuterol has been a serious threat to human health due to its illegal usage in live-stock feeding. Peng Miao et al. has reported on a highly sensitive electrochemical sensor for clenbuterol using melanine-functionalized silver nanoparticles. The sandwich structure permits sensitive and selective detection of clenbuterol, and melanine-AgNPs provide a couple of well-defined sharp silver stripping peaks, which stands for a highly characteristic solid-state Ag/AgCl reaction; a rather low detection limit of 10 pM was achieved with a detection range of 10 pM to 100 nM [105].

5. Metal and metal-oxide based nanoparticles in catalysis

Nanostructured metal and metal oxides play an important role as cost-effective, selective, and sensitive catalysts in electrochemical sensing [106]. The metal oxide nanoparticles (MO NPs) are prepared with different morphologies through versatile methods. The size, stability, and high surface area of the MO NPs, along with electrical and photo-chemical properties, make them suitable for various applications. The main function of MNPs in electroanalysis involves the architecting of the conductive sensing interface in which the catalytic properties of nanoparticles are expanded leading to the electrical contact of redox-centers with the surface of the sensing materials [107].

Because of nanosize and increased surface area with higher surface energy, NPs exhibit fast electron transfer between the surface of sensor and analyte molecule, and NPs behave as efficient mediators in electrocatalysis. The biocompatibilities of MO NPs are exploited in the immobilization of biomolecules for the fabrication of immuno-sensors and DNA sensors. On the other hand, semiconducting nanoparticles are being mainly used as markers and tracers in the electrochemical study. Various techniques including physical adsorption, electrodeposition, chemical covalent bonding, and elec-tropolymerization are employed to facilitate the strong affinity of MO NPs with the surface of the working electrode [108].

Major transition MO NPs applied for the electrochemical devices are oxides of iron, copper, cobalt, nickel, manganese, titanium, silver, vanadium, zirconium, zinc, and tungsten. However, some disadvantages of these MO NPs are their wide bandgap, which exhibits semiconducting or even insulating properties resulting in poor electron/ion transport kinetics [109] and volume expansion and contraction during the reversible reactions. These unpleasant limitations can be overcome by the functionalization with carbonaceous materials or other MNPs, and conducting polymers.

5.1 Iron oxide nanoparticles

Hematite (Fe_2O_3) and magnetite (Fe_3O_4) nanoparticles have attracted a lot of interest for electrochemical sensing. In Fe_2O_3, the oxidation state of iron is Fe^{2+}, and in Fe_3O_4, the iron has mixed oxidation states such as Fe^{2+} and Fe^{3+}. The electron hopping process between the Fe^{2+} and Fe^{3+} induces Fe_3O_4 as better electrically conducting nanoparticle.

Various analytes like heavy metals (Pb, Cd, Hg, Zn), glucose, H_2O_2, nitrites, nitrates, and many organic species (dopamine, bisphenol A, urea) have been extensively detected by exploiting iron oxide nanoparticles as electrocatalysts [110]. In many cases, virgin iron oxide (Fe_2O_3/Fe_3O_4) nanoparticles with proper shape and size are used for electrode modification, which exhibits improvement in the sensitivity and detection limit of the sensors. The introduction of iron oxide nanoparticles into the core-shell bimetallic structures has resulted in the enhancement of the efficiency of catalytic activity due to the synergistic effect. Metals like Au, Zr, Ag and cobalt hexacyanoferrate, etc. when incorporated with iron oxide yields core-shell modifications. Doping of some metals (Pd-Pt, Ba, Co, and Ni) with iron oxide also enhanced the activity of electrochemical sensors [111].

The functionalization of iron oxide with organic species and carbonaceous entities like CNTs and graphene oxide (GO) form multicomponent composites and exhibit improved electrocatalytic activity toward many analytes. In most of these cases, carbonaceous materials act as a support to ensure size and shape selectivity of nanoparticles and provide high surface-to-volume ratio for electrocatalytic increment by limiting the agglomeration. Carbonaceous materials also exhibit large surface area, high mechanical strength, fast electron transfer rate, and excellent thermal and electrical conductivities that, in turn, enhance the electrocatalytic activity due to the synergic effect of carbonaceous material and iron oxide NPs [112].

5.2 Titanium dioxide nanoparticles

Titanium oxide (TiO_2) nanoparticles are ceramic nanomaterials and find important applications in the area of electrochemical sensing because of their reliability, biocompatibility, high conductivity, and low cost. TiO_2 nanoparticles are chemically stable in both acidic as well as alkaline solutions and exhibit an efficient catalytic activity for the reduction of some small organic molecules [113]. TiO_2 NPs can be prepared in various forms like nanoparticles, nanotubes, nanofibers, and nanoneedles and have become a prospective environmental-friendly electrode material for electrochemical sensing and biosensing applications (Table 1) [114]. However, low solubility of TiO_2 nanoparticles and the unstable nature of TiO_2 film on the surface of electrode results in low sensitivity, and this is a major limitation involved in the fabrication and commercialization of electrochemical sensors using pure TiO_2 nanomaterials. In an effort to overcome these limitations, TiO_2 combined with carbonaceous materials, polymers, or enzymes has shown improved activity as well as better stability.

Electrodes immobilized with proteins such as cytochrome-c (Cyt-c) and hemoglobin on the surface of nanoporous TiO_2 film acts as the working electrode for the detection of dissolved CO in an aqueous solutions [86]. Nanocrystalline TiO_2 film-modified graphite electrode provides a promising platform for the immobilization of protein molecules, and

Table 1 Nanomaterial-based electrochemical sensors for the biomolecule analytes.

Nanomaterial	Analyte	Detection method	Linear range/ detection limit	Ref.
SWNT-Nafion-GOx	Glucose	Amperometry	2–20 mM	[19]
SWNT-GOx	Glucose	Amperometry	Up to 40 mM	[20]
SWNT-mineral-oil paste	Lactate	Amperometry	Up to 7.0 mM/ 3×10^{-4} M	[21]
Nafion-SWNT	Dopamine	DPV	0.02–6.0 mM/ 5.00 nM	[22]
SWNT polymer composite	Dopamine	CV	1.6×10^{-8} to 6×10^{-4} M/8 nM	[23]
SWNT	Rutin	CV	20 nM–5.0 mM/ 10 nM	[24]
SWNT	DNA	CV	5–30 mM/1.43 mM	[26]
SWNT	Human serum albumin	CV	0.075–7.5 nM/75 pM	[27]
MWNT	DNA	DPV	0.2–50 nM/ 1.0×10^{-11} M	[31]
MWNT-ZrO_2	DNA	DPV	0.149–93.2 nM/ 0.075 nM	[32]
MWNT	DNA	DPV	2–40 mg mL^{-1}/ 16 nM	[33]
MWNT-Nafion	Epinephrine	CV and DPV	0.06–0.24 mM/ 0.02 mM	[34]
MWNT nanocomposite	Epinephrine	LSV	50 nM–10 mM/ 10 nM	[35]
MWNT	Cholesterol	Amperometry	Up to 6.0 mM/ 0.2 mM	[36]
MWNT mat	Cholesterol	Amperometry	100–400 mg dL^{-1}	[37]
MWNT	Methimazole	Amperometry	0.074–63.5 mM/ 0.056 mM	[38]
MWNT-silver nanoparticles	Sumatriptan	CV	80 nM–100 mM/ 40 nM	[39]
MWNT	Paracetamol	Adsorptive-stripping voltammetry	0.01–20 mM/ 10 nM	[40]
Buckypaper	Glucose	CV	Up to 9 mM/ 0.01 mM	[41]
Buckypaper	H_2O_2	CV	0.1–500 mM/ 75 nM	[42]
Pt nanoparticle-MWNT	DNA	DPV	2.25×10^{-7} to 2.25×10^{-11} M/ 1.0×10^{-11} M	[43]

Continued

Table 1 Nanomaterial-based electrochemical sensors for the biomolecule analytes—cont'd

Nanomaterial	Analyte	Detection method	Linear range/ detection limit	Ref.
Pt–Nafion–SWNT-GOx	Glucose	Amperometry	0.5 mM to 5 mM/ 0.5 mM	[44]
Glutathione derivatized Au Nanoparticle	Cancer biomarker	Amperometry	1.0–500 fg mL^{-1}/ 1.0 fg mL^{-1}	[45]
Gold nanoparticle	Cancer biomarker	Amperometry	20–400 pg mL^{-1}/ 20 pg mL^{-1}	[46]
Au nanoparticle-MWNT	*Plasmodium falciparum*	Amperometry	Up to 120 ng mL^{-1}/ 8 ng mL^{-1}	[47]
Gold nanoparticle	DNA	Chronocoulometry	8.0×10^{-17} to 1.6×10^{-12} M/28 aM	[48]
Gold nanoparticle	Human chorionic Gonadotrophin	Amperometry	1.0–100.0 mIU mL^{-1}/ 0.3 mIU mL^{-1}	[49]
Silver nanoparticle	H$_2$O$_2$	Amperometry	2.0 μM–353 mM/ 0.6 mM	[50]
Silver nanoparticle	H$_2$O$_2$	Amperometry	5.0×10^{-5} to 6.5×10^{-3} M/ 2.7×10^{-5} M	[51]
Silver-DNA hybrid nanoparticles	H$_2$O$_2$	Amperometry and CV	2.0 μM to 2.5 mM/ 0.6 mM	[52]
Ag nanoparticle-MWNT-COOH	DNA	Amperometry and DPV	9.0×10^{-12} to 9.0 nM/ 3.2×10^{-12} M	[53]
Silver nanoparticle	Tumor marker	Anodic-stripping voltammetry	5.0 pg mL^{-1} to 5.0 ng mL^{-1}/ 3.5 pg mL^{-1}	[54]
Silver nanoparticle	Sudan	CV	40 nM–4.0 μM/ 8.0 nM	[55]
TiO$_2$ nanotube array	H$_2$O$_2$	Amperometry	11 mM–11 mM/ 1.2 mM	[56]
Au nanoparticle-TiO$_2$ nanotube array	H$_2$O$_2$	CV and chronoamperometry	5.0 mM–0.4 mM/ 2.0 mM	[57]
Pt–Au nanoparticle-TiO$_2$ nanotube array	Glucose	Amperometry	Up to 1.8 mM/ 0.1 mM	[58]

Table 1 Nanomaterial-based electrochemical sensors for the biomolecule analytes—cont'd

Nanomaterial	Analyte	Detection method	Linear range/detection limit	Ref.
TiO_2 nanotube array-Ni composite	Glucose	Amperometry and CV	0.1–1.7 mM/4 mM	[59]
TiO_2-MWNT	Glucose	CV and chronoamperometry	0.01–15.2 mM/2 mM	[60]
Au nanoparticle-TiO_2 nanotube array	Ascorbic acid	CV, DPV, and amperometry	1.0 mM–5.0 mM/0.1 mM	[61]

it enhances the electron transfer process between biomolecules and the electrode [115]. Palladium nanoparticles fabricated on the surface of titanium dioxide-silicon carbide nanohybrid (TiO_2-SiC) via a chemical reduction method (Pd@TiO_2–SiC) on GCE exhibits good electrochemical performance for the simultaneous detection and quantification of hydroquinone and bisphenol A with detection limits ($S/N = 3$) of 5.5 and 4.3 nM, respectively.

Hu et al. have demonstrated a photoenhanced electrochemical-sensing method based on Au nanoparticles decorated carbon-doped TiO_2 nanotube arrays (TiO_2/Au NTAs) for the detection of bisphenol A (BPA). The efficacy of the electrochemical detection of BPA on TiO_2/Au NTA was enriched under UV irradiation. The electrode generated a fresh surface continuously and exhibits an increased photocurrent due to the consumption of holes by BPA.

5.3 Cobalt oxide nanoparticles

Cobalt oxide nanoparticles play an inevitable role in electrochemical sensing. Cobalt oxide NPs have attracted wide attention due to their superior activity and higher stability. Cobalt oxide nanoparticles can be synthesized via electrodeposition method, heating cobalt foil, chemical method, calcination, hydrothermal, and green synthesis. The detection of humidity and temperature has been successfully carried out with cobalt oxide-modified electrode [116, 117].

Electrodeposited cobalt oxide has been a promising material for nitrite detection over a linear range of 1–30 μM with limit of detection of 0.20 μM [118]. Single crystal and vertically aligned cobalt oxide (Co_3O_4) nanowalls coated on GCE using conductive silver paint displayed phenomenal activity for the electrocatalytic oxidation and reduction of H_2O_2 up to 10 μM concentration in 0.01 M phosphate buffer pH 7.4 [119]. Salimi et al. electrodeposited CoO nanoparticles on the surface of GCE and, on top of it, hemoglobin was immobilized. One electron redox reaction through the CoO/hemoglobin film resulted in the reduction of H_2O_2, and this sensor exhibited a quick amperometric response with linear response in the wide concentration range from 5 to 700 μM with a detection limit of 0.5 μM [120]. The same group also employed cobalt oxide nanoparticle-based sensor for the detection of arsenic(III) to deliver promising results [121].

CoO nanoflakes were synthesized via electrochemical method using cyclic voltammetry in a pH-controlled solution containing sodium potassium tartarate on the surface of GCE. Electrocatalytic oxidation of glucose on cobalt oxide nanoflakes modified glassy carbon (CoONF/GC) in alkaline solution yielded fruitful results, and even the kinetics of the reaction has been investigated [122, 123]. Furthermore, cobalt oxide nanoparticles with an average size of 70 nm were prepared on the surface of carbon ceramic electrode (CCE), which exhibited high electrocatalytic activity toward acetaminophen in a wide range of pH and has been used to develop a flow injection analysis (FIA) system for its detection [124, 125]. Saghatforoush et al. have synthesized cobalt hydroxide nanoparticle using cobalt chloride in ammonia solution at room temperature, which in turn, upon calcination at 500°C for 2 h yielded cobalt oxide nanoparticles. This Co_3O_4/GC-modified electrode displayed high catalytic activity for the oxidation of levodopa and serotonin with one pair of redox peaks [126, 127].

5.4 Nanomaterials based on conducting polymers

Conducting polymers (CPs) are organic semiconductors consisting of conjugated chemical double bonds along the polymer chain. The electrical conduction of these polymers can be improved by doping, which induces the delocalization of electrons along the chain. Electrical conductivity of CPs can be tuned by the degree of doping and addition of electroactive species.

The doping process can be classified into two types, namely redox doping and nonredox doping [128]. The former process commonly applies to the case of polypyrrole (PPy) and poly(3,4-ethylenedioxy thiophene (PEDOT). Under ambient conditions, PPy's and PEDOT's backbone are bearing a positive charge on every three to four monomeric units and are nonconductive and prone to oxidation. As a result, they can attract negatively charged dopants, which promote electrical conduction as well as balance the charges on the polymer backbone. On the contrary, there is no change in the number of electrons along the polymer backbone during the nonredox doping, but the energy levels are rearranged instead. The nonredox doping process applies to the production of conductive poly aniline (PANI). In this case, nonconductive PANI is exposed to protonic acids, either inorganic acid (e.g., HCl) or organic (e.g., camphorsulfonic acid). The imine nitrogen atoms on the PANI backbone are protonated, and the degree of protonation is dependent on its oxidation state and pH. Upon complete protonation of the emeraldine base (half-oxidized polymer) with aqueous HCl, for example, an increase in conductivity of about 10 orders of magnitude can be obtained due to the formation of delocalized polysemiquinone radical cation [129]. The doping with organic acids is favorable for the development of electrochemical biosensors because the hydrophobic moieties of the acid dopant stabilize the protonating form under physiological pH conditions. As the doping process is reversible and the dopant stability is highly dependent on the surrounding conditions, these polymers can be exploited as transducers for monitoring small electrical conductivity changes. For example, PANI has been widely applied for NH_3 sensing by determining the change in resistivity [130]. Furthermore, electrochemical detection techniques like potentiometry [131], conductometry, and amperometry employs CPs in the development of chemical sensor/biosensor systems.

The merit of sensing mechanism in CPs depends on the dopant concentration and charge on the surrounding environment. For example, negatively charged biomolecules such as nucleic acids can be electrostatically adsorbed on the positively charged conducting polymers, facilitating the filtration process or assisting the immobilization process [132]. For sample preparation purpose, effective desorption of the adsorbed species is accomplished by simply changing the surrounding environment that promotes the opposite charge of the CPs [133].

Finally, CPs with important functional groups, e.g., –CN, –NH$_2$, or, most importantly, –SH, facilitate the formation of CP-MNP hybrids through self-assembly. Fig. 4 illustrates the structure of different conducting polymers and their conductivity ranges. Furthermore, the high affinity between noble MNPs and CPs containing sulfur groups is interesting in the fabrication of sensing elements, which enhance electrocatalytic activity and also serve as immobilization surfaces for biorecognition elements (BREs) [134]. Moreover, metal ions can be spontaneously or electrochemically reduced on the CPs, forming MNP decorated CPs [135].

6. Macrocycle-stabilized metal nanoparticles

To have better chemical properties, it is very important to control the size and shape of the nanoparticles. There is a lot of research work that has been carried out worldwide to synthesize nanoparticles of different shapes and sizes [136]. Researchers have made much effort to prepare nanoparticles with a narrow size distribution and to organize these nanoparticles into highly ordered superlattices that have unique and specific properties that are different from those of single nanoparticles and random aggregates. Self-assembly offers control over the size, shape, and properties of the individual nanoparticle building blocks [137]. In addition, the properties introduced by the functionalization while stabilizing the nanoparticle lead to the enhancement of device properties [138]. Macromolecules such as polymers, dendrimers, porphyrins, and phthalocyanines have also been used to stabilize and functionalize the nanoparticles but to a very limited extent [139]. Chromophoric porphyrin structure is highly versatile and can be tuned based on chemical modifications on the ring as well as the central core [140]. The porphyrin template serves to control the

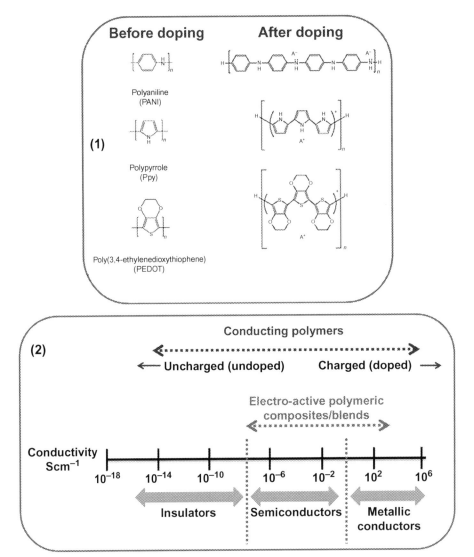

Fig. 4 Example of conducting polymers (CPs). (1) Structure of undoped and doped (A) polyaniline (PANI), (B) polypyrrole (PPy), and (C) poly(3,4-ethylenedioxythiophene) (PEDOT) (A* is an arbitrary dopant species). (2) Conductivity range of CPs and conductive polymeric composites.

shape and size of the nanoparticles, which in turn, affects not only the activity of the catalyst but also the selectivity [141]. In this context, porphyrin-protected metallic nanoclusters are promising. Porphyrin monolayer-protected gold and silver nanoparticles are potential components for the development of the artificial photosynthetic materials, photocatalysts, chemical sensors, etc. [142]. MNPs protected with macromolecules are attractive catalysts not only due to their high surface-to-volume ratio and surface energy but also the effective functionalization that yields better selectivity during catalysis and molecular recognition [143]. So far, very few reports present the potential application of these macromolecule monolayer-protected gold and silver nanoparticles in catalysis of small molecules [144].

The cobalt tetra-amino phenyl porphyrin has been used to stabilize gold and silver nanoparticles for the first time [145]. The four amino groups present at the periphery of the benzene ring of the porphyrin macrocycle and the steric factor of the macrocycle are expected to stabilize the nanoparticle in a controlled fashion to prevent the aggregation. As the porphyrin molecule is a well-known catalyst, when used for the stabilization of nanoparticles, it is expected that the catalytic property will be enhanced to a larger extent due to synergic effect [146].

One pot approach has been used to synthesize Au and AgNPs by sonicating $HAuCl_4$ or $AgNO_3$ in DMSO with aminophenyl porphyrin at 5°C, and sodium borohydride ($NaBH_4$) was used for reduction to obtain stable Au and Ag nanoparticles [147]. The particles were homogeneously dispersed with nearly spherical shape, and the size varies from 4 to 5 nm. The homogeneity and monodispersity of the nanoparticles indicate the effective capping of porphyrin moiety with the MNPs, which resists the aggregation of nanoparticles [148]. The porphyrin-capped gold nanoparticles exhibited superior efficiency for H_2O_2 reduction in terms of catalytic current and overpotential compared with silver nanoparticles [149]. Proposed mechanism for the reduction of H_2O_2 has been presented as:

$$2M^0 + H_2O_2 \rightarrow 2M^{2+} - OH + H_2O$$

$$2M^{2+} - OH + 2e^- + 2H^+ \rightarrow 2M^0 + H_2O$$

$$[M - CoTAPP] + nH^+ + ne^- \rightarrow [M - CoTAPP]_{red}$$

$$[M - CoTAPP]_{red} + nH_2O_2 \rightarrow [M - CoTAPP] + nH_2O$$

where M = Au or Ag nanoparticles.

Cobalt is a highly oxidizing metal, and even after stabilizing it using an appropriate capping agent, it undergoes oxidation. Although thiol, carboxylic acid, and amine-based capping agents play an important role in stabilization of Co nanoparticles, the stabilized nanoparticles still undergo oxidation with time. [150] Cobalt nanoparticles were prepared by mixng cobalt teramine phthalocyanine in DMSO with $CoCl_2$ at 283 K in inert atmosphere and reducing with $NaBH_4$ (Fig. 5).

The $NaBH_4$ reduces Co^{+2} ions to produce metallic cobalt. The nitrogen atoms present in the amine group of the peripheral benzene and delocalized π-electrons of the phthalocyanine ring interact with cobalt nanoparticles and stabilize them in an efficient way [151]. Covalent-like strong interactions between the amine and metal and physical

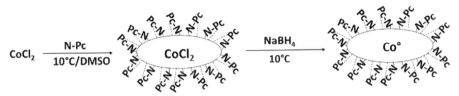

Fig. 5 Scheme for preparation of cobalt nanoparticles capped with CoTAPc. Where N-Pc = CoTAPc and - - - bond between Pc and Co indicates the interaction of delocalized π-electrons of CoTAPc with Co.

and Van der Waals interactions due to the delocalized π-electrons of macrocycle with metal and also steric effects due to macromolecule parts in the phthalocyanine have combined efficiently to stabilize the cobalt nanoparticles [152].

The stabilized cobalt nanoparticles on dispersion in ethanol solvent showed nanoparticle-interconnected pearl necklace-like structure due to the magnetic dipolar attraction between the neighboring particles in ethanol solvent [153].

The cobalt nanoparticles and composite of CNT and Co NPs (CNT-Co NP) were employed for ORR [154], and an increase in peak current of almost four times that of the nanoparticle electrode was noticed with a shift in the onset potential to −0.05 V for CNT-dispersed nanoparticle composite [155]. The enhancement in the activity for the composite was due to the increase in the conductivity of the nanoparticles on mixing with CNT, increased surface area, and porous structure of CNT, which influences the reactant-product mass transport and, in turn, promotes the charge transfer [156]. This results in higher catalytically active sites that allow the adsorption of oxygen and simultaneous reduction (Fig. 6) [157].

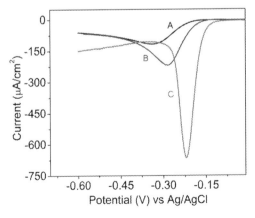

Fig. 6 LSV curves for ORR of (A) pristine GC, modified GC with (B) Co nanoparticles, (C) composite of Co nanoparticles and CNT in O_2 saturated 0.3 M KOH.

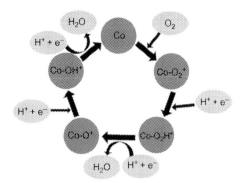

Fig. 7 Mechanism proposed for ORR at cobalt nanoparticles.

The catalytic activity and mechanism of oxygen reduction depends on the orientation and packing of the nanoparticles on the electrode surface. The possible ORR mechanism has been depicted in Fig. 7.

Ding et al. have reported the fabrication of ultrathin films of TiO_2 nanoparticle capped with tetrasulfonated copper phthalocyanine (CuTsPc) by layer-by-layer (LBL) self-assembly technique. Alternating bilayer structures were formed by consecutive adsorption of CuTsPc-capped TiO_2 nanoparticles with poly(diallyldimethylammonium chloride). These ordered ultrathin films have been used for electrocatalysis, photocatalysis, and photovolotaics [158]. Zhang et al. reported the stable organic sol of solvent-stabilized oxovanadium phthalocynine (VOPc) nanoparticles with excellent photoconductivity. The nanoscopic VOPc particles were well-dispersed in an insulating polycarbonate (PC) resin, resulting in single-layered photoreceptors with high surface charge durability in the dark and excellent photoconductivity. [159].

Paul et al. has reported the detection of lead and copper ions by calix-functionalized gold nanoparticles, CFAuNPs [160]. The aggregation of gold nanoparticles in case of lead ion detection is due to interparticle plasmon coupling, and in the case of copper due to alloy formation of gold-copper by antigalvanic exchange.

Gold nanoparticles functionalized with dithiacarbamide derivative of calix are insoluble in water but displayed excellent selective response toward cobalt metal ion in THF [161]. No significant interference from other metal ions was noticed in amperometric method in aqueous medium. The selectivity for cobalt ions is due to the aggregation of AuNPs as a result of coupling of the plasmon resonances of equidistantly placed neighboring AuNPs.

Paul et al. have also reported the detection of mercury ions using water-soluble calix-functionalized gold nanoparticles [162]. Hg(II) ions were selectively and sensitively detected with remarkable detection limit of 40 ppb. The efficiency and multivalency of nanoparticles increase when they are bound by multiple ligating units compared with a monovalent ligand unit. Avvakumova et al. prepared glycocalixarene-functionalized water-dispersible gold nanoparticles for the first time and explained their efficiency for targeting cancerous cells [163].

Qian et al. reported a novel, facile, and clean synthesis of monodispersed AuNPs with an average diameter of 5 nm by reducing $HAuCl_4$ with dihydroxylatopillar[5]arene (2HP5) in a basic solution without the use of harsh reagents and/or external energy. The electrochemcial indicator, toluidine blue, could enter into the cavity of 2HP5 and forms host–guest complex that significantly enhanced the loading quantity of TB and effectively suppressed the leaking of TB resulting in an ultrasensitive and robust electrochemical response. Furthermore, the construction of a sandwich-type electrochemical immunosensor using TB-2HP5@Au-Pd/MnO_2 nanocomposite improved the robustness and ultrasensitive detection of cardiac troponin I (cTnI), a significant biomarker of acute myocardial infarction. This immunosensor had remarkable stability, specificity, and linearly response observed in the concentration range from 0.005 to 20 ng mL^{-1} with a low detection limit of 2 pg mL^{-1} ($S/N = 3$). This electrochemical immunosensor was employed for human serum analysis, and it exhibited excellent recovery values indicating a promising emerging transducer material for the detection of biological markers [164].

7. Bimetallic nanoparticles

Bimetallic nanoparticles have attracted significant attention as electrocatalysts, because the addition of the second metal brings about the variations in particle shape, size, surface morphology, and chemical and physical properties [165]. Compared with the corresponding monometal nanoparticles, bimetallic particles have exhibited higher catalytic activity, better resistance to deactivation, and greater catalytic selectivity [166]. These bimetallic nanostructures are usually in the form of alloys, core/shell structures, and mixed monometallic nanoparticles [167]. These bimetallic nanoparticles have been applied for constructing electrochemical sensors.

Different forms of Au–Ag bimetallic nanoparticles have been prepared to catalyze H_2O_2. Wang et al. [168] and Tsai et al. [169] fabricated Au–Ag alloy nanoparticles through the reduction of Ag on the Au nanoparticles modified film and electrodeposition process, respectively, for the detection of H_2O_2. Manivannan and Ramaraj [169] synthesized the core-shell Au_{100}–xAg_x bimetallic nanoparticles embedded in methyl functionalized silicate, i.e., methyltrimethoxysilane (MTMOS) network and investigated their electrocatalytic activity toward the reduction of H_2O_2.

Besides Au–Ag bimetallic nanoparticles, Au–Pt alloy nanoparticles are also extensively studied for electrocatalytic applications [170]. Yu et al. [113] prepared a series of room-temperature ionic liquids (RTILs) containing different functional groups, which were utilized as electrodeposition substrate for depositing Au/Pt nanostructures.

Meanwhile, Au/Pt nanoparticles have provided a facile way to construct a third-generation H_2O_2 sensor. Niu et al. [171] proposed novel snowflake-like Pt-Pd bimetallic cluster modified screen-printed gold nanofilm electrode for H_2O_2 and glucose sensing.

Two different reactivities are generally used in the case of mixed or alloyed bimetallic nanoparticles, e.g., $RuCl_3$ and H_2PtCl_6 or $\{Na_6Pt (SO_3)_4, Na_6Ru (SO_3)_4\}$ or $\{PtCl_2$ and $RuCl_3)\}$ and even various complex mixed precursors have also been used. Either radiolytic, sonochemical, or electrochemical method is employed for the chemical activation of the physical mixing.

In addition, bimetallic nanoparticles such as Pt/Ir, Pt/Ag and Pt/Cu, Pt/Pd, Pd/Cu, and Rh/Pd and Ru/Rh were employed for constructing an electrochemical-sensing platform for the determination of various analytes [165, 172, 173].

8. Conclusion

This chapter provides a glimpse into some simpler MNPs that are being currently used for their potential applications in electrochemical and biosensors. However, the field of nanoscience has blossomed over the last two decades, and the need for nanotechnology to explore beyond the sensor applications has become more important. MNPs have successfully integrated in detection of various analytes, but the advances in sensitivity and detection limit depend largely on the shape, size, and selectivity of the nanoparticle to the analyte molecules. Moreover, the type of the nanoparticle synthesized also governs the sensing modality to be used and thus the cost of sensor devices. Even though current investigations have demonstrated that multivalent composite materials can provide significant advantages, the ambiguity in developing them for a particular target with high specificity is still challenging. Fortunately, the field of nanotechnology continues to grow in the chemical research community with major discoveries as well as new scientific challenges. Nevertheless, future studies should also aim to address safety and biocompatibility of these MNPs, in particular long-term durability and on-sight applicability.

References

[1] M. Reibold, P. Paufler, A.A. Levin, W. Kochmann, N. Pätzke, D.C. Meyer, Carbon nanotubes in an ancient damascus sabre, Nature 444 (2006) 286.
[2] M. Krukemeyer, V. Krenn, F. Huebner, W. Wagner, R. Resch, History and possible uses of nano-medicine based on nanoparticles and nanotechnological progress, J. Nanomed. Nanotechnol. 6 (2015) 336.
[3] M.C. Daniel, D. Astruc, Gold nanoparticles: assembly, supramolecular chemistry, quantum-size-related properties, and applications toward biology, catalysis, and nanotechnology, Chem. Rev. 104 (2004) 293–346.
[4] A.N. Goldstein, C.M. Echer, A.P. Alivisatos, Melting in semiconductor nanocrystals, Science 256 (1992) 1425.

[5] S.E.F. Kleijn, S.C.S. Lai, M.T.M. Koper, P.R. Unwin, Electrochemistry of nanoparticles, Angew. Chem. Int. Ed. 53 (2014) 3558.

[6] V.J. Nagaraj, M. Jacobs, K.M. Vattipalli, V.P. Annam, S. Prasad, Nanochannel-based electrochemical sensor for the detection of pharmaceutical contaminants in water, Environ Sci Process Impacts 16 (2014) 135.

[7] L.D. Rampino, F.F. Nord, Preparation of palladium and platinum synthetic high polymer catalysts and the relationship between particle size and rate of hydrogenation, J. Am. Chem. Soc. 63 (1941) 2745.

[8] A. Sinha, J. Manjhi, Silver nanoparticles: green route of synthesis and antimicrobial profile, In. J. Nanoparticles 8 (2015) 30.

[9] M.A. Hayat, Colloidal Gold: Principles, Methods, and Applications, Elsevier, 2012.

[10] J.A. Khan, R.A. Kudgus, A. Szabolcs, S. Dutta, E. Wang, et al., Designing nanoconjugates to effectively target pancreatic cancer cells in vitro and in vivo, PLoS One 6 (6) (2011) 20347.

[11] J.M. Simard, Synthesis of Gold Nanoparticles for Biomacromolecular Recognition, University of Massachusetts Amherst, 2017.

[12] J. Turkevich, P.C. Stevenson, J. Hillier, A study of the nucleation and growth processes in the synthesis of colloidal gold, Discuss. Faraday Soc. 11 (1951) 55.

[13] M. Brust, M. Walker, D. Bethell, D.J. Schiffrin, R. Whyman, Synthesis of thiol-derivatised gold nanoparticles in a two-phase liquid-liquid system, J. Chem. Soc. Chem. Commun. 1994 (1994) 801.

[14] S. Song, Y. Qin, Y. He, Q. Huang, C. Fan, H.Y. Chen, Functional nanoprobes for ultrasensitive detection of biomolecules, Chem. Soc. Rev. 39 (2010) 4234.

[15] Y. Meng, L. Aldous, S.R. Belding, R.G. Compton, The formal potentials and electrode kinetics of the proton/hydrogen couple in various room temperature ionic liquids, Chem. Commun. 48 (2012) 5572.

[16] B. Shah, A. Chen, Novel electrochemical approach for the monitoring of biodegradation of phenolic pollutants and determination of enzyme activity, Electrochem. Commun. 25 (2012) 79.

[17] S. Chatterjee, A. Chen, Voltammetric detection of the alpha-dicarbonyl compound: methylglyoxal as a flavouring agent in wine and beer, Anal. Chim. Acta 751 (2012) 66.

[18] R.N. Goyal, S. Chatterjee, A.R.S. Rana, The effect of modifying an edge-plane pyrolytic graphite electrode with single-wall carbon nanotubes on its use for sensing diclofenac, Carbon 48 (2010) 4136.

[19] S.B. Revin, S.A. John, Electrochemical sensor for neurotransmitters at physiological pH using a heterocyclic conducting polymer modified electrode, Analyst 137 (2012) 209.

[20] J. Lei, H. Ju, Nanotubes in biosensing, Wiley Interdiscip. Rev. Nanomed. Nanobiotechnol. 2 (2010) 496.

[21] L. Wu, X. Zhang, H. Ju, Detection of NADH and ethanol based on catalytic activity of soluble carbon nanofibre with low overpotential, Anal. Chem. 79 (2007) 453.

[22] S. Wu, H. Ju, Y. Liu, Conductive mesocellular silica-carbon nanocomposite foams for immobilization, direct electrochemistry and biosensing of proteins, Adv. Funct. Mater. 17 (2007) 585.

[23] L. Shi, X. Liu, W. Niu, H. Li, S. Han, J. Chen, G. Xu, Hydrogen peroxide biosensor based on direct electrochemistry of soybean peroxide immobilized on single-walled carbon nanohorn modified electrode, Biosens. Bioelectron. 24 (2009) 1159.

[24] Y. Tang, B.L. Allen, D.R. Kauffman, A. Star, Electrocatalytic activity of nitrogen-doped nanotube cups, J. Am. Chem. Soc. 131 (2009) 13200.

[25] X. Xu, S. Jiang, Z. Hu, S. Liu, Nitrogen doped carbon nanotubes: high electrocatlytic activity toward the oxidation of hydrogen peroxide and its application for biosensing, ACS Nano 4 (2010) 4292.

[26] D.A.C. Brownson, D.K. Kampouris, C.E. Banks, Graphene electrochemistry: fundamental concepts through to prominent applications, Chem. Soc. Rev. 41 (2012) 6944.

[27] Y. Zhang, Y. Bai, B. Yan, Functionalized carbon nanotubes for potential medicinal applications, Drug Discov. Today 15 (2010) 428.

[28] J. Wang, M. Musameh, Y. Lin, Solubilization of carbon nanotubes by nafion toward the preparation of amperometric biosensors, J. Am. Chem. Soc. 125 (2003) 2408.

[29] J. Wang, M. Musameh, Enzyme-dispersed carbon-nanotube electrodes: a needle microsensor for monitoring glucose, Analyst 128 (2003) 1382.

[30] M.D. Rubianes, G.A. Rivas, Enzymatic biosensor based on carbon nanotubes paste electrodes, Electroanalysis 17 (2005) 73.

[31] H.S. Wang, T.H. Li, W.L. Jia, H.Y. Xu, Highly selective and sensitive determination of dopamine using a nafion/carbon nanotubes coated poly(3-methylthiophene) modified electrode, Biosens. Bioelectron. 22 (2006) 664.

[32] H. Horvath, Gustav Mie and the scattering and absorption of light by particles: historic developments and basics, J. Quant. Spectrosc. Radiat. Transf. 110 (2009) 787.

[33] Y. Zhang, Y. Cai, S. Su, Determination of dopamine in the presence of ascorbic acid by poly(styrene sulfonic acid) sodium salt/single-wall carbon nanotube film modified glassy carbon electrode, Anal. Biochem. 350 (2006) 285.

[34] B. Zeng, S. Wei, F. Xiao, F. Zhao, Voltammetric behaviour and determination of rutin at a single-walled carbon nanotubes modified gold electrode, Sensors Actuators B Chem. 115 (2006) 240.

[35] M. Silvestrini, L. Fruk, P. Ugo, Functionalized ensembles of nanoelectrodes as affinity biosensors for DNA hybridization detection, Biosens. Bioelectron. 40 (2013) 265.

[36] J. Li, Y. Zhang, T. Yang, H. Zhang, Y. Yang, P. Xiao, DNA biosensor by self-assembly of carbon nanotubes and DNA to detect riboflavin, Mater. Sci. Eng. C 29 (2009) 2360.

[37] T.C. Deivaraj, J.Y. Lee, Preparation of carbon-supported PtRu nanoparticles for direct methanol fuel cell applications—a comparative study, J. Power Sources 142 (2005) 43.

[38] J. Tang, J. Liu, N.L. Torad, T. Kimura, Tailored design of functional nanoporous carbon materials toward fuel cell applications, NanoToday 9 (2014) 305.

[39] C.H. Cui, S.H. Yu, Engineering interface and surface of noble metal nanoparticle nanotubes toward enhanced catalytic activity for fuel cell applications, Acc. Chem. Res. 46 (2013) 1427.

[40] N. Jung, D.Y. Chungb, J. Ryu, S.J. Yoo, Y.E. Sung, Pt-based nanoarchitecture and catalystdesign for fuel cell applications, NanoToday 9 (2014) 433.

[41] P. Scodeller, V. Flexer, R. Szamocki, E.J. Calvo, N. Tognalli, H. Troiani, A. Fainstein, Wired-enzyme core-shell Au nanoparticles biosensor, J. Am. Chem. Soc. 130 (2008) 12690.

[42] K.A. Mahmoud, J.H. Luong, Impedance method for detecting HIV-1 protease and screening for its inhibitors usning ferrocene-peptide conjugate/Au nanoparticles/single-walled carbon nanotube modified electrode, Anal. Chem. 80 (2008) 7056.

[43] D. Tian, C. Duan, W. Wang, H. Cui, Ultrasensitive electrochemiluminescence immunosensor based on luminal functionalized gold nanoparticles labelling, Biosens. Bioelectron. 25 (2010) 2290.

[44] J. Wang, Electrochemical detection for microscale analytical systems: a review, Talanta 56 (2002) 223.

[45] R.S. Martin, A.J. Gawron, B.A. Fogarty, F.B. Regan, E. Dempsey, S.M. Lunte, Carbon paste-based electrochemical detectors for microchip capillary electrophoresis/electrochemistry, Analyst 126 (2001) 277.

[46] J.S. Rossier, M.A. Roberts, R. Ferrigno, H.H. Girault, Electrochemical detection in polymer microchannels, Anal. Chem. 71 (1999) 4294.

[47] H.Y. Tan, W.K. Loke, N.T. Nguyen, S.N. Tan, N.B. Tay, W. Wang, S.H. Ng, Lab-on-a-chip for rapid electrochemical detection of nerve agent sarin, Biomed. Microdevices 16 (2014) 269.

[48] S. Schumacher, J. Nestler, T. Otto, M. Wegener, E.E. Forster, D. Michel, K. Wunderlich, S. Palzer, K. Sohn, A. Weber, et al., Highly-integrated lab-on-chip system for point-of-care multiparameter analysis, Lab Chip 12 (2012) 464.

[49] A.T. Woolley, K. Lao, A.N. Glazer, R.A. Mathies, Capillary electrophoresis chips with integrated electrochemical detection, Anal. Chem. 70 (1998) 684.

[50] J. Min, A.J. Baeumner, Characterization and optimization of interdigitated ultramicroelectrode arrays as electrochemical biosensor transducers, Electroanalysis 16 (2004) 724.

[51] Z. Zou, A. Jang, E. MacKnight, P.M. Wu, J. Do, P.L. Bishop, C.H. Ahn, Environmentally friendly disposable sensors with microfabricated on-chip planar bismuth electrode for in situ heavy metal ions measurement, Sensors Actuators B Chem. 134 (2008) 18.

[52] P. Jothimuthu, R.A. Wilson, J. Herren, E.N. Haynes, W.R. Heineman, I. Papautsky, Lab-on-a-chip sensor for detection of highly electronegative heavy metals by anodic stripping voltammetry, Biomed. Microdevices 13 (2011) 695.

[53] D. Desmond, B. Lane, J. Alderman, J.D. Glennon, D. Diamond, D.W.M. Arrigan, Evaluation of miniaturised solid state reference electrodes on a silicon based component, Sensors Actuators B Chem. 44 (1997) 389.

[54] C.C. Shih, C.M. Shih, K.Y. Chou, S.J. Lin, Y.Y. Su, R.A. Gerhardt, Mechanism of degradation of AgCl coating on biopotential sensors, J. Biomed. Mater. Res. A 82 (2007) 872.

[55] R. Mamińska, A. Dybko, W. Wróblewski, All-solid-state miniaturised planar reference electrodes based on ionic liquids, Sensors Actuators B Chem. 115 (2006) 552.

[56] M. Ciobanu, J.P. Wilburn, N.I. Buss, P. Ditavong, D. Lowy, Miniaturized reference electrodes based on Ag/AgiX internal reference elements. I. Manufacturing and performance, Electroanalysis 14 (2002) 989.

[57] M. Ciobanu, J.P. Wilburn, D.A. Lowy, Miniaturized reference electrodes. II. Use in corrosive, biological, and organic media, Electroanalysis 16 (2004) 1351.

[58] N. Wongkaew, S.E.K. Kirschbaum, W. Surareungchai, R.A. Durst, A.J. Baeumner, A novel three-electrode system fabricated on polymethyl methacrylate for on-chip electrochemical detection, Electroanalysis 24 (2012) 1903.

[59] X. Yu, S.N. Kim, F. Papadimitrakopoulos, J.F. Rusling, Protein immunosensor using single-wall carbon nanotube forests with electrochemical detection of enzyme labels, Mol. BioSyst. 1 (2005) 70.

[60] S. Eustis, H.Y. Hsu, M.A. El-Sayed, Gold nanoparticle formation from photochemical reduction of Au^{3+} by continuous excitation in colloidal solutions. A proposed molecular mechanism, J. Phys. Chem. B 109 (11) (2005) 4811.

[61] S. Iravani, Green synthesis of metal nanoparticles using plants, Green Chem. 13 (2011) 2638.

[62] A.D.E. Muniz, M.M. Costa, C.S. Espinel, B.D. Freitas, J.F. Suarez, A.G. Fernandez, A. Merkoci, Gold nanoparticle-based electrochemical magnetoimmunosensor for rapid detection of anti-hepatitis B visus antibodies in human serum, Biosens. Bioelectron. 26 (2010) 1710.

[63] G.J. Yang, J.L. Huang, W.J. Meng, M. Shen, X.A. Jiao, A reusable capacitive immunosensor for detection of salmonella spp. Based on grafted ethylene diamine and self-assembled gold nanoparticle monolayers, Anal. Chim. Acta 647 (2009) 159.

[64] K. Glynou, P.C. Ioannou, T.K. Christopoulos, V. Syriopoulou, Oligonucleotide-functionalized gold nanoparticles as probes in dry-reagent strip biosensor for DNA analysis by hybridization, Anal. Chem. 75 (2003) 4155.

[65] K. Hu, D. Lan, X. Li, S. Zhang, Electrochemical DNA biosensor based on nanoporous gold electrode and multifunctional encoded DNA-Au bar codes, Anal. Chem. 80 (2008) 9124.

[66] J. Wang, Portable electrochemical systems, TrAC Trends Anal. Chem. 21 (2002) 226.

[67] J. Wang, B. Tian, E. Sahlin, Micromachined electrophoresis chips with thick-film electrochemical detectors, Anal. Chem. 71 (1999) 5436.

[68] A. Chałupniak, A. Merkoçi, Graphene oxide − poly-(dimethylsiloxane)-based lab-on-a-chip platform for heavy-metals preconcentration and electrochemical detection, ACS Appl. Mater. Interfaces 9 (2017) 44766.

[69] J. Wang, M. Pumera, M.P. Chatrathi, A. Escarpa, R. Konrad, A. Griebel, W. Dorner, H. Lowe, Toward disposable lab-on-a-chip: poly(methylmethacrylate) microchip electrophoresis device with electrochemical detection, Electrophoresis 23 (2002) 596.

[70] N. Ruecha, J. Lee, H. Chae, H. Cheong, V. Soum, P. Preechakasedkit, O. Chailapakul, G. Tanev, J. Madsen, N. Rodthongkum, et al., Paper-based digital microfluidic chip for multiple electrochemical assay operated by a wireless portable control system, Adv. Mater. Technol. 2 (2017) 1600267.

[71] J. Wu, R. Wang, H. Yu, G. Li, K. Xu, N.C. Tien, R.C. Roberts, D. Li, Inkjet-printed microelectrodes on PDMS as biosensors for functionalized microfluidic systems, Lab Chip 15 (2015) 690.

[72] P. Sjöberg, A. Määttänen, U. Vanamo, M. Novell, P. Ihalainen, F.J. Andrade, J. Bobacka, J. Peltonen, Paper-based potentiometric ion sensors constructed on ink-jet printed gold electrodes, Sensors Actuators B Chem. 224 (2016) 325.

[73] W. Dungchai, O. Chailapakul, C.S. Henry, Electrochemical detection for paper-based microfluidics, Anal. Chem. 81 (2009) 5821.

[74] H. Ko, J. Lee, Y. Kim, B. Lee, C.H. Jung, J.H. Choi, O.S. Kwon, K. Shin, Active digital microfluidic paper chips with inkjet-printed patterned electrodes, Adv. Mater. 26 (2014) 2335.

[75] P. Ginet, K. Montagne, S. Akiyama, A. Rajabpour, A. Taniguchi, T. Fujii, Y. Sakai, B. Kim, D. Fourny, S. Volz, Toward single cell heat shock response by accurate control on thermal confinement with an on-chip microwire electrode, Lab Chip 11 (2011) 1513.

[76] J. Wang, B. Tian, E. Sahlin, Integrated electrophoresis chips/amperometric detection with sputtered gold working electrodes, Anal. Chem. 71 (1999) 3901.

[77] A. Hilmi, J.H.T. Luong, Electrochemical detectors prepared by electroless deposition for microfabricated electrophoresis chips, Anal. Chem. 72 (2000) 4677.

[78] S. Saem, Y. Zhu, H. Luu, J.M. Mirabal, Bench-top fabrication of an all-PDMS microfluidic electrochemical cell sensor integrating micro/nanostructured electrodes, Sensors 17 (2017) 732.

[79] A. Wisitsoraat, P. Sritongkham, C. Karuwan, D. Phokharatkul, T. Maturos, A. Tuantranont, Fast cholesterol detection using flow injection microfluidic device with functionalized carbon nanotubes based electrochemical sensor, Biosens. Bioelectron. 26 (2010) 1514.

[80] S.K. Kim, H. Lim, T.D. Chung, H.C. Kim, A miniaturized electrochemical system with a novel polyelectrolyte reference electrode and its application to thin layer electroanalysis, Sensors Actuators B Chem. 115 (2006) 212.

[81] E.W. Keefer, B.R. Botterman, M.I. Romero, A.F. Rossi, G.W. Gross, Carbon nanotube coating improves neuronal recordings, Nat. Nanotechnol. 3 (2008) 434.

[82] V. Parpura, Instrumentation: carbon nanotubes on the brain, Nat. Nanotechnol. 3 (2008) 384.

[83] H. Zhou, Z. Zhang, P. Yu, L. Su, T. Ohsaka, L. Mao, Noncovalent attachment of NAD + cofactor onto carbon nanotubes for preparation of integrated dehydrogenase-based electrochemical biosensors, Langmuir 26 (2010) 6028.

[84] S. Cheemalapati, S. Palanisamy, V. Mani, S.M. Chen, Simultaneous electrochemical determination of dopamine and paracetamol on multiwalled carbon nanotubes/graphene oxide nanocomposite-modified glassy carbon electrode, Talanta 117 (2013) 297.

[85] X. Sun, J. Wu, Z. Chen, X. Su, B.J. Hinds, Fouling characteristics and electrochemical recovery of carbon nanotube membranes, Adv. Funct. Mater. 23 (2013) 1500.

[86] L.G. Carrascosa, M. Moreno, M. Alvarez, L.M. Lechuga, Nanomechanical biosensors: a new sensing tool, Trends Anal. Chem. 25 (2006) 196.

[87] E. Lahiff, C. Lynam, N. Gilmartin, R. Kennedy, D. Diamond, The increasing importance of carbon nanotubes and nanostructured conducting polymers in biosensors, Anal. Bioanal. Chem. 398 (2010) 1575.

[88] M. Comotti, C.D. Pina, R. Matarrese, M. Rossi, The catalytic activity of "naked" gold particles, Angew. Chem. Int. Ed. 43 (2004) 5812.

[89] K.R. Brown, D.G. Walter, M.J. Natan, Seeding of colloidal Au nanoparticle solutions. 2. Improved control of particle size and shape, Chem. Mater. 12 (2000) 306.

[90] G.C. Jensen, C.E. Krause, G.A. Sotzing, J.F. Rusling, Inkjet-printed gold nanoparticle electrochemical arrays on plastics. Application to immunodetection of a cancer biomarker protein, Phys. Chem. Chem. Phys. 13 (2011) 4888.

[91] Z. Yin, Y. Liu, L.P. Jiang, J.J. Zhu, Electrochemical immunosensor of tumor necrosis α based on alkaline phosphate functionalized nanospheres, Biosens. Bioelectron. 26 (2011) 1890.

[92] R.T. Kachoosangi, G.G. Wildgoose, R.G. Compton, Sensitive adsorptive stripping voltammetric determination of paracetamol at multiwalled carbon nanotube modified basal plane pyrolytic graphite electrode, Anal. Chim. Acta 618 (2008) 54.

[93] M. Endo, H. Muramatsu, T. Hayashi, Y.A. Kim, M. Terrones, M.S. Dresselhaus, Buckypaper from coaxial nanotubes, Nature 433 (2005) 476.

[94] A. Ahmadalinezhad, G. Wu, A. Chen, Mediator-free electrochemical biosensor based on buckypaper on with enhanced stability and sensitivity for glucose detection, Biosens. Bioelectron. 30 (2011) 287.

[95] S. Chatterjee, A. Chen, Functionalization of carbon buckypaer for sensitive determination of hydrogen peroxide in human urine, Biosens. Bioelectron. 35 (2012) 302.

[96] E.C. Dreaden, A.M. Alkilany, X. Huang, C.J. Murphy, M.A. El-Sayed, The golden age: gold nanoparticles for biomedicine, Chem. Soc. Rev. 41 (2012) 2740.

[97] N. Zhu, Z. Chang, P. He, Y. Fang, Electrochemical DNA biosensors based on platinum nanoparticles combined carbon nanotubes, Anal. Chim. Acta 545 (2005) 21.

[98] A. Chen, P. Holt-Hindle, Platinum-based nanostructured materials: synthesis, properties and applications, Chem. Rev. 110 (2010) 3767.

[99] S. Hrapovic, Y. Liu, K. Male, J.H. Luong, Electrochemcial biosensing platforms using platinum nanoparticles and carbon nanotubes, Anal. Chem. 76 (2004) 1083.

[100] J.A. Ho, H.C. Chang, N.Y. Shih, L.C. Wu, Y.F. Chang, C.C. Chen, C. Chou, Diagnostic detection of human lung cancer-associated antigen using gold nanoparticle-based electrochemical immunosensor, Anal. Chem. 82 (2010) 5944.

[101] B.S. Munge, A.L. Coffey, J.M. Doucette, B.K. Somba, R. Malhotra, V. Patel, J.S. Gutkind, J.F. Rusling, Nanostructured immunosensor for attomolar detection of cancer biomarker interleukin-8 using massively labelled superparamagnetic particles, Angew. Chem. Int. Ed. 50 (2011) 7915.

[102] J. Wang, D.F. Thomas, A. Chen, Nonezymatic electrochemical glucose sensor based on nanoporous PtPb networks, Anal. Chem. 80 (2008) 997.

[103] P.H. Hindle, S. Nigro, M. Asmussen, A. Chen, Amperometric glucose sensor based on platinum-iridium nanomaterials, Electrochem. Commun. 10 (2008) 1438.

[104] P.K. Rastogi, V. Ganesan, S. Krishnamoorthi, A promising electrochemical sensing platform based on a silver nanoparticles decorated copolymer for sensitive nitrite determination, J. Mater. Chem. A 2 (2014) 933.

[105] P. Miao, K. Han, H. Sun, J. Yin, J. Zhao, B. Wang, Y. Tang, Melamine functionalized silver nanoparticles as the probe for electrochemical sensing of clenbuterol, ACS Appl. Mater. Interfaces 6 (2014) 8667.

[106] Y. Zhang, K. Zhang, H. Ma, Electrochemical DNA biosensor based on silver nanoparticles/poly(3-(3-pyridyl)acrylic acid)/carbon nanotubes modified electrode, Anal. Biochem. 387 (2009) 13.

[107] G. Lai, F. Yan, J. Wu, C. Leng, H. Ju, Ultrasensitive multiplexed immunoassay with electrochemical stripping analysis of silver nanoapticles catalytically deposited by gold nanoparticles and enzymatic reaction, Anal. Chem. 83 (2011) 2726.

[108] G. Lai, L. Wang, J. Wu, H. Ju, F. Yan, Electrochemcial stripping analysis of nanogold label-induced silver deposition for ultrasensitive multiplexed detection of tumor markers, Anal. Chim. Acta 721 (2012) 1.

[109] M. Chao, X. Ma, Electrochemical determination of sudan I at a silver nanoparticles/poly(aminosulfonic acid) modified glassy carbon electrode, Int. J. Electrochem. Sci. 7 (2012) 6331.

[110] Z. Su, W. Zhou, Formation, morphology control and applications of anodic TiO_2 nanotube arrays, J. Mater. Chem. 21 (2011) 8955.

[111] S. Liu, A. Chen, Coadsorption of horseradish peroxidase with thionine on TiO_2 nanotubes for biosensing, Langmuir 21 (2005) 8409.

[112] A.K.M. Kafi, G. Wu, A. Chen, A novel hydrogen peroxide biosensor based on the immobilization of horseradish onto Au-modified titanium dioxide nanotube arrays, Biosens. Bioelectron. 24 (2008) 566.

[113] Q. Kang, L. Yang, Q. Cai, An electro-catalytic biosensor fabricated with Pt-Au nanoparticle-decorated titania nanotube array, Bioelectrochemistry 74 (2008) 62.

[114] C. Wang, L. Yin, L. Zhang, R. Gao, Ti/TiO_2 nanotube array/Ni composite electrodes for nonenzymatic amperometric glucose sensing, J. Phys. Chem. C 114 (2010) 4408.

[115] J. Li, D. Kuang, Y. Feng, F. Zhang, M. Liu, Glucose biosensor based on glucose oxidase immobilized on a nanofilm composed of mesoporous hydroxyapatite, titanium dioxide, and modified with multi-walled carbon nanotubes, Microchim. Acta 176 (2012) 73.

[116] C. Kung, C. Lin, Y. Lai, et al., Cobalt oxide acicular nanorods with high sensitivity for the non-enzymatic detection of glucose, Biosens. Bioelectron. 27 (2011) 125.

[117] K. Keat, P. Yee, C. Haur, W. Shong, CoOOH nanosheet electrodes: simple fabrication for sensitive electrochemical sensing of hydrogen peroxide and hydrazine, Biosens. Bioelectron. 39 (2013) 255.

[118] A. Salimi, H. Mamkhezri, R. Hallaj, S. Soltanian, Electrochemical detection of trace amount of arsenic (III) at glassy carbon electrode modified with cobalt oxide nanoparticles, Sensors Actuators B Chem. 129 (2008) 246.

[119] A. Salimi, R. Hallaj, H. Mamkhezri, et al., Electrochemical properties and electrocatalytic activity of FAD immobilized onto cobalt oxide nanoparticles: application to nitrite detection, J. Electroanal. Chem. 620 (2008) 31.

[120] W. Jia, M. Guo, Z. Zheng, et al., Electrocatalytic oxidation and reduction of H_2O_2 on vertically aligned Co_3O_4 nanowalls electrode: toward H_2O_2 detection, J. Electroanal. Chem. 625 (2009) 27.

[121] S.M. Chen, C. Karuppiah, S. Palanisamy, V. Veeramani, A novel enzymatic glucose biosensor and sensitive non-enzymatic hydrogen peroxide sensor based on graphene and cobalt oxide nanoparticles composite modified glassy carbon electrode, Sensors Actuators B Chem. 196 (2014) 450.

[122] M.M. Shahid, N.M. Huang, P. Rameshkumar, A. Pandikumar, An electrochemical sensing platform based on reduced grapheme oxide-cobalt oxide nanocubes@platinum nanocomposite for nitric oxide detection, J. Mater. Chem. A 3 (2015) 14458.

[123] A.L. Saghatforoush, S. Sanati, M. Hasanzadeh, Synthesis, characterization and electrochemical properties of Co_3O_4 nanostructures by using cobalt hydroxide as a precursor Lotf, Res. Chem. Intermed. 41 (2014) 4361.

[124] S. Buratti, B. Brunetti, S. Mannino, Amperometric detection of carbohydrates and thiols by using a glassy carbon electrode coated with Co oxide/multi-wall carbon nanotubes catalytic system, Talanta 76 (2008) 454.

[125] T. Toan, V.H. Nguyen, R. Kumar, Facile synthesis of cobalt oxide/reduced graphene oxide composites for electrochemical capacitor and sensor applications, Solid State Sci. 53 (2016) 71.

[126] F. Chekin, S.M. Vahdat, M.J. Asadi, Green synthesis and characterization of cobalt oxide nanoparticles and its electrocatalytic behavior, Russ. J. Appl. Chem. 89 (2016) 816.

[127] A. Salimi, R. Hallaj, S. Soltanian, Immobilization of hemoglobin on electrodeposited cobalt-oxide nanoparticles: direct voltammetry and electrocatalytic activity, Biophys. Chem. 130 (2007) 122.

[128] P.C. Wang, L.H. Liu, D.A. Mengistie, K.H. Li, B.J. Wen, T.S. Liu, C.W. Chu, Transparent electrodes based on conducting polymers for display applications, Displays 34 (2013) 301.

[129] A.G. MacDiarmid, "Synthetic metals": a novel role for organic polymers (nobel lecture), Angew. Chem. Int. Ed. 40 (2001) 2581.

[130] J. Janata, M. Josowicz, Conducting polymers in electronic chemical sensors, Nat. Mater. 2 (2003) 19.

[131] N.R. Tanguy, M. Thompson, N. Yan, A review on advances in application of polyaniline for ammonia detection, Sensors Actuators B Chem. 257 (2018) 1044.

[132] M. Trojanowicz, M.L. Hitchman, A potentiometric polypyrrole-based glucose biosensor, Electroanalysis 8 (1996) 263.

[133] W.Q. Brandão, J.C.M. Llamas, J.J.A. Espinoza, A.E.C. Guajardo, C.P. Melo, Polyaniline—polystyrene membrane for simple and efficient retrieval of double-stranded DNA from aqueous media, RSC Adv. 6 (2016) 104566.

[134] T. Ahuja, I.A. Mir, D. Kumar, Rajesh, Biomolecular immobilization on conducting polymers for biosensing applications, Biomaterials 28 (2007) 791.

[135] V.C. Ferreira, A.I. Melato, A.F. Silva, L.M. Abrantes, Attachment of noble metal nanoparticles to conducting polymers containing sulphur—preparation conditions for enhanced electrocatalytic activity, Electrochim. Acta 56 (2011) 3567.

[136] R. Elghanian, J.J. Storhoff, R.C. Mucic, R.L. Letsinger, C.A. Mirkin, Selective colorimetric detection of polynucleotides based on the distance-dependent optical properties of gold nanoparticles, Science 277 (1997) 1078.

[137] B.L. Cushing, V.L. Kolesnichenko, C.J. O'Connor, Recent advances in the liquid-phase synthesis of inorganic nanoparticles, Chem. Rev. 104 (2004) 3893.

[138] J.B. Reitz, E.I. Solomon, Polyaniline oxidation of copper oxide surfaces: electronic and geometric contributions to reactivity and selectivity, J. Am. Chem. Soc. 120 (1998) 11467.

[139] J.M. Tarascon, M. Arand, Issues and challenges facing rechargeable lithium batteries, Nature 414 (2001) 359.

[140] S. Chen, J.M. Sommers, Alkanethiolate-protected copper nanoparticles: spectroscopy, electrochemistry, and solid-state morphological evolution, J. Phys. Chem. B 105 (2001) 8816.

[141] K. Akamatsu, S. Ikeda, H. Nawafune, H. Yanagimoto, Direct patterning of copper on polyimide using ion exchangeable surface templates generated by site-selective surface modification, J. Am. Chem. Soc. 126 (2004) 10822.

[142] Y.H. Kim, D.K. Lee, B.G. Jo, J.H. Jeong, Y.S. Kang, Synthesis of oleate Cu nanoparticles by thermal decomposition, Colloids Surf. A Physicochem. Eng. Asp. 285 (2006) 364.

[143] E.K. Athanassiou, R.N. Grass, W.J. Stark, Large-scale production of carbon-coated copper nanoparticles for sensor applications, Nanotechnology 17 (2006) 1668.

[144] K.S. Lokesh, Y. Shivaraj, B.P. Dayananda, S. Chandra, Synthesis of phthalocyanine stabilized rhodium nanoparticles and their application in biosensing of cytochrome c, Bioelectrochemistry 75 (2009) 104.

[145] C. Sudeshna, K.S. Lokesh, A. Nicolai, H. Lang, Dendrimer-rhodium nanoparticle modified glassy carbon electrode for amperometric detection of hydrogen peroxide, Anal. Chim. Acta 632 (2009) 63.

[146] M. Imadadulla, M. Nemakal, K.S. Lokesh, Solvent dependent dispersion behaviour of macrocycle stabilized cobalt nanoparticles and their applications, New J. Chem. 42 (2018) 11364.

[147] L.K. Sannegowda, S. Aralekallu, M. Nemakal, I. Mohammed, H. Mirabbos, Porphyrin macrocycle-stabilized gold and silver nanoparticles and their application in catalysis of hydrogen peroxide, Dyes Pigments 120 (2015) 155.

[148] C.P.K. Prabhu, M. Nemakal, S. Aralekallu, I. Mohammed, M. Palanna, V.A. Sajjan, D. Akshitha, L.K. Sannegowda, Ni foam-supported azo linkage cobalt phthalocyanine as an efficient electrocatalyst for oxygen evolution reaction, J. Electroanal. Chem. 847 (2019) 113262.

[149] C.P. Keshavananda Prabhu, N. Manjunatha, A. Shambhulinga, M. Imadadulla, K.H. Shivaprasad, M.K. Amshumali, K.S. Lokesh, Synthesis and characterization of novel imine substituted phthalocyanine for sensing of L-cysteine, J. Electroanal. Chem. 834 (2019) 130.

[150] V.A. Sajjan, I. Mohammed, M. Nemakal, S. Aralekallu, K.R.H. Kumar, S. Swamy, L.K. Sannegowda, Synthesis and electropolymerization of cobalt tetraaminebenzmidephthalocyanine macrocycle for the amperometric sensing of dopamine, J. Electroanal. Chem. 838 (2019) 33.

[151] S. Aralekallu, G. Kuntoji, M. Nemakal, I. Mohammed, L.K. Sannegowda, Self-assembly of reactive difunctional molecules on nickel electrode, Surf. Interfaces 15 (2019) 19.

[152] S. Aralekallu, I. Mohammed, N. Manjunatha, M. Palanna, Dhanjai, L.K. Sannegowda, Synthesis of novel azo group substituted polymeric phthalocyanine for amperometric sensing of nitrite, Sensors Actuators B Chem. 282 (2019) 417.

[153] M. Nemakal, S. Aralekallu, I. Mohammed, M. Pari, K.R.V. Reddy, L.K. Sannegowda, Nanomolar detection of 4-aminophenol using amperometric sensor based on a novel phthalocyanine, Electrochim. Acta 318 (2019) 342.

[154] N. Manjunatha, A. Shambhulinga, M. Imadadulla, C.P.K. Prabhu, K.S. Lokesh, Chemisorbed palladium phthalocyanine for simultaneous determination of biomolecueles, Microchem. J. 143 (2018) 82.

[155] N. Manjunatha, A. Shambhulinga, M. Imadadulla, S. Swamy, K.S. Lokesh, Electropolymerized octabenzimidazole phthalocyanine as an amperometric sensor for hydrazine, J. Electroanal. Chem. 839 (2019) 238.

[156] A. Shambhulinga, V.A. Sajjan, C.P. Keshavananda Prabhu, P.M. Palanna, M. Hajamberdiev, K.S. Lokesh, Ni foam-supported azo linkage cobalt phthalocyanine as an efficient electrocatalyst for oxygen evolution reaction, J. Power Sources 449 (2019) 227516.

[157] M. Imadadulla, N. Manjunatha, A. Shambhulinga, S.A. Veeresh, T.R. Divakara, P. Manjunatha, C.P. Keshavananda Prabu, K.S. Lokesh, Phthalocyanine sheet polymer based amperometric sensor for the selective detection of 2,4-dichlorophenol, J. Electroanal. Chem. 871 (2020) 114292.

[158] X. Zhang, Y. Wang, Y. Ma, Y. Ye, Y. Wang, K. Wu, Solvent-stabilized oxovanadium phthalocyanine nanoparticles and their application in xerographic photoreceptors, Langmuir 22 (2006) 344–348.

[159] D. Hanming, X. Zhang, M.K. Ram, C. Nicolini, Ultrathin films of tetrasulfonated copper phthalocyanine-capped titanium dioxide nanoparticles: fabrication, characterization, and photovoltaic effect, J. Colloid Interface Sci. 290 (2005) 166–171.

[160] R. Gunupuru, D. Maity, G.R. Bhadu, A. Chakraborty, D.N. Srivastava, P. Paul, Colorimetric detection of Cu^{2+} and Pb^{2+} ions using calix[4]arene functionalized gold nanoparticles, J. Chem. Sci. 126 (3) (2014) 627–635.

[161] D. Maity, R. Gupta, R. Gunupuru, D.N. Srivastava, P. Paul, Calix[4]arene functionalized goldnanoparticles: application in colorimetric and electrochemical sensing of cobalt ion in organic

andaqueous medium, Sensors Actuators B Chem. 191 (2014) 757–764, https://doi.org/10.1016/j.snb.2013.10.066.

[162] D. Maity, A. Kumar, R. Gunupuru, P. Paul, Colorimetric detection of mercury(II) in aqueous media with high selectivity using calixarene functionalized gold nanoparticles, Colloids Surf. A Physicochem. Eng. Asp. 455 (2014) 122–128, https://doi.org/10.1016/j.colsurfa.2014.04.047.

[163] S. Avvakumova, P. Fezzardi, L. Pandolfi, M. Colombo, F. Sansone, A. Casnati, D. Prosperi, Gold nanoparticles decorated by clustered multivalent cone-glycocalixarenes actively improve the targeting efficiency toward cancer cells, Chem. Commun. 50 (75) (2014) 11029–11032.

[164] X. Qiana, X. Zhoua, X. Rana, H. Nic, Z. Lia, Q. Qub, J. Lia, G. Dua, L. Yanga, Facile and clean synthesis of dihydroxylatopillar[5]arene-stabilized gold nanoparticles integrated Pd/MnO$_2$ nanocomposites for robust and ultrasensitive detection of cardiac troponin I, Biosens. Bioelectron. 130 (2019) 214–224.

[165] K.J. Chen, K.C. Pillai, J. Rick, C.J. Pan, S.H. Wang, C.C. Liu, B.J. Hwang, Bimetallic PtM (M 0 Pd, Ir) nanoparticle decorated multi-walled carbon nanotube enzyme-free, mediator-less amperometric sensor for H$_2$O$_2$, Biosens. Bioelectron. 33 (2012) 120–127.

[166] Y.Y. Yu, Q. Sun, X.Q. Liu, H.H. Wu, T.S. Zhou, G.Y. Shi, Sizecontrollable gold–platinum alloy nanoparticles on nine functionalized ionic-liquid surfaces and their application as electrocatalysts for hydrogen peroxide reduction, Chem. Eur. J. 17 (2011) 11314–11323.

[167] K.J. Chen, C.F. Lee, J. Rick, S.H. Wang, C.C. Liu, B.J. Hwang, Fabrication and application of amperometric glucose biosensor based on a novel PtPd bimetallic nanoparticle decorated multiwalled carbon nanotube catalyst, Biosens. Bioelectron. 33 (2012) 75–81.

[168] L. Wang, F. Wang, L. Shang, C.Z. Zhu, W. Ren, S.J. Dong, AuAg bimetallic nanoparticles film fabricated based on H$_2$O$_2$-mediated silver reduction and its application, Talanta 82 (2010) 113–117.

[169] T.H. Tsai, S. Thiagarajan, S.M. Chen, Green synthesized Au–Ag bimetallic nanoparticles modified electrodes for the amperometric detection of hydrogen peroxide, J. Appl. Electrochem. 40 (2010) 2071–2076.

[170] S. Manivannan, R. Ramaraj, Core–shell Au/Ag nanoparticles embedded in silicate sol–gel network for sensor application toward hydrogen peroxide, J. Chem. Sci. 121 (2009) 735–743.

[171] X.H. Niu, C. Chen, H.L. Zhao, Y. Chai, M.B. Lan, Novel snowflake-like Pt-Pd bimetallic clusters on screen-printed gold nanofilm electrode for H$_2$O$_2$ and glucose sensing, Biosens. Bioelectron. 36 (2012) 262–266.

[172] M. Rajkumar, S. Thiagarajan, S.M. Chen, Electrochemical fabrication of Rh–Pd particles and electrocatalytic applications, J. Appl. Electrochem. 41 (2011) 663–668.

[173] D. Janasek, W. Vastarella, U.S.N. Teuscher, A. Heilmann, Ruthenium/rhodium modified gold electrodes for the amperometric detection of hydrogen peroxide at low potentials, Anal. Bioanal. Chem. 374 (2002) 1267–1273.

Index

Note: Page numbers followed by *f* indicate figures, *t* indicate tables, and *s* indicate schemes.

Printed in the United States
by Baker & Taylor Publisher Services